Campbell Biology in Focus, 2/E

for Anderson University

| PEARSON COLLECTIONS |

PEARSON

Cover Art: Courtesy of Getty Images

Attention bookstores: For permission to return any unsold stock, contact us at pe-uscustomreturns@pearson.com

Pearson Learning Solutions, 501 Boylston Street, Suite 900, Boston, MA 02116

A Pearson Education Company
www.pearsoned.com

PEARSON

ISBN 10: 132366985X

ISBN 13: 9781323669853

Printed in the USA
1 17

Table of Contents

Introduction: Evolution and the Foundations of Biology

From Chapter 1 of *Campbell Biology in Focus*, Second Edition. Lisa A. Urry, Michael L. Cain, Steven A. Wasserman, Peter V. Minorsky, and Jane B. Reece.

Introduction: Evolution and the Foundations of Biology

KEY CONCEPTS

1 The study of life reveals common themes
2 The Core Theme: Evolution accounts for the unity and diversity of life
3 In studying nature, scientists form and test hypotheses

▲ **Figure 1 What can this beach mouse (*Peromyscus polionotus*) teach us about biology?**

Inquiring About Life

There are few hiding places for a mouse among the sparse clumps of beach grass that dot the brilliant white sand dunes along the Florida seashore. However, the beach mice that live there have light, dappled fur, allowing them to blend into their surroundings **(Figure 1)**. Mice of the same species (*Peromyscus polionotus*) also inhabit nearby inland areas. These mice are much darker in color, as are the soil and vegetation where they live **(Figure 2)**. For both beach mice and inland mice, the close color match of coat (fur) and environment is vital for survival, since hawks, herons, and other sharp-eyed predators periodically scan the landscape for prey. How has the color of each group of mice come to be so well matched, or *adapted*, to the local background?

▲ **Figure 2 An inland mouse of the species *Peromyscus polionotus*.** This mouse has a much darker back, side, and face than mice of the same species that inhabit sand dunes.

An organism's adaptations to its environment, such as the mouse's protective camouflage, are the result of **evolution**, the process of change over time that has resulted in the astounding array of organisms found on Earth. Evolution is the fundamental principle of biology and the core theme of this text.

Posing questions about the living world and seeking answers through scientific inquiry are the central activities of **biology**, the scientific study of life. Biologists' questions can be ambitious. They may ask how a single tiny cell becomes a tree or a dog, how the human mind works, or how the different forms of life in a forest interact. When questions occur to you as you observe the living world, you are thinking like a biologist.

How do biologists make sense of life's diversity and complexity? This opening chapter sets up a framework for answering this question. We begin with a panoramic view of the biological "landscape," organized around a set of unifying themes. We'll then focus on biology's core theme, evolution. Finally, we'll examine the process of scientific inquiry—how scientists ask and attempt to answer questions about the natural world.

CONCEPT 1

The study of life reveals common themes

Biology is a subject of enormous scope, and exciting new biological discoveries are being made every day. How can you organize and make sense of all the information you'll encounter as you study biology? Focusing on a few big ideas will help. Here are five unifying themes—ways of thinking about life that will still hold true decades from now:

- Organization
- Information
- Energy and Matter
- Interactions
- Evolution

In this chapter, we'll briefly define and explore each theme.

Theme: New Properties Emerge at Successive Levels of Biological Organization

ORGANIZATION The study of life on Earth extends from the microscopic scale of the molecules and cells that make up organisms to the global scale of the entire living planet. As biologists, we can divide this enormous range into different levels of biological organization.

In **Figure 3**, we zoom in from space to take a closer and closer look at life in a mountain meadow. This journey, depicted in the figure as a series of numbered steps, highlights the hierarchy of biological organization.

Zooming in at ever-finer resolution illustrates the principle that underlies *reductionism*, an approach that reduces complex systems to simpler components that are more manageable to study. Reductionism is a powerful strategy in biology. For example, by studying the molecular structure of DNA that had been extracted from cells, James Watson and Francis Crick inferred the chemical basis of biological inheritance. Despite its importance, reductionism provides an incomplete view of life, as we'll discuss next.

Emergent Properties

Let's reexamine Figure 3, beginning this time at the molecular level and then zooming out. Viewed this way, we see that novel properties emerge at each level that are absent from the preceding one. These **emergent properties** are due to the arrangement and interactions of parts as complexity increases. For example, although photosynthesis occurs in an intact chloroplast, it will not take place if chlorophyll and other chloroplast molecules are simply mixed in a test tube. The coordinated processes of photosynthesis require a specific organization of these molecules in the chloroplast. In general, isolated components of living systems—the objects of study in a reductionist approach—lack a number of significant properties that emerge at higher levels of organization.

Emergent properties are not unique to life. A box of bicycle parts won't transport you anywhere, but if they are arranged in a certain way, you can pedal to your chosen destination. Compared with such nonliving examples, however, biological systems are far more complex, making the emergent properties of life especially challenging to study.

To fully explore emergent properties, biologists complement reductionism with **systems biology**, the exploration of the network of interactions that underlie the emergent properties of a system. A single leaf cell can be considered a system, as can a frog, an ant colony, or a desert ecosystem. By examining and modeling the dynamic behavior of an integrated network of components, systems biology enables us to pose new kinds of questions. For example, how do networks of genes in our cells produce oscillations in the activity of the molecules that generate our 24-hour cycle of wakefulness and sleep? At a larger scale, how does a gradual increase in atmospheric carbon dioxide alter ecosystems and the entire biosphere? Systems biology can be used to study life at all levels.

Structure and Function

At each level of biological organization, we find a correlation between structure and function. Consider a leaf in Figure 3: Its thin, flat shape maximizes the capture of sunlight by chloroplasts. Because such correlations of structure and function are common in all forms of life, analyzing a biological structure gives us clues about what it does and how it works. A good example from the animal kingdom is the hummingbird. The hummingbird's anatomy allows its wings to rotate at the shoulder, so hummingbirds have the ability, unique among birds, to fly backward or hover in place. While hovering, the birds can extend their long slender beaks into flowers and feed on nectar. Such an elegant match of form and function in the structures of life is explained by natural selection, as we'll explore shortly.

The Cell: An Organism's Basic Unit of Structure and Function

The cell is the smallest unit of organization that can perform all activities required for life. In fact, the actions of an organism are all based on the activities of its cells. For instance, the movement of your eyes as you read this sentence results from the activities of muscle and nerve cells. Even a process that occurs on a global scale, such as the recycling of carbon atoms, is the cumulative product of cellular functions,

▼ Figure 3 **Exploring Levels of Biological Organization**

◄ 1 The Biosphere

Even from space, we can see signs of Earth's life—in the green mosaic of the forests, for example. We can also see the entire **biosphere**, which consists of all life on Earth and all the places where life exists: most regions of land, most bodies of water, the atmosphere to an altitude of several kilometers, and even sediments far below the ocean floor.

◄ 2 Ecosystems

Our first scale change brings us to a North American mountain meadow, which is an example of an ecosystem, as are tropical forests, grasslands, deserts, and coral reefs. An **ecosystem** consists of all the living things in a particular area, along with all the nonliving components of the environment with which life interacts, such as soil, water, atmospheric gases, and light.

► 3 Communities

The array of organisms inhabiting a particular ecosystem is called a biological **community**. The community in our meadow ecosystem includes many kinds of plants, various animals, mushrooms and other fungi, and enormous numbers of diverse microorganisms, such as bacteria, that are too small to see without a microscope. Each of these forms of life belongs to a *species*—a group whose members can only reproduce with other members of the group.

► 4 Populations

A **population** consists of all the individuals of a species living within the bounds of a specified area. For example, our meadow includes a population of lupine (some of which are shown here) and a population of mule deer. A community is therefore the set of populations that inhabit a particular area.

▲ 5 Organisms

Individual living things are called **organisms**. Each plant in the meadow is an organism, and so is each animal, fungus, and bacterium.

including the photosynthetic activity of chloroplasts in leaf cells.

All cells share certain characteristics, such as being enclosed by a membrane that regulates the passage of materials between the cell and its surroundings. Nevertheless, we distinguish two main forms of cells: prokaryotic and eukaryotic. The cells of two groups of single-celled microorganisms—bacteria and archaea—are prokaryotic. All other forms of life, including plants and animals, are composed of eukaryotic cells.

A **eukaryotic cell** contains membrane-enclosed organelles (**Figure 4**). Some organelles, such as the DNA-containing nucleus, are found in the cells of all eukaryotes; other organelles are specific to particular cell types. For example, the chloroplast in Figure 3 is an organelle found only in eukaryotic cells that carry out photosynthesis. In contrast to eukaryotic cells, a **prokaryotic cell** lacks a nucleus or other membrane-enclosed organelles. Furthermore, prokaryotic cells are generally smaller than eukaryotic cells, as shown in Figure 4.

▼6 Organs

The structural hierarchy of life continues to unfold as we explore the architecture of a complex organism. A leaf is an example of an **organ**, a body part that is made up of multiple tissues and has specific functions in the body. Leaves, stems, and roots are the major organs of plants. Within an organ, each tissue has a distinct arrangement and contributes particular properties to organ function.

▼7 Tissues

Viewing the tissues of a leaf requires a microscope. Each **tissue** is a group of cells that work together, performing a specialized function. The leaf shown here has been cut on an angle. The honeycombed tissue in the interior of the leaf (left side of photo) is the main location of photosynthesis, the process that converts light energy to the chemical energy of sugar. The jigsaw puzzle–like "skin" on the surface of the leaf is a tissue called epidermis (right side of photo). The pores through the epidermis allow entry of the gas CO_2, a raw material for sugar production.

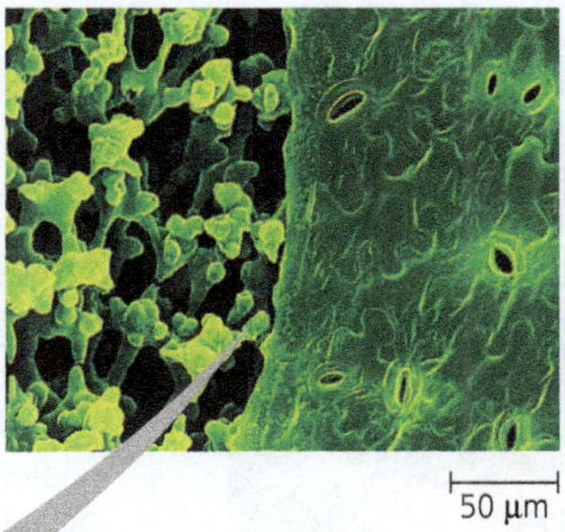

►8 Cells

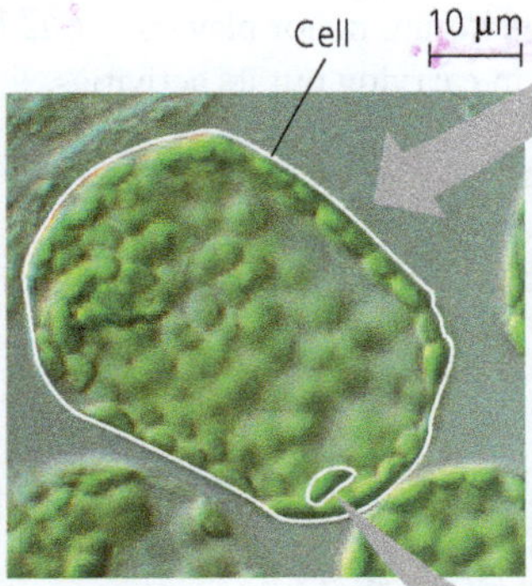

The **cell** is life's fundamental unit of structure and function. Some organisms consist of a single cell, which performs all the functions of life. Other organisms are multicellular and feature a division of labor among specialized cells. Here we see a magnified view of a cell in a leaf tissue. This cell is about 40 micrometers (µm) across—about 500 of them would reach across a small coin. Within these tiny cells are even smaller green structures called chloroplasts, which are responsible for photosynthesis.

▼9 Organelles

Chloroplasts are examples of **organelles**, the various functional components present in cells. The image below, taken by a powerful microscope, shows a single chloroplast.

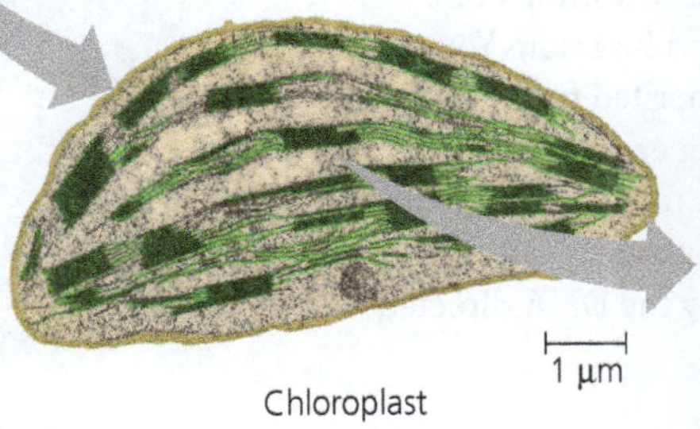

▼10 Molecules

Our last scale change drops us into a chloroplast for a view of life at the molecular level. A **molecule** is a chemical structure consisting of two or more units called atoms, represented as balls in this computer graphic of a chlorophyll molecule. Chlorophyll is the pigment that makes a leaf green, and it absorbs sunlight during photosynthesis. Within each chloroplast, millions of chlorophyll molecules are organized into systems that convert light energy to the chemical energy of food.

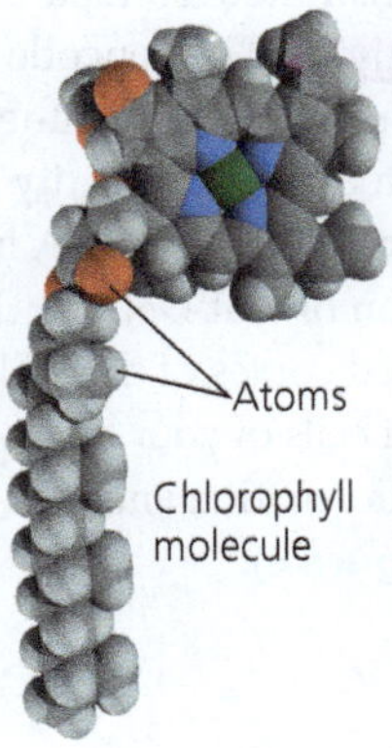

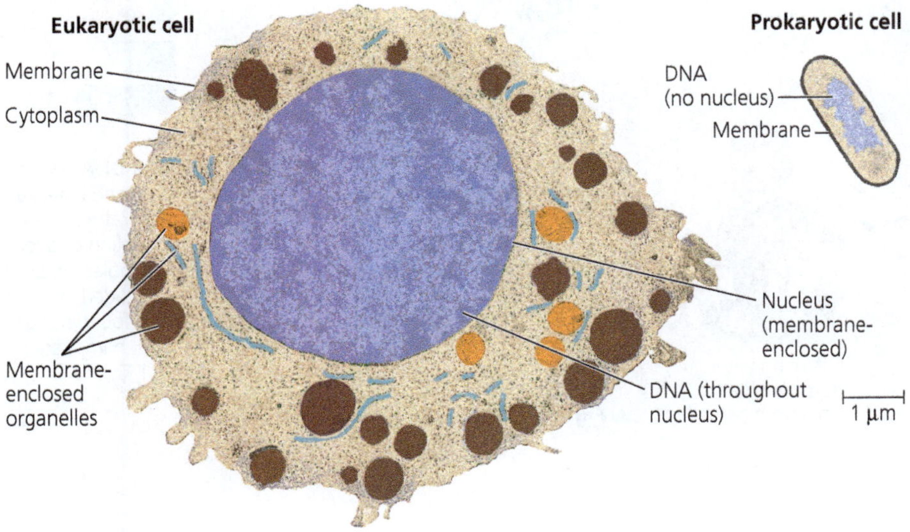

◄**Figure 4 Contrasting eukaryotic and prokaryotic cells in size and complexity.** Cells vary in size, but eukaryotic cells are generally much larger than prokaryotic cells.

Theme: Life's Processes Involve the Expression and Transmission of Genetic Information

INFORMATION Within cells, structures called chromosomes contain genetic material in the form of **DNA (deoxyribonucleic acid)**. In cells that are preparing to divide, the chromosomes may be made visible using a dye that appears blue when bound to the DNA **(Figure 5)**.

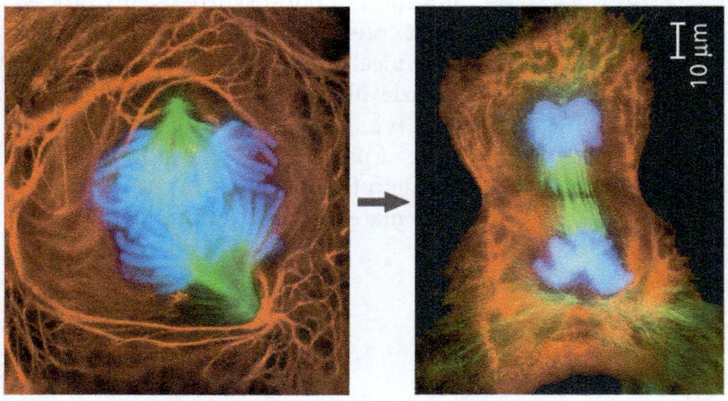

▲ **Figure 5 A lung cell from a newt divides into two smaller cells that will grow and divide again.**

DNA, the Genetic Material

Each chromosome contains one very long DNA molecule with hundreds or thousands of **genes**, each a section of the DNA of the chromosome. Transmitted from parents to offspring, genes are the units of inheritance. They encode the information necessary to build all of the molecules synthesized within a cell, which in turn establish that cell's identity and function. You began as a single cell stocked with DNA inherited from your parents. The replication of that DNA during each round of cell division transmitted copies of the DNA to what eventually became the trillions of cells of your body. As the cells grew and divided, the genetic information encoded by the DNA directed your development **(Figure 6)**.

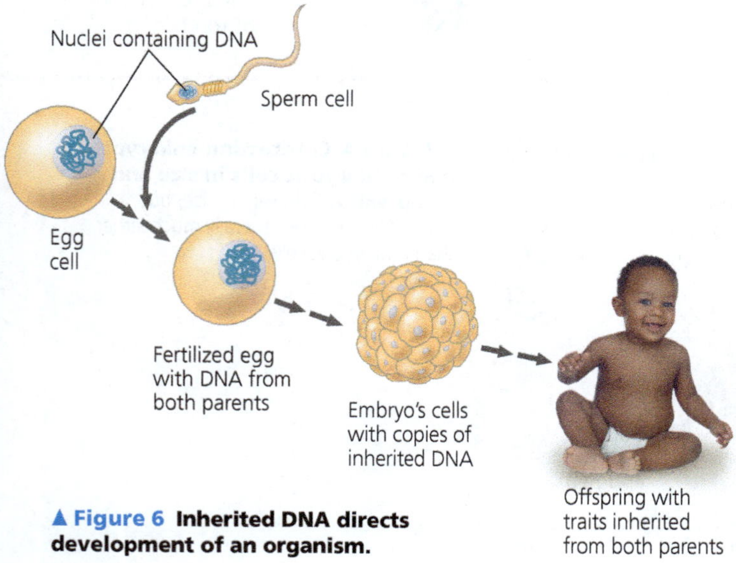

▲ **Figure 6 Inherited DNA directs development of an organism.**

The molecular structure of DNA accounts for its ability to store information. A DNA molecule is made up of two long chains, called strands, arranged in a double helix. Each chain is made up of four kinds of chemical building blocks called nucleotides, abbreviated A, T, C, and G **(Figure 7)**. Specific sequences of these four nucleotides encode the information in genes. The way DNA encodes information is analogous to how we arrange the letters of the alphabet into words and phrases with specific meanings. The word *rat*, for example, evokes a rodent; the words *tar* and *art*, which contain the same letters, mean very different things. We can think of the set of nucleotides as a four-letter alphabet.

For many genes, the sequence provides the blueprint for making a protein. For instance, a given bacterial gene may specify a particular protein (an enzyme) required to assemble the cell membrane, while a certain human gene may denote a different protein (an antibody) that helps fight off infection. Overall, proteins are major players in building and maintaining the cell and in carrying out its activities.

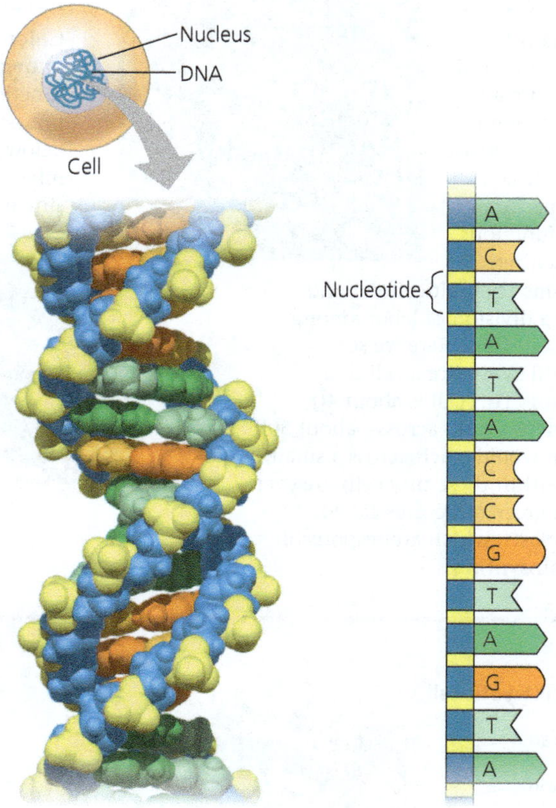

(a) DNA double helix. This model shows the atoms in a segment of DNA. Made up of two long chains (strands) of building blocks called nucleotides, a DNA molecule takes the three-dimensional form of a double helix.

(b) Single strand of DNA. These geometric shapes and letters are simple symbols for the nucleotides in a small section of one strand of a DNA molecule. Genetic information is encoded in specific sequences of the four types of nucleotides. Their names are abbreviated A, T, C, and G.

▲ **Figure 7 DNA: The genetic material.**

Genes control protein production indirectly, using a related molecule called mRNA as an intermediary **(Figure 8)**. The sequence of nucleotides along a gene is transcribed into mRNA, which is then translated into a chain of protein building blocks called amino acids. Once completed, this chain forms a specific protein with a unique shape and function. The entire process by which the information in a gene directs the production of a cellular product is called **gene expression**.

In carrying out gene expression, all forms of life employ essentially the same genetic code: A particular sequence of nucleotides says the same thing in one organism as it does in another. Differences between organisms reflect differences between their nucleotide sequences rather than between their genetic codes. This universality of the genetic code is a strong piece of evidence that all life is related. Comparing the sequences in several species for a gene that codes for a particular protein can provide valuable information both about the protein and about the evolutionary relationship of the species to each other.

The mRNA molecule in Figure 8 is translated into a protein, but other cellular RNAs function differently. For example, we have known for decades that some types of RNA are actually components of the cellular machinery that manufactures proteins. Recently, scientists have discovered whole new classes of RNA that play other roles in the cell, such as regulating the function of protein-coding genes. Genes also specify all of these RNAs, and their production is also referred to as gene expression. By carrying the instructions for making proteins and RNAs and by replicating with each cell division, DNA ensures faithful inheritance of genetic information from generation to generation.

Genomics: Large-Scale Analysis of DNA Sequences

The entire "library" of genetic instructions that an organism inherits is called its **genome**. A typical human cell has two similar sets of chromosomes, and each set has approximately 3 billion nucleotide pairs of DNA. If the one-letter abbreviations for the nucleotides of one strand in a set were written in letters the size of those you are now reading, the genomic text would fill about 700 biology textbooks.

Since the early 1990s, the pace at which researchers can determine the sequence of a genome has accelerated at an astounding rate, enabled by a revolution in technology. The genome sequence—the entire sequence of nucleotides for a representative member of a species—is now known for humans and many other animals, as well as numerous plants, fungi, bacteria, and archaea. To make sense of the deluge of data from genome-sequencing projects and the growing catalog of known gene functions, scientists are applying a systems biology approach at the cellular and molecular levels. Rather than investigating a single gene at a time, researchers study whole sets of genes in one or more species—an approach called **genomics**. Likewise, the term **proteomics** refers to the study of sets of proteins and their properties. (The entire set of proteins expressed by a given cell or group of cells is called a **proteome**.)

▼ Figure 8 Gene expression: Cells use information encoded in a gene to synthesize a functional protein.

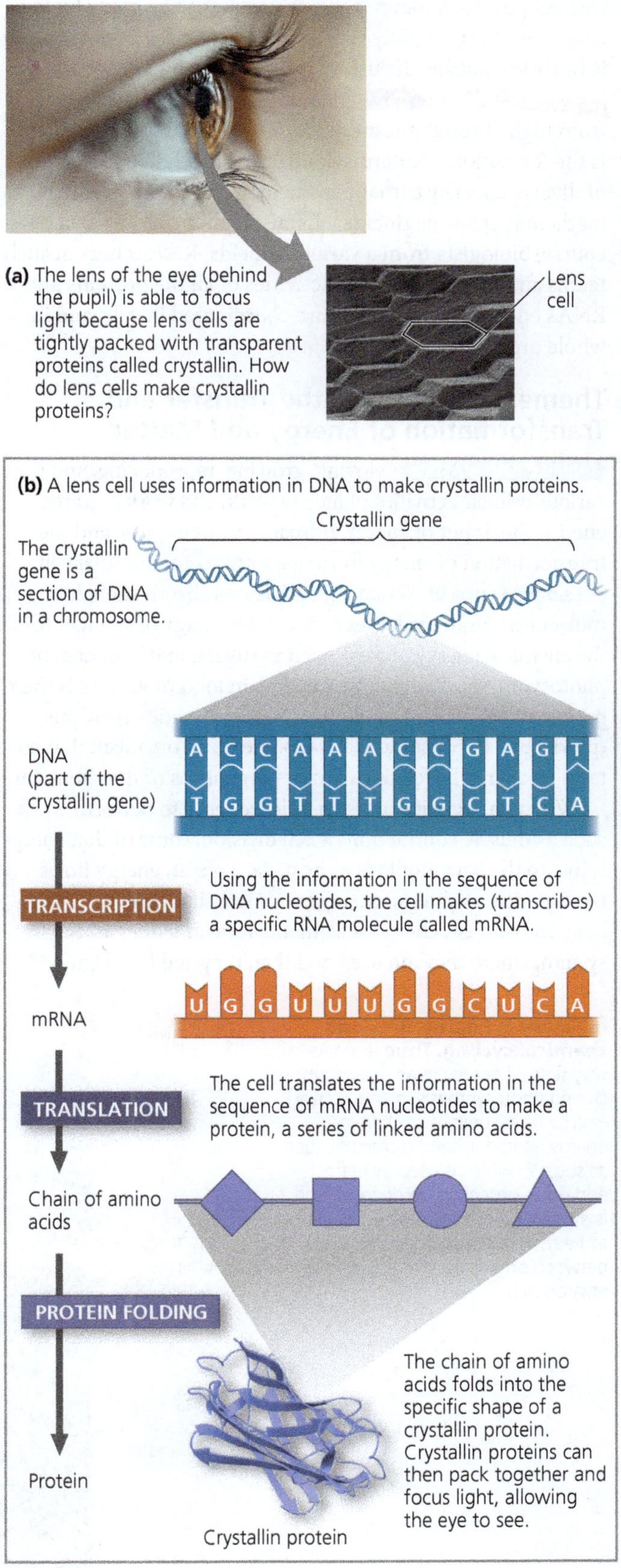

Three important research developments have made the genomic and proteomic approaches possible. One is "high-throughput" technology, tools that can analyze many biological samples very rapidly. The second major development is **bioinformatics**, the use of computational tools to store, organize, and analyze the huge volume of data that results from high-throughput methods. The third key development is the formation of interdisciplinary research teams—groups of diverse specialists that may include computer scientists, mathematicians, engineers, chemists, physicists, and, of course, biologists from a variety of fields. Researchers in such teams aim to learn how the activities of all the proteins and RNAs encoded by the DNA are coordinated in cells and in whole organisms.

Theme: Life Requires the Transfer and Transformation of Energy and Matter

ENERGY AND MATTER Moving, growing, reproducing, and the various cellular activities of life are work, and work requires energy. The input of energy, primarily from the sun, and the transformation of energy from one form to another make life possible **(Figure 9)**. When a plant's leaves absorb sunlight, molecules within the leaves convert the energy of sunlight to the chemical energy of food, such as sugars, in the process of photosynthesis. The chemical energy in food molecules is then passed along by plants and other photosynthetic organisms (producers) to consumers. A consumer is an organism that obtains its energy by feeding on other organisms or their remains.

When an organism uses chemical energy to perform work, such as muscle contraction or cell division, some of that energy is lost to the surroundings as heat. As a result, energy flows *through* an ecosystem, usually entering as light and exiting as heat. In contrast, chemical elements remain *within* an ecosystem, where they are used and then recycled (see Figure 9). Chemicals that a plant absorbs from the air or soil may be incorporated into the plant's body and then passed to an animal that eats the plant. Eventually, these chemicals will be returned to the environment by decomposers, such as bacteria and fungi, that break down waste products, organic debris, and the bodies of dead organisms. The chemicals are then available to be taken up by plants again, thereby completing the cycle.

Theme: Organisms Interact with Other Organisms and the Physical Environment

INTERACTIONS Every organism in an ecosystem interacts with other organisms. A flowering plant, for example, interacts with soil microorganisms associated with its roots, insects that pollinate its flowers, and animals that eat its leaves and petals. Interactions between organisms include those that are mutually beneficial (as when fish eat small parasites on a turtle, shown in **Figure 10**), and those in which one species benefits and the other is harmed (as when a lion kills and eats a zebra). In some interactions between species both are harmed (as when two plants compete for a soil resource that is in short supply).

Each organism in an ecosystem also interacts continuously with physical factors in its environment. The leaves of a flowering plant, for example, absorb light from the sun, take in carbon dioxide from the air, and release oxygen to the air. The environment is also affected by the organisms living there. For example, a plant takes up water and minerals from the soil through its roots, and its roots break up rocks, thereby contributing to the formation of soil. On a global scale, plants and other photosynthetic organisms have generated all the oxygen in the atmosphere.

Like other organisms, we humans interact with our environment. Unfortunately, our interactions sometimes have dire consequences. For example, over the past 150 years, humans have greatly increased the burning of fossil fuels (coal, oil, and gas). This practice releases large amounts of carbon dioxide

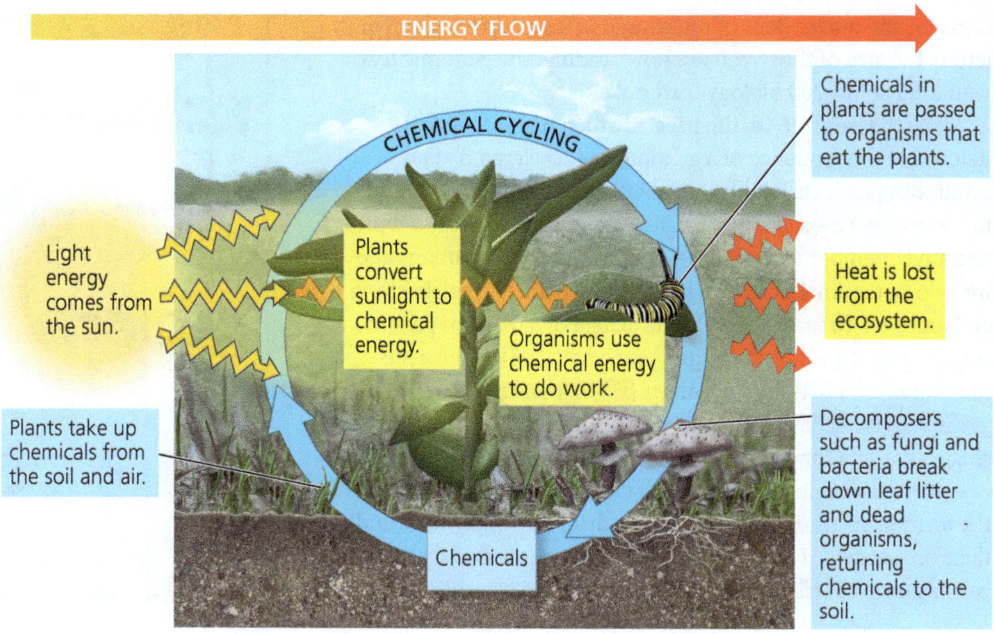

▶ **Figure 9 Energy flow and chemical cycling.** There is a one-way flow of energy in an ecosystem: During photosynthesis, plants convert energy from sunlight to chemical energy (stored in food molecules such as sugars), which is used by plants and other organisms to do work and is eventually lost from the ecosystem as heat. In contrast, chemicals cycle between organisms and the physical environment.

(CO_2) and other gases into the atmosphere. About half of this CO_2 stays in the atmosphere, causing heat to be trapped close to Earth's surface. Scientists calculate that the CO_2 that human activities have added to the atmosphere has increased the average temperature of the planet by about 1°C since 1900. At the current rates that CO_2 and other gases are being added to the atmosphere, global models predict an additional rise of at least 3°C before the end of this century.

This ongoing global warming is a major aspect of **climate change**, a directional change to the global climate that lasts for three decades or more (as opposed to short-term changes in the weather). But global warming is not the only way the climate is changing: wind and precipitation patterns are also shifting, and extreme weather events such as storms and droughts are occurring more often. Climate change has already affected organisms and their habitats all over the planet. For example, polar bears have lost much of the ice platform from which they hunt, leading to food shortages and increased mortality rates. As habitats deteriorate, hundreds of plant and animal species are shifting their ranges to more suitable locations—but for some, there is insufficient suitable habitat, or they may not be able to migrate quickly enough. As a result, the populations of many species are shrinking in size or even disappearing (**Figure 11**). This trend can ultimately result in extinction, the permanent loss of a species. The consequences of these changes for humans and other organisms may be profound.

▶ **Figure 10 A mutually beneficial interaction between species.** These fish feed on small organisms living on the sea turtle's skin and shell. The sea turtle benefits from the removal of parasites, and the fish gain a meal and protection from enemies.

Evolution, the Core Theme of Biology

Having considered four of the unifying themes that run through this text (organization, information, energy and matter, and interactions), let's now turn to biology's core theme—evolution. Evolution makes sense of everything we know about living organisms. As the fossil record clearly shows, life has been evolving on Earth for billions of years, resulting in a vast diversity of past and present organisms. But along with the diversity there is also unity. For example, while sea horses, jackrabbits, hummingbirds, crocodiles, and giraffes all look very different, their skeletons are organized in the same basic way.

The scientific explanation for the unity and diversity of organisms—as well as for the adaptation of organisms to their particular environments—is **evolution**: the concept that the organisms living on Earth today are the modified descendants of common ancestors. As a result of descent with modification, two species share certain traits (unity) simply because they have descended from a common ancestor. Furthermore, we can account for differences between two species (diversity) with the idea that certain heritable changes occurred after the two species diverged from their common ancestor. An abundance of evidence of different types supports the occurrence of evolution and the theory that describes how it takes place. Meanwhile, in the next section, we'll continue our introduction to the fundamental concept of evolution.

▶ **Figure 11 Threatened by global warming.** A warmer environment causes lizards in the genus *Sceloporus* to spend more time in refuges from the heat, reducing the time available for foraging. The lizards' food intake drops, decreasing their reproductive success. Indeed, surveys of 200 populations of *Sceloporus* species in Mexico show that 12% of these populations have disappeared since 1975.

CONCEPT CHECK 1

1. Starting with the molecular level in Figure 3, write a sentence that includes components from the previous (lower) level of biological organization, for example, "A molecule consists of *atoms* bonded together." Continue with organelles, moving up the biological hierarchy.
2. Identify the theme or themes exemplified by (a) the sharp quills of a porcupine, (b) the development of a multicellular organism from a single fertilized egg, and (c) a hummingbird using sugar to power its flight.
3. **WHAT IF?** For each theme discussed in this section, give an example not mentioned in the text.

CONCEPT 2

The Core Theme: Evolution accounts for the unity and diversity of life

EVOLUTION Diversity is a hallmark of life. Biologists have identified and named about 1.8 million species of organisms, and estimates of the number of living species range from about 10 million to over 100 million. These remarkably diverse forms of life arose by evolutionary processes. Before exploring evolution further, however, let's first consider how biologists organize the enormous variety of life forms on this planet into manageable and informative groupings.

Classifying the Diversity of Life

Humans have a tendency to group diverse items according to their similarities and relationships to each other. Following this inclination, biologists have long used careful comparisons of form and function to classify life-forms into a hierarchy of increasingly inclusive groups. Consider, for example, the species known as the leopard (*Panthera pardus*). Leopards belong to the same genus (*Panthera*) as tigers and lions. Bringing together several similar genera forms a family, which in turn is a component of an order and then a class. For the leopard, this means being grouped with cougars, cheetahs, and others in the family Felidae, with wolves in the order Carnivora, and with dolphins (and us) in the class Mammalia. These animals can be classified into still broader groupings: the phylum Chordata and the kingdom Animalia.

In the last few decades, new methods of assessing species relationships, especially comparisons of DNA sequences, have led to a reevaluation of the larger groupings. Although this reevaluation is ongoing, there is consensus among biologists that the kingdoms of life, whatever their number, can be further grouped into three higher levels of classification called domains: Bacteria, Archaea, and Eukarya **(Figure 12)**.

▼ Figure 12 The three domains of life.

(a) Domain Bacteria

Bacteria are the most diverse and widespread prokaryotes and are now classified into multiple kingdoms. Each rod-shaped structure in this photo is a bacterial cell.

(b) Domain Archaea

Domain Archaea includes multiple kingdoms. Some of the prokaryotes known as **archaea** live in Earth's extreme environments, such as salty lakes and boiling hot springs. Each round structure in this photo is an archaeal cell.

(c) Domain Eukarya

▲ **Kingdom Plantae** (land plants) consists of terrestrial multicellular eukaryotes that carry out photosynthesis, the conversion of light energy to the chemical energy in food.

► **Kingdom Fungi** is defined in part by the nutritional mode of its members (such as this mushroom), which absorb nutrients from outside their bodies.

◄ **Kingdom Animalia** consists of multicellular eukaryotes that ingest other organisms.

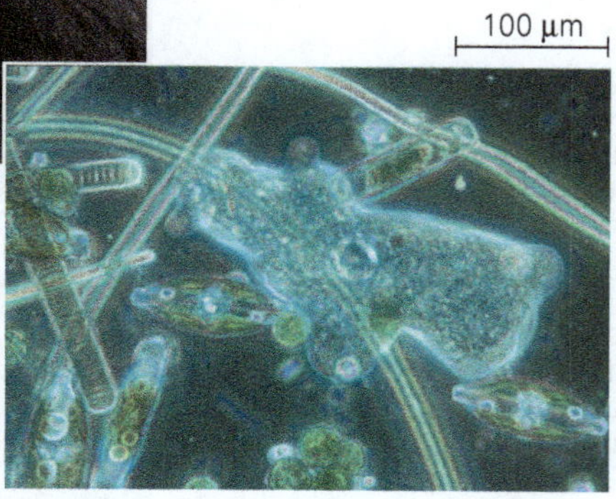

► **Protists** are mostly unicellular eukaryotes and some relatively simple multicellular relatives. Pictured here is an assortment of protists inhabiting pond water. Scientists are currently debating how to classify protists in a way that accurately reflects their evolutionary relationships.

B + A = P

The organisms making up two of the three domains—**Bacteria** and **Archaea**—are prokaryotic. All the eukaryotes (organisms with eukaryotic cells) are grouped in domain **Eukarya**. This domain includes three kingdoms of multicellular eukaryotes: Plantae, Fungi, and Animalia. These three kingdoms are distinguished partly by their modes of nutrition. Plants produce their own sugars and other food molecules by photosynthesis; fungi absorb dissolved nutrients from their surroundings; and animals obtain food by eating and digesting other organisms. Animalia is, of course, our own kingdom.

The most numerous and diverse eukaryotes are the mostly single-celled protists. Although protists once were placed in a single kingdom, they are now classified into several groups. One major reason for this change is the recent DNA evidence showing that some protists are less closely related to other protists than they are to plants, animals, or fungi.

Unity in the Diversity of Life

Although diversity is apparent in the many forms of life, there is also remarkable unity. Consider, for example, the similar skeletons of different animals and the universal genetic language of DNA (the genetic code), both mentioned earlier. In fact, similarities between organisms are evident at all levels of the biological hierarchy.

How can we account for life's dual nature of unity and diversity? The process of evolution, explained next, illuminates both the similarities and differences in the world of life. It also introduces another dimension of biology: the passage of time. The history of life, as documented by fossils and other evidence, is the saga of an ever-changing Earth billions of years old, inhabited by an evolving cast of living forms **(Figure 13)**.

▲ **Figure 13 Digging into the past.** Paleontologists carefully excavate the hind leg of a long-necked dinosaur (*Rapetosaurus krausei*) from rocks in Madagascar.

▶ **Figure 14 Charles Darwin as a young man.** His revolutionary book, which is commonly referred to as *The Origin of Species*, was first published in 1859.

Charles Darwin and the Theory of Natural Selection

An evolutionary view of life came into sharp focus in 1859, when Charles Darwin published one of the most important and influential books ever written, *On the Origin of Species by Means of Natural Selection* **(Figure 14)**. *The Origin of Species* articulated two main points. The first was that species arise from a succession of ancestors that were different from them. Darwin called this process "descent with modification." This insightful phrase captured the duality of life's unity and diversity—unity in the kinship among species that descended from common ancestors and diversity in the modifications that evolved as species branched from their common ancestors **(Figure 15)**. Darwin's second main point was his proposal that "natural selection" is a primary cause of descent with modification.

▲ **Figure 15 Unity and diversity among birds.** These three birds are variations on a common body plan. For example, each has feathers, a beak, and wings, although these features are highly specialized for the birds' diverse lifestyles.

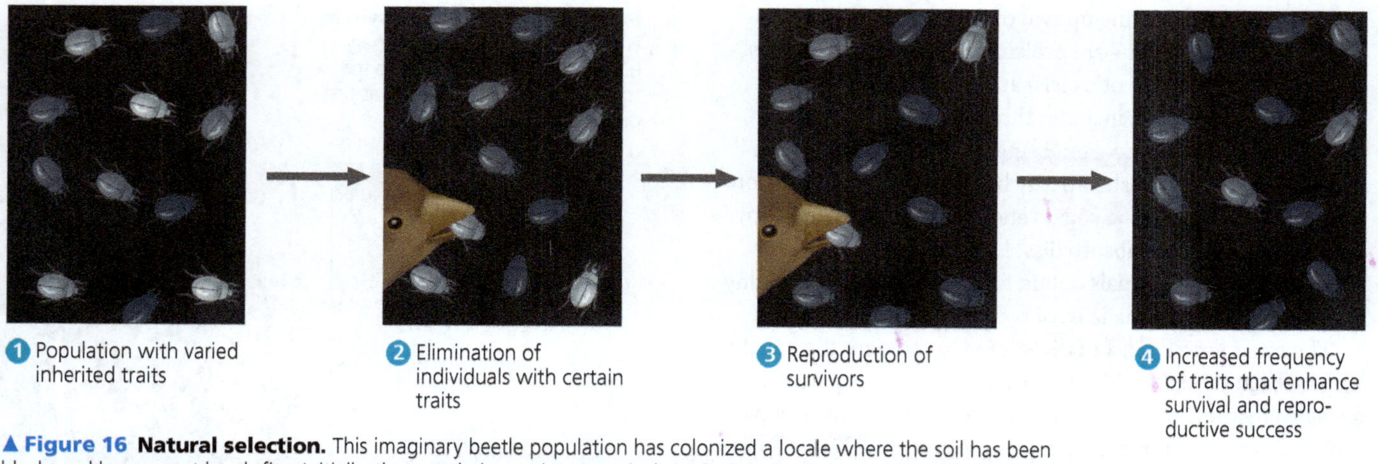

▲ **Figure 16 Natural selection.** This imaginary beetle population has colonized a locale where the soil has been blackened by a recent brush fire. Initially, the population varies extensively in the inherited coloration of the individuals, from very light gray to charcoal. For hungry birds that prey on the beetles, it is easiest to spot the beetles that are lightest in color.

Darwin developed his theory of natural selection from observations that by themselves were neither new nor profound. However, although others had described the pieces of the puzzle, it was Darwin who saw how they fit together. His three essential observations were the following: First, individuals in a population vary in their traits, many of which seem to be heritable (passed on from parents to offspring). Second, a population can produce far more offspring than can survive to produce offspring of their own. Competition is thus inevitable. Third, species generally are suited to their environments—in other words, they are adapted to their circumstances. For instance, a common adaptation among birds that eat mostly hard seeds is an especially strong beak.

By making inferences from these three observations, Darwin arrived at his theory of evolution. He reasoned that individuals with inherited traits that are better suited to the local environment are more likely to survive and reproduce than are less well-suited individuals. Over many generations, a higher and higher proportion of individuals in a population will have the advantageous traits. Darwin called this mechanism of evolutionary adaptation **natural selection** because the natural environment consistently "selects" for the propagation of certain traits among naturally occurring variant traits in the population **(Figure 16)**.

The Tree of Life

For another example of unity and diversity, consider the human arm. The bones, joints, nerves, and blood vessels in your forelimb are very similar to those in the foreleg of a horse, the flipper of a whale, and the wing of a bat. Indeed, all mammalian forelimbs are anatomical variations of a common architecture. According to the Darwinian concept of descent with modification, the shared anatomy of mammalian limbs reflects inheritance of the limb structure from a common ancestor—the "prototype" mammal from which all other mammals descended. The diversity of mammalian forelimbs results from modification by natural selection operating over millions of years in different environmental contexts.

Darwin proposed that natural selection, by its cumulative effects over long periods of time, could cause an ancestral species to give rise to two or more descendant species. This could occur, for example, if one population of organisms fragmented into several subpopulations isolated in different environments. In these separate arenas of natural selection, one species could gradually radiate into multiple species as the geographically isolated populations adapted over many generations to different environmental conditions.

The "family tree" of six finch species shown in **Figure 17** illustrates a famous example of the process of radiation. Darwin collected specimens of finches during his 1835 visit to the remote Galápagos Islands, 900 kilometers (km) west of South America. The Galápagos finches are believed to have descended from an ancestral finch species that reached the archipelago from South America or the Caribbean. Over time, the Galápagos finches diversified from their ancestor as populations became adapted to different food sources on their particular islands. Years after Darwin collected the finches, researchers began to sort out their evolutionary relationships, first from anatomical and geographic data and more recently using DNA sequence comparisons.

Biologists' diagrams of such evolutionary relationships generally take treelike forms, though the trees are often turned sideways, as in Figure 17. Tree diagrams make sense: Just as an individual has a genealogy that can be diagrammed as a family tree, each species is one twig of a branching tree of life extending back in time through ancestral species more and more remote. Species that are very similar, such as the Galápagos finches, share a relatively recent common ancestor. Through an ancestor that lived much farther back in time, finches are related to sparrows, hawks, penguins, and all other birds. Furthermore, finches and other birds are related to us through a common ancestor even more ancient. Trace life back far enough, and we reach the early prokaryotes that inhabited Earth 3.5 billion years ago. We can recognize their vestiges in our own cells—in the universal genetic code, for example. Indeed, all of life is connected through its long evolutionary history.

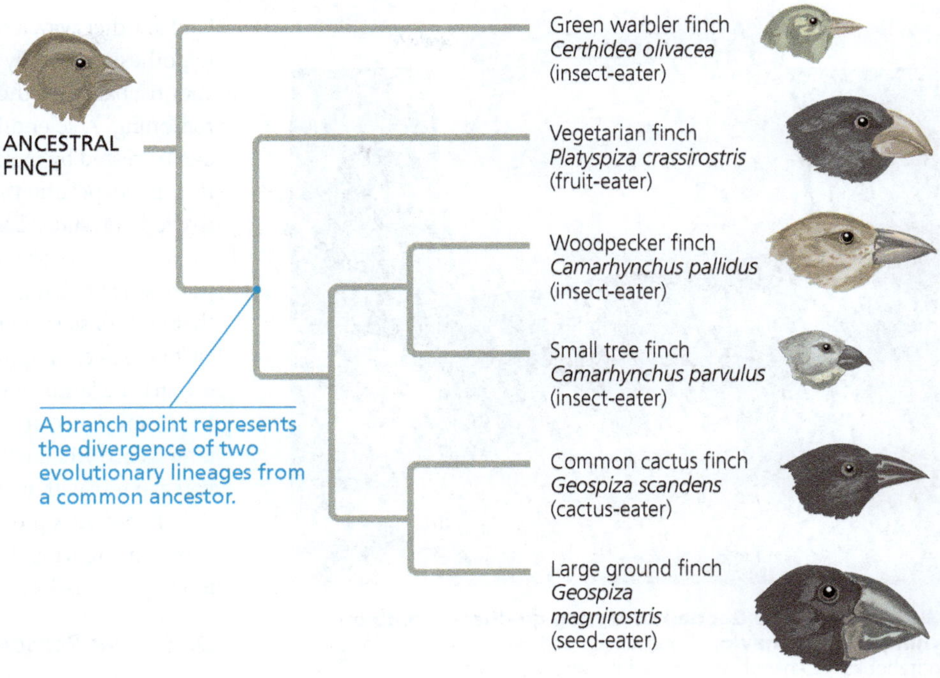

▶ **Figure 17 Descent with modification: finches on the Galápagos Islands.** This "tree" diagram illustrates a current model for the evolutionary relationships of finches on the Galápagos. Note the various beaks, which are adapted to particular food sources. For example, heavier, thicker beaks are better at cracking seeds, while more slender beaks are better at grasping insects.

CONCEPT CHECK 2

1. How is a mailing address analogous to biology's hierarchical classification system?
2. Explain why "editing" is an appropriate metaphor for how natural selection acts on a population's heritable variation.
3. **DRAW IT** Recent evidence indicates that fungi and animals are more closely related to each other than either of these kingdoms is to plants. Draw a simple branching pattern that symbolizes the proposed relationship between these three kingdoms of multicellular eukaryotes.

CONCEPT 3

In studying nature, scientists form and test hypotheses

Science is a way of knowing—an approach to understanding the natural world. It developed out of our curiosity about ourselves, other life forms, our planet, and the universe.

At the heart of science is **inquiry**, a search for information and explanations of natural phenomena. There is no formula for successful scientific inquiry, no single scientific method that researchers must rigidly follow. As in all quests, science includes elements of challenge, adventure, and luck, along with careful planning, reasoning, creativity, patience, and the persistence to overcome setbacks. Such diverse elements of inquiry make science far less structured than most people realize. That said, it is possible to highlight certain characteristics that help to distinguish science from other ways of describing and explaining nature.

Scientists use a process of inquiry that includes making observations, forming logical explanations (*hypotheses*), and testing them. The process is necessarily repetitive: In testing a hypothesis, our observations may inspire revision of the original hypothesis or formation of a new one, thus leading to further testing. In this way, scientists circle closer and closer to their best estimation of the laws governing nature.

Exploration and Discovery

Biology, like other sciences, begins with careful observation. In gathering information, biologists often use tools, such as microscopes, precision thermometers, or high-speed cameras, that extend their senses or facilitate careful measurement. Observations can reveal valuable information about the natural world. For example, a series of detailed observations have shaped our understanding of cell structure. Another set of observations is currently expanding our databases of genome sequences from diverse species and of genes whose expression is altered in diseases.

In exploring nature, biologists also rely heavily on the scientific literature, the published contributions of fellow scientists. By reading about and understanding past studies, scientists can build on the foundation of existing knowledge, focusing their investigations on observations that are original and on hypotheses that are consistent with previous findings. Identifying publications relevant to a new line of research is now easier than at any point in the past, thanks to indexed and searchable electronic databases.

Gathering and Analyzing Data

Recorded observations are called **data**. Put another way, data are items of information on which scientific inquiry is based. Some data are *qualitative*, such as descriptions of what is observed. For example, British primate researcher Jane Goodall spent decades recording her observations of chimpanzee behavior during field research in a Tanzanian jungle

qualitative = words

▲ **Figure 18 Jane Goodall collecting qualitative data on chimpanzee behavior.** Goodall recorded her observations in field notebooks, often with sketches of the animals' behavior.

quantitative = numbers

(**Figure 18**). She also documented her observations with photographs and movies.

In her studies, Goodall also gathered and recorded volumes of *quantitative* data, such as the frequency and duration of specific behaviors for different members of a group of chimpanzees in a variety of situations. Quantitative data are generally expressed as numerical measurements and often organized into tables or graphs. Scientists analyze their data using a type of mathematics called *statistics* to test whether their results are significant or merely due to random fluctuations. All results presented in this text have been shown to be statistically significant.

Collecting and analyzing observations can lead to important conclusions based on a type of logic called **inductive reasoning**. Through induction, we derive generalizations from a large number of specific observations. The generalization "All organisms are made of cells" was based on two centuries of microscopic observations made by biologists examining cells in diverse biological specimens. Careful observations and data analyses, along with the generalizations reached by induction, are fundamental to our understanding of nature.

Forming and Testing Hypotheses

Our innate curiosity often stimulates us to pose questions about the natural basis for the phenomena we observe in the world. What *caused* the different chimpanzee behaviors observed in the wild? What *explains* the variation in coat color among the mice of a single species, shown in Figures 1 and 2? In science, answering such questions usually involves forming and testing logical explanations—that is, hypotheses.

In science, a **hypothesis** is an explanation, based on observations and assumptions, that leads to a testable prediction. Said another way, a hypothesis is an explanation on trial. The hypothesis is usually a rational accounting for a set of observations, based on the available data and guided by inductive reasoning. A scientific hypothesis must lead to predictions that can be tested by making additional observations or by performing experiments. An **experiment** is a scientific test, often carried out under controlled conditions.

We all make observations and develop questions and hypotheses in solving everyday problems. Let's say, for example, that your desk lamp is plugged in and turned on but the bulb isn't lit. That's an observation. The question is obvious: Why doesn't the lamp work? Two reasonable hypotheses based on your experience are that (1) the bulb is burnt out or (2) the lamp is broken. Each of these hypotheses leads to predictions you can test with experiments. For example, the burnt-out bulb hypothesis predicts that replacing the bulb will fix the problem. Figuring things out in this way by trial and error is a hypothesis-based approach.

Deductive Reasoning

A type of logic called deduction is also built into the use of hypotheses in science. While induction entails reasoning from a set of specific observations to reach a general conclusion, **deductive reasoning** involves logic that flows in the opposite direction, from the general to the specific. From general premises, we extrapolate to the specific results we should expect if the premises are true. In the scientific process, deductions usually take the form of predictions of results that will be found if a particular hypothesis (premise) is correct. We then test the hypothesis by carrying out experiments or observations to see whether or not the results are as predicted. This deductive testing takes the form of "*If . . . then*" logic. In the case of the desk lamp example: *If* the burnt-out bulb hypothesis is correct, *then* the lamp should work when you replace the bulb with a new one.

We can use the desk lamp example to illustrate two other key points about the use of hypotheses in science. First, one can always devise additional hypotheses to explain a set of observations. For instance, another of the many possible alternative hypotheses to explain our dead desk lamp is that the wall outlet is faulty. Although you could design an experiment to test this hypothesis, we can never test all possible explanations. Second, we can never *prove* that a hypothesis is true. The burnt-out bulb hypothesis is the most likely explanation, but testing supports that hypothesis *not* by proving that it is correct, but rather by not finding that it is false. For example, if replacing the bulb fixed the desk lamp, it might have been because the bulb we replaced was good but not screwed in properly.

Although a hypothesis can never be proved beyond the shadow of a doubt, testing it in various ways can significantly increase our confidence in its validity. Often, rounds of hypothesis formulation and testing lead to a scientific consensus—the shared conclusion of many scientists that a particular hypothesis explains the known data well and stands up to experimental testing.

Questions That Can and Cannot Be Addressed by Science

Scientific inquiry is a powerful way to learn about nature, but there are limitations to the kinds of questions it can answer. A scientific hypothesis must be *testable*; there must be some observation or experiment that could reveal if such an idea is more likely to be true or false. For example, the hypothesis that a burnt-out bulb is the sole reason the lamp doesn't work would not be supported if replacing the bulb with a new one didn't fix the lamp.

Not all hypotheses meet the criteria of science: You wouldn't be able to test the hypothesis that invisible ghosts are fooling with your desk lamp! Because science only deals with natural, testable explanations for natural phenomena, it can neither support nor contradict the invisible ghost hypothesis, nor whether spirits or elves cause storms, rainbows, or illnesses. Such supernatural explanations are simply outside the bounds of science, as are religious matters, which are issues of personal faith.

The Flexibility of the Scientific Process

The way that researchers answer questions about the natural and physical world is often idealized as the scientific method. However, very few scientific studies adhere rigidly to the sequence of steps that are typically used to describe this approach. For example, a scientist may start to design an experiment, but then backtrack after realizing that more preliminary observations are necessary. In other cases, observations remain too puzzling to prompt well-defined questions until further study provides a new context in which to view those observations. For example, scientists could not unravel the details of how genes encode proteins until *after* the discovery of the structure of DNA (an event that took place in 1953).

A more realistic model of the scientific process is shown in **Figure 19**. The focus of this model, shown in the central circle in the figure, is the forming and testing of hypotheses. This core set of activities is the reason that science does so well in explaining phenomena in the natural world. These activities, however, are shaped by exploration and discovery (upper circle)

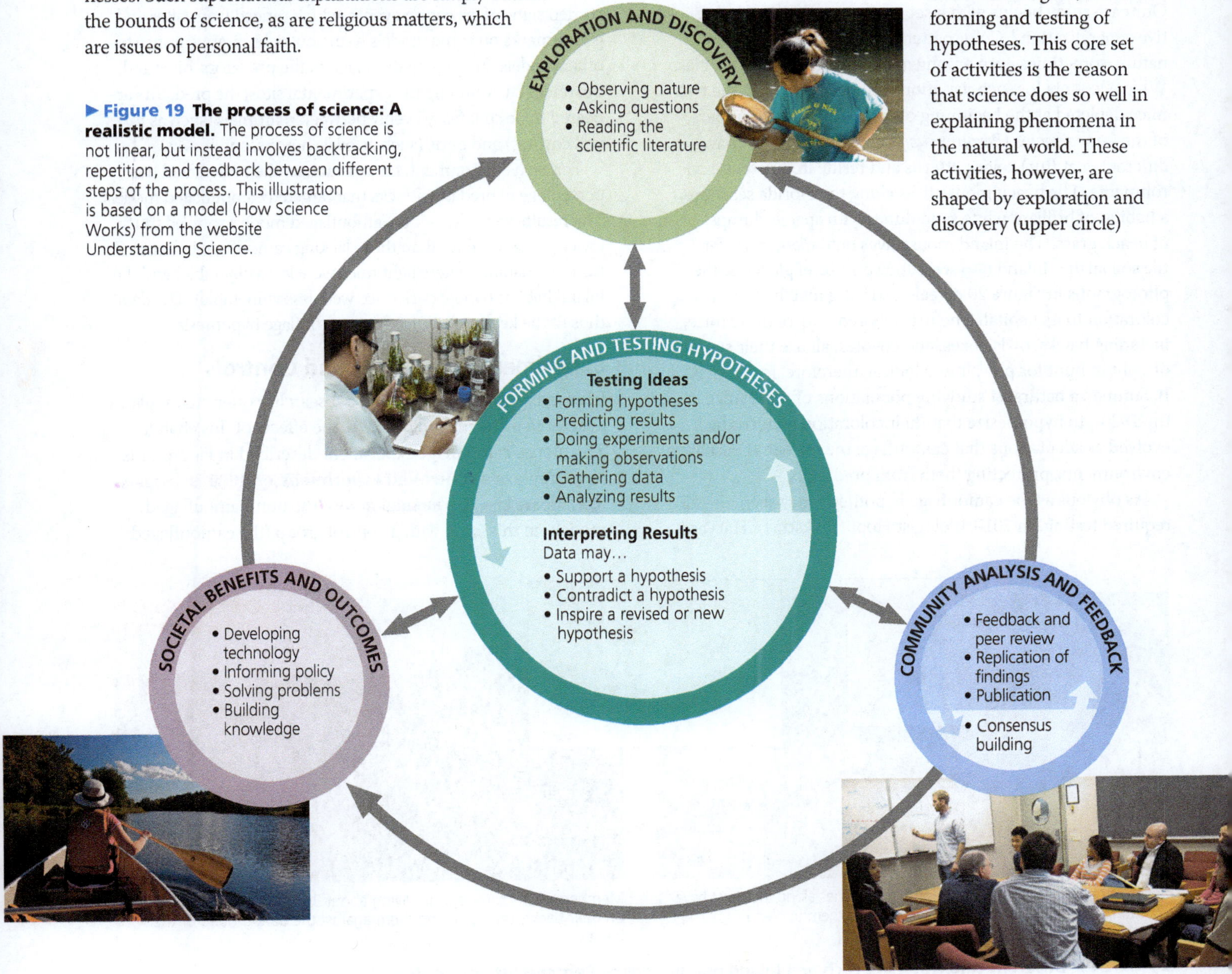

▶ **Figure 19 The process of science: A realistic model.** The process of science is not linear, but instead involves backtracking, repetition, and feedback between different steps of the process. This illustration is based on a model (How Science Works) from the website Understanding Science.

and influenced by interactions with other scientists and with society more generally (lower circles). For example, the community of scientists influences which hypotheses are tested, how test results are interpreted, and what value is placed on the findings. Similarly, societal needs—such as the push to cure cancer or understand the process of climate change—may help shape what research projects are funded and how extensively the results are discussed.

Now that we have highlighted the key features of scientific inquiry—making observations and forming and testing hypotheses—you should be able to recognize these features in a case study of actual scientific research.

A Case Study in Scientific Inquiry: Investigating Coat Coloration in Mouse Populations

Our case study begins with a set of observations and inductive generalizations. Color patterns of animals vary widely in nature, sometimes even among members of the same species. What accounts for such variation? As you may recall, the two mice depicted at the beginning of this chapter are members of the same species (*Peromyscus polionotus*), but they have different coat (fur) color patterns and reside in different environments. The beach mouse lives along the Florida seashore, a habitat of brilliant white sand dunes with sparse clumps of beach grass. The inland mouse lives on darker, more fertile soil farther inland (**Figure 20**). Even a brief glance at the photographs in Figure 20 reveals a striking match of mouse coloration to its habitat. The natural predators of these mice, including hawks, owls, foxes, and coyotes, all use their sense of sight to hunt for prey. It was logical, therefore, for Francis B. Sumner, a naturalist studying populations of these mice in the 1920s, to hypothesize that their coloration patterns had evolved as adaptations that camouflage the mice in their native environments, protecting them from predation.

As obvious as the camouflage hypothesis may seem, it still required testing. In 2010, biologist Hopi Hoekstra of Harvard University and a group of her students headed to Florida to test the prediction that mice with coloration that did not match their habitat would be preyed on more heavily than the native, well-matched mice. **Figure 21** summarizes this field experiment, introducing a format we will use throughout the text to walk through other examples of biological inquiry.

The researchers built hundreds of models of mice and spray-painted them to resemble either beach or inland mice, so that the models differed only in their color patterns. The researchers placed equal numbers of these model mice randomly in both habitats and left them overnight. The mouse models resembling the native mice in the habitat were the *control* group (for instance, light-colored mouse models in the beach habitat), while the models with the non-native coloration were the *experimental* group (for example, darker models in the beach habitat). The following morning, the team counted and recorded signs of predation events, which ranged from bites and gouge marks on some models to the outright disappearance of other models. Judging by the shape of the predators' bites and the tracks surrounding the experimental sites, the predators appeared to be split fairly evenly between mammals (such as foxes and coyotes) and birds (such as owls, herons, and hawks).

For each environment, the researchers then calculated the percentage of predator attacks that targeted camouflaged models. The results were clear-cut: Camouflaged models showed much lower predation rates than those lacking camouflage in both the dune habitat (where light mice were less vulnerable) and the inland habitat (where dark mice were less vulnerable). The data thus fit the key prediction of the camouflage hypothesis.

Experimental Variables and Controls

In carrying out an experiment, a researcher often manipulates a factor in a system and observes the effects of this change. The mouse camouflage experiment described in Figure 21 is an example of a **controlled experiment**, one that is designed to compare an experimental group (the non-camouflaged models, in this case) with a control group (the camouflaged

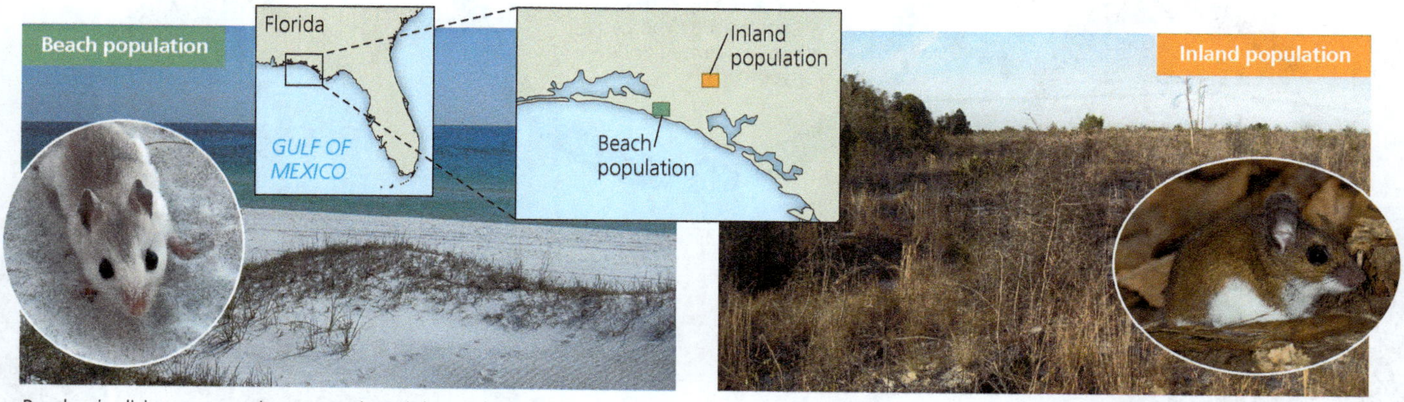

Beach mice living on sparsely vegetated sand dunes along the coast have light tan, dappled fur on their backs that allows them to blend into their surroundings, providing camouflage.

Members of the same species living about 30 km inland have dark fur on their backs, camouflaging them against the dark ground of their habitat.

▲ **Figure 20 Different coloration in beach and inland populations of *Peromyscus polionotus*.**

▼ Figure 21 Inquiry

Does camouflage affect predation rates on two populations of mice?

Experiment Hopi Hoekstra and colleagues tested the hypothesis that coat coloration provides camouflage that protects beach and inland populations of *Peromyscus polionotus* mice from predation in their habitats. The researchers spray-painted mouse models with light or dark color patterns that matched those of the beach and inland mice and placed models with each of the patterns in both habitats. The next morning, they counted damaged or missing models.

Results For each habitat, the researchers calculated the percentage of attacked models that were camouflaged or non-camouflaged. In both cases, the models whose pattern did not match their surroundings suffered much higher "predation" than did the camouflaged models.

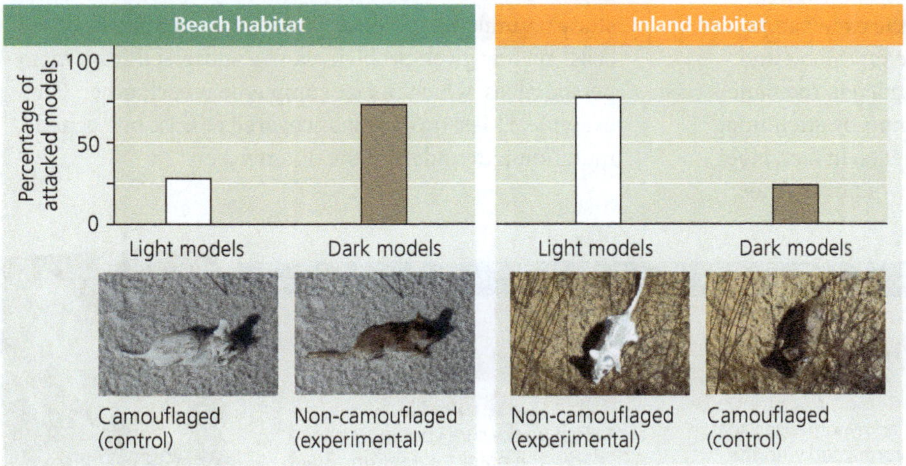

Conclusion The results are consistent with the researchers' prediction that mouse models with camouflage coloration would be preyed on less often than non-camouflaged mouse models. Thus, the experiment supports the camouflage hypothesis.

Data from S. N. Vignieri, J. G. Larson, and H. E. Hoekstra, The selective advantage of crypsis in mice, *Evolution* 64:2153–2158 (2010).

INTERPRET THE DATA *The bars indicate the percentage of the attacked models that were either light or dark. Assume 100 mouse models were attacked in each habitat. For the beach habitat, how many were light models? Dark models? Answer the same questions for the inland habitat.*

models). Both the factor that is manipulated and the factor that is subsequently measured are experimental **variables**—a feature or quantity that varies in an experiment. In our example, the color of the mouse model was the **independent variable**—the factor manipulated by the researchers. The **dependent variable** is the factor being measured that is predicted to be affected by the independent variable; in this case, the researchers measured the amount of predation in response to variation in color of the mouse model. Ideally, the experimental and control groups differ in only one independent variable—in the mouse experiment, color.

Without the control group of camouflaged models, the researchers would not have been able to rule out other factors as causes of the more frequent attacks on the non-camouflaged models—such as different numbers of predators or different temperatures in the various test areas. The clever experimental design left coloration as the only factor that could account for the low predation rate on models camouflaged with respect to the surrounding environment.

A common misconception is that the term *controlled experiment* means that scientists control all features of the experimental environment. But that's impossible in field research and can be very difficult even in a highly regulated laboratory setting. Researchers usually "control" unwanted variables not by *eliminating* them but by *canceling out* their effects using control groups.

Theories in Science

"It's just a theory!" Our everyday use of the term *theory* often implies an untested speculation. But the term *theory* has a different meaning in science. What is a scientific theory, and how is it different from a hypothesis or from mere speculation?

First, a scientific **theory** is much broader in scope than a hypothesis. *This* is a hypothesis: "Coat coloration well-matched to their habitat is an adaptation that protects mice from predators." But *this* is a theory: "Evolutionary adaptations arise by natural selection." This theory proposes that natural selection accounts for an enormous variety of adaptations, of which coat color in mice is but one example.

Second, a theory is general enough to spin off many new, testable hypotheses. For example, the theory of natural selection motivated two researchers at Princeton University, Peter and Rosemary Grant, to test the specific hypothesis that the beaks of Galápagos finches evolve in response to changes in the types of available food.

And third, compared to any one hypothesis, a theory is generally supported by a much greater body of evidence. Those theories that become widely adopted in science (such as the theory of natural selection or the theory of gravity) explain a great diversity of observations and are supported by a vast accumulation of evidence.

In spite of the body of evidence supporting a widely accepted theory, scientists will sometimes modify or even reject theories when new research produces results that don't fit. For example, biologists once lumped bacteria and archaea together as a kingdom of prokaryotes. When new methods for comparing cells and molecules could be used to test such relationships, the evidence led scientists to reject the theory that bacteria and archaea are members of the same kingdom. If there is "truth" in science, it is conditional, based on the weight of available evidence.

Science as a Social Process

The great scientist Sir Isaac Newton once said: "To explain all nature is too difficult a task for any one man or even for any one age. 'Tis much better to do a little with certainty, and leave the rest for others that come after you." Anyone who becomes a scientist, driven by curiosity about nature, is sure to benefit from the rich storehouse of discoveries by others who have come before. In fact, while movies and cartoons sometimes portray scientists as loners working in isolated labs, science is an intensely social activity. Most scientists work in teams, which often include graduate and undergraduate students.

Science is continuously vetted through the expectation that observations and experiments must be repeatable and hypotheses must be testable. Scientists working in the same research field often check one another's claims by attempting to confirm observations or repeat experiments. In fact, Hopi Hoekstra's experiment benefited from the work of another researcher, D. W. Kaufman, four decades earlier. You can study the design of Kaufman's experiment and interpret the results in the **Scientific Skills Exercise**.

If scientific colleagues cannot repeat experimental findings, this failure may reflect an underlying weakness in the original claim, which will then have to be revised. In this sense, science polices itself. Adherence to high professional standards in reporting results is central to the scientific endeavor, since the validity of experimental data is key to designing further inquiry.

Biologists may approach questions from different angles. Some biologists focus on ecosystems, while others study natural phenomena at the level of organisms or cells. Yet any given problem can be addressed from many perspectives, which in fact complement each other. For example, Hoekstra's work uncovered at least one genetic mutation that underlies the differences

Scientific Skills Exercise

Interpreting a Pair of Bar Graphs

How Much Does Camouflage Affect Predation on Mice by Owls with and without Moonlight? D. W. Kaufman hypothesized that the extent to which the coat color of a mouse contrasted with the color of its surroundings would affect the rate of nighttime predation by owls. He also hypothesized that contrast would be affected by the amount of moonlight. In this exercise, you will analyze data from his studies of owl-mouse predation that tested these hypotheses.

How the Experiment Was Done Pairs of mice (*Peromyscus polionotus*) with different coat colors, one light brown and one dark brown, were released simultaneously into an enclosure that contained a hungry owl. The researcher recorded the color of the mouse that was first caught by the owl. If the owl did not catch either mouse within 15 minutes, the test was recorded as a zero. The release trials were repeated multiple times in enclosures with either a dark-colored soil surface or a light-colored soil surface. The presence or absence of moonlight during each assay was recorded.

Data from the Experiment

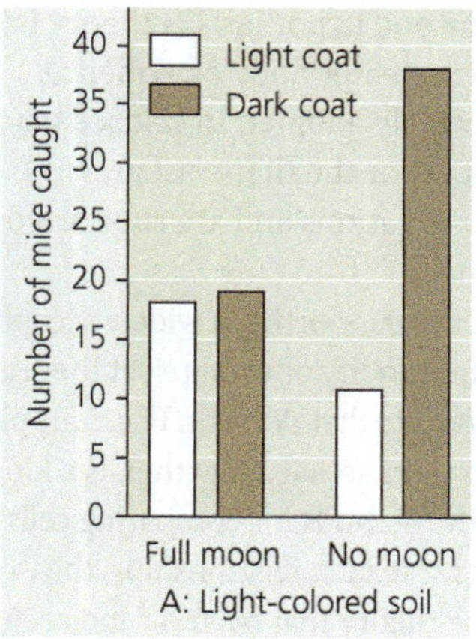

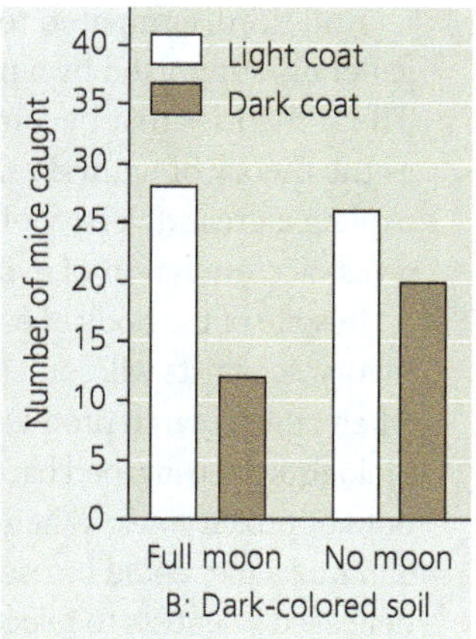

Data from D. W. Kaufman, Adaptive coloration in *Peromyscus polionotus*: Experimental selection by owls, *Journal of Mammalogy* 55:271–283 (1974).

INTERPRET THE DATA

1. First, make sure you understand how the graphs are set up. Graph A shows data from the light-colored soil enclosure and graph B from the dark-colored enclosure, but in all other respects the graphs are the same. (a) There is more than one independent variable in these graphs. What are the independent variables, the variables that were tested by the researcher? Which axis of the graphs has the independent variables? (b) What is the dependent variable, the response to the variables being tested? Which axis of the graphs has the dependent variable? (For additional information about graphs, see the Scientific Skills Review in the Study Area in MasteringBiology.)
2. (a) How many dark brown mice were caught in the light-colored soil enclosure on a moonlit night? (b) How many dark brown mice were caught in the dark-colored soil enclosure on a moonlit night? (c) On a moonlit night, would a dark brown mouse be more likely to escape predation by owls on dark- or light-colored soil? Explain your answer.
3. (a) Is a dark brown mouse on dark-colored soil more likely to escape predation under a full moon or with no moon? (b) What about a light brown mouse on light-colored soil? Explain.
4. (a) Under which conditions would a dark brown mouse be most likely to escape predation at night? (b) A light brown mouse?
5. (a) What combination of independent variables led to the highest predation level in enclosures with light-colored soil? (b) What combination of independent variables led to the highest predation level in enclosures with dark-colored soil?
6. Thinking about your answers to question 5, provide a simple statement describing conditions that are especially deadly for either color of mouse.
7. Combining the data from both graphs, estimate the number of mice caught in moonlight versus no-moonlight conditions. Which condition is optimal for predation by the owl? Explain.

MB A version of this Scientific Skills Exercise can be assigned in MasteringBiology.

between beach and inland mouse coloration. Because the biologists in her lab have different specialties, her research group has been able not only to characterize evolutionary adaptations, but also to define the molecular basis for particular adaptations in the DNA sequence of the mouse genome.

The research community is part of society at large. The relationship of science to society becomes clearer when we add technology to the picture. The goal of **technology** is to *apply* scientific knowledge for some specific purpose. Because scientists put new technology to work in their research, science and technology are interdependent.

In centuries past, many major technological innovations originated along trade routes, where a rich mix of different cultures ignited new ideas. For example, the printing press was invented around 1440 by Johannes Gutenberg, living in what is now Germany. This invention relied on several innovations from China, including paper and ink, and from Iraq, where technology was developed for the mass production of paper. Like technology, science stands to gain much from embracing a diversity of backgrounds and viewpoints among its practitioners.

The scientific community reflects the customs and behaviors of society at large. It is therefore not surprising that until recently, women and certain racial and ethnic groups have faced huge obstacles in their pursuit to become professional scientists. Over the past 50 years, changing attitudes about career choices have increased the proportion of women in biology, and women now constitute roughly half of undergraduate majors and Ph.D. students in the field. The pace of change has been slow at higher levels in the profession, however, and women and many racial and ethnic groups are still significantly underrepresented in many branches of science. This lack of diversity hampers the progress of science. The more voices that are heard at the table, the more robust and productive the scientific conversation will be.

CONCEPT CHECK 3

1. Contrast inductive reasoning with deductive reasoning.
2. What qualitative observation led to the quantitative study outlined in Figure 21?
3. Why is natural selection called a theory?
4. How does science differ from technology?

Go to **MasteringBiology**® for Assignments, the eText, and the Study Area with Animations, Activities, Vocab Self-Quiz, and Practice Tests.

Chapter Review

SUMMARY OF KEY CONCEPTS

VOCAB SELF-QUIZ

goo.gl/gbai8v

CONCEPT 1

The study of life reveals common themes

Theme: Organization

- The hierarchy of life unfolds as follows: biosphere > ecosystem > community > population > organism > organ system > organ > tissue > cell > organelle > molecule > atom. With each step up, new properties emerge (**emergent properties**) as a result of interactions among components at the lower levels.
- Structure and function are correlated at all levels of biological organization. The cell is the lowest level of organization that can perform all activities required for life. Cells are either prokaryotic or eukaryotic. **Eukaryotic cells** have a DNA-containing nucleus and other membrane-enclosed organelles. **Prokaryotic cells** lack such organelles.

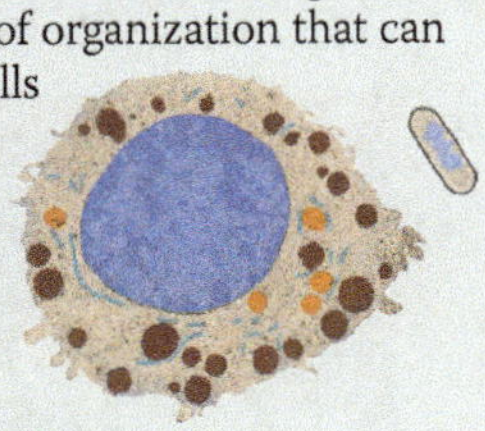

Theme: Information

- Genetic information is encoded in the nucleotide sequences of **DNA**. It is DNA that transmits heritable information from parents to offspring. DNA sequences (called **genes**) program a cell's protein production by being transcribed into mRNA and then translated into specific proteins, a process called **gene expression**. Gene expression also produces RNAs that are not translated into proteins but serve other important functions.

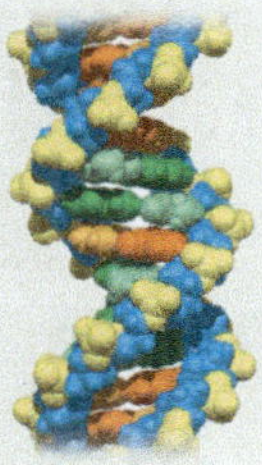

Theme: Energy and Matter

- Energy flows through an ecosystem. All organisms must perform work, which requires energy. Producers convert energy from sunlight to chemical energy, some of which is then passed on to consumers (the rest is lost from the ecosystem as heat). Chemicals cycle between organisms and the environment.

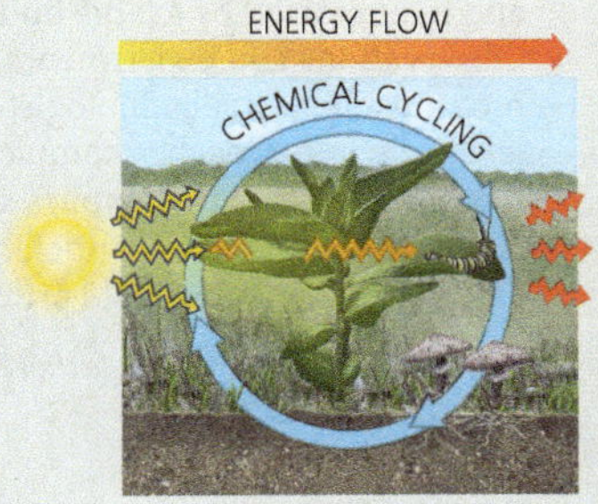

Theme: Interactions

- Organisms interact continuously with physical factors. Plants take up nutrients from the soil and chemicals from the air and use energy from the sun. Interactions among plants, animals, and other organisms affect the participants in varying ways.

Core Theme: Evolution

- Evolution accounts for the unity and diversity of life and also for the match of organisms to their environments.

? *Thinking about the muscles and nerves in your hand, how does the activity of text messaging reflect the five unifying themes of biology described in this chapter?*

CONCEPT 2

The Core Theme: Evolution accounts for the unity and diversity of life

- Biologists classify species according to a system of broader and broader groups. Domain **Bacteria** and domain **Archaea** consist of prokaryotes. Domain **Eukarya**, the eukaryotes, includes various groups of protists as well as plants, fungi, and animals. As diverse as life is, there is also evidence of remarkable unity, which is revealed in the similarities between different kinds of organisms.
- Darwin proposed **natural selection** as the mechanism for evolutionary adaptation of populations to their environments. Natural selection is the evolutionary process that occurs when a population is exposed to environmental factors that consistently cause individuals with certain heritable traits to have greater reproductive success than do individuals with other heritable traits.
- Each species is one twig of a branching tree of life extending back in time through more and more remote ancestral species. All of life is connected through its long evolutionary history.

? *How could natural selection have led to the evolution of adaptations such as camouflaging coat color in beach mice?*

CONCEPT 3

In studying nature, scientists form and test hypotheses

- In scientific **inquiry**, scientists make and record observations (collect **data**) and use **inductive reasoning** to draw a general conclusion, which can be developed into a testable **hypothesis**. **Deductive reasoning** makes predictions that can be used to test hypotheses. Scientific hypotheses must be testable.
- **Controlled experiments**, such as the investigation of coat color in mouse populations, are designed to demonstrate the effect of one **variable** by testing control groups and experimental groups differing in only that one variable.
- A scientific **theory** is broad in scope, generates new hypotheses, and is supported by a large body of evidence.
- Observations and experiments must be repeatable, and hypotheses must be testable. Biologists approach questions at different levels; their approaches complement each other. **Technology** is a method or device that applies scientific knowledge for some specific purpose that affects society as well as for scientific research. Diversity among scientists promotes progress in science.

? *What are the roles of gathering and interpreting data in scientific inquiry?*

TEST YOUR UNDERSTANDING

PRACTICE TEST

goo.gl/CRZjvS

Level 1: Knowledge/Comprehension

1. All the organisms on your campus make up
(A) an ecosystem. (C) a population.
(B) a community. (D) a taxonomic domain.

2. Which of the following best demonstrates the unity among all organisms?
(A) emergent properties (C) DNA structure and function
(B) descent with modification (D) natural selection

3. A controlled experiment is one that
(A) proceeds slowly enough that a scientist can make careful records of the results.
(B) tests experimental and control groups in parallel.
(C) is repeated many times to make sure the results are accurate.
(D) keeps all variables constant.

4. Which of the following statements best distinguishes hypotheses from theories in science?
(A) Theories are hypotheses that have been proved.
(B) Hypotheses are guesses; theories are correct answers.
(C) Hypotheses usually are relatively narrow in scope; theories have broad explanatory power.
(D) Theories are proved true; hypotheses are often contradicted by experimental results.

Level 2: Application/Analysis

5. Which of the following best describes the logic of scientific inquiry?
(A) If I generate a testable hypothesis, tests and observations will support it.
(B) If my prediction is correct, it will lead to a testable hypothesis.
(C) If my observations are accurate, they will support my hypothesis.
(D) If my hypothesis is correct, I can expect certain test results.

6. DRAW IT With rough sketches, draw a biological hierarchy similar to the one in Figure 3 but using a coral reef as the ecosystem, a fish as the organism, its stomach as the organ, and DNA as the molecule. Include all levels in the hierarchy.

Level 3: Synthesis/Evaluation

7. **SCIENTIFIC INQUIRY**
Based on the results of the mouse coloration case study, propose a hypothesis researchers might use to further study the role of predators in the natural selection process.

8. **SCIENTIFIC INQUIRY**
Scientists search the scientific literature by means of electronic databases such as PubMed, a free online database maintained by the National Center for Biotechnology Information. Use PubMed to find the abstract of a scientific article that Hopi Hoekstra published in 2014 or later.

9. **FOCUS ON EVOLUTION**
In a short essay (100–150 words), describe Darwin's view of how natural selection resulted in both unity and diversity of life on Earth. Include in your discussion some of his evidence. (See tips for writing good essays and a suggested grading rubric in the Study Area of MasteringBiology under "Writing Tips and Rubric.")

10. **FOCUS ON INFORMATION**
A typical prokaryotic cell has about 3,000 genes in its DNA, while a human cell has almost 21,000 genes. About 1,000 of these genes are present in both types of cells. (a) Based on your understanding of evolution, explain how such different organisms could have this same subset of 1,000 genes. (b) What sorts of functions might these shared genes have? Justify your choices.

11. **SYNTHESIZE YOUR KNOWLEDGE**

Can you pick out the mossy leaf-tailed gecko lying against the tree trunk in this photo? How is the appearance of the gecko a benefit in terms of survival? Given what you learned about evolution, natural selection, and genetic information in this chapter, describe how the gecko's coloration might have evolved.

Answers

Concept Check 1

1. Examples: A molecule consists of *atoms* bonded together. Each organelle has an orderly arrangement of *molecules*. Photosynthetic plant cells contain *organelles* called chloroplasts. A tissue consists of a group of similar *cells*. Organs such as the heart are constructed from several *tissues*. A complex multicellular organism, such as a plant, has several types of *organs*, such as leaves and roots. A population is a set of *organisms* of the same species. A community consists of *populations* of the various species inhabiting a specific area. An ecosystem consists of a biological *community* along with the nonliving factors important to life, such as air, soil, and water. The biosphere is made up of all of Earth's *ecosystems*. **2.** (a) New properties emerge at successive levels of biological organization: Structure and function are correlated. (b) Life's processes involve the expression and transmission of genetic information. (c) Life requires the transfer and transformation of energy and matter. **3.** Some possible answers: *Organization (emergent properties)*: The ability of a human heart to pump blood requires an intact heart; it is not a capability of any of the heart's tissues or cells working alone. *Organization (structure and function)*: The strong, sharp teeth of a wolf are well suited to grasping and dismembering its prey. *Information*: Human eye color is determined by the combination of genes inherited from the two parents. *Energy and matter*: A plant, such as a grass, absorbs energy from the sun and transforms it into molecules that act as stored fuel. Animals can eat parts of the plant and use the food for energy to carry out their activities. *Interactions*: A mouse eats food, such as nuts or grasses, and deposits some of the food material as wastes (feces and urine). Construction of a nest rearranges the physical environment and may hasten degradation of some of its components. The mouse may also act as food for a predator. *Evolution*: All plants have chloroplasts, indicating their descent from a common ancestor.

Concept Check 2

1. An address pinpoints a location by tracking from broader to narrower categories—a state, city, zip code, street, and building number. This is analogous to the groups-subordinate-to-groups structure of biological taxonomy. **2.** The naturally occurring heritable variation in a population is "edited" by natural selection because individuals with heritable traits better suited to the environment survive and reproduce more successfully than others. Over time, better-suited individuals persist and their percentage in the population increases, while less well-suited individuals become less prevalent—a type of population editing.
3.

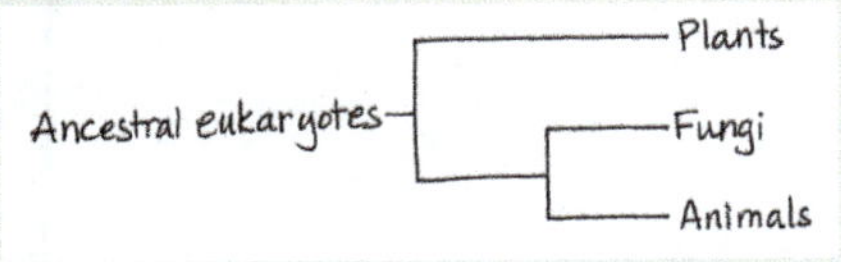

Concept Check 3

1. Inductive reasoning derives generalizations from specific cases; deductive reasoning predicts specific outcomes from general premises. **2.** Mouse coat color matches the environment for both beach and inland populations. **3.** Compared to a hypothesis, a scientific theory is usually more general and substantiated by a much greater amount of evidence. Natural selection is an explanatory idea that applies to all kinds of organisms and is supported by vast amounts of evidence. **4.** Science aims to understand natural phenomena and how they work, while technology involves application of scientific discoveries for a particular purpose or to solve a specific problem.

Summary of Key Concepts Questions

1 Finger movements rely on the coordination of the many structural components of the hand (muscles, nerves, bones, etc.), each of which is composed of elements from lower levels of biological *organization* (cells, molecules). The development of the hand relies on the genetic *information* encoded in chromosomes found in cells throughout the body. To power the finger movements that result in a text message, muscle and nerve cells require chemical *energy* that they transform in powering muscle contraction or in propagating nerve impulses. Texting is in essence communication, an *interaction* that conveys information between organisms, in this case of the same species. Finally, all of the anatomical and physiological features that allow the activity of texting are the outcome of a process of natural selection that resulted in the *evolution* of hands and of the mental facilities for use of language. **2** Ancestors of the beach mouse may have exhibited variations in their coat color. Because of the prevalence of visual predators, the better camouflaged (lighter) mice may have survived longer and been able to produce more offspring. Over time, a higher and higher proportion of individuals in the population would have had the adaptation of lighter fur that acted to camouflage the mouse. **3** Gathering and interpreting data are core activities in the scientific process, and they are affected by, and affect in turn, three other arenas of the scientific process: exploration and discovery, community analysis and feedback, and societal benefits and outcomes.

Test Your Understanding

1. B **2.** C **3.** B **4.** C **5.** D
6. Your figure should show the following: (1) For the biosphere, Earth with an arrow coming out of a tropical ocean; (2) for the ecosystem, a distant view of a coral reef; (3) for the community, a collection of reef animals and algae, with corals, fishes, some seaweed, and any other organisms you can think of; (4) for the population, a group of fish of the same species; (5) for the organism, one fish from your population; (6) for the organ, the fish's stomach; (7) for a tissue, a group of similar cells from the stomach; (8) for a cell, one cell from the tissue, showing its nucleus and a few other organelles; (9) for an organelle, the nucleus, where most of the cell's DNA is located; and (10) for a molecule, a DNA double helix. Your sketches can be very rough!

Credits

Credits are listed in the order of appearance.

Photo Credits

1 J.B. Miller/Florida Park Service; **2** Shawn P. Carey, Migration Productions; **UN1 hummingbird** Jim Zipp/Science Source; **3.1** Leonello Calvetti/Stocktrek Images/Getty Images; **3.2** Terry Donnelly/Alamy; **3.3, 3.4** Floris van Breugel/Nature Picture Library; **3.5** Greg Vaughn/Alamy; **3.6** Pat Burner, Pearson Education; **3.7** Jeremy Burgess/Science Source; **3.8** Texas A & M University; **3.9** Jeremy Burgess/Science Source; **4a** Steve Gschmeissner/Science Source; **4b** A. Barry Dowsett/Science Source; **5** Conly Rieder; **6** Gelpi/fotolia; **7a** Photodisc/Getty Images; **8a.1** Carol Yepes/Moment/Getty Images; **8a.2** Ralf Dahm/Max Planck Institute of Neurobiology; **10** WaterFrame/Alamy; **11** Rod Williams/Nature Picture Library; **12a** Oliver Meckes and Nicole Ottawa/Eye of Science/Science Source; **12b** Eye of Science/Science Source; **12c1** Kunst & Scheidulin/age fotostock; **12c2** daksel/fotolia; **12c3** Anup Shah/Nature Picture Library; **12c4** M. I. Walker/Science Source; **13** Dede Randrianarisata/Macalester College; **14a** From: *Origin of Species* by Charles Darwin, 1859. Murray edition; **14b** ARCHIV/Science Source; **15a** zhaoyan/Shutterstock; **15b** Sebastian Knight/Shutterstock; **15c** Volodymyr Goinyk/Shutterstock; **18 inset** Karl Ammann/Corbis; **18** Tim Ridley/Dorling Kindersley; **19 clockwise from top** Martin Shields/Alamy; Maureen Spuhler, Pearson Education; Rolf Hicker Photography/All Canada Photos/Alamy; xPacifica/The Image Bank/Getty Images; **20a** From: Darwin to DNA: The Genetic Basis of Color Adaptations. From *In the Light of Evolution: Essays from the Laboratory and Field*, by J. Losos, Roberts and Co. Photo by Sacha Vignieri; **20a inset** Courtesy of Hopi Hoekstra, Harvard University; **20b** Sacha Vignieri; **20b inset** Shawn P. Carey, Migration Productions; **21** From: The selective advantage of cryptic coloration in mice. S. N. Vignieri, J. Larson, and H. E. Hoekstra. 2010. *Evolution* 64:2153–2158. Fig. 1; **1 SSE** Imagebroker/FLPA; **Chapter Review eukaryotic cell** Steve Gschmeissner/Science Source; **prokaryotic cell** A. Barry Dowsett/Science Source; **turtle** WaterFrame/Alamy; **SYK** Chris Mattison/Alamy.

Illustration and Text Credits

Chapter 1 **Quote** ("To explain all . . .") Sir Isaac Newton, as quoted in Richard S. Westfall, *The Life of Isaac Newton* (1994), 256; **21** Data from S. N. Vignieri, J. G. Larson, and H. E. Hoekstra, The selective advantage of crypsis in mice, *Evolution* 64:2153–2158 (2010); **SSE.2** Data from D. W. Kaufman, Adaptive coloration in *Peromyscus polionotus*: Experimental selection by owls, *Journal of Mammalogy* 55:271–283 (1974).

Glossary

Archaea (ar′-kē′-uh) One of two prokaryotic domains, the other being Bacteria.

Bacteria One of two prokaryotic domains, the other being Archaea.

bioinformatics The use of computers, software, and mathematical models to process and integrate biological information from large data sets.

biology The scientific study of life.

climate change A directional change in temperature, precipitation, or other aspect of the global climate that lasts for three decades or more.

controlled experiment An experiment designed to compare an experimental group with a control group; ideally, the two groups differ only in the factor being tested.

data Recorded observations.

deductive reasoning A type of logic in which specific results are predicted from a general premise.

DNA (deoxyribonucleic acid) (dē-ok′-sē-rī′-bō-nū-klā′-ik) A nucleic acid molecule, usually a double-stranded helix, in which each polynucleotide strand consists of nucleotide monomers with a deoxyribose sugar and the nitrogenous bases adenine (A), cytosine (C), guanine (G), and thymine (T); capable of being replicated and determining the inherited structure of a cell's proteins.

emergent properties New properties that arise with each step upward in the hierarchy of life, owing to the arrangement and interactions of parts as complexity increases.

Eukarya (yū-kar′-ē-uh) The domain that includes all eukaryotic organisms.

eukaryotic cell (yū′-ker-ē-ot′-ik) A type of cell with a membrane-enclosed nucleus and membrane-enclosed organelles. Organisms with eukaryotic cells (protists, plants, fungi, and animals) are called eukaryotes.

experiment A scientific test. Often carried out under controlled conditions that involve manipulating one factor in a system in order to see the effects of changing that factor.

gene A discrete unit of hereditary information consisting of a specific nucleotide sequence in DNA (or RNA, in some viruses).

genome (jē′-nōm) The genetic material of an organism or virus; the complete complement of an organism's or virus's genes along with its noncoding nucleic acid sequences.

genomics (juh-nō′-miks) The study of whole sets of genes and their interactions within a species, as well as genome comparisons between species.

hypothesis (hī-poth′-uh-sis) A testable explanation for a set of observations based on the available data and guided by inductive reasoning. A hypothesis is narrower in scope than a theory.

inductive reasoning A type of logic in which generalizations are based on a large number of specific observations.

inquiry The search for information and explanation, often focusing on specific questions.

proteome The entire set of proteins expressed by a given cell, tissue, or organism.

science An approach to understanding the natural world.

technology The application of scientific knowledge for a specific purpose, often involving industry or commerce but also including uses in basic research.

theory An explanation that is broader in scope than a hypothesis, generates new hypotheses, and is supported by a large body of evidence.

variable A factor that varies during an experiment.

The Chemical Context of Life

From Chapter 2 of *Campbell Biology in Focus*, Second Edition. Lisa A. Urry, Michael L. Cain, Steven A. Wasserman, Peter V. Minorsky, and Jane B. Reece.

The Chemical Context of Life

KEY CONCEPTS

1. Matter consists of chemical elements in pure form and in combinations called compounds
2. An element's properties depend on the structure of its atoms
3. The formation and function of molecules depend on chemical bonding between atoms
4. Chemical reactions make and break chemical bonds
5. Hydrogen bonding gives water properties that help make life possible on Earth

▲ **Figure 1 What weapon are these wood ants shooting into the air?**

A Chemical Connection to Biology

Like other animals, ants have mechanisms that defend them from attack. Wood ants live in colonies of hundreds or thousands, and the colony as a whole has a particularly effective way of dealing with enemies. When threatened from above, the ants shoot volleys of formic acid into the air from their abdomens, and the acid bombards the potential predator, such as a hungry bird **(Figure 1)**. Formic acid is produced by many species of ants and got its name from the Latin word for ant, *formica*. In quite a few ant species, the formic acid isn't shot out, but probably serves as a disinfectant that protects the ants against microbial parasites. Scientists have long known that chemicals play a major role in insect communication, the attraction of mates, and defense against predators.

Research on ants and other insects is a good example of how relevant chemistry is to the study of life. Unlike college courses, nature is not neatly packaged into individual sciences—biology, chemistry, physics, and so forth. Biologists specialize in the study of life, but organisms and their environments are natural systems to which the concepts of chemistry and physics apply. Biology is multidisciplinary.

We'll start with some basic concepts of chemistry that apply to the study of life. Somewhere in the transition from molecules to cells, we will cross the blurry boundary between nonlife and life. This chapter focuses on the chemical components that make up all matter, with a final section on the substance that supports all of life—water.

CONCEPT 1

Matter consists of chemical elements in pure form and in combinations called compounds

Organisms are composed of **matter**, which is anything that takes up space and has mass. Matter exists in many forms. Rocks, metals, oils, gases, and living organisms are a few examples of what seems to be an endless assortment of matter.

Elements and Compounds

Matter is made up of elements. An **element** is a substance that cannot be broken down to other substances by chemical reactions. Today, chemists recognize 92 elements occurring in nature; gold, copper, carbon, and oxygen are examples. Each element has a symbol, usually the first letter or two of its name. Some symbols are derived from Latin or German; for instance, the symbol for sodium is Na, from the Latin word *natrium*.

A **compound** is a substance consisting of two or more different elements combined in a fixed ratio. Table salt, for example, is sodium chloride (NaCl), a compound composed of the elements sodium (Na) and chlorine (Cl) in a 1:1 ratio. Pure sodium is a metal, and pure chlorine is a poisonous gas. When chemically combined, however, sodium and chlorine form an edible compound. Water (H_2O), another compound, consists of the elements hydrogen (H) and oxygen (O) in a 2:1 ratio. These are simple examples of organized matter having *emergent properties*: A compound has chemical and physical characteristics different from those of its constituent elements (**Figure 2**).

The Elements of Life

Of the 92 natural elements, about 20–25% are **essential elements** that an organism needs to live a healthy life and reproduce. The essential elements are similar among organisms, but there is some variation—for example, humans need 25 elements, but plants need only 17.

Relative amounts of all the elements in the human body are listed in **Table 1**. Just four elements—oxygen (O), carbon (C), hydrogen (H), and nitrogen (N)—make up approximately 96% of living matter. Calcium (Ca), phosphorus (P), potassium (K), sulfur (S), and a few other elements account for most of the remaining 4% or so of an organism's mass. **Trace elements** are required by an organism in only minute quantities. Some trace elements, such as iron (Fe), are needed by all forms of life; others are required only by certain species. For example, in vertebrates (animals with backbones), the element iodine (I) is an essential ingredient of a hormone produced by the thyroid gland. A daily intake of only 0.15 milligram (mg) of iodine is adequate for normal activity of the human thyroid. An iodine deficiency in the diet causes the thyroid gland to grow to abnormal size, a condition called goiter. Consuming seafood or iodized salt reduces the incidence of goiter.

▲ **Figure 2 The emergent properties of a compound.** The metal sodium combines with the poisonous gas chlorine, forming the edible compound sodium chloride, or table salt.

Table 1 Elements in the Human Body

Element	Symbol	Percentage of Body Mass (including water)	
Oxygen	O	65.0%	96.3%
Carbon	C	18.5%	
Hydrogen	H	9.5%	
Nitrogen	N	3.3%	
Calcium	Ca	1.5%	3.7%
Phosphorus	P	1.0%	
Potassium	K	0.4%	
Sulfur	S	0.3%	
Sodium	Na	0.2%	
Chlorine	Cl	0.2%	
Magnesium	Mg	0.1%	

Trace elements (less than 0.01% of mass): Boron (B), chromium (Cr), cobalt (Co), copper (Cu), fluorine (F), iodine (I), iron (Fe), manganese (Mn), molybdenum (Mo), selenium (Se), silicon (Si), tin (Sn), vanadium (V), zinc (Zn)

INTERPRET THE DATA *Given the makeup of the human body, what compound do you think accounts for the high percentage of oxygen?*

Evolution of Tolerance to Toxic Elements

EVOLUTION Some naturally occurring elements are toxic to organisms. In humans, for instance, the element arsenic has been linked to numerous diseases and can be lethal. Some species, however, have become adapted to environments containing elements that are usually toxic. For example, sunflower plants can take up lead, zinc, and other heavy metals in concentrations that would kill most organisms. This capability enabled sunflowers to be used to detoxify contaminated soils after Hurricane Katrina. Presumably, variants of ancestral sunflower species were able to grow in soils with heavy metals, and subsequent natural selection resulted in their survival and reproduction.

CONCEPT CHECK 1

1. Is a trace element an essential element? Explain.
2. **WHAT IF?** In humans, iron is a trace element required for the proper functioning of hemoglobin, the molecule that carries oxygen in red blood cells. What might be the effects of an iron deficiency?

CONCEPT 2

An element's properties depend on the structure of its atoms

Each element consists of a certain type of atom that is different from the atoms of any other element. An **atom** is the smallest unit of matter that still retains the properties of an element. Atoms are so small that it would take about a million of them to stretch across the period at the end of this sentence. We symbolize atoms with the same abbreviation used for the element that is made up of those atoms. For example, the symbol C stands for both the element carbon and a single carbon atom.

Subatomic Particles

Although the atom is the smallest unit having the properties of an element, these tiny bits of matter are composed of even smaller parts, called *subatomic particles*. Using high-energy collisions, physicists have produced more than a hundred types of particles from the atom, but only three kinds of particles are relevant here: **neutrons**, **protons**, and **electrons**. Protons and electrons are electrically charged. Each proton has one unit of positive charge, and each electron has one unit of negative charge. A neutron, as its name implies, is electrically neutral.

Protons and neutrons are packed together tightly in a dense core, or **atomic nucleus**, at the center of an atom; protons give the nucleus a positive charge. The rapidly moving electrons form a "cloud" of negative charge around the nucleus, and it is the attraction between opposite charges that keeps the electrons in the vicinity of the nucleus. **Figure 3** shows two commonly used models of the structure of the helium atom as an example.

The neutron and proton are almost identical in mass, each about 1.7×10^{-24} gram (g). Grams and other conventional units are not very useful for describing the mass of objects that are so minuscule. Thus, for atoms and subatomic particles (and for molecules, too), we use a unit of measurement called the **dalton** (the same as the *atomic mass unit*, or *amu*). Neutrons and protons have masses close to 1 dalton. Because the mass of an electron is only about 1/2,000 that of a neutron or proton, we can ignore electrons when computing the total mass of an atom.

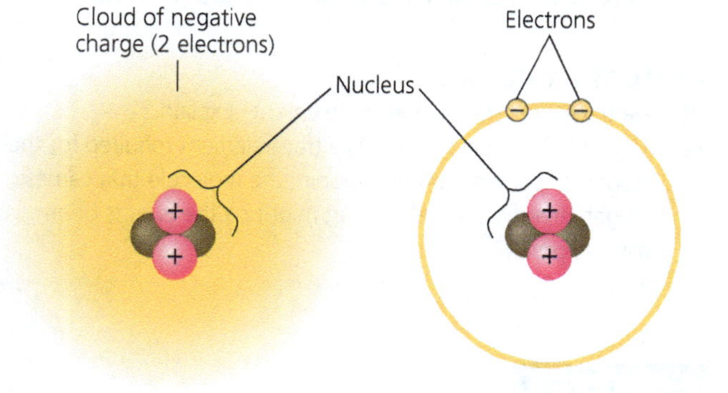

(a) This model shows the two electrons as a cloud of negative charge, a result of their motion around the nucleus.

(b) In this more simplified model, the electrons are shown as two small yellow spheres on a circle around the nucleus.

▲ **Figure 3 Simplified models of a helium (He) atom.** The helium nucleus consists of 2 neutrons (brown) and 2 protons (pink). Two electrons (yellow) exist outside the nucleus. These models are not to scale; they greatly overestimate the size of the nucleus in relation to the electron cloud.

Atomic Number and Atomic Mass

Atoms of the various elements differ in their number of subatomic particles. All atoms of a particular element have the same number of protons in their nuclei. This number of protons, which is unique to that element, is called the **atomic number** and is written as a subscript to the left of the symbol for the element. The abbreviation $_2$He, for example, tells us that an atom of the element helium has 2 protons in its nucleus. Unless otherwise indicated, an atom is neutral in electrical charge, which means that its protons must be balanced by an equal number of electrons. Therefore, the atomic number tells us the number of protons and also the number of electrons in an electrically neutral atom.

We can deduce the number of neutrons from a second quantity, the **mass number**, which is the total number of protons and neutrons in the nucleus of an atom. The mass number is written as a superscript to the left of an element's symbol. For example, we can use this shorthand to write an atom of helium as $^{4}_{2}$He. Because the atomic number indicates how many protons there are, we can determine the number of neutrons by subtracting the atomic number from the mass number: The helium atom $^{4}_{2}$He has 2 neutrons. For sodium (Na):

$^{23}_{11}$Na

Mass number = number of protons + neutrons
= 23 for sodium

Atomic number = number of protons
= number of electrons in a neutral atom
= 11 for sodium

Number of neutrons = mass number − atomic number
= 23 − 11 = 12 for sodium

The simplest atom is hydrogen $^{1}_{1}$H, which has no neutrons; it consists of a single proton with a single electron.

Because the contribution of electrons to mass is negligible, almost all of an atom's mass is concentrated in its nucleus. Neutrons and protons each have a mass very close to 1 dalton, so the mass number is close to, but slightly different from, the total mass of an atom, called its **atomic mass**. For example, the mass number of sodium ($^{23}_{11}$Na) is 23, but its atomic mass is 22.9898 daltons.

Isotopes

All atoms of a given element have the same number of protons, but some atoms have more neutrons than other atoms of the same element and thus have greater mass. These different atomic forms of the same element are called **isotopes** of the element. In nature, an element may occur as a mixture of its isotopes. As an example, the element carbon, which has the atomic number 6, has three naturally occurring isotopes. The most common isotope is carbon-12, $^{12}_{6}$C, which accounts for about 99% of the carbon in nature. The isotope $^{12}_{6}$C has 6 neutrons. Most of the remaining 1% of carbon consists of atoms of the isotope $^{13}_{6}$C, with 7 neutrons. A third, even rarer isotope, $^{14}_{6}C$, has 8 neutrons. Notice that all three isotopes of carbon have 6 protons; otherwise, they would not be carbon. Although the isotopes of an element have slightly different masses, they behave identically in chemical reactions. (For an element with more than one naturally occurring isotope, the atomic mass is an average of those isotopes, weighted by their abundance. Thus carbon has an atomic mass of 12.01 daltons.)

Both ^{12}C and ^{13}C are stable isotopes, meaning that their nuclei do not have a tendency to lose subatomic particles, a process called decay. The isotope ^{14}C, however, is unstable, or radioactive. A **radioactive isotope** is one in which the nucleus decays spontaneously, giving off particles and energy. When the radioactive decay leads to a change in the number of protons, it transforms the atom to an atom of a different element. For example, when an atom of ^{14}C decays, it becomes an atom of nitrogen.

Radioactive isotopes have many useful applications in biology. For example, researchers use measurements of radioactivity in fossils to date these relics of past life. Radioactive isotopes are also useful as tracers to follow atoms through metabolism, the chemical processes of an organism. Cells can use radioactive atoms just as they would use nonradioactive isotopes of the same element. The radioactive isotopes are incorporated into biologically active molecules, which can then be tracked by monitoring the radioactivity.

Radioactive tracers are important diagnostic tools in medicine. For example, certain kidney disorders can be diagnosed by injecting small doses of substances containing radioactive isotopes into the blood and then measuring the amount of tracer excreted in the urine. Radioactive tracers are also used in combination with sophisticated imaging instruments, such as PET scanners, that can monitor the growth and metabolism of cancers in the body **(Figure 4)**.

Although radioactive isotopes are very useful in biological research and medicine, radiation from decaying isotopes also poses a hazard to life by damaging cellular molecules. The severity of this damage depends on the type and amount of radiation an organism absorbs. One of the most serious environmental threats is radioactive fallout from nuclear accidents. The doses of most isotopes used in medical diagnosis, however, are relatively safe.

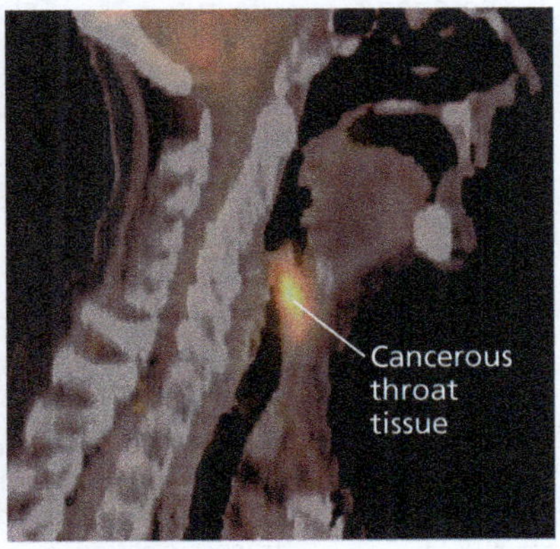

▶ **Figure 4 A PET scan, a medical use for radioactive isotopes.** PET (positron-emission tomography) detects locations of intense chemical activity in the body. The bright yellow spot marks an area with an elevated level of radioactively labeled glucose, which in turn indicates high metabolic activity, a hallmark of cancerous tissue.

The Energy Levels of Electrons

The simplified models of the atom in Figure 3 greatly exaggerate the size of the nucleus relative to that of the whole atom. If an atom of helium were the size of a typical football stadium, the nucleus would be the size of a pencil eraser in the center of the field. Moreover, the electrons would be like two tiny gnats buzzing around the stadium. Atoms are mostly empty space.

When two atoms approach each other during a chemical reaction, their nuclei do not come close enough to interact. Of the three kinds of subatomic particles we have discussed, only electrons are directly involved in the chemical reactions between atoms.

An atom's electrons vary in the amount of energy they possess. **Energy** is defined as the capacity to cause change—for instance, by doing work. **Potential energy** is the energy that matter possesses because of its location or structure. For example, water in a reservoir on a hill has potential energy because of its altitude. When the gates of the reservoir's dam are opened and the water runs downhill, the energy can be used to do work, such as moving the blades of turbines to generate electricity. Because energy has been expended, the water has less energy at the bottom of the hill than it did in the reservoir. Matter has a natural tendency to move toward the lowest possible state of potential energy; in our example, the water runs downhill. To restore the potential energy of a reservoir, work must be done to elevate the water against gravity.

The electrons of an atom have potential energy due to their distance from the nucleus **(Figure 5)**. The negatively charged

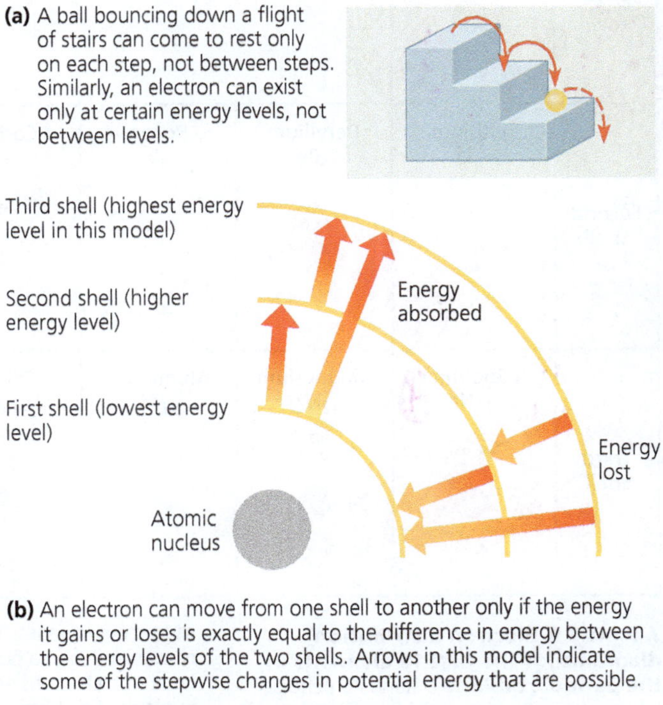

▲ **Figure 5 Energy levels of an atom's electrons.** Electrons exist only at fixed levels of potential energy called electron shells.

electrons are attracted to the positively charged nucleus. It takes work to move a given electron farther away from the nucleus, so the more distant an electron is from the nucleus, the greater its potential energy. Unlike the continuous flow of water downhill, changes in the potential energy of electrons can occur only in steps of fixed amounts. An electron having a certain amount of energy is something like a ball on a staircase (Figure 5a). The ball can have different amounts of potential energy, depending on which step it is on, but it cannot spend much time between the steps. Similarly, an electron's potential energy is determined by its energy level. An electron can exist only at certain energy levels, not between them.

An electron's energy level is correlated with its average distance from the nucleus. Electrons are found in different **electron shells**, each with a characteristic average distance and energy level. In diagrams, shells can be represented by concentric circles (Figure 5b). The first shell is closest to the nucleus, and electrons in this shell have the lowest potential energy. Electrons in the second shell have more energy, and electrons in the third shell even more energy. An electron can move from one shell to another, but only by absorbing or losing an amount of energy equal to the difference in potential energy between its position in the old shell and that in the new shell. When an electron absorbs energy, it moves to a shell farther out from the nucleus. For example, light energy can excite an electron to a higher energy level. (Indeed, this is the first step taken when plants harness the energy of sunlight for photosynthesis, the process that produces food from carbon dioxide and water.) When an electron loses energy, it "falls back" to a shell closer to the nucleus, and the lost energy is usually released to the environment as heat. For example, sunlight excites electrons in the surface of a car to higher energy levels. When the electrons fall back to their original levels, the car's surface heats up. This thermal energy can be transferred to the air or to your hand if you touch the car.

Electron Distribution and Chemical Properties

The chemical behavior of an atom is determined by the distribution of electrons in the atom's electron shells. Beginning with hydrogen, the simplest atom, we can imagine building the atoms of the other elements by adding 1 proton and 1 electron at a time (along with an appropriate number of neutrons). **Figure 6**, a modified version of what is called the *periodic table of the elements*, shows this distribution of electrons for the first 18 elements, from hydrogen ($_1H$) to argon ($_{18}Ar$). The elements are arranged in three rows, or *periods*, corresponding to the number of electron shells in their atoms. The

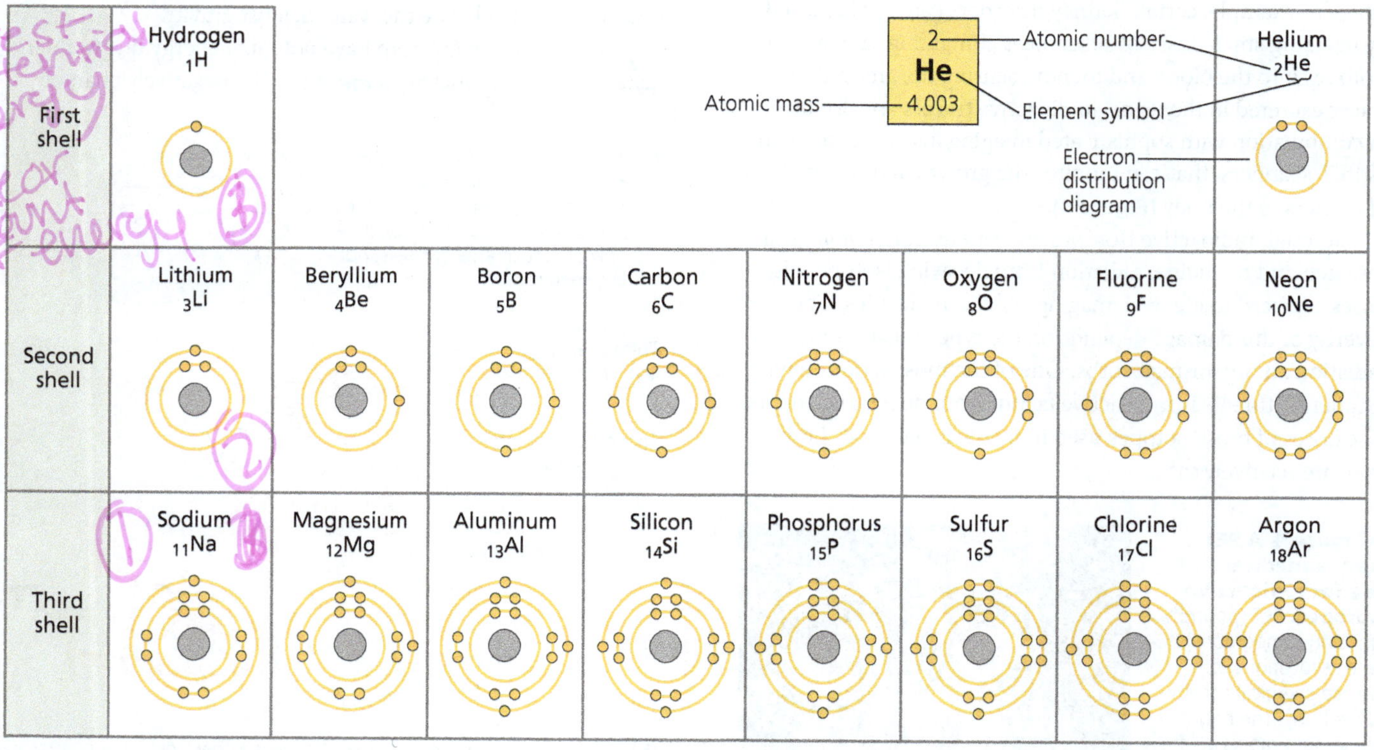

▲ **Figure 6 Electron distribution diagrams for the first 18 elements in the periodic table.** In a standard periodic table, information for each element is presented as shown for helium in the inset. In the diagrams in this table, electrons are represented as yellow dots and electron shells as concentric circles. These diagrams are a convenient way to picture the distribution of an atom's electrons among its electron shells, but these simplified models do not accurately represent the shape of the atom or the location of its electrons. The elements are arranged in rows, each representing the filling of an electron shell. As electrons are added, they occupy the lowest available shell.

? *What is the atomic number of magnesium? How many protons and electrons does it have? How many electron shells? How many valence electrons?*

left-to-right sequence of elements in each row corresponds to the sequential addition of electrons and protons.

Hydrogen's 1 electron and helium's 2 electrons are located in the first shell. Electrons, like all matter, tend to exist in the lowest available state of potential energy. In an atom, this state is in the first shell. However, the first shell can hold no more than 2 electrons; thus, hydrogen and helium are the only elements in the first row of the table. In an atom with more than 2 electrons, the additional electrons must occupy higher shells because the first shell is full. The next element, lithium, has 3 electrons. Two of these electrons fill the first shell, while the third electron occupies the second shell. The second shell holds a maximum of 8 electrons. Neon, at the end of the second row, has 8 electrons in the second shell, giving it a total of 10 electrons.

The chemical behavior of an atom depends mostly on the number of electrons in its *outermost* shell. We call those outer electrons **valence electrons** and the outermost electron shell the **valence shell**. In the case of lithium, there is only 1 valence electron, and the second shell is the valence shell. Atoms with the same number of electrons in their valence shells exhibit similar chemical behavior. For example, fluorine (F) and chlorine (Cl) both have 7 valence electrons, and both form compounds when combined with the element sodium (Na): Sodium fluoride (NaF) is commonly added to toothpaste to prevent tooth decay, and, as described earlier, NaCl is table salt (see Figure 2). An atom with a completed valence shell is unreactive; that is, it will not interact readily with other atoms. At the far right of the periodic table are helium, neon, and argon, the only three elements shown in Figure 6 that have full valence shells. These elements are said to be *inert*, meaning chemically unreactive. All the other atoms in Figure 6 are chemically reactive because they have incomplete valence shells.

Notice that as we "build" the atoms in Figure 6, the first 4 electrons added to the second and third shells are not shown in pairs; only after 4 electrons are present do the next electrons complete pairs. The reactivity of an atom arises from the presence of one or more unpaired electrons in its valence shell. As you will see in the next section, atoms interact in a way that completes their valence shells. When they do so, it is the *unpaired* electrons that are involved.

CONCEPT CHECK 2

1. A nitrogen atom has 7 protons, and the most common isotope of nitrogen has 7 neutrons. A radioactive isotope of nitrogen has 8 neutrons. Write the atomic number and mass number of this radioactive nitrogen as a chemical symbol with a subscript and superscript.
2. How many electrons does fluorine have? How many electron shells? How many electrons are needed to fill the valence shell?
3. **WHAT IF?** In Figure 6, if two or more elements are in the same row, what do they have in common? If two or more elements are in the same column, what do they have in common?

CONCEPT 3

The formation and function of molecules depend on chemical bonding between atoms

Now that we have looked at the structure of atoms, we can move up the hierarchy of organization and see how atoms combine to form molecules and ionic compounds. Atoms with incomplete valence shells can interact with certain other atoms in such a way that each partner completes its valence shell: The atoms either share or transfer valence electrons. These interactions usually result in atoms staying close together, held by attractions called **chemical bonds**. The strongest kinds of chemical bonds are covalent bonds and ionic bonds (when in dry ionic compounds; ionic bonds are weak when in aqueous solutions).

Covalent Bonds

A **covalent bond** is the sharing of a pair of valence electrons by two atoms. For example, let's consider what happens when two hydrogen atoms approach each other. Recall that hydrogen has 1 valence electron in the first shell, but the shell's capacity is 2 electrons. When the two hydrogen atoms come close enough for their electron shells to overlap, they can share their electrons **(Figure 7)**. Each hydrogen atom is now associated with 2 electrons in what amounts to a completed valence shell. Two or more atoms held together by covalent bonds constitute a **molecule**, in this case a hydrogen molecule.

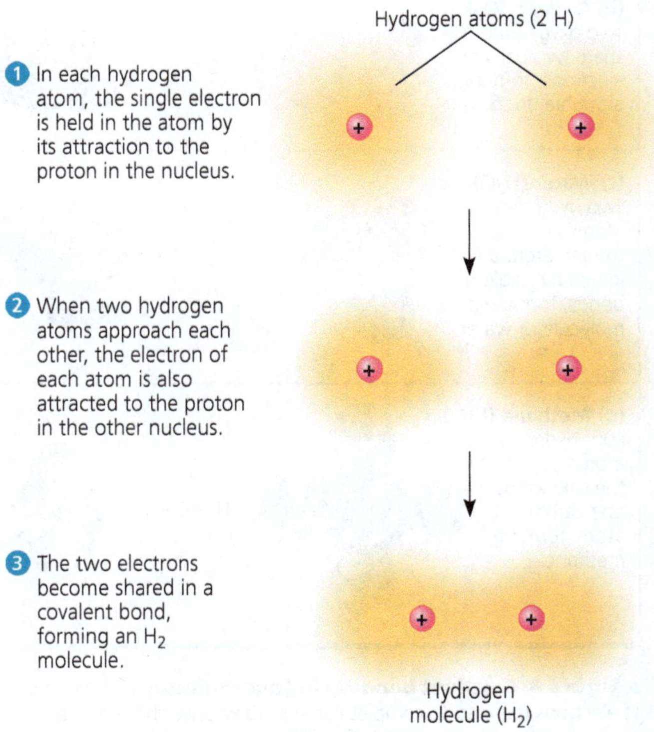

▲ **Figure 7 Formation of a covalent bond.**

Figure 8a shows several ways of representing a hydrogen molecule. Its *molecular formula*, H_2, simply indicates that the molecule consists of two atoms of hydrogen. Electron sharing can be depicted by an electron distribution diagram or by a *structural formula*, H—H, where the line represents a **single bond**, a pair of shared electrons. A space-filling model comes closest to representing the actual shape of the molecule.

Oxygen has 6 electrons in its second electron shell and therefore needs 2 more electrons to complete its valence shell. Two oxygen atoms form a molecule by sharing *two* pairs of valence electrons **(Figure 8b)**. The atoms are thus joined by a **double bond** (O═O).

Each atom that can share valence electrons has a bonding capacity corresponding to the number of covalent bonds the atom can form. When the bonds form, they give the atom a full complement of electrons in the valence shell. The bonding capacity of oxygen, for example, is 2. This bonding capacity is called the atom's **valence** and usually equals the number of electrons required to complete the atom's outermost (valence) shell. See if you can determine the valences of hydrogen, oxygen, nitrogen, and carbon by studying the electron distribution diagrams in Figure 6. You can see that the valence of hydrogen is 1; oxygen, 2; nitrogen, 3; and carbon, 4. (The situation is more complicated for phosphorus, in the third row of the periodic table, which can have a valence of 3 or 5 depending on the combination of single and double bonds it makes.)

The molecules H_2 and O_2 are pure elements rather than compounds because a compound is a combination of two or more *different* elements. Water, with the molecular formula H_2O, is a compound. Two atoms of hydrogen are needed to satisfy the valence of one oxygen atom. **Figure 8c** shows the structure of a water molecule. Water is so important to life that the last section of this chapter, Concept 5, is devoted to its structure and behavior.

Methane, the main component of natural gas, is a compound with the molecular formula CH_4. It takes four hydrogen atoms, each with a valence of 1, to complement one atom of carbon, with its valence of 4 **(Figure 8d)**.

Atoms in a molecule attract shared bonding electrons to varying degrees, depending on the element. The attraction of a particular atom for the electrons of a covalent bond is called its **electronegativity**. The more electronegative an atom is, the more strongly it pulls shared electrons toward itself. In a covalent bond between two atoms of the same element, the electrons are shared equally because the two atoms have the same electronegativity—the tug-of-war is at a standoff. Such a bond is called a **nonpolar covalent bond**. For example, the single bond of H_2 is nonpolar, as is the double bond of O_2. However, when an atom is bonded to a more electronegative atom, the electrons of the bond are not shared equally. This type of bond is called a **polar covalent bond**. Such bonds vary in their polarity, depending on the relative electronegativity of the two atoms. For example, the bonds between the oxygen and hydrogen atoms of a water molecule are quite polar **(Figure 9)**. Oxygen is one of the most electronegative elements, attracting shared electrons much more strongly than hydrogen does. In a covalent bond between oxygen and hydrogen, the electrons spend more time near the oxygen nucleus than they do near the hydrogen nucleus. Because electrons have a negative charge and are pulled toward oxygen in a water molecule, the oxygen atom has a partial negative charge (indicated by the Greek letter δ with a

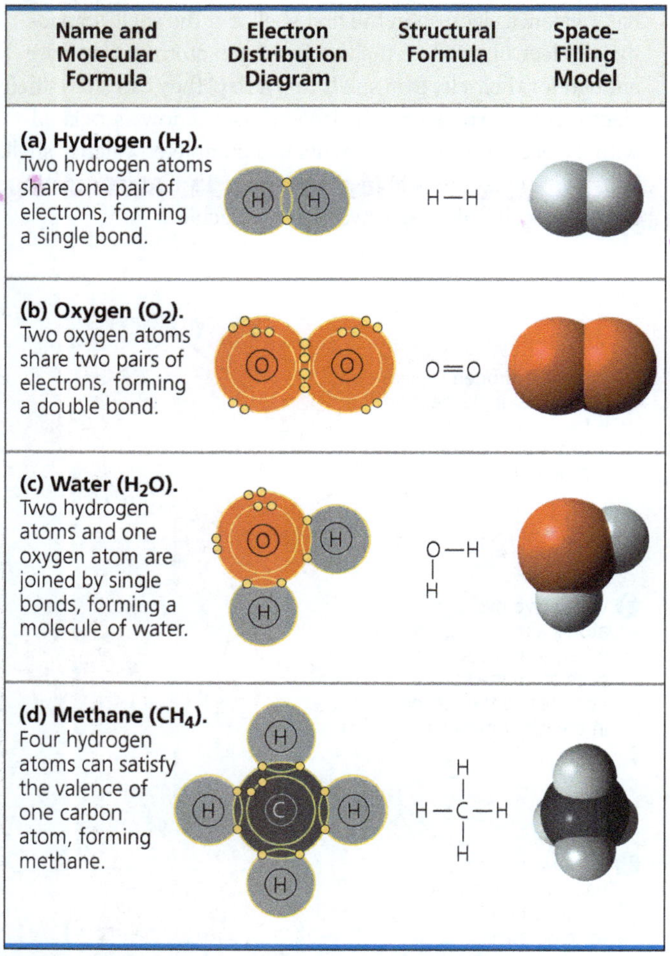

▲ **Figure 8 Covalent bonding in four molecules.** The number of electrons required to complete an atom's valence shell generally determines how many covalent bonds that atom will form. This figure shows several ways of indicating covalent bonds.

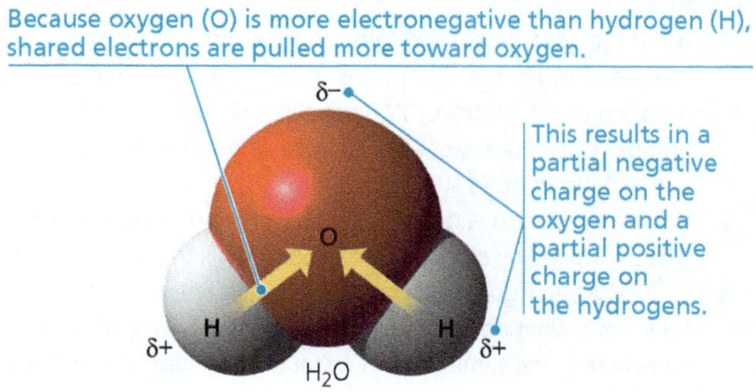

▲ **Figure 9 Polar covalent bonds in a water molecule.**

minus sign, δ–, or "delta minus"), and each hydrogen atom has a partial positive charge (δ+, or "delta plus"). In contrast, the individual bonds of methane (CH_4) are much less polar because the electronegativities of carbon and hydrogen are similar.

Ionic Bonds

In some cases, two atoms are so unequal in their attraction for valence electrons that the more electronegative atom strips an electron completely away from its partner. The two resulting oppositely charged atoms (or molecules) are called **ions**. A positively charged ion is called a **cation**, while a negatively charged ion is called an **anion**. Because of their opposite charges, cations and anions attract each other; this attraction is called an ***ionic bond***. Note that the transfer of an electron is not, by itself, the formation of a bond; rather, it allows a bond to form because it results in two ions of opposite charge. Any two such ions can form an ionic bond—the ions do not need to have acquired their charge by an electron transfer with each other.

This is what happens when an atom of sodium ($_{11}Na$) encounters an atom of chlorine ($_{17}Cl$) **(Figure 10)**. A sodium atom has a total of 11 electrons, with its single valence electron in the third electron shell. A chlorine atom has a total of 17 electrons, with 7 electrons in its valence shell. When these two atoms meet, the lone valence electron of sodium is transferred to the chlorine atom, and both atoms end up with their valence shells complete. (Because sodium no longer has an electron in the third shell, the second shell is now the valence shell.)

The electron transfer between the two atoms moves one unit of negative charge from sodium to chlorine. Sodium, now with 11 protons but only 10 electrons, has a net electrical charge of 1+; the sodium atom has become a cation. Conversely, the chlorine atom, having gained an extra electron, now has 17 protons and 18 electrons, giving it a net electrical charge of 1–; it has become a chloride ion—an anion.

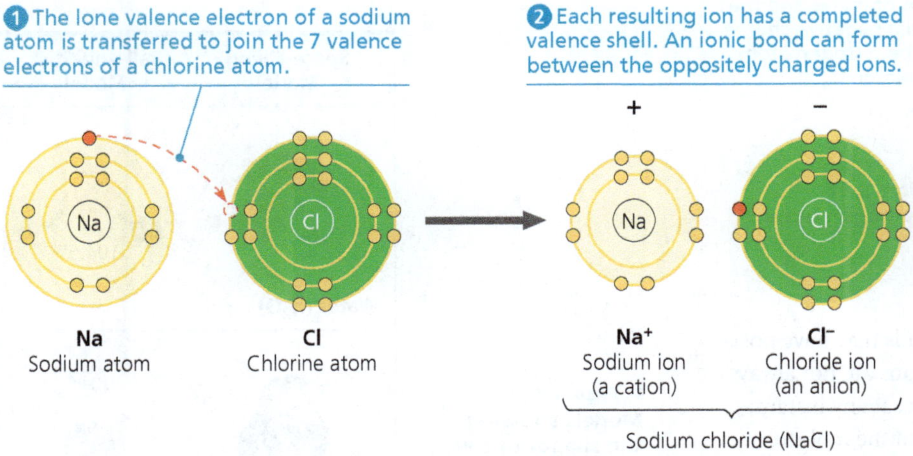

▲ **Figure 10 Electron transfer and ionic bonding.** The attraction between oppositely charged atoms, or ions, is an ionic bond. An ionic bond can form between any two oppositely charged ions, even if they have not been formed by transfer of an electron from one to the other.

▲ **Figure 11 A sodium chloride (NaCl) crystal.** The sodium ions (Na^+) and chloride ions (Cl^-) are held together by ionic bonds. The formula NaCl tells us that the ratio of Na^+ to Cl^- is 1:1.

Compounds formed by ionic bonds are called **ionic compounds**, or **salts**. We know the ionic compound sodium chloride (NaCl) as table salt **(Figure 11)**. Salts are often found in nature as crystals of various sizes and shapes. Each salt crystal is an aggregate of vast numbers of cations and anions bonded by their electrical attraction and arranged in a three-dimensional lattice. Unlike a covalent compound, which consists of molecules having a definite size and number of atoms, an ionic compound does not consist of molecules. The formula for an ionic compound, such as NaCl, indicates only the ratio of elements in a crystal of the salt. "NaCl" by itself is not a molecule.

Not all salts have equal numbers of cations and anions. For example, the ionic compound magnesium chloride ($MgCl_2$) has two chloride ions for each magnesium ion. Magnesium ($_{12}Mg$) must lose 2 outer electrons if the atom is to have a complete valence shell, so it has a tendency to become a cation with a net charge of 2+ (Mg^{2+}). One magnesium cation can therefore form ionic bonds with two chloride anions (Cl^-).

The term *ion* also applies to entire molecules that are electrically charged. In the salt ammonium chloride (NH_4Cl), for instance, the anion is a single chloride ion (Cl^-), but the cation is ammonium (NH_4^+), a nitrogen atom covalently bonded to four hydrogen atoms. The whole ammonium ion has an electrical charge of 1+ because it has given up 1 electron and thus is 1 electron short.

Environment affects the strength of ionic bonds. In a dry salt crystal, the bonds are so strong that it takes a hammer and chisel to break enough of them to crack the crystal in two. If the same salt crystal is dissolved in water, however, the ionic bonds are much weaker because each ion is partially shielded by its interactions with water molecules. Most drugs are manufactured as salts because they are quite stable when dry but can dissociate (come apart) easily in water.

Weak Chemical Bonds

In organisms, most of the strongest chemical bonds are covalent bonds, which link atoms to form a cell's molecules. But weaker bonding within and between molecules is also indispensable in the cell, contributing greatly to the emergent properties of life. Many large biological molecules are held in their functional form by weak bonds. In addition, when two molecules in the cell make contact, they may adhere temporarily by weak bonds. The reversibility of weak bonding can be an advantage: Two molecules can come together, respond to one another in some way, and then separate.

Several types of weak chemical bonds are important in organisms. One is the ionic bond as it exists between ions dissociated in water, which we just discussed. Hydrogen bonds and van der Waals interactions are also crucial to life.

Hydrogen Bonds

Among the various kinds of weak chemical bonds, hydrogen bonds are so central to the chemistry of life that they deserve special attention. When a hydrogen atom is covalently bonded to an electronegative atom, the hydrogen atom has a partial positive charge that allows it to be attracted to a different electronegative atom nearby. This noncovalent attraction between a hydrogen and an electronegative atom is called a **hydrogen bond**. In living cells, the electronegative partners are usually oxygen or nitrogen atoms. Refer to **Figure 12** to examine the simple case of hydrogen bonding between water (H_2O) and ammonia (NH_3).

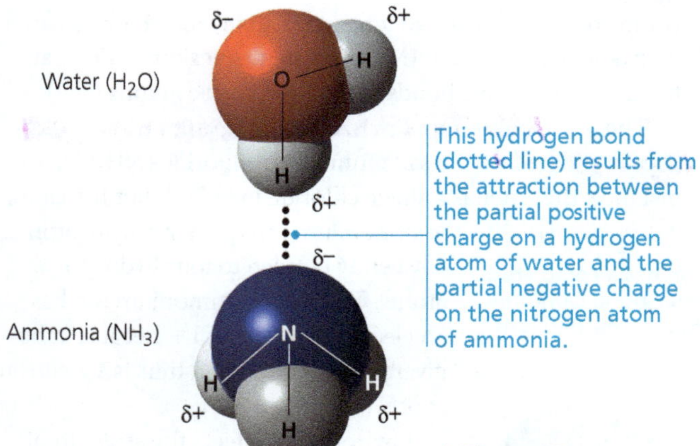

▲ **Figure 12**
A hydrogen bond.

Van der Waals Interactions

Even a molecule with nonpolar covalent bonds may have positively and negatively charged regions. Electrons are not always symmetrically distributed in such a molecule; at any instant, they may accumulate by chance in one part of the molecule or another. The results are ever-changing regions of positive and negative charge that enable all atoms and molecules to stick to one another. These **van der Waals interactions** are individually weak and occur only when atoms and molecules are very close together. When many such interactions occur simultaneously, however, they can be powerful: Van der Waals interactions allow the gecko lizard, shown here, to walk straight up a wall! A gecko toe has hundreds of thousands of tiny hairs with multiple projections on each, which help to maximize surface contact with the wall. The van der Waals interactions between the molecules of the foot and those of the wall's surface are so numerous that despite their individual weakness, together they can support the gecko's body weight. This discovery has inspired development of an artificial adhesive called Geckskin: A patch the size of an index card can hold a 700-pound weight to a wall!

Van der Waals interactions, hydrogen bonds, ionic bonds in water, and other weak bonds may form not only between molecules but also between parts of a large molecule, such as a protein. The cumulative effect of weak bonds is to reinforce the three-dimensional shape of the molecule.

Molecular Shape and Function

A molecule has a characteristic size and shape, which are key to its function in the living cell. A molecule consisting of two atoms, such as H_2 or O_2, is always linear, but most molecules with more than two atoms have more complicated shapes. To take a very simple example, a water molecule (H_2O) is shaped roughly like a V, with its two covalent bonds spread apart at an angle of 104.5° **(Figure 13)**. A methane molecule (CH_4) has a geometric shape called a tetrahedron, a pyramid with a triangular base. The carbon nucleus is inside, at the center, with its four covalent bonds radiating to hydrogen nuclei at the

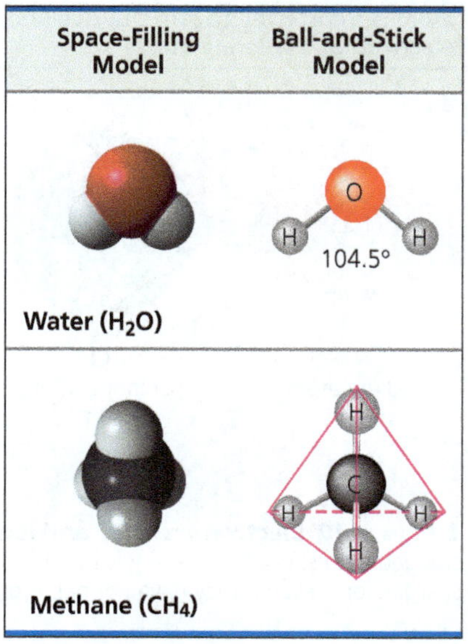

► **Figure 13**
Models showing the shapes of two small molecules.
Each of the molecules, water and methane, is represented in two different ways.

corners of the tetrahedron. Larger molecules containing multiple carbon atoms, including many of the molecules that make up living matter, have more complex overall shapes. However, the tetrahedral shape of a carbon atom bonded to four other atoms is often a repeating motif within such molecules.

Molecular shape is crucial in biology: It determines how biological molecules recognize and respond to one another with specificity. Biological molecules often bind temporarily to each other by forming weak bonds, but only if their shapes are complementary. Consider the effects of opiates (drugs derived from opium); morphine and heroin are two examples. Opiates relieve pain and alter mood by weakly binding to specific receptor molecules on the surfaces of brain cells. Why would brain cells carry receptors for opiates, compounds that are not made by the body? In 1975, the discovery of endorphins answered this question. Endorphins are signaling molecules made by the pituitary gland that bind to the receptors, relieving pain and producing euphoria during times of stress, such as intense exercise. Opiates have shapes similar to endorphins and mimic them by binding to endorphin receptors in the brain. That is why opiates and endorphins have similar effects **(Figure 14)**.

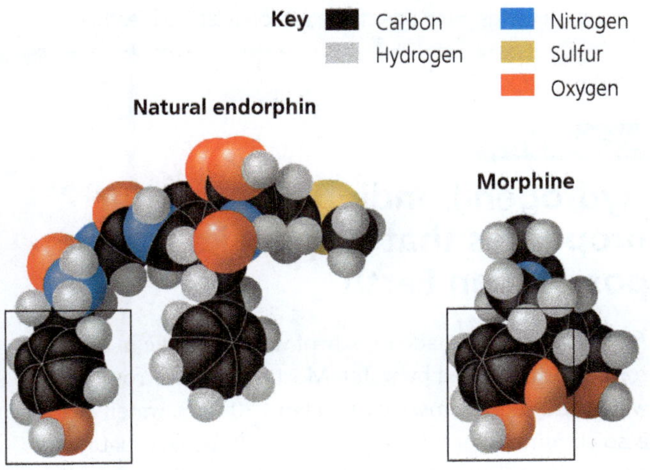

(a) Structures of endorphin and morphine. The boxed portion of the endorphin molecule (left) binds to receptor molecules on target cells in the brain. The boxed portion of the morphine molecule (right) is a close match.

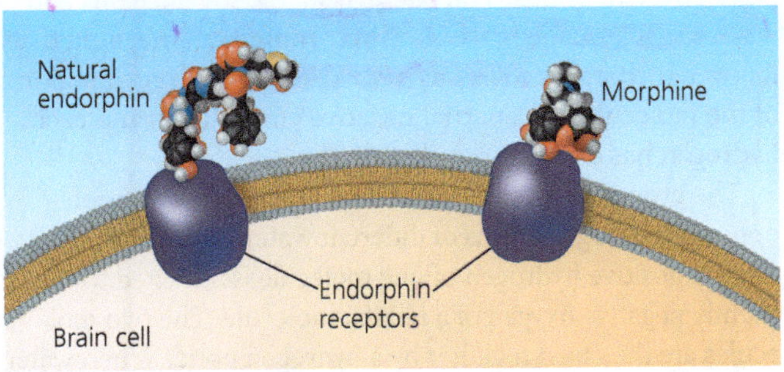

(b) Binding to endorphin receptors. Both endorphin and morphine can bind to endorphin receptors on the surface of a brain cell.

▲ **Figure 14 A molecular mimic.** Morphine affects pain perception and emotional state by mimicking the brain's natural endorphins.

The role of molecular shape in brain chemistry illustrates the match between structure and function in biological organization, one of biology's unifying themes.

CONCEPT CHECK 3

1. Why does the structure H—C=C—H fail to make sense chemically?
2. What holds the atoms together in a crystal of magnesium chloride ($MgCl_2$)?
3. **WHAT IF?** If you were a pharmaceutical researcher, why would you want to learn the three-dimensional shapes of naturally occurring signaling molecules?

CONCEPT 4

Chemical reactions make and break chemical bonds

The making and breaking of chemical bonds, leading to changes in the composition of matter, are called **chemical reactions**. An example is the reaction between hydrogen and oxygen molecules that forms water:

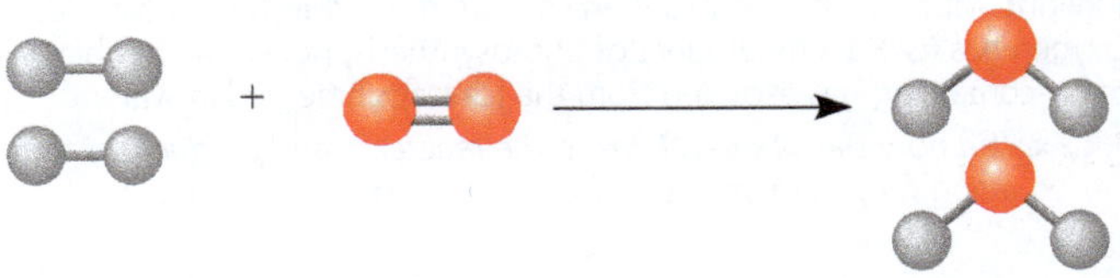

$$2\,H_2 + O_2 \longrightarrow 2\,H_2O$$

Reactants **Reaction** **Products**

This reaction breaks the covalent bonds of H_2 and O_2 and forms the new bonds of H_2O. When we write the equation for a chemical reaction, we use an arrow to indicate the conversion of the starting materials, called the **reactants**, to the resulting materials, or **products**. The coefficients indicate the number of molecules involved; for example, the coefficient 2 in front of H_2 means that the reaction starts with two molecules of hydrogen. Notice that all atoms of the reactants must be accounted for in the products. Matter is conserved in a chemical reaction: Reactions cannot create or destroy atoms but can only rearrange (redistribute) the electrons among them.

Photosynthesis, which takes place within the cells of green plant tissues, is an important biological example of how chemical reactions rearrange matter. Humans and other animals ultimately depend on photosynthesis for food and oxygen, and this process is at the foundation of almost all ecosystems. The following chemical shorthand summarizes the process of photosynthesis:

$$6\,CO_2 + 6\,H_2O \rightarrow C_6H_{12}O_6 + 6\,O_2$$

The raw materials of photosynthesis are carbon dioxide (CO_2), which is taken from the air, and water (H_2O), which is

▲ **Figure 15 Photosynthesis: a solar-powered rearrangement of matter.** *Elodea*, a freshwater plant, produces sugar by rearranging the atoms of carbon dioxide and water in the chemical process known as photosynthesis, which is powered by sunlight. Much of the sugar is then converted to other food molecules. Oxygen gas (O_2) is a by-product of photosynthesis; notice the bubbles of O_2-containing gas escaping from the leaves submerged in water.

? *Explain how this photo relates to the reactants and products in the equation for photosynthesis given in the text.*

absorbed from the soil. Within the plant cells, sunlight powers the conversion of these ingredients to a sugar called glucose ($C_6H_{12}O_6$) and oxygen molecules (O_2), a by-product that the plant releases into the surroundings **(Figure 15)**. Although photosynthesis is actually a sequence of many chemical reactions, we still end up with the same number and types of atoms that we had when we started. Matter has simply been rearranged, with an input of energy provided by sunlight.

All chemical reactions are reversible, with the products of the forward reaction becoming the reactants for the reverse reaction. For example, hydrogen and nitrogen molecules can combine to form ammonia, but ammonia can also decompose to regenerate hydrogen and nitrogen:

$$3\,H_2 + N_2 \rightleftharpoons 2\,NH_3$$

The two opposite-headed arrows indicate that the reaction is reversible.

One of the factors affecting the rate of a reaction is the concentration of reactants. The greater the concentration of reactant molecules, the more frequently they collide with one another and have an opportunity to react and form products. The same holds true for products. As products accumulate, collisions resulting in the reverse reaction become more frequent. Eventually, the forward and reverse reactions occur at the same rate, and the relative concentrations of products and reactants stop changing. The point at which the reactions offset one another exactly is called **chemical equilibrium**. This is a dynamic equilibrium; reactions are still going on, but with no net effect on the concentrations of reactants and products. Equilibrium does *not* mean that the reactants and products are equal in concentration, but only that their concentrations have stabilized at a particular ratio. The reaction involving ammonia reaches equilibrium when ammonia decomposes as rapidly as it forms. In some chemical reactions, the equilibrium point may lie so far to the right that these reactions go essentially to completion; that is, virtually all the reactants are converted to products.

To conclude this chapter, we focus on water, the substance in which all the chemical processes of organisms occur.

CONCEPT CHECK 4

1. Which type of chemical reaction occurs faster at equilibrium, the formation of products from reactants or that of reactants from products?
2. WHAT IF? Write an equation that uses the products of photosynthesis as reactants and the reactants of photosynthesis as products. Add energy as another product. This new equation describes a process that occurs in your cells. Describe this equation in words. How does this equation relate to breathing?

CONCEPT 5

Hydrogen bonding gives water properties that help make life possible on Earth

All organisms are made mostly of water and live in an environment dominated by water. Most cells are surrounded by water, and cells themselves are about 70–95% water. Water is so common that it is easy to overlook the fact that it is an exceptional substance with many extraordinary qualities. We can trace water's unique behavior to the structure and interactions of its molecules. As you saw in Figure 9, the connections between the atoms of a water molecule are polar covalent bonds. The unequal sharing of electrons and water's V-like shape make it a **polar molecule**, meaning that its overall charge is unevenly distributed: The oxygen region of the molecule has a partial negative charge ($\delta-$), and each hydrogen has a partial positive charge ($\delta+$).

The properties of water arise from attractions between oppositely charged atoms of different water molecules: The slightly positive hydrogen of one molecule is attracted to the slightly negative oxygen of a nearby molecule. The two molecules are thus held together by a hydrogen bond. When water is in its liquid form, its hydrogen bonds are very fragile, each only about 1/20 as strong as a covalent bond. The hydrogen bonds form, break, and re-form with great frequency. Each lasts only a few trillionths of a second, but the molecules are

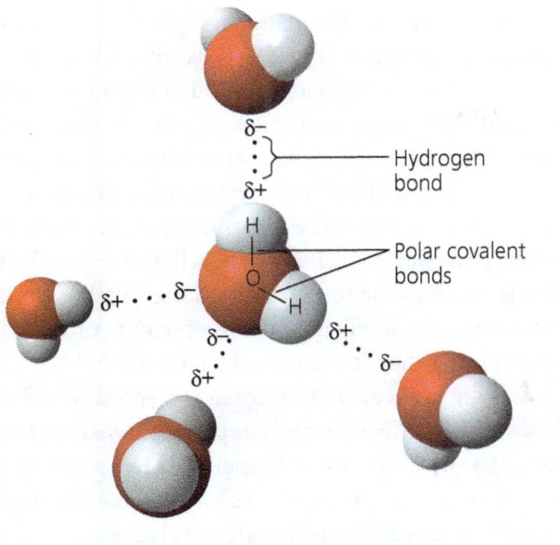

▲ **Figure 16 Hydrogen bonds between water molecules.** The charged regions in a water molecule are due to its polar covalent bonds. Oppositely charged regions of neighboring water molecules are attracted to each other, forming hydrogen bonds. Each molecule can hydrogen-bond to multiple partners, and these associations are constantly changing.

DRAW IT *Draw partial charges on all the atoms of the water molecule on the far left, and draw two more water molecules hydrogen-bonded to it.*

constantly forming new hydrogen bonds with a succession of partners. Therefore, at any instant, most of the water molecules are hydrogen-bonded to their neighbors **(Figure 16)**. The extraordinary properties of water emerge from this hydrogen bonding, which organizes water molecules into a higher level of structural order. We will examine four emergent properties of water that contribute to Earth's suitability as an environment for life: cohesive behavior, ability to moderate temperature, expansion upon freezing, and versatility as a solvent. After that, we'll discuss a critical aspect of water chemistry—acids and bases.

Cohesion of Water Molecules

Water molecules stay close to each other as a result of hydrogen bonding. At any given moment, many of the molecules in liquid water are linked by multiple hydrogen bonds. These linkages make water more structured than most other liquids. Collectively, the hydrogen bonds hold the substance together, a phenomenon called **cohesion**.

Cohesion due to hydrogen bonding contributes to the transport of water and dissolved nutrients against gravity in plants **(Figure 17)**. Water from the roots reaches the leaves through a network of water-conducting cells. As water evaporates from a leaf, hydrogen bonds cause water molecules leaving the veins to tug on molecules farther down, and the upward pull is transmitted through the water-conducting cells all the way to the roots. **Adhesion**, the clinging of one substance to another, also plays a role. Adhesion of water to cell walls by hydrogen bonds helps counter the downward pull of gravity.

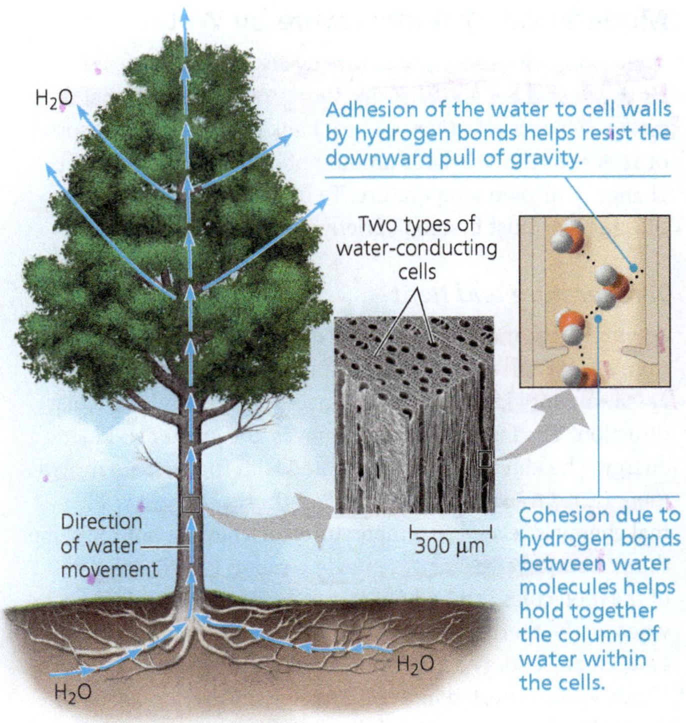

▲ **Figure 17 Water transport in plants.** Evaporation from leaves pulls water upward from the roots through water-conducting cells. Because of the properties of cohesion and adhesion, the tallest trees can transport water more than 100 m upward—approximately one-quarter the height of the Empire State Building in New York City.

Visit the Study Area in **MasteringBiology** for the BioFlix® 3-D Animation on Water Transport in Plants.

water droplets on penny

Related to cohesion is **surface tension**, a measure of how difficult it is to stretch or break the surface of a liquid. The hydrogen bonds in water give it an unusually high surface tension, making it behave as though it were coated with an invisible film. You can observe the surface tension of water by slightly overfilling a drinking glass; the water will stand above the rim. The spider in **Figure 18** takes advantage of the surface tension of water to walk across a pond without breaking the surface.

► **Figure 18 Walking on water.** The high surface tension of water, resulting from the collective strength of its hydrogen bonds, allows this raft spider to walk on the surface of a pond.

Moderation of Temperature by Water

Water moderates air temperature by absorbing heat from air that is warmer and releasing the stored heat to air that is cooler. Water is effective as a heat bank because it can absorb or release a relatively large amount of heat with only a slight change in its own temperature. To understand this capability of water, we must first look briefly at temperature and heat.

Temperature and Heat

Anything that moves has **kinetic energy**, the energy of motion. Atoms and molecules have kinetic energy because they are always moving, although not necessarily in any particular direction. The faster a molecule moves, the greater its kinetic energy. The kinetic energy associated with the random movement of atoms or molecules is called **thermal energy**. Thermal energy is related to temperature, but they are not the same thing. **Temperature** represents the *average* kinetic energy of the molecules in a body of matter, regardless of volume, whereas the thermal energy of a body of matter reflects the *total* kinetic energy and thus depends on the matter's volume. When water is heated in a coffeemaker, the average speed of the molecules increases, and the thermometer records this as a rise in temperature of the liquid. The total amount of thermal energy also increases in this case. Note, however, that although the pot of coffee has a much higher temperature than, say, the water in a swimming pool, the swimming pool contains more thermal energy because of its much greater volume.

Whenever two objects of different temperature are brought together, thermal energy passes from the warmer to the cooler object until the two are the same temperature. Molecules in the cooler object speed up at the expense of the thermal energy of the warmer object. An ice cube cools a drink not by adding coldness to the liquid, but by absorbing thermal energy from the liquid as the ice itself melts. Thermal energy in transfer from one body of matter to another is defined as **heat**.

One convenient unit of heat used in this text is the **calorie (cal)**. A calorie is the amount of heat it takes to raise the temperature of 1 g of water by 1°C. Conversely, a calorie is also the amount of heat that 1 g of water releases when it cools by 1°C. A **kilocalorie (kcal)**, 1,000 cal, is the quantity of heat required to raise the temperature of 1 kilogram (kg) of water by 1°C. (The "calories" on food packages are actually kilocalories.) Another energy unit used in this text is the **joule (J)**. One joule equals 0.239 cal; one calorie equals 4.184 J.

Water's High Specific Heat

The ability of water to stabilize temperature stems from its relatively high specific heat. The **specific heat** of a substance is defined as the amount of heat that must be absorbed or lost for 1 g of that substance to change its temperature by 1°C. We already know water's specific heat because we have defined a calorie as the amount of heat that causes 1 g of water to change its temperature by 1°C. Therefore, the specific heat of water is 1 calorie per gram per degree Celsius, abbreviated as 1 cal/(g • °C). Compared with most other substances, water has an unusually high specific heat. As a result, water will change its temperature less than other liquids when it absorbs or loses a given amount of heat. The reason you can burn your fingers by touching the side of an iron pot on the stove when the water in the pot is still lukewarm is that the specific heat of water is ten times greater than that of iron. In other words, the same amount of heat will raise the temperature of 1 g of the iron much faster than it will raise the temperature of 1 g of the water. Specific heat can be thought of as a measure of how well a substance resists changing its temperature when it absorbs or releases heat. Water resists changing its temperature; when it does change its temperature, it absorbs or loses a relatively large quantity of heat for each degree of change.

We can trace water's high specific heat, like many of its other properties, to hydrogen bonding. Heat must be absorbed in order to break hydrogen bonds; by the same token, heat is released when hydrogen bonds form. A calorie of heat causes a relatively small change in the temperature of water because much of the heat is used to disrupt hydrogen bonds before the water molecules can begin moving faster. And when the temperature of water drops slightly, many additional hydrogen bonds form, releasing a considerable amount of energy in the form of heat.

What is the relevance of water's high specific heat to life on Earth? A large body of water can absorb and store a huge amount of heat from the sun in the daytime and during summer while warming up only a few degrees. At night and during winter, the gradually cooling water can warm the air. This capability of water serves to moderate air temperatures in coastal areas (**Figure 19**). The high specific heat of water also tends to stabilize ocean temperatures, creating a favorable environment for marine life. Thus, because of its high specific heat, the water that covers most of Earth keeps temperature fluctuations on land and in water within limits that permit life. Also, because organisms are made primarily of water, they are better able to resist changes in their own temperature than if they were made of a liquid with a lower specific heat.

▲ **Figure 19 Temperatures for the Pacific Ocean and Southern California on an August day.**

INTERPRET THE DATA *Explain the pattern of temperatures shown in this diagram.*

Evaporative Cooling

Molecules of any liquid stay close together because they are attracted to one another. Molecules moving fast enough to overcome these attractions can depart the liquid and enter the air as a gas (vapor). This transformation from a liquid to a gas is called vaporization, or *evaporation*. Recall that the speed of molecular movement varies and that temperature is the *average* kinetic energy of molecules. Even at low temperatures, the speediest molecules can escape into the air. Some evaporation occurs at any temperature; a glass of water at room temperature, for example, will eventually evaporate completely. If a liquid is heated, the average kinetic energy of molecules increases and the liquid evaporates more rapidly.

Heat of vaporization is the quantity of heat a liquid must absorb for 1 g of it to be converted from the liquid to the gaseous state. For the same reason that water has a high specific heat, it also has a high heat of vaporization relative to most other liquids. To evaporate 1 g of water at 25°C, about 580 cal of heat is needed—nearly double the amount needed to vaporize a gram of alcohol, for example. Water's high heat of vaporization is another emergent property resulting from the strength of its hydrogen bonds, which must be broken before the molecules can exit from the liquid in the form of water vapor.

The high amount of energy required to vaporize water has a wide range of effects. On a global scale, for example, it helps moderate Earth's climate. A considerable amount of solar heat absorbed by tropical seas is consumed during the evaporation of surface water. Then, as moist tropical air circulates poleward, it releases heat as it condenses and forms rain. On an organismal level, water's high heat of vaporization accounts for the severity of steam burns. These burns are caused by the heat energy released when steam condenses into liquid on the skin.

As a liquid evaporates, the surface of the liquid that remains behind cools down (its temperature decreases). This **evaporative cooling** occurs because the "hottest" molecules, those with the greatest kinetic energy, are the ones most likely to leave as gas. It is as if the hundred fastest runners at a college transferred to another school; the average speed of the remaining students would decline.

Evaporative cooling of water contributes to the stability of temperature in lakes and ponds and also provides a mechanism that prevents terrestrial organisms from overheating. For example, evaporation of water from the leaves of a plant helps keep the tissues in the leaves from becoming too warm in the sunlight. Evaporation of sweat from human skin dissipates body heat and helps prevent overheating on a hot day or when excess heat is generated by strenuous activity. High humidity on a hot day increases discomfort because the high concentration of water vapor in the air inhibits the evaporation of sweat from the body.

Floating of Ice on Liquid Water

Water is one of the few substances that are less dense as a solid than as a liquid. In other words, ice floats on liquid water. While other materials contract and become denser when they solidify, water expands. The cause of this exotic behavior is, once again, hydrogen bonding. At temperatures above 4°C, water behaves like other liquids, expanding as it warms and contracting as it cools. As the temperature falls from 4°C to 0°C, water begins to freeze because more and more of its molecules are moving too slowly to break hydrogen bonds. At 0°C, the molecules become locked into a crystalline lattice, each water molecule hydrogen-bonded to four partners **(Figure 20)**. The hydrogen bonds keep the molecules at "arm's length," far enough apart to make ice about 10% less dense than liquid water at 4°C. When ice absorbs enough heat for its temperature to rise above 0°C, hydrogen bonds between molecules are disrupted. As the crystal collapses, the ice melts, and molecules are free to slip closer together. Water reaches its greatest density at 4°C and then begins to expand as the molecules move faster.

The ability of ice to float due to its lower density is an important factor in the suitability of the environment for life. If ice sank, then eventually all ponds, lakes, and even oceans

◀ **Figure 20 Ice: crystalline structure and floating barrier.** In ice, each molecule is hydrogen-bonded to four neighbors in a three-dimensional crystal. Because the crystal is spacious, ice has fewer molecules than an equal volume of liquid water. In other words, ice is less dense than liquid water. Floating ice becomes a barrier that insulates the liquid water below from the colder air. The marine organism shown here is a type of shrimp called krill; it was photographed beneath floating ice in the Southern Ocean near Antarctica.

WHAT IF? *If water did not form hydrogen bonds, what would happen to the shrimp's habitat, shown here?*

would freeze solid, making life as we know it impossible on Earth. During summer, only the upper few inches of the ocean would thaw. Instead, when a deep body of water cools, the floating ice insulates the liquid water below, preventing it from freezing and allowing life to exist under the frozen surface, as shown in the photo in Figure 20.

Water: The Solvent of Life

A sugar cube placed in a glass of water will dissolve. Eventually, the glass will contain a uniform mixture of sugar and water; the concentration of dissolved sugar will be the same everywhere in the mixture. A liquid that is a completely homogeneous mixture of two or more substances is called a **solution**. The dissolving agent of a solution is the **solvent**, and the substance that is dissolved is the **solute**. In this case, water is the solvent and sugar is the solute. An **aqueous solution** is one in which the solute is dissolved in water; water is the solvent.

Water is a very versatile solvent, a quality we can trace to the polarity of the water molecule. Suppose, for example, that a spoonful of table salt, the ionic compound sodium chloride (NaCl), is placed in water **(Figure 21)**. At the surface of each grain, or crystal, of salt, the sodium and chloride ions are exposed to the solvent. These ions and regions of the water molecules are attracted to each other due to their opposite charges. The oxygen regions of the water molecules are negatively

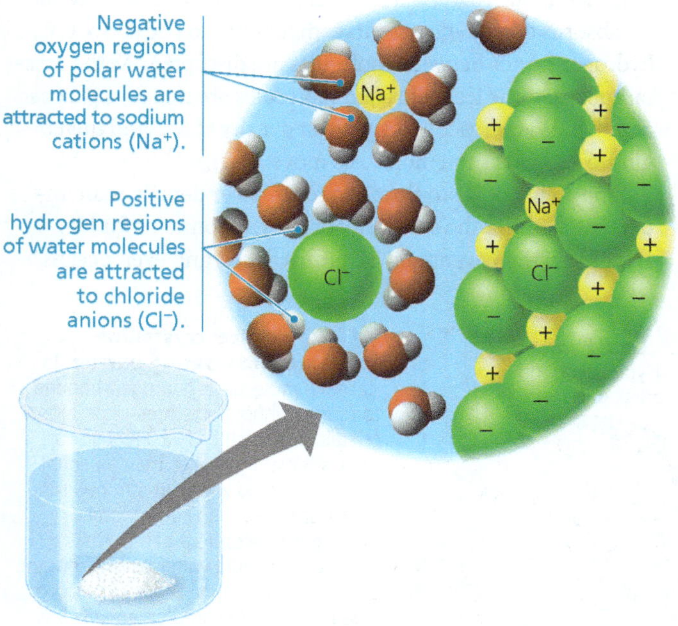

▲ **Figure 21 Table salt dissolving in water.** A sphere of water molecules, called a hydration shell, surrounds each solute ion.

WHAT IF? *What would happen if you heated this solution for a long time?*

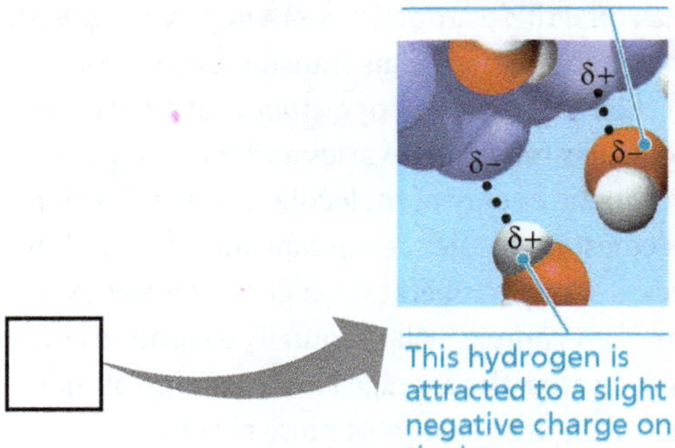

▲ **Figure 22 A water-soluble protein.** Human lysozyme is a protein found in tears and saliva that has antibacterial action. This model shows the lysozyme molecule (purple) in an aqueous environment. Ionic and polar regions on the protein's surface attract the slightly charged regions of water molecules.

charged and are attracted to sodium cations. The hydrogen regions are positively charged and are attracted to chloride anions. As a result, water molecules surround the individual sodium and chloride ions, separating and shielding them from one another. The sphere of water molecules around each dissolved ion is called a **hydration shell**. Working inward from the surface of each salt crystal, water eventually dissolves all the ions. The result is a solution of two solutes, sodium cations and chloride anions, homogeneously mixed with water, the solvent. Other ionic compounds also dissolve in water. Seawater, for instance, contains a great variety of dissolved ions, as do living cells.

A compound does not need to be ionic to dissolve in water; many compounds made up of nonionic polar molecules, such as sugars, are also water-soluble. Such compounds dissolve when water molecules surround each of the solute molecules, forming hydrogen bonds with them. Even molecules as large as proteins can dissolve in water if they have ionic and polar regions on their surface **(Figure 22)**. Many different kinds of polar compounds are dissolved (along with ions) in the water of such biological fluids as blood, the sap of plants, and the liquid within all cells. Water is the solvent of life.

Hydrophilic and Hydrophobic Substances

Any substance that has an affinity for water is said to be **hydrophilic** (from the Greek *hydro*, water, and *philos*, loving). In some cases, substances can be hydrophilic without actually dissolving. For example, some molecules in cells are so large that they do not dissolve. Another example of a hydrophilic substance that does not dissolve is cotton, a plant product. Cotton consists of giant molecules of cellulose, a compound with numerous regions of partial positive and partial negative charges that can form hydrogen bonds with water. Water adheres to the cellulose fibers. Thus, a cotton towel does a great job of drying the body, yet it does not dissolve in the washing machine. Cellulose is also present in the walls of plant cells that conduct water; you read earlier how the adhesion of water to these hydrophilic walls helps water move up the plant against gravity.

There are, of course, substances that do not have an affinity for water. Substances that are nonionic and nonpolar (or otherwise cannot form hydrogen bonds) actually seem to repel water; these substances are said to be **hydrophobic** (from the Greek *phobos*, fearing). An example from the kitchen is vegetable oil, which, as you know, does not mix stably with water-based substances such as vinegar. The hydrophobic behavior of the oil molecules results from a prevalence of relatively nonpolar covalent bonds, in this case bonds between carbon and hydrogen, which share electrons almost equally. Hydrophobic molecules related to oils are major ingredients of cell membranes. (Imagine what would happen to a cell if its membrane dissolved!)

Solute Concentration in Aqueous Solutions

Most of the chemical reactions in organisms involve solutes dissolved in water. To understand such reactions, we must know how many atoms and molecules are involved and be able to calculate the concentration of solutes in an aqueous solution (the number of solute molecules in a volume of solution).

When carrying out experiments, we use mass to calculate the number of molecules. We first calculate the **molecular mass**, which is simply the sum of the masses of all the atoms in a molecule. As an example, let's calculate the molecular mass of table sugar (sucrose), $C_{12}H_{22}O_{11}$, by multiplying the number of atoms by the atomic mass of each element. In round numbers, sucrose has a molecular mass of $(12 \times 12) + (22 \times 1) + (11 \times 16) = 342$ daltons. Because we can't measure out small numbers of molecules, we usually measure substances in units called moles. Just as a dozen always means 12 objects, a **mole (mol)** represents an exact number of objects: 6.02×10^{23}, which is called Avogadro's number. There are 6.02×10^{23} daltons in 1 g. Once we determine the molecular mass of a molecule such as sucrose, we can use the same number (342), but with the unit *gram*, to represent the mass of 6.02×10^{23} molecules of sucrose, or 1 mol of sucrose. To obtain 1 mol of sucrose in the lab, therefore, we weigh out 342 g.

The practical advantage of measuring a quantity of chemicals in moles is that a mole of one substance has exactly the same number of molecules as a mole of any other substance. Measuring in moles makes it convenient for scientists working in the laboratory to combine substances in fixed ratios of molecules.

How would we make a liter (L) of solution consisting of 1 mol of sucrose dissolved in water? We would measure out 342 g of sucrose and then add enough water to bring the total volume of the solution up to 1 L. At that point, we would have a 1-molar (1 *M*) solution of sucrose. **Molarity**—the number of moles of solute per liter of solution—is the unit of concentration most often used by biologists for aqueous solutions.

Acids and Bases

Occasionally, a hydrogen atom participating in a hydrogen bond between two water molecules shifts from one molecule to the other. When this happens, the hydrogen atom leaves its electron behind, and what is actually transferred is a **hydrogen ion** (H^+), a single proton with a charge of 1+. The water molecule that lost a proton is now a **hydroxide ion** (OH^-), which has a charge of 1−. The proton binds to the other water molecule, making that molecule a **hydronium ion** (H_3O^+):

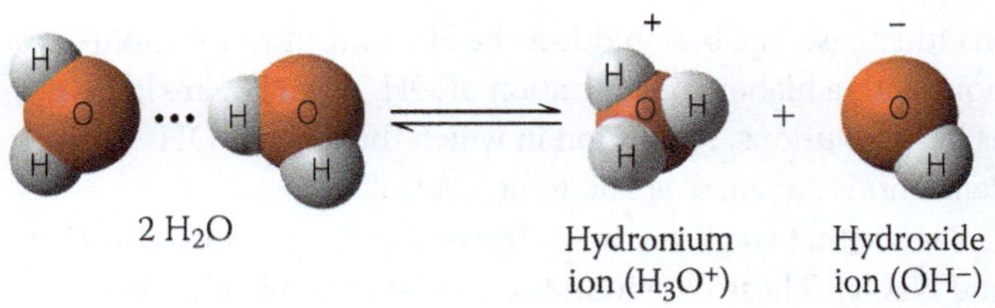

By convention, H^+ (the hydrogen ion) is used to represent H_3O^+ (the hydronium ion), and we follow that practice here. Keep in mind, though, that H^+ does not exist on its own in an aqueous solution. It is always associated with a water molecule in the form of H_3O^+.

As indicated by the double arrows, this is a reversible reaction that reaches a state of dynamic equilibrium when water molecules dissociate at the same rate that they are being reformed from H^+ and OH^-. At this equilibrium point, the concentration of water molecules greatly exceeds the concentrations of H^+ and OH^-. In pure water, only one water molecule in every 554 million is dissociated; the concentration of each ion in pure water is 10^{-7} *M* (at 25°C). This means there is only one ten-millionth of a mole of hydrogen ions per liter of pure water and an equal number of hydroxide ions.

Although the dissociation of water is reversible and statistically rare, it is exceedingly important in the chemistry of life. H^+ and OH^- are very reactive. Changes in their concentrations can drastically affect a cell's proteins and other complex molecules. As we have seen, the concentrations of H^+ and OH^- are equal in pure water, but adding certain kinds of solutes, called acids and bases, disrupts this balance.

What would cause an aqueous solution to have an imbalance in H^+ and OH^- concentrations? When acids dissolve in water, they donate additional H^+ to the solution. An **acid** is a substance that increases the hydrogen ion concentration of a solution. For example, when hydrochloric acid (HCl) is added to water, hydrogen ions dissociate from chloride ions:

$$HCl \rightarrow H^+ + Cl^-$$

This source of H^+ (dissociation of water is the other source) results in an acidic solution—one having more H^+ than OH^-.

A substance that *reduces* the hydrogen ion concentration of a solution is called a **base**. Some bases reduce the H^+ concentration directly by accepting hydrogen ions. Ammonia (NH_3), for instance, acts as a base when the unshared electron pair in nitrogen's valence shell attracts a hydrogen ion from the solution, resulting in an ammonium ion (NH_4^+):

$$NH_3 + H^+ \rightleftharpoons NH_4^+$$

Other bases reduce the H^+ concentration indirectly by dissociating to form hydroxide ions, which combine with hydrogen ions and form water. One such base is sodium hydroxide (NaOH), which in water dissociates into its ions:

$$NaOH \rightarrow Na^+ + OH^-$$

In either case, the base reduces the H^+ concentration. Solutions with a higher concentration of OH^- than H^+ are known as basic solutions. A solution in which the H^+ and OH^- concentrations are equal is said to be neutral.

Notice that single arrows were used in the reactions for HCl and NaOH. These compounds dissociate completely when mixed with water, so hydrochloric acid is called a strong acid and sodium hydroxide a strong base. In contrast, ammonia is a relatively weak base. The double arrows in the reaction for ammonia indicate that the binding and release of hydrogen ions are reversible reactions, although at equilibrium there will be a fixed ratio of NH_4^+ to NH_3.

Weak acids are acids that reversibly release and accept back hydrogen ions. An example is carbonic acid:

$$\underset{\text{Carbonic acid}}{H_2CO_3} \rightleftharpoons \underset{\text{Bicarbonate ion}}{HCO_3^-} + \underset{\text{Hydrogen ion}}{H^+}$$

Here the equilibrium so favors the reaction in the left direction that when carbonic acid is added to pure water, only 1% of the molecules are dissociated at any particular time. Still, that is enough to shift the balance of H^+ and OH^- from neutrality.

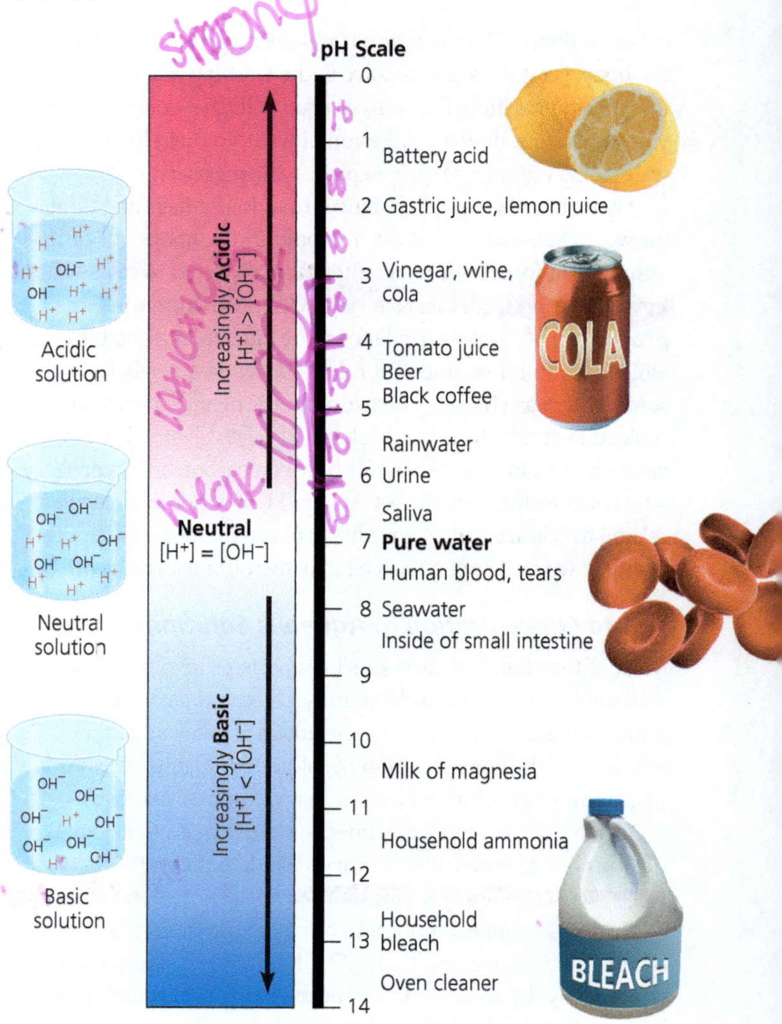

▲ **Figure 23 The pH scale and pH values of some aqueous solutions.**

The pH Scale

In any aqueous solution at 25°C, the *product* of the H^+ and OH^- concentrations is constant at 10^{-14}. This can be written

$$[H^+][OH^-] = 10^{-14}$$

In such an equation, brackets indicate molar concentration. In a neutral solution at room temperature (25°C), $[H^+] = 10^{-7}$ and $[OH^-] = 10^{-7}$. In this case, $10^{-7} \times 10^{-7} = 10^{-14}$. If enough acid is added to a solution to increase $[H^+]$ to $10^{-5}\,M$, then $[OH^-]$ will decline by an equivalent factor to $10^{-9}\,M$ (note that $10^{-5} \times 10^{-9} = 10^{-14}$). This constant relationship expresses the behavior of acids and bases in an aqueous solution. An acid not only adds hydrogen ions to a solution, but also removes hydroxide ions because of the tendency for H^+ to combine with OH^-, forming water. A base has the opposite effect, increasing OH^- concentration but also reducing H^+ concentration by the formation of water. If enough of a base is added to raise the OH^- concentration to $10^{-4}\,M$, it will cause the H^+ concentration to drop to $10^{-10}\,M$. Whenever we know the concentration of either H^+ or OH^- in an aqueous solution, we can deduce the concentration of the other ion.

Because the H^+ and OH^- concentrations of solutions can vary by a factor of 100 trillion or more, scientists have developed a way to express this variation more conveniently than in moles per liter. The pH scale **(Figure 23)** compresses the range of H^+ and OH^- concentrations by employing logarithms. The **pH** of a solution is defined as the negative logarithm (base 10) of the hydrogen ion concentration:

$$pH = -\log [H^+]$$

For a neutral aqueous solution, $[H^+]$ is $10^{-7}\,M$, giving us

$$-\log 10^{-7} = -(-7) = 7$$

Notice that pH *declines* as H^+ concentration *increases*. Notice, too, that although the pH scale is based on H^+ concentration, it also implies OH^- concentration. A solution of pH 10 has a hydrogen ion concentration of $10^{-10}\,M$ and a hydroxide ion concentration of $10^{-4}\,M$.

The pH of a neutral aqueous solution at 25°C is 7, the midpoint of the pH scale. A pH value less than 7 denotes an acidic solution; the lower the number, the more acidic the solution. The pH for basic solutions is above 7. Most biological fluids, such as blood and saliva, are within the range of pH 6–8. There are a few exceptions, however, including the strongly acidic digestive juice of the human stomach, which has a pH of about 2.

Remember that each pH unit represents a tenfold difference in H^+ and OH^- concentrations. It is this mathematical feature that makes the pH scale so compact. A solution of pH 3 is not twice as acidic as a solution of pH 6, but a thousand times ($10 \times 10 \times 10$) more acidic. When the pH of a solution changes slightly, the actual concentrations of H^+ and OH^- in the solution change substantially.

Buffers

The internal pH of most living cells is close to 7. Even a slight change in pH can be harmful because the chemical processes of the cell are very sensitive to the concentrations of hydrogen and hydroxide ions. The pH of human blood is very close to 7.4, which is slightly basic. A person cannot survive for more than a few minutes if the blood pH drops to 7 or rises to 7.8, and a chemical system exists in the blood that maintains a stable pH. If 0.01 mol of a strong acid is added to a liter of pure water, the pH drops from 7.0 to 2.0. If the same amount of acid is added to a liter of blood, however, the pH decrease is only from 7.4 to 7.3. Why does the addition of acid have so much less of an effect on the pH of blood than it does on the pH of water?

The presence of substances called buffers allows biological fluids to maintain a relatively constant pH despite the addition of acids or bases. A **buffer** is a substance that minimizes changes in the concentrations of H^+ and OH^- in a solution. It does so by accepting hydrogen ions from the solution when they are in excess and donating hydrogen ions to the solution when they have been depleted. Most buffer solutions contain a weak acid and its corresponding base, which combine reversibly with hydrogen ions.

Several buffers contribute to pH stability in human blood and many other biological solutions. One of these is carbonic acid (H_2CO_3), which is formed when CO_2 reacts with water in blood plasma. As mentioned earlier, carbonic acid dissociates to yield a bicarbonate ion (HCO_3^-) and a hydrogen ion (H^+):

$$\underset{\substack{H^+ \text{ donor} \\ \text{(acid)}}}{H_2CO_3} \underset{\substack{\text{Response to} \\ \text{a drop in pH}}}{\overset{\substack{\text{Response to} \\ \text{a rise in pH}}}{\rightleftharpoons}} \underset{\substack{H^+ \text{ acceptor} \\ \text{(base)}}}{HCO_3^-} + \underset{\substack{\text{Hydrogen} \\ \text{ion}}}{H^+}$$

The chemical equilibrium between carbonic acid and bicarbonate acts as a pH regulator, the reaction shifting left or right as other processes in the solution add or remove hydrogen ions. If the H^+ concentration in blood begins to fall (that is, if pH rises), the reaction proceeds to the right and more carbonic acid dissociates, replenishing hydrogen ions. But when the H^+ concentration in blood begins to rise (when pH drops), the reaction proceeds to the left, with HCO_3^- (the base) removing the hydrogen ions from the solution and forming H_2CO_3. Thus, the carbonic acid–bicarbonate buffering system consists of an acid and a base in equilibrium with each other. Most other buffers are also acid-base pairs.

Acidification: A Threat to Our Oceans

Among the many threats to water quality posed by human activities is the burning of fossil fuels, which releases CO_2 into the atmosphere. The resulting increase in atmospheric CO_2 levels has caused global warming. In addition, about 25% of human-generated CO_2 is absorbed by the oceans. In spite of the huge volume of water in the oceans, scientists worry that the absorption of so much CO_2 will harm marine ecosystems.

Recent data have shown that such fears are well founded. When CO_2 dissolves in seawater, it reacts with water to form carbonic acid, which lowers ocean pH. This process, known as **ocean acidification**, alters the delicate balance of conditions for life in the oceans **(Figure 24)**. Based on measurements of CO_2 levels in air bubbles trapped in ice over thousands of years, scientists calculate that the pH of the oceans is 0.1 pH unit lower now than at any time in the past 420,000 years. Recent studies predict that it will drop another 0.3–0.5 pH unit by the end of this century.

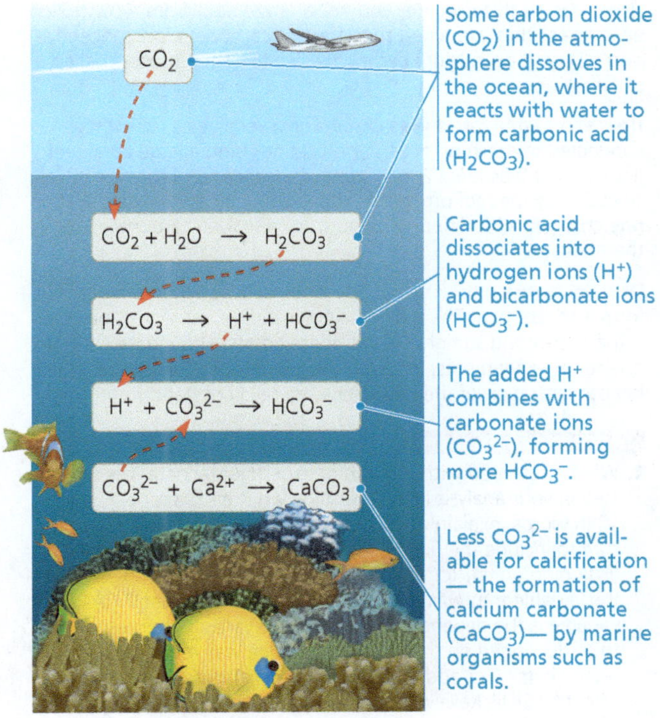

▲ **Figure 24 Atmospheric CO_2 from human activities and its fate in the ocean.**

WHAT IF? *Would lowering the ocean's carbonate concentration have any effect, even indirectly, on organisms that don't form $CaCO_3$? Explain.*

As seawater acidifies, the extra hydrogen ions combine with carbonate ions (CO_3^{2-}) to form bicarbonate ions (HCO_3^-), thereby reducing the carbonate ion concentration (see Figure 24). Scientists predict that ocean acidification will cause the carbonate ion concentration to decrease by 40% by the year 2100. This is of great concern because carbonate ions are required for calcification, the production of calcium carbonate ($CaCO_3$), by many marine organisms, including reef-building corals and animals that build shells. The **Scientific Skills Exercise** allows you to work with data from an experiment examining the effect of carbonate ion concentration on coral reefs. Coral reefs are sensitive ecosystems that act as havens for a great diversity of marine life. The disappearance of coral reef ecosystems would be a tragic loss of biological diversity.

CONCEPT CHECK 5

1. Describe how properties of water contribute to the upward movement of water in a tree.
2. How can the freezing of water crack boulders?
3. Compared with a basic solution at pH 9, the same volume of an acidic solution at pH 4 has _______ times as many hydrogen ions (H^+).
4. **WHAT IF?** What would be the effect on the properties of the water molecule if oxygen and hydrogen had equal electronegativity?
5. **INTERPRET THE DATA** The concentration of the appetite-regulating hormone ghrelin is about 1.3×10^{-10} *M* in the blood of a fasting person. How many molecules of ghrelin are in 1 L of blood?

Scientific Skills Exercise

Interpreting a Scatter Plot with a Regression Line

How Does the Carbonate Ion Concentration of Seawater Affect the Calcification Rate of a Coral Reef? Scientists predict that acidification of the ocean due to higher levels of atmospheric CO_2 will lower the concentration of dissolved carbonate ions, which living corals use to build calcium carbonate reef structures. In this exercise, you will analyze data from a controlled experiment that examined the effect of carbonate ion concentration ($[CO_3^{2-}]$) on calcium carbonate deposition, a process called calcification.

How the Experiment Was Done For several years, scientists conducted research on ocean acidification using a large coral reef aquarium at Biosphere 2 in Arizona. They measured the rate of calcification by the reef organisms and examined how the calcification rate changed with differing amounts of dissolved carbonate ions in the seawater.

Data from the Experiment The black data points in the graph form a scatter plot. The red line, known as a linear regression line, is the best-fitting straight line for these points. These data are from one set of experiments, in which the pH, temperature, and calcium ion concentration of the seawater were held constant.

Data from C. Langdon et al., Effect of calcium carbonate saturation state on the calcification rate of an experimental coral reef, *Global Biogeochemical Cycles* 14:639–654 (2000).

INTERPRET THE DATA

1. When presented with a graph of experimental data, the first step in your analysis is to determine what each axis represents. (a) In words, explain what is being shown on the *x*-axis. Be sure to include the units. (b) What is being shown on the *y*-axis (including units)? (c) Which variable is the independent variable—the variable that was *manipulated* by the researchers? (d) Which variable is the dependent variable—the variable that responded to or depended on the treatment, which was *measured* by the researchers? (For additional information about graphs, see the Scientific Skills Review in the Study Area in MasteringBiology.)
2. Based on the data shown in the graph, describe in words the relationship between carbonate ion concentration and calcification rate.
3. (a) If the seawater carbonate ion concentration is 270 μmol/kg, what is the approximate rate of calcification, and approximately how many days would it take 1 square meter of reef to accumulate 30 mmol of calcium carbonate ($CaCO_3$)? (b) If the seawater carbonate ion concentration is 250 μmol/kg, what is the approximate rate of calcification, and approximately how many days would it take 1 square meter of reef to accumulate 30 mmol of calcium carbonate? (c) If carbonate ion concentration decreases, how does the calcification rate change, and how does that affect the time it takes coral to grow?
4. (a) Referring to the reactions in Figure 24, determine which step of the process is measured in this experiment. (b) Are the results of this experiment consistent with the hypothesis that increased atmospheric $[CO_2]$ will slow the growth of coral reefs? Why or why not?

MB A version of this Scientific Skills Exercise can be assigned in MasteringBiology.

Chapter Review

Go to **MasteringBiology®** for Assignments, the eText, and the Study Area with Animations, Activities, Vocab Self-Quiz, and Practice Tests.

SUMMARY OF KEY CONCEPTS

VOCAB SELF-QUIZ
goo.gl/gbai8v

CONCEPT 1

Matter consists of chemical elements in pure form and in combinations called compounds

- **Elements** cannot be broken down chemically to other substances. A **compound** contains two or more different elements in a fixed ratio. Oxygen, carbon, hydrogen, and nitrogen make up approximately 96% of living matter.

? *In what way does the need for iodine or iron in your diet differ from your need for calcium or phosphorus?*

CONCEPT 2

An element's properties depend on the structure of its atoms

- An **atom**, the smallest unit of an element, has the following components:

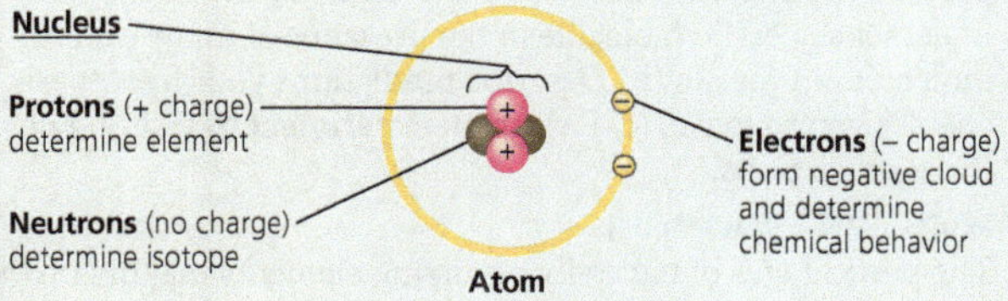

- An electrically neutral atom has equal numbers of electrons and protons; the number of protons determines the **atomic number**. **Isotopes** of an element differ from each other in neutron number and therefore mass. Unstable isotopes give off particles and energy as radioactivity.
- In an atom, electrons occupy specific **electron shells**; the electrons in a shell have a characteristic energy level. Electron distribution in shells determines the chemical behavior of an atom. An atom that has an incomplete outer shell, the **valence shell**, is reactive.

DRAW IT *Draw the electron distribution diagrams for neon ($_{10}$Ne) and argon ($_{18}$Ar). Why are they chemically unreactive?*

CONCEPT 3

The formation and function of molecules depend on chemical bonding between atoms

- **Chemical bonds** form when atoms interact and complete their valence shells. **Covalent bonds** form when pairs of electrons are shared. H_2 has a **single bond**: H—H. A **double bond** is the sharing of two pairs of electrons, as in O=O.
- **Molecules** consist of two or more covalently bonded atoms. The attraction of an atom for the electrons of a covalent bond is its **electronegativity**. Electrons of a **polar covalent bond** are pulled closer to the more electronegative atom, such as the oxygen in H_2O.
- An **ion** forms when an atom or molecule gains or loses an electron and becomes charged. An **ionic bond** is the attraction between two oppositely charged ions, such as Na^+ and Cl^-.
- Weak bonds reinforce the shapes of large molecules and help molecules adhere to each other. A **hydrogen bond** is an attraction between a hydrogen atom carrying a partial positive charge (δ+) and an electronegative atom (δ–). **Van der Waals interactions** occur between transiently positive and negative regions of molecules.
- Molecular shape is usually the basis for the recognition of one biological molecule by another.

? *In terms of electron sharing between atoms, compare nonpolar covalent bonds, polar covalent bonds, and the formation of ions.*

CONCEPT 4

Chemical reactions make and break chemical bonds

- **Chemical reactions** change **reactants** into **products** while conserving matter. All chemical reactions are theoretically reversible. **Chemical equilibrium** is reached when the forward and reverse reaction rates are equal.

? *What would happen to the concentration of products if more reactants were added to a reaction that was in chemical equilibrium? How would this addition affect the equilibrium?*

CONCEPT 5

Hydrogen bonding gives water properties that help make life possible on Earth

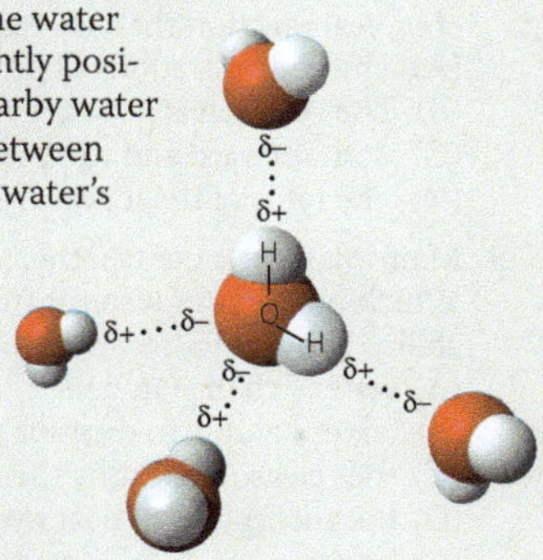

- A **hydrogen bond** forms when the slightly negatively charged oxygen of one water molecule is attracted to the slightly positively charged hydrogen of a nearby water molecule. Hydrogen bonding between water molecules is the basis for water's properties.
- Hydrogen bonding keeps water molecules close to each other, giving water **cohesion**. Hydrogen bonding is also responsible for water's **surface tension**.
- Water has a high **specific heat**: Heat is absorbed when hydrogen bonds break and is released when hydrogen bonds form. This helps keep temperatures relatively steady, within limits that permit life. **Evaporative cooling** is based on water's high **heat of vaporization**. The evaporative loss of the most energetic water molecules cools a surface.
- Ice floats because it is less dense than liquid water. This property allows life to exist under the frozen surfaces of lakes and seas.

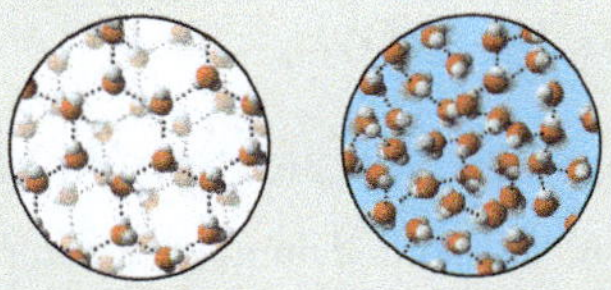

Ice: stable hydrogen bonds
Liquid water: transient hydrogen bonds

- Water is an unusually versatile **solvent** because its polar molecules are attracted to ions and polar substances that can form hydrogen bonds. **Hydrophilic** substances have an affinity for water; **hydrophobic** substances do not. **Molarity**, the number of moles of **solute** per liter of **solution**, is used as a measure of solute concentration in solutions. A **mole** is a certain number of molecules of a substance. The mass of a mole of a substance in grams is the same as the **molecular mass** in daltons.

- A water molecule can transfer an H^+ to another water molecule to form H_3O^+ (represented simply by H^+) and OH^-.
- The concentration of H^+ is expressed as **pH**; pH = $-\log [H^+]$. A **buffer** consists of an acid-base pair that combines reversibly with hydrogen ions, allowing it to resist pH changes.

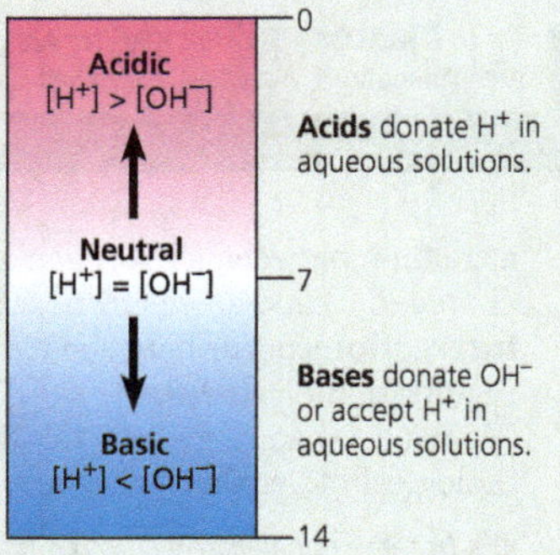

- The burning of fossil fuels increases the amount of CO_2 in the atmosphere. Some CO_2 dissolves in the oceans, causing **ocean acidification**, which has potentially grave consequences for coral reefs.

? *Describe how the properties of water result from the molecule's polar covalent bonds and how these properties contribute to Earth's suitability for life.*

TEST YOUR UNDERSTANDING

PRACTICE TEST

goo.gl/CRZjvS

Level 1: Knowledge/Comprehension

1. The reactivity of an atom arises from
(A) the average distance of the outermost electron shell from the nucleus.
(B) the existence of unpaired electrons in the valence shell.
(C) the sum of the potential energies of all the electron shells.
(D) the potential energy of the valence shell.

2. Which of the following statements correctly describes any chemical reaction that has reached equilibrium?
(A) The concentrations of products and reactants are equal.
(B) The reaction is now irreversible.
(C) Both forward and reverse reactions have halted.
(D) The rates of the forward and reverse reactions are equal.

3. Many mammals control their body temperature by sweating. Which property of water is most directly responsible for the ability of sweat to lower body temperature?
(A) water's change in density when it condenses
(B) water's ability to dissolve molecules in the air
(C) the release of heat by the formation of hydrogen bonds
(D) the absorption of heat by the breaking of hydrogen bonds

4. We can be sure that a mole of table sugar and a mole of vitamin C are equal in their
(A) mass. (C) number of atoms.
(B) volume. (D) number of molecules.

5. Measurements show that the pH of a particular lake is 4.0. What is the hydrogen ion concentration of the lake?
(A) 4.0 M (B) 10^{-4} M (C) 10^4 M (D) 10^{-10} M

Level 2: Application/Analysis

6. The atomic number of sulfur is 16. Sulfur combines with hydrogen by covalent bonding to form a compound, hydrogen sulfide. Based on the number of valence electrons in a sulfur atom, predict the molecular formula of the compound.
(A) HS (B) HS_2 (C) H_2S (D) H_3S_2

7. What coefficients must be placed in the following blanks so that all atoms are accounted for in the products?

$$C_6H_{12}O_6 \rightarrow \_\_\_\_ C_2H_6O + \_\_\_\_ CO_2$$

(A) 1; 2 (B) 3; 1 (C) 1; 3 (D) 2; 2

8. A slice of pizza has 500 kcal. If we could burn the pizza and use all the heat to warm a 50-L container of cold water, what would be the approximate increase in the temperature of the water? (*Note*: A liter of cold water weighs about 1 kg.)
(A) 50°C (B) 5°C (C) 1°C (D) 10°C

9. DRAW IT Draw the hydration shells that form around a potassium ion and a chloride ion when potassium chloride (KCl) dissolves in water. Label the positive, negative, and partial charges on the atoms.

10. MAKE CONNECTIONS What do climate change and ocean acidification have in common?

Level 3: Synthesis/Evaluation

11. **SCIENTIFIC INQUIRY**

Female luna moths (*Actias luna*) attract males by emitting chemical signals that spread through the air. A male hundreds of meters away can detect these molecules and fly toward their source. The sensory organs responsible for this behavior are the comblike antennae visible in the photograph shown here. Each filament of an antenna is equipped with thousands of receptor cells that detect the sex attractant. (a) Based on what you learned in this chapter, propose a hypothesis to account for the ability of the male moth to detect a specific molecule in the presence of many other molecules in the air. (b) Describe predictions your hypothesis enables you to make. (c) Design an experiment to test one of these predictions.

12. **FOCUS ON EVOLUTION**
The percentages of naturally occurring elements making up the human body are similar to the percentages of these elements found in other organisms. How could you account for this similarity among organisms? Explain your thinking.

13. **FOCUS ON ORGANIZATION**
Several emergent properties of water contribute to the suitability of the environment for life. In a short essay (100–150 words), describe how the ability of water to function as a versatile solvent arises from the structure of water molecules.

14. **SYNTHESIZE YOUR KNOWLEDGE**

How do cats drink? Scientists using high-speed video have shown that cats use an interesting technique to drink aqueous substances like water and milk. Four times a second, the cat touches the tip of its tongue to the water and draws a column of water up into its mouth (as you can see in the photo), which then shuts before gravity can pull the water back down. Describe how the properties of water allow cats to drink in this fashion, including how water's molecular structure contributes to the process.

Answers

Figure Questions

Figure 6 Atomic number = 12; 12 protons, 12 electrons; 3 electron shells; 2 valence electrons **Figure 15** The plant is submerged in water (H_2O), in which the CO_2 is dissolved. The sun's energy is used to make sugar, which is found in the plant and can act as food for the plant itself, as well as for animals that eat the plant. The oxygen (O_2) is present in the bubbles. **Figure 16** One possible answer is the following:

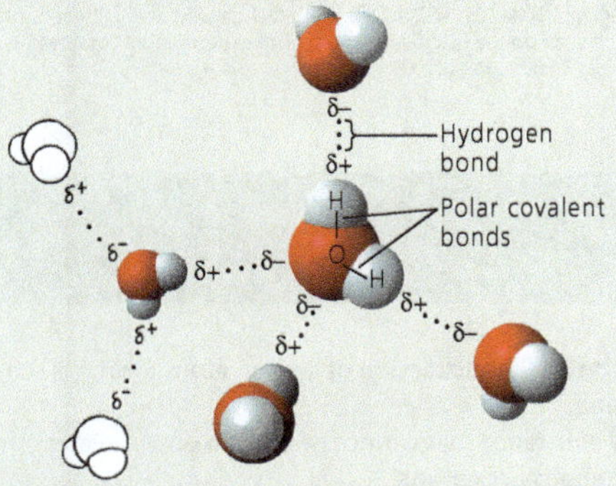

Figure 20 Without hydrogen bonds, water would behave like other small molecules, and the solid phase (ice) would be denser than liquid water. The ice would sink to the bottom and would no longer insulate the body of water. All the water would eventually freeze because the average annual temperature at the South Pole is –50°C. The krill could not survive. **Figure 21** Heating the solution would cause the water to evaporate faster than it is evaporating at room temperature. At a certain point, there wouldn't be enough water molecules to dissolve the salt ions. The salt would start coming out of solution and re-forming crystals. Eventually, all the water would evaporate, leaving behind a pile of salt like the original pile. **Figure 24** By causing the loss of coral reefs, a decrease in the ocean's carbonate concentration would have a ripple effect on noncalcifying organisms. Some of these organisms depend on the reef structure for protection, while others feed on species associated with reefs.

Concept Check 1

1. Yes, because an organism requires trace elements, even though only in small amounts **2.** A person with an iron deficiency will probably show fatigue and other effects of a low oxygen level in the blood. (The condition is called anemia and can also result from too few red blood cells or abnormal hemoglobin.)

Concept Check 2

1. $^{15}_{7}N$ **2.** 9 electrons; two electron shells; 1 electron is needed to fill the valence shell. **3.** The elements in a row all have the same number of electron shells. In a column, all the elements have the same number of electrons in their valence shells.

Concept Check 3

1. Each carbon atom has only three covalent bonds instead of the required four. **2.** The attraction between oppositely charged ions, forming ionic bonds **3.** If you could synthesize molecules that mimic these shapes, you might be able to treat diseases or conditions caused by the inability of affected individuals to synthesize such molecules.

Concept Check 4

1. At equilibrium, the forward and reverse reactions occur at the same rate. **2.** $C_6H_{12}O_6 + 6\,O_2 \rightarrow 6\,CO_2 + 6\,H_2O$ + Energy. Glucose and oxygen react to form carbon dioxide and water, releasing energy. We breathe in oxygen because we need it for this reaction to occur, and we breathe out carbon dioxide because it is a product of this reaction. (This reaction is called cellular respiration.)

Concept Check 5

1. Hydrogen bonds hold neighboring water molecules together. This cohesion helps the chains of water molecules move upward against gravity in water-conducting cells as water evaporates from the leaves. Adhesion between water molecules and the walls of the water-conducting cells also helps counter gravity. **2.** As water freezes, it expands because water molecules move farther apart in forming ice crystals. When there is water in a crevice of a boulder, expansion due to freezing may crack the boulder. **3.** 10^5, or 100,000 **4.** The covalent bonds of water molecules would not be polar, and water molecules would not form hydrogen bonds with each other. Water would therefore not have the unusual properties described in this chapter—such as cohesion, surface tension, high specific heat, high heat of vaporization, and versatility as a solvent.

Summary of Key Concepts Questions

1 Iodine (part of a thyroid hormone) and iron (part of hemoglobin in blood) are both trace elements, required in minute quantities. Calcium and phosphorus (components of bones and teeth) are needed by the body in much greater quantities.
2

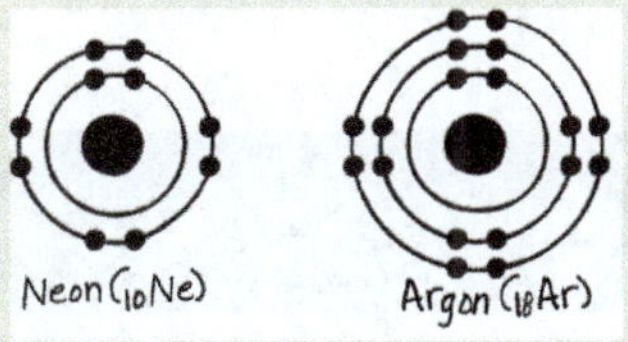

Both neon and argon are unreactive because they have completed valence shells. They do not have unpaired electrons that could participate in chemical bonds. **3** Electrons are shared equally between the two atoms in a nonpolar covalent bond. In a polar covalent bond, the electrons are drawn closer to the more electronegative atom. In the formation of ions, one or more electrons are completely transferred from one atom to a much more electronegative atom. **4** The concentration of products would increase as the added reactants were converted to products. Eventually, an equilibrium would again be reached in which the forward and reverse reactions were proceeding at the same rate and the relative concentrations of reactants and products returned to where they were before the addition of more reactants. **5** The polar covalent bonds of a water molecule allow it to form hydrogen bonds with other water molecules and other polar molecules as well. The sticking together of water molecules, called cohesion, and the sticking of water to other molecules, called adhesion, help water rise from the roots of plants to their leaves, among other biological benefits. Hydrogen bonding between water molecules is responsible for water's high specific heat (resistance to temperature change), which helps moderate temperature on Earth. Hydrogen bonding is also responsible for water's high heat of vaporization, which makes water useful for evaporative cooling. A lattice of stable hydrogen bonds in ice makes it less dense than liquid water, so that it floats, creating an insulating surface on bodies of water that allows organisms to live underneath. Finally, the polarity of water molecules resulting from their polar covalent bonds makes water an excellent solvent; polar and ionic atoms and molecules that are needed for life can exist in a dissolved state and participate in chemical reactions.

Test Your Understanding

1. B **2.** D **3.** D **4.** D **5.** B **6.** C **7.** D **8.** D
9.

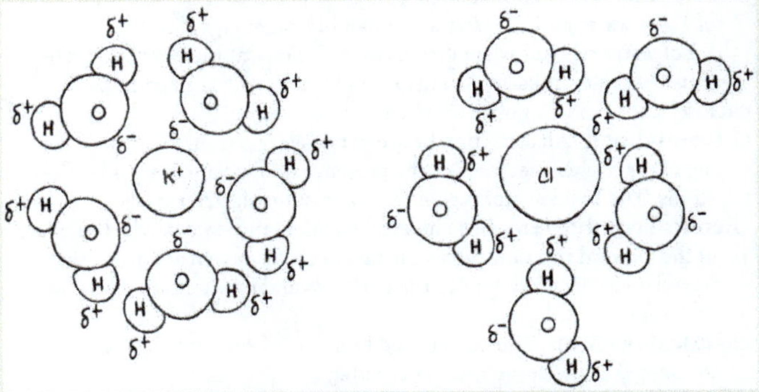

10. Both global warming and ocean acidification are caused by increasing levels of carbon dioxide in the atmosphere, the result of burning fossil fuels.

Credits

Credits are listed in the order of appearance.

Photo Credits

1 Paul Quagliana/BNPS; **2a** Chip Clark; **2bc** Stephen Frisch/Pearson Education; **4** National Library of Medicine; **11** Stephen Frisch/Pearson Education; **UN01 gecko** Martin Harvey/Photolibrary/Getty Images; **15** Nigel Cattlin/Science Source; **17** N.C. Brown Center for Ultrastructure Studies, SUNY, Syracuse, NY; **18** Alasdair Thomson/E+/Getty Images; **20** Jan van Franeker, Alfred-Wegener Institute fur Polar-und-Meeresforschung; **23a** Paulista/fotolia; **23b** Fotofermer/fotolia; **23c** SciePro/SPL/age fotostock; **23d** Beth Van Trees/Shutterstock; **SSE** From: Coral reefs under rapid climate change and ocean acidification. O. Hoegh-Guldberg et al. *Science*. 2007 Dec 14; 318(5857):1737–42. Fig. 5a; **Chapter Review moth** Rolf Nussbaumer/Nature Picture Library; **SYK** Eric Guilloret/Biosphoto/Science Source.

Illustration and Text Credits

22a Based on M. Gerstein and M. Levitt. Simulating water and the molecules of life, *Scientific American* Nov. 1998, pp. 100–105; **SSE.1** Data from C. Langdon et al., Effect of calcium carbonate saturation state on the calcification rate of an experimental coral reef, *Global Biogeochemical Cycles* 14:639–654 (2000).

Glossary

acid A substance that increases the hydrogen ion concentration of a solution.

adhesion The clinging of one substance to another, such as water to plant cell walls, by means of hydrogen bonds.

anion (an′-ī-on) A negatively charged ion.

aqueous solution (ā′-kwē-us) A solution in which water is the solvent.

atom The smallest unit of matter that retains the properties of an element.

atomic mass The total mass of an atom, numerically equivalent to the mass in grams of 1 mole of the atom. (For an element with more than one isotope, the atomic mass is the average mass of the naturally occurring isotopes, weighted by their abundance.)

atomic nucleus An atom's dense central core, containing protons and neutrons.

atomic number The number of protons in the nucleus of an atom, unique for each element and designated by a subscript.

base A substance that reduces the hydrogen ion concentration of a solution.

buffer A solution that contains a weak acid and its corresponding base. A buffer minimizes changes in pH when acids or bases are added to the solution.

calorie (cal) The amount of heat energy required to raise the temperature of 1 g of water by 1°C; also the amount of heat energy that 1 g of water releases when it cools by 1°C. The Calorie (with a capital C), usually used to indicate the energy content of food, is a kilocalorie.

cation (cat′-ī′-on) A positively charged ion.

chemical bond An attraction between two atoms, resulting from a sharing of outer-shell electrons or the presence of opposite charges on the atoms. The bonded atoms gain complete outer electron shells.

chemical equilibrium In a chemical reaction, the state in which the rate of the forward reaction equals the rate of the reverse reaction, so that the relative concentrations of the reactants and products do not change with time.

chemical reaction The making and breaking of chemical bonds, leading to changes in the composition of matter.

cohesion The linking together of like molecules, often by hydrogen bonds.

compound A substance consisting of two or more different elements combined in a fixed ratio.

covalent bond (kō-vā′-lent) A type of strong chemical bond in which two atoms share one or more pairs of valence electrons.

dalton A measure of mass for atoms and subatomic particles; the same as the atomic mass unit, or amu.

double bond A double covalent bond; the sharing of two pairs of valence electrons by two atoms.

electron A subatomic particle with a single negative electrical charge and a mass about 1/2,000 that of a neutron or proton. One or more electrons move around the nucleus of an atom.

electron shell An energy level of electrons at a characteristic average distance from the nucleus of an atom.

electronegativity The attraction of a given atom for the electrons of a covalent bond.

element Any substance that cannot be broken down to any other substance by chemical reactions.

energy The capacity to cause change, especially to do work (to move matter against an opposing force).

essential element A chemical element required for an organism to survive, grow, and reproduce.

evaporative cooling The process in which the surface of an object becomes cooler during evaporation, a result of the molecules with the greatest kinetic energy changing from the liquid to the gaseous state.

heat Thermal energy in transfer from one body of matter to another.

heat of vaporization The quantity of heat a liquid must absorb for 1 g of it to be converted from the liquid to the gaseous state.

hydration shell The sphere of water molecules around a dissolved ion.

hydrogen bond A type of weak chemical bond that is formed when the slightly positive hydrogen atom of a polar covalent bond in one molecule is attracted to the slightly negative atom of a polar covalent bond in another molecule or in another region of the same molecule.

hydrogen ion A single proton with a charge of 1+. The dissociation of a water molecule (H_2O) leads to the generation of a hydroxide ion (OH^-) and a hydrogen ion (H^+); in water, H^+ is not found alone but associates with a water molecule to form a hydronium ion.

hydronium ion A water molecule that has an extra proton bound to it; H_3O^+, commonly represented as H^+.

hydrophilic (hī′-drō-fil′-ik) Having an affinity for water.

hydrophobic (hī′-drō-fō′-bik) Having no affinity for water; tending to coalesce and form droplets in water.

hydroxide ion A water molecule that has lost a proton; OH^-.

ion (ī′-on) An atom or group of atoms that has gained or lost one or more electrons, thus acquiring a charge.

ionic bond (ī-on′-ik) A chemical bond resulting from the attraction between oppositely charged ions.

ionic compound A compound resulting from the formation of an ionic bond; also called a salt.

isotope (ī′-sō-tōp′) One of several atomic forms of an element, each with the same number of protons but a different number of neutrons, thus differing in atomic mass.

kilocalorie (kcal) A thousand calories; the amount of heat energy required to raise the temperature of 1 kg of water by 1°C.

mass number The total number of protons and neutrons in an atom's nucleus.

matter Anything that takes up space and has mass.

molarity A common measure of solute concentration, referring to the number of moles of solute per liter of solution.

mole (mol) The number of grams of a substance that equals its molecular weight in daltons and contains Avogadro's number of molecules.

molecular mass The sum of the masses of all the atoms in a molecule; sometimes called molecular weight.
molecule Two or more atoms held together by covalent bonds.
neutron A subatomic particle having no electrical charge (electrically neutral), with a mass of about 1.7×10^{-24} g, found in the nucleus of an atom.
nonpolar covalent bond A type of covalent bond in which electrons are shared equally between two atoms of similar electronegativity.
ocean acidification The process by which the pH of the ocean is lowered (made more acidic) when excess CO_2 dissolves in seawater and forms carbonic acid (H_2CO_3).
pH A measure of hydrogen ion concentration equal to $-\log [H^+]$ and ranging in value from 0 to 14.
polar covalent bond A covalent bond between atoms that differ in electronegativity. The shared electrons are pulled closer to the more electronegative atom, making it slightly negative and the other atom slightly positive.
polar molecule A molecule (such as water) with an uneven distribution of charges in different regions of the molecule.
product A material resulting from a chemical reaction.
proton (prō′-ton) A subatomic particle with a single positive electrical charge, with a mass of about 1.7×10^{-24} g, found in the nucleus of an atom.
radioactive isotope An isotope (an atomic form of a chemical element) that is unstable; the nucleus decays spontaneously, giving off detectable particles and energy.
reactant A starting material in a chemical reaction.
salt A compound resulting from the formation of an ionic bond; also called an ionic compound.
single bond A single covalent bond; the sharing of a pair of valence electrons by two atoms.
solute (sol′-yūt) A substance that is dissolved in a solution.
solution A liquid that is a homogeneous mixture of two or more substances.
solvent The dissolving agent of a solution. Water is the most versatile solvent known.
specific heat The amount of heat that must be absorbed or lost for 1 g of a substance to change its temperature by 1°C.
surface tension A measure of how difficult it is to stretch or break the surface of a liquid.
temperature A measure in degrees of the average kinetic energy (thermal energy) of the atoms and molecules in a body of matter.
trace element An element indispensable for life but required in extremely minute amounts.
valence electron An electron in the outermost electron shell.
valence shell The outermost energy shell of an atom, containing the valence electrons involved in the chemical reactions of that atom.
van der Waals interactions Weak attractions between molecules or parts of molecules that result from transient local partial charges.

Carbon and the Molecular Diversity of Life

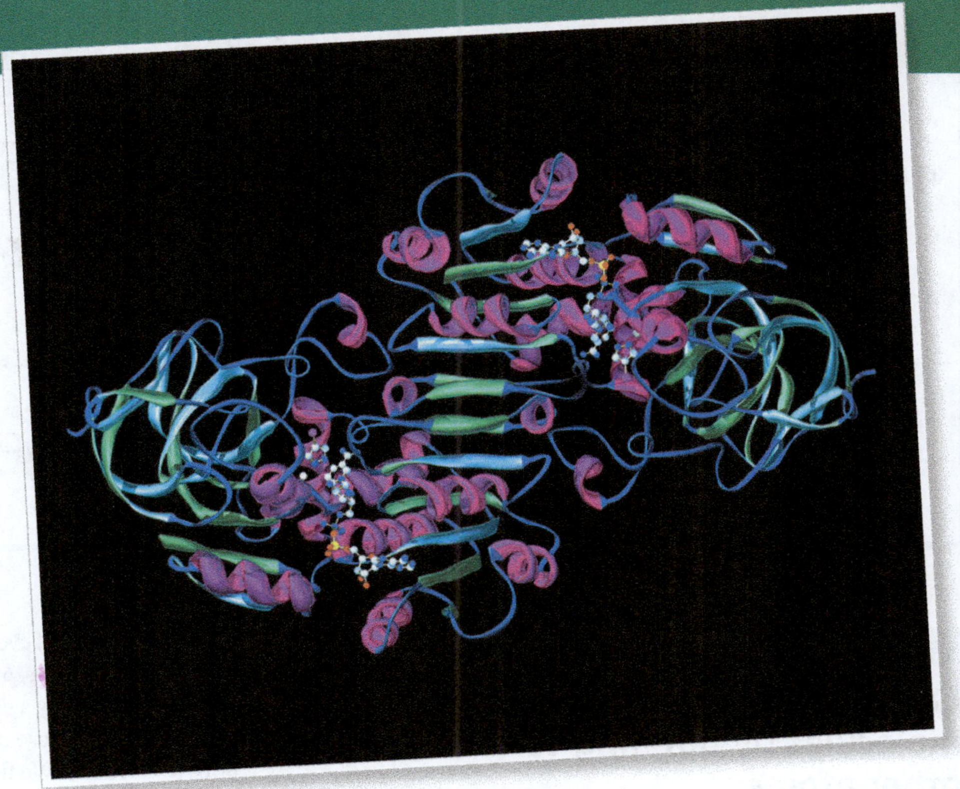

▲ **Figure 1 Why is the structure of a protein important for its function?**

KEY CONCEPTS

1. Carbon atoms can form diverse molecules by bonding to four other atoms
2. Macromolecules are polymers, built from monomers
3. Carbohydrates serve as fuel and building material
4. Lipids are a diverse group of hydrophobic molecules
5. Proteins include a diversity of structures, resulting in a wide range of functions
6. Nucleic acids store, transmit, and help express hereditary information
7. Genomics and proteomics have transformed biological inquiry and applications

Carbon Compounds and Life

Water is the universal medium for life on Earth, but water aside, living organisms are made up of chemicals based mostly on the element carbon. Of all chemical elements, carbon is unparalleled in its ability to form molecules that are large, complex, and varied. Hydrogen (H), oxygen (O), nitrogen (N), sulfur (S), and phosphorus (P) are other common ingredients of these compounds, but it is the element carbon (C) that accounts for the enormous variety of biological molecules. For historical reasons, a compound containing carbon is said to be an **organic compound**; furthermore, almost all organic compounds associated with life contain hydrogen atoms in addition to carbon atoms. Different species of organisms and even different individuals within a species are distinguished by variations in their large organic compounds.

Given the rich complexity of life on Earth, it may surprise you to learn that the critically important large molecules of all living things—from bacteria to elephants—fall into just four main classes: carbohydrates, lipids, proteins, and nucleic acids. On the molecular scale, members of three of these classes—carbohydrates, proteins, and nucleic acids—are huge and are therefore called **macromolecules**. For example, a protein may consist of thousands of atoms that form a molecular colossus with a mass well over 100,000 daltons. Considering the size and complexity of macromolecules, it is noteworthy that biochemists have determined the detailed structure of so many of them. The image in **Figure 1** is a molecular model of a protein called alcohol dehydrogenase, which breaks down alcohol in the body. The structures of macromolecules can provide important information about their functions.

In this chapter, we'll first investigate the properties of small organic molecules and then go on to discuss the larger biological molecules. After considering how macromolecules are built, we'll examine the structure and function of all four classes of large biological molecules. Like small molecules, large biological molecules exhibit unique emergent properties arising from the orderly arrangement of their atoms.

From Chapter 3 of *Campbell Biology in Focus*, Second Edition. Lisa A. Urry, Michael L. Cain, Steven A. Wasserman, Peter V. Minorsky, and Jane B. Reece.

▼ **Figure 2 The shapes of three simple organic molecules.**

Molecule and Molecular Shape	Molecular Formula	Structural Formula	Ball-and-Stick Model (molecular shape in pink)	Space-Filling Model
(a) Methane. When a carbon atom has four single bonds to other atoms, the molecule is tetrahedral.	CH_4	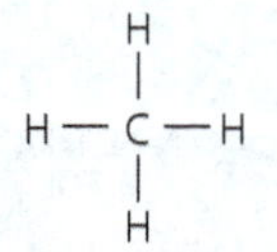	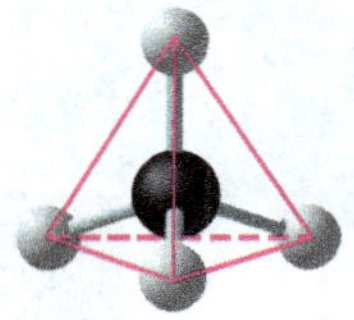	
(b) Ethane. A molecule may have more than one tetrahedral group of single-bonded atoms. (Ethane consists of two such groups.)	C_2H_6	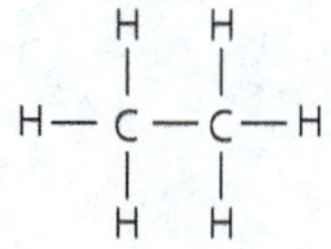	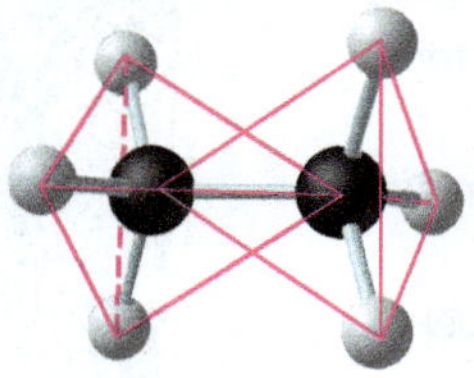	
(c) Ethene (ethylene). When two carbon atoms are joined by a double bond, all atoms attached to those carbons are in the same plane, and the molecule is flat.	C_2H_4	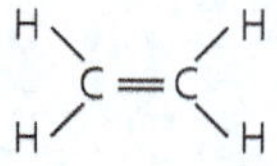	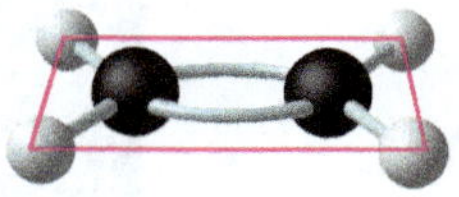	

CONCEPT 1

Carbon atoms can form diverse molecules by bonding to four other atoms

The key to an atom's chemical characteristics is its electron configuration. This configuration determines the kinds and number of bonds an atom will form with other atoms, and it is the source of carbon's versatility.

The Formation of Bonds with Carbon

Carbon has 6 electrons, with 2 in the first electron shell and 4 in the second shell; thus, it has 4 valence electrons in a shell that can hold up to 8 electrons. A carbon atom usually completes its valence shell by sharing its 4 electrons with other atoms so that 8 electrons are present. Each pair of shared electrons constitutes a covalent bond. In organic molecules, carbon usually forms single or double covalent bonds. Each carbon atom acts as an intersection point from which a molecule can branch off in as many as four directions. This enables carbon to form large, complex molecules.

When a carbon atom forms four single covalent bonds, the bonds angle toward the corners of an imaginary tetrahedron. The bond angles in methane (CH_4) are 109.5° **(Figure 2a)**, and they are roughly the same in any group of atoms where carbon has four single bonds. For example, ethane (C_2H_6) is shaped like two overlapping tetrahedrons **(Figure 2b)**. In molecules with more carbons, every grouping of a carbon bonded to four other atoms has a tetrahedral shape. But when two carbon atoms are joined by a double bond, as in ethene (C_2H_4), the atoms joined to those carbons are in the same plane as the carbons **(Figure 2c)**. We find it convenient to write molecules as structural formulas, as if the molecules being represented are two-dimensional, but keep in mind that molecules are three-dimensional and that the shape of a molecule is central to its function.

covalent = shared

The electron configuration of carbon gives it covalent compatibility with many different elements. **Figure 3** shows electron distribution diagrams for carbon and its most frequent partners—hydrogen, oxygen, and nitrogen. These are the four main atoms in organic molecules. The number of unpaired electrons in the valence shell of an atom is generally equal to the atom's **valence**, the number of covalent bonds it can form. Let's consider how valence and the rules of covalent bonding

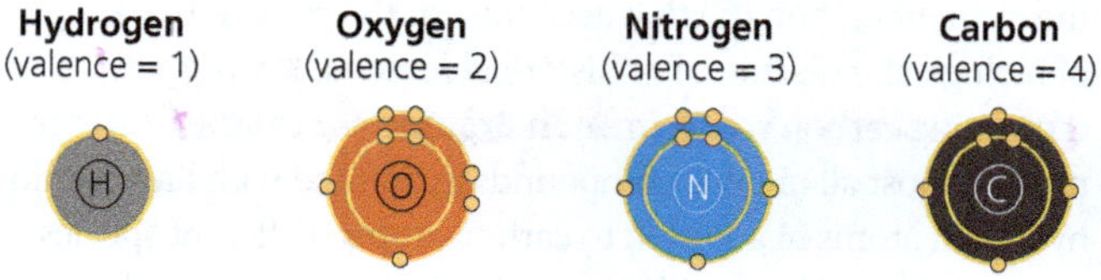

▲ **Figure 3 Valences of the major elements of organic molecules.** Valence is the number of covalent bonds an atom can form. It is generally equal to the number of electrons required to complete the valence (outermost) shell. Note that carbon can form four bonds.

apply to carbon atoms with partners other than hydrogen. We'll first look at the simple example of carbon dioxide.

In the carbon dioxide molecule (CO_2), a single carbon atom is joined to two atoms of oxygen by double covalent bonds. The structural formula for CO_2 is shown here:

$$O{=}C{=}O$$

Each line in a structural formula represents a pair of shared electrons. Thus, the two double bonds in CO_2 have the same number of shared electrons as four single bonds. The arrangement completes the valence shells of all atoms in the molecule:

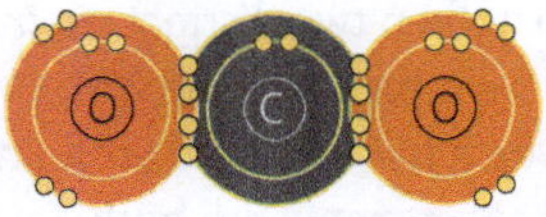

Because CO_2 is a very simple molecule and lacks hydrogen, it is often considered inorganic, even though it contains carbon. Whether we call CO_2 organic or inorganic, however, it is clearly important to the living world as the source of carbon for all organic molecules in organisms.

Carbon dioxide is a molecule with only one carbon atom. But a carbon atom can also use one or more valence electrons to form covalent bonds to other carbon atoms, linking the atoms into chains, as shown here for C_3H_8:

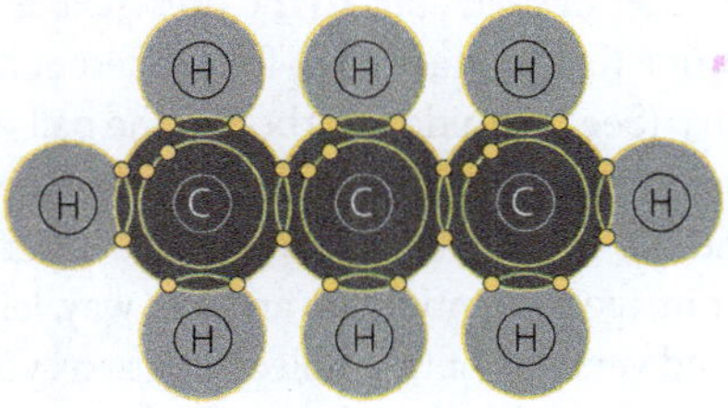

Molecular Diversity Arising from Variation in Carbon Skeletons

Carbon chains form the skeletons of most organic molecules. The skeletons vary in length and may be straight, branched, or arranged in closed rings (**Figure 4**). Some carbon skeletons have double bonds, which vary in number and location. Such variation in carbon skeletons is one important source of the molecular complexity and diversity that characterize living matter. In addition, atoms of other elements can be bonded to the skeletons at available sites.

Hydrocarbons

All of the molecules shown in Figures 2 and 4 are **hydrocarbons**, organic molecules consisting of only carbon and hydrogen. Atoms of hydrogen are attached to the carbon skeleton wherever electrons are available for covalent bonding. Hydrocarbons are the major components of petroleum, which is called a fossil fuel because it consists of the partially decomposed remains of organisms that lived millions of years ago. Although hydrocarbons are not prevalent in most living organisms, many of a cell's organic molecules have regions consisting of only carbon and hydrogen. For example, the molecules known as fats have long hydrocarbon tails attached to a nonhydrocarbon component (as you will see in Figure 13). Neither petroleum nor fat dissolves in water; both are hydrophobic compounds because the great majority of their bonds are relatively nonpolar carbon-to-hydrogen linkages. Another characteristic of hydrocarbons is that they can undergo reactions that release a relatively large amount of energy. The gasoline that fuels a car consists of hydrocarbons, and the hydrocarbon tails of fats serve as stored fuel for plant embryos (seeds) and animals.

▼ Figure 4 Four ways that carbon skeletons can vary.

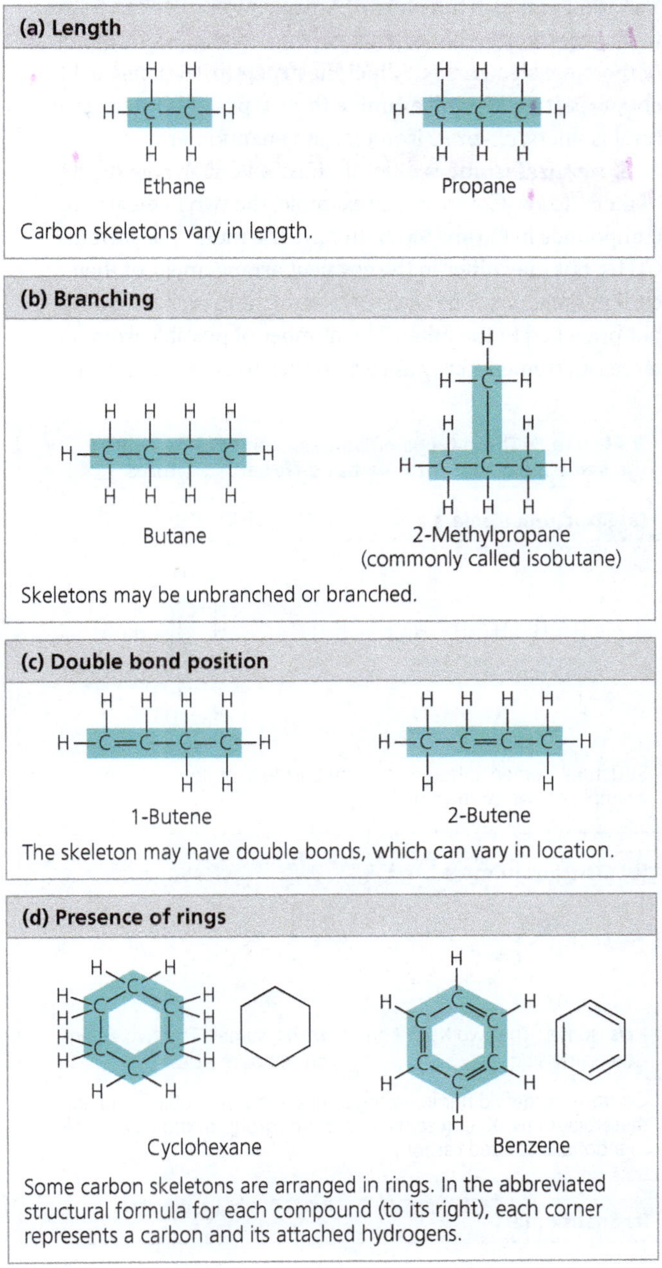

Isomers

Variation in the architecture of organic molecules can be seen in **isomers**, compounds that have the same numbers of atoms of the same elements but different structures and hence different properties. We will examine three types of isomers: structural isomers, *cis-trans* isomers, and enantiomers.

Structural isomers differ in the covalent arrangements of their atoms. Compare, for example, the two five-carbon compounds in **Figure 5a.** Both have the molecular formula C_5H_{12}, but they differ in the covalent arrangement of their carbon skeletons. The skeleton is straight in one compound but branched in the other. The number of possible isomers increases tremendously as carbon skeletons increase in size.

▼ Figure 5 Three types of isomers, compounds with the same molecular formula but different structures.

(a) Structural isomers

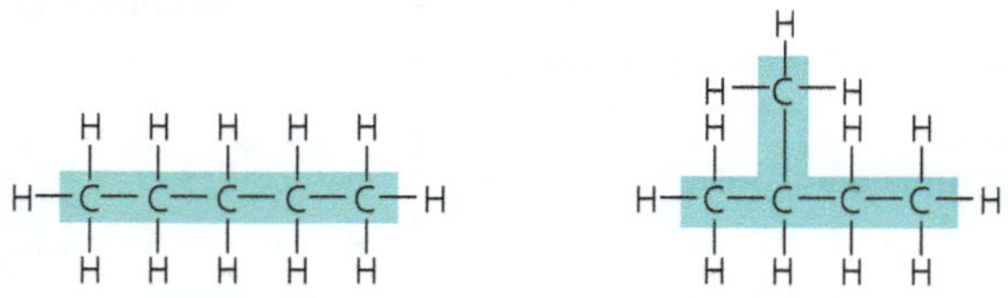

Pentane 2-Methylbutane

Structural isomers differ in covalent partners, as shown in this example of two isomers of C_5H_{12}.

(b) *Cis-trans* isomers

cis isomer: The two Xs are on the same side.

trans isomer: The two Xs are on opposite sides.

Cis-trans isomers differ in arrangement about a double bond. In these diagrams, X represents an atom or group of atoms attached to a double-bonded carbon.

(c) Enantiomers

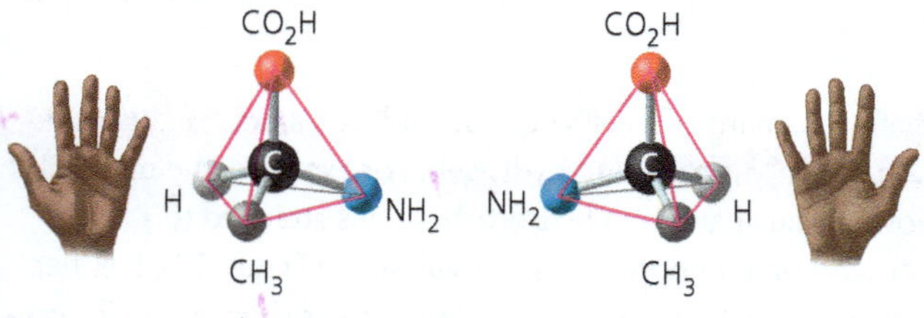

L isomer D isomer

Enantiomers differ in spatial arrangement around an asymmetric carbon, resulting in molecules that are mirror images, like left and right hands. The two isomers here are designated the L and D isomers from the Latin for "left" and "right" (*levo* and *dextro*). Enantiomers cannot be superimposed on each other.

DRAW IT *There are three structural isomers of C_5H_{12} (two of which are shown in Figure 5a); draw the one not shown in (a).*

There are only three forms of C_5H_{12} (two of which are shown in Figure 5a), but there are 18 variants of C_8H_{18} and 366,319 possible structural isomers of $C_{20}H_{42}$. Structural isomers may also differ in the location of double bonds.

In ***cis-trans*** **isomers**, carbons have covalent bonds to the same atoms, but these atoms differ in their spatial arrangements due to the inflexibility of double bonds. Single bonds allow the atoms they join to rotate freely about the bond axis without changing the compound. In contrast, double bonds do not permit such rotation. If a double bond joins two carbon atoms, and each C also has two different atoms (or groups of atoms) attached to it, then two distinct *cis-trans* isomers are possible. Consider a simple molecule with two double-bonded carbons, each of which has an H and an X attached to it **(Figure 5b)**. The arrangement with both Xs on the same side of the double bond is called a *cis isomer*, and that with the Xs on opposite sides is called a *trans isomer*. The subtle difference in shape between such isomers can dramatically affect the biological activities of organic molecules. For example, the biochemistry of vision involves a light-induced change of retinal, a chemical compound in the eye, from the *cis* isomer to the *trans* isomer. Another example involves *trans* fats, which are discussed in later in this chapter.

Enantiomers are isomers that are mirror images of each other and that differ in shape due to the presence of an *asymmetric carbon*, one that is attached to four different atoms or groups of atoms. (See the middle carbon in the ball-and-stick models shown in **Figure 5c**.) The four groups can be arranged in space around the asymmetric carbon in two different ways that are mirror images. Enantiomers are, in a way, left-handed and right-handed versions of the molecule. Just as your right hand won't fit into a left-handed glove, a "right-handed" molecule won't fit into the same space as the "left-handed" version. Usually, only one isomer is biologically active because only that form can bind to specific molecules in an organism.

The concept of enantiomers is important in the pharmaceutical industry because the two enantiomers of a drug may not be equally effective, as is the case for both ibuprofen and the asthma medication albuterol. Methamphetamine also occurs in two enantiomers that have very different effects. One enantiomer is the highly addictive stimulant drug known as "crank," sold illegally in the street drug trade. The other has a much weaker effect and is the active ingredient in an over-the-counter vapor inhaler for treatment of nasal congestion. The differing effects of enantiomers in the body demonstrate that organisms are sensitive to even the most subtle variations in molecular architecture. Once again, we see that molecules have emergent properties that depend on the specific arrangement of their atoms.

The Chemical Groups Most Important to Life

The distinctive properties of an organic molecule depend not only on the arrangement of its carbon skeleton but also on the various chemical groups attached to that skeleton **(Figure 6)**.

▼ **Figure 6** **Some biologically important chemical groups.**

Chemical Group	Compound Name	Examples
Hydroxyl group (— OH)  (may be written HO—)	**Alcohol** (The specific name usually ends in *-ol*.)	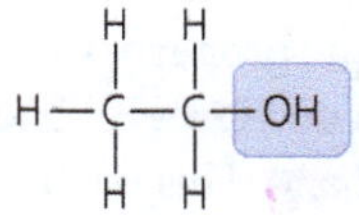 **Ethanol**, the alcohol present in alcoholic beverages
Carbonyl group (>C=O) 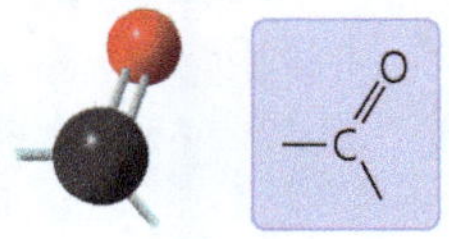	**Ketone** if the carbonyl group is within a carbon skeleton **Aldehyde** if the carbonyl group is at the end of a carbon skeleton	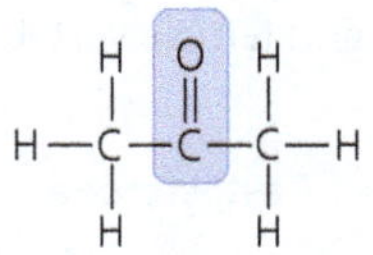 **Acetone**, the simplest ketone 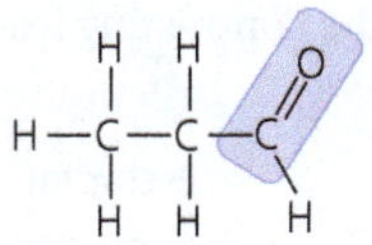 **Propanal**, an aldehyde
Carboxyl group (— COOH) 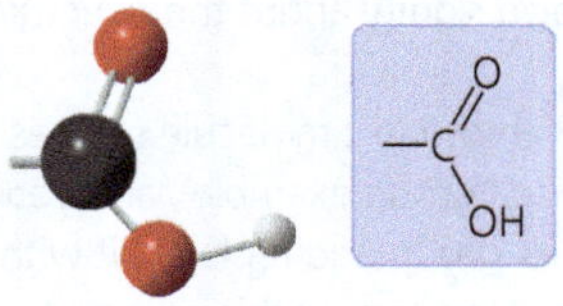	**Carboxylic acid**, or **organic acid**	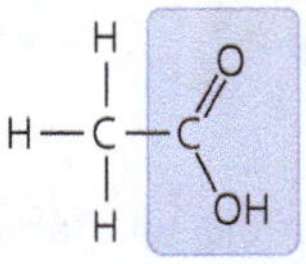 **Acetic acid**, which gives vinegar its sour taste 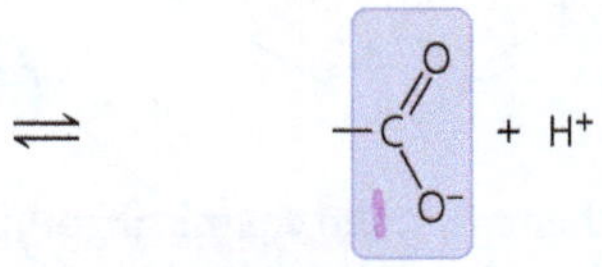 Ionized form of —COOH (carboxylate ion), found in cells
Amino group (—NH_2) 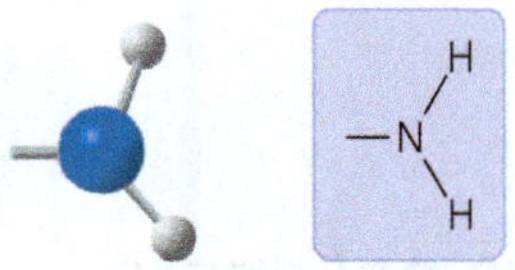	**Amine**	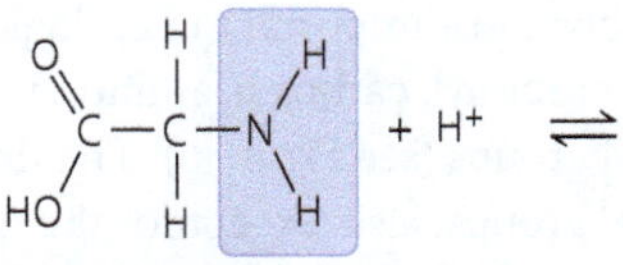 **Glycine**, an amino acid (note its carboxyl group) 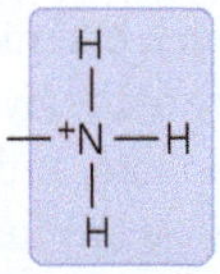 Ionized form of —NH_2, found in cells
Sulfhydryl group (—SH) 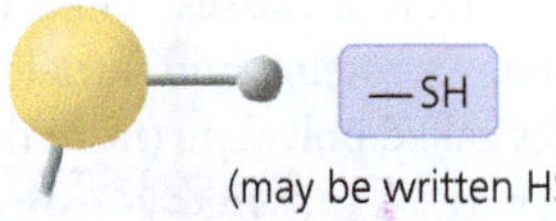 (may be written HS —)	**Thiol**	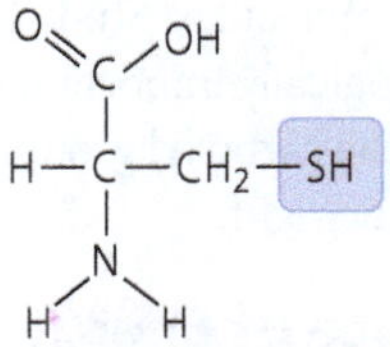 **Cysteine**, a sulfur-containing amino acid
Phosphate group (—OPO_3^{2-}) 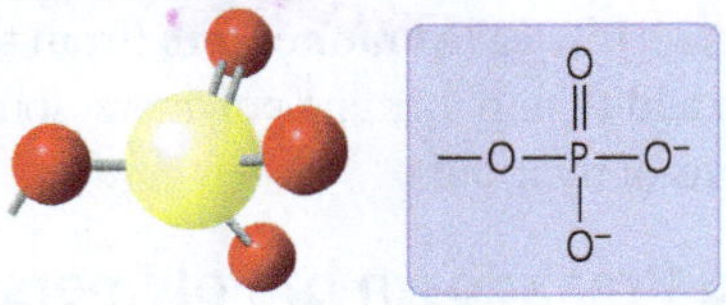	**Organic phosphate**	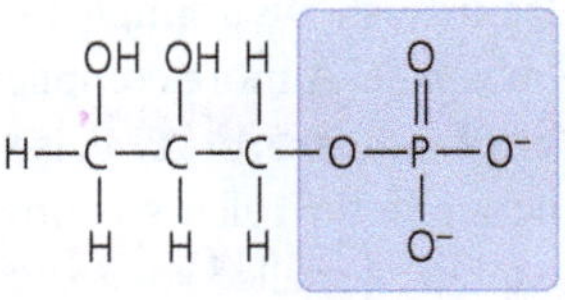 **Glycerol phosphate**, which takes part in many important chemical reactions in cells
Methyl group (—CH_3) 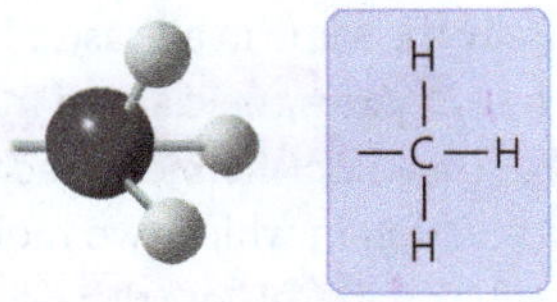	**Methylated compound**	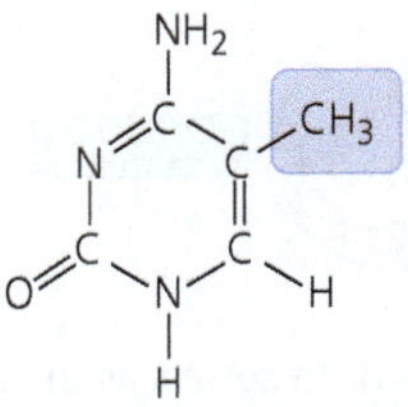 **5-Methylcytosine**, a component of DNA that has been modified by addition of a methyl group

We can think of hydrocarbons, the simplest organic molecules, as the underlying framework for more complex organic molecules. A number of chemical groups can replace one or more of the hydrogens bonded to the carbon skeleton of the hydrocarbon. The number and arrangement of chemical groups help give each organic molecule its unique properties.

In some cases, chemical groups contribute to function primarily by affecting the molecule's shape. This is true for the steroid sex hormones estradiol (a type of estrogen) and testosterone, which differ in attached chemical groups and act to produce the contrasting features of male and female vertebrates.

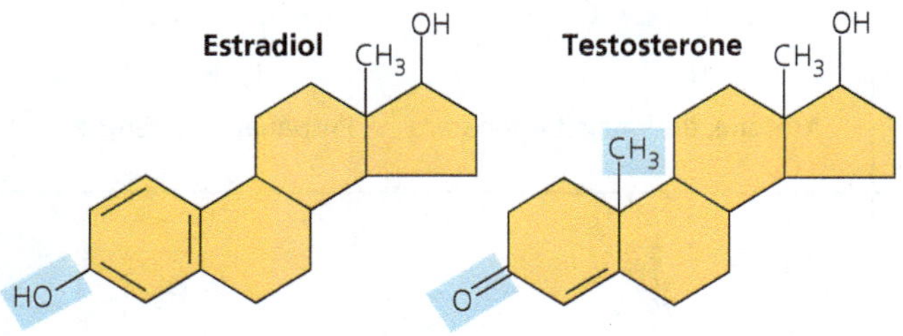

In other cases, the chemical groups affect molecular function by being directly involved in chemical reactions; these important chemical groups are known as **functional groups**. Each functional group participates in chemical reactions in a characteristic way.

The seven chemical groups most important in biological processes are the hydroxyl, carbonyl, carboxyl, amino, sulfhydryl, phosphate, and methyl groups (see Figure 6). The first six groups can act as functional groups; also, except for the sulfhydryl group, they are hydrophilic and thus increase the solubility of organic compounds in water. The last group, the methyl group, is not reactive, but instead often serves as a recognizable tag on biological molecules. Study Figure 6 to become familiar with these biologically important chemical groups. At normal cellular pH, the carboxyl group and amino group are ionized, as shown at the right.

ATP: An Important Source of Energy for Cellular Processes

The "Phosphate group" row in Figure 6 shows a simple example of an organic phosphate molecule. A more complicated organic phosphate, **adenosine triphosphate**, or **ATP**, is worth mentioning here because its function in the cell is so important. ATP consists of an organic molecule called adenosine attached to a string of three phosphate groups:

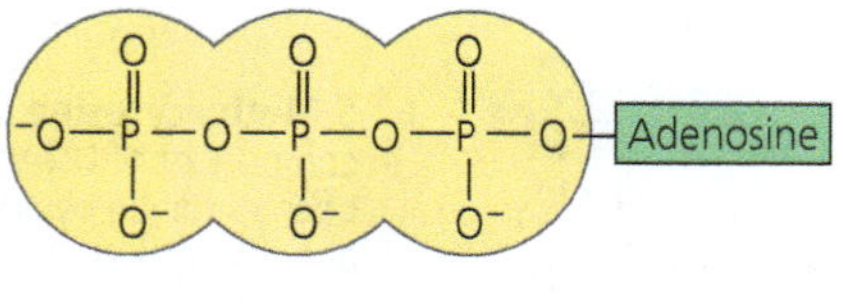

Where three phosphates are present in series, as in ATP, one phosphate may be split off as a result of a reaction with water. This inorganic phosphate ion, $HOPO_3^{2-}$, is often abbreviated $Ⓟ_i$ in this text, and a phosphate group in an organic molecule is often written as Ⓟ. Having lost one phosphate, ATP becomes adenosine *di*phosphate, or ADP. Although ATP is sometimes said to store energy, it is more accurate to think of it as storing the potential to react with water. This reaction releases energy that can be used by the cell.

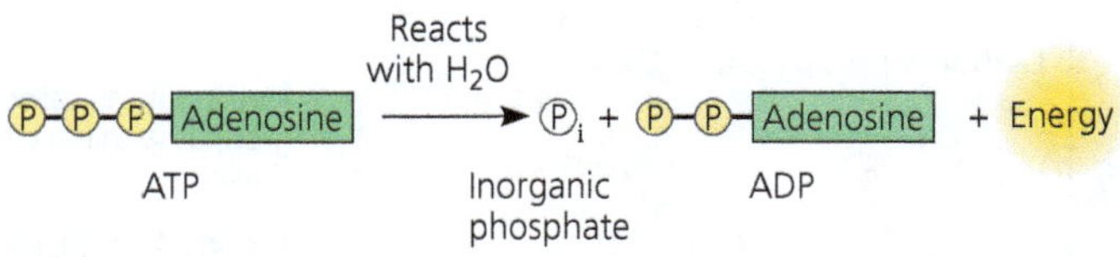

CONCEPT CHECK 1

1. How are gasoline and fat chemically similar?
2. Which molecules in Figure 4 are isomers? For each pair, identify the type of isomer.
3. What does the term *amino acid* signify about the structure of such a molecule?
4. **DRAW IT** Suppose you had an organic molecule such as cysteine (see Figure 6, sulfhydryl group example), and you chemically removed the —NH_2 group and replaced it with —COOH. Draw this structure. How would this change the chemical properties of the molecule? Is the central carbon asymmetric before the change? After?

CONCEPT 2

Macromolecules are polymers, built from monomers

The macromolecules in three of the four classes of life's organic compounds—carbohydrates, proteins, and nucleic acids—are chain-like molecules called polymers (from the Greek *polys*, many, and *meros*, part). A **polymer** is a long molecule consisting of many similar or identical building blocks linked by covalent bonds, much as a train consists of a chain of cars. The repeating units that serve as the building blocks of a polymer are smaller molecules called **monomers** (from the Greek *monos*, single). In addition to forming polymers, some monomers have functions of their own.

The Synthesis and Breakdown of Polymers

Although each class of polymer is made up of a different type of monomer, the chemical mechanisms by which cells make and break down polymers are basically the same in all cases. In cells, these processes are facilitated by **enzymes**, specialized macromolecules (usually proteins) that speed up chemical reactions. Monomers are connected by a reaction in which two molecules are covalently bonded to each other, with the loss of a water molecule; this is known as a **dehydration reaction** (Figure 7a).

When a bond forms between two monomers, each monomer contributes part of the water molecule that is released during the reaction: One monomer provides a hydroxyl group (—OH), while the other provides a hydrogen (—H). This reaction is repeated as monomers are added to the chain one by one, making a polymer (also called polymerization).

Polymers are disassembled to monomers by **hydrolysis**, a process that is essentially the reverse of the dehydration reaction **(Figure 7b)**. Hydrolysis means water breakage (from the Greek *hydro*, water, and *lysis*, break). The bond between the monomers is broken by the addition of a water molecule, with a hydrogen from water attaching to one monomer and the hydroxyl group attaching to the other. An example of hydrolysis working within our bodies is the process of digestion. The bulk of the organic material in our food is in the form of polymers that are much too large to enter our cells. Within the digestive tract, various enzymes attack the polymers, speeding up hydrolysis. Released monomers are then absorbed into the bloodstream for distribution to all body cells. Those cells can then use dehydration reactions to assemble the monomers into new, different polymers that can perform specific functions required by the cell.

The Diversity of Polymers

A cell has thousands of different macromolecules; the collection varies from one type of cell to another even in the same organism. The inherited differences between close relatives, such as human siblings, reflect small variations in polymers, particularly DNA and proteins. Molecular differences between unrelated individuals are more extensive, and those between species greater still. The diversity of macromolecules in the living world is vast, and the possible variety is effectively limitless.

What is the basis for such diversity in life's polymers? These molecules are constructed from only 40 to 50 common monomers and some others that occur rarely. Building a huge variety of polymers from such a limited number of monomers is analogous to constructing hundreds of thousands of words from only 26 letters of the alphabet. The key is arrangement—the particular linear sequence that the units follow. However, this analogy falls far short of describing the great diversity of macromolecules because most biological polymers have many more monomers than the number of letters in even the longest word. Proteins, for example, are built from 20 kinds of amino acids arranged in chains that are typically hundreds of amino acids long. The molecular logic of life is simple but elegant: Small molecules common to all organisms are ordered into unique macromolecules.

Despite this immense diversity, molecular structure and function can still be grouped roughly by class. Let's examine each of the four major classes of large biological molecules. For each class, the large molecules have emergent properties not found in their individual building blocks.

▼ **Figure 7 The synthesis and breakdown of polymers.**

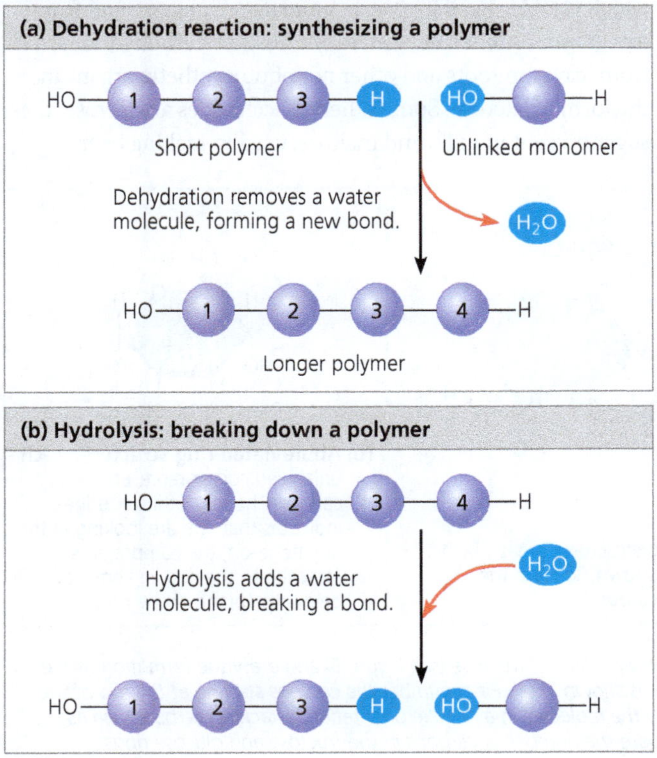

CONCEPT CHECK 2

1. How many molecules of water are needed to completely hydrolyze a polymer that is ten monomers long?
2. **WHAT IF?** Suppose you eat a serving of fish. What reactions must occur for the amino acid monomers in the protein of the fish to be converted to new proteins in your body?

CONCEPT 3

Carbohydrates serve as fuel and building material

Carbohydrates include both sugars and polymers of sugars. The simplest carbohydrates are the monosaccharides, or simple sugars; these are the monomers from which more complex carbohydrates are built. Disaccharides are double sugars, consisting of two monosaccharides joined by a covalent bond. Carbohydrates also include macromolecules called polysaccharides, polymers composed of many sugar building blocks joined together by dehydration reactions.

Sugars

Monosaccharides (from the Greek *monos*, single, and *sacchar*, sugar) generally have molecular formulas that are some multiple of the unit CH_2O. Glucose ($C_6H_{12}O_6$), the most common

▼ **Figure 8 Examples of monosaccharides.** Sugars vary in the location of their carbonyl groups (orange) and the length of their carbon skeletons.

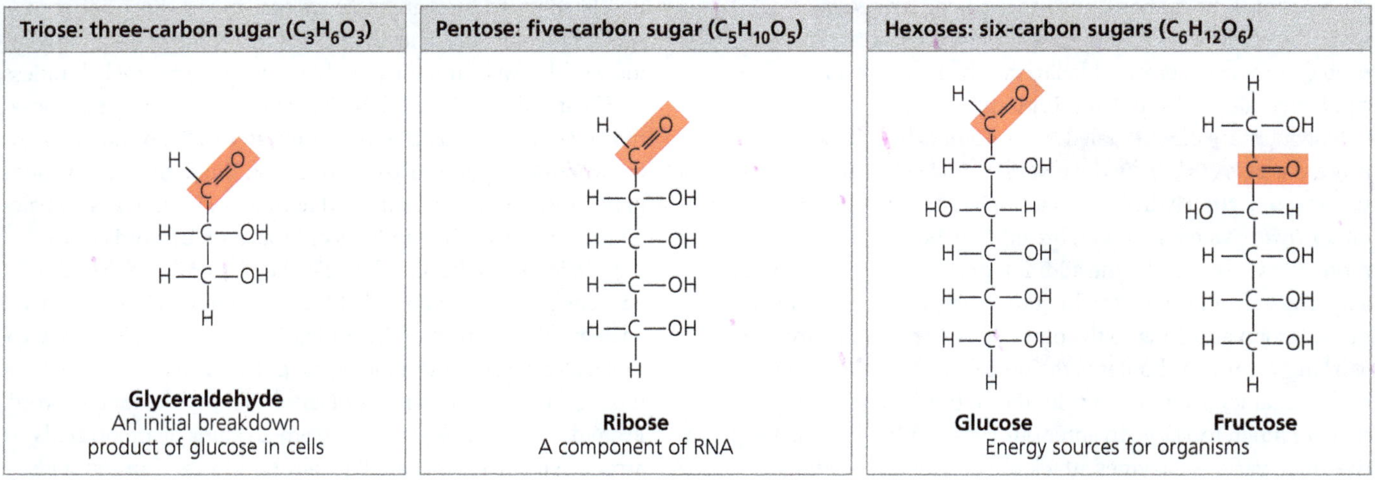

Sugars end in -ose

monosaccharide, is of central importance in the chemistry of life. In the structure of glucose, we can see the trademarks of a sugar: The molecule has a carbonyl group (>C=O) and multiple hydroxyl groups (—OH) **(Figure 8)**. The carbonyl group can be on the end of the linear sugar molecule, as in glucose, or attached to an interior carbon, as in fructose. (Thus, sugars are either aldehydes or ketones; see Figure 6.) The carbon skeleton of a sugar molecule ranges from three to seven carbons long. Glucose, fructose, and other sugars that have six carbons are called hexoses. Trioses (three-carbon sugars) and pentoses (five-carbon sugars) are also common. Note that most names for sugars end in *-ose*.

Although it is convenient to draw glucose with a linear carbon skeleton, this representation is not completely accurate. In aqueous solutions, glucose molecules (as well as most other five- and six-carbon sugars) form rings, because they are the most stable form of these sugars under physiological conditions **(Figure 9)**.

Monosaccharides, particularly glucose, are major nutrients for cells. In the process known as cellular respiration, cells extract energy from glucose molecules by breaking them down in a series of reactions. Also, the carbon skeletons of sugars serve as raw material for the synthesis of other types of small organic molecules, such as amino acids. Sugar molecules that are not immediately used in these ways are generally incorporated as monomers into disaccharides or polysaccharides.

A **disaccharide** consists of two monosaccharides joined by a **glycosidic linkage**, a covalent bond formed between two monosaccharides by a dehydration reaction (*glyco* refers to carbohydrate). The most prevalent disaccharide is sucrose, which is table sugar. Its two monomers are glucose and fructose **(Figure 10)**. Plants generally transport carbohydrates from leaves to roots and other nonphotosynthetic organs in the form of sucrose. Some other disaccharides are lactose, the sugar present in milk, and maltose, used in making beer.

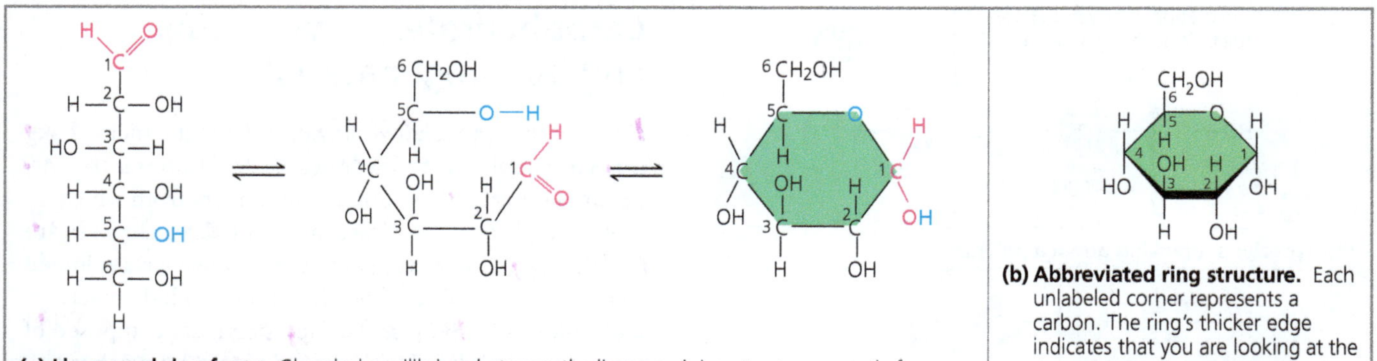

(a) Linear and ring forms. Chemical equilibrium between the linear and ring structures greatly favors the formation of rings. The carbons of the sugar are numbered 1 to 6, as shown. To form the glucose ring, carbon 1 (magenta) bonds to the oxygen (blue) attached to carbon 5.

(b) Abbreviated ring structure. Each unlabeled corner represents a carbon. The ring's thicker edge indicates that you are looking at the ring edge-on; the components attached to the ring lie above or below the plane of the ring.

▲ **Figure 9 Linear and ring forms of glucose.**

DRAW IT *Start with the linear form of fructose (see Figure 8) and draw the formation of the fructose ring in two steps, as shown in (a). First, number the carbons starting at the top of the linear structure. Then draw the molecule in a ringlike orientation, attaching carbon 5 via its oxygen to carbon 2. Compare the number of carbons in the fructose and glucose rings.*

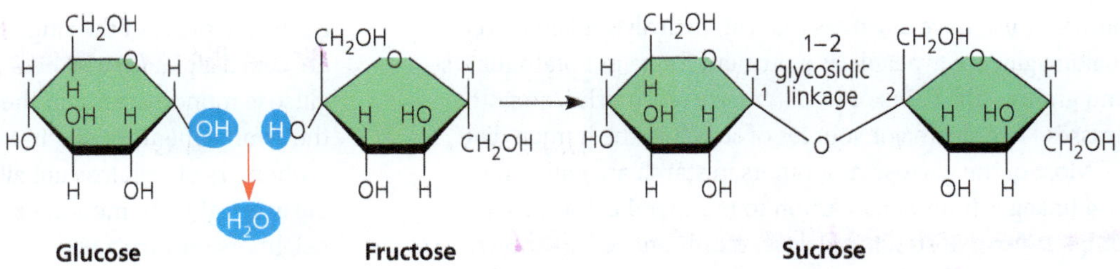

▶ **Figure 10 Disaccharide synthesis.** Sucrose is a disaccharide formed from glucose and fructose by a dehydration reaction. Notice that fructose, though a hexose like glucose, forms a five-sided ring.

DRAW IT *Number the carbons in each sugar (see Figure 9). How does the name of the linkage relate to the numbers?*

Polysaccharides

Polysaccharides are macromolecules, polymers with a few hundred to a few thousand monosaccharides joined by glycosidic linkages. Some polysaccharides serve as storage material, hydrolyzed as needed to provide sugar for cells. Other polysaccharides serve as building material for structures that protect the cell or the whole organism. The structure and function of a polysaccharide are determined by its sugar monomers and by the positions of its glycosidic linkages.

Storage Polysaccharides

Both plants and animals store sugars for later use in the form of storage polysaccharides (**Figure 11**). Plants store **starch**, a polymer of glucose monomers, as granules within cells. Synthesizing starch enables the plant to stockpile surplus glucose. Because glucose is a major cellular fuel, starch represents stored energy. The sugar can later be withdrawn from this carbohydrate "bank" by hydrolysis, which breaks the bonds between the glucose monomers. Most animals, including

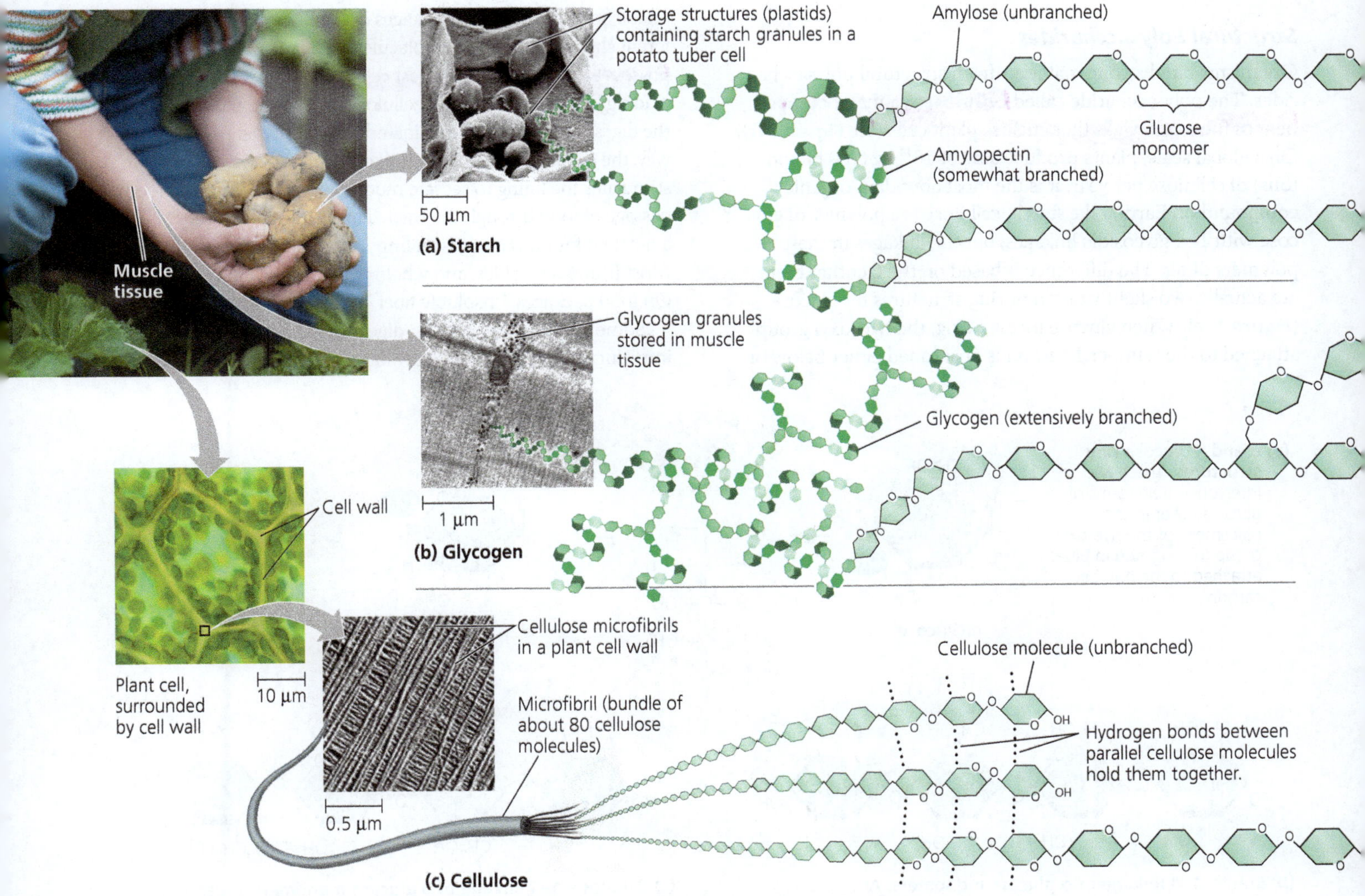

▲ **Figure 11 Polysaccharides of plants and animals. (a)** Starch stored in plant cells, **(b)** glycogen stored in muscle cells, and **(c)** structural cellulose fibers in plant cell walls are all polysaccharides composed entirely of glucose monomers (green hexagons). In starch and glycogen, the polymer chains tend to form helices in unbranched regions because of the angle of the 1–4 linkage between the glucose monomers. Cellulose, with a different kind of 1–4 linkage, is always unbranched.

humans, also have enzymes that can hydrolyze plant starch, making glucose available as a nutrient for cells. Potato tubers and grains—the fruits of wheat, maize (corn), rice, and other grasses—are the major sources of starch in the human diet.

Most of the glucose monomers in starch are joined by 1–4 linkages (number 1 carbon to number 4 carbon). The simplest form of starch, amylose, is unbranched, as shown in **Figure 11a**. Amylopectin, a more complex starch, is a branched polymer with 1–6 linkages at the branch points.

Animals store a polysaccharide called **glycogen,** a polymer of glucose that is like amylopectin but more extensively branched (see **Figure 11b**). Humans and other vertebrates store glycogen mainly in liver and muscle cells. Hydrolysis of glycogen in these cells releases glucose when the demand for sugar increases. This stored fuel cannot sustain an animal for long, however. In humans, for example, glycogen stores are depleted in about a day unless they are replenished by eating. This is an issue of concern in ultra-low-carbohydrate diets, which can result in weakness and fatigue.

Structural Polysaccharides

Organisms build strong materials from structural polysaccharides. The polysaccharide called **cellulose** is a major component of the tough walls that enclose plant cells (see **Figure 11c**). On a global scale, plants produce almost 10^{14} kg (100 billion tons) of cellulose per year; it is the most abundant organic compound on Earth. Like starch, cellulose is a polymer of glucose with 1–4 glycosidic linkages, but the linkages in these two polymers differ. The difference is based on the fact that there are actually two slightly different ring structures for glucose **(Figure 12a)**. When glucose forms a ring, the hydroxyl group attached to the number 1 carbon is positioned either below or above the plane of the ring. These two ring forms for glucose are called alpha (α) and beta (β), respectively. In starch, all the glucose monomers are in the α configuration **(Figure 12b)**, the arrangement we saw in Figure 9. In contrast, the glucose monomers of cellulose are all in the β configuration, making every glucose monomer "upside down" with respect to its neighbors **(Figure 12c)**.

The differing glycosidic linkages in starch and cellulose give the two molecules distinct three-dimensional shapes. Whereas certain starch molecules are largely helical, a cellulose molecule is straight. Cellulose is never branched, and some hydroxyl groups on its glucose monomers are free to hydrogen-bond with the hydroxyls of other cellulose molecules lying parallel to it. In plant cell walls, parallel cellulose molecules held together in this way are grouped into units called microfibrils (see Figure 11c). These cable-like microfibrils are a strong building material for plants and an important substance for humans because cellulose is the major constituent of paper and the only component of cotton.

Enzymes that digest starch by hydrolyzing its α linkages are unable to hydrolyze the β linkages of cellulose due to the different shapes of these two molecules. In fact, few organisms possess enzymes that can digest cellulose. Almost all animals, including humans, do not; the cellulose in our food passes through the digestive tract and is eliminated with the feces. Along the way, the cellulose abrades the wall of the digestive tract and stimulates the lining to secrete mucus, which aids in the smooth passage of food through the tract. Thus, although cellulose is not a nutrient for humans, it is an important part of a healthful diet. Most fruits, vegetables, and whole grains are rich in cellulose. On food packages, "insoluble fiber" refers mainly to cellulose.

Some microorganisms can digest cellulose, breaking it down into glucose monomers. A cow harbors cellulose-digesting

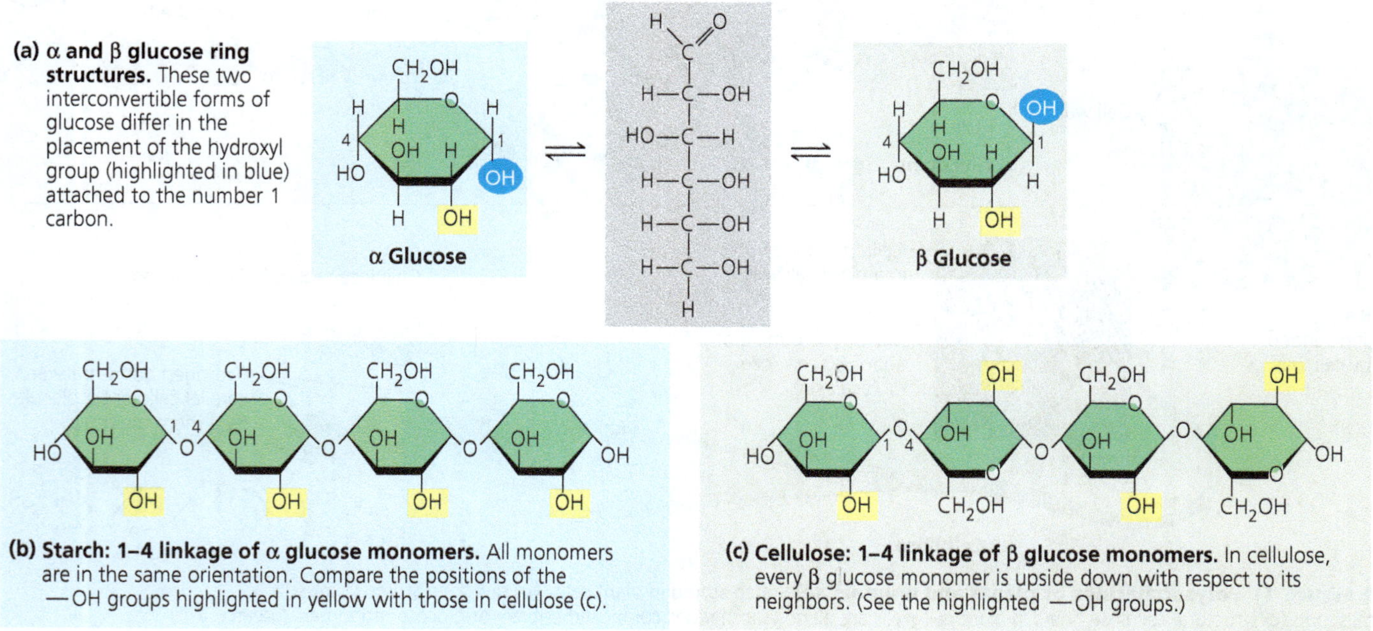

(a) α and β glucose ring structures. These two interconvertible forms of glucose differ in the placement of the hydroxyl group (highlighted in blue) attached to the number 1 carbon.

(b) Starch: 1–4 linkage of α glucose monomers. All monomers are in the same orientation. Compare the positions of the —OH groups highlighted in yellow with those in cellulose (c).

(c) Cellulose: 1–4 linkage of β glucose monomers. In cellulose, every β glucose monomer is upside down with respect to its neighbors. (See the highlighted —OH groups.)

▲ **Figure 12 Starch and cellulose structures.**

prokaryotes and protists in its gut. These microbes hydrolyze the cellulose of hay and grass and convert the glucose to other compounds that nourish the cow. Similarly, a termite, which is unable to digest cellulose by itself, has prokaryotes or protists living in its gut that can make a meal of wood. Some fungi can also digest cellulose in soil and elsewhere, thereby helping recycle chemical elements within Earth's ecosystems.

Another important structural polysaccharide is **chitin**, the carbohydrate used by arthropods (insects, spiders, crustaceans, and related animals) to build their exoskeletons—hard cases that surround the soft parts of these animals. Chitin is also found in many fungi, which use this polysaccharide as the building material for their cell walls. Chitin is similar to cellulose except that the glucose monomer of chitin has a nitrogen-containing attachment.

CONCEPT CHECK 3

1. Write the formula for a monosaccharide that has three carbons.
2. A dehydration reaction joins two glucose molecules to form maltose. The formula for glucose is $C_6H_{12}O_6$. What is the formula for maltose?
3. **WHAT IF?** After a cow is given antibiotics to treat an infection, a vet gives the animal a drink of "gut culture" containing various prokaryotes. Why is this necessary?

CONCEPT 4

Lipids are a diverse group of hydrophobic molecules

Lipids are the one class of large biological molecules that does not include true polymers, and they are generally not big enough to be considered macromolecules. The compounds called **lipids** are grouped together because they share one important trait: They mix poorly, if at all, with water. The hydrophobic behavior of lipids is based on their molecular structure. Although they may have some polar bonds associated with oxygen, lipids consist mostly of hydrocarbon regions. Lipids are varied in form and function. They include waxes and certain pigments, but we will focus on the types of lipids that are most biologically important: fats, phospholipids, and steroids.

Fats

Although fats are not polymers, they are large molecules assembled from smaller molecules by dehydration reactions. A **fat** is constructed from two kinds of smaller molecules: glycerol and fatty acids (Figure 13a). Glycerol is an alcohol; each of its three carbons bears a hydroxyl group. A **fatty acid** has a long carbon skeleton, usually 16 or 18 carbon atoms in length. The carbon at one end of the skeleton is part of a carboxyl group, the functional group that gives these molecules the name fatty *acid*. The rest of the skeleton consists of a hydrocarbon chain. The relatively nonpolar C—H bonds in the hydrocarbon chains of fatty acids are the reason fats are hydrophobic. Fats separate from water because the water molecules hydrogen-bond to one another and exclude the fats. This is the reason that vegetable oil (a liquid fat) separates from the aqueous vinegar solution in a bottle of salad dressing.

In making a fat, three fatty acid molecules are each joined to glycerol by an ester linkage, a bond between a hydroxyl group and a carboxyl group. The resulting fat, also called a **triacylglycerol**, thus consists of three fatty acids linked to one glycerol molecule. (Still another name for a fat is *triglyceride*, a word often found in the list of ingredients on packaged foods.) The fatty acids in a fat can be the same, or they can be of two or three different kinds, as in Figure 13b.

The terms *saturated fats* and *unsaturated fats* are commonly used in the context of nutrition. These terms refer to the structure of the hydrocarbon chains of the fatty acids. If there are no double bonds between carbon atoms composing a chain, then as many hydrogen atoms as possible are bonded to the carbon skeleton. Such a structure is said to be *saturated* with hydrogen, and the resulting fatty acid is called a **saturated fatty acid**. An **unsaturated fatty acid** has one or more double bonds, with one fewer hydrogen atom on each double-bonded carbon. Nearly every double bond in naturally occurring fatty acids is a *cis* double bond, which creates a kink

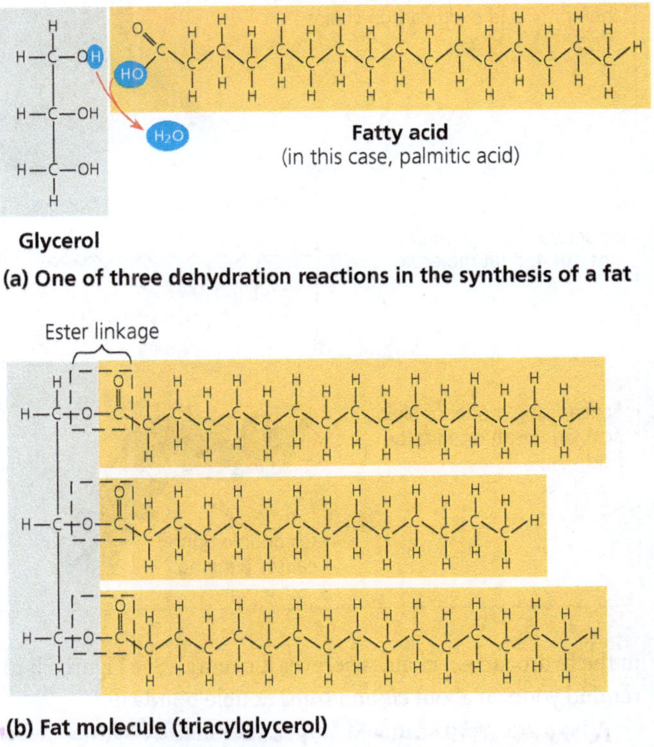

▲ **Figure 13 The synthesis and structure of a fat, or triacylglycerol.** The molecular building blocks of a fat are one molecule of glycerol and three molecules of fatty acids. **(a)** One water molecule is removed for each fatty acid joined to the glycerol. **(b)** A fat molecule with three fatty acid units, two of them identical. The carbons of the fatty acids are arranged zigzag to suggest the actual orientations of the four single bonds extending from each carbon (see Figure 2a).

▼ Figure 14 Saturated and unsaturated fats and fatty acids.

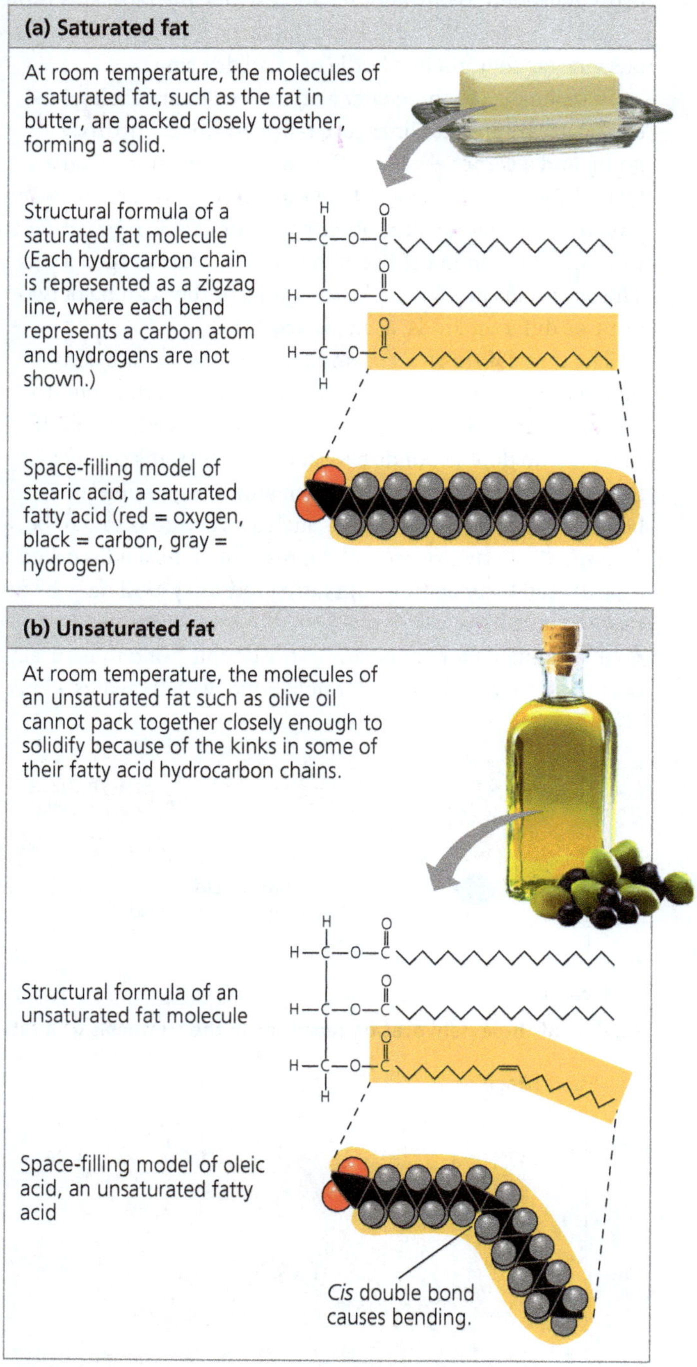

in the hydrocarbon chain wherever it occurs. (See Figure 5b to remind yourself about *cis* and *trans* double bonds.)

A fat made from saturated fatty acids is called a saturated fat. Most animal fats are saturated: The hydrocarbon chains of their fatty acids—the "tails" of the fat molecules—lack double bonds, and their flexibility allows the fat molecules to pack together tightly. Saturated animal fats—such as lard and butter—are solid at room temperature (**Figure 14a**). In contrast, the fats of plants and fishes are generally unsaturated, meaning that they are built of one or more types of unsaturated fatty acids. Usually liquid at room temperature, plant and fish fats are referred to as oils—olive oil and cod liver oil are examples (**Figure 14b**). The kinks where the *cis* double bonds are located prevent the molecules from packing together closely enough to solidify at room temperature. The phrase "hydrogenated vegetable oils" on food labels means that unsaturated fats have been synthetically converted to saturated fats by adding hydrogen, allowing them to solidify. This process also produces unsaturated fats with *trans* double bonds, known as ***trans* fats**. It appears that *trans* fats can contribute to coronary heart disease. Because *trans* fats are especially common in baked goods and processed foods, the U.S. Food and Drug Administration (FDA) requires nutritional labels to include information on *trans* fat content. The FDA has proposed a ban on *trans* fats in the U.S. food supply; some countries, such as Denmark and Switzerland, have already implemented restrictions on the level of *trans* fats in foods.

The major function of fats is energy storage. The hydrocarbon chains of fats are similar to gasoline molecules and just as rich in energy. A gram of fat stores more than twice as much energy as a gram of a polysaccharide, such as starch. Because plants are relatively immobile, they can function with bulky energy storage in the form of starch. (Vegetable oils are generally obtained from seeds, where more compact storage is an asset to the plant.) Animals, however, must carry their energy stores with them, so there is an advantage to having a more compact reservoir of fuel—fat.

Phospholipids

Cells could not exist without another type of lipid, called phospholipids. Phospholipids are essential for cells because they are major constituents of cell membranes. Their structure provides a classic example of how form fits function at the molecular level. As shown in **Figure 15**, a **phospholipid** is similar to a fat molecule but has only two fatty acids attached to glycerol rather than three. The third hydroxyl group of glycerol is joined to a phosphate group, which has a negative electrical charge in the cell. Additional small molecules, which are usually charged or polar, can be linked to the phosphate group to form a variety of phospholipids.

The two ends of a phospholipid exhibit different behavior toward water. The hydrocarbon tails are hydrophobic and are excluded from water. However, the phosphate group and its attachments form a hydrophilic head that has an affinity for water. When phospholipids are added to water, they self-assemble into double-layered structures called "bilayers," shielding their hydrophobic portions from water (see Figure 15d).

At the surface of a cell, phospholipids are arranged in a similar bilayer. The hydrophilic heads of the molecules are on the outside of the bilayer, in contact with the aqueous solutions inside and outside of the cell. The hydrophobic tails point toward the interior of the bilayer, away from the water. The phospholipid bilayer forms a boundary between the cell and its external environment; in fact, the existence of cells depends on the properties of phospholipids.

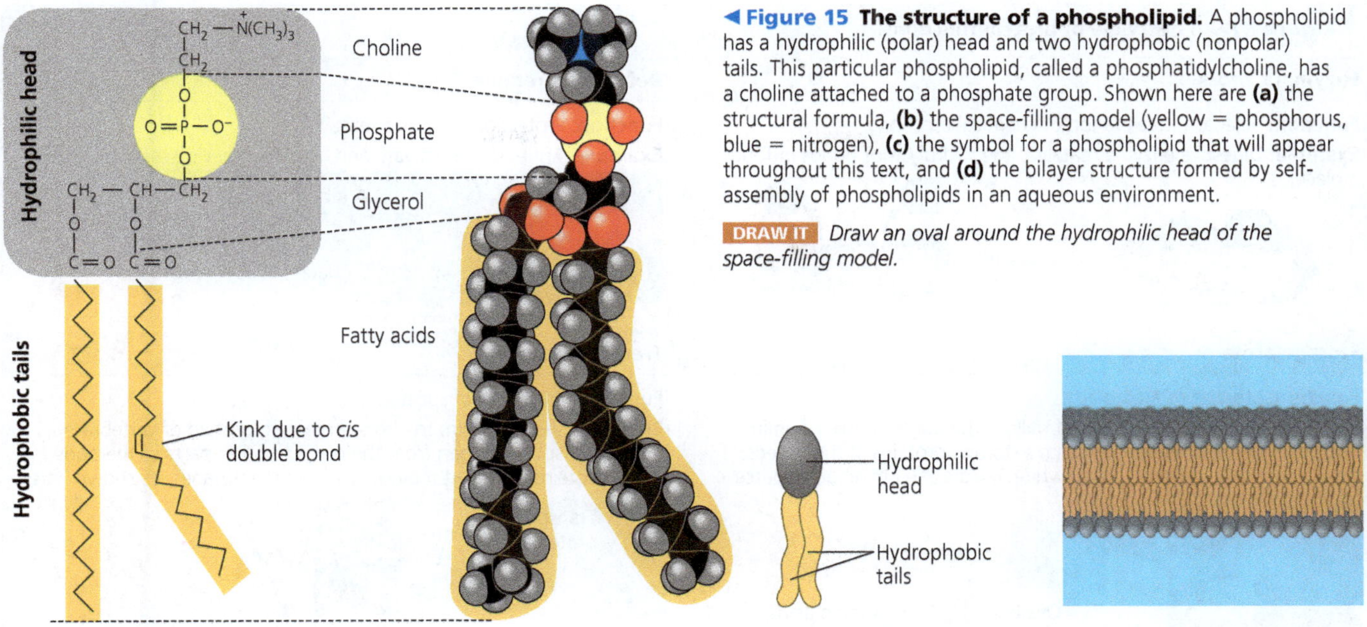

(a) Structural formula **(b) Space-filling model** **(c) Phospholipid symbol** **(d) Phospholipid bilayer**

◀ **Figure 15 The structure of a phospholipid.** A phospholipid has a hydrophilic (polar) head and two hydrophobic (nonpolar) tails. This particular phospholipid, called a phosphatidylcholine, has a choline attached to a phosphate group. Shown here are **(a)** the structural formula, **(b)** the space-filling model (yellow = phosphorus, blue = nitrogen), **(c)** the symbol for a phospholipid that will appear throughout this text, and **(d)** the bilayer structure formed by self-assembly of phospholipids in an aqueous environment.

DRAW IT *Draw an oval around the hydrophilic head of the space-filling model.*

Steroids

Steroids are lipids characterized by a carbon skeleton consisting of four fused rings. Different steroids are distinguished by the particular chemical groups attached to this ensemble of rings. Shown in **Figure 16**, **cholesterol** is a crucial steroid in animals. It is a common component of animal cell membranes and is also the precursor from which other steroids are synthesized, such as the vertebrate sex hormones estrogen and testosterone (see Concept 1). In vertebrates, cholesterol is synthesized in the liver and is also obtained from the diet. A high level of cholesterol in the blood may contribute to atherosclerosis, although some researchers are questioning the roles of cholesterol and saturated fats in development of this condition.

▶ **Figure 16 Cholesterol, a steroid.**

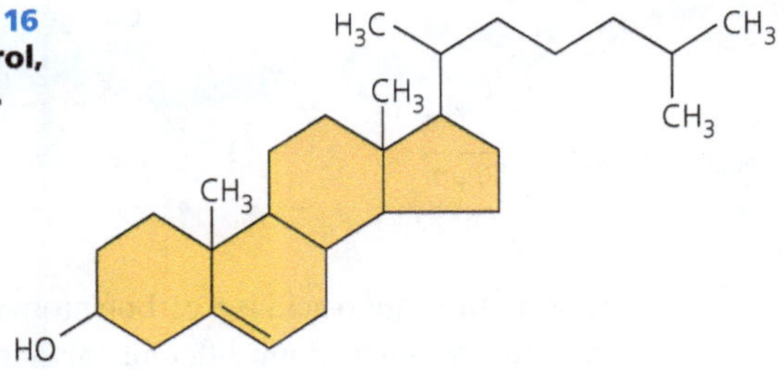

CONCEPT CHECK 4

1. Compare the structure of a fat (triglyceride) with that of a phospholipid.
2. Why are human sex hormones considered lipids?
3. **WHAT IF?** Suppose a membrane surrounded an oil droplet, as it does in the cells of plant seeds. Describe and explain the form it might take.

CONCEPT 5

Proteins include a diversity of structures, resulting in a wide range of functions

Nearly every dynamic function of a living being depends on proteins. In fact, the importance of proteins is underscored by their name, which comes from the Greek word *proteios*, meaning "first," or "primary." Proteins account for more than 50% of the dry mass of most cells, and they are instrumental in almost everything organisms do. Some proteins speed up chemical reactions, while others play a role in defense, storage, transport, cellular communication, movement, or structural support. **Figure 17** shows examples of proteins with these functions.

Life would not be possible without enzymes, most of which are proteins. Enzymatic proteins regulate metabolism by acting as **catalysts**, chemical agents that selectively speed up chemical reactions without being consumed in the reaction. Because an enzyme can perform its function over and over again, these molecules can be thought of as workhorses that keep cells running by carrying out the processes of life.

A human has tens of thousands of different proteins, each with a specific structure and function; proteins, in fact, are the most structurally sophisticated molecules known. Consistent with their diverse functions, they vary extensively in structure, each type of protein having a unique three-dimensional shape.

Proteins are all constructed from the same set of 20 amino acids, linked in unbranched polymers. The bond between amino acids is called a peptide bond, so a polymer of amino

▼ Figure 17 An overview of protein functions.

Enzymatic proteins

Function: Selective acceleration of chemical reactions

Example: Digestive enzymes catalyze the hydrolysis of bonds in food molecules.

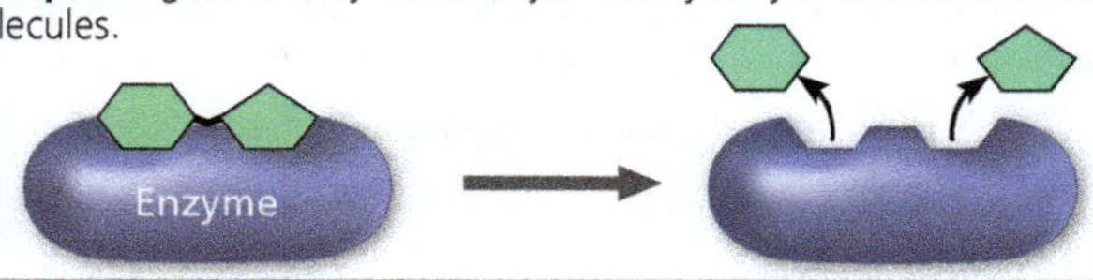

Defensive proteins

Function: Protection against disease

Example: Antibodies inactivate and help destroy viruses and bacteria.

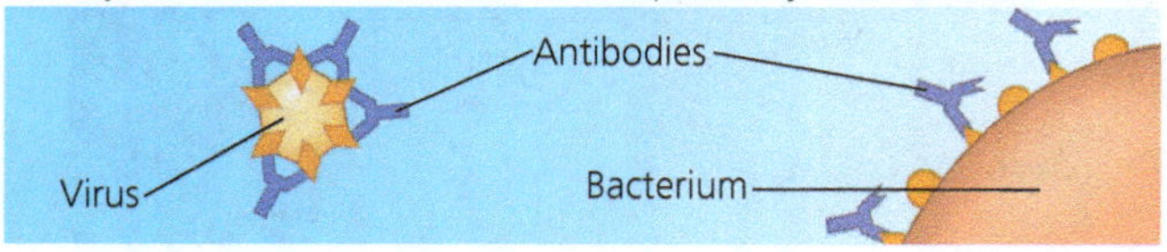

Storage proteins

Function: Storage of amino acids

Examples: Casein, the protein of milk, is the major source of amino acids for baby mammals. Plants have storage proteins in their seeds. Ovalbumin is the protein of egg white, used as an amino acid source for the developing embryo.

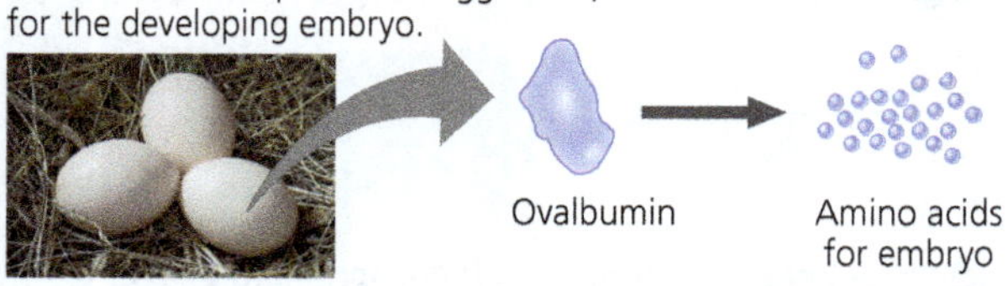

Transport proteins

Function: Transport of substances

Examples: Hemoglobin, the iron-containing protein of vertebrate blood, transports oxygen from the lungs to other parts of the body. Other proteins transport molecules across membranes, as shown here.

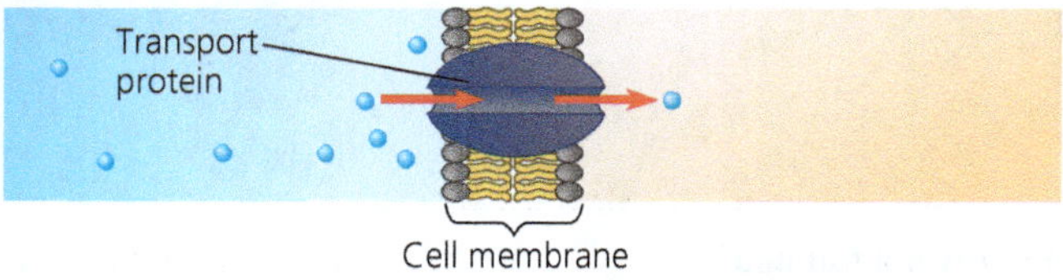

Hormonal proteins

Function: Coordination of an organism's activities

Example: Insulin, a hormone secreted by the pancreas, causes other tissues to take up glucose, thus regulating blood sugar concentration.

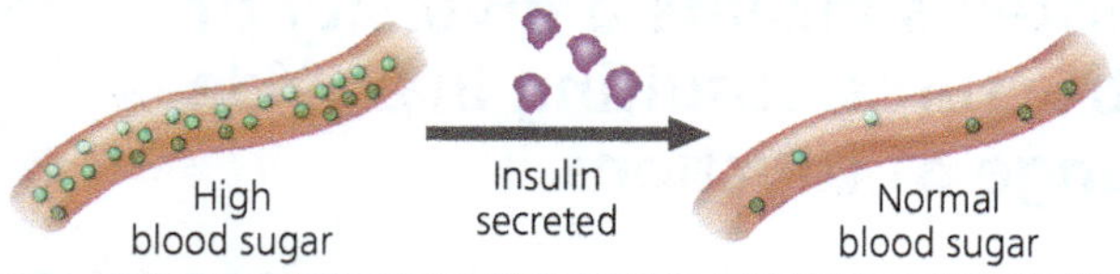

Receptor proteins

Function: Response of cell to chemical stimuli

Example: Receptors built into the membrane of a nerve cell detect signaling molecules released by other nerve cells.

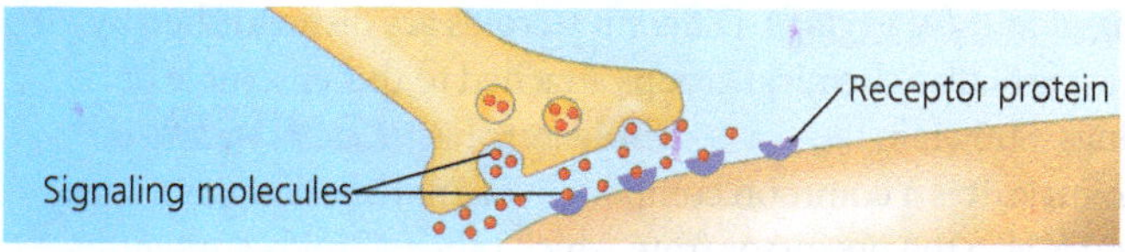

Contractile and motor proteins

Function: Movement

Examples: Motor proteins are responsible for the undulations of cilia and flagella. Actin and myosin proteins are responsible for the contraction of muscles.

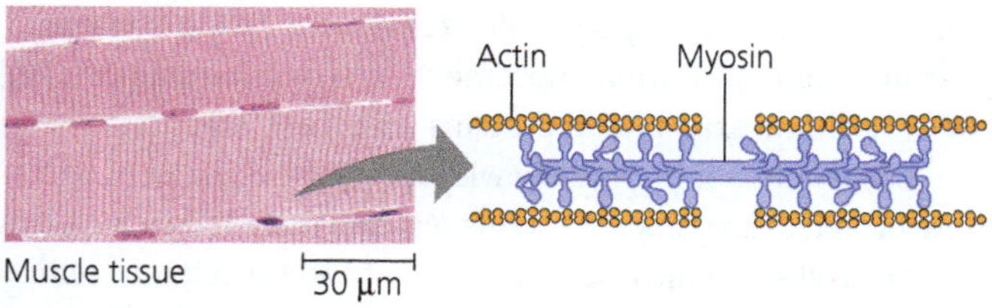

Structural proteins

Function: Support

Examples: Keratin is the protein of hair, horns, feathers, and other skin appendages. Insects and spiders use silk fibers to make their cocoons and webs, respectively. Collagen and elastin proteins provide a fibrous framework in animal connective tissues.

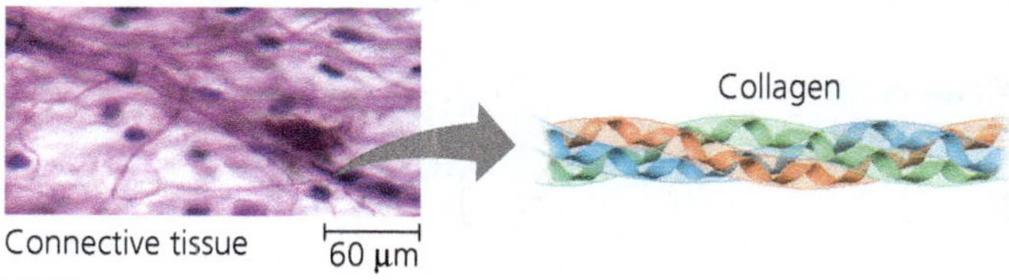

acids is called a **polypeptide**. A **protein** is a biologically functional molecule made up of one or more polypeptides folded and coiled into a specific three-dimensional structure.

Amino Acid Monomers

All amino acids share a common structure. An **amino acid** is an organic molecule with both an amino group and a carboxyl group; the small figure shows the general formula for an amino acid. At the center of the amino acid is a carbon atom called the *alpha* (α) *carbon*. Its four different partners are an amino group, a carboxyl group, a hydrogen atom, and a variable group symbolized by R. The R group, also called the side chain, differs with each amino acid **(Figure 18)**.

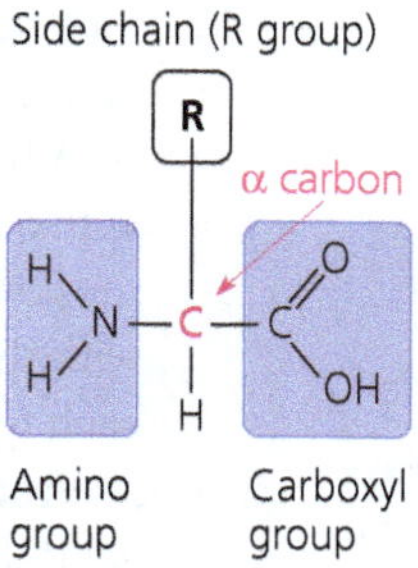

The 20 amino acids in Figure 18 are the ones cells use to build their proteins. Here the amino groups and carboxyl groups are all depicted in ionized form, the way they usually exist at the pH found in a cell. The

▼ Figure 18 The 20 amino acids of proteins. The amino acids are grouped here according to the properties of their side chains (R groups) and shown in their prevailing ionic forms at pH 7.2, the pH within a cell. The three-letter and one-letter abbreviations for the amino acids are in parentheses.

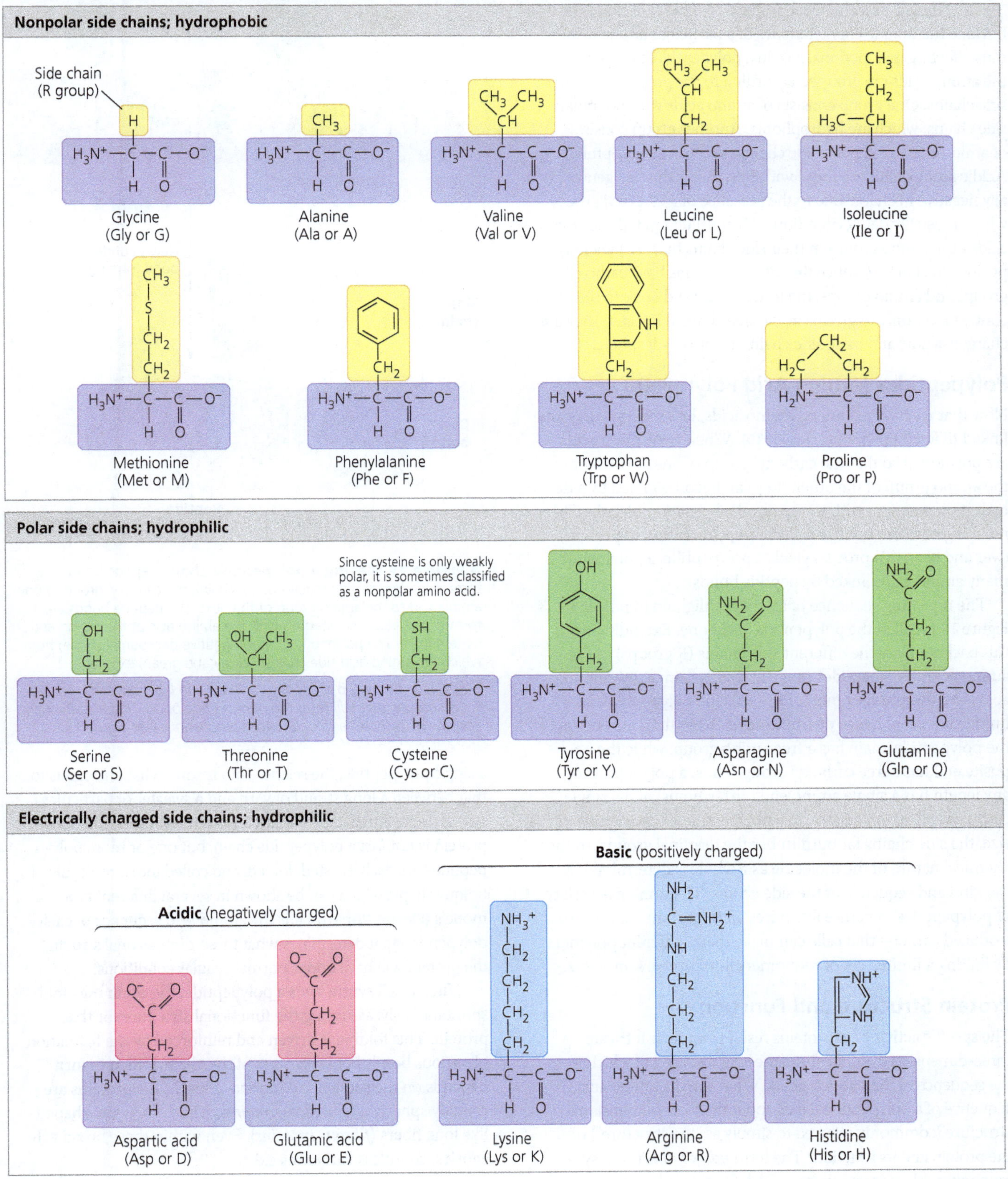

side chain (R group) may be as simple as a hydrogen atom, as in the amino acid glycine, or it may be a carbon skeleton with various functional groups attached, as in glutamine.

The physical and chemical properties of the side chain determine the unique characteristics of a particular amino acid, thus affecting its functional role in a polypeptide. In Figure 18, the amino acids are grouped according to the properties of their side chains. One group consists of amino acids with nonpolar side chains, which are hydrophobic. Another group consists of amino acids with polar side chains, which are hydrophilic. Acidic amino acids are those with side chains that are generally negative in charge due to the presence of a carboxyl group, which is usually dissociated (ionized) at cellular pH. Basic amino acids have amino groups in their side chains that are generally positive in charge. (Notice that *all* amino acids have carboxyl groups and amino groups; the terms *acidic* and *basic* in this context refer only to groups in the side chains.) Because they are charged, acidic and basic side chains are also hydrophilic.

Polypeptides (Amino Acid Polymers)

Now that we have examined amino acids, let's see how they are linked to form polymers (**Figure 19**). When two amino acids are positioned so that the carboxyl group of one is adjacent to the amino group of the other, they can become joined by a dehydration reaction, with the removal of a water molecule. The resulting covalent bond is called a **peptide bond**. Repeated over and over, this process yields a polypeptide, a polymer of many amino acids linked by peptide bonds.

The repeating sequence of atoms highlighted in purple in Figure 19 is called the polypeptide backbone. Extending from this backbone are the different side chains (R groups) of the amino acids. Polypeptides range in length from a few amino acids to a thousand or more. Each specific polypeptide has a unique linear sequence of amino acids. Note that one end of the polypeptide chain has a free amino group, while the opposite end has a free carboxyl group. Thus, a polypeptide of any length has a single amino end (N-terminus) and a single carboxyl end (C-terminus). In a polypeptide of any significant size, the side chains far outnumber the terminal groups, so the chemical nature of the molecule as a whole is determined by the kind and sequence of the side chains. The immense variety of polypeptides in nature illustrates an important concept introduced earlier—that cells can make many different polymers by linking a limited set of monomers into diverse sequences.

Protein Structure and Function

The specific activities of proteins result from their intricate three-dimensional architecture, the simplest level of which is the sequence of their amino acids. What can the amino acid sequence of a polypeptide tell us about the three-dimensional structure (commonly referred to simply as "the structure") of the protein and its function? The term *polypeptide* is not synonymous with the term *protein*. Even for a protein consisting of a single polypeptide, the relationship is somewhat analogous to that between a long strand of yarn and a sweater of particular size and shape that can be knitted from the yarn. A functional protein is not *just* a polypeptide chain, but one or more polypeptides precisely twisted, folded, and coiled into a molecule of unique shape, which can be shown in several different types of models (**Figure 20**). And it is the amino acid sequence of each polypeptide that determines what three-dimensional structure the protein will have under normal cellular conditions.

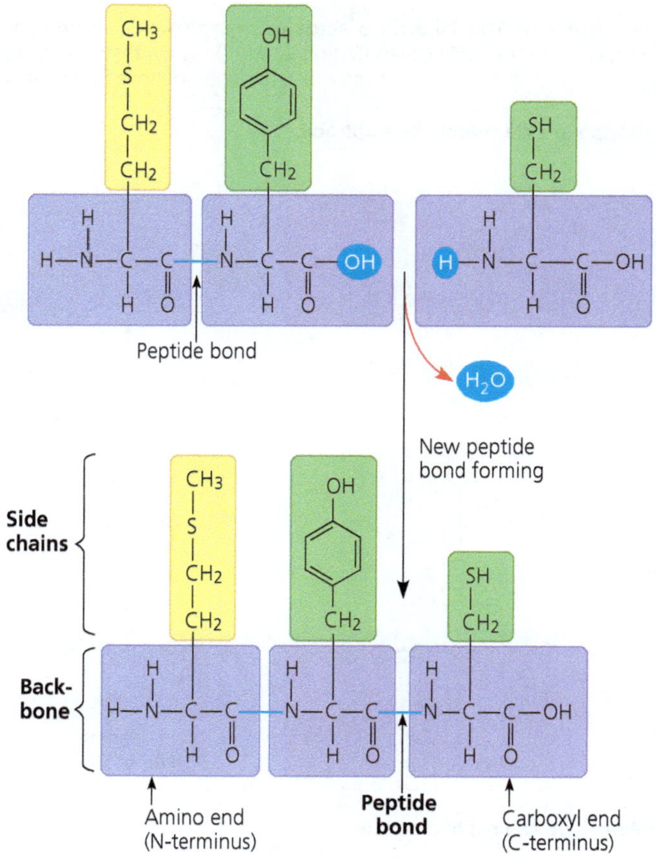

▲ **Figure 19 Making a polypeptide chain.** Peptide bonds are formed by dehydration reactions, which link the carboxyl group of one amino acid to the amino group of the next. The peptide bonds are formed one at a time, starting with the amino acid at the amino end (N-terminus). The polypeptide has a repetitive backbone (purple) from which the amino acid side chains (yellow and green) extend.

DRAW IT *At the top of the figure, circle and label the carboxyl and amino groups that will form the new peptide bond. Under each amino acid, write its three- and one-letter abbreviations (see Figure 18).*

When a cell synthesizes a polypeptide, the chain may fold spontaneously, assuming the functional structure for that protein. This folding is driven and reinforced by the formation of various bonds between parts of the chain, which in turn depends on the sequence of amino acids. Many proteins are roughly spherical (*globular proteins*), while others are shaped like long fibers (*fibrous proteins*). Even within these broad categories, countless variations exist.

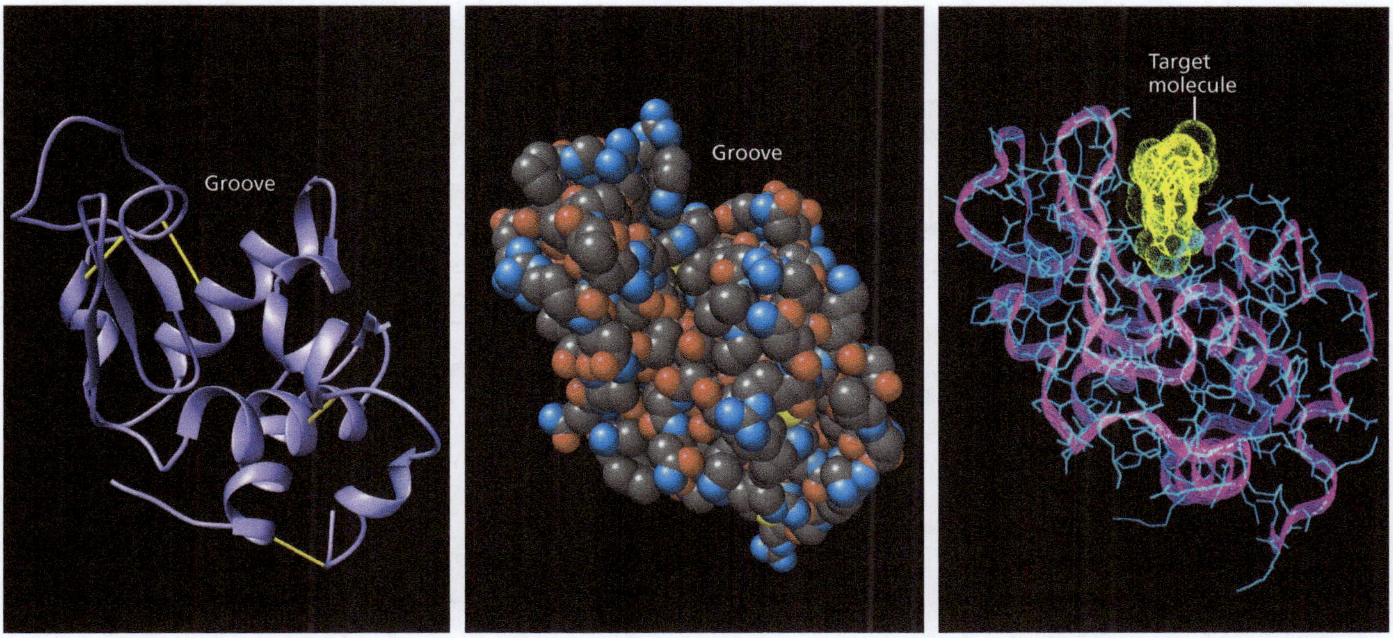

(a) A **ribbon model** shows how the single polypeptide chain folds and coils to form the functional protein. (The yellow lines represent disulfide bridges that stabilize the protein's shape.)

(b) A **space-filling model** shows more clearly the globular shape seen in many proteins, as well as the specific three-dimensional structure unique to lysozyme. The groove is the site that will bind to the target bacterial molecule.

(c) In this view, a ribbon model is superimposed on a **wireframe model**, which shows the backbone with the side chains extending from it. The yellow structure is the target molecule on the bacterial cell surface.

▲ **Figure 20 Structure of a protein, the enzyme lysozyme.** Present in our sweat, tears, and saliva, lysozyme is an enzyme that helps prevent infection by binding to and catalyzing the destruction of specific (target) molecules on the surface of many kinds of bacteria. The groove is the part of the protein that recognizes and binds to the target molecules on the surface of bacterial cell walls.

A protein's specific structure determines how it works. In almost every case, the function of a protein depends on its ability to recognize and bind to some other molecule. In an especially striking example of the marriage of form and function, **Figure 21** shows the exact match of shape between an antibody (a protein in the body) and the particular foreign substance on a flu virus that the antibody binds to and marks for destruction.

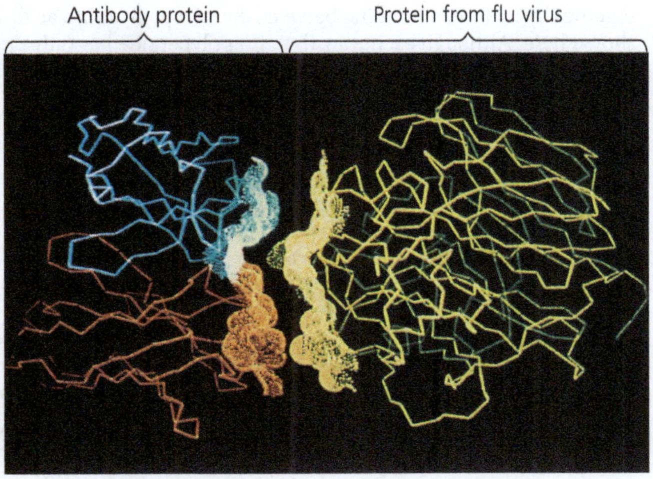

▲ **Figure 21 An antibody binding to a protein from a flu virus.** A technique called X-ray crystallography was used to generate a computer model of an antibody protein (blue and orange, left) bound to a flu virus protein (green and yellow, right). Computer software was then used to back the images away from each other, revealing the exact complementarity of shape between the two protein surfaces.

Another example of molecules with matching shapes is endorphin molecules (produced by the body) and morphine molecules (a manufactured drug), both of which fit into receptor molecules on the surface of brain cells in humans, producing euphoria and relieving pain. Morphine, heroin, and other opiate drugs are able to mimic endorphins because they all share a similar shape with endorphins and can thus fit into and bind to endorphin receptors. This fit is very specific, something like a lock and key. The endorphin receptor, like other receptor molecules, is a protein. The function of a protein—for instance, the ability of a receptor protein to bind to a particular pain-relieving signaling molecule—is an emergent property resulting from exquisite molecular order.

Four Levels of Protein Structure

In spite of their great diversity, all proteins share three superimposed levels of structure, known as primary, secondary, and tertiary structure. A fourth level, quaternary structure, arises when a protein consists of two or more polypeptide chains. **Figure 22** describes these four levels of protein structure. Be sure to study this figure thoroughly before going on to the next section.

▼ Figure 22 **Exploring Levels of Protein Structure**

Primary Structure

Linear chain of amino acids

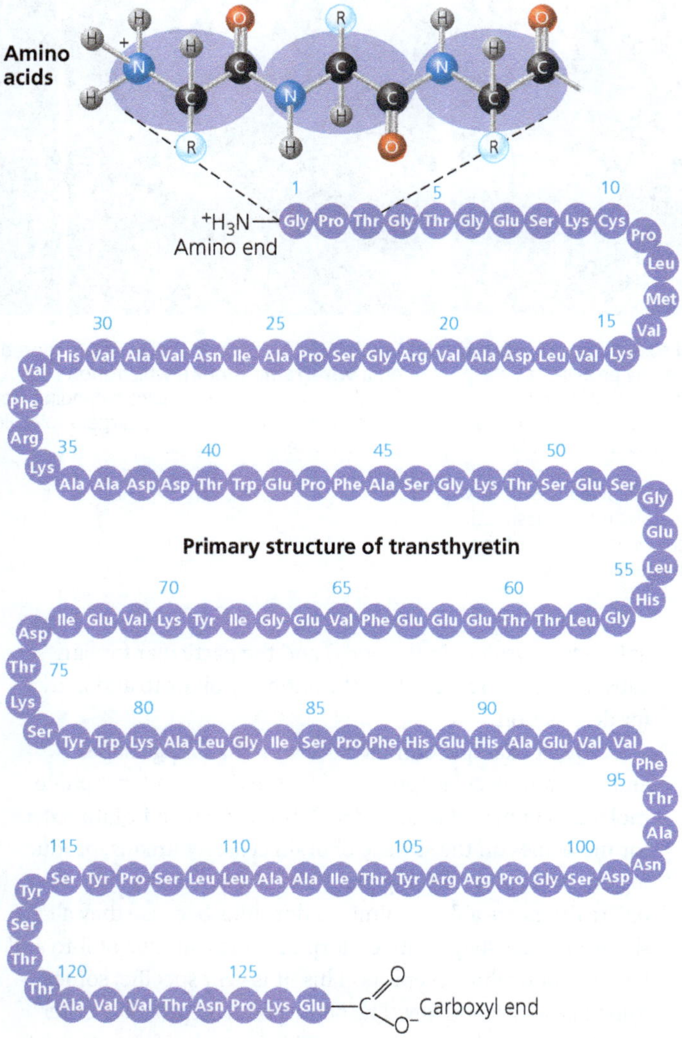

The **primary structure** of a protein is its sequence of amino acids. As an example, let's consider transthyretin, a globular blood protein that transports vitamin A and one of the thyroid hormones throughout the body. Transthyretin is made up of four identical polypeptide chains, each composed of 127 amino acids. Shown here is one of these chains unraveled for a closer look at its primary structure. Each of the 127 positions along the chain is occupied by one of the 20 amino acids, indicated here by its three-letter abbreviation.

The primary structure is like the order of letters in a very long word. If left to chance, there would be 20^{127} different ways of making a polypeptide chain 127 amino acids long. However, the precise primary structure of a protein is determined not by the random linking of amino acids, but by inherited genetic information. The primary structure in turn dictates secondary and tertiary structure, due to the chemical nature of the backbone and the side chains (R groups) of the amino acids along the polypeptide.

Secondary Structure

Regions stabilized by hydrogen bonds between atoms of the polypeptide backbone

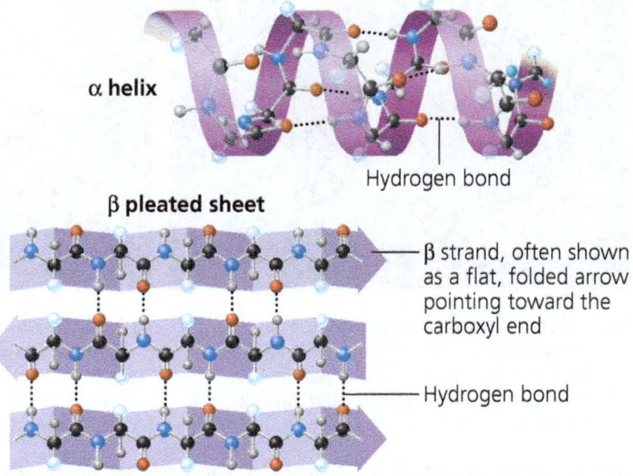

Most proteins have segments of their polypeptide chains repeatedly coiled or folded in patterns that contribute to the protein's overall shape. These coils and folds, collectively referred to as **secondary structure**, are the result of hydrogen bonds between the repeating constituents of the polypeptide backbone (not the amino acid side chains). Within the backbone, the oxygen atoms have a partial negative charge, and the hydrogen atoms attached to the nitrogens have a partial positive charge (see Figure 12); therefore, hydrogen bonds can form between these atoms. Individually, these hydrogen bonds are weak, but because they are repeated many times over a relatively long region of the polypeptide chain, they can support a particular shape for that part of the protein.

One such secondary structure is the **α helix**, a delicate coil held together by hydrogen bonding between every fourth amino acid, as shown here. Although each transthyretin polypeptide has only one α helix region (see the Tertiary Structure section), other globular proteins have multiple stretches of α helix separated by nonhelical regions (see hemoglobin in the Quaternary Structure section). Some fibrous proteins, such as α-keratin, the structural protein of hair, have the α helix formation over most of their length.

The other main type of secondary structure is the **β pleated sheet**. As shown here, in this structure two or more segments of the polypeptide chain lying side by side (called β strands) are connected by hydrogen bonds between parts of the two parallel segments of polypeptide backbone. β pleated sheets make up the core of many globular proteins, as is the case for transthyretin (see Tertiary Structure), and dominate some fibrous proteins, including the silk protein of a spider's web. The teamwork of so many hydrogen bonds makes each spider silk fiber stronger than a steel strand of the same weight.

► Spiders secrete silk fibers made of a structural protein containing β pleated sheets, which allow the spider web to stretch and recoil.

Tertiary Structure

Three-dimensional shape stabilized by interactions between side chains

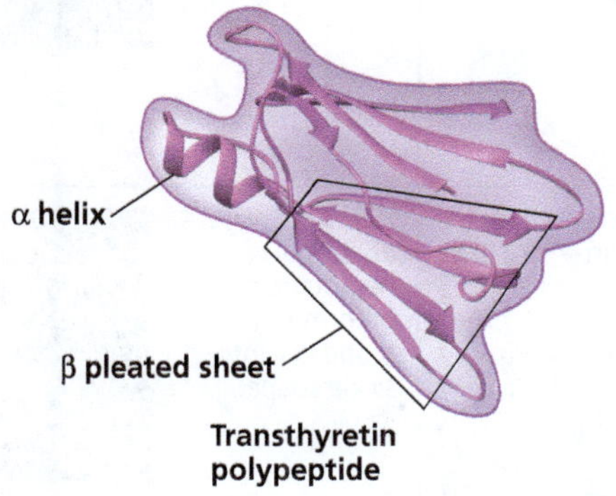

Superimposed on the patterns of secondary structure is a protein's tertiary structure, shown here in a ribbon model of the transthyretin polypeptide. While secondary structure involves interactions between backbone constituents, **tertiary structure** is the overall shape of a polypeptide resulting from interactions between the side chains (R groups) of the various amino acids. One type of interaction that contributes to tertiary structure is called—somewhat misleadingly—a **hydrophobic interaction**. As a polypeptide folds into its functional shape, amino acids with hydrophobic (nonpolar) side chains usually end up in clusters at the core of the protein, out of contact with water. Thus, a "hydrophobic interaction" is actually caused by the exclusion of nonpolar substances by water molecules. Once nonpolar amino acid side chains are close together, van der Waals interactions help hold them together. Meanwhile, hydrogen bonds between polar side chains and ionic bonds between positively and negatively charged side chains also help stabilize tertiary structure. These are all weak interactions in the aqueous cellular environment, but their cumulative effect helps give the protein a unique shape.

Covalent bonds called **disulfide bridges** may further reinforce the shape of a protein. Disulfide bridges form where two cysteine monomers, which have sulfhydryl groups (—SH) on their side chains (see Figure 6), are brought close together by the folding of the protein. The sulfur of one cysteine bonds to the sulfur of the second, and the disulfide bridge (—S—S—) rivets parts of the protein together (see yellow lines in Figure 20a). All of these different kinds of interactions can contribute to the tertiary structure of a protein, as shown here in a small part of a hypothetical protein:

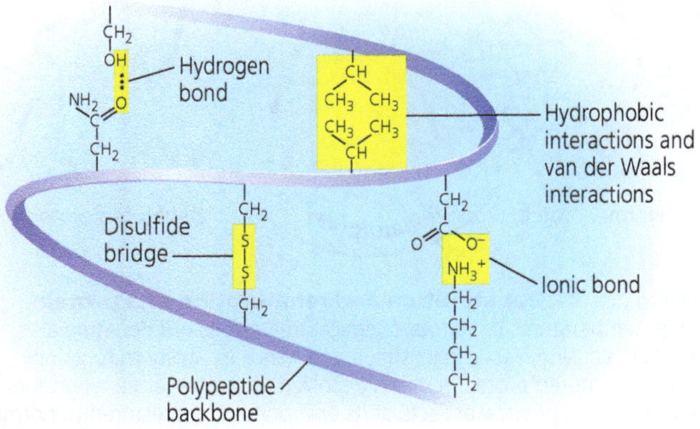

Quaternary Structure

Association of two or more polypeptides (some proteins only)

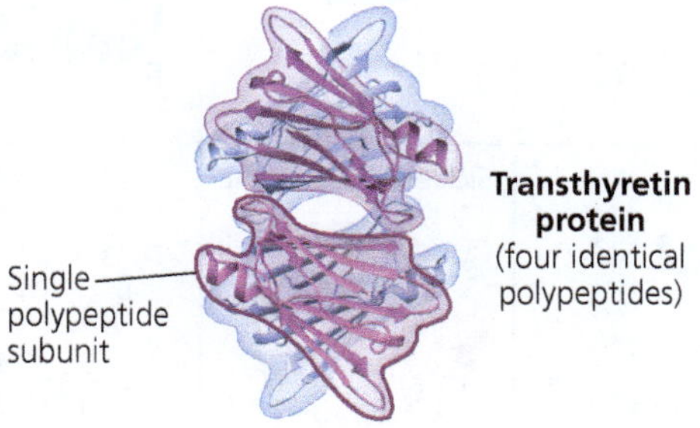

Some proteins consist of two or more polypeptide chains aggregated into one functional macromolecule. **Quaternary structure** is the overall protein structure that results from the aggregation of these polypeptide subunits. For example, shown here is the complete globular transthyretin protein, made up of its four polypeptides.

Another example is collagen, which is a fibrous protein that has three identical helical polypeptides intertwined into a larger triple helix, giving the long fibers great strength. This suits collagen fibers to their function as the girders of connective tissue in skin, bone, tendons, ligaments, and other body parts. (Collagen accounts for 40% of the protein in a human body.)

Hemoglobin, the oxygen-binding protein of red blood cells, is another example of a globular protein with quaternary structure. It consists of four polypeptide subunits, two of one kind (α) and two of another kind (β). Both α and β subunits consist primarily of α-helical secondary structure. Each subunit has a nonpolypeptide component, called heme, with an iron atom that binds oxygen.

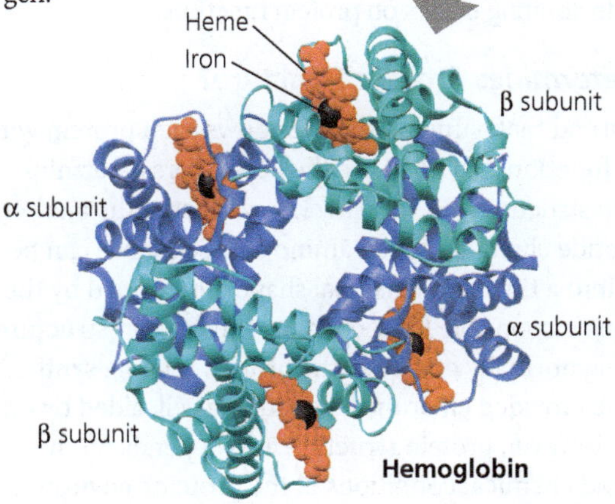

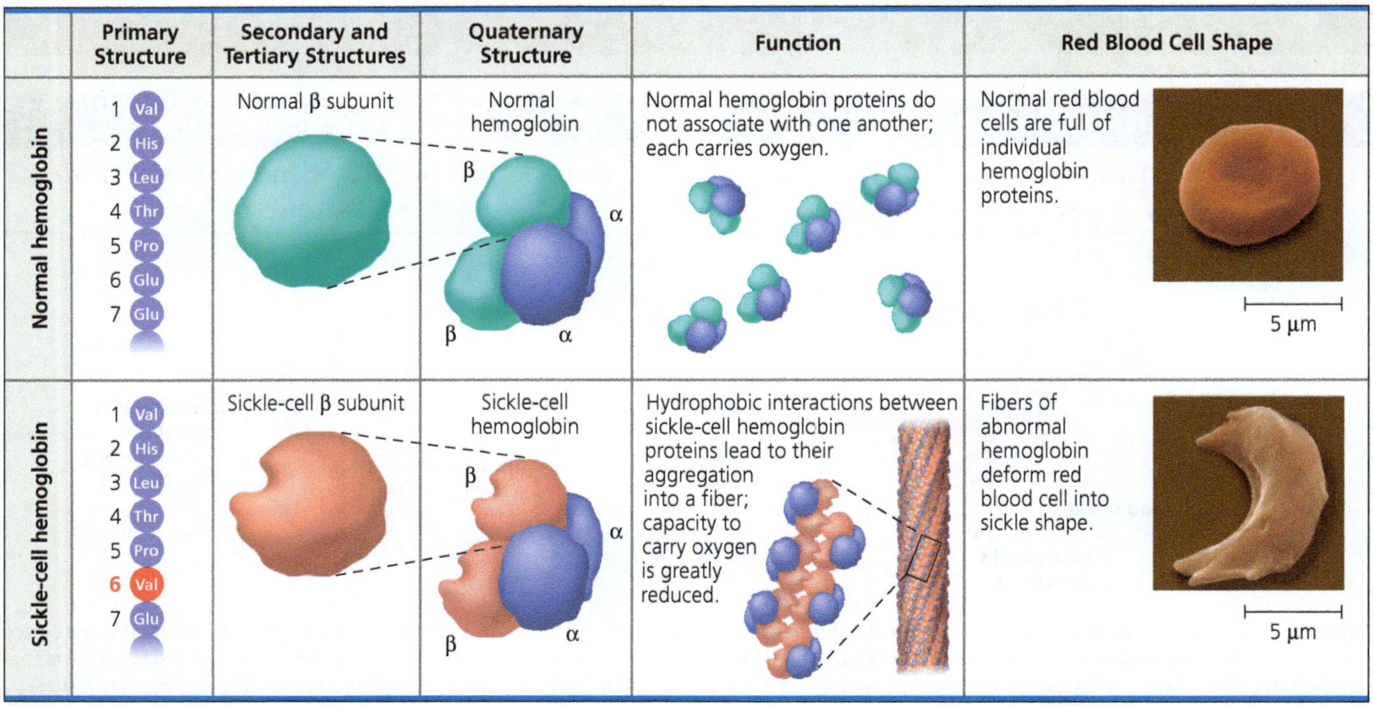

▲ **Figure 23** A single amino acid substitution in a protein causes sickle-cell disease.

MAKE CONNECTIONS *Considering the chemical characteristics of the amino acids valine and glutamic acid (see Figure 18), propose a possible explanation for the dramatic effect on protein function that occurs when valine is substituted for glutamic acid.*

Sickle-Cell Disease: A Change in Primary Structure

Even a slight change in primary structure can affect a protein's shape and ability to function. For instance, **sickle-cell disease**, an inherited blood disorder, is caused by the substitution of one amino acid (valine) for the normal one (glutamic acid) at a particular position in the primary structure of hemoglobin, the protein that carries oxygen in red blood cells. Normal red blood cells are disk-shaped, but in sickle-cell disease, the abnormal hemoglobin molecules tend to aggregate into fibers, deforming some of the cells into a sickle shape **(Figure 23)**. A person with the disease has periodic "sickle-cell crises" when the angular cells clog tiny blood vessels, impeding blood flow. The toll taken on such patients is a dramatic example of how a simple change in protein structure can have devastating effects on protein function.

What Determines Protein Structure?

You've learned that a unique shape endows each protein with a specific function. But what are the key factors determining protein structure? You already know most of the answer: A polypeptide chain of a given amino acid sequence can be arranged into a three-dimensional shape determined by the interactions responsible for secondary and tertiary structure. This folding normally occurs as the protein is being synthesized in the crowded environment within a cell, aided by other proteins. However, protein structure also depends on the physical and chemical conditions of the protein's environment. If the pH, salt concentration, temperature, or other aspects of its environment are altered, the weak chemical bonds and interactions within a protein may be destroyed, causing the protein to unravel and lose its native shape, a change called **denaturation** **(Figure 24)**. Because it is misshapen, the denatured protein is biologically inactive.

Most proteins become denatured if they are transferred from an aqueous environment to a nonpolar solvent, such as ether or chloroform; the polypeptide chain refolds so that its hydrophobic regions face outward toward the solvent.

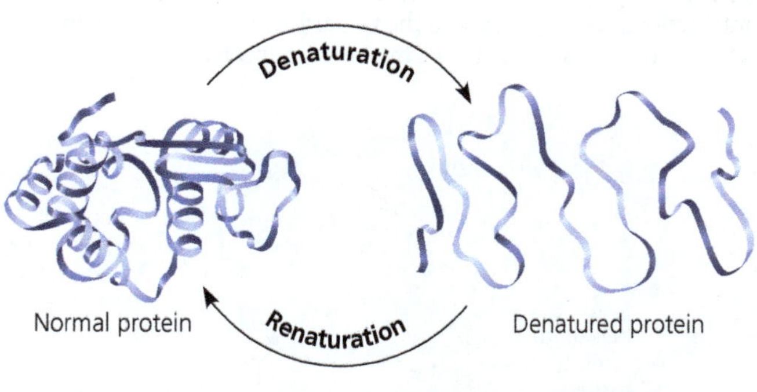

▲ **Figure 24 Denaturation and renaturation of a protein.** High temperatures or various chemical treatments will denature a protein, causing it to lose its shape and hence its ability to function. If the denatured protein remains dissolved, it may renature when the chemical and physical aspects of its environment are restored to normal.

Other denaturation agents include chemicals that disrupt the hydrogen bonds, ionic bonds, and disulfide bridges that maintain a protein's shape. Denaturation can also result from excessive heat, which agitates the polypeptide chain enough to overpower the weak interactions that stabilize the structure. The white of an egg becomes opaque during cooking because the denatured proteins are insoluble and solidify. This also explains why excessively high fevers can be fatal: Proteins in the blood tend to denature at very high body temperatures.

When a protein in a test-tube solution has been denatured by heat or chemicals, it can sometimes return to its functional shape when the denaturing agent is removed. (Sometimes this is not possible: For example, a fried egg will not become liquefied when placed back into the refrigerator!) We can conclude that the information for building a specific shape is intrinsic to the protein's primary structure. The sequence of amino acids determines the protein's shape—where an α helix can form, where β pleated sheets can exist, where disulfide bridges are located, where ionic bonds can form, and so on. But how does protein folding occur in the cell?

Protein Folding in the Cell

Biochemists now know the amino acid sequence for nearly 50 million proteins, with about 1.7 million added each month, and the three-dimensional shape for more than 31,000. Researchers have tried to correlate the primary structure of many proteins with their three-dimensional structure to discover the rules of protein folding. Unfortunately, however, the protein-folding process is not that simple. Most proteins probably go through several intermediate structures on their way to a stable shape, and looking at the mature structure does not reveal the stages of folding required to achieve that form. However, biochemists have developed methods for tracking a protein through such stages. They are still working to develop computer programs that can predict the 3-D structure of a polypeptide from its primary structure alone.

Misfolding of polypeptides is a serious problem in cells. Many diseases, such as Alzheimer's, Parkinson's, and mad cow disease, are associated with an accumulation of misfolded proteins. In fact, misfolded versions of the transthyretin protein featured in Figure 22 have been implicated in several diseases, including one form of senile dementia.

Even when scientists have a correctly folded protein in hand, determining its exact three-dimensional structure is not simple, for a single protein molecule has thousands of atoms. The method most commonly used to determine the 3-D shape of a protein is **X-ray crystallography**, which depends on the diffraction of an X-ray beam by the atoms of a crystallized molecule. Using this technique, scientists can build a 3-D model that shows the exact position of every atom in a protein molecule **(Figure 25)**. Nuclear magnetic resonance (NMR) spectroscopy and bioinformatics are complementary approaches to understanding protein structure and function.

▼ Figure 25 Research Method

X-Ray Crystallography

Application Scientists use X-ray crystallography to determine the three-dimensional (3-D) structure of macromolecules such as nucleic acids and proteins. As an example, we show how this method was used to determine the 3-D shape of transthyretin, the blood transport protein whose levels of structure are described in Figure 22.

Technique Researchers aimed an X-ray beam through the crystallized protein. The atoms of the crystal diffracted (bent) the X-rays into an orderly array that a digital detector recorded as a pattern of spots called an X-ray diffraction pattern, an example of which is shown below.

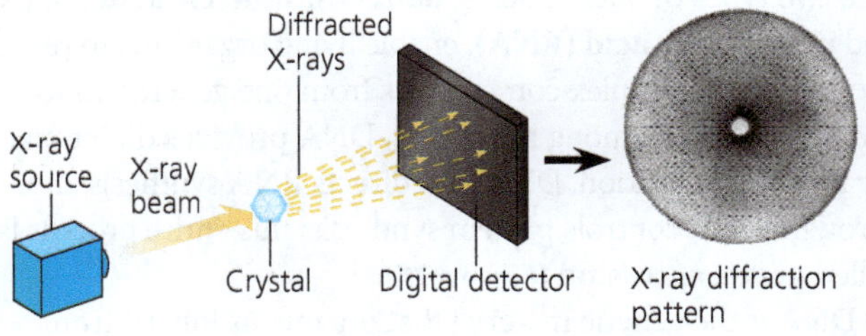

Results Using data from X-ray diffraction patterns, as well as the amino acid sequence determined by chemical methods, researchers built a 3-D model of the four-subunit transthyretin protein with the help of computer software.

CONCEPT CHECK 5

1. Why does a denatured protein no longer function normally?
2. What parts of a polypeptide participate in the bonds that hold together secondary structure? Tertiary structure?
3. **WHAT IF?** Where would you expect a polypeptide region rich in the amino acids valine, leucine, and isoleucine to be located in a folded polypeptide? Explain.

CONCEPT 6

Nucleic acids store, transmit, and help express hereditary information

If the primary structure of polypeptides determines a protein's shape, what determines primary structure? The amino acid sequence of a polypeptide is programmed by a discrete unit of inheritance known as a **gene**. Genes consist of DNA, which belongs to the class of compounds called nucleic acids. **Nucleic acids** are polymers made of monomers called nucleotides.

The Roles of Nucleic Acids

The two types of nucleic acids, **deoxyribonucleic acid (DNA)** and **ribonucleic acid (RNA)**, enable living organisms to reproduce their complex components from one generation to the next. Unique among molecules, DNA provides directions for its own replication. DNA also directs RNA synthesis and, through RNA, controls protein synthesis; this entire process is called **gene expression** (Figure 26).

DNA is the genetic material that organisms inherit from their parents. Each chromosome contains one long DNA molecule, usually carrying several hundred or more genes. When a cell reproduces itself by dividing, its DNA molecules are copied and passed along from one generation of cells to the next. Encoded in the structure of DNA is the information that programs all the cell's activities. The DNA, however, is not directly involved in running the operations of the cell, any more than computer software by itself can print a bank statement or read the bar code on a box of cereal. Just as a printer is needed to print out a statement and a scanner is needed to read a bar code, proteins are required to implement genetic programs. The molecular hardware of the cell—the tools for biological functions—consists mostly of proteins. For example, the oxygen carrier in red blood cells is the protein hemoglobin (see Figure 22), not the DNA that specifies its structure.

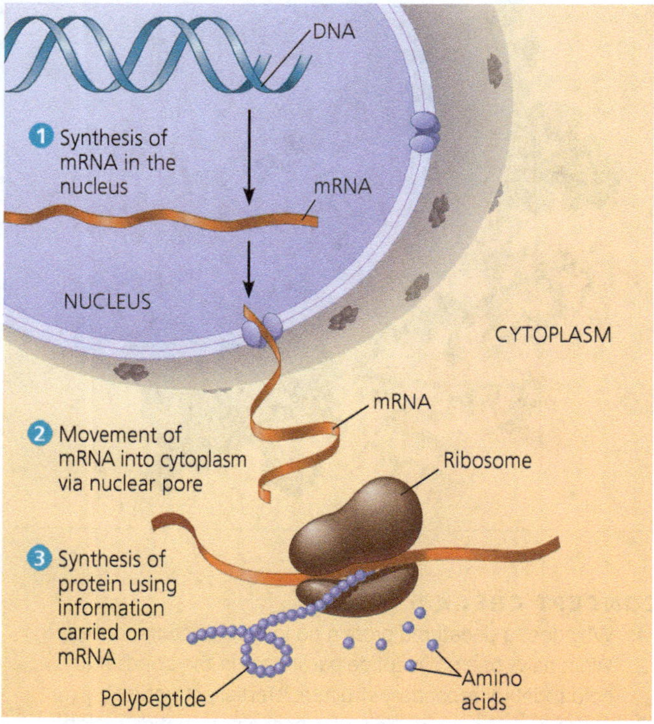

▲ **Figure 26 Gene expression: DNA → RNA → protein.** In a eukaryotic cell, DNA in the nucleus programs protein production in the cytoplasm by dictating synthesis of messenger RNA (mRNA).

How does RNA, the other type of nucleic acid, fit into gene expression, the flow of genetic information from DNA to proteins? A given gene along a DNA molecule can direct synthesis of a type of RNA called *messenger RNA* (*mRNA*). The mRNA molecule interacts with the cell's protein-synthesizing machinery to direct production of a polypeptide, which folds into all or part of a protein. We can summarize the flow of genetic information as DNA → RNA → protein (see Figure 26). The sites of protein synthesis are cellular structures called ribosomes. In a eukaryotic cell, ribosomes are in the cytoplasm—the region between the nucleus and the cell's outer boundary, the plasma membrane—but DNA resides in the nucleus. Messenger RNA conveys genetic instructions for building proteins from the nucleus to the cytoplasm. Prokaryotic cells lack nuclei but still use mRNA to convey a message from the DNA to ribosomes and other cellular equipment that translate the coded information into amino acid sequences. Some recently discovered RNA molecules have other functions; the stretches of DNA that direct synthesis of these RNAs are also considered genes.

The Components of Nucleic Acids

Nucleic acids are macromolecules that exist as polymers called **polynucleotides** (Figure 27a). As indicated by the name, each polynucleotide consists of monomers called **nucleotides**. A nucleotide, in general, is composed of three parts: a nitrogen-containing (nitrogenous) base, a five-carbon sugar (a pentose), and one to three phosphate groups (Figure 27b). The beginning monomer used to build a polynucleotide has three phosphate groups, but two are lost during the polymerization process. The portion of a nucleotide without any phosphate groups is called a *nucleoside*.

To understand the structure of a single nucleotide, let's first consider the nitrogenous bases (Figure 27c). Each nitrogenous base has one or two rings that include nitrogen atoms. (They are called nitrogenous *bases* because the nitrogen atoms tend to take up H^+ from solution, thus acting as bases.) There are two families of nitrogenous bases: pyrimidines and purines. A **pyrimidine** has one six-membered ring of carbon and nitrogen atoms. The members of the pyrimidine family are cytosine (C), thymine (T), and uracil (U). **Purines** are larger, with a

▼ Figure 27 Components of nucleic acids. (a) A polynucleotide has a sugar-phosphate backbone with variable appendages, the nitrogenous bases. **(b)** A nucleotide monomer includes a nitrogenous base, a sugar, and a phosphate group. Note that carbon numbers in the sugar include primes ('). **(c)** A nucleoside includes a nitrogenous base (purine or pyrimidine) and a five-carbon sugar (deoxyribose or ribose).

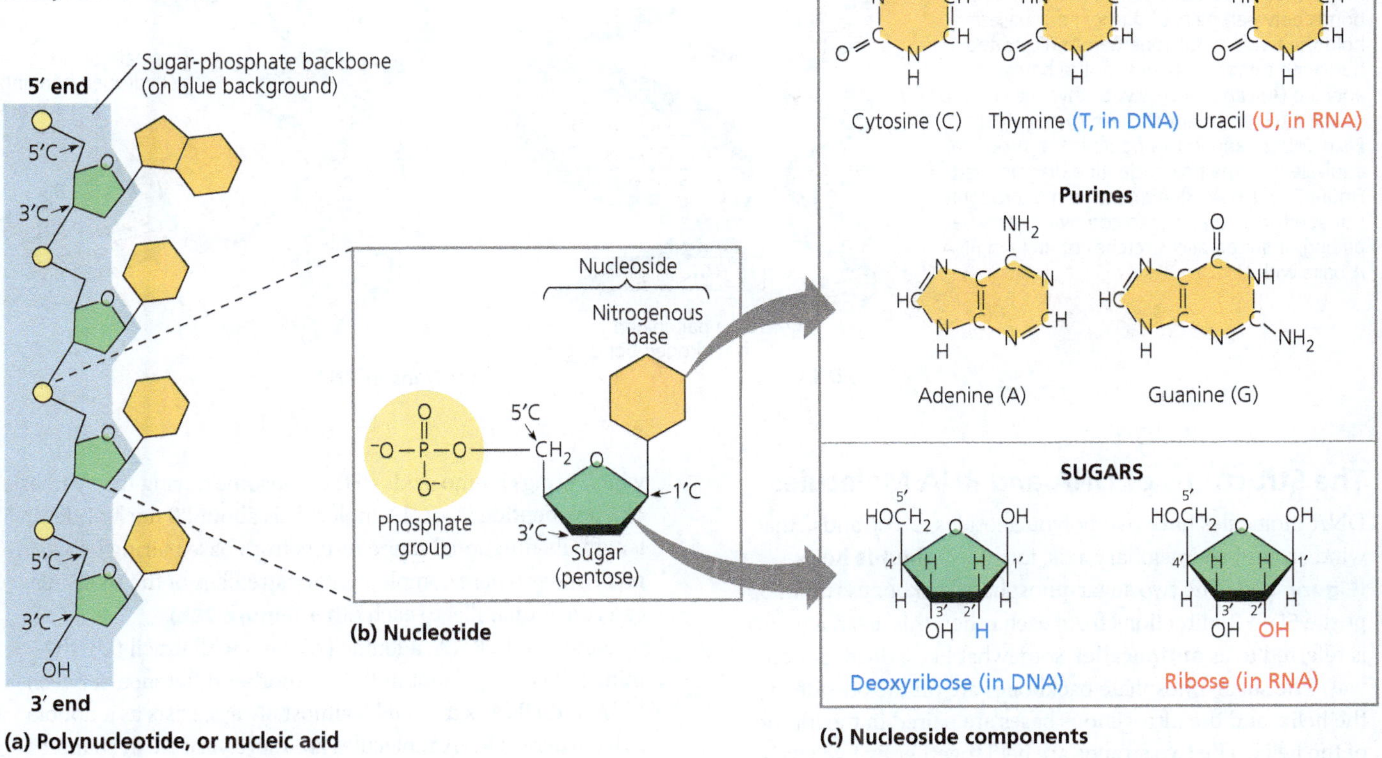

six-membered ring fused to a five-membered ring. The purines are adenine (A) and guanine (G). The specific pyrimidines and purines differ in the chemical groups attached to the rings. Adenine, guanine, and cytosine are found in both DNA and RNA; thymine is found only in DNA and uracil only in RNA.

Now let's add the sugar to which the nitrogenous base is attached. In DNA the sugar is **deoxyribose**; in RNA it is **ribose** (see Figure 27c). The only difference between these two sugars is that deoxyribose lacks an oxygen atom on the second carbon in the ring, hence the name *deoxy*ribose.

So far, we have built a nucleoside (nitrogenous base plus sugar). To complete the construction of a nucleotide, we attach a phosphate group to the 5′ carbon of the sugar (the carbon numbers in the sugar include ′, the prime symbol; see Figure 27b). The molecule is now a nucleoside monophosphate, more often called a nucleotide.

Nucleotide Polymers

The linkage of nucleotides into a polynucleotide involves a dehydration reaction. In the polynucleotide, adjacent nucleotides are joined by a phosphodiester linkage, which consists of a phosphate group that covalently links the sugars of two nucleotides. This bonding results in a backbone with a repeating pattern of sugar-phosphate units called the *sugar-phosphate backbone* (see Figure 27a). (Note that the nitrogenous bases are not part of the backbone.) The two free ends of the polymer are distinctly different from each other. One end has a phosphate attached to a 5′ carbon, and the other end has a hydroxyl group on a 3′ carbon; we refer to these as the *5′ end* and the *3′ end*, respectively. We can say that a polynucleotide has a built-in directionality along its sugar-phosphate backbone, from 5′ to 3′, somewhat like a one-way street. All along this sugar-phosphate backbone are appendages consisting of the nitrogenous bases.

The sequence of bases along a DNA (or mRNA) polymer is unique for each gene and provides very specific information to the cell. Because genes are hundreds to thousands of nucleotides long, the number of possible base sequences is effectively limitless. A gene's meaning to the cell is encoded in its specific sequence of the four DNA bases. For example, the sequence 5′-AGGTAACTT-3′ means one thing, whereas the sequence 5′-CGCTTTAAC-3′ has a different meaning. (Entire genes, of course, are much longer.) The linear order of bases in a gene specifies the amino acid sequence—the primary structure—of a protein, which in turn specifies that protein's three-dimensional structure, thus enabling its function in the cell.

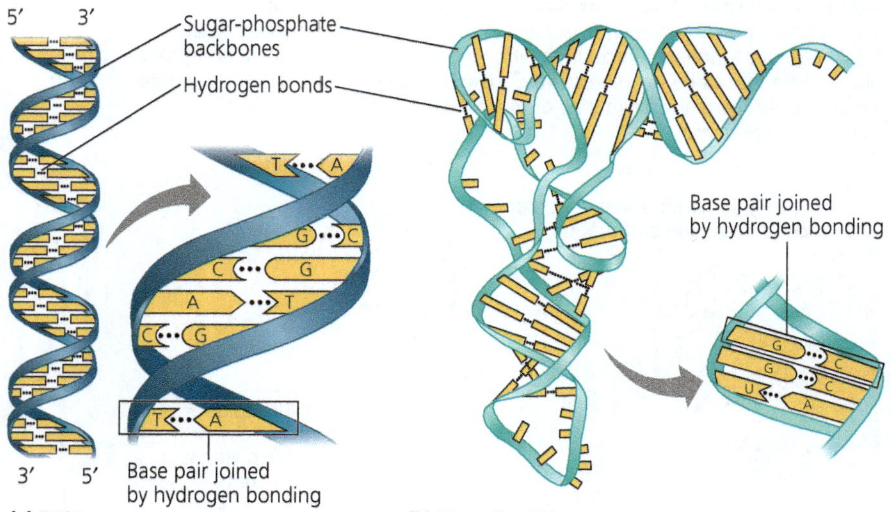

(a) DNA

(b) Transfer RNA

▶ **Figure 28 The structures of DNA and tRNA molecules. (a)** The DNA molecule is usually a double helix, with the sugar-phosphate backbones of the antiparallel polynucleotide strands (symbolized here by blue ribbons) on the outside of the helix. Hydrogen bonds between pairs of nitrogenous bases hold the two strands together. As illustrated here with symbolic shapes for the bases, adenine (A) can pair only with thymine (T), and guanine (G) can pair only with cytosine (C). Each DNA strand in this figure is the structural equivalent of the polynucleotide diagrammed in Figure 27a. **(b)** A tRNA molecule has a roughly L-shaped structure, due to complementary base pairing of antiparallel stretches of RNA. In RNA, A pairs with U.

The Structures of DNA and RNA Molecules

DNA molecules have two polynucleotides, or "strands," that wind around an imaginary axis, forming a **double helix** **(Figure 28a)**. The two sugar-phosphate backbones run in opposite 5′ → 3′ directions from each other; this arrangement is referred to as **antiparallel**, somewhat like a divided highway. The sugar-phosphate backbones are on the outside of the helix, and the nitrogenous bases are paired in the interior of the helix. The two strands are held together by hydrogen bonds between the paired bases (see Figure 28a). Most DNA molecules are very long, with thousands or even millions of base pairs. The one long DNA double helix in a eukaryotic chromosome includes many genes, each one a particular segment of the molecule.

In base pairing, only certain bases in the double helix are compatible with each other. Adenine (A) in one strand always pairs with thymine (T) in the other, and guanine (G) always pairs with cytosine (C). The two strands of the double helix are said to be *complementary*, each the predictable counterpart of the other. It is this feature of DNA that makes it possible to generate two identical copies of each DNA molecule in a cell that is preparing to divide. When the cell divides, the copies are distributed to the daughter cells, making them genetically identical to the parent cell. Thus, the structure of DNA accounts for its function of transmitting genetic information whenever a cell reproduces.

RNA molecules, by contrast, exist as single strands. Complementary base pairing can occur, however, between regions of two RNA molecules or even between two stretches of nucleotides in the *same* RNA molecule. In fact, base pairing within an RNA molecule allows it to take on the particular three-dimensional shape necessary for its function. Consider, for example, the type of RNA called *transfer RNA* (*tRNA*), which brings amino acids to the ribosome during the synthesis of a polypeptide. A tRNA molecule is about 80 nucleotides in length. Its functional shape results from base pairing between nucleotides where complementary stretches of the molecule can run antiparallel to each other **(Figure 28b)**.

Note that in RNA, adenine (A) pairs with uracil (U); thymine (T) is not present in RNA. Another difference between RNA and DNA is that DNA almost always exists as a double helix, whereas RNA molecules are more variable in shape. RNAs are versatile molecules, and many biologists believe RNA may have preceded DNA as the carrier of genetic information in early forms of life.

CONCEPT CHECK 6

1. DRAW IT Go to Figure 27a and, for the top three nucleotides, number all the carbons in the sugars (don't forget the primes), circle the nitrogenous bases, and star the phosphates.
2. DRAW IT In a DNA double helix, a region along one DNA strand has the following sequence of nitrogenous bases: 5′-TAGGCCT-3′. Copy this sequence, and write down its complementary strand, clearly indicating the 5′ and 3′ ends of the complementary strand.

CONCEPT 7

Genomics and proteomics have transformed biological inquiry and applications

Experimental work in the first half of the 20th century established the role of DNA as the bearer of genetic information, passed from generation to generation, that specified the functioning of living cells and organisms. Once the structure of the DNA molecule was described in 1953, and the linear sequence of nucleotide bases was understood to specify the amino acid sequence of proteins, biologists sought to "decode" genes by learning their base sequences.

The first chemical techniques for *DNA sequencing*, or determining the sequence of nucleotides along a DNA strand, one by one, were developed in the 1970s. Researchers began to study gene sequences, gene by gene, and the more they learned, the more questions they had: How was expression of genes regulated? Genes and their protein products clearly interacted with each other, but how? What was the function, if any, of the DNA that is not part of genes? To fully understand the genetic functioning of a living organism, the entire sequence of the full complement of DNA, the organism's *genome*, would be most enlightening. In spite of the apparent impracticality of this idea, in the late 1980s several prominent biologists put forth an audacious proposal to launch a project that would sequence the entire human genome—all 3 billion bases of it! This endeavor began in 1990 and was effectively completed in the early 2000s.

An unplanned but profound side benefit of this project—the Human Genome Project—was the rapid development of faster and less expensive methods of sequencing. This trend has continued apace: The cost for sequencing 1 million bases in 2001, well over $5,000, has decreased to about $0.08 in 2014. And a human genome, the first of which took over 10 years to sequence, could be completed at today's pace in just a few days **(Figure 29)**. The number of genomes that have been fully sequenced has burgeoned, generating reams of data and prompting development of *bioinformatics*, the use of computer software and other computational tools that can handle and analyze these large data sets.

The reverberations of these developments have transformed the study of biology and related fields. Biologists often look at problems by analyzing large sets of genes or even comparing whole genomes of different species, an approach called **genomics**. A similar analysis of large sets of proteins, including their sequences, is called **proteomics**. (Protein sequences can be determined either by using biochemical techniques or by translating the DNA sequences that code for them.) These approaches permeate all fields of biology, some examples of which are shown in **Figure 30**.

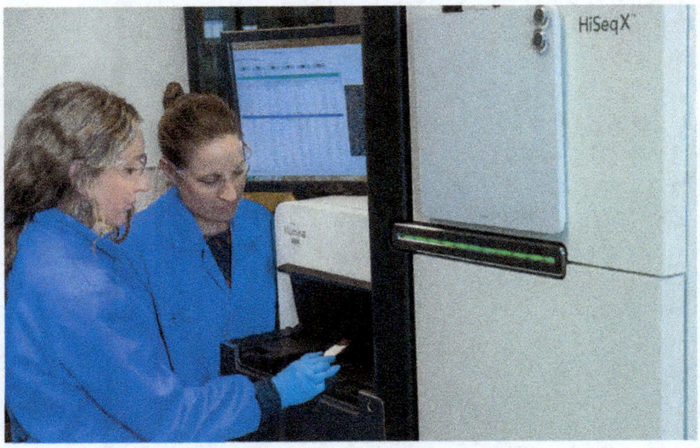

▲ **Figure 29 Automatic DNA sequencing machines and abundant computing power enable rapid sequencing of genes and genomes.**

DNA and Proteins as Tape Measures of Evolution

EVOLUTION We are accustomed to thinking of shared traits, such as hair and milk production in mammals, as evidence of shared ancestry. Because DNA carries heritable information in the form of genes, sequences of genes and their protein products document the hereditary background of an organism. The linear sequences of nucleotides in DNA molecules are passed from parents to offspring; these sequences determine the amino acid sequences of proteins. As a result, siblings have greater similarity in their DNA and proteins than do unrelated individuals of the same species.

Given our evolutionary view of life, we can extend this concept of "molecular genealogy" to relationships between species: We would expect two species that appear to be closely related based on anatomical evidence (and possibly fossil evidence) to also share a greater proportion of their DNA and protein sequences than do less closely related species. In fact, that is the case. An example is the comparison of the β polypeptide chain of human hemoglobin with the corresponding hemoglobin polypeptide in other vertebrates. In this chain of 146 amino acids, humans and gorillas differ in just 1 amino acid, while humans and frogs, more distantly related, differ in 67 amino acids. In the **Scientific Skills Exercise**, you can apply this sort of reasoning to additional species. The relative sequence similarity also holds true when comparing whole genomes: The human genome is 95–98% identical to that of the chimpanzee, but only roughly 85% identical to that of the mouse, a more distant evolutionary relative. Molecular biology has added a new tape measure to the toolkit biologists use to assess evolutionary kinship.

Perhaps the most significant impact of genomics and proteomics on the field of biology has been their contributions to our understanding of evolution. To quote one of the founders of modern evolutionary theory, Theodosius Dobzhansky, "Nothing in biology makes sense except in the light of evolution." In addition to confirming evidence for evolution from the study of fossils and characteristics of currently existing species, genomics has helped us tease out relationships among different groups of organisms that had not been resolved by previous types of evidence, and thus infer their evolutionary history.

CONCEPT CHECK 7

1. How would sequencing the entire genome of an organism help scientists to understand how that organism functioned?
2. Given the function of DNA, why would you expect two species with very similar traits to also have very similar genomes?

▼ **Figure 30**

MAKE CONNECTIONS

Contributions of Genomics and Proteomics to Biology

Nucleotide sequencing and the analysis of large sets of genes and proteins can be done rapidly and inexpensively due to advances in technology and information processing. Taken together, genomics and proteomics have advanced our understanding of biology across many different fields.

Paleontology

New DNA sequencing techniques have allowed decoding of minute quantities of DNA found in ancient tissues from our extinct relatives, the Neanderthals (*Homo neanderthalensis*). Sequencing the Neanderthal genome has informed our understanding of their physical appearance as well as their relationship with modern humans.

Medical Science

Identifying the genetic basis for human diseases like cancer helps researchers focus their search for potential future treatments. Currently, sequencing the sets of genes expressed in an individual's tumor can allow a more targeted approach to treating the cancer, a type of "personalized medicine."

Evolution

A major aim of evolutionary biology is to understand the relationships among species, both living and extinct. For example, genome sequence comparisons have identified the hippopotamus as the land mammal sharing the most recent common ancestor with whales.

Hippopotamus

Short-finned pilot whale

Conservation Biology

The tools of molecular genetics and genomics are increasingly used by forensic ecologists to identify which species of animals and plants are killed illegally. In one case, genomic sequences of DNA from illegal shipments of elephant tusks were used to track down poachers and pinpoint the territory where they were operating.

Species Interactions

Most plant species exist in a mutually beneficial partnership with fungi (right) and bacteria associated with the plants' roots; these interactions improve plant growth. Genome sequencing and analysis of gene expression have allowed characterization of plant-associated communities. Such studies will help advance our understanding of such interactions and may improve agricultural practices.

MAKE CONNECTIONS *Considering the examples provided here, describe how the approaches of genomics and proteomics help us to address a variety of biological questions.*

Scientific Skills Exercise

Analyzing Polypeptide Sequence Data

▶ Human

▶ Rhesus monkey

▶ Gibbon

Are Rhesus Monkeys or Gibbons More Closely Related to Humans? DNA and polypeptide sequences from closely related species are more similar to each other than are sequences from more distantly related species. In this exercise, you will look at amino acid sequence data for the β polypeptide chain of hemoglobin, often called β-globin. You will then interpret the data to hypothesize whether the monkey or the gibbon is more closely related to humans.

How Such Experiments Are Done Researchers can isolate the polypeptide of interest from an organism and then determine the amino acid sequence. More frequently, the DNA of the relevant gene is sequenced, and the amino acid sequence of the polypeptide is deduced from the DNA sequence of its gene.

Data from the Experiments In the data below, the letters give the sequence of the 146 amino acids in β-globin from humans, rhesus monkeys, and gibbons. Because a complete sequence would not fit on one line here, the sequences are broken into three segments.

The sequences for the three different species are aligned so that you can compare them easily. For example, you can see that for all three species, the first amino acid is V (valine; see Figure 18) and the 146th amino acid is H (histidine).

INTERPRET THE DATA

1. Scan along the monkey and gibbon sequences, letter by letter, circling any amino acids that do not match the human sequence. (a) How many amino acids differ between the monkey and the human sequences? (b) Between the gibbon and human?
2. For each nonhuman species, what percent of its amino acids is identical to the human sequence of β-globin?
3. Based on these data alone, state a hypothesis for which of these two species is more closely related to humans. What is your reasoning?
4. What other evidence could you use to support your hypothesis?

Species	Alignment of Amino Acid Sequences of β-globin
Human	1 VHLTPEEKSA VTALWGKVNV DEVGGEALGR LLVVYPWTQR FFESFGDLST
Monkey	1 VHLTPEEKNA VTTLWGKVNV DEVGGEALGR LLLVYPWTQR FFESFGDLSS
Gibbon	1 VHLTPEEKSA VTALWGKVNV DEVGGEALGR LLVVYPWTQR FFESFGDLST
Human	51 PDAVMGNPKV KAHGKKVLGA FSDGLAHLDN LKGTFATLSE LHCDKLHVDP
Monkey	51 PDAVMGNPKV KAHGKKVLGA FSDGLNHLDN LKGTFAQLSE LHCDKLHVDP
Gibbon	51 PDAVMGNPKV KAHGKKVLGA FSDGLAHLDN LKGTFAQLSE LHCDKLHVDP
Human	101 ENFRLLGNVL VCVLAHHFGK EFTPPVQAAY QKVVAGVANA LAHKYH
Monkey	101 ENFKLLGNVL VCVLAHHFGK EFTPQVQAAY QKVVAGVANA LAHKYH
Gibbon	101 ENFRLLGNVL VCVLAHHFGK EFTPQVQAAY QKVVAGVANA LAHKYH

Data from Human: http://www.ncbi.nlm.nih.gov/protein/AAA21113.1; rhesus monkey: http://www.ncbi.nlm.nih.gov/protein/122634; gibbon: http://www.ncbi.nlm.nih.gov/protein/122616

(MB) A version of this Scientific Skills Exercise can be assigned in MasteringBiology.

Chapter Review

Go to **MasteringBiology**® for Assignments, the eText, and the Study Area with Animations, Activities, Vocab Self-Quiz, and Practice Tests.

SUMMARY OF KEY CONCEPTS

VOCAB SELF-QUIZ
goo.gl/gbai8v

CONCEPT 1

Carbon atoms can form diverse molecules by bonding to four other atoms

- Carbon, with a **valence** of 4, can bond to various other atoms, including O, H, and N. Carbon can also bond to other carbon atoms, forming the carbon skeletons of **organic compounds**. These skeletons vary in length and shape. **Hydrocarbons** consist of carbon and hydrogen. **Isomers** have the same molecular formula but different structures and properties.
- Chemical groups attached to the carbon skeletons of organic molecules participate in chemical reactions (as **functional groups**) or contribute to function by affecting molecular shape.
- **ATP (adenosine triphosphate)** can react with water, releasing energy that can be used by the cell.

? *In what ways does a methyl group differ chemically from the other six important chemical groups shown in Figure 6?*

CONCEPT 2

Macromolecules are polymers, built from monomers

- Large carbohydrates (polysaccharides), proteins, and nucleic acids are **polymers**, which are chains of **monomers**. The components of lipids vary. Monomers form larger molecules by **dehydration reactions**, in which water molecules are released. Polymers can disassemble by the reverse process, **hydrolysis**. An immense variety of polymers can be built from a small set of monomers.

? *What is the fundamental basis for the differences between large carbohydrates, proteins, and nucleic acids?*

Large Biological Molecules	Components	Examples	Functions
CONCEPT 3 **Carbohydrates serve as fuel and building material** ? *Compare the composition, structure, and function of starch and cellulose. What role do starch and cellulose play in the human body?*	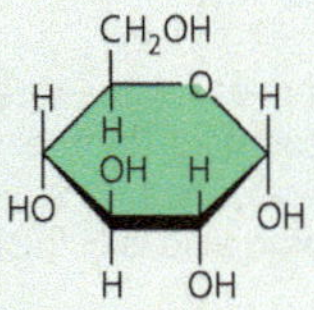 Monosaccharide monomer	**Monosaccharides:** glucose, fructose	Fuel; carbon sources that can be converted to other molecules or combined into polymers
		Disaccharides: lactose, sucrose	
		Polysaccharides: • Cellulose (plants) • Starch (plants) • Glycogen (animals) • Chitin (animals and fungi)	 • Strengthens plant cell walls • Stores glucose for energy • Stores glucose for energy • Strengthens exoskeletons and fungal cell walls
CONCEPT 4 **Lipids are a diverse group of hydrophobic molecules** ? *Why are lipids not considered to be polymers or macromolecules?*	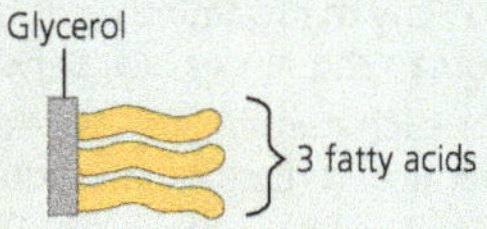	**Triacylglycerols** (fats or oils): glycerol + three fatty acids	Important energy source 
	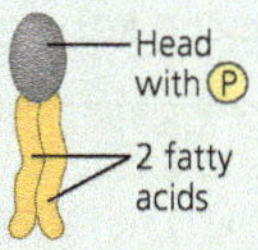	**Phospholipids:** glycerol + phosphate group + two fatty acids	Lipid bilayers of membranes 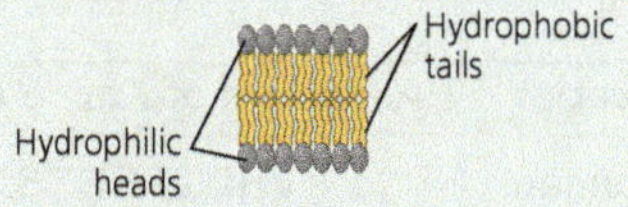
	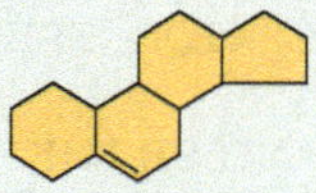 Steroid backbone	**Steroids:** four fused rings with attached chemical groups	• Component of cell membranes (cholesterol) • Signaling molecules that travel through the body (hormones)
CONCEPT 5 **Proteins include a diversity of structures, resulting in a wide range of functions** ? *Explain the basis for the great diversity of proteins.*	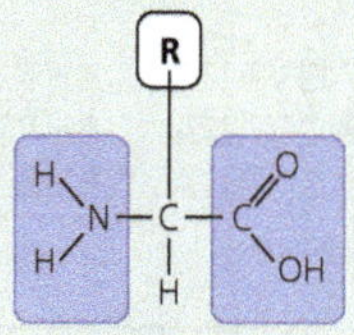 Amino acid monomer (20 types)	• Enzymes • Structural proteins • Storage proteins • Transport proteins • Hormones • Receptor proteins • Motor proteins • Defensive proteins	• Catalyze chemical reactions • Provide structural support • Store amino acids • Transport substances • Coordinate organismal responses • Receive signals from outside cell • Function in cell movement • Protect against disease
CONCEPT 6 **Nucleic acids store, transmit, and help express hereditary information** ? *What role does complementary base pairing play in the functions of nucleic acids?*	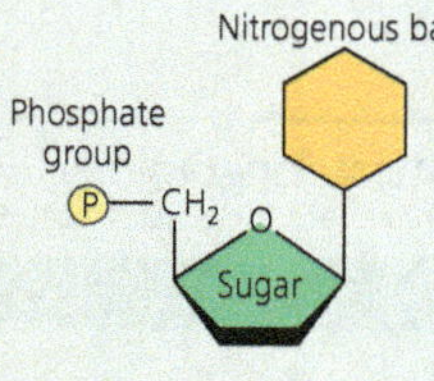 Nucleotide monomer	**DNA:**  • Sugar = deoxyribose • Nitrogenous bases = C, G, A, T • Usually double-stranded	Stores hereditary information
		RNA:  • Sugar = ribose • Nitrogenous bases = C, G, A, U • Usually single-stranded	Various functions in gene expression, including carrying instructions from DNA to ribosomes

CONCEPT 7

Genomics and proteomics have transformed biological inquiry and applications

- Recent technological advances in DNA sequencing have given rise to **genomics**, an approach that analyzes large sets of genes or whole genomes, and **proteomics,** a similar approach for large sets of proteins. Bioinformatics is the use of computational tools and computer software to analyze these large data sets.
- The more closely two species are related evolutionarily, the more similar their DNA sequences are. DNA sequence data confirm models of evolution based on fossils and anatomical evidence.

? *Given the sequences of a particular gene in fruit flies, fish, mice, and humans, predict the relative similarity of the human sequence to that of each of the other species.*

TEST YOUR UNDERSTANDING

PRACTICE TEST

goo.gl/CRZjvS

Level 1: Knowledge/Comprehension

1. Choose the term that correctly describes the relationship between these two sugar molecules:

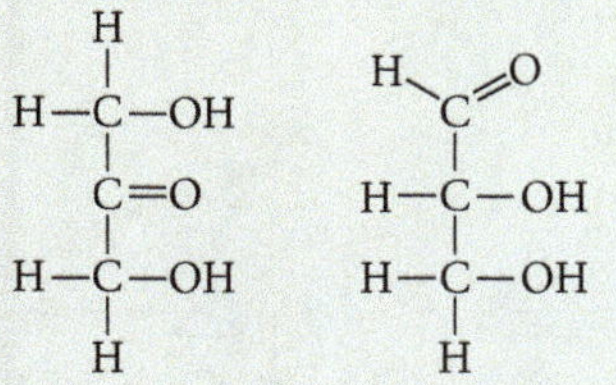

(A) structural isomers
(B) *cis-trans* isomers
(C) enantiomers
(D) isotopes

2. Which functional group is *not* present in this molecule?

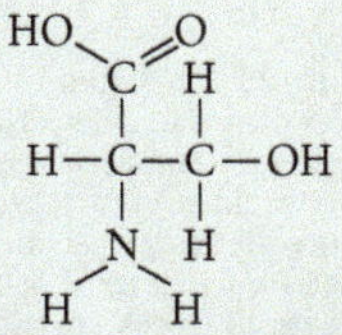

(A) carboxyl
(B) sulfhydryl
(C) hydroxyl
(D) amino

3. MAKE CONNECTIONS Which chemical group is most likely to be responsible for an organic molecule behaving as a base?
(A) hydroxyl
(B) carbonyl
(C) amino
(D) phosphate

4. Which of the following categories includes all others in the list?
(A) disaccharide
(B) starch
(C) carbohydrate
(D) polysaccharide

5. Which of the following statements concerning *unsaturated* fats is true?
(A) They are more common in animals than in plants.
(B) They have double bonds in their fatty acid chains.
(C) They generally solidify at room temperature.
(D) They contain more hydrogen than do saturated fats having the same number of carbon atoms.

6. The structural level of a protein *least* affected by a disruption in hydrogen bonding is the
(A) primary level.
(B) secondary level.
(C) tertiary level.
(D) quaternary level.

7. Enzymes that break down DNA catalyze the hydrolysis of the covalent bonds that join nucleotides together. What would happen to DNA molecules treated with these enzymes?
(A) The two strands of the double helix would separate.
(B) The phosphodiester linkages of the polynucleotide backbone would be broken.
(C) The pyrimidines would be separated from the deoxyribose sugars.
(D) All bases would be separated from the deoxyribose sugars.

Level 2: Application/Analysis

8. Which of the following hydrocarbons has a double bond in its carbon skeleton?
(A) C_3H_8
(B) C_2H_6
(C) C_2H_4
(D) C_2H_2

9. The molecular formula for glucose is $C_6H_{12}O_6$. What would be the molecular formula for a polymer made by linking ten glucose molecules together by dehydration reactions?
(A) $C_{60}H_{120}O_{60}$
(B) $C_{60}H_{102}O_{51}$
(C) $C_{60}H_{100}O_{50}$
(D) $C_{60}H_{111}O_{51}$

10. Construct a table that organizes the following terms, and label the columns and rows.

Monosaccharides	Polypeptides	Phosphodiester linkages
Fatty acids	Triacylglycerols	Peptide bonds
Amino acids	Polynucleotides	Glycosidic linkages
Nucleotides	Polysaccharides	Ester linkages

11. DRAW IT Copy the polynucleotide strand in Figure 27a and label the bases G, T, C, and T, starting from the 5′ end. Assuming this is a DNA polynucleotide, now draw the complementary strand, using the same symbols for phosphates (circles), sugars (pentagons), and bases. Label the bases. Draw arrows showing the 5′ → 3′ direction of each strand. Use the arrows to make sure the second strand is antiparallel to the first. *Hint*: After you draw the first strand vertically, turn the paper upside down; it is easier to draw the second strand from the 5′ toward the 3′ direction as you go from top to bottom.

Level 3: Synthesis/Evaluation

12. **SCIENTIFIC INQUIRY**
Suppose you are a research assistant in a lab studying DNA-binding proteins. You have been given the amino acid sequences of all the proteins encoded by the genome of a certain species and have been asked to find candidate proteins that could bind DNA. What type of amino acids would you expect to see in the DNA-binding regions of such proteins? Explain your thinking.

13. **FOCUS ON EVOLUTION**
Comparisons of amino acid sequences can shed light on the evolutionary divergence of related species. If you were comparing two living species, would you expect all proteins to show the same degree of divergence? Why or why not? Justify your answer.

14. **FOCUS ON ORGANIZATION**
Proteins, which have diverse functions in a cell, are all polymers of the same kinds of monomers—amino acids. Write a short essay (100–150 words) that discusses how the structure of amino acids allows this one type of polymer to perform so many functions.

15. **SYNTHESIZE YOUR KNOWLEDGE**

Given that the function of egg yolk is to nourish and support the developing chick, explain why egg yolks are so high in fat, protein, and cholesterol.

Answers

Figure Questions

Figure 5

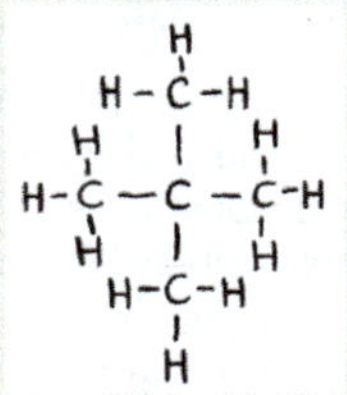

Figure 9

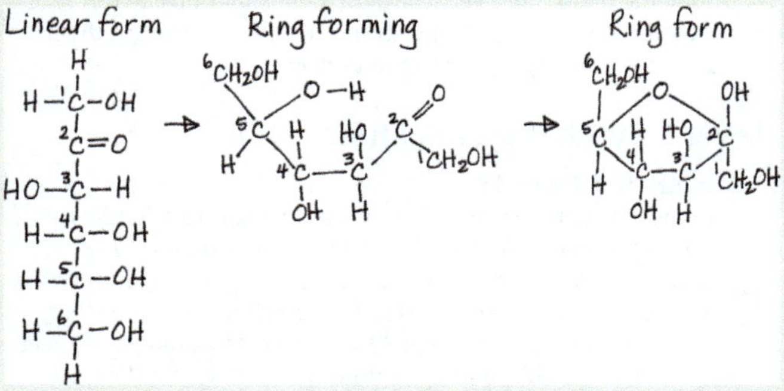

Note that the oxygen on carbon 5 lost its proton and that the oxygen on carbon 2, which used to be the carbonyl oxygen, gained a proton. Four carbons are in the fructose ring, and two are not. (The latter two carbons are attached to carbons 2 and 5, which are in the ring.) The fructose ring differs from the glucose ring, which has five carbons in the ring and one that is not. (Note that the orientation of this fructose molecule is flipped horizontally relative to that of the one in Figure 10.)

Figure 10

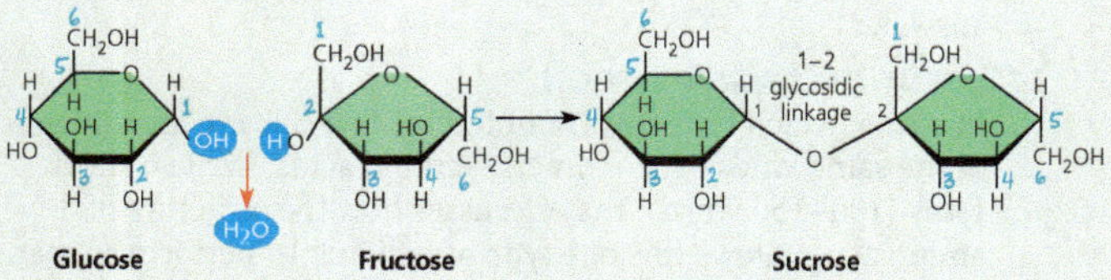

In sucrose, the linkage is called a 1–2 glycosidic linkage because the number 1 carbon in the left monosaccharide (glucose) is linked to the number 2 carbon in the right monosaccharide (fructose). (Note that the fructose molecule is oriented differently from the glucose molecules in Figures 9 and 10, and from the fructose shown in the answer for Figure 9, above. In Figure 10 and here, carbon 2 of fructose is close to carbon 1 of glucose.)

Figure 15

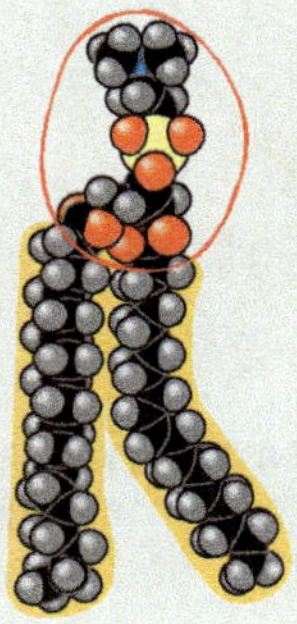

Figure 19

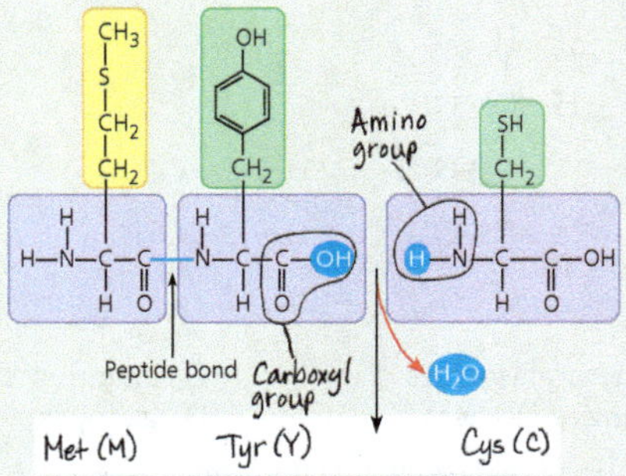

Figure 23 The R group of glutamic acid is acidic and hydrophilic, whereas that of valine is nonpolar and hydrophobic. Therefore, it is unlikely that valine can participate in the same intramolecular interactions that glutamic acid can. A change in these interactions would be expected to (and does) cause a disruption of molecular structure.
Figure 30 Using a genomics approach allows us to use gene sequences to identify species and to learn about evolutionary relationships among any two species. This is because all species are related by their evolutionary history, and the evidence is in the DNA sequences. Proteomics—looking at proteins that are expressed—allows us to learn about how organisms or cells are functioning at a given time or in an association with another species.

Concept Check 1

1. Both consist largely of hydrocarbon chains, which provide fuel—gasoline for engines and fats for plant embryos and animals. Reactions of both types of molecule release energy. **2.** The forms of C_4H_{10} in (b) are structural isomers, as are the butenes (forms of C_4H_8) in (c). **3.** It has both an amino group ($—NH_2$), which makes it an amine, and a carboxyl group (—COOH), which makes it a carboxylic acid. **4.** A chemical group that can act as a base (by picking up H^+) has been replaced with a group that can act as an acid, increasing the acidic properties of the molecule. The shape of the molecule would also change, likely changing the molecules with which it can interact.

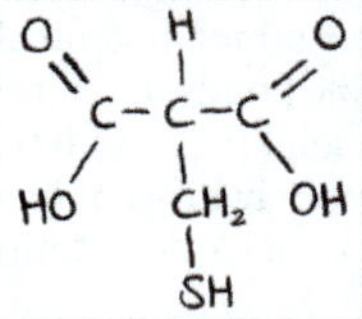

The original cysteine molecule has an asymmetric carbon in the center. After replacement of the amino group with a carboxyl group, this carbon is no longer asymmetric.

Concept Check 2

1. Nine, with one water molecule required to hydrolyze each connected pair of monomers **2.** The amino acids in the fish protein must be released in hydrolysis reactions and incorporated into other proteins in dehydration reactions.

Concept Check 3

1. $C_3H_6O_3$ **2.** $C_{12}H_{22}O_{11}$ **3.** The antibiotic treatment is likely to have killed the cellulose-digesting prokaryotes in the cow's stomach. The absence of these prokaryotes would hamper the cow's ability to obtain energy from food and could lead to weight loss and possibly death. Thus, prokaryotic species are reintroduced, in appropriate combinations, in the gut culture given to treated cows.

Concept Check 4

1. Both have a glycerol molecule attached to fatty acids. The glycerol of a fat has three fatty acids attached, whereas the glycerol of a phospholipid is attached to two fatty acids and one phosphate group. **2.** Human sex hormones are steroids, a type of compound that is hydrophobic and thus classified as a lipid. **3.** The oil droplet membrane could consist of a single layer of phospholipids rather than a bilayer, because an arrangement in which the hydrophobic tails of the membrane phospholipids were in contact with the hydrocarbon regions of the oil molecules would be more stable.

Concept Check 5

1. The function of a protein is a consequence of its specific shape, which is lost when a protein becomes denatured. **2.** Secondary structure involves hydrogen bonds between atoms of the polypeptide backbone. Tertiary structure involves interactions between atoms of the side chains of the amino acid monomers. **3.** These are all nonpolar, hydrophobic amino acids, so you would expect this region to be located in the interior of the folded polypeptide, where it would not contact the aqueous environment inside the cell.

Concept Check 6

1.

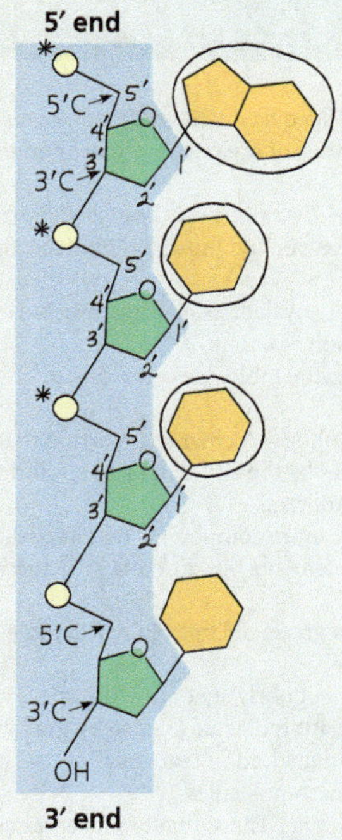

2.

5′–T A G G C C T–3′
3′–A T C C G G A–5′

Concept Check 7

1. The DNA of an organism encodes all of its proteins, and proteins are the molecules that carry out the work of cells, whether an organism is unicellular or multicellular. By knowing the DNA sequence of an organism, scientists would be able to catalog the protein sequences as well. **2.** Ultimately, the DNA sequence carries the information necessary to make the proteins that determine the traits of a particular species. Because the traits of the two species are similar, you would expect the proteins to be similar as well, and therefore the gene sequences should also have a high degree of similarity.

Summary of Key Concepts Questions

1 The methyl group is nonpolar and not reactive. The other six groups are called functional groups because they can participate in chemical reactions. Except for the sulfhydryl group, these functional groups are hydrophilic; they increase the solubility of organic compounds in water. **2** The polymers of large carbohydrates, proteins, and nucleic acids are built from three different types of monomers (monosaccharides, amino acids, and nucleotides, respectively). **3** Both starch and cellulose are polymers of glucose, but the glucose monomers are in the α configuration in starch and the β configuration in cellulose. The glycosidic linkages thus have different geometries, giving the polymers different shapes and thus different properties. Starch is an energy-storage compound in plants; cellulose is a structural component of plant cell walls. Humans can hydrolyze starch to provide energy but cannot hydrolyze cellulose. Cellulose aids in the passage of food through the digestive tract. **4** Lipids are not polymers because they do not exist as a chain of linked monomers. They are not considered macromolecules because they do not reach the giant size of many polysaccharides, proteins, and nucleic acids. **5** A polypeptide, which may consist of hundreds of amino acids in a specific sequence (primary structure), has regions of coils and pleats (secondary structure), which are then folded into irregular contortions (tertiary structure) and may be noncovalently associated with other polypeptides (quaternary structure). The linear order of amino acids, with the varying properties of their side chains (R groups), determines what secondary and tertiary structures will form to produce a protein. The resulting unique three-dimensional shapes of proteins are key to their specific and diverse functions. **6** The complementary base pairing of the two strands of DNA makes possible the precise replication of DNA every time a cell divides, ensuring that genetic information is faithfully transmitted. In some types of RNA, complementary base pairing enables RNA molecules to assume specific three-dimensional shapes that facilitate diverse functions. **7** You would expect the human gene sequence to be most similar to that of the mouse (another mammal), and then to that of the fish (another vertebrate), and least similar to that of the fruit fly (an invertebrate).

Test Your Understanding

1. A **2.** B **3.** C **4.** C **5.** B **6.** A **7.** B **8.** C **9.** B

10.

	Monomers or Components	Polymer or larger molecule	Type of linkage
Carbohydrates	Monosaccharides	Polysaccharides	Glycosidic linkages
Fats	Fatty acids	Triacylglycerols	Ester linkages
Proteins	Amino acids	Polypeptides	Peptide bonds
Nucleic acids	Nucleotides	Polynucleotides	Phosphodiester linkages

11.

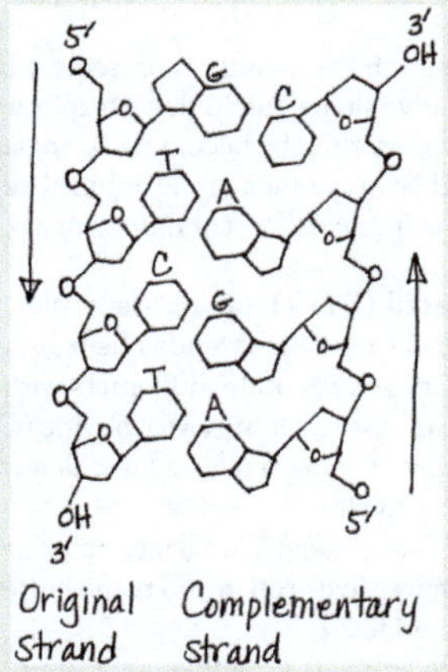

Credits

Credits are listed in the order of appearance.

Photo Credits

1 Mark J. Winter/Science Source; **11 clockwise from top left** Dougal Waters/Getty Images; John N.A. Lott/Biological Photo Service; Paul Lazarow; Biophoto Associates/Science Source; John Durham/Science Source; **14a** Jim Forrest/Dorling Kindersley; **14b** David Murray/Dorling Kindersley; **17a** Andrey Stratilatov/Shutterstock; **17bc** Nina Zanetti/Pearson Education; **20ab** PDB ID 2LYZ, From: Real-space refinement of the structure of hen egg-white lysozyme. R. Diamond. *Journal of Molecular Biology* 82(3):371–91 (Jan 25, 1974); **20c** Clive Freeman, The Royal Institution/Science Source; **21** Peter Colman; **22 spider** Dieter Hopf/Imagebroker/age fotostock; **blood cells** SciePro/SPL/age fotostock; **23** Eye of Science/Science Source; **25 top** CC-BY-3.0 Photo: Dsrjsr; **25 bottom** Laguna Design/Science Source; **29** P. Morris/Garvan Institute of Medical Research; **30 DNA** Alfred Pasieka/Science Source; **hippo** Frontline Photography/Alamy; **whale** WaterFrame/Alamy; **elephants** Imagebroker/FLPA; **Neanderthal** Illustration by Viktor Deak; **doctor** Chassenet/BSIP SA/Alamy; **fungus** David Read, Department of Animal and Plant Sciences, University of Sheffield; **SSE.1** lanych/Shutterstock; **SSE.2** David Bagnall/Alamy; **SSE.3** Eric Isselée/Shutterstock; **Chapter Review butter** Jim Forrest/Dorling Kindersley; **oil** David Murray/Dorling Kindersley; **SYK** Africa Studio/Shutterstock.

Illustration and Text Credits

5 © Pearson Education, Inc.; **15** Adapted from *Biology: The Science of Life*, 3rd ed., by R. Wallace, G. Sanders and R. Ferl, 1991. Copyright © 1991 by Pearson Education, Inc. Reprinted and electronically reproduced by permission of Pearson Education, Inc. Upper Saddle River, NJ. **17h** PDB ID 1CGD: J. Bella, B. Brodsky, and H. M. Berman. Hydration structure of a collagen peptide, *Structure* 3:893–906 (1995); **20** PDB ID 2LYZ: Based on R. Diamond. Real-space refinement of the structure of hen egg-white lysozyme. *Journal of Molecular Biology* 82(3): 371–91 (Jan 25, 1974); **22** (part 4): PDB ID 3GS0: S.K. Palaninathan, N.N. Mohamedmohaideen, E. Orlandini, G. Ortore, S. Nencetti, A. Lapucci, A. Rossello, J.S. Freundlich, J.C. Sacchettini. Novel transthyretin amyloid fibril formation inhibitors: Synthesis, biological evaluation, and X-ray structural analysis. *Public Library of Science ONE* 4(7): e6290. doi:10.1371/journal.pone.0006290 (2009); **22** (part 6): PDB ID 1CGD: J. Bella, B. Brodsky, and H.M. Berman. Hydration structure of a collagen peptide, *Structure* 3:893–906 (1995); **22** (part 7): PDB ID 2HHB: G. Fermi, M.F. Perutz, B. Shaanan, and R. Fourme. The crystal structure of human deoxyhaemoglobin at 1.74 Å resolution. *J. Mol. Biol.* 175:159–174 (1984); **SSE Table** Data from human: http://www.ncbi.nlm.nih.gov/protein/AAA21113.1; Data from rhesus monkey: http://www.ncbi.nlm.nih.gov/protein/122634; Data from gibbon: http://www.ncbi.nlm.nih.gov/protein/122616.

Glossary

adenosine triphosphate (a-den′-ō-sēn trī-fos′-fāt) An adenine-containing nucleoside triphosphate that releases free energy when its phosphate bonds are hydrolyzed. This energy is used to drive endergonic reactions in cells.

amino acid (uh-mēn′-ō) An organic molecule possessing both a carboxyl and an amino group. Amino acids serve as the monomers of polypeptides.

antiparallel Referring to the arrangement of the sugar-phosphate backbones in a DNA double helix (they run in opposite 5′ → 3′ directions).

carbohydrate (kar′-bō-hī′-drāt) A sugar (monosaccharide) or one of its dimers (disaccharides) or polymers (polysaccharides).

catalyst (kat′-uh-list) A chemical agent that selectively increases the rate of a reaction without being consumed by the reaction.

chitin (kī′-tin) A structural polysaccharide, consisting of amino sugar monomers, found in many fungal cell walls and in the exoskeletons of all arthropods.

cholesterol (kō-les′-tuh-rol) A steroid that forms an essential component of animal cell membranes and acts as a precursor molecule for the synthesis of other biologically important steroids, such as many hormones.

***cis-trans* isomer** One of two or more compounds that have the same molecular formula and covalent bonds between atoms but differ in the spatial arrangements of their atoms owing to the inflexibility of double bonds;
formerly called a geometric isomer.

dehydration reaction A chemical reaction in which two molecules become covalently bonded to each other with the removal of a water molecule.

denaturation (dē-nā′-chur-ā′-shun) In proteins, a process in which a protein loses its native shape due to the disruption of weak chemical bonds and interactions, thereby becoming biologically inactive; in DNA, the separation of the two strands of the double helix. Denaturation occurs under extreme (noncellular) conditions of pH, salt concentration, or temperature.

deoxyribonucleic acid (DNA) (dē-ok′-sē-rī′-bō-nū-klā′-ik) A nucleic acid molecule, usually a double-stranded helix, in which each polynucleotide strand consists of nucleotide monomers with a deoxyribose sugar and the nitrogenous bases adenine (A), cytosine (C), guanine (G), and thymine (T); capable of being replicated and determining the inherited structure of a cell's proteins.

deoxyribose (dē-ok′-si-rī′-bōs) The sugar component of DNA nucleotides, having one fewer hydroxyl group than ribose, the sugar component of RNA nucleotides.

disaccharide (dī-sak′-uh-rīd) A double sugar, consisting of two monosaccharides joined by a glycosidic linkage formed by a dehydration reaction.

double helix The form of native DNA, referring to its two adjacent antiparallel polynucleotide strands wound around an imaginary axis into a spiral shape.

enantiomer One of two compounds that are mirror images of each other and that differ in shape due to the presence of an asymmetric carbon.

enzyme (en′-zīm) A macromolecule serving as a catalyst, a chemical agent that increases the rate of a reaction without being consumed by the reaction. Most enzymes are proteins.

fat A lipid consisting of three fatty acids linked to one glycerol molecule; also called a triacylglycerol or triglyceride.

fatty acid A carboxylic acid with a long carbon chain. Fatty acids vary in length and in the number and location of double bonds; three fatty acids linked to a glycerol molecule form a fat molecule, also known as a triacylglycerol or triglyceride.

functional group A specific configuration of atoms commonly attached to the carbon skeletons of organic molecules and involved in chemical reactions.

glycogen (glī′-kō-jen) An extensively branched glucose storage polysaccharide found in the liver and muscle of animals; the animal equivalent of starch.

glycosidic linkage A covalent bond formed between two monosaccharides by a dehydration reaction.

hydrocarbon An organic molecule consisting of only carbon and hydrogen.

hydrolysis (hī-drol′-uh-sis) A chemical reaction that breaks bonds between two molecules by the addition of water; functions in disassembly of polymers to monomers.

isomer One of two or more compounds that have the same numbers of atoms of the same elements but different structures and hence different properties.

lipid (lip′-id) Any of a group of large biological molecules, including fats, phospholipids,
and steroids, that mix poorly, if at all, with water.

macromolecule A giant molecule formed by the joining of smaller molecules, usually by a dehydration reaction. Polysaccharides, proteins, and nucleic acids are macromolecules.

monomer (mon′-uh-mer) The subunit that serves as the building block of a polymer.

monosaccharide (mon′-ō-sak′-uh-rīd) The simplest carbohydrate, active alone or serving as a monomer for disaccharides and polysaccharides. Also known as simple sugars, monosaccharides have molecular formulas that are generally some multiple of CH_2O.

nucleic acid (nū-klā′-ik) A polymer (polynucleotide) consisting of many nucleotide monomers; serves as a blueprint for proteins and, through the actions of proteins, for all cellular activities. The two types are DNA and RNA.

nucleotide (nū′-klē-ō-tīd′) The building block of a nucleic acid, consisting of a five-carbon sugar covalently bonded to a nitrogenous base and one to three phosphate groups.

organic compound A chemical compound containing carbon.

peptide bond The covalent bond between the carboxyl group on one amino acid and the amino group on another, formed by a dehydration reaction.

phospholipid (fos′-fō-lip′-id) A lipid made up of glycerol joined to two fatty acids and a phosphate group. The hydrocarbon chains of the fatty acids act as nonpolar, hydrophobic tails, while the rest of the molecule acts as a polar, hydrophilic head. Phospholipids form bilayers that function as biological membranes.

polymer (pol′-uh-mer) A long molecule consisting of many similar or identical monomers linked together by covalent bonds.

polynucleotide (pol′-ē-nū′-klē-ō-tīd) A polymer consisting of many nucleotide monomers in a chain. The nucleotides can be those of DNA or RNA.

polypeptide (pol′-ē-pep′-tīd) A polymer of many amino acids linked together by peptide bonds.

polysaccharide (pol′-ē-sak′-uh-rīd) A polymer of many monosaccharides, formed by dehydration reactions.

protein (prō′-tēn) A biologically functional molecule consisting of one or more polypeptides folded and coiled into a specific three-dimensional structure.

purine (pyū′-rēn) One of two types of nitrogenous bases found in nucleotides, characterized by a six-membered ring fused to a five-membered ring. Adenine (A) and guanine (G) are purines.

pyrimidine (puh-rim′-uh-dēn) One of two types of nitrogenous bases found in nucleotides, characterized by a six-membered ring. Cytosine (C), thymine (T), and uracil (U) are pyrimidines.

ribonucleic acid (RNA) (rī′-bō-nū-klā′-ik) A type of nucleic acid consisting of a polynucleotide made up of nucleotide monomers with a ribose sugar and the nitrogenous bases adenine (A), cytosine (C), guanine (G), and uracil (U); usually single-stranded; functions in protein synthesis, gene regulation, and as the genome of some viruses.

ribose The sugar component of RNA nucleotides.

saturated fatty acid A fatty acid in which all carbons in the hydrocarbon tail are connected by single bonds, thus maximizing the number of hydrogen atoms attached to the carbon skeleton.

starch A storage polysaccharide in plants, consisting entirely of glucose monomers joined by α glycosidic linkages.

steroid A type of lipid characterized by a carbon skeleton consisting of four fused rings with various chemical groups attached.

structural isomer One of two or more compounds that have the same molecular formula but differ in the covalent arrangements of their atoms.

***trans* fats** An unsaturated fat, formed artificially during hydrogenation of oils, containing one or more *trans* double bonds.

triacylglycerol (trī-as′-ul-glis′-uh-rol) A lipid consisting of three fatty acids linked to one glycerol molecule; also called a fat or triglyceride.

unsaturated fatty acid A fatty acid that has one or more double bonds between carbons in the hydrocarbon tail. Such bonding reduces the number of hydrogen atoms attached to the carbon skeleton.

valence The bonding capacity of a given atom; the number of covalent bonds an atom can form, which usually equals the number of unpaired electrons in its outermost (valence) shell.

X-ray crystallography A technique used to study the three-dimensional structure of molecules. It depends on the diffraction of an X-ray beam by the individual atoms of a crystallized molecule.

A Tour of the Cell

From Chapter 4 of *Campbell Biology in Focus*, Second Edition. Lisa A. Urry, Michael L. Cain, Steven A. Wasserman, Peter V. Minorsky, and Jane B. Reece.

A Tour of the Cell

KEY CONCEPTS

1. Biologists use microscopes and the tools of biochemistry to study cells
2. Eukaryotic cells have internal membranes that compartmentalize their functions
3. The eukaryotic cell's genetic instructions are housed in the nucleus and carried out by the ribosomes
4. The endomembrane system regulates protein traffic and performs metabolic functions in the cell
5. Mitochondria and chloroplasts change energy from one form to another
6. The cytoskeleton is a network of fibers that organizes structures and activities in the cell
7. Extracellular components and connections between cells help coordinate cellular activities

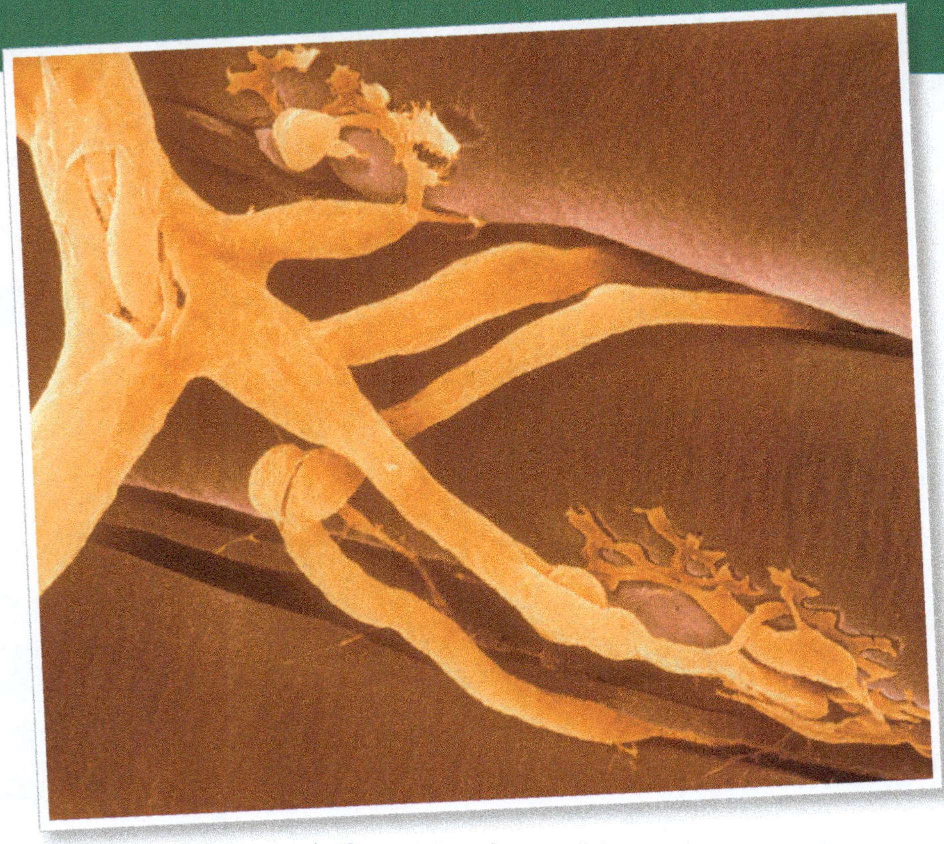

▲ **Figure 1** How do your cells help you learn about biology?

The Fundamental Units of Life

Given the scope of biology, you may wonder sometimes how you will ever learn all the material in this course! The answer involves cells, which are as fundamental to the living systems of biology as the atom is to chemistry. The contraction of muscle cells moves your eyes as you read this sentence. **Figure 1** shows extensions from a nerve cell (orange) making contact with muscle cells (red). The words on the page are translated into signals that nerve cells carry to your brain, where they are passed on to other nerve cells. As you study, you are making cell connections like these that solidify memories and permit learning to occur.

All organisms are made of cells. In the hierarchy of biological organization, the cell is the simplest collection of matter that can be alive. Indeed, many forms of life exist as single-celled organisms. (You may be familiar with single-celled eukaryotic organisms that live in pond water, such as paramecia.) Larger, more complex organisms, including plants and animals, are multicellular; their bodies are cooperatives of many kinds of specialized cells that could not survive for long on their own. Even when cells are arranged into higher levels of organization, such as tissues and organs, the cell remains the organism's basic unit of structure and function.

All cells are related by their descent from earlier cells. During the long evolutionary history of life on Earth, cells have been modified in many different ways. But although cells can differ substantially from one another, they share common features. In this chapter, we'll first examine the tools and techniques that allow us to understand cells, and then tour the cell and become acquainted with its components.

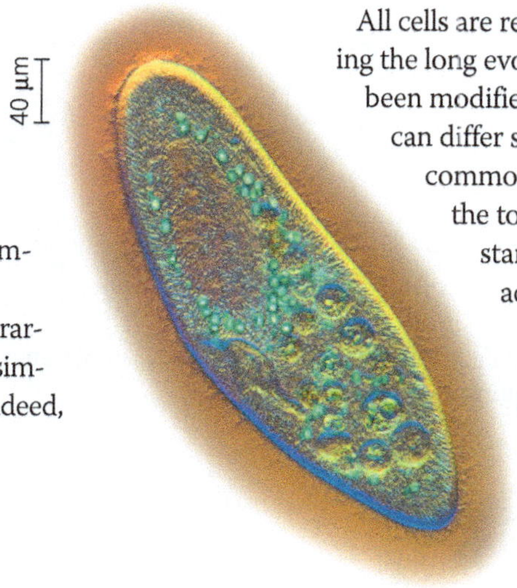

◀ *Paramecium caudatum*

CONCEPT 1

Biologists use microscopes and the tools of biochemistry to study cells

How can cell biologists investigate the inner workings of a cell, usually too small to be seen by the unaided eye? Before we tour the cell, it will be helpful to learn how cells are studied.

Microscopy

The development of instruments that extend the human senses has gone hand in hand with the advance of science. Microscopes were invented in 1590 and further refined during the 1600s. Cell walls were first seen by Robert Hooke in 1665 as he looked through a microscope at dead cells from the bark of an oak tree. But it took the wonderfully crafted lenses of Antoni van Leeuwenhoek to visualize living cells. Imagine Hooke's awe when he visited van Leeuwenhoek in 1674 and the world of microorganisms—what his host called "very little animalcules"—was revealed to him.

The microscopes first used by Renaissance scientists, as well as the microscopes you are likely to use in the laboratory, are all light microscopes. In a **light microscope (LM)**, visible light is passed through the specimen and then through glass lenses. The lenses refract (bend) the light in such a way that the image of the specimen is magnified as it is projected into the eye or into a camera.

Three important parameters in microscopy are magnification, resolution, and contrast. *Magnification* is the ratio of an object's image size to its real size. Light microscopes can magnify effectively to about 1,000 times the actual size of the specimen; at greater magnifications, additional details cannot be seen clearly. *Resolution* is a measure of the clarity of the image; it is the minimum distance two points can be separated and still be distinguished as separate points. For example, what appears to the unaided eye as one star in the sky may be resolved as twin stars with a telescope, which has a higher resolving ability than the eye. Similarly, using standard techniques, the light microscope cannot resolve detail finer than about 0.2 micrometer (μm), or 200 nanometers (nm), regardless of the magnification **(Figure 2)**. The third parameter, *contrast*, is the difference in brightness between the light and dark areas of an image. Methods for enhancing contrast in light microscopy include staining or labeling cell components to stand out visually. **Figure 3** shows some different types of microscopy; study this figure as you read the rest of this section.

Until recently, the resolution barrier prevented cell biologists from using standard light microscopy when studying **organelles**, the membrane-enclosed structures within eukaryotic cells. To see these structures in any detail required the development of a new instrument. In the 1950s, the electron microscope was introduced to biology. Rather than focusing light, the **electron microscope (EM)** focuses a beam of electrons through the specimen or onto its surface. Resolution is inversely related to the wavelength of the light (or electrons) a microscope uses for imaging, and electron beams have much shorter wavelengths than visible light. Modern electron microscopes can theoretically achieve a resolution of about 0.002 nm, though in practice they usually cannot resolve structures smaller than about 2 nm across. Still, this is a 100-fold improvement over the standard light microscope.

The **scanning electron microscope (SEM)** is especially useful for detailed study of the topography of a specimen

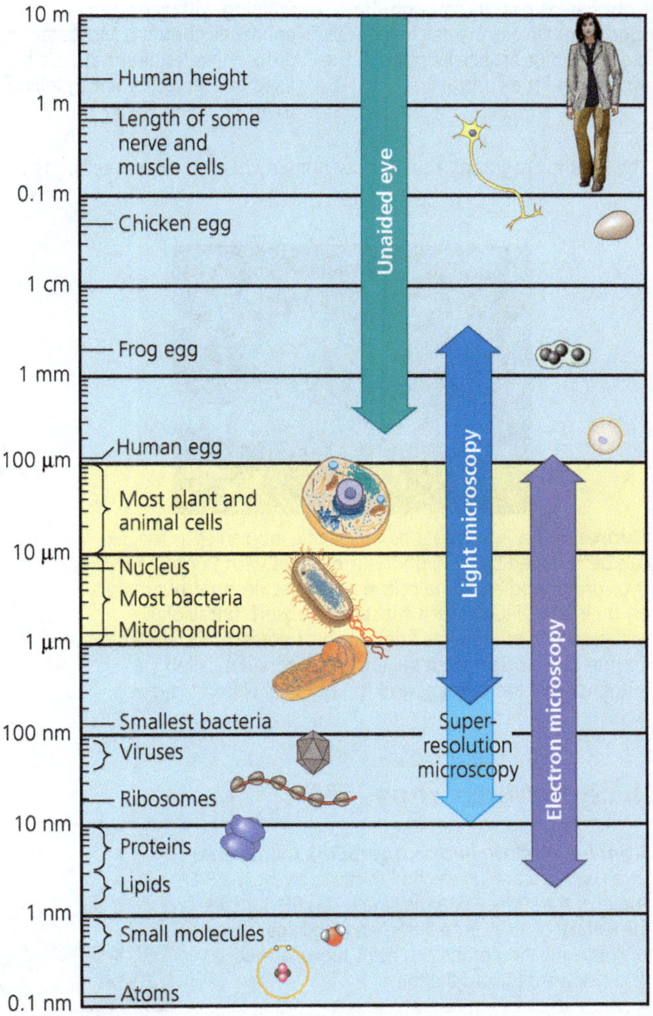

1 centimeter (cm) = 10^{-2} meter (m) = 0.4 inch
1 millimeter (mm) = 10^{-3} m
1 micrometer (μm) = 10^{-3} mm = 10^{-6} m
1 nanometer (nm) = 10^{-3} μm = 10^{-9} m

▲ **Figure 2 The size range of cells.** Most cells are between 1 and 100 μm in diameter (yellow region of chart), and their components are even smaller, as are viruses. Notice that the scale along the left side is logarithmic to accommodate the range of sizes shown. Starting at the top of the scale with 10 m and going down, each reference measurement marks a tenfold decrease in diameter or length.

▼ **Figure 3** **Exploring Microscopy**

Light Microscopy (LM)

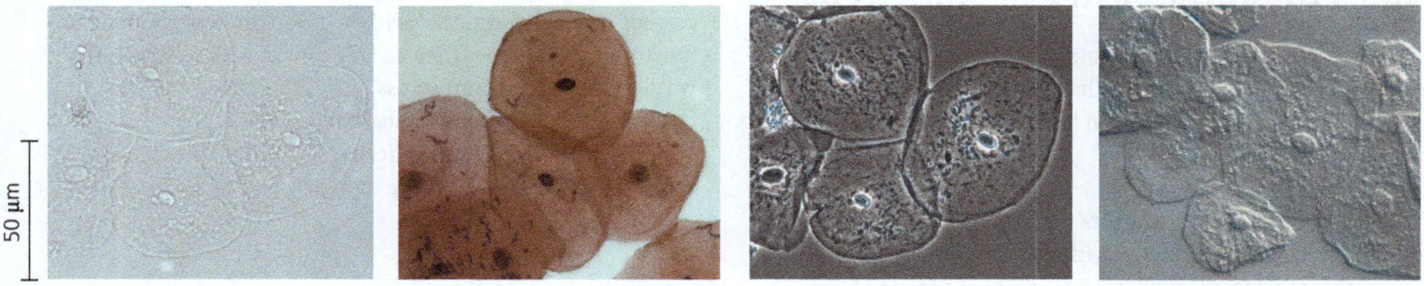

Brightfield (unstained specimen). Light passes directly through the specimen. Unless the cell is naturally pigmented or artificially stained, the image has little contrast.

Brightfield (stained specimen). Staining with various dyes enhances contrast. Most staining procedures require that cells be fixed (preserved), thereby killing them.

Phase-contrast. Variations in density within the specimen are amplified to enhance contrast in unstained cells; this is especially useful for examining living, unpigmented cells.

Differential-interference contrast (Nomarski). As in phase-contrast microscopy, optical modifications are used to exaggerate differences in density; the image appears almost 3-D.

The light micrographs above show human cheek epithelial cells; the scale bar pertains to all four micrographs.

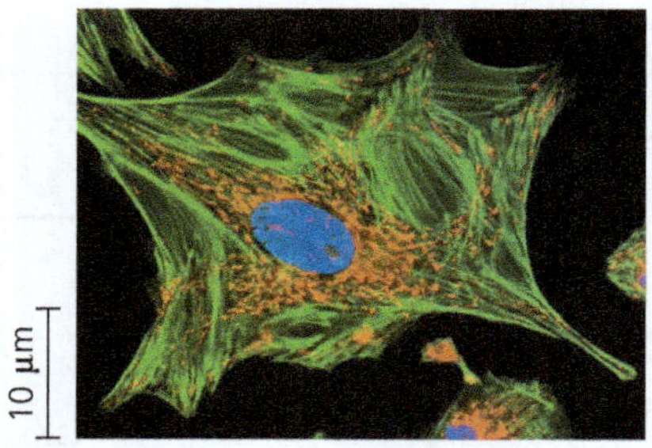

Fluorescence. The locations of specific molecules in the cell can be revealed by labeling the molecules with fluorescent dyes or antibodies; some cells have molecules that fluoresce on their own. Fluorescent substances absorb ultraviolet radiation and emit visible light. In this fluorescently labeled uterine cell, nuclear material is blue, organelles called mitochondria are orange, and the cell's "skeleton" is green.

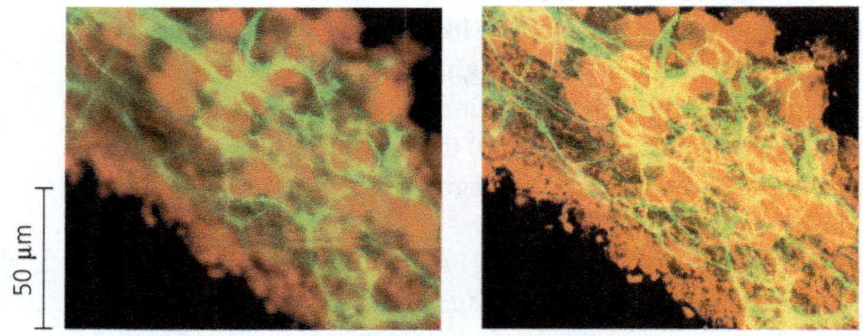

Confocal. The left image is a standard fluorescence micrograph of fluorescently labeled nervous tissue (nerve cells are green, support cells are orange, and regions of overlap are yellow); the right image is a confocal image of the same tissue. Using a laser, this "optical sectioning" technique eliminates out-of-focus light from a thick sample, creating a single plane of fluorescence in the image. By capturing sharp images at many different planes, a 3-D reconstruction can be created. The standard image is blurry because out-of-focus light is not excluded.

Electron Microscopy (EM)

Scanning electron microscopy (SEM). Micrographs taken with a scanning electron microscope show a 3-D image of the surface of a specimen. This SEM shows the surface of a cell from a trachea (windpipe) covered with cilia. (Beating of the cilia helps move inhaled debris upward toward the throat.)

Electron micrographs are black and white but are often artificially colorized to highlight particular structures, as has been done with both electron micrographs (SEM and TEM) shown here.

Abbreviations used in figure legends throughout this text:
LM = Light Micrograph
SEM = Scanning Electron Micrograph
TEM = Transmission Electron Micrograph

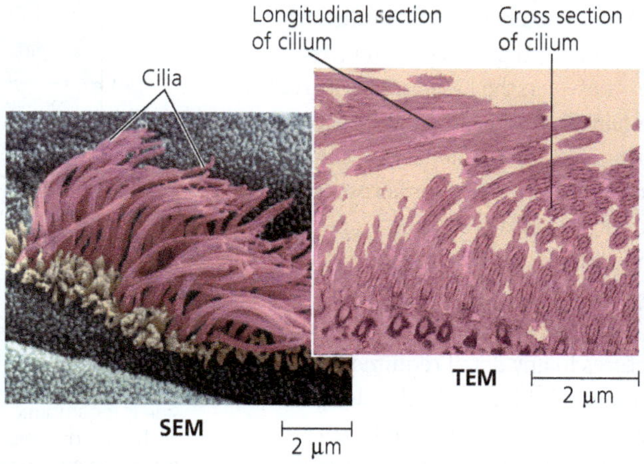

Transmission electron microscopy (TEM). A transmission electron microscope profiles a thin section of a specimen. This TEM shows a section through a tracheal cell, revealing its internal structure. In preparing the specimen, some cilia were cut along their lengths, creating longitudinal sections, while other cilia were cut straight across, creating cross sections.

(see Figure 3). The electron beam scans the surface of the sample, usually coated with a thin film of gold. The beam excites electrons on the surface, and these secondary electrons are detected by a device that translates the pattern of electrons into an electronic signal sent to a video screen. The result is an image of the specimen's surface that appears three-dimensional.

The **transmission electron microscope (TEM)** is used to study the internal structure of cells (see Figure 3). The TEM aims an electron beam through a very thin section of the specimen, much as a light microscope aims light through a sample on a slide. For the TEM, the specimen has been stained with atoms of heavy metals, which attach to certain cellular structures, thus enhancing the electron density of some parts of the cell more than others. The electrons passing through the specimen are scattered more in the denser regions, so fewer are transmitted. The image displays the pattern of transmitted electrons. Instead of using glass lenses, both the SEM and TEM use electromagnets as lenses to bend the paths of the electrons, ultimately focusing the image onto a monitor for viewing.

Electron microscopes have revealed many subcellular structures that were impossible to resolve with the light microscope. But the light microscope offers advantages, especially in studying living cells. A disadvantage of electron microscopy is that the methods used to prepare the specimen kill the cells. Specimen preparation for any type of microscopy can introduce artifacts, structural features seen in micrographs that do not exist in the living cell.

In the past several decades, light microscopy has been revitalized by major technical advances. Labeling individual cellular molecules or structures with fluorescent markers has made it possible to see such structures with increasing detail. In addition, confocal and other newer types of fluorescent light microscopy have produced sharper images of three-dimensional tissues and cells. Finally, a group of new techniques and labeling molecules developed in recent years have allowed researchers to "break" the resolution barrier and distinguish subcellular structures as small as 10–20 nm across. As this "super-resolution microscopy" becomes more widespread, the images we see of living cells are proving as awe-inspiring to us as van Leeuwenhoek's were to Robert Hooke 350 years ago.

Microscopes are the most important tools of *cytology*, the study of cell structure. Understanding the function of each structure, however, required the integration of cytology and *biochemistry*, the study of the chemical processes (metabolism) of cells.

Cell Fractionation

A useful technique for studying cell structure and function is **cell fractionation**. Broken-up cells are placed in a tube that is spun in a centrifuge. The resulting force causes the largest cell components to settle to the bottom of the tube, forming a pellet. The liquid above the pellet is poured into a new tube and centrifuged at a higher speed for a longer time. This process is repeated several times, resulting in a series of pellets that consist of nuclei, mitochondria (and chloroplasts if the cells are from a photosynthetic organism), pieces of membrane, and ribosomes, the smallest components.

Cell fractionation enables researchers to prepare specific cell components in bulk and identify their functions, a task not usually possible with intact cells. For example, in one of the cell fractions resulting from centrifugation, biochemical tests showed the presence of enzymes involved in cellular respiration, while electron microscopy revealed large numbers of the organelles called mitochondria. Together, these data helped biologists determine that mitochondria are the sites of cellular respiration. Biochemistry and cytology thus complement each other in correlating cell function with structure.

CONCEPT CHECK 1

1. How do stains used for light microscopy compare with those used for electron microscopy?
2. **WHAT IF?** Which type of microscope would you use to study (a) the changes in shape of a living white blood cell and (b) the details of surface texture of a hair?

CONCEPT 2

Eukaryotic cells have internal membranes that compartmentalize their functions

Cells—the basic structural and functional units of every organism—are of two distinct types: prokaryotic and eukaryotic. Organisms of the domains Bacteria and Archaea consist of prokaryotic cells. Protists, fungi, animals, and plants all consist of eukaryotic cells. ("Protist" is an informal term referring to a group of mostly unicellular eukaryotes.)

Comparing Prokaryotic and Eukaryotic Cells

All cells share certain basic features: They are all bounded by a selective barrier, called the *plasma membrane*. Inside all cells is a semifluid, jellylike substance called **cytosol**, in which subcellular components are suspended. All cells contain *chromosomes*, which carry genes in the form of DNA. And all cells have *ribosomes*, tiny complexes that make proteins according to instructions from the genes.

A major difference between prokaryotic and eukaryotic cells is the location of their DNA. In a **eukaryotic cell**, most of the DNA is in an organelle called the *nucleus*, which is bounded by a double membrane (see Figure 7). In a **prokaryotic cell**, the DNA is concentrated in a region that is not membrane-enclosed, called the **nucleoid** **(Figure 4)**. *Eukaryotic* means "true nucleus" (from the Greek *eu*, true, and *karyon*, kernel, here referring to the nucleus), and *prokaryotic* means "before nucleus" (from the Greek *pro*, before), reflecting the earlier evolution of prokaryotic cells.

The interior of either type of cell is called the **cytoplasm**; in eukaryotic cells, this term refers only to the region between the nucleus and the plasma membrane. Within the cytoplasm of a eukaryotic cell, suspended in cytosol, are a variety of organelles of specialized form and function. These membrane-bounded structures are absent in prokaryotic cells, another distinction between prokaryotic and eukaryotic cells. However, in spite of the absence of organelles, the prokaryotic cytoplasm is not a formless soup. For example, some prokaryotes contain regions surrounded by proteins (not membranes), within which specific reactions take place.

Eukaryotic cells are generally much larger than prokaryotic cells (see Figure 2). Size is a general feature of cell structure that relates to function. The logistics of carrying out cellular metabolism sets limits on cell size. At the lower limit, the smallest cells known are bacteria called mycoplasmas, which have diameters between 0.1 and 1.0 μm. These are perhaps the smallest packages with enough DNA to program metabolism and enough enzymes and other cellular equipment to carry out the activities necessary for a cell to sustain itself and reproduce. Typical bacteria are 1–5 μm in diameter, about ten times the size of mycoplasmas. Eukaryotic cells are typically 10–100 μm in diameter.

Metabolic requirements also impose theoretical upper limits on the size that is practical for a single cell. At the boundary of every cell, the **plasma membrane** functions as a selective barrier that allows passage of enough oxygen, nutrients, and wastes to service the entire cell **(Figure 5)**. For each square micrometer of membrane, only a limited amount of a particular substance can cross per second, so the ratio of surface area to volume is critical. As a cell (or any other object) increases in size, its surface area grows proportionately less than its volume. (Area is proportional to a linear dimension squared, whereas volume is proportional to the linear dimension cubed.) Thus, a smaller object has a greater ratio of surface area to volume **(Figure 6)**. The **Scientific Skills Exercise** for this chapter gives you a chance to calculate the volumes and surface areas of two actual cells—a mature yeast cell and a cell budding from it.

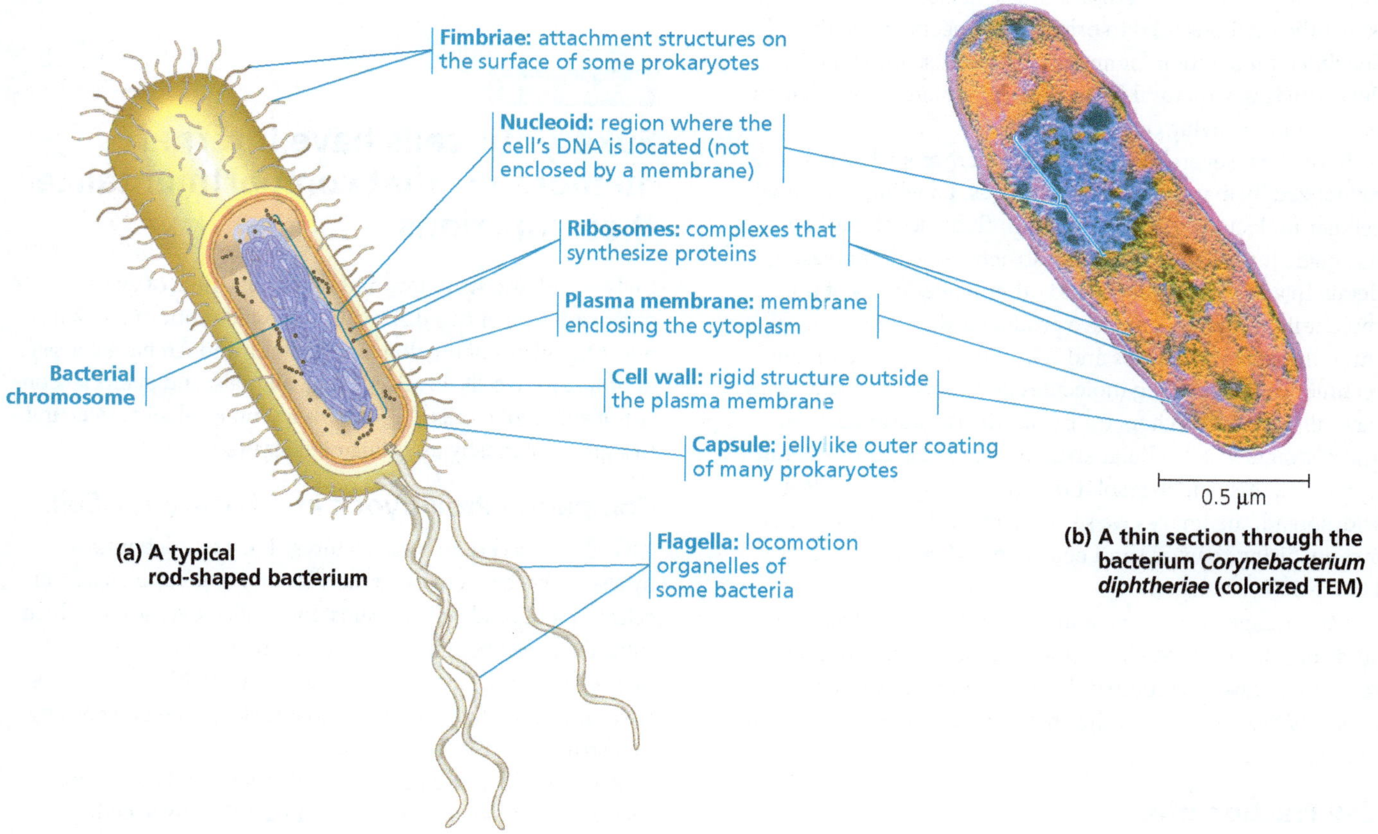

▲ **Figure 4 A prokaryotic cell.** Lacking a true nucleus and the other membrane-enclosed organelles of the eukaryotic cell, the prokaryotic cell appears much simpler in internal structure. Prokaryotes include bacteria and archaea; the general cell structure of the two domains is quite similar.

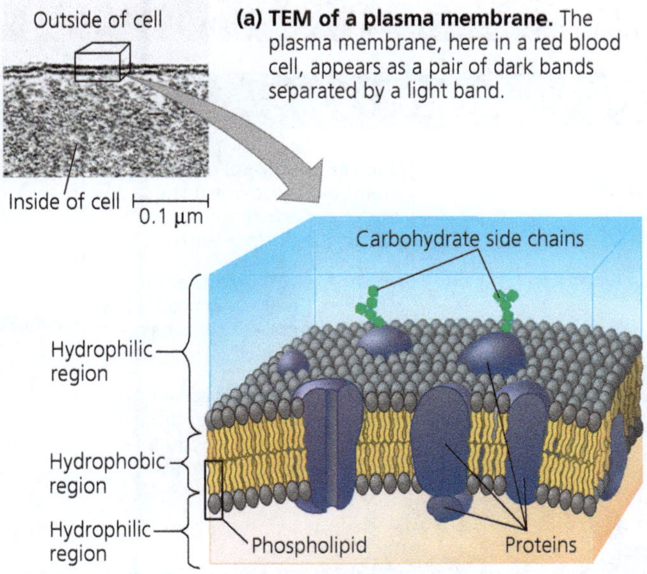

(a) TEM of a plasma membrane. The plasma membrane, here in a red blood cell, appears as a pair of dark bands separated by a light band.

(b) Structure of the plasma membrane

▲ **Figure 5 The plasma membrane.** The plasma membrane and the membranes of organelles consist of a double layer (bilayer) of phospholipids with various proteins attached to or embedded in it. The hydrophobic parts of phospholipids and membrane proteins are found in the interior of the membrane, while the hydrophilic parts are in contact with aqueous solutions on either side. Carbohydrate side chains may be attached to proteins or lipids on the outer surface of the plasma membrane.

MAKE CONNECTIONS *Describe the characteristics of phospholipids that allow them to function as the major components of the plasma membrane.*

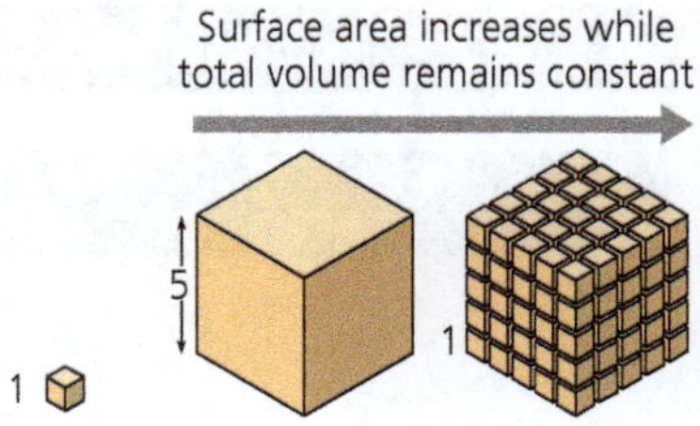

Total surface area [sum of the surface areas (height × width) of all box sides × number of boxes]	6	150	750
Total volume [height × width × length × number of boxes]	1	125	125
Surface-to-volume (S-to-V) ratio [surface area ÷ volume]	6	1.2	6

▲ **Figure 6 Geometric relationships between surface area and volume.** In this diagram, cells are represented as boxes. Using arbitrary units of length, we can calculate the cell's surface area (in square units, or units^2), volume (in cubic units, or units^3), and ratio of surface area to volume. A high surface-to-volume ratio facilitates the exchange of materials between a cell and its environment.

The need for a surface area sufficiently large to accommodate the volume helps explain the microscopic size of most cells and the narrow, elongated shapes of others, such as nerve cells. Larger organisms do not generally have *larger* cells than smaller organisms—they simply have *more* cells. A sufficiently high ratio of surface area to volume is especially important in cells that exchange a lot of material with their surroundings, such as intestinal cells. Such cells may have many long, thin projections from their surface called *microvilli*, which increase surface area without an appreciable increase in volume.

The evolutionary relationships between prokaryotic and eukaryotic cells will be discussed later in this chapter. Most of the discussion of cell structure that follows in this chapter applies to eukaryotic cells.

A Panoramic View of the Eukaryotic Cell

In addition to the plasma membrane at its outer surface, a eukaryotic cell has extensive, elaborately arranged internal membranes that divide the cell into compartments—the organelles mentioned earlier. The cell's compartments provide different local environments that support specific metabolic functions, so incompatible processes can go on simultaneously inside a single cell. The plasma membrane and organelle membranes also participate directly in the cell's metabolism, because many enzymes are built right into the membranes.

The basic fabric of most biological membranes is a double layer of phospholipids and other lipids. Embedded in this lipid bilayer or attached to its surfaces are diverse proteins (see Figure 5). However, each type of membrane has a unique composition of lipids and proteins suited to that membrane's specific functions. For example, enzymes embedded in the membranes of the organelles called mitochondria function in cellular respiration.

Before continuing with this chapter, examine the eukaryotic cells in **Figure 7**. The generalized diagrams of an animal cell and a plant cell introduce the various organelles and show the key differences between animal and plant cells. The micrographs at the bottom of the figure give you a glimpse of cells from different types of eukaryotic organisms.

CONCEPT CHECK 2

1. Briefly describe the structure and function of the nucleus, the mitochondrion, the chloroplast, and the endoplasmic reticulum.
2. **WHAT IF?** Imagine an elongated cell (such as a nerve cell) that measures 125 × 1 × 1 arbitrary units. Predict how its surface-to-volume ratio would compare with those in Figure 6. Then calculate the ratio and check your prediction.

▼ Figure 7 **Exploring Eukaryotic Cells**

Animal Cell (cutaway view of generalized cell)

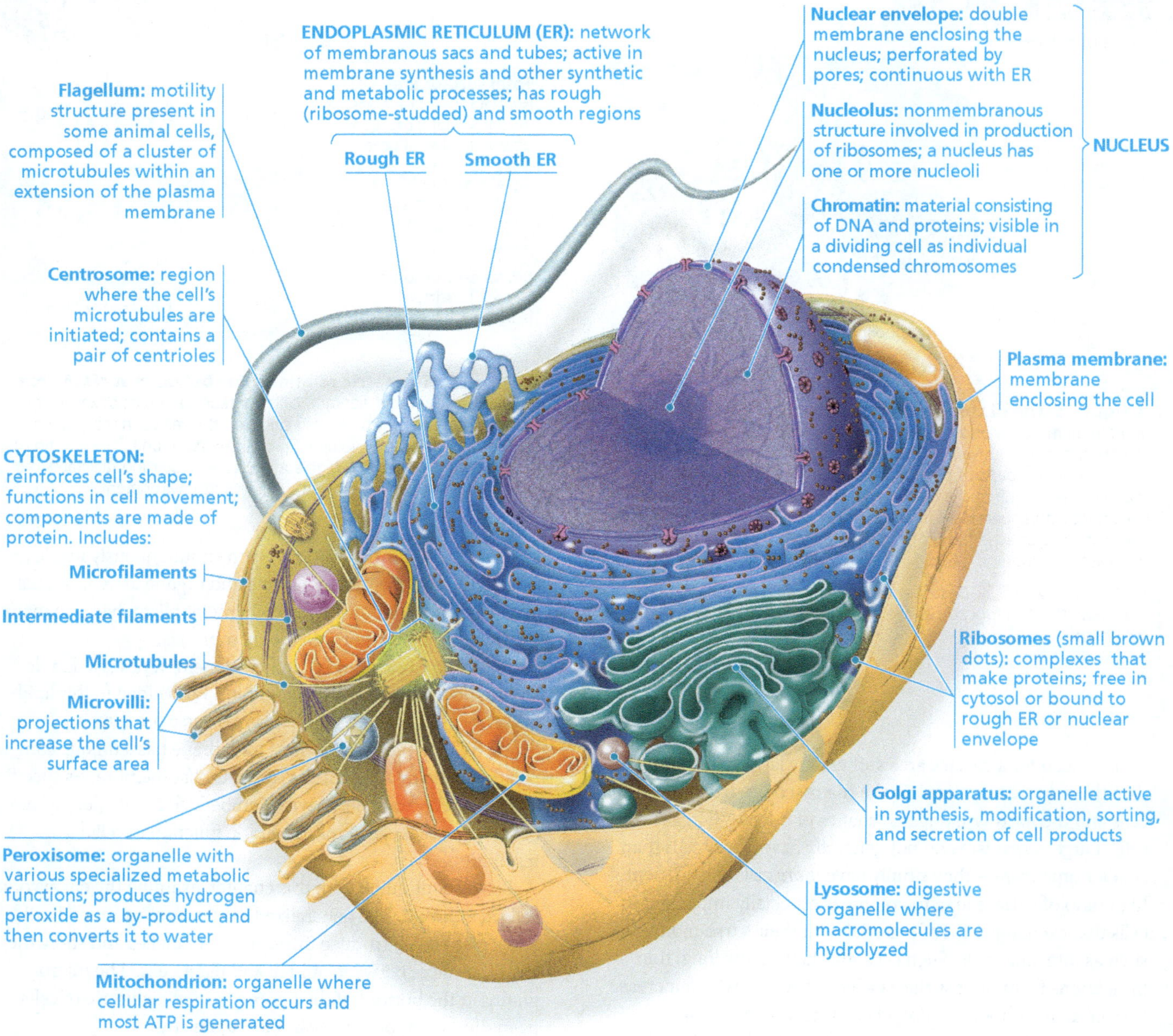

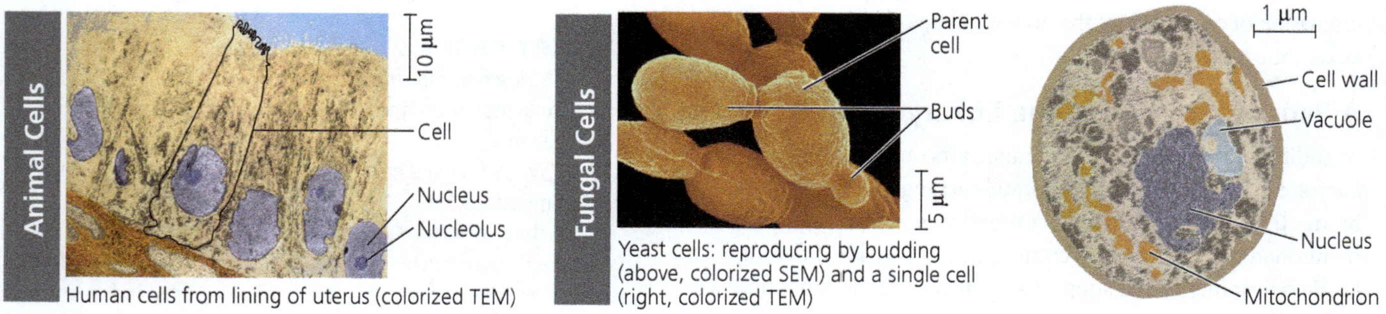

Human cells from lining of uterus (colorized TEM)

Yeast cells: reproducing by budding (above, colorized SEM) and a single cell (right, colorized TEM)

Plant Cell (cutaway view of generalized cell)

ANIMATION

Visit the Study Area in **MasteringBiology** for the BioFlix® 3-D Animations Tour of an Animal Cell and Tour of a Plant Cell.

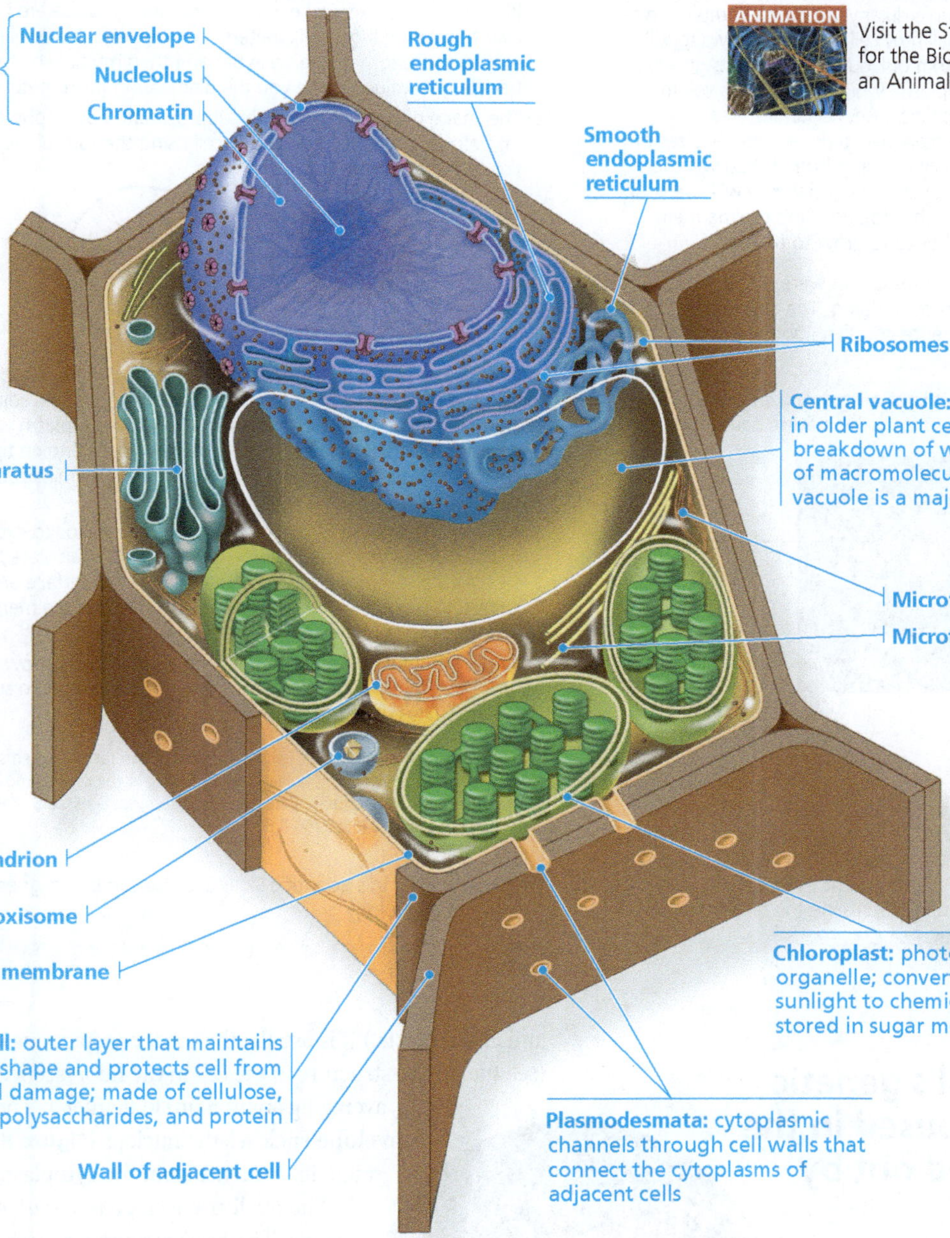

Plant Cells

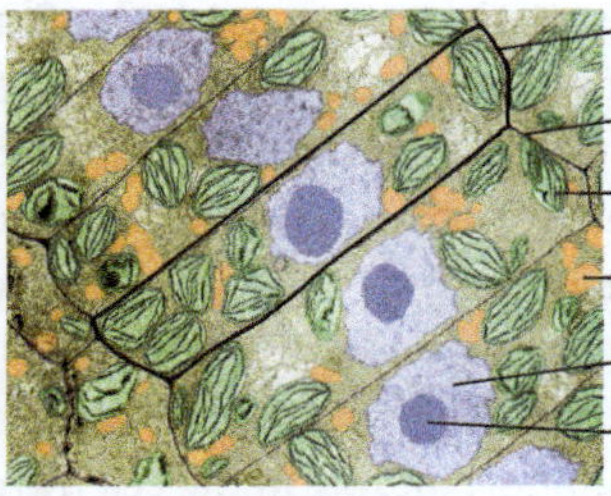

Cells from duckweed (*Spirodela oligorrhiza*), a floating plant (colorized TEM)

Fungal Cells

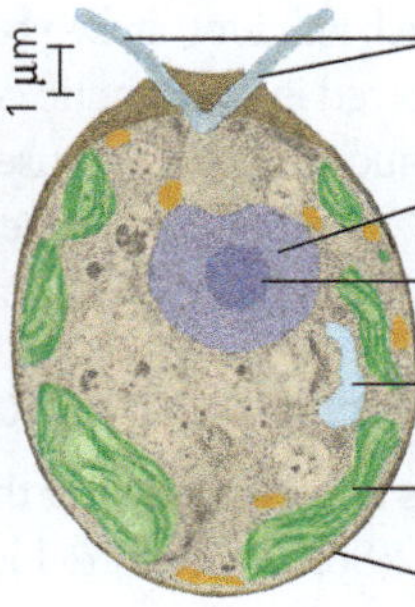

Unicellular green alga *Chlamydomonas* (above, colorized SEM; right, colorized TEM)

Scientific Skills Exercise

Using a Scale Bar to Calculate Volume and Surface Area of a Cell

How Much New Cytoplasm and Plasma Membrane Are Made by a Growing Yeast Cell? The unicellular yeast *Saccharomyces cerevisiae* divides by budding off a small new cell that then grows to full size (see the yeast cells at the bottom of Figure 7). During its growth, the new cell synthesizes new cytoplasm, which increases its volume, and new plasma membrane, which increases its surface area.

In this exercise, you will use a scale bar to determine the sizes of a mature parent yeast cell and a cell budding from it. You will then calculate the volume and surface area of each cell. You will use your calculations to determine how much cytoplasm and plasma membrane the new cell needs to synthesize to grow to full size.

How the Experiment Was Done Yeast cells were grown under conditions that promoted division by budding. The cells were then viewed with a differential interference contrast light microscope and photographed.

Data from the Experiment This light micrograph shows a budding yeast cell about to be released from the mature parent cell:

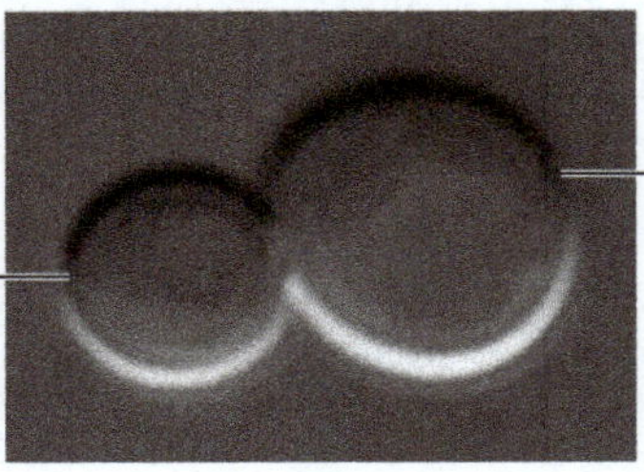

INTERPRET THE DATA

1. Examine the micrograph of the yeast cells. The scale bar under the photo is labeled 1 µm. The scale bar works in the same way as a scale on a map, where, for example, 1 inch equals 1 mile. In this case the bar represents one thousandth of a millimeter. Using the scale bar as a basic unit, determine the diameter of the mature parent cell and the new cell. Start by measuring the scale bar and the diameter of each cell. The units you use are irrelevant, but working in millimeters is convenient. Divide each diameter by the length of the scale bar and then multiply by the scale bar's length value to give you the diameter in micrometers.
2. The shape of a yeast cell can be approximated by a sphere. (a) Calculate the volume of each cell using the formula for the volume of a sphere:

$$V = \frac{4}{3}\pi r^3$$

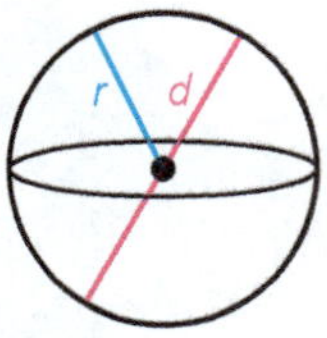

Note that π (the Greek letter pi) is a constant with an approximate value of 3.14, *d* stands for diameter, and *r* stands for radius, which is half the diameter. (b) What volume of new cytoplasm will the new cell have to synthesize as it matures? To determine this, calculate the difference between the volume of the full-sized cell and the volume of the new cell.
3. As the new cell grows, its plasma membrane needs to expand to contain the increased volume of the cell. (a) Calculate the surface area of each cell using the formula for the surface area of a sphere: $A = 4\pi r^2$. (b) How much area of new plasma membrane will the new cell have to synthesize as it matures?
4. When the new cell matures, it will be approximately how many times greater in volume and how many times greater in surface area than its current size?

Micrograph from Kelly Tatchell, using yeast cells grown for experiments described in L. Kozubowski et al., Role of the septin ring in the asymmetric localization of proteins at the mother-bud neck in *Saccharomyces cerevisiae*, *Molecular Biology of the Cell* 16:3455–3466 (2005).

MB A version of this Scientific Skills Exercise can be assigned in MasteringBiology.

CONCEPT 3

The eukaryotic cell's genetic instructions are housed in the nucleus and carried out by the ribosomes

On the first stop of our detailed tour of the eukaryotic cell, let's look at two cellular components involved in the genetic control of the cell: the nucleus, which houses most of the cell's DNA, and the ribosomes, which use information from the DNA to make proteins.

The Nucleus: Information Central

The **nucleus** contains most of the genes in the eukaryotic cell. (Some genes are located in mitochondria and chloroplasts.) It is usually the most conspicuous organelle (see the purple structure in the cell in the fluorescence micrograph), averaging about 5 µm in diameter. The **nuclear envelope** encloses the nucleus **(Figure 8)**, separating its contents from the cytoplasm.

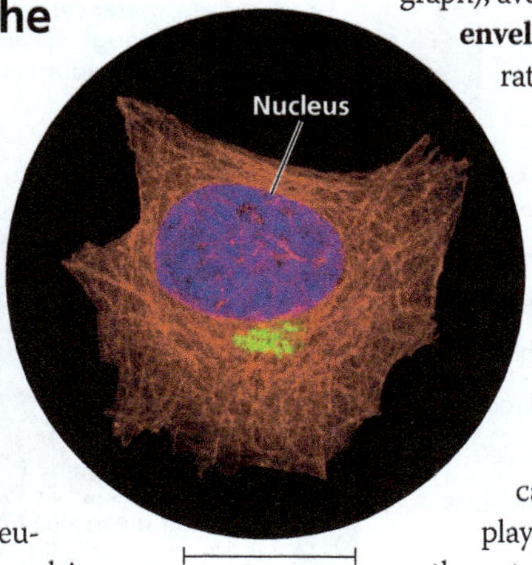

The nuclear envelope is a *double* membrane. The two membranes, each a lipid bilayer with associated proteins, are separated by a space of 20–40 nm. The envelope is perforated by pore structures that are about 100 nm in diameter. At the lip of each pore, the inner and outer membranes of the nuclear envelope are continuous. An intricate protein structure called a *pore complex* lines each pore and plays an important role in the cell by regulating the entry and exit of proteins and RNAs, as well as

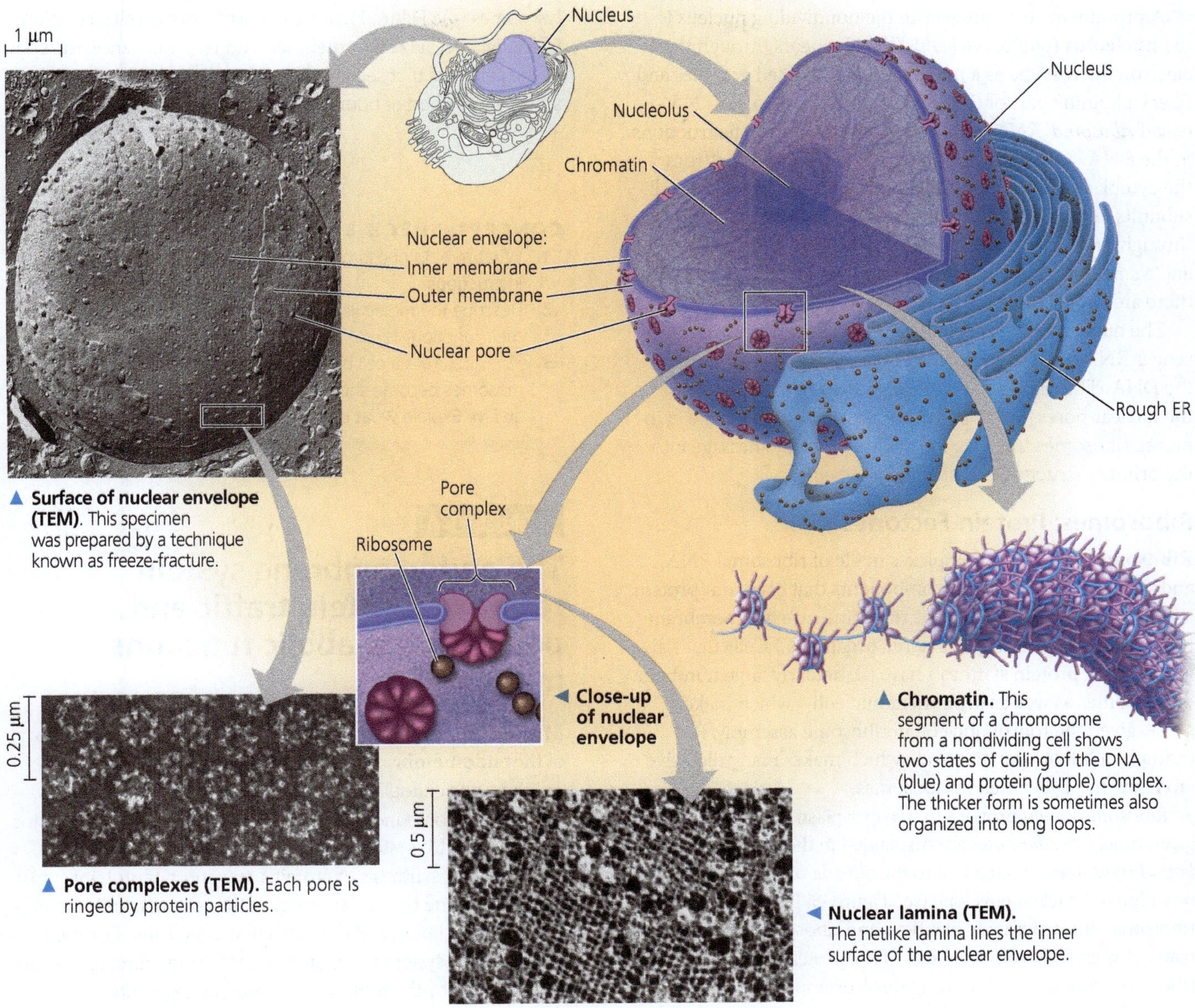

▲ **Surface of nuclear envelope (TEM).** This specimen was prepared by a technique known as freeze-fracture.

◄ **Close-up of nuclear envelope**

▲ **Chromatin.** This segment of a chromosome from a nondividing cell shows two states of coiling of the DNA (blue) and protein (purple) complex. The thicker form is sometimes also organized into long loops.

▲ **Pore complexes (TEM).** Each pore is ringed by protein particles.

◄ **Nuclear lamina (TEM).** The netlike lamina lines the inner surface of the nuclear envelope.

▲ **Figure 8 The nucleus and its envelope.** Within the nucleus are the chromosomes, which appear as a mass of chromatin (DNA and associated proteins), and one or more nucleoli (singular, *nucleolus*), which function in ribosome synthesis. The nuclear envelope, which consists of two membranes separated by a narrow space, is perforated with pores and lined by the nuclear lamina.

MAKE CONNECTIONS *Since the chromosomes contain the genetic material and reside in the nucleus, how does the rest of the cell get access to the information they carry?*

large complexes of macromolecules. Except at the pores, the nuclear side of the envelope is lined by the **nuclear lamina**, a netlike array of protein filaments that maintains the shape of the nucleus by mechanically supporting the nuclear envelope.

Within the nucleus, the DNA is organized into discrete units called **chromosomes**, structures that carry the genetic information. Each chromosome contains one long DNA molecule associated with many proteins. Some of the proteins help coil the DNA molecule of each chromosome, reducing its length and allowing it to fit into the nucleus. The complex of DNA and proteins making up chromosomes is called **chromatin**. When a cell is not dividing, stained chromatin appears as a diffuse mass in micrographs, and the chromosomes cannot be distinguished from one another, even though discrete chromosomes are present. As a cell prepares to divide, however, the chromosomes coil (condense) further, becoming thick enough to be distinguished under a microscope as separate structures. Each eukaryotic species has a characteristic number of chromosomes. For example, a typical human cell has 46 chromosomes in its nucleus; the exceptions are the sex cells (eggs and sperm), which have only 23 chromosomes in humans.

A prominent structure within the nondividing nucleus is the **nucleolus** (plural, *nucleoli*), which appears through the electron microscope as a mass of densely stained granules and fibers adjoining part of the chromatin. Here a type of RNA called *ribosomal RNA* (*rRNA*) is synthesized from instructions in the DNA. Also in the nucleolus, proteins imported from the cytoplasm are assembled with rRNA into large and small subunits of ribosomes. These subunits then exit the nucleus through the nuclear pores to the cytoplasm, where a large and a small subunit can assemble into a ribosome. Sometimes there are two or more nucleoli.

The nucleus directs protein synthesis by synthesizing messenger RNA (mRNA) according to instructions provided by the DNA. The mRNA is then transported to the cytoplasm via the nuclear pores. Once an mRNA molecule reaches the cytoplasm, ribosomes translate the mRNA's genetic message into the primary structure of a specific polypeptide.

Ribosomes: Protein Factories

Ribosomes, which are complexes made of ribosomal RNA and protein, are the cellular components that carry out protein synthesis **(Figure 9)**. (Note that ribosomes are not membrane bound and thus are not considered organelles.) Cells that have high rates of protein synthesis have particularly large numbers of ribosomes as well as prominent nucleoli—which makes sense, given the role of nucleoli in ribosome assembly. For example, a human pancreas cell, which makes many digestive enzymes, has a few million ribosomes.

Ribosomes build proteins in two cytoplasmic locales. At any given time, *free ribosomes* are suspended in the cytosol, while *bound ribosomes* are attached to the outside of the endoplasmic reticulum or nuclear envelope (see Figure 9). Bound and free ribosomes are structurally identical, and ribosomes can alternate between the two roles. Most of the proteins made on free ribosomes function within the cytosol; examples are enzymes that catalyze the first steps of sugar breakdown. Bound ribosomes generally make proteins that are destined for insertion into membranes, for packaging within certain organelles such as lysosomes (see Figure 7), or for export from the cell (secretion). Cells that specialize in protein secretion—for instance, the cells of the pancreas that secrete digestive enzymes—frequently have a high proportion of bound ribosomes.

CONCEPT CHECK 3

1. What role do ribosomes play in carrying out genetic instructions?
2. Describe the molecular composition of nucleoli, and explain their function.
3. **WHAT IF?** As a cell begins the process of dividing, its chromosomes become shorter, thicker, and individually visible in an LM. Explain what is happening at the molecular level.

CONCEPT 4

The endomembrane system regulates protein traffic and performs metabolic functions in the cell

Many of the different membranes of the eukaryotic cell are part of the **endomembrane system**, which includes the nuclear envelope, the endoplasmic reticulum, the Golgi apparatus, lysosomes, various kinds of vesicles and vacuoles, and the plasma membrane. This system carries out a variety of tasks in the cell, including synthesis of proteins, transport of proteins into membranes and organelles or out of the cell, metabolism and movement of lipids, and detoxification of poisons. The membranes of this system are related either through direct physical continuity or by the transfer of membrane segments as tiny **vesicles** (sacs made of membrane). Despite these relationships, the various membranes are not identical in structure and function. Moreover, the thickness, molecular composition, and types

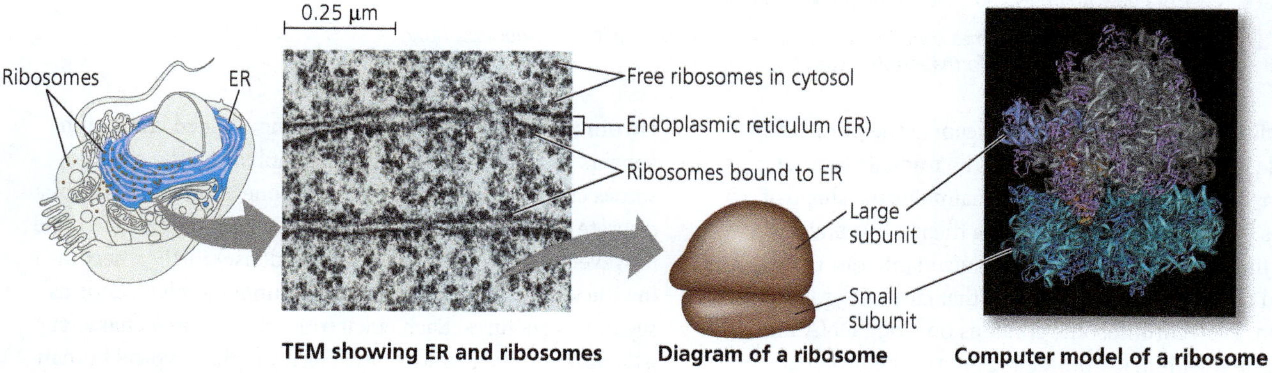

▲ **Figure 9 Ribosomes.** This electron micrograph of part of a pancreas cell shows both free and bound ribosomes. The simplified diagram and computer model show the two subunits of a ribosome.

DRAW IT *After you have read the section on ribosomes, circle a ribosome in the micrograph that might be making a protein that will be secreted.*

of chemical reactions carried out in a given membrane are not fixed, but may be modified several times during the membrane's life. Having already discussed the nuclear envelope, we will now focus on the endoplasmic reticulum and the other endomembranes to which the endoplasmic reticulum gives rise.

The Endoplasmic Reticulum: Biosynthetic Factory

The **endoplasmic reticulum (ER)** is such an extensive network of membranes that it accounts for more than half the total membrane in many eukaryotic cells. (The word *endoplasmic* means "within the cytoplasm," and *reticulum* is Latin for "little net.") The ER consists of a network of membranous tubules and sacs called cisternae (from the Latin *cisterna*, a reservoir for a liquid). The ER membrane separates the internal compartment of the ER, called the ER lumen (cavity) or cisternal space, from the cytosol. And because the ER membrane is continuous with the nuclear envelope, the space between the two membranes of the envelope is continuous with the lumen of the ER **(Figure 10)**.

There are two distinct, though connected, regions of the ER that differ in structure and function: smooth ER and rough ER. **Smooth ER** is so named because its outer surface lacks ribosomes. **Rough ER** is studded with ribosomes on the outer surface of the membrane and thus appears rough through the electron microscope. As already mentioned, ribosomes are also attached to the cytoplasmic side of the nuclear envelope's outer membrane, which is continuous with rough ER.

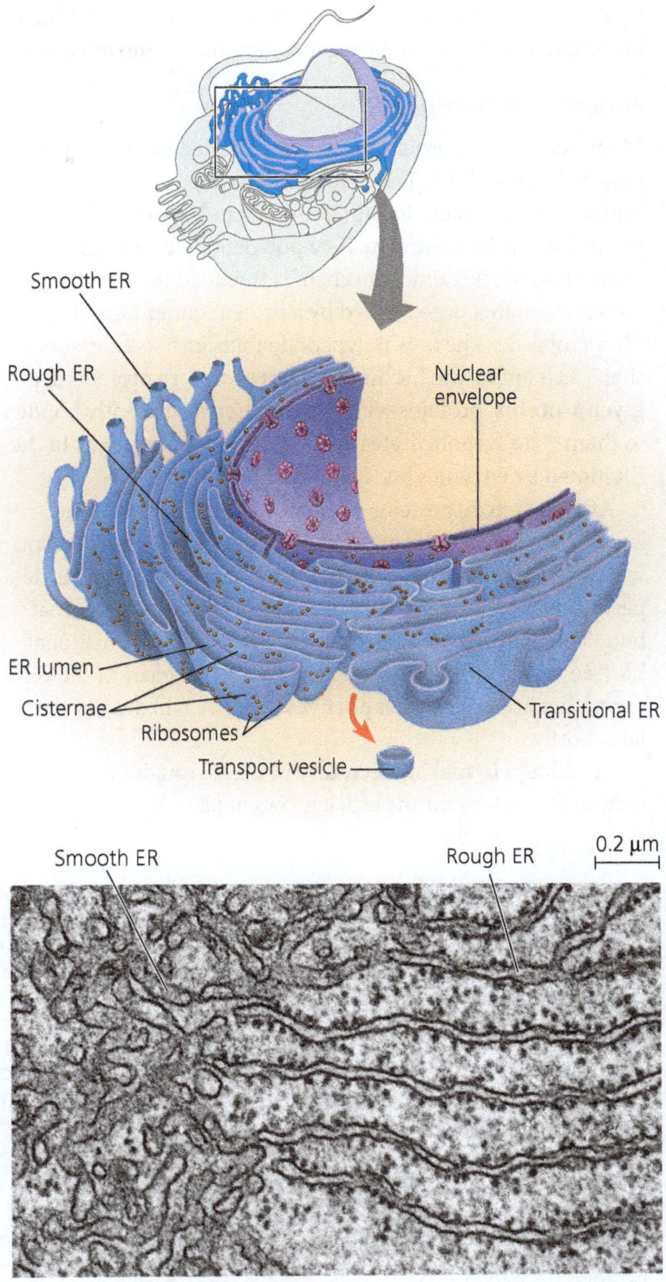

▲ **Figure 10 Endoplasmic reticulum (ER).** A membranous system of interconnected tubules and flattened sacs called cisternae, the ER is also continuous with the nuclear envelope, as shown in the cutaway diagram at the top. The membrane of the ER encloses a continuous compartment called the ER lumen (or cisternal space). Rough ER, which is studded on its outer surface with ribosomes, can be distinguished from smooth ER in the electron micrograph (TEM). Transport vesicles bud off from a region of the rough ER called transitional ER and travel to the Golgi apparatus and other destinations.

Functions of Smooth ER

The smooth ER functions in diverse metabolic processes, which vary with cell type. These processes include synthesis of lipids, metabolism of carbohydrates, detoxification of drugs and poisons, and storage of calcium ions.

Enzymes of the smooth ER are important in the synthesis of lipids, including oils, steroids, and new membrane phospholipids. Among the steroids produced by the smooth ER in animal cells are the sex hormones of vertebrates and the various steroid hormones secreted by the adrenal glands. The cells that synthesize and secrete these hormones—in the testes and ovaries, for example—are rich in smooth ER, a structural feature that fits the function of these cells.

Other enzymes of the smooth ER help detoxify drugs and poisons, especially in liver cells. Detoxification usually involves adding hydroxyl groups to drug molecules, making them more soluble and easier to flush from the body. The sedative phenobarbital and other barbiturates are examples of drugs metabolized in this manner by smooth ER in liver cells. In fact, barbiturates, alcohol, and many other drugs induce the proliferation of smooth ER and its associated detoxification enzymes, thus increasing the rate of detoxification. This, in turn, increases tolerance to the drugs, meaning that higher doses are required to achieve a particular effect, such as sedation. Also, because some of the detoxification enzymes have relatively broad action, the proliferation of smooth ER in response to one drug can increase the need for higher dosages of other drugs as well. Barbiturate abuse, for example, can decrease the effectiveness of certain antibiotics and other useful drugs.

The smooth ER also stores calcium ions. In muscle cells, for example, the smooth ER membrane pumps calcium ions from the cytosol into the ER lumen. When a muscle cell is stimulated by a nerve impulse, calcium ions rush back across the ER membrane into the cytosol and trigger contraction of the muscle cell.

Functions of Rough ER

Many cells secrete proteins that are produced by ribosomes attached to rough ER. For example, certain pancreatic cells synthesize the protein insulin in the ER and secrete this hormone into the bloodstream. As a polypeptide chain grows from a bound ribosome, the chain is threaded into the ER lumen through a pore formed by a protein complex in the ER membrane. The new polypeptide folds into its functional shape as it enters the ER lumen. Most secretory proteins are **glycoproteins**, proteins with carbohydrates covalently bonded to them. The carbohydrates are attached to the proteins in the ER lumen by enzymes built into the ER membrane.

After secretory proteins are formed, the ER membrane keeps them separate from proteins that remain in the cytosol, which are produced by free ribosomes. Secretory proteins depart from the ER wrapped in the membranes of vesicles that bud like bubbles from a specialized region called transitional ER (see Figure 10). Vesicles in transit from one part of the cell to another are called **transport vesicles**; we will discuss their fate shortly.

In addition to making secretory proteins, rough ER is a membrane factory for the cell; it grows in place by adding membrane proteins and phospholipids to its own membrane. As polypeptides destined to be membrane proteins grow from the ribosomes, they are inserted into the ER membrane itself and anchored there by their hydrophobic portions. Like the smooth ER, the rough ER also makes membrane phospholipids; enzymes built into the ER membrane assemble phospholipids from precursors in the cytosol. The ER membrane expands, and portions of it are transferred in the form of transport vesicles to other components of the endomembrane system.

The Golgi Apparatus: Shipping and Receiving Center

After leaving the ER, many transport vesicles travel to the **Golgi apparatus**. We can think of the Golgi as a warehouse for receiving, sorting, shipping, and even some manufacturing. Here, products of the ER, such as proteins, are modified and stored and then sent to other destinations. Not surprisingly, the Golgi apparatus is especially extensive in cells specialized for secretion.

The Golgi apparatus consists of flattened membranous sacs—cisternae—looking like a stack of pita bread **(Figure 11)**. A cell may have many, even hundreds, of these stacks. The membrane of each cisterna in a stack separates its internal space from the cytosol. Vesicles concentrated in the vicinity of the Golgi apparatus are engaged in the transfer of material between parts of the Golgi and other structures.

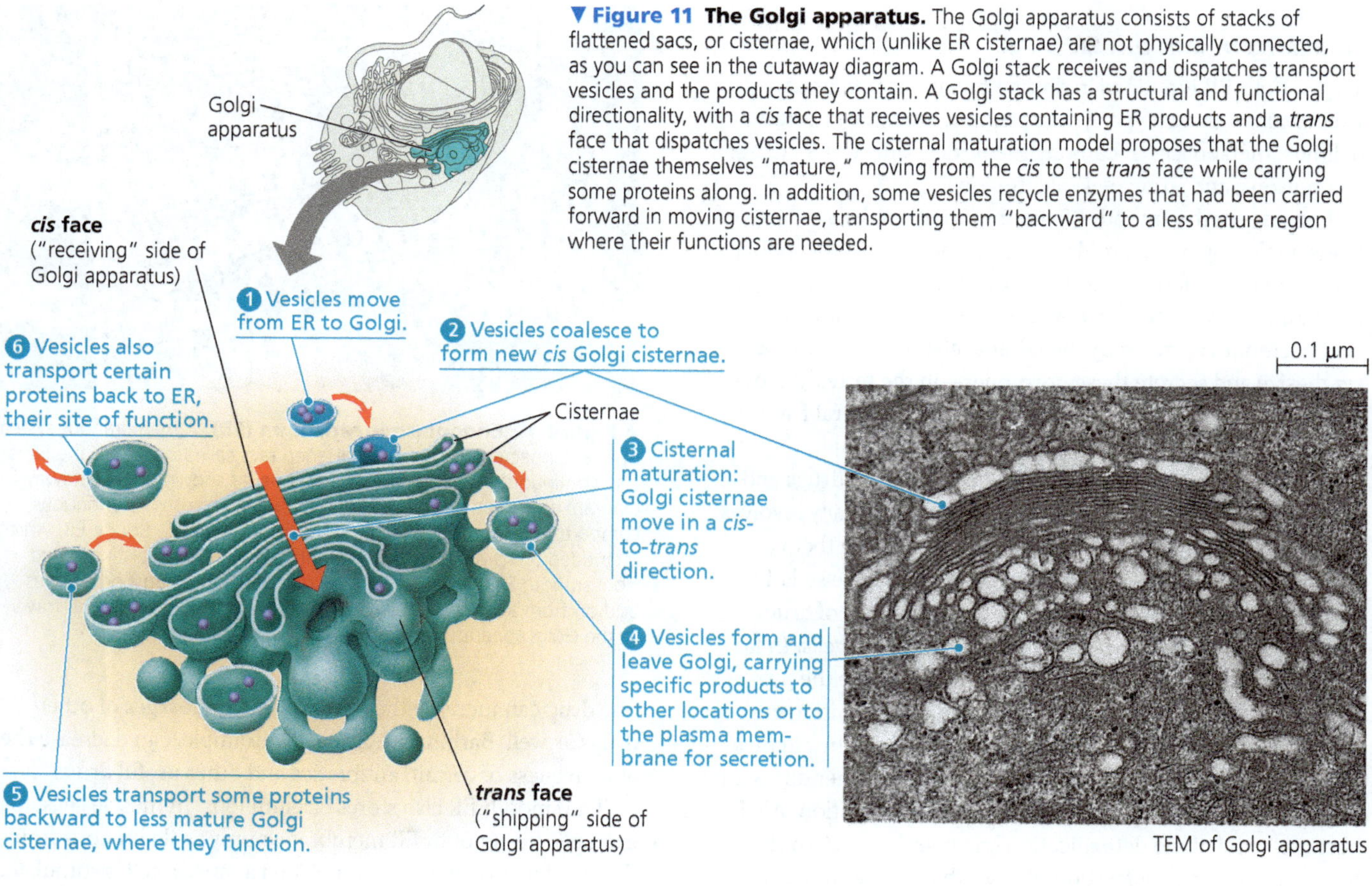

▼ Figure 11 The Golgi apparatus. The Golgi apparatus consists of stacks of flattened sacs, or cisternae, which (unlike ER cisternae) are not physically connected, as you can see in the cutaway diagram. A Golgi stack receives and dispatches transport vesicles and the products they contain. A Golgi stack has a structural and functional directionality, with a *cis* face that receives vesicles containing ER products and a *trans* face that dispatches vesicles. The cisternal maturation model proposes that the Golgi cisternae themselves "mature," moving from the *cis* to the *trans* face while carrying some proteins along. In addition, some vesicles recycle enzymes that had been carried forward in moving cisternae, transporting them "backward" to a less mature region where their functions are needed.

A Golgi stack has a distinct structural directionality, with the membranes of cisternae on opposite sides of the stack differing in thickness and molecular composition. The two sides of a Golgi stack are referred to as the *cis* face and the *trans* face; these act, respectively, as the receiving and shipping departments of the Golgi apparatus. The term *cis* means "on the same side," and the *cis* face is usually located near the ER. Transport vesicles move material from the ER to the Golgi apparatus. A vesicle that buds from the ER can add its membrane and the contents of its lumen to the *cis* face by fusing with a Golgi membrane. The *trans* face ("on the opposite side") gives rise to vesicles that pinch off and travel to other sites.

Products of the endoplasmic reticulum are usually modified during their transit from the *cis* region to the *trans* region of the Golgi apparatus. For example, glycoproteins formed in the ER have their carbohydrates modified, first in the ER itself, and then as they pass through the Golgi. The Golgi removes some sugar monomers and substitutes others, producing a large variety of carbohydrates. Membrane phospholipids may also be altered in the Golgi.

In addition to its finishing work, the Golgi apparatus also manufactures some macromolecules. Many polysaccharides secreted by cells are Golgi products. For example, pectins and certain other noncellulose polysaccharides are made in the Golgi of plant cells and then incorporated along with cellulose into their cell walls. Like secretory proteins, nonprotein Golgi products that will be secreted depart from the *trans* face of the Golgi inside transport vesicles that eventually fuse with the plasma membrane.

The Golgi manufactures and refines its products in stages, with different cisternae containing unique teams of enzymes. Until recently, biologists viewed the Golgi as a static structure, with products in various stages of processing transferred from one cisterna to the next by vesicles. While this may occur, research from several labs has given rise to a new model of the Golgi as a more dynamic structure. According to the *cisternal maturation model*, the cisternae of the Golgi actually progress forward from the *cis* to the *trans* face, carrying and modifying their cargo as they move. Figure 11 shows the details of this model.

Before a Golgi stack dispatches its products by budding vesicles from the *trans* face, it sorts these products and targets them for various parts of the cell. Molecular identification tags, such as phosphate groups added to the Golgi products, aid in sorting by acting like zip codes on mailing labels. Finally, transport vesicles budded from the Golgi may have external molecules on their membranes that recognize "docking sites" on the surface of specific organelles or on the plasma membrane, thus targeting the vesicles appropriately.

Lysosomes: Digestive Compartments

A **lysosome** is a membranous sac of hydrolytic enzymes that many eukaryotic cells use to digest (hydrolyze) macromolecules (**Figure 12**). Lysosomal enzymes work best in the acidic environment found in lysosomes. If a lysosome breaks open or leaks its contents, the released enzymes are not very active because the cytosol has a near-neutral pH. However, excessive leakage from a large number of lysosomes can destroy a cell by self-digestion.

Hydrolytic enzymes and lysosomal membrane are made by rough ER and then transferred to the Golgi apparatus for further processing. At least some lysosomes probably arise by budding from the *trans* face of the Golgi apparatus (see Figure 11). How are the proteins of the inner surface of the lysosomal membrane and the digestive enzymes themselves spared from destruction? Apparently, the three-dimensional shapes of these lysosomal proteins protect vulnerable bonds from enzymatic attack.

Lysosomes carry out intracellular digestion in a variety of circumstances. Amoebas and many other unicellular eukaryotes eat by engulfing smaller organisms or food particles, a process called **phagocytosis** (from the Greek *phagein*, to eat, and *kytos*, vessel, referring here to the cell). The *food vacuole* formed in this way then fuses with a lysosome, whose enzymes digest the food (see Figure 12, bottom). Digestion products, including simple sugars, amino acids, and other monomers,

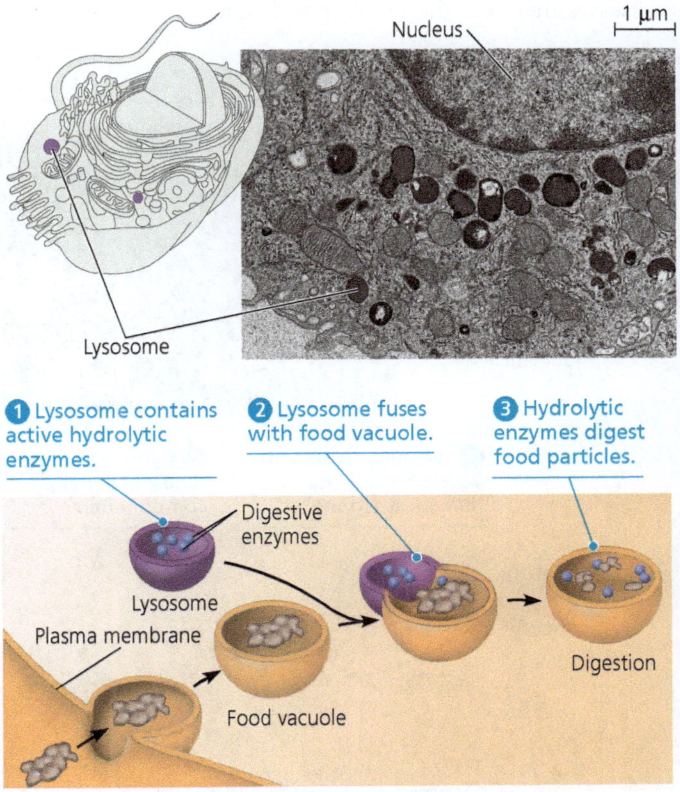

▲ **Figure 12 Lysosomes: Phagocytosis.** In phagocytosis, lysosomes digest (hydrolyze) materials taken into the cell and recycle intracellular materials. *Top*: In this macrophage (a type of white blood cell) from a rat, the lysosomes are very dark because of a stain that reacts with one of the products of digestion inside the lysosome (TEM). Macrophages ingest bacteria and viruses and destroy them using lysosomes. *Bottom*: This diagram shows a lysosome fusing with a food vacuole during the process of phagocytosis by a unicellular eukaryote.

pass into the cytosol and become nutrients for the cell. Some human cells also carry out phagocytosis. Among them are macrophages, a type of white blood cell that helps defend the body by engulfing and destroying bacteria and other invaders (see Figure 12, top, and Figure 28).

Lysosomes also use their hydrolytic enzymes to recycle the cell's own organic material, a process called *autophagy*. During autophagy, a damaged organelle or small amount of cytosol becomes surrounded by a double membrane, and a lysosome fuses with the outer membrane of this vesicle **(Figure 13)**. The lysosomal enzymes dismantle the enclosed material, and the resulting small organic compounds are released to the cytosol for reuse. With the help of lysosomes, the cell continually renews itself. A human liver cell, for example, recycles half of its macromolecules each week.

The cells of people with inherited lysosomal storage diseases lack a functioning hydrolytic enzyme normally present in lysosomes. The lysosomes become engorged with indigestible material, which begins to interfere with other cellular activities. In Tay-Sachs disease, for example, a lipid-digesting enzyme is missing or inactive, and the brain becomes impaired by an accumulation of lipids in the cells. Fortunately, lysosomal storage diseases are rare in the general population.

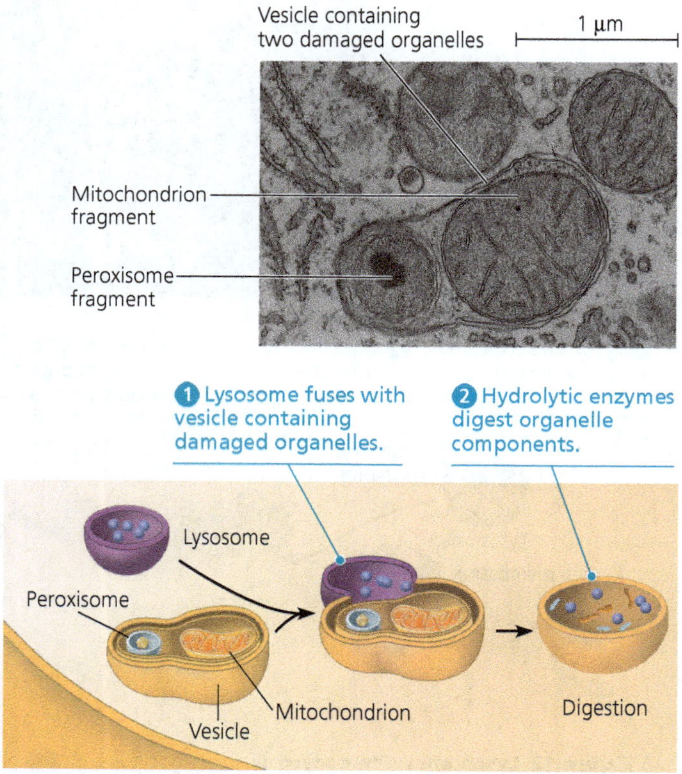

▲ **Figure 13 Lysosomes: Autophagy.** In autophagy, lysosomes recycle intracellular materials. *Top*: In the cytoplasm of this rat liver cell is a vesicle containing two disabled organelles; the vesicle will fuse with a lysosome in the process of autophagy (TEM). *Bottom*: This diagram shows fusion of such a vesicle with a lysosome and the subsequent digestion of the damaged organelles.

Vacuoles: Diverse Maintenance Compartments

Vacuoles are large vesicles derived from the endoplasmic reticulum and Golgi apparatus. Thus, vacuoles are an integral part of a cell's endomembrane system. Like all cellular membranes, the vacuolar membrane is selective in transporting solutes; as a result, the solution inside a vacuole differs in composition from the cytosol.

Vacuoles perform a variety of functions in different kinds of cells. **Food vacuoles**, formed by phagocytosis, have already been mentioned (see Figure 12). Many unicellular eukaryotes living in fresh water have **contractile vacuoles** that pump excess water out of the cell, thereby maintaining a suitable concentration of ions and molecules inside the cell. In plants and fungi, certain vacuoles carry out enzymatic hydrolysis, a function shared by lysosomes in animal cells. (In fact, some biologists consider these hydrolytic vacuoles to be a type of lysosome.) In plants, small vacuoles can hold reserves of important organic compounds, such as the proteins stockpiled in the storage cells in seeds. Vacuoles may also help protect the plant against herbivores by storing compounds that are poisonous or unpalatable to animals. Some plant vacuoles contain pigments, such as the red and blue pigments of petals that help attract pollinating insects to flowers.

Mature plant cells generally contain a large **central vacuole (Figure 14)**, which develops by the coalescence of smaller vacuoles. The solution inside the central vacuole, called cell sap, is the plant cell's main repository of inorganic ions, including potassium and chloride. The central vacuole plays a major role in the growth of plant cells, which enlarge as the vacuole absorbs water, enabling the cell to become larger with a minimal investment in new cytoplasm. The cytosol often occupies only a thin layer between the central vacuole and the plasma

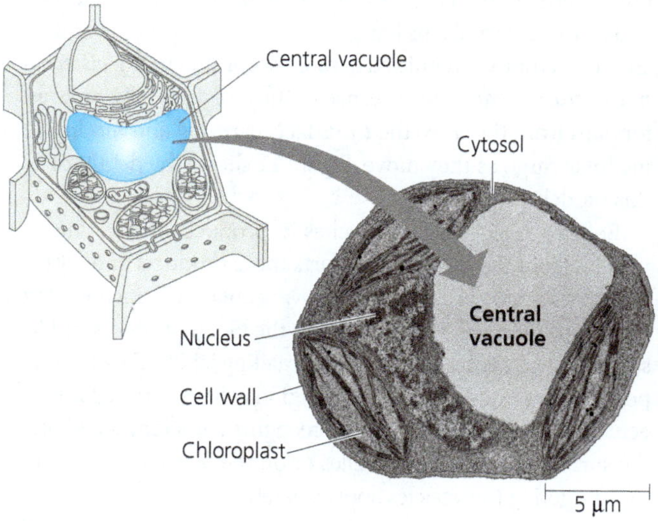

▲ **Figure 14 The plant cell vacuole.** The central vacuole is usually the largest compartment in a plant cell; the rest of the cytoplasm is often confined to a narrow zone between the vacuolar membrane and the plasma membrane (TEM).

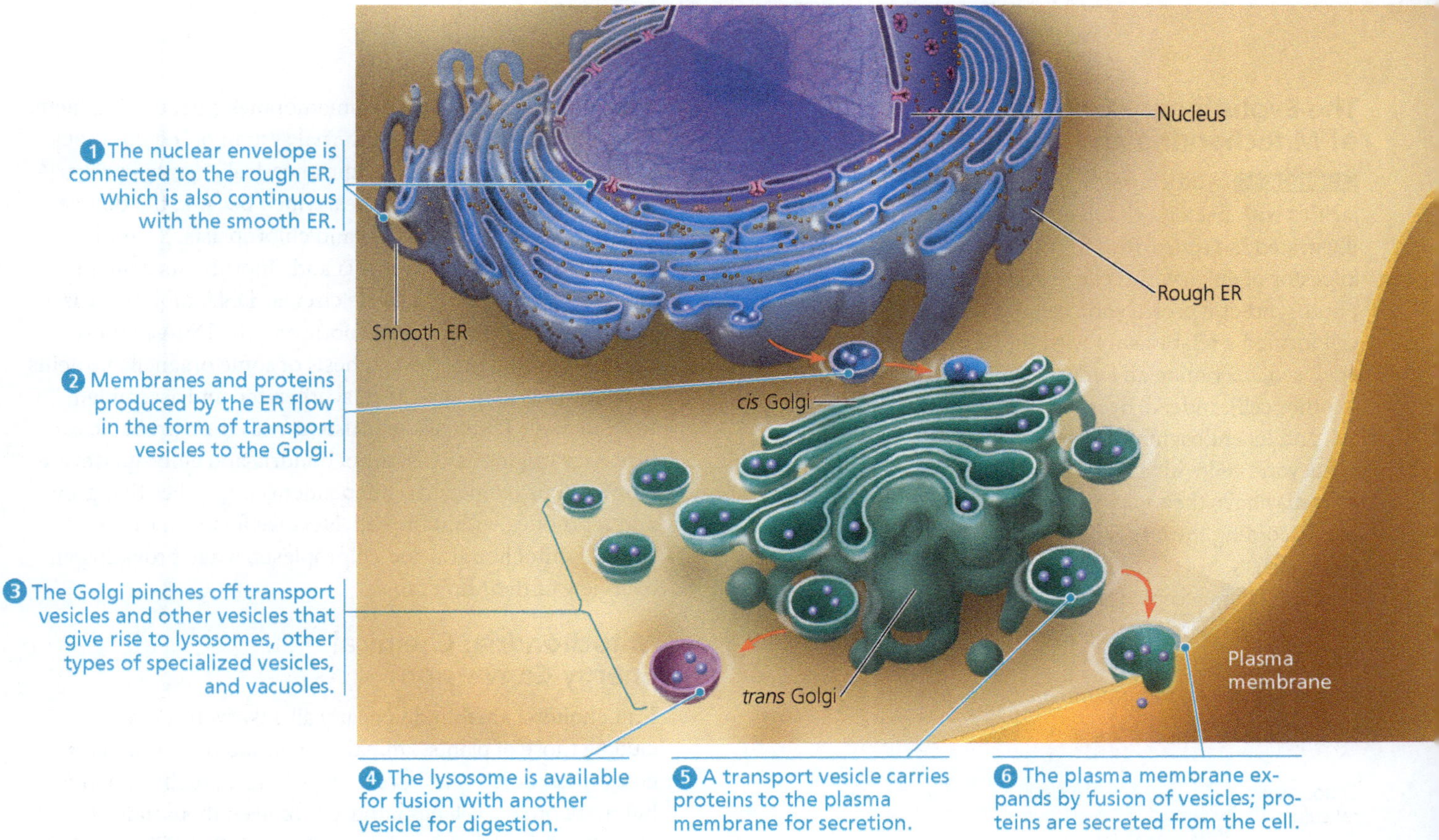

▲ **Figure 15 Review: relationships among organelles of the endomembrane system.** The red arrows show some of the migration pathways for membranes and the materials they enclose.

membrane, so the ratio of plasma membrane surface to cytosolic volume is sufficient, even for a large plant cell.

The Endomembrane System: *A Review*

Figure 15 reviews the endomembrane system, showing the flow of membrane lipids and proteins through the various organelles. As the membrane moves from the ER to the Golgi and then elsewhere, its molecular composition and metabolic functions are modified, along with those of its contents. The endomembrane system is a complex and dynamic player in the cell's compartmental organization.

We'll continue our tour of the cell with some organelles that are not closely related to the endomembrane system but play crucial roles in the energy transformations carried out by cells.

CONCEPT CHECK 4

1. Describe the structural and functional distinctions between rough and smooth ER.
2. Describe how transport vesicles integrate the endomembrane system.
3. WHAT IF? Imagine a protein that functions in the ER but requires modification in the Golgi apparatus before it can achieve that function. Describe the protein's path through the cell, starting with the mRNA molecule that specifies the protein.

CONCEPT 5

Mitochondria and chloroplasts change energy from one form to another

Organisms transform the energy they acquire from their surroundings. In eukaryotic cells, mitochondria and chloroplasts are the organelles that convert energy to forms that cells can use for work. **Mitochondria** (singular, *mitochondrion*) are the sites of cellular respiration, the metabolic process that uses oxygen to drive the generation of ATP by extracting energy from sugars, fats, and other fuels. **Chloroplasts**, found in plants and algae, are the sites of photosynthesis. This process in chloroplasts converts solar energy to chemical energy by absorbing sunlight and using it to drive the synthesis of organic compounds such as sugars from carbon dioxide and water.

In addition to having related functions, mitochondria and chloroplasts share similar evolutionary origins, which we'll discuss briefly before describing their structures. In this section, we will also consider the peroxisome, an oxidative organelle. The evolutionary origin of the peroxisome, as well as its relation to other organelles, is still under debate.

The Evolutionary Origins of Mitochondria and Chloroplasts

EVOLUTION Mitochondria and chloroplasts display similarities with bacteria that led to the **endosymbiont theory**, illustrated in **Figure 16**. This theory states that an early ancestor of eukaryotic cells engulfed an oxygen-using nonphotosynthetic prokaryotic cell. Eventually, the engulfed cell formed a relationship with the host cell in which it was enclosed, becoming an *endosymbiont* (a cell living within another cell). Indeed, over the course of evolution, the host cell and its endosymbiont merged into a single organism, a eukaryotic cell with a mitochondrion. At least one of these cells may have then taken up a photosynthetic prokaryote, becoming the ancestor of eukaryotic cells that contain chloroplasts.

This widely accepted theory is consistent with many structural features of mitochondria and chloroplasts. First, rather than being bounded by a single membrane like organelles of the endomembrane system, mitochondria and typical chloroplasts have two membranes surrounding them. (Chloroplasts also have an internal system of membranous sacs.) There is evidence that the ancestral engulfed prokaryotes had two outer membranes, which became the double membranes of mitochondria and chloroplasts. Second, like prokaryotes, mitochondria and chloroplasts contain ribosomes, as well as multiple circular DNA molecules associated with their inner membranes. The DNA in these organelles programs the synthesis of some organelle proteins on ribosomes that have been synthesized and assembled there as well. Third, also consistent with their probable evolutionary origins as cells, mitochondria and chloroplasts are autonomous (somewhat independent) organelles that grow and reproduce within the cell. Next we focus on the structures of mitochondria and chloroplasts, while providing an overview of their functions.

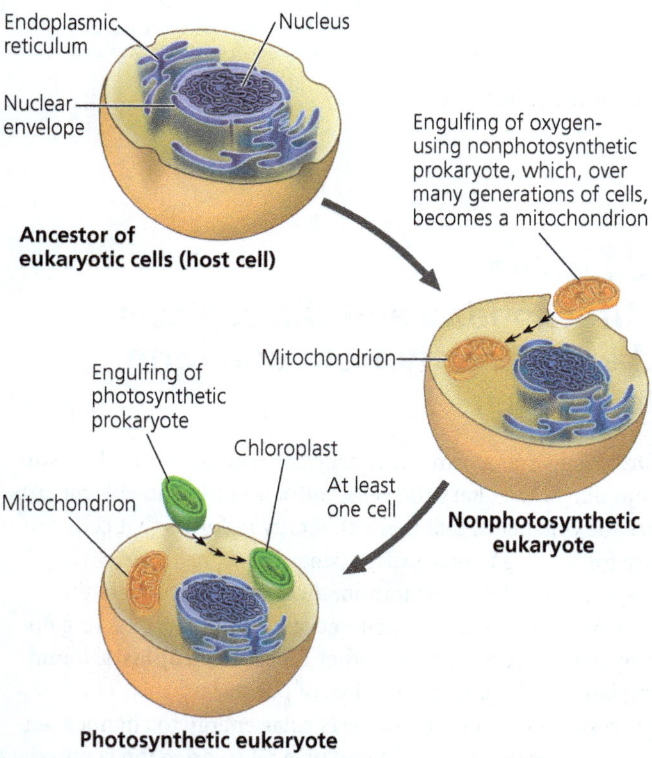

▲ **Figure 16 The endosymbiont theory of the origins of mitochondria and chloroplasts in eukaryotic cells.** According to this theory, the proposed ancestors of mitochondria were oxygen-using nonphotosynthetic prokaryotes, while the proposed ancestors of chloroplasts were photosynthetic prokaryotes. The large arrows represent change over evolutionary time; the small arrows inside the cells show the process of the endosymbiont becoming an organelle, also over long periods of time.

Mitochondria: Chemical Energy Conversion

Mitochondria are found in nearly all eukaryotic cells, including those of plants, animals, fungi, and most unicellular eukaryotes. Some cells have a single large mitochondrion, but more often a cell has hundreds or even thousands of mitochondria; the number correlates with the cell's level of metabolic activity. For example, cells that move or contract have proportionally more mitochondria per volume than less active cells.

Each of the two membranes enclosing the mitochondrion is a phospholipid bilayer with a unique collection of embedded proteins **(Figure 17)**. The outer membrane is smooth, but the inner membrane is convoluted, with infoldings called **cristae**. The inner membrane divides the mitochondrion into two internal compartments. The first is the intermembrane space, the narrow region between the inner and outer membranes. The second compartment, the **mitochondrial matrix**, is enclosed by the inner membrane. The matrix contains many different enzymes as well as the mitochondrial DNA and ribosomes. Enzymes in the matrix catalyze some of the steps of cellular respiration. Other proteins that function in respiration, including the enzyme that makes ATP, are built into the inner membrane. As highly folded surfaces, the cristae give the inner mitochondrial membrane a large surface area, thus enhancing the productivity of cellular respiration. This is another example of structure fitting function.

Mitochondria are generally in the range of 1–10 μm long. Time-lapse films of living cells reveal mitochondria moving around, changing their shapes, and fusing or dividing in two, unlike the static structures seen in electron micrographs of dead cells. These studies helped biologists understand that mitochondria form a branched tubular network (see Figure 17b) in a dynamic state of flux.

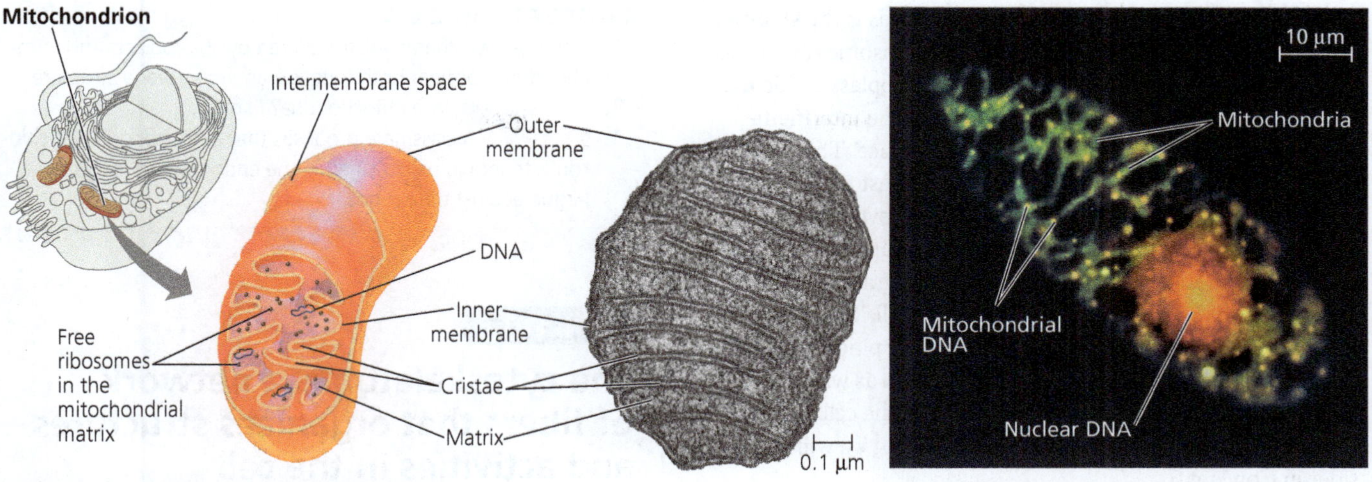

(a) Diagram and TEM of mitochondrion

(b) Network of mitochondria in *Euglena* (LM)

▲ **Figure 17 The mitochondrion, site of cellular respiration. (a)** The inner and outer membranes of the mitochondrion are evident in the drawing and electron micrograph (TEM). The cristae are infoldings of the inner membrane, which increase its surface area. The cutaway drawing shows the two compartments bounded by the membranes: the intermembrane space and the mitochondrial matrix. Many respiratory enzymes are found in the inner membrane and the matrix. Free ribosomes are also present in the matrix. The circular DNA molecules are associated with the inner mitochondrial membrane. **(b)** The light micrograph shows an entire unicellular eukaryote (*Euglena gracilis*) at a much lower magnification than the TEM. The mitochondrial matrix has been stained green. The mitochondria form a branched tubular network. The nuclear DNA is stained red; molecules of mitochondrial DNA appear as bright yellow spots.

Chloroplasts: Capture of Light Energy

Chloroplasts contain the green pigment chlorophyll, along with enzymes and other molecules that function in the photosynthetic production of sugar. These lens-shaped organelles, about 3–6 µm in length, are found in leaves and other green organs of plants and in algae **(Figure 18)**.

The contents of a chloroplast are partitioned from the cytosol by an envelope consisting of two membranes separated by a very narrow intermembrane space. Inside the chloroplast is another membranous system in the form of flattened, interconnected sacs called **thylakoids**. In some regions, thylakoids are stacked like poker chips; each stack is called a **granum**

▼ **Figure 18 The chloroplast, site of photosynthesis. (a)** Many plants have lens-shaped chloroplasts, as shown here in a diagram and a TEM. A chloroplast has three compartments: the intermembrane space, the stroma, and the thylakoid space. Free ribosomes are present in the stroma, as are copies of chloroplast DNA molecules. **(b)** This fluorescence micrograph, at much lower magnification than the TEM, shows a whole cell of the green alga *Spirogyra crassa*, which is named for its spiral chloroplasts. Under natural light the chloroplasts appear green, but under ultraviolet light they naturally fluoresce red, as shown here.

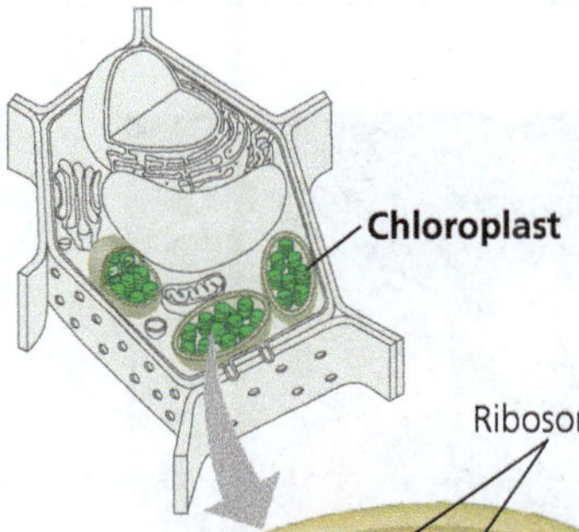

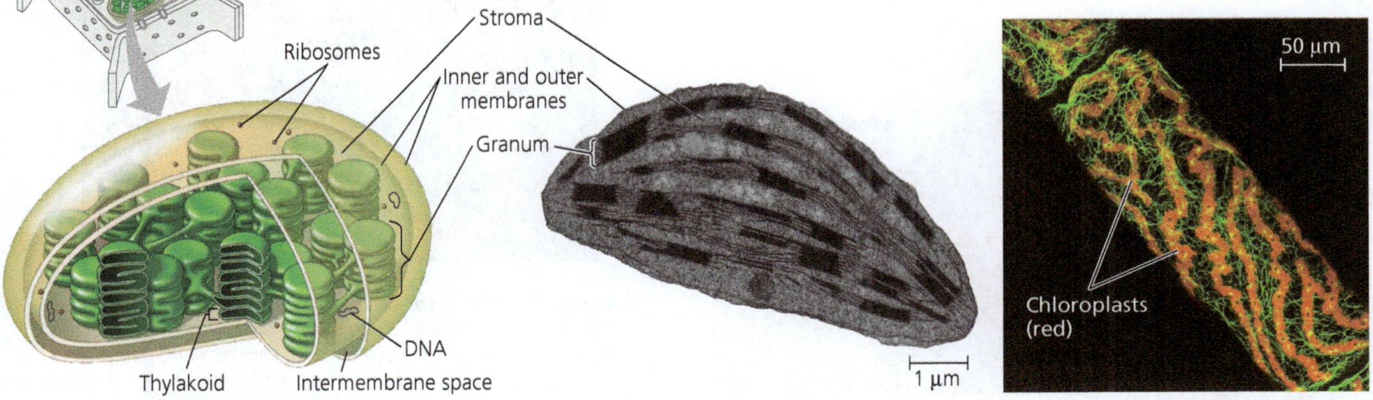

(a) Diagram and TEM of chloroplast

(b) Chloroplasts in an algal cell

(plural, *grana*). The fluid outside the thylakoids is the **stroma**, which contains the chloroplast DNA and ribosomes as well as many enzymes. The membranes of the chloroplast divide the chloroplast space into three compartments: the intermembrane space, the stroma, and the thylakoid space. This compartmental organization enables the chloroplast to convert light energy to chemical energy during photosynthesis.

As with mitochondria, the static and rigid appearance of chloroplasts in micrographs or schematic diagrams is not true to their dynamic behavior in the living cell. Their shape is changeable, and they grow and occasionally pinch in two, reproducing themselves. They are mobile and, as with mitochondria and other organelles, move around the cell along tracks of the cytoskeleton, a structural network we will consider in Concept 6.

The chloroplast is a specialized member of a family of closely related plant organelles called **plastids**. One type of plastid, the *amyloplast*, is a colorless organelle that stores starch (amylose), particularly in roots and tubers. Another is the *chromoplast*, which has pigments that give fruits and flowers their orange and yellow hues.

Peroxisomes: Oxidation

The **peroxisome** is a specialized metabolic compartment bounded by a single membrane **(Figure 19)**. Peroxisomes contain enzymes that remove hydrogen atoms from certain molecules and transfer them to oxygen (O_2), producing hydrogen peroxide (H_2O_2). These reactions have many different functions. For example, peroxisomes in the liver detoxify alcohol and other harmful compounds by transferring hydrogen from the poisons to oxygen. The H_2O_2 formed by peroxisomes is itself toxic, but the organelle also contains an enzyme that converts H_2O_2 to water. This is an excellent example of how the cell's compartmental structure is crucial to its functions: The enzymes that produce H_2O_2 and those that dispose of this toxic compound are sequestered from other cellular components that could be damaged.

Peroxisomes grow larger by incorporating proteins made in the cytosol and ER, as well as lipids made in the ER and within the peroxisome itself. But how peroxisomes increase in number and how they arose in evolution are still open questions.

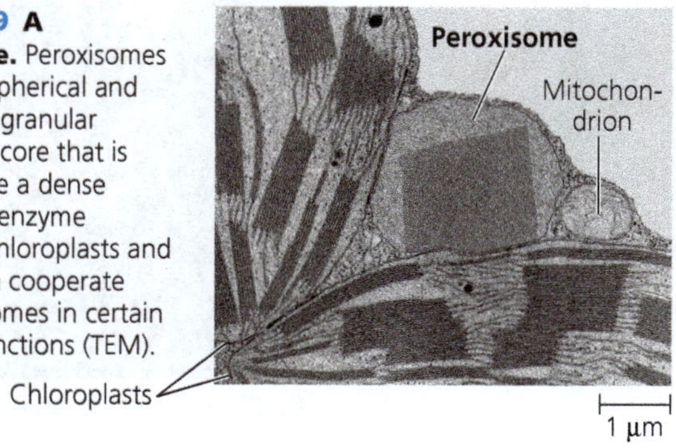

▶ **Figure 19 A peroxisome.** Peroxisomes are roughly spherical and often have a granular or crystalline core that is thought to be a dense collection of enzyme molecules. Chloroplasts and mitochondria cooperate with peroxisomes in certain metabolic functions (TEM).

CONCEPT CHECK 5

1. Describe two characteristics shared by chloroplasts and mitochondria. Consider both function and membrane structure.
2. Do plant cells have mitochondria? Explain.
3. **WHAT IF?** A classmate proposes that mitochondria and chloroplasts should be classified in the endomembrane system. Argue against the proposal.

CONCEPT 6

The cytoskeleton is a network of fibers that organizes structures and activities in the cell

In the early days of electron microscopy, biologists thought that the organelles of a eukaryotic cell floated freely in the cytosol. But improvements in both light microscopy and electron microscopy have revealed the **cytoskeleton**, a network of fibers extending throughout the cytoplasm **(Figure 20)**. The cytoskeleton plays a major role in organizing the structures and activities of the cell.

Roles of the Cytoskeleton: Support and Motility

The most obvious function of the cytoskeleton is to give mechanical support to the cell and maintain its shape. This is especially important for animal cells, which lack walls. The remarkable strength and resilience of the cytoskeleton as a whole are based on its architecture. Like a dome tent, the cytoskeleton is stabilized by a balance between opposing forces exerted by its elements. And just as the skeleton of an animal helps fix the positions of other body parts, the cytoskeleton provides anchorage for many organelles and even cytosolic enzyme molecules. The cytoskeleton is more dynamic than an animal skeleton,

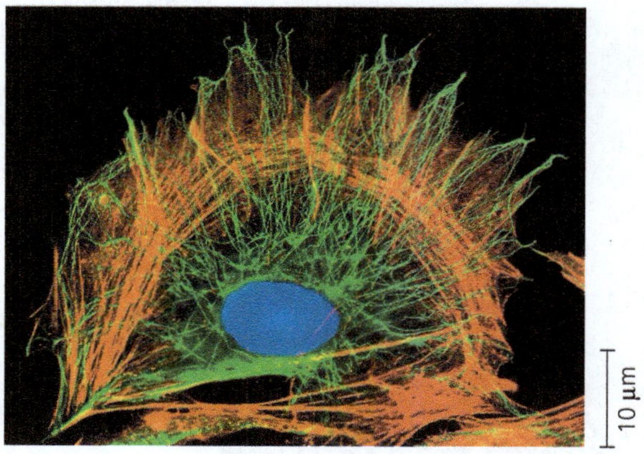

▲ **Figure 20 The cytoskeleton.** As shown in this fluorescence micrograph, the cytoskeleton extends throughout the cell. The cytoskeletal elements have been tagged with different fluorescent molecules: green for microtubules and reddish-orange for microfilaments. A third component of the cytoskeleton, intermediate filaments, is not evident here. (The blue color tags the DNA in the nucleus.)

however. It can be quickly dismantled in one part of the cell and reassembled in a new location, changing the shape of the cell.

Some types of cell motility (movement) also involve the cytoskeleton. The term *cell motility* includes both changes in cell location and more limited movements of cell parts. Cell motility generally requires the interaction of the cytoskeleton with **motor proteins**. Examples of such cell motility abound. Cytoskeletal elements and motor proteins work together with plasma membrane molecules to allow whole cells to move along fibers outside the cell. Inside the cell, vesicles and other organelles often use motor protein "feet" to "walk" to their destinations along a track provided by the cytoskeleton. For instance, this is how vesicles containing neurotransmitter molecules migrate to the tips of axons, the long extensions of nerve cells that release these molecules as chemical signals to adjacent nerve cells **(Figure 21)**. The cytoskeleton also manipulates the plasma membrane, bending it inward to form food vacuoles or other phagocytic vesicles.

Components of the Cytoskeleton

Let's look more closely at the three main types of fibers that make up the cytoskeleton: *Microtubules* are the thickest, *microfilaments* (actin filaments) are the thinnest, and *intermediate filaments* are fibers with diameters in a middle range. **Table 1** summarizes the properties of these fibers.

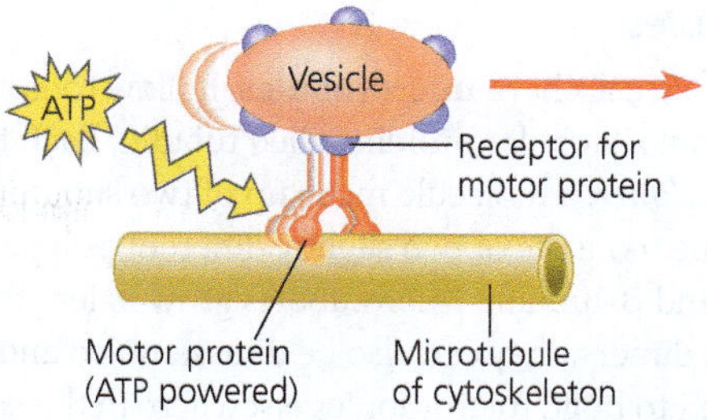

(a) Motor proteins that attach to receptors on vesicles can "walk" the vesicles along microtubules or, in some cases, along microfilaments.

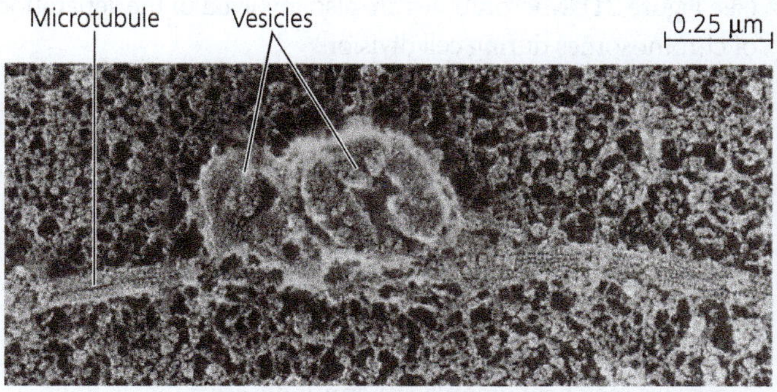

(b) Two vesicles containing neurotransmitters move along a microtubule toward the tip of a nerve cell extension called an axon (SEM).

▲ **Figure 21 Motor proteins and the cytoskeleton.**

Table 1 The Structure and Function of the Cytoskeleton

Property	Microtubules (Tubulin Polymers)	Microfilaments (Actin Filaments)	Intermediate Filaments
Structure	Hollow tubes	Two intertwined strands of actin	Fibrous proteins coiled into cables
Diameter	25 nm with 15-nm lumen	7 nm	8–12 nm
Protein subunits	Tubulin, a dimer consisting of α-tubulin and β-tubulin	Actin	One of several different proteins (such as keratins)
Main functions	Maintenance of cell shape; cell motility; chromosome movements in cell division; organelle movements	Maintenance of cell shape; changes in cell shape; muscle contraction; cytoplasmic streaming (plant cells); cell motility; cell division (animal cells)	Maintenance of cell shape; anchorage of nucleus and certain other organelles; formation of nuclear lamina
Fluorescence micrographs of fibroblasts. Fibroblasts are a favorite cell type for cell biology studies because they spread out flat and their internal structures are easy to see. In each, the structure of interest has been tagged with fluorescent molecules. The DNA in the nucleus has also been tagged in the first micrograph (blue) and third micrograph (orange).	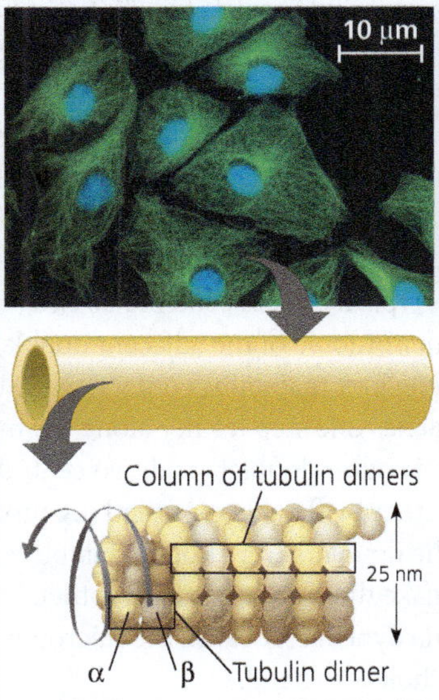	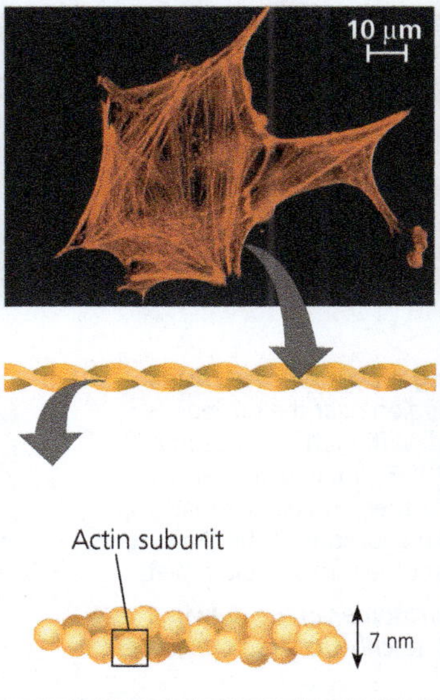	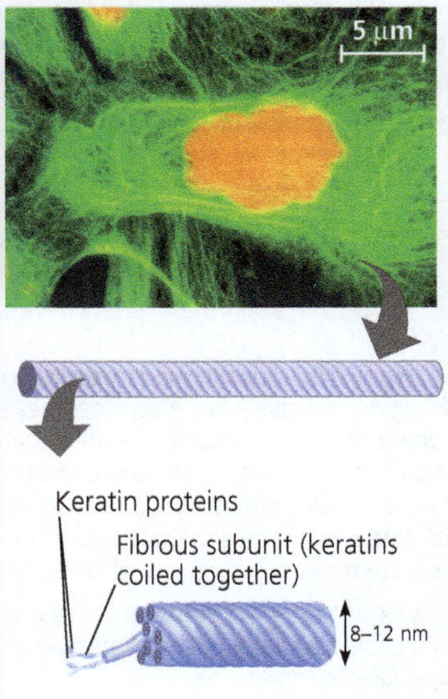

Microtubules

All eukaryotic cells have **microtubules**, hollow rods constructed from a globular protein called tubulin. Each tubulin protein is a *dimer*, a molecule made up of two subunits. A tubulin dimer consists of two slightly different polypeptides, α-tubulin and β-tubulin. Microtubules grow in length by adding tubulin dimers; they can also be disassembled and their tubulin used to build microtubules elsewhere in the cell.

Microtubules shape and support the cell and serve as tracks along which organelles equipped with motor proteins can move (see Figure 21). Microtubules are also involved in the separation of chromosomes during cell division.

Centrosomes and Centrioles In animal cells, microtubules grow out from a **centrosome**, a region that is often located near the nucleus and is considered a "microtubule-organizing center." These microtubules function as compression-resisting girders of the cytoskeleton. Within the centrosome is a pair of **centrioles**, each composed of nine sets of triplet microtubules arranged in a ring **(Figure 22)**. Although centrosomes with centrioles may help organize microtubule assembly in animal cells, many other eukaryotic cells lack centrosomes with centrioles and instead organize microtubules by other means.

Cilia and Flagella In eukaryotes, a specialized arrangement of microtubules is responsible for the beating of **flagella** (singular, *flagellum*) and **cilia** (singular, *cilium*), microtubule-containing extensions that project from some cells. (The bacterial flagellum, shown in Figure 4, has a completely different structure.) Many unicellular eukaryotes are propelled through water by cilia or flagella that act as locomotor appendages, and the sperm of animals, algae, and some plants have flagella. When cilia or flagella extend from cells that are held in place as part of a tissue layer, they can move fluid over the surface of the tissue. For example, the ciliated lining of the trachea (windpipe) sweeps mucus containing trapped debris out of the lungs (see the EMs in Figure 3). In a woman's reproductive tract, the cilia lining the oviducts help move an egg toward the uterus.

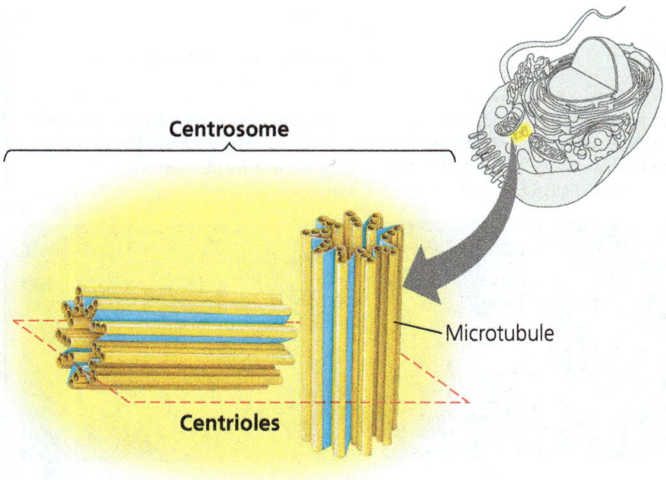

▲ Figure 22 Centrosome containing a pair of centrioles. Most animal cells have a centrosome, a region near the nucleus where the cell's microtubules are initiated. Within the centrosome is a pair of centrioles, each about 250 nm (0.25 μm) in diameter. The two centrioles are at right angles to each other, and each is made up of nine sets of three microtubules. The blue portions of the drawing represent nontubulin proteins that connect the microtubule triplets.

? *How many microtubules are in a centrosome? In the drawing, circle and label one microtubule and describe its structure. Circle and label a triplet.*

Motile cilia usually occur in large numbers on the cell surface. Flagella are usually limited to just one or a few per cell, and they are longer than cilia. Flagella and cilia also differ in their beating patterns. A flagellum has an undulating motion like the tail of a fish. In contrast, cilia have alternating power and recovery strokes, like the oars of a racing crew boat.

A cilium may also act as a signal-receiving antenna for the cell. Cilia that have this function are generally nonmotile, and there is only one per cell. (In fact, in vertebrate animals, it appears that almost all cells have such a cilium, which is called a *primary cilium*.) Membrane proteins on this kind of cilium transmit molecular signals from the cell's environment to its interior, triggering signaling pathways that may lead to changes in the cell's activities. Cilium-based signaling appears to be crucial to brain function and to embryonic development.

Though different in length, number per cell, and beating pattern, motile cilia and flagella share a common structure. Each motile cilium or flagellum has a group of microtubules sheathed in an extension of the plasma membrane **(Figure 23a)**. Nine doublets of microtubules are arranged in a ring with two single microtubules in its center **(Figure 23b)**. This arrangement, referred to as the "9 + 2" pattern, is found in nearly all eukaryotic flagella and motile cilia. (Nonmotile primary cilia have a "9 + 0" pattern, lacking the central pair of microtubules.) The microtubule assembly of a cilium or flagellum is anchored in the cell by a **basal body**, which is structurally similar to a centriole, with microtubule triplets in a "9 + 0" pattern **(Figure 23c)**. In fact, in many animals (including humans), the basal body of the fertilizing sperm's flagellum enters the egg and becomes a centriole.

How does the microtubule assembly produce the bending movements of flagella and motile cilia? Bending involves large motor proteins called **dyneins** (red in the diagram in Figure 23b) that are attached along each outer microtubule doublet. A typical dynein protein has two "feet" that "walk" along the microtubule of the adjacent doublet, using ATP for energy. One foot maintains contact, while the other releases and reattaches one step farther along the microtubule (see Figure 21). The outer doublets and two central microtubules are held together by flexible cross-linking proteins (blue in Figure 23b). If the doublets were not held in place, the walking action would make them slide past each other. Instead, the movements of the dynein feet cause the microtubules—and the organelle as a whole—to bend.

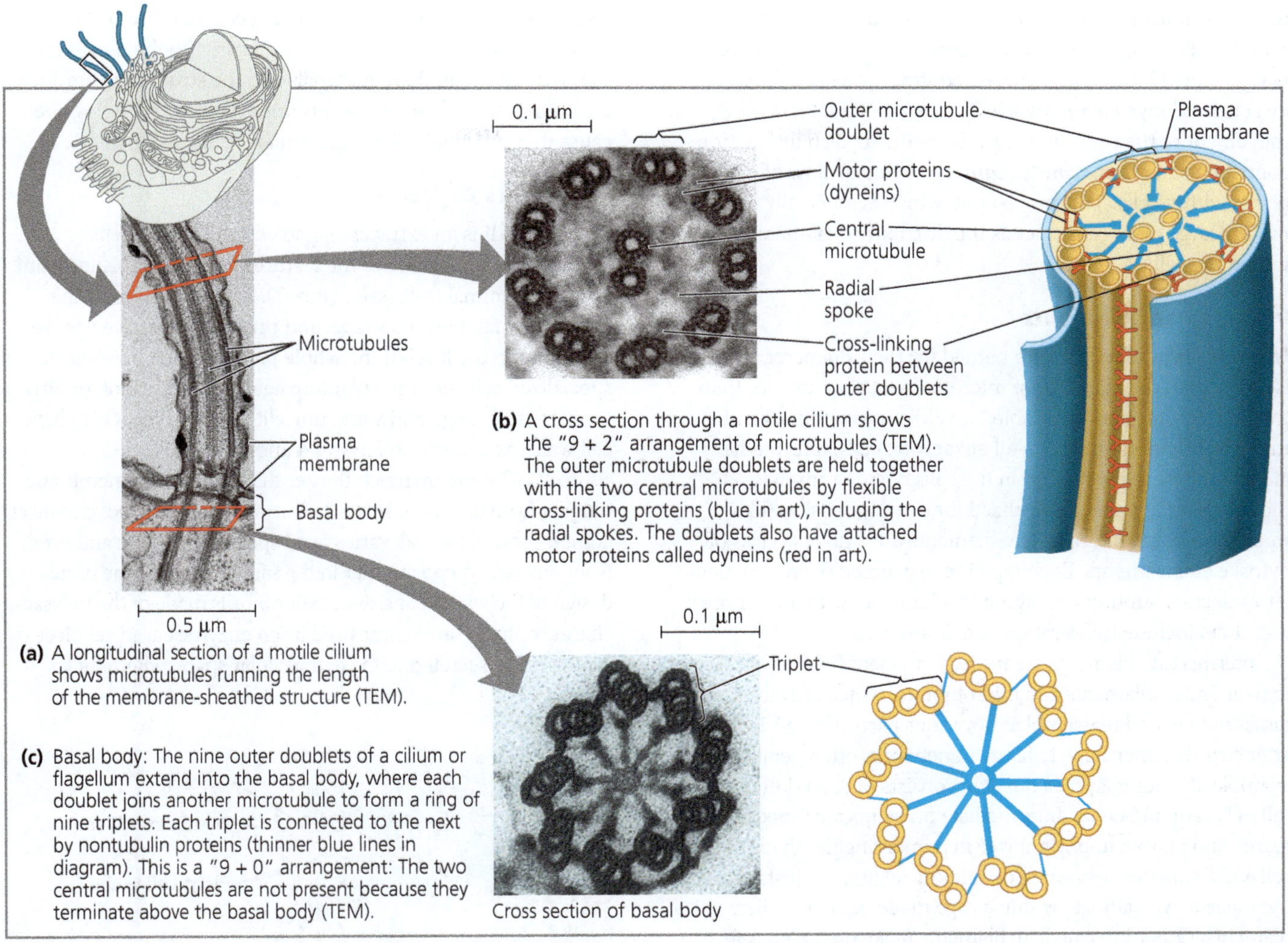

(a) A longitudinal section of a motile cilium shows microtubules running the length of the membrane-sheathed structure (TEM).

(b) A cross section through a motile cilium shows the "9 + 2" arrangement of microtubules (TEM). The outer microtubule doublets are held together with the two central microtubules by flexible cross-linking proteins (blue in art), including the radial spokes. The doublets also have attached motor proteins called dyneins (red in art).

(c) Basal body: The nine outer doublets of a cilium or flagellum extend into the basal body, where each doublet joins another microtubule to form a ring of nine triplets. Each triplet is connected to the next by nontubulin proteins (thinner blue lines in diagram). This is a "9 + 0" arrangement: The two central microtubules are not present because they terminate above the basal body (TEM).

Cross section of basal body

▲ **Figure 23 Structure of a flagellum or motile cilium.**

DRAW IT *In* ***(a)****, circle and label the central pair of microtubules. Show where they terminate, and explain why they aren't seen in the cross section of the basal body in* ***(c)****.*

Microfilaments (Actin Filaments)

Microfilaments are thin solid rods. They are also called actin filaments because they are built from molecules of **actin**, a globular protein. A microfilament is a twisted double chain of actin subunits (see Table 1). Besides occurring as linear filaments, microfilaments can form structural networks when certain proteins bind along the side of such a filament and allow a new filament to extend as a branch.

The structural role of microfilaments in the cytoskeleton is to bear tension (pulling forces). A three-dimensional network formed by microfilaments just inside the plasma membrane helps support the cell's shape. In some kinds of animal cells, such as nutrient-absorbing intestinal cells, bundles of microfilaments make up the core of microvilli, delicate projections that increase the cell's surface area **(Figure 24)**.

Microfilaments are well known for their role in cell motility. Thousands of actin filaments and thicker filaments of a

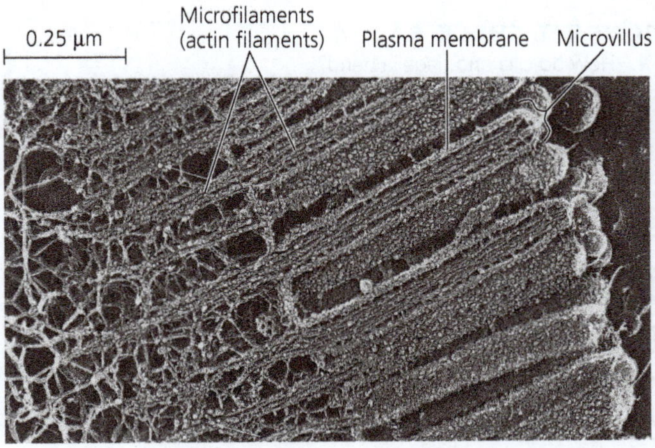

▲ **Figure 24 A structural role of microfilaments.** The surface area of this intestinal cell is increased by its many microvilli (singular, *microvillus*), cellular extensions reinforced by bundles of microfilaments (TEM).

motor protein called **myosin** interact to cause contraction of muscle cells. In the unicellular eukaryote *Amoeba* and some of our white blood cells, localized contractions brought about by actin and myosin are involved in the amoeboid (crawling) movement of the cells. In plant cells, actin-myosin interaction contributes to *cytoplasmic streaming*, a circular flow of cytoplasm within cells. This movement, which is especially common in large plant cells, speeds the distribution of materials within the cell.

Intermediate Filaments

Intermediate filaments are named for their diameter, which is larger than the diameter of microfilaments but smaller than that of microtubules (see Table 1). While microtubules and microfilaments are found in all eukaryotic cells, intermediate filaments are only found in the cells of some animals, including vertebrates. Specialized for bearing tension (like microfilaments), intermediate filaments are a diverse class of cytoskeletal elements. Each type is constructed from a particular molecular subunit belonging to a family of proteins whose members include the keratins in hair and nails.

Intermediate filaments are more permanent fixtures of cells than are microfilaments and microtubules, which are often disassembled and reassembled in various parts of a cell. Even after cells die, intermediate filament networks often persist; for example, the outer layer of our skin consists of dead skin cells full of keratin filaments. Intermediate filaments are especially sturdy and play an important role in reinforcing the shape of a cell and fixing the position of certain organelles. For instance, the nucleus typically sits within a cage made of intermediate filaments. Other intermediate filaments make up the nuclear lamina, which lines the interior of the nuclear envelope (see Figure 8). In general, the various kinds of intermediate filaments seem to function together as the permanent framework of the entire cell.

CONCEPT CHECK 6

1. How do cilia and flagella bend?
2. WHAT IF? Males afflicted with Kartagener's syndrome are sterile because of immotile sperm, and they tend to suffer from lung infections. This disorder has a genetic basis. Suggest what the underlying defect might be.

CONCEPT 7

Extracellular components and connections between cells help coordinate cellular activities

Having crisscrossed the cell to explore its interior components, we complete our tour of the cell by returning to the surface of this microscopic world, where there are additional structures with important functions. The plasma membrane is usually regarded as the boundary of the living cell, but most cells synthesize and secrete materials extracellularly (to the outside of the cell). Although these materials and the structures they form are outside the cell, their study is important to cell biology because they are involved in a great many cellular functions.

Cell Walls of Plants

The **cell wall** is an extracellular structure of plant cells **(Figure 25)**. This is one of the features that distinguishes plant cells from animal cells (see Figure 7). The wall protects the plant cell, maintains its shape, and prevents excessive uptake of water. On the level of the whole plant, the strong walls of specialized cells hold the plant up against the force of gravity. Prokaryotes, fungi, and some unicellular eukaryotes also have cell walls, as you saw in Figures 4 and 7.

Plant cell walls are much thicker than the plasma membrane, ranging from 0.1 μm to several micrometers. The exact chemical composition of the wall varies from species to species and even from one cell type to another in the same plant, but the basic design of the wall is consistent. Microfibrils made of the polysaccharide cellulose are synthesized by an enzyme called cellulose synthase and secreted to the extracellular space, where they

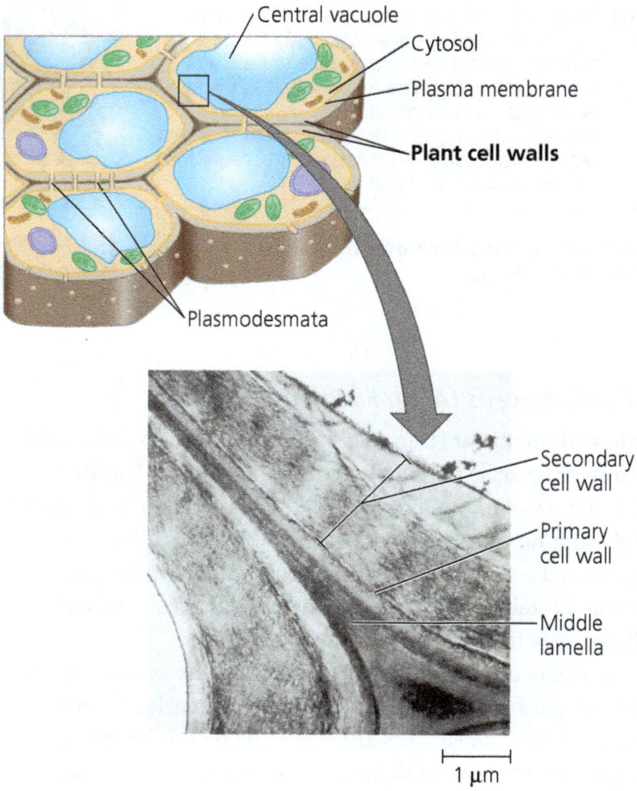

▲ **Figure 25 Plant cell walls.** The drawing shows several cells, each with a large vacuole, a nucleus, and several chloroplasts and mitochondria. The TEM shows the cell walls where two cells come together. The multilayered partition between plant cells consists of adjoining walls individually secreted by the cells. Plasmodesmata are channels through cell walls that connect the cytoplasm of adjacent plant cells.

become embedded in a matrix of other polysaccharides and proteins. This combination of materials, strong fibers in a "ground substance" (matrix), is the same basic architectural design found in steel-reinforced concrete and in fiberglass.

A young plant cell first secretes a relatively thin and flexible wall called the **primary cell wall** (see Figure 25). Between primary walls of adjacent cells is the **middle lamella**, a thin layer rich in sticky polysaccharides called pectins. The middle lamella glues adjacent cells together. (Pectin is used as a thickening agent in jams and jellies.) When the cell matures and stops growing, it strengthens its wall. Some plant cells do this simply by secreting hardening substances into the primary wall. Other cells add a **secondary cell wall** between the plasma membrane and the primary wall. The secondary wall, often deposited in several laminated layers, has a strong and durable matrix that affords the cell protection and support. Wood, for example, consists mainly of secondary walls. Plant cell walls are usually perforated by channels between adjacent cells called plasmodesmata, which will be discussed shortly.

The Extracellular Matrix (ECM) of Animal Cells

Although animal cells lack walls akin to those of plant cells, they do have an elaborate **extracellular matrix (ECM)**. The main ingredients of the ECM are glycoproteins and other carbohydrate-containing molecules secreted by the cells. (Recall that glycoproteins are proteins with covalently bonded carbohydrates.) The most abundant glycoprotein in the ECM of most animal cells is **collagen**, which forms strong fibers outside the cells. In fact, collagen accounts for about 40% of the total protein in the human body. The collagen fibers are embedded in a network woven of secreted **proteoglycans** (**Figure 26**). A proteoglycan molecule consists of a small core protein with many carbohydrate chains covalently attached; it may be up to 95% carbohydrate. Large proteoglycan complexes can form when hundreds of proteoglycan molecules become noncovalently attached to a single long polysaccharide molecule, as shown in Figure 26. Some cells are attached to the ECM by ECM glycoproteins such as **fibronectin**. Fibronectin and other ECM proteins bind to cell-surface receptor proteins called **integrins** that are built into the plasma membrane. Integrins span the membrane and bind on their cytoplasmic side to associated proteins attached to microfilaments of the cytoskeleton. The name *integrin* is based on the word *integrate*: Integrins are in a position to transmit signals between the ECM and the cytoskeleton and thus to integrate changes occurring outside and inside the cell.

Current research is revealing the influential role of the ECM in the lives of cells. By communicating with a cell through integrins, the ECM can regulate a cell's behavior. For example, some cells in a developing embryo migrate along specific pathways by matching the orientation of their microfilaments to the "grain" of fibers in the extracellular matrix. Researchers have also learned that the extracellular matrix around a cell can influence the activity of genes in the nucleus. Information about the ECM probably reaches the nucleus by a combination of mechanical and chemical signaling pathways. Mechanical signaling involves fibronectin, integrins, and microfilaments of the cytoskeleton. Changes in the cytoskeleton may in turn trigger chemical signaling pathways inside the cell, leading to changes in the set of proteins being made by the cell and therefore changes in the cell's function. In this way, the extracellular

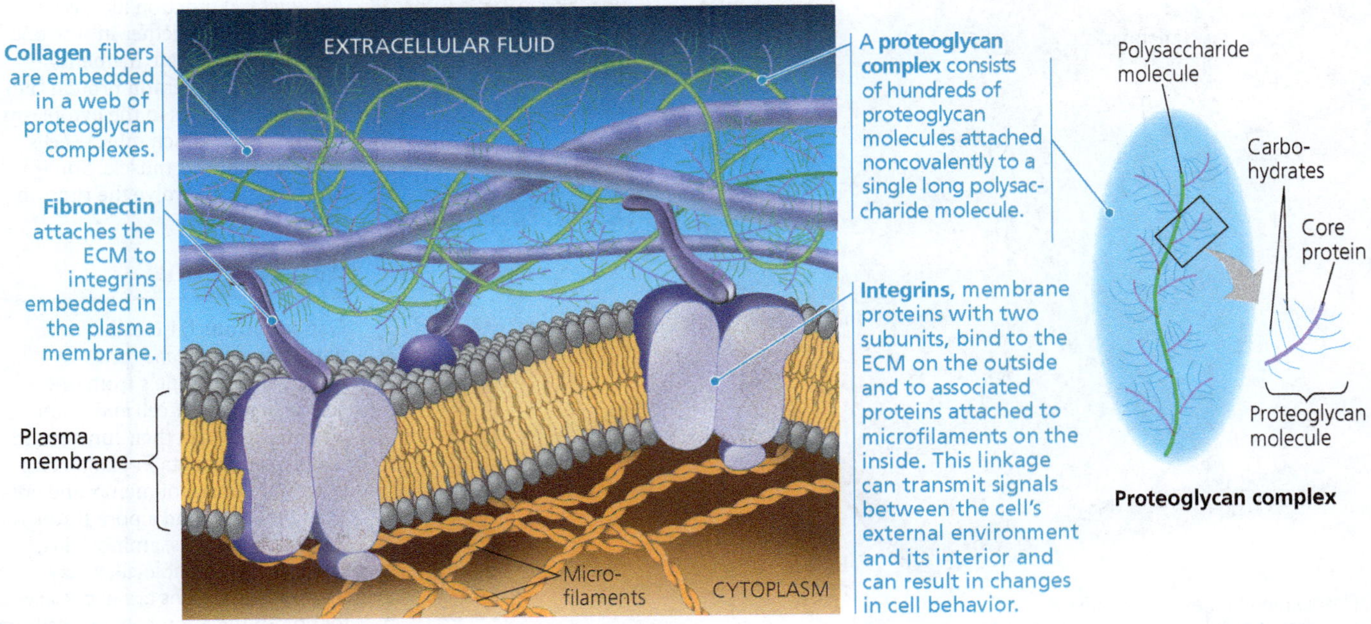

▲ **Figure 26 Extracellular matrix (ECM) of an animal cell.** The molecular composition and structure of the ECM vary from one cell type to another. In this example, three different types of ECM molecules are present: proteoglycans, collagen, and fibronectin.

matrix of a particular tissue may help coordinate the behavior of all the cells of that tissue. Direct connections between cells also function in this coordination, as we discuss next.

Cell Junctions

Neighboring cells in an animal or plant often adhere, interact, and communicate via sites of direct physical contact.

Plasmodesmata in Plant Cells

It might seem that the nonliving cell walls of plants would isolate plant cells from one another. But in fact, as shown in Figure 25, cell walls are perforated with **plasmodesmata** (singular, *plasmodesma*; from the Greek *desma*, bond), membrane-lined channels filled with cytosol. By joining adjacent cells, plasmodesmata unify most of a plant into one living continuum. The plasma membranes of adjacent cells line the channel of each plasmodesma and thus are continuous. Water and small solutes can pass freely from cell to cell, and experiments have shown that in some circumstances, certain proteins and RNA molecules can do this as well. The macromolecules transported to neighboring cells appear to reach the plasmodesmata by moving along fibers of the cytoskeleton.

Tight Junctions, Desmosomes, and Gap Junctions in Animal Cells

In animals, there are three main types of cell junctions: *tight junctions*, *desmosomes*, and *gap junctions* (**Figure 27**). All three types are especially common in epithelial tissue, which

▼ Figure 27 Exploring Cell Junctions in Animal Tissues

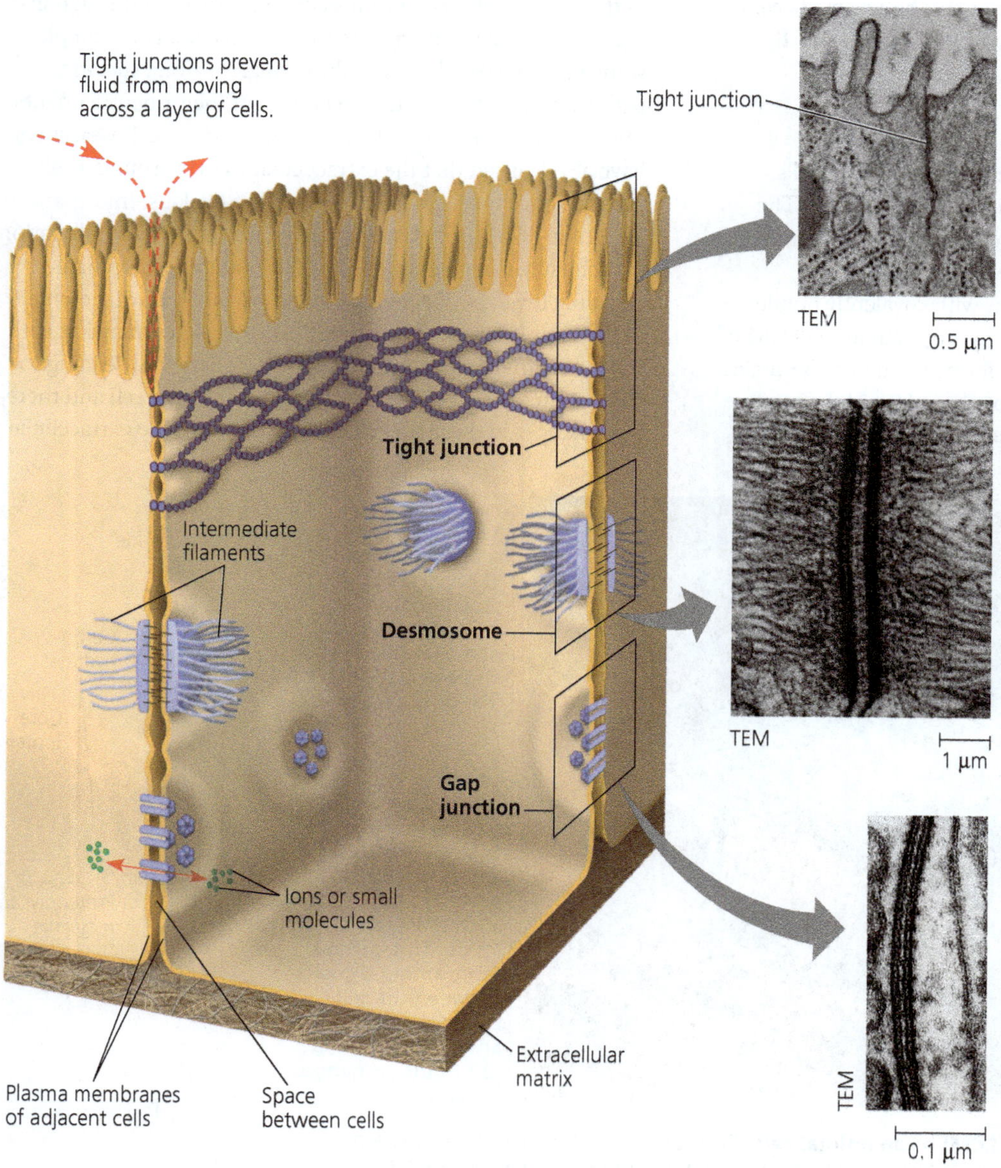

Tight Junctions

At **tight junctions**, the plasma membranes of neighboring cells are very tightly pressed against each other, bound together by specific proteins (purple). Forming continuous seals around the cells, tight junctions establish a barrier that prevents leakage of extracellular fluid across a layer of epithelial cells (see red dashed arrow). For example, tight junctions between skin cells make us watertight.

Desmosomes

Desmosomes (one type of *anchoring junction*) function like rivets, fastening cells together into strong sheets. Intermediate filaments made of sturdy keratin proteins anchor desmosomes in the cytoplasm. Desmosomes attach muscle cells to each other in a muscle. Some "muscle tears" involve the rupture of desmosomes.

Gap Junctions

Gap junctions (also called *communicating junctions*) provide cytoplasmic channels from one cell to an adjacent cell and in this way are similar in their function to the plasmodesmata in plants. Gap junctions consist of membrane proteins that surround a pore through which ions, sugars, amino acids, and other small molecules may pass. Gap junctions are necessary for communication between cells in many types of tissues, such as heart muscle, and in animal embryos.

lines the external and internal surfaces of the body. Figure 27 uses epithelial cells of the intestinal lining to illustrate these junctions. (Gap junctions are most like the plasmodesmata of plants, although gap junction pores are not lined with membrane.)

CONCEPT CHECK 7

1. In what way are the cells of plants and animals structurally different from single-celled eukaryotes?
2. **WHAT IF?** If the plant cell wall or the animal extracellular matrix were impermeable, what effect would this have on cell function?
3. **MAKE CONNECTIONS** The polypeptide chain that makes up a tight junction weaves back and forth through the membrane four times, with two extracellular loops, and one loop plus short C-terminal and N-terminal tails in the cytoplasm. What would you predict about the amino acids making up the tight junction protein?

The Cell: A Living Unit Greater Than the Sum of Its Parts

From our panoramic view of the cell's compartmental organization to our close-up inspection of each organelle's architecture, this tour of the cell has provided many opportunities to correlate structure with function. But even as we dissect the cell, remember that none of its components works alone. As an example of cellular integration, consider the microscopic scene in **Figure 28**. The large cell is a macrophage (see Figure 12). It helps defend the mammalian body against infections by ingesting bacteria (the smaller cells) into phagocytic vesicles. The macrophage crawls along a surface and reaches out to the bacteria with thin cell extensions called pseudopodia (specifically, filopodia). Actin filaments interact with other elements of the cytoskeleton in these movements. After the macrophage engulfs the bacteria, they are destroyed by lysosomes. The elaborate endomembrane system produces the lysosomes. The digestive enzymes of the lysosomes and the proteins of the cytoskeleton are all made on ribosomes. And the synthesis of these proteins is programmed by genetic messages dispatched from the DNA in the nucleus. All these processes require energy, which mitochondria supply in the form of ATP. Cellular functions arise from cellular order: The cell is a living unit greater than the sum of its parts.

▲ **Figure 28 The emergence of cellular functions.** The ability of this macrophage (brown) to recognize, apprehend, and destroy *Staphylococcus* bacteria (orange) is a coordinated activity of the whole cell. Its cytoskeleton, lysosomes, and plasma membrane are among the components that function in phagocytosis (colorized SEM).

Chapter Review

Go to **MasteringBiology**® for Assignments, the eText, and the Study Area with Animations, Activities, Vocab Self-Quiz, and Practice Tests.

SUMMARY OF KEY CONCEPTS

VOCAB SELF-QUIZ
goo.gl/gbai8v

CONCEPT 1

Biologists use microscopes and the tools of biochemistry to study cells

- Improvements in microscopy that affect the parameters of magnification, resolution, and contrast have catalyzed progress in the study of cell structure. **Light microscopy** (LM) and **electron microscopy** (EM), as well as other types, remain important tools.
- Cell biologists can obtain pellets enriched in particular cellular components by centrifuging disrupted cells at sequential speeds, a process known as **cell fractionation**. Larger cellular components are in the pellet after lower-speed centrifugation, and smaller components are in the pellet after higher-speed centrifugation.

? *How do microscopy and biochemistry complement each other to reveal cell structure and function?*

CONCEPT 2

Eukaryotic cells have internal membranes that compartmentalize their functions

- All cells are bounded by a **plasma membrane**, a bilayer of phospholipids with their hydrophobic tails on in the interior of the membrane and their hydrophilic heads in contact with the aqueous solutions on either side.
- **Prokaryotic cells** lack nuclei and other membrane-enclosed **organelles**, while **eukaryotic cells** have internal membranes that compartmentalize cellular functions.
- The surface-to-volume ratio is an important parameter affecting cell size and shape.
- Plant and animal cells have most of the same organelles: a nucleus, endoplasmic reticulum, Golgi apparatus, and mitochondria. Some organelles are found only in plant or in animal cells. Chloroplasts are present only in cells of photosynthetic eukaryotes.

? *Explain how the compartmental organization of a eukaryotic cell contributes to its biochemical functioning.*

	Cell Component	Structure	Function
CONCEPT 3 **The eukaryotic cell's genetic instructions are housed in the nucleus and carried out by the ribosomes** ? *Describe the relationship between the nucleus and ribosomes.*	Nucleus 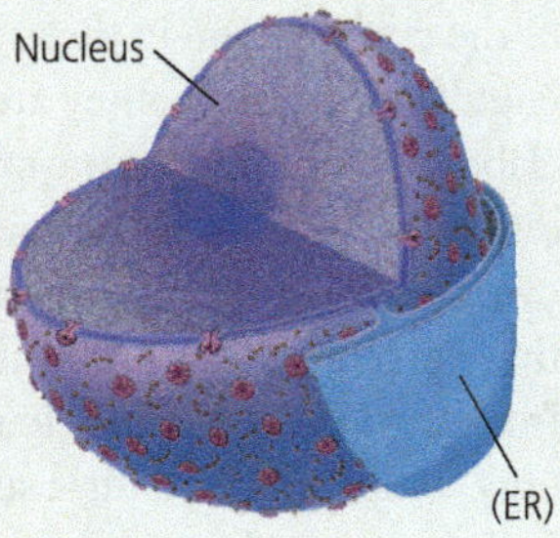	Surrounded by nuclear envelope (double membrane) perforated by nuclear pores; nuclear envelope continuous with endoplasmic reticulum (ER)	Houses chromosomes, which are made of chromatin (DNA and proteins); contains nucleoli, where ribosomal subunits are made; pores regulate entry and exit of materials
	Ribosome 	Two subunits made of ribosomal RNA and proteins; can be free in cytosol or bound to ER	Protein synthesis
CONCEPT 4 **The endomembrane system regulates protein traffic and performs metabolic functions in the cell** ? *Describe the key role played by transport vesicles in the endomembrane system.*	Endoplasmic reticulum 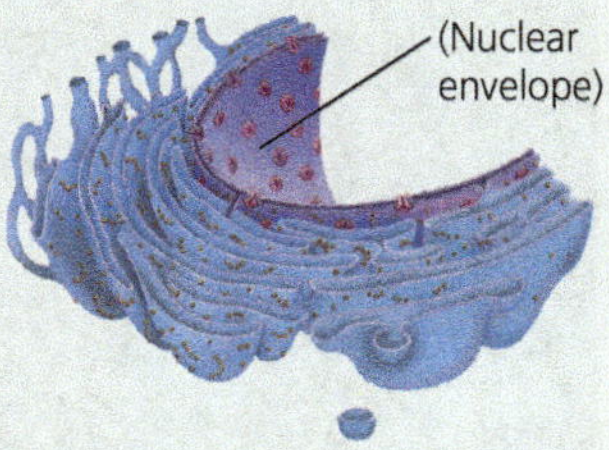	Extensive network of membrane-bounded tubules and sacs; membrane separates lumen from cytosol; continuous with nuclear envelope	Smooth ER: synthesis of lipids, metabolism of carbohydrates, Ca^{2+} storage, detoxification of drugs and poisons Rough ER: aids in synthesis of secretory and other proteins from bound ribosomes; adds carbohydrates to proteins to make glycoproteins; produces new membrane
	Golgi apparatus 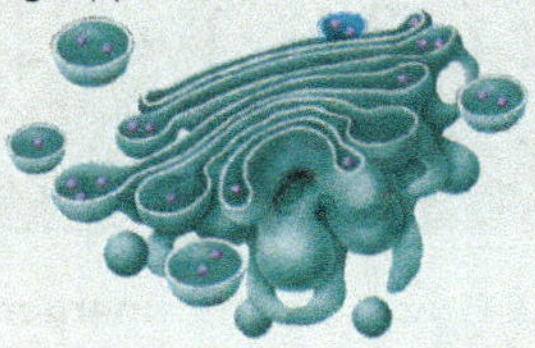	Stacks of flattened membranous sacs; has polarity (*cis* and *trans* faces)	Modification of proteins, carbohydrates on proteins, and phospholipids; synthesis of many polysaccharides; sorting of Golgi products, which are then released in vesicles
	Lysosome 	Membranous sac of hydrolytic enzymes (in animal cells)	Breakdown of ingested substances, cell macromolecules, and damaged organelles for recycling
	Vacuole 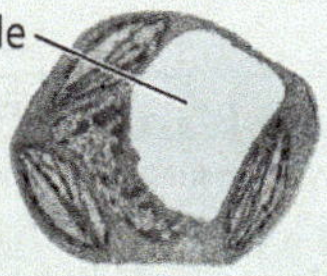	Large membrane-bounded vesicle	Digestion, storage, waste disposal, water balance, plant cell growth and protection
CONCEPT 5 **Mitochondria and chloroplasts change energy from one form to another** ? *What is the endosymbiont theory?*	Mitochondrion 	Bounded by double membrane; inner membrane has infoldings (cristae)	Cellular respiration
	Chloroplast 	Typically two membranes around fluid stroma, which contains thylakoids stacked into grana (in cells of photosynthetic eukaryotes, including plants)	Photosynthesis
	Peroxisome 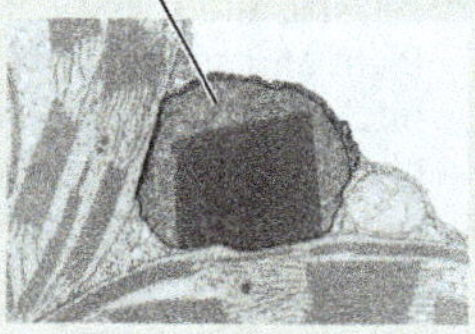	Specialized metabolic compartment bounded by a single membrane	Contains enzymes that transfer hydrogen atoms from certain molecules to oxygen, producing hydrogen peroxide (H_2O_2) as a by-product; H_2O_2 is converted to water by another enzyme

CONCEPT 6

The cytoskeleton is a network of fibers that organizes structures and activities in the cell

- The **cytoskeleton** functions in structural support for the cell and in motility and signal transmission.
- **Microtubules** shape the cell, guide organelle movement, and separate chromosomes in dividing cells. **Cilia** and **flagella** are motile appendages containing microtubules. *Primary cilia* play sensory and signaling roles. **Microfilaments** are thin rods that function in muscle contraction, amoeboid movement, cytoplasmic streaming, and support of microvilli. **Intermediate filaments** support cell shape and fix organelles in place.

? *Describe the role of motor proteins inside the eukaryotic cell and in whole-cell movement.*

CONCEPT 7

Extracellular components and connections between cells help coordinate cellular activities

- Plant **cell walls** are made of cellulose fibers embedded in other polysaccharides and proteins.
- Animal cells secrete glycoproteins and proteoglycans that form the **extracellular matrix (ECM)**, which functions in support, adhesion, movement, and regulation.
- Cell junctions connect neighboring cells in plants and animals. Plants have **plasmodesmata** that pass through adjoining cell walls. Animal cells have **tight junctions**, **desmosomes**, and **gap junctions**.

? *Compare the structure and functions of a plant cell wall and the extracellular matrix of an animal cell.*

TEST YOUR UNDERSTANDING

PRACTICE TEST

goo.gl/CRZjvS

Level 1: Knowledge/Comprehension

1. Which structure is *not* part of the endomembrane system?
(A) nuclear envelope
(B) chloroplast
(C) Golgi apparatus
(D) plasma membrane

2. Which structure is common to plant *and* animal cells?
(A) chloroplast
(B) wall made of cellulose
(C) mitochondrion
(D) centriole

3. Which of the following is present in a prokaryotic cell?
(A) mitochondrion
(B) ribosome
(C) nuclear envelope
(D) chloroplast

4. Which structure-function pair is *mismatched*?
(A) microtubule; muscle contraction
(B) ribosome; protein synthesis
(C) Golgi; protein trafficking
(D) nucleolus; production of ribosomal subunits

Level 2: Application/Analysis

5. Cyanide binds to at least one molecule involved in producing ATP. If a cell is exposed to cyanide, most of the cyanide will be found within the
(A) mitochondria.
(B) ribosomes.
(C) peroxisomes.
(D) lysosomes.

6. What is the most likely pathway taken by a newly synthesized protein that will be secreted by a cell?
(A) ER → Golgi → nucleus
(B) nucleus → ER → Golgi
(C) ER → Golgi → vesicles that fuse with plasma membrane
(D) ER → lysosomes → vesicles that fuse with plasma membrane

7. Which cell would be best for studying lysosomes?
(A) muscle cell
(B) nerve cell
(C) phagocytic white blood cell
(D) bacterial cell

8. DRAW IT From memory, draw two eukaryotic cells. Label the structures listed here and show any physical connections between the internal structures of each cell: nucleus, rough ER, smooth ER, mitochondrion, centrosome, chloroplast, vacuole, lysosome, microtubule, cell wall, ECM, microfilament, Golgi apparatus, intermediate filament, plasma membrane, peroxisome, ribosome, nucleolus, nuclear pore, vesicle, flagellum, microvilli, plasmodesma.

Level 3: Synthesis/Evaluation

9. SCIENTIFIC INQUIRY
In studying micrographs of an unusual protist (single-celled eukaryote) that you found in a sample of pond water, you spot an organelle that you can't recognize. You successfully develop a method for growing this organism in liquid in the laboratory. Describe how you would go about finding out what this organelle is and what it does in the cell. Assume that you would make use of additional microscopy, cell fractionation, and biochemical tests.

10. FOCUS ON EVOLUTION
Compare different aspects of cell structure. (a) What structures best reveal evolutionary unity? (b) Provide an example of diversity related to specialized modifications.

11. FOCUS ON ORGANIZATION
Considering some of the characteristics that define life and drawing on your new knowledge of cellular structures and functions, write a short essay (100–150 words) that discusses this statement: Life is an emergent property that appears at the level of the cell.

12. SYNTHESIZE YOUR KNOWLEDGE

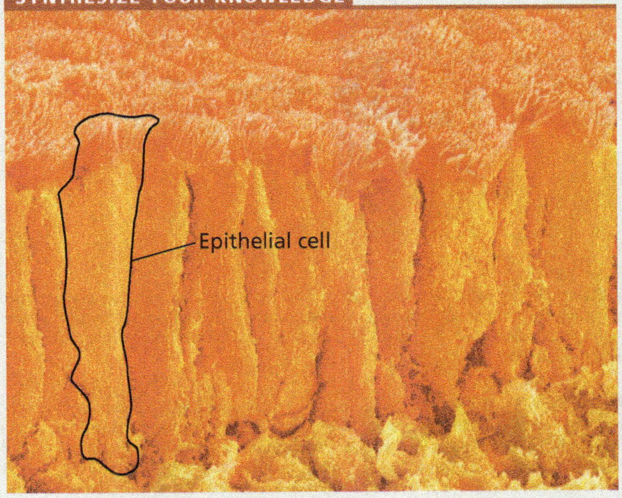

The cells in the SEM are epithelial cells from the small intestine. Discuss how their cellular structure contributes to their specialized functions of nutrient absorption and as a barrier between intestinal contents and the blood supply on the other side of the cell sheet.

Answers

Figure Questions

Figure 5 A phospholipid is a lipid consisting of a glycerol molecule joined to two fatty acids and one phosphate group. Together, the glycerol and phosphate end of the phospholipid form the "head," which is hydrophilic, while the hydrocarbon chains on the fatty acids form hydrophobic "tails." The presence in a single molecule of both a hydrophilic and a hydrophobic region makes the molecule ideal as the main building block of a cell or organelle membrane, which is a phospholipid bilayer. In this bilayer, the hydrophobic regions can associate with each other on the inside of the membrane, while the hydrophilic region of each can be in contact with the aqueous solution on either side. **Figure 8** The DNA in a chromosome dictates synthesis of a messenger RNA (mRNA) molecule, which then moves out to the cytoplasm. There, the information is used for the production, on ribosomes, of proteins that carry out cellular functions. **Figure 9** Any of the bound ribosomes (attached to the endoplasmic reticulum) could be circled, because any could be making a protein that will be secreted.
Figure 22 Each centriole has 9 sets of 3 microtubules, so the entire centrosome (two centrioles) has 54 microtubules. Each microtubule consists of a helical array of tubulin dimers (as shown in Table 1).

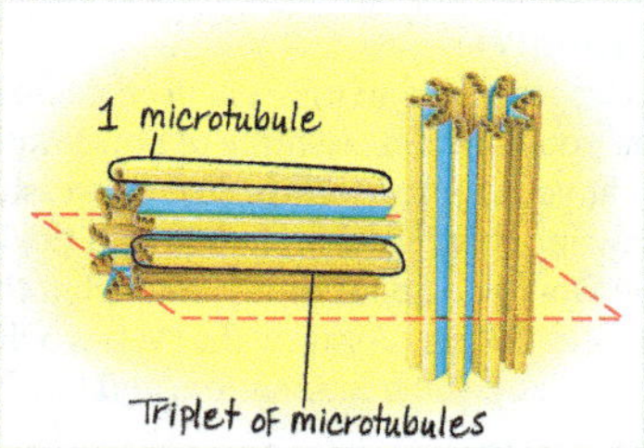

Figure 23 The two central microtubules terminate above the basal body, so they aren't present at the level of the cross section through the basal body, indicated by the lower red rectangle in **(a)**.

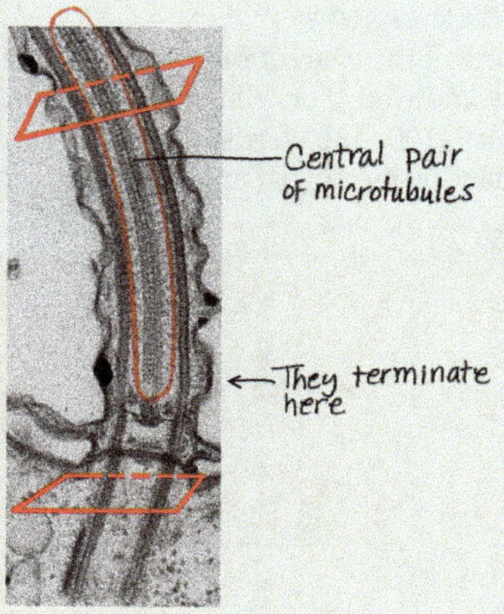

Concept Check 1

1. Stains used for light microscopy are colored molecules that bind to cell components, affecting the light passing through, while stains used for electron microscopy involve heavy metals that affect the beams of electrons. **2.** (a) Light microscope, (b) scanning electron microscope

Concept Check 2

1. See Figure 7. **2.** This cell would have the same volume as the cells in columns 2 and 3 but proportionally more surface area than that in column 2 and less than that in column 3. Thus, the surface-to-volume ratio should be greater than 1.2 but less than 6. To obtain the surface area, you would add the area of the six sides (the top, bottom, sides, and ends): $125 + 125 + 125 + 125 + 1 + 1 = 502$. The surface-to-volume ratio equals 502 divided by a volume of 125, or 4.0.

Concept Check 3

1. Ribosomes in the cytoplasm translate the genetic message, carried from the DNA in the nucleus by mRNA, into a polypeptide chain. **2.** Nucleoli consist of DNA and the ribosomal RNA (rRNA) made according to its instructions, as well as proteins imported from the cytoplasm. Together, the rRNA and proteins are assembled into large and small ribosomal subunits. (These are exported through nuclear pores to the cytoplasm, where they will participate in polypeptide synthesis.) **3.** Each chromosome consists of one long DNA molecule attached to numerous protein molecules, a combination called chromatin. As a cell begins division, each chromosome becomes "condensed" as its diffuse mass of chromatin coils up.

Concept Check 4

1. The primary distinction between rough and smooth ER is the presence of bound ribosomes on the rough ER. Both types of ER make phospholipids, but membrane proteins and secretory proteins are all produced by the ribosomes on the rough ER. The smooth ER also functions in detoxification, carbohydrate metabolism, and storage of calcium ions. **2.** Transport vesicles move membranes and the substances they enclose between other components of the endomembrane system. **3.** The mRNA is synthesized in the nucleus and then passes out through a nuclear pore to be translated on a bound ribosome, attached to the rough ER. The protein is synthesized into the lumen of the ER and perhaps modified there. A transport vesicle carries the protein to the Golgi apparatus. After further modification in the Golgi, another transport vesicle carries it back to the ER, where it will perform its cellular function.

Concept Check 5

1. Both organelles are involved in energy transformation, mitochondria in cellular respiration and chloroplasts in photosynthesis. They both have multiple membranes that separate their interiors into compartments. In both organelles, the innermost membranes—cristae, or infoldings of the inner membrane, in mitochondria, and the thylakoid membranes in chloroplasts—have large surface areas with embedded enzymes that carry out their main functions. **2.** Yes. Plant cells are able to make their own sugar by photosynthesis, but mitochondria in these eukaryotic cells are the organelles that are able to generate energy from sugars, a function required in all cells.
3. Mitochondria and chloroplasts are not derived from the ER, nor are they connected physically or via transport vesicles to organelles of the endomembrane system. Mitochondria and chloroplasts are structurally quite different from vesicles derived from the ER, which are bounded by a single membrane.

Concept Check 6

1. Dynein arms, powered by ATP, move neighboring doublets of microtubules relative to each other. Because they are anchored within the organelle and with respect to one another, the doublets bend instead of sliding past each other. Synchronized bending of the nine microtubule doublets brings about bending of both cilia and flagella.
2. Such individuals have defects in the microtubule-based movement of cilia and flagella. Thus, the sperm can't move because of malfunctioning or nonexistent flagella, and the airways are compromised because cilia that line the trachea malfunction or don't exist, and so mucus cannot be cleared from the lungs.

Concept Check 7

1. The most obvious difference is the presence of direct cytoplasmic connections between cells of plants (plasmodesmata) and animals (gap junctions). These connections result in the cytoplasm being continuous between adjacent cells. **2.** The cell would not be able to function properly and would probably soon die, as the cell wall or ECM must be permeable to allow the exchange of matter between the cell and its external environment. Molecules involved in energy production and use must be allowed entry, as well as those that provide information about the cell's environment. Other molecules, such as products synthesized by the cell for export and the by-products of cellular respiration, must be allowed to exit. **3.** The parts of the protein that face aqueous regions would be expected to have polar or charged (hydrophilic) amino acids, while the parts that go through the membrane would be expected to have nonpolar (hydrophobic) amino acids. You would predict polar or charged amino acids at each end (tail), in the region of the cytoplasmic loop, and in the regions of the two extracellular loops. You would predict nonpolar amino acids in the four regions that go through the membrane between the tails and loops.

Summary of Key Concepts Questions

1 Both light and electron microscopy allow cells to be studied visually, thus helping us understand internal cellular structure and the arrangement of cell components. Cell fractionation techniques separate out different groups of cell components, which can then be analyzed biochemically to determine their function. Performing microscopy on the same cell fraction helps to correlate the biochemical function of the cell with the cell component responsible. **2** The separation of different functions in different organelles has several advantages. Reactants and enzymes can be concentrated in one area instead of spread throughout the cell. Reactions that require specific conditions, such as a lower pH, can be compartmentalized. And enzymes for specific reactions are often embedded in the membranes that enclose or partition an organelle.
3 The nucleus contains the genetic material of the cell in the form of DNA, which codes for messenger RNA, which in turn provides instructions for the synthesis of proteins (including the proteins that make up part of the ribosomes). DNA also codes for ribosomal RNA, which is combined with proteins in the nucleolus into the subunits of ribosomes. Within the cytoplasm, ribosomes join with mRNA to build polypeptides, using the genetic information in the mRNA. **4** Transport vesicles move proteins and membrane synthesized by the rough ER to the Golgi for further processing and then to the plasma membrane, lysosomes, or other locations in the cell, including back to the ER. **5** According to the endosymbiont theory, mitochondria originated from an oxygen-using prokaryotic cell that was engulfed by an ancestral eukaryotic cell. Over time, the host and endosymbiont evolved into a single unicellular organism.

Chloroplasts originated when at least one of these eukaryotic cells containing mitochondria engulfed and then retained a photosynthetic prokaryote. **6** Inside the cell, motor proteins interact with components of the cytoskeleton to move cellular parts. Motor proteins may "walk" vesicles along microtubules. The movement of cytoplasm within a cell involves interactions of the motor protein myosin and microfilaments (actin filaments). Whole cells can be moved by the rapid bending of flagella or cilia, which is caused by the motor-protein-powered sliding of microtubules within these structures. Some cells move by amoeboid movement, which involves interactions of microfilaments with myosin. Interactions of motor proteins and microfilaments in muscle cells cause muscle contraction that can propel whole organisms (for example, by walking or swimming). **7** A plant cell wall is primarily composed of microfibrils of cellulose embedded in other polysaccharides and proteins. The ECM of animal cells is primarily composed of collagen and other protein fibers, such as fibronectin and other glycoproteins. These fibers are embedded in a network of carbohydrate-rich proteoglycans. A plant cell wall provides structural support for the cell and, collectively, for the plant body. In addition to giving support, the ECM of an animal cell allows for communication of environmental changes into the cell.

Test Your Understanding

1. B **2.** C **3.** B **4.** A **5.** A **6.** C **7.** C **8.** See Figure 7.

Credits

Credits are listed in the order of appearance.

Photo Credits

1 Don W. Fawcett/Science Source; **UNCO paramecium** M.I. Walker/Science Source; **3.1–3.4** Elisabeth Pierson, Pearson Education; **3.5** Michael W. Davidson/Molecular Expressions; **3.6** Karl Garsha; **3.7a** J.L. Carson/Custom Medical Stock Photo; **3.7b** William Dentler/Biological Photo Service; **4** CNRI/Science Source; **5** Daniel S. Friend; **7 left to right** S. Cinti/Science Source; SPL/Science Source; A. Barry Dowsett/Science Source; Biophoto Associates/Science Source; SPL/Science Source; The University of Kansas Center for Bioinformatics; **SSE** Kelly Tatchell; **UN01 nucleus** Thomas Deerinck and Mark Ellisman, NCMIR; **8.1** Reproduced with permission from Freeze-Etch Histology, by L. Orci and A. Perrelet, Springer-Verlag, Heidelberg, 1975, Plate 25, page 53. Copyright © 1975 by Springer-Verlag GmbH & Co KG; **8.2** Don W. Fawcett/Science Source; **8.3** Ueli Aebi; **9.1** Don W. Fawcett/Science Source; **9.2** Harry Noller; **10** Don Fawcett; R. Bolender/Science Source; **11** Don W. Fawcett/Science Source; **12–13** Daniel S. Friend; **14** Eldon H. Newcomb; **17a** Daniel S. Friend; **17b** From: The shape of mitochondria and the number of mitochondrial nucleoids during the cell cycle of *Euglena gracilis*. Y. Hayashi and K. Ueda. *Journal of Cell Science*, 93:565–570, Copyright © 1989 by Company of Biologists; **18a** Jeremy Burgess/Science Source; **18b** Franz Golig, Philipps-University, Marburg, Germany; **19** Eldon H. Newcomb; **20** Albert Tousson; **21b** Bruce J. Schnapp; **Table 1 left to right** Mary Osborn; Frank Solomon; Mark Ladinsky; **23a** Omikron/Science Source; **23b** Dartmouth Electron Microscope Facility, Dartmouth College; **23c** Richard W. Linck; **24** From: Cross-linker system between neurofilaments, microtubules, and membranous organelles in frog axons revealed by the quick-freeze, deep-etching method. Hirokawa Nobutaka. *Journal of Cell Biology* 94(1): 129–142, 1982. Reproduced by permission of Rockefeller University Press; **25** G. F. Leedale/Science Source; **27.1** Reproduced with permission from *Freeze-Etch Histology*, by L. Orci and A. Perrelet, Springer-Verlag, Heidelberg, 1975. Plate 32. Page 68. Copyright © 1975 by Springer-Verlag GmbH & Co KG; **27.2** From: Fine structure of desmosomes, hemidesmosomes, and an adepidermal globular layer in developing newt epidermis. D. E. Kelly. *J Cell Biol.* 1966 Jan; 28(1):51–72. Fig. 7; **27.3** From: Low resistance junctions in crayfish. Structural changes with functional uncoupling. C. Peracchia and A. F. Dulhunty. *The Journal of Cell Biology*. 1976 Aug; 70(2 pt 1):419–39. Fig. 6. Reproduced by permission of Rockefeller University Press; **28** Eye of Science/Science Source; **Chapter Review vacuole, peroxisome** Eldon H. Newcomb; **SYK** Susumu Nishinaga/Science Source.

Illustration and Text Credits

5b Adapted from *The World of the Cell*, 3rd Edition, by W.M. Becker, J.B. Reece, and M.F. Poenie, 1996. Copyright © 1996 by Pearson Education, Inc. Reprinted and electronically reproduced by permission of Pearson Education, Inc. Upper Saddle River, NJ; **7** Adapted from illustrations by Tomo Narashima in *Human Anatomy and Physiology*, 5th ed., by E.N. Marieb, 2001. Copyright © 2001 by E.N. Marieb. Reprinted and electronically reproduced with permission by Pearson Education, Inc. Upper Saddle River, NJ; **8** Adapted from *Human Anatomy and Physiology*, 5th ed., by E.N. Marieb, 2001. Copyright © 2001 by E.N. Marieb. Reprinted and electronically reproduced with permission by Pearson Education, Inc. Upper Saddle River, NJ; **10** Adapted from illustrations by Tomo Narashima in *Human Anatomy and Physiology*, 5th ed., by E.N. Marieb, 2001. Copyright © 2001 by E.N. Marieb. Reprinted and electronically reproduced with permission by Pearson Education, Inc. Upper Saddle River, NJ; **11** Adapted from illustrations by Tomo Narashima in *Human Anatomy and Physiology*, 5th ed., by E.N. Marieb, 2001. Copyright © 2001 by E.N. Marieb. Reprinted and electronically reproduced with permission by Pearson Education, Inc. Upper Saddle River, NJ; **Table 1** Adapted from *The World of the Cell*, 4th ed., by W.M. Becker, L.J. Kleinsmith, and J. Hardin, 2000. Copyright © 2000 by Pearson Education, Inc. Reprinted and electronically reproduced by permission of Pearson Education, Inc. Upper Saddle River, NJ; **15** © Pearson Education, Inc.

Glossary

actin (ak′-tin) A globular protein that links into chains, two of which twist helically about each other, forming microfilaments (actin filaments) in muscle and other kinds of cells.

basal body (bā′-sul) A eukaryotic cell structure consisting of a "9 + 0" arrangement of microtubule triplets. The basal body may organize the microtubule assembly of a cilium or flagellum and is structurally very similar to a centriole.

cell fractionation The disruption of a cell and separation of its parts by centrifugation at successively higher speeds.

cell wall A protective layer external to the plasma membrane in the cells of plants, prokaryotes, fungi, and some protists. Polysaccharides such as cellulose (in plants and some protists), chitin (in fungi), and peptidoglycan (in bacteria) are important structural components of cell walls.

central vacuole In a mature plant cell, a large membranous sac with diverse roles in growth, storage, and sequestration of toxic substances.

centriole (sen′-trē-ōl) A structure in the centrosome of an animal cell composed of a cylinder of microtubule triplets arranged in a 9 + 0 pattern. A centrosome has a pair of centrioles.

centrosome (sen′-trō-sōm) A structure present in the cytoplasm of animal cells that functions as a microtubule-organizing center and is important during cell division. A centrosome has two centrioles.

chloroplast (klōr′-ō-plast) An organelle found in plants and photosynthetic protists that absorbs sunlight and uses it to drive the synthesis of organic compounds from carbon dioxide and water.

chromatin (krō′-muh-tin) The complex of DNA and proteins that makes up eukaryotic chromosomes. When the cell is not dividing, chromatin exists in its dispersed form, as a mass of very long, thin fibers that are not visible with a light microscope.

chromosome (krō′-muh-sōm) A cellular structure consisting of one DNA molecule and associated protein molecules. (In some contexts, such as genome sequencing, the term may refer to the DNA alone.) A eukaryotic cell typically has multiple, linear chromosomes, which are located in the nucleus. A prokaryotic cell often has a single, circular chromosome, which is found in the nucleoid, a region that is not enclosed by a membrane. *See also* chromatin.

collagen A glycoprotein in the extracellular matrix of animal cells that forms strong fibers, found extensively in connective tissue and bone; the most abundant protein in the animal kingdom.

contractile vacuole A membranous sac that helps move excess water out of certain freshwater protists.

crista (plural, **cristae**) (kris′-tuh, kris′-tē) An infolding of the inner membrane of a mitochondrion. The inner membrane houses electron transport chains and molecules of the enzyme catalyzing the synthesis of ATP (ATP synthase).

cytoplasm (sī′-tō-plaz′-um) The contents of the cell enclosed by the plasma membrane; in eukaryotes, the portion exclusive of the nucleus.

cytoskeleton A network of microtubules, microfilaments, and intermediate filaments that extends throughout the cytoplasm and serves a variety of mechanical, transport, and signaling functions.

cytosol (sī′-tō-sol) The semifluid portion of the cytoplasm.

dynein (dī′-nē-un) In cilia and flagella, a large motor protein extending from one microtubule doublet to the adjacent doublet. ATP hydrolysis drives changes in dynein shape that lead to bending of cilia and flagella.

electron microscope (EM) A microscope that uses magnets to focus an electron beam on or through a specimen, resulting in a practical resolution 100-fold greater than that of a light microscope using standard techniques. A transmission electron microscope (TEM) is used to study the internal structure of thin sections of cells. A scanning electron microscope (SEM) is used to study the fine details of cell surfaces.

endomembrane system The collection of membranes inside and surrounding a eukaryotic cell, related either through direct physical contact or by the transfer of membranous vesicles; includes the plasma membrane, the nuclear envelope, the smooth and rough endoplasmic reticulum, the Golgi apparatus, lysosomes, vesicles, and vacuoles.

endoplasmic reticulum (ER) (en′-dō-plaz′-mik ruh-tik′-yū-lum) An extensive membranous network in eukaryotic cells, continuous with the outer nuclear membrane and composed of ribosome-studded (rough) and ribosome-free (smooth) regions.

extracellular matrix (ECM) The meshwork surrounding animal cells, consisting of glycoproteins, polysaccharides, and proteoglycans synthesized and secreted by the cells.

fibronectin An extracellular glycoprotein secreted by animal cells that helps them attach to the extracellular matrix.

filter feeder An animal that feeds by using a filtration mechanism to strain small organisms or food particles from its surroundings.

food vacuole A membranous sac formed by phagocytosis of microorganisms or particles to be used as food by the cell.

glycoprotein A protein with one or more covalently attached carbohydrates.

Golgi apparatus (gol′-jē) An organelle in eukaryotic cells consisting of stacks of flat membranous sacs that modify, store, and route products of the endoplasmic reticulum and synthesize some products, notably noncellulose carbohydrates.

granum (gran′-um) (plural, **grana**) A stack of membrane-bounded thylakoids in the chloroplast. Grana function in the light reactions of photosynthesis.

integrin In animal cells, a transmembrane receptor protein with two subunits that interconnects the extracellular matrix and the cytoskeleton.

intermediate filament A component of the cytoskeleton that includes filaments intermediate in size between microtubules and microfilaments.

light microscope (LM) An optical instrument with lenses that refract (bend) visible light to magnify images of specimens.

lysosome (lī′-suh-sōm) A membrane-enclosed sac of hydrolytic enzymes found in the cytoplasm of animal cells and some protists.

microfilament A cable composed of actin proteins in the cytoplasm of almost every eukaryotic cell, making up part of the cytoskeleton and acting alone or with myosin to cause cell contraction; also known as an actin filament.

microtubule A hollow rod composed of tubulin proteins that makes up part of the cytoskeleton in all eukaryotic cells and is found in cilia and flagella.

middle lamella (luh-mel′-uh) In plants, a thin layer of adhesive extracellular material, primarily pectins, found between the primary walls of adjacent young cells.

mitochondrial matrix The compartment of the mitochondrion enclosed by the inner membrane and containing enzymes and substrates for the citric acid cycle, as well as ribosomes and DNA.

motor protein A protein that interacts with cytoskeletal elements and other cell components, producing movement of the whole cell or parts of the cell.

myosin (mī′-uh-sin) A type of motor protein that associates into filaments that interact with actin filaments, causing cell contraction.

nuclear envelope In a eukaryotic cell, the double membrane that surrounds the nucleus, perforated with pores that regulate traffic with the cytoplasm. The outer membrane is continuous with the endoplasmic reticulum.

nuclear lamina A netlike array of protein filaments that lines the inner surface of the nuclear envelope and helps maintain the shape of the nucleus.

nucleolus (nū-klē′-ō-lus) (plural, **nucleoli**) A specialized structure in the nucleus consisting of chromosomal regions containing ribosomal RNA (rRNA) genes along with ribosomal proteins imported from the cytoplasm; site of rRNA synthesis and ribosomal subunit assembly. *See also* ribosome.

nucleus (1) An atom's central core, containing protons and neutrons. (2) The organelle of a eukaryotic cell that contains the genetic material in the form of chromosomes, made up of chromatin. (3) A cluster of neurons.

organelle (ōr-guh-nel′) Any of several kinds of membrane-enclosed structures with specialized functions, suspended in the cytosol of eukaryotic cells.

peroxisome (puh-rok′-suh-sōm′) An organelle containing enzymes that transfer hydrogen atoms from various substrates to oxygen (O_2), producing and then degrading hydrogen peroxide (H_2O_2).

plasma membrane The membrane at the boundary of every cell that acts as a selective barrier, regulating the cell's chemical composition.

plastid One of a family of closely related organelles that includes chloroplasts, chromoplasts, and amyloplasts. Plastids are found in the cells of photosynthetic eukaryotes.

primary cell wall In plants, a relatively thin and flexible layer that surrounds the plasma membrane of a young cell.

prokaryotic cell (prō′-kār′-ē-ot′-ik) A type of cell lacking a membrane-enclosed nucleus and membrane-enclosed organelles. Organisms with prokaryotic cells (bacteria and archaea) are called prokaryotes.

proteoglycan (prō′-tē-ō-glī′-kan) A large molecule consisting of a small core protein with many carbohydrate chains attached, found in the extracellular matrix of animal cells. A proteoglycan may consist of up to 95% carbohydrate.

ribosome (rī′-buh-sōm′) A complex of rRNA and protein molecules that functions as a site of protein synthesis in the cytoplasm; consists of a large subunit and a small subunit. In eukaryotic cells, each subunit is assembled in the nucleolus. *See also* nucleolus.

rough ER That portion of the endoplasmic reticulum with ribosomes attached.

scanning electron microscope (SEM) A microscope that uses an electron beam to scan the surface of a sample, coated with metal atoms, to study details of its topography.

secondary cell wall In plant cells, a strong and durable matrix that is often deposited in several laminated layers around the plasma membrane and that provides protection and support.

smooth ER That portion of the endoplasmic reticulum that is free of ribosomes.

thylakoid (thī′-luh-koyd) A flattened, membranous sac inside a chloroplast. Thylakoids often exist in stacks called grana that are interconnected; their membranes contain molecular "machinery" used to convert light energy to chemical energy.

transmission electron microscope (TEM) A microscope that passes an electron beam through very thin sections stained with metal atoms and is primarily used to study the internal ultrastructure of cells.

transport vesicle A small membranous sac in a eukaryotic cell's cytoplasm carrying molecules produced by the cell.

vacuole (vak′-yū-ōl′) A membrane-bounded vesicle whose specialized function varies in different kinds of cells.

vesicle (ves′-i-kul) A membranous sac in the cytoplasm of a eukaryotic cell.

Membrane Transport and Cell Signaling

From Chapter 5 of *Campbell Biology in Focus*, Second Edition. Lisa A. Urry, Michael L. Cain, Steven A. Wasserman, Peter V. Minorsky, and Jane B. Reece.

Membrane Transport and Cell Signaling

KEY CONCEPTS

1 Cellular membranes are fluid mosaics of lipids and proteins
2 Membrane structure results in selective permeability
3 Passive transport is diffusion of a substance across a membrane with no energy investment
4 Active transport uses energy to move solutes against their gradients
5 Bulk transport across the plasma membrane occurs by exocytosis and endocytosis
6 The plasma membrane plays a key role in most cell signaling

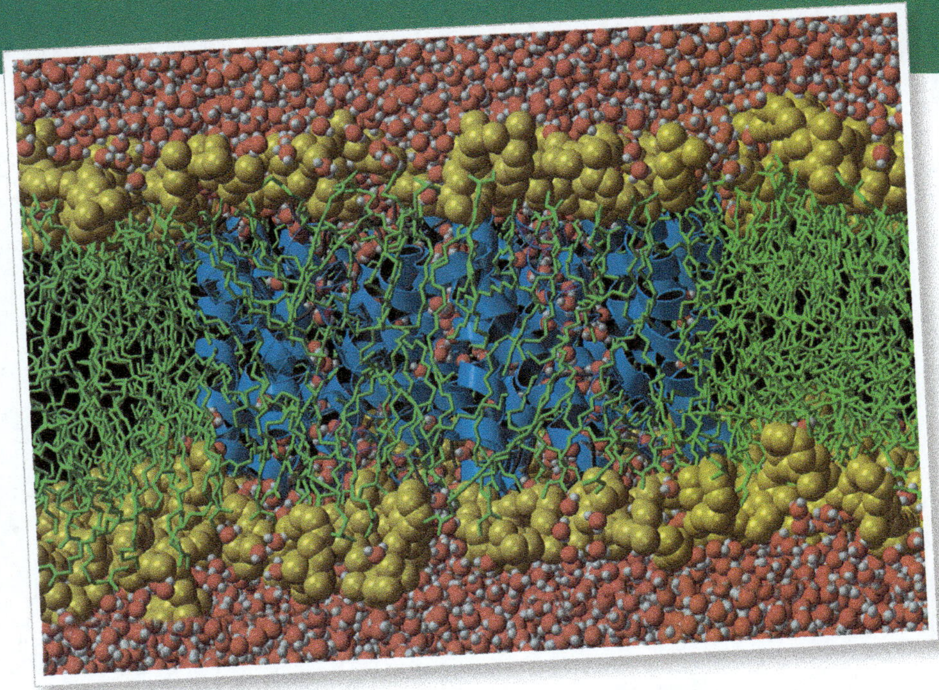

▲ **Figure 1** How do cell membrane proteins help regulate chemical traffic?

Life at the Edge

The plasma membrane is the edge of life, the boundary that separates the living cell from its surroundings. A remarkable film only about 8 nm thick—it would take over 8,000 plasma membranes to equal the thickness of a piece of paper—the plasma membrane controls traffic into and out of the cell it surrounds. Like all biological membranes, the plasma membrane exhibits **selective permeability**; that is, it allows some substances to cross it more easily than others. The resulting ability of the cell to discriminate in its chemical exchanges with its environment is fundamental to life.

Most of this chapter is devoted to how cellular membranes control the passage of substances through them. **Figure 1** shows a computer model of water molecules (red and gray) passing through a short section of a membrane, a phospholipid bilayer (phosphates are yellow, and hydrocarbon tails are green). The blue ribbons within the lipid bilayer represent helical regions of a membrane protein called an aquaporin. One molecule of this protein enables billions of water molecules to pass through the membrane every second, many more than could cross on their own. Found in many kinds of cells, aquaporins are but one example of how the plasma membrane and its proteins enable cells to survive and function.

To understand how membranes work, we'll begin by examining their molecular structure. Then we'll describe in some detail how plasma membranes control transport into and out of cells. Finally, we'll discuss cell signaling, emphasizing the role of the plasma membrane in cell communication.

CONCEPT 1

Cellular membranes are fluid mosaics of lipids and proteins

Figure 2 shows the currently accepted model of the arrangement of molecules in the plasma membrane. Lipids and proteins are the staple ingredients of membranes, although carbohydrates are also important. The most abundant lipids in most membranes are phospholipids. The ability of phospholipids to form membranes is inherent in their molecular structure. A phospholipid is an **amphipathic** molecule, meaning it has both a hydrophilic region and a hydrophobic region. A phospholipid bilayer can exist as a stable boundary between two aqueous compartments because the molecular arrangement shelters the hydrophobic tails of the phospholipids from water while exposing the hydrophilic heads to water **(Figure 3)**.

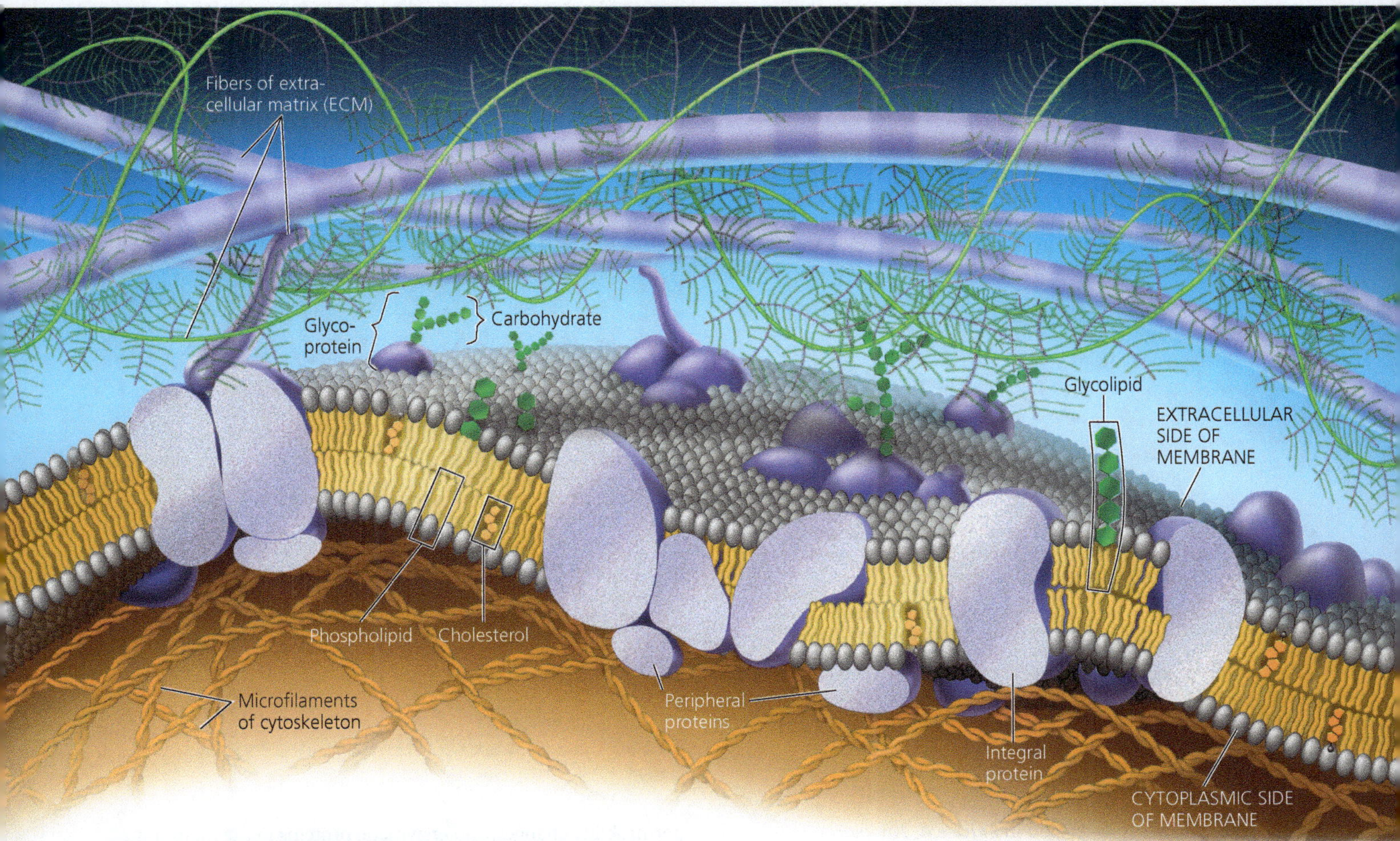

▲ **Figure 2 Current model of an animal cell's plasma membrane (cutaway view).** Lipids are colored gray and gold, proteins purple, and carbohydrates green.

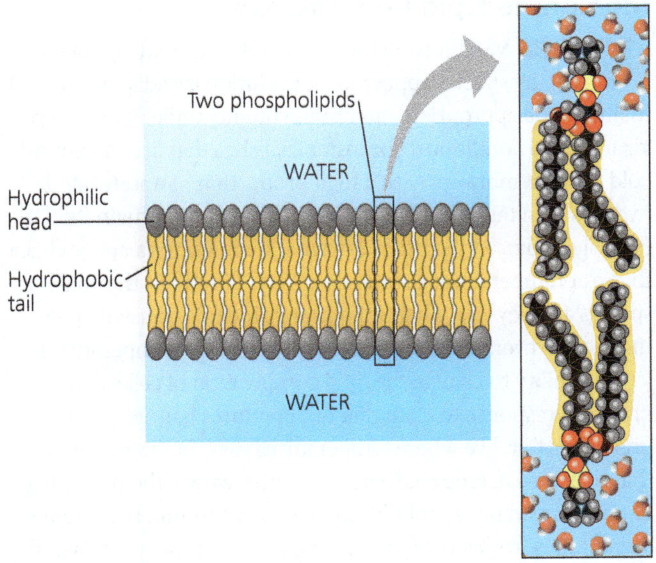

▲ **Figure 3 Phospholipid bilayer (cross section).**

MAKE CONNECTIONS *Circle the hydrophilic and hydrophobic portions of the upper phospholipid on the right side of Figure 3. Explain what each portion contacts when the phospholipid is in the plasma membrane.*

Like membrane lipids, most membrane proteins are amphipathic. Such proteins can reside in the phospholipid bilayer with their hydrophilic regions protruding. This molecular orientation maximizes contact of the hydrophilic regions of a protein with water in the cytosol and extracellular fluid, while providing its hydrophobic parts with a nonaqueous environment.

In the **fluid mosaic model** in Figure 2, the membrane is a mosaic of protein molecules bobbing in a fluid bilayer of phospholipids. The proteins are not randomly distributed in the membrane, however. Groups of proteins are often associated in long-lasting, specialized patches, as are certain lipids. In some regions, the membrane may be much more packed with proteins than shown in Figure 2. Like all models, the fluid mosaic model is continually being refined as new research reveals more about membrane structure.

The Fluidity of Membranes

Membranes are not static sheets of molecules locked rigidly in place. A membrane is held together primarily by hydrophobic interactions, which are much weaker than covalent bonds. Most of the lipids and some of the proteins can shift about laterally—that is, in the plane of the membrane—like partygoers elbowing their way through a crowded room.

The lateral movement of phospholipids within the membrane is rapid. Proteins are much larger than lipids and move more slowly, but some membrane proteins do drift, as shown

▼ Figure 4 Inquiry

Do membrane proteins move?

Experiment Larry Frye and Michael Edidin, at Johns Hopkins University, labeled the plasma membrane proteins of a mouse cell and a human cell with two different markers and fused the cells. Using a microscope, they observed the markers on the hybrid cell.

Results

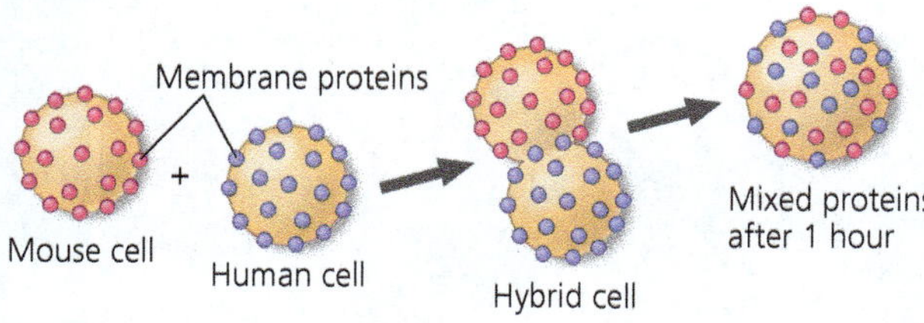

Conclusion The mixing of the mouse and human membrane proteins indicates that at least some membrane proteins move sideways within the plane of the plasma membrane.

Data from L. D. Frye and M. Edidin, The rapid intermixing of cell surface antigens after formation of mouse-human heterokaryons, *Journal of Cell Science* 7:319 (1970).

WHAT IF? Suppose the proteins did not mix in the hybrid cell, even many hours after fusion. Would you be able to conclude that proteins don't move within the membrane? What other explanation could there be?

▼ Figure 5 Factors that affect membrane fluidity.

(a) Unsaturated versus saturated hydrocarbon tails.

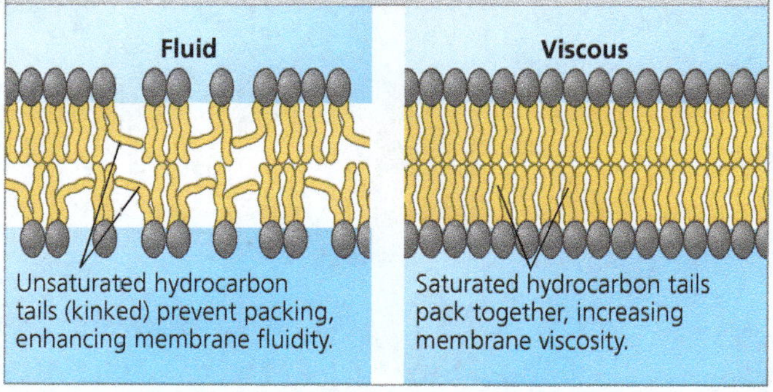

(b) Cholesterol within the animal cell membrane.

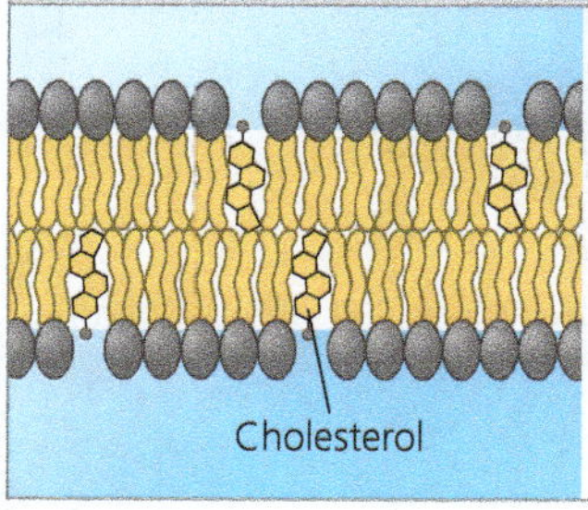

Cholesterol reduces membrane fluidity at moderate temperatures by reducing phospholipid movement, but at low temperatures it hinders solidification by disrupting the regular packing of phospholipids.

in a classic experiment described in **Figure 4**. And some membrane proteins seem to move in a highly directed manner, perhaps driven along cytoskeletal fibers by motor proteins. However, many other membrane proteins seem to be held immobile by their attachment to the cytoskeleton or to the extracellular matrix (see Figure 2).

A membrane remains fluid as temperature decreases until the phospholipids settle into a closely packed arrangement and the membrane solidifies, much as bacon grease forms lard when it cools. The temperature at which a membrane solidifies depends on the types of lipids it is made of. The membrane remains fluid to a lower temperature if it is rich in phospholipids with unsaturated hydrocarbon tails. Because of kinks in the tails where double bonds are located, unsaturated hydrocarbon tails cannot pack together as closely as saturated hydrocarbon tails, and this looseness makes the membrane more fluid **(Figure 5a)**.

The steroid cholesterol, which is wedged between phospholipid molecules in the plasma membranes of animal cells, has different effects on membrane fluidity at different temperatures **(Figure 5b)**. At relatively high temperatures—at 37°C, the body temperature of humans, for example—cholesterol makes the membrane less fluid by restraining phospholipid movement. However, because cholesterol also hinders the close packing of phospholipids, it lowers the temperature required for the membrane to solidify. Thus, cholesterol helps membranes resist changes in fluidity when the temperature changes.

Membranes must be fluid to work properly; they are usually about as fluid as salad oil. When a membrane solidifies, its permeability changes, and enzymatic proteins in the membrane may become inactive. However, membranes that are too fluid cannot support protein function either. Therefore, extreme environments pose a challenge for life, resulting in evolutionary adaptations that include differences in membrane lipid composition.

Evolution of Differences in Membrane Lipid Composition

EVOLUTION Variations in the cell membrane lipid compositions of many species appear to be evolutionary adaptations that maintain the appropriate membrane fluidity under specific environmental conditions. For instance, fishes that live in extreme cold have membranes with a high proportion of unsaturated hydrocarbon tails, enabling their membranes to remain fluid (see Figure 5a). At the other extreme, some bacteria and archaea thrive at temperatures greater than 90°C (194°F) in thermal hot springs and geysers. Their membranes include unusual lipids that help prevent excessive fluidity at such high temperatures.

The ability to change the lipid composition of cell membranes in response to changing temperatures has evolved in organisms that live where temperatures vary. In many plants that tolerate extreme cold, such as winter wheat, the percentage of unsaturated phospholipids increases in autumn, keeping the membranes from solidifying during winter. Some bacteria and archaea can also change the proportion of unsaturated phospholipids in their cell membranes, depending on the temperature at which they are growing. Overall, natural selection has apparently favored organisms whose mix of membrane lipids ensures an appropriate level of membrane fluidity for their environment.

Membrane Proteins and Their Functions

Now we return to the *mosaic* aspect of the fluid mosaic model. Somewhat like a tile mosaic, a membrane is a collage of different proteins embedded in the fluid matrix of the lipid bilayer (see Figure 2). More than 50 kinds of proteins have been found so far in the plasma membrane of red blood cells, for example. Phospholipids form the main fabric of the membrane, but proteins determine most of the membrane's functions. Different types of cells contain different sets of membrane proteins, and the various membranes within a cell each have a unique collection of proteins.

Notice in Figure 2 that there are two major populations of membrane proteins: integral proteins and peripheral proteins. **Integral proteins** penetrate the hydrophobic interior of the lipid bilayer. The majority are *transmembrane proteins*, which span the membrane; other integral proteins extend only partway into the hydrophobic interior. The hydrophobic regions of an integral protein consist of one or more stretches of nonpolar amino acids, usually coiled into α helices **(Figure 6)**. The hydrophilic parts of the molecule are exposed to the aqueous solutions on either side of the membrane. Some proteins also have one or more hydrophilic channels that allow passage of hydrophilic substances (even water itself; see Figure 1). **Peripheral proteins** are not embedded in the lipid bilayer at all; they are loosely bound to the surface of the membrane, often to exposed parts of integral proteins (see Figure 2).

On the cytoplasmic side of the plasma membrane, some membrane proteins are held in place by attachment to the cytoskeleton. And on the extracellular side, certain membrane proteins are attached to fibers of the extracellular matrix. These attachments combine to give animal cells a stronger framework than the plasma membrane alone could provide.

Figure 7 gives an overview of six major functions performed by proteins of the plasma membrane. A single cell may have membrane proteins carrying out several of these functions, and a single membrane protein may have multiple

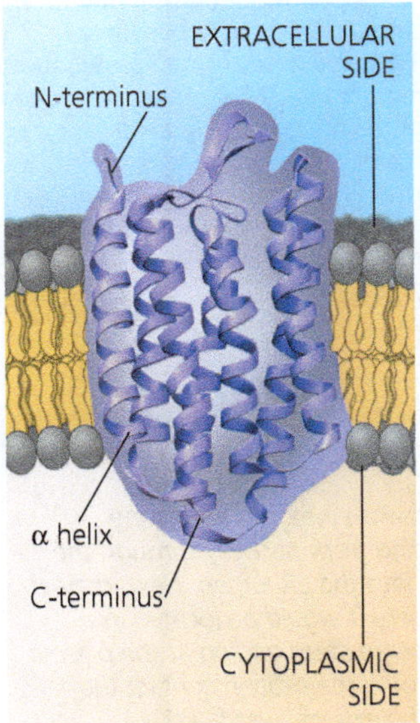

◀ **Figure 6 The structure of a transmembrane protein.** Bacteriorhodopsin (a bacterial transport protein) has a distinct orientation in the membrane, with its N-terminus outside the cell and its C-terminus inside. This ribbon model highlights the α-helical secondary structure of the hydrophobic parts, which lie mostly within the hydrophobic interior of the membrane. The protein includes seven transmembrane helices. The nonhelical hydrophilic segments are in contact with the aqueous solutions on the extracellular and cytoplasmic sides of the membrane.

(a) Transport. *Left:* A protein that spans the membrane may provide a hydrophilic channel across the membrane that is selective for a particular solute. *Right:* Other transport proteins shuttle a substance from one side to the other by changing shape. Some of these proteins hydrolyze ATP as an energy source to actively pump substances across the membrane.

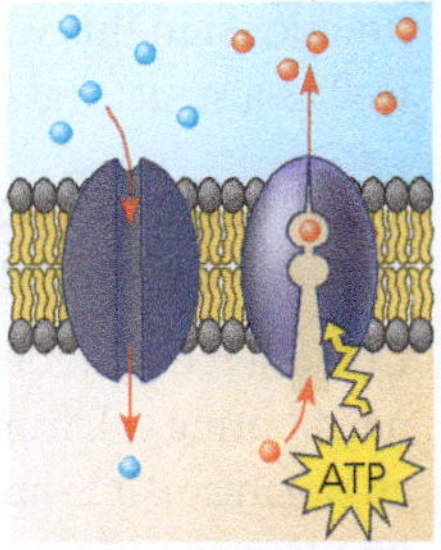

(b) Enzymatic activity. A protein built into the membrane may be an enzyme with its active site (where the target molecule binds) exposed to substances in the adjacent solution. In some cases, several enzymes in a membrane are organized as a team that carries out sequential steps of a metabolic pathway.

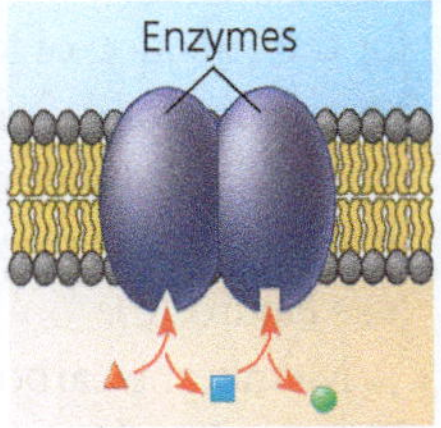

(c) Signal transduction. A membrane protein (receptor) may have a binding site with a specific shape that fits the shape of a chemical messenger, such as a hormone. The external messenger (signaling molecule) may cause the protein to change shape, allowing it to relay the message to the inside of the cell, usually by binding to a cytoplasmic protein.

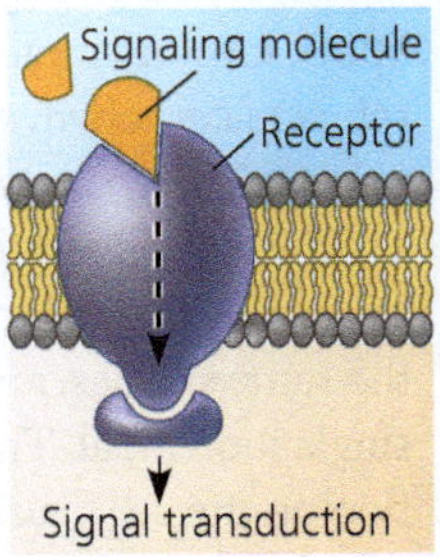

(d) Cell-cell recognition. Some glycoproteins serve as identification tags that are specifically recognized by membrane proteins of other cells. This type of cell-cell binding is usually short-lived compared to that shown in (e).

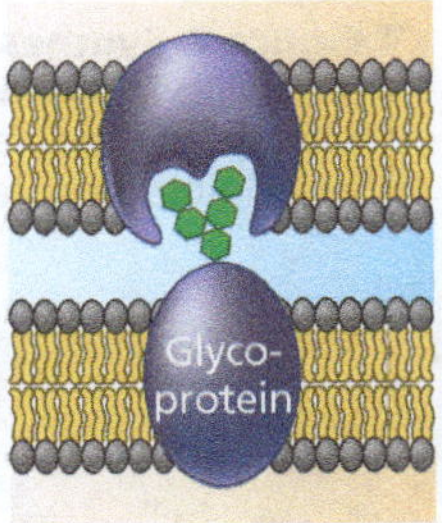

(e) Intercellular joining. Membrane proteins of adjacent cells may hook together in various kinds of junctions, such as gap junctions or tight junctions. This type of binding is more long-lasting than that shown in (d).

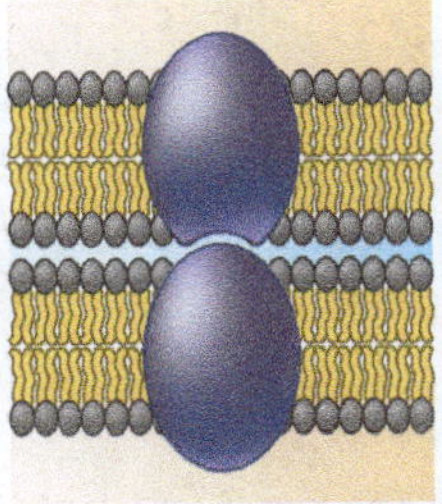

(f) Attachment to the cytoskeleton and extracellular matrix (ECM). Microfilaments or other elements of the cytoskeleton may be noncovalently bound to membrane proteins, a function that helps maintain cell shape and stabilizes the location of certain membrane proteins. Proteins that can bind to ECM molecules can coordinate extracellular and intracellular changes.

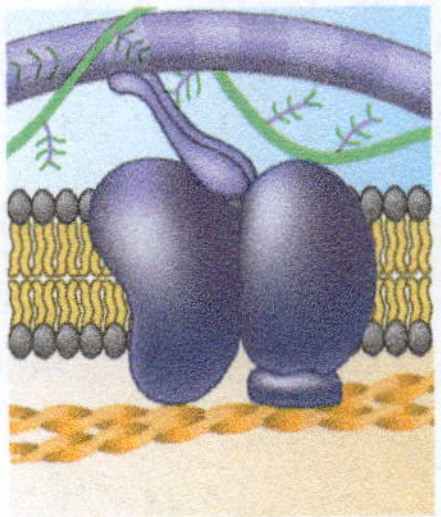

▲ **Figure 7 Some functions of membrane proteins.** In many cases, a single protein performs multiple tasks.

? *Some transmembrane proteins can bind to a particular ECM molecule and, when bound, transmit a signal into the cell. Use the proteins shown in (c) and (f) to explain how this might occur.*

functions. In this way, the membrane is a functional mosaic as well as a structural one.

The Role of Membrane Carbohydrates in Cell-Cell Recognition

Cell-cell recognition, a cell's ability to distinguish one type of neighboring cell from another, is crucial to the functioning of an organism. It is important, for example, in the sorting of cells into tissues and organs in an animal embryo. It is also the basis for the rejection of foreign cells by the immune system, an important line of defense in vertebrate animals. Cells recognize other cells by binding to molecules, often containing carbohydrates, on the extracellular surface of the plasma membrane (see Figure 7d).

Membrane carbohydrates are usually short, branched chains of fewer than 15 sugar units. Some are covalently bonded to lipids, forming molecules called **glycolipids**. (Recall that *glyco* refers to carbohydrate.) However, most are covalently bonded to proteins, which are thereby **glycoproteins**.

The carbohydrates on the extracellular side of the plasma membrane vary from species to species, among individuals of the same species, and even from one cell type to another in a single individual. The diversity of the molecules and their location on the cell's surface enable membrane carbohydrates to function as markers that distinguish one cell from another. For example, the four human blood types designated A, B, AB, and O reflect variation in the carbohydrate part of glycoproteins on the surface of red blood cells.

Synthesis and Sidedness of Membranes

Membranes have distinct inside and outside faces. The two lipid layers may differ in lipid composition, and each protein has directional orientation in the membrane (see Figure 6, for example). **Figure 8** shows how membrane sidedness arises: The asymmetric arrangement of proteins, lipids, and their associated carbohydrates in the plasma membrane is determined as the membrane is being built by the endoplasmic reticulum (ER) and Golgi apparatus.

CONCEPT CHECK 1

1. Plasma membrane proteins have carbohydrates attached to them in the ER and Golgi apparatus and then are transported in vesicles to the cell surface. On which side of the vesicle membrane are the carbohydrates?
2. **WHAT IF?** How would the membrane lipid composition of a native grass found in very warm soil around hot springs compare with that of a native grass found in cooler soil? Explain.

▼ **Figure 8 Synthesis of membrane components and their orientation in the membrane.** The cytoplasmic (orange) face of the plasma membrane differs from the extracellular (aqua) face. The latter arises from the inside face of ER, Golgi, and vesicle membranes.

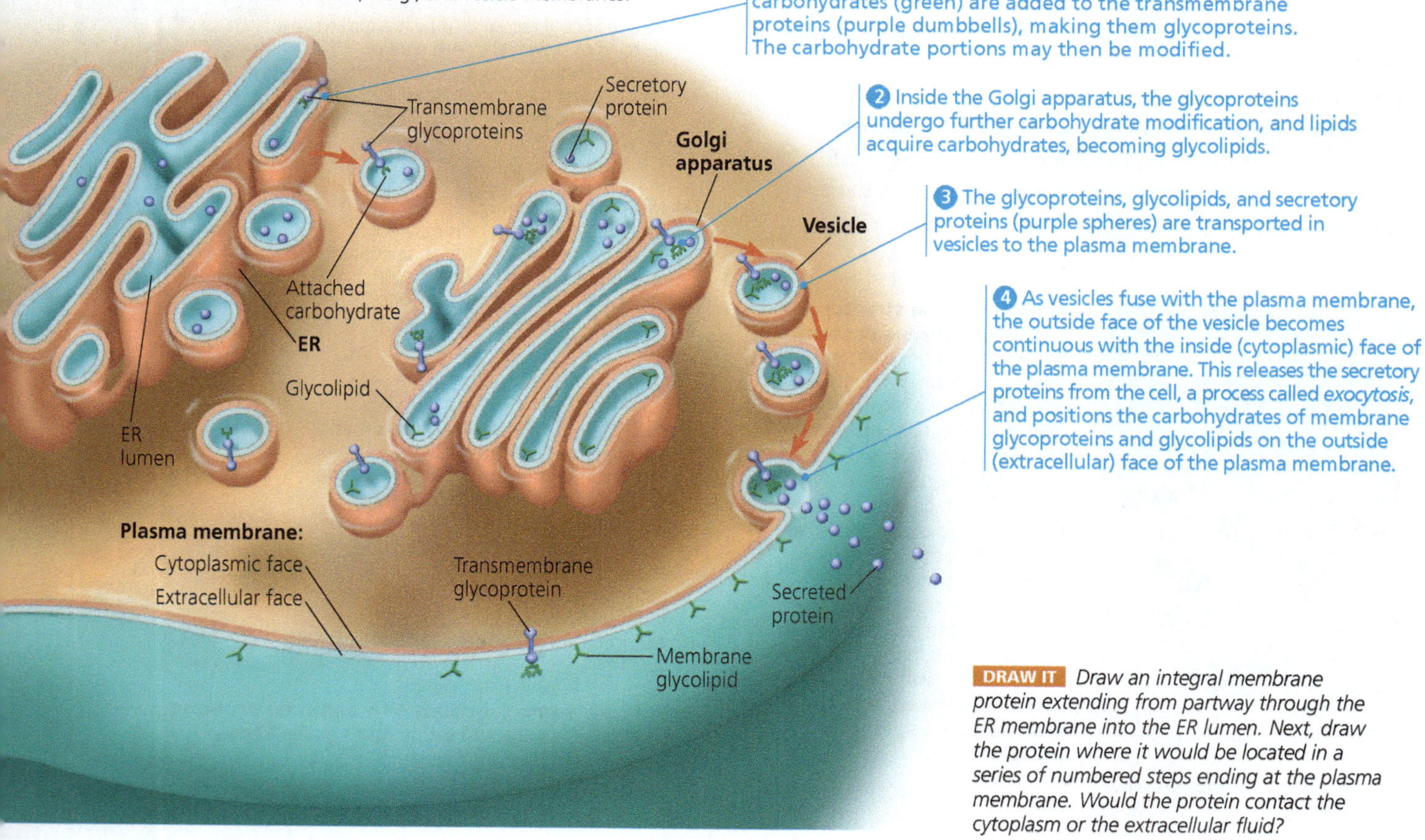

DRAW IT *Draw an integral membrane protein extending from partway through the ER membrane into the ER lumen. Next, draw the protein where it would be located in a series of numbered steps ending at the plasma membrane. Would the protein contact the cytoplasm or the extracellular fluid?*

CONCEPT 2

Membrane structure results in selective permeability

The biological membrane is an exquisite example of a supramolecular structure—many molecules ordered into a higher level of organization—with emergent properties beyond those of the individual molecules. We now focus on one of the most important of those properties: the ability to regulate transport across cellular boundaries, a function essential to the cell's existence. We will see once again that form fits function: The fluid mosaic model helps explain how membranes regulate the cell's molecular traffic.

A steady traffic of small molecules and ions moves across the plasma membrane in both directions. Consider the chemical exchanges between a muscle cell and the extracellular fluid that bathes it. Sugars, amino acids, and other nutrients enter the cell, and metabolic waste products leave it. The cell takes in O_2 for use in cellular respiration and expels CO_2. Also, the cell regulates its concentrations of inorganic ions, such as Na^+, K^+, Ca^{2+}, and Cl^-, by shuttling them one way or the other across the plasma membrane. In spite of heavy traffic through them, cell membranes are selectively permeable, and substances do not cross the barrier indiscriminately. The cell is able to take up some small molecules and ions and exclude others.

The Permeability of the Lipid Bilayer

Nonpolar molecules, such as hydrocarbons, CO_2, and O_2, are hydrophobic. They can therefore dissolve in the lipid bilayer of the membrane and cross it easily, without the aid of membrane proteins. However, the hydrophobic interior of the membrane impedes the direct passage through the membrane of ions and polar molecules, which are hydrophilic. Polar molecules such as glucose and other sugars pass only slowly through a lipid bilayer, and even water, a very small polar molecule, does not cross rapidly. A charged atom or molecule and its surrounding shell of water are even less likely to penetrate the hydrophobic interior of the membrane. Furthermore, the lipid bilayer is only one aspect of the gatekeeper system responsible for a cell's selective permeability. Proteins built into the membrane play key roles in regulating transport.

Transport Proteins

Specific ions and a variety of polar molecules can't move through cell membranes on their own. However, these hydrophilic substances can avoid contact with the lipid bilayer by passing through **transport proteins** that span the membrane.

Some transport proteins, called *channel proteins*, function by having a hydrophilic channel that certain molecules or atomic ions use as a tunnel through the membrane (see Figure 7a, left). For example, as you read earlier, the passage of water molecules through the plasma membrane of certain cells is greatly facilitated by channel proteins called **aquaporins** (see Figure 1). Most aquaporin proteins consist of four identical subunits. The polypeptide making up each subunit forms a channel that allows single-file passage of up to *3 billion* (3×10^9) water molecules per second, many more than would cross the membrane without aquaporin. Other transport proteins, called *carrier proteins*, hold onto their passengers and change shape in a way that shuttles them across the membrane
(see Figure 7a, right).

A transport protein is specific for the substance it translocates (moves), allowing only a certain substance (or a small group of related substances) to cross the membrane. For example, a specific carrier protein in the plasma membrane of red blood cells transports glucose across the membrane 50,000 times faster than glucose can pass through on its own. This "glucose transporter" is so selective that it even rejects fructose, a structural isomer of glucose.

Thus, the selective permeability of a membrane depends on both the discriminating barrier of the lipid bilayer and the specific transport proteins built into the membrane. But what establishes the *direction* of traffic across a membrane? At a given time, what determines whether a particular substance will enter the cell or leave the cell? And what mechanisms actually drive molecules across membranes? We will address these questions next as we explore two modes of membrane traffic: passive transport and active transport.

CONCEPT CHECK 2

1. What property allows O_2 and CO_2 to cross a lipid bilayer without the help of membrane proteins?
2. Why is a transport protein needed to move many water molecules rapidly across a membrane?
3. MAKE CONNECTIONS Aquaporins exclude passage of hydronium ions (H_3O^+), but some aquaporins allow passage of glycerol, a three-carbon alcohol, as well as H_2O. Since H_3O^+ is closer in size to water than glycerol is, yet cannot pass through, what might be the basis of this selectivity?

CONCEPT 3

Passive transport is diffusion of a substance across a membrane with no energy investment

Molecules have a type of energy called thermal energy, due to their constant motion. One result of this motion is **diffusion**, the movement of particles of any substance so that they tend to spread out into the available space. Each molecule moves randomly, yet diffusion of a *population* of molecules may be directional. To understand this process, let's imagine a synthetic membrane separating pure water from

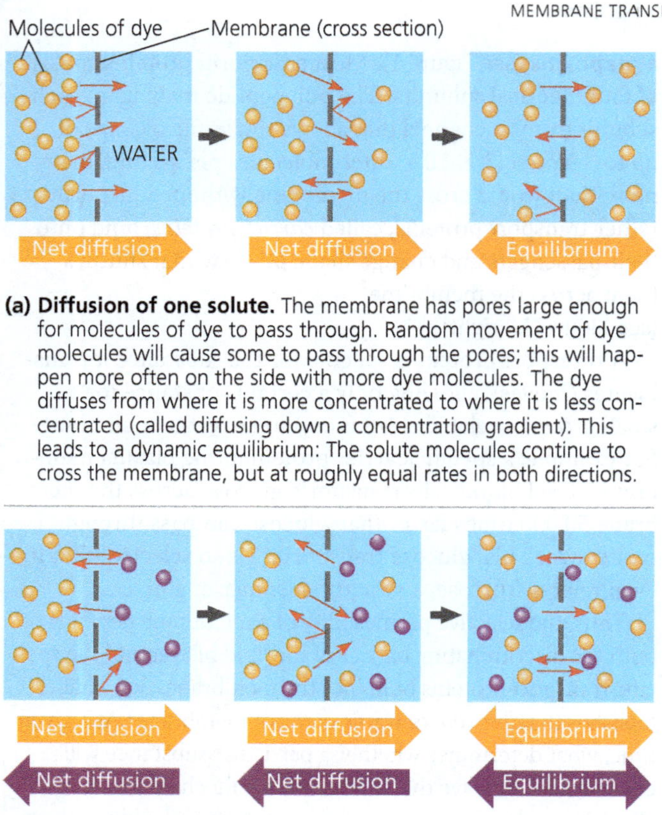

(a) Diffusion of one solute. The membrane has pores large enough for molecules of dye to pass through. Random movement of dye molecules will cause some to pass through the pores; this will happen more often on the side with more dye molecules. The dye diffuses from where it is more concentrated to where it is less concentrated (called diffusing down a concentration gradient). This leads to a dynamic equilibrium: The solute molecules continue to cross the membrane, but at roughly equal rates in both directions.

(b) Diffusion of two solutes. Solutions of two different dyes are separated by a membrane that is permeable to both. Each dye diffuses down its own concentration gradient. There will be a net diffusion of the purple dye toward the left, even though the *total* solute concentration was initially greater on the left side.

▲ Figure 9 The diffusion of solutes across a synthetic membrane. Each of the large arrows under the diagrams shows the net diffusion of the dye molecules of that color.

a solution of a dye in water. Study **Figure 9a** to appreciate how diffusion would result in both solutions having equal concentrations of the dye molecules. Once that point is reached, there will be a dynamic equilibrium, with roughly as many dye molecules crossing the membrane each second in one direction as in the other.

We can now state a simple rule of diffusion: In the absence of other forces, a substance will diffuse from where it is more concentrated to where it is less concentrated. Put another way, any substance will diffuse down its **concentration gradient**, the region along which the density of a substance increases or decreases (in this case, decreases). No work must be done to make this happen; diffusion is a spontaneous process, needing no input of energy. Note that each substance diffuses down its *own* concentration gradient, unaffected by the concentration gradients of other substances **(Figure 9b)**.

Much of the traffic across cell membranes occurs by diffusion. When a substance is more concentrated on one side of a membrane than on the other, there is a tendency for the substance to diffuse across the membrane down its concentration gradient (assuming that the membrane is permeable to that substance). One important example is the uptake of oxygen by a cell performing cellular respiration. Dissolved oxygen diffuses into the cell across the plasma membrane. As long as cellular respiration consumes the O_2 as it enters, diffusion into the cell will continue because the concentration gradient favors movement in that direction.

The diffusion of a substance across a biological membrane is called **passive transport** because the cell does not have to expend energy to make it happen. The concentration gradient itself represents potential energy and drives diffusion. Remember, however, that membranes are selectively permeable and therefore have different effects on the rates of diffusion of various molecules. In the case of water, aquaporins allow water to diffuse very rapidly across the membranes of certain cells. As we'll see next, the movement of water across the plasma membrane has important consequences for cells.

Effects of Osmosis on Water Balance

To see how two solutions with different solute concentrations interact, picture a U-shaped glass tube with a selectively permeable artificial membrane separating two sugar solutions **(Figure 10)**. Pores in this synthetic membrane are too small

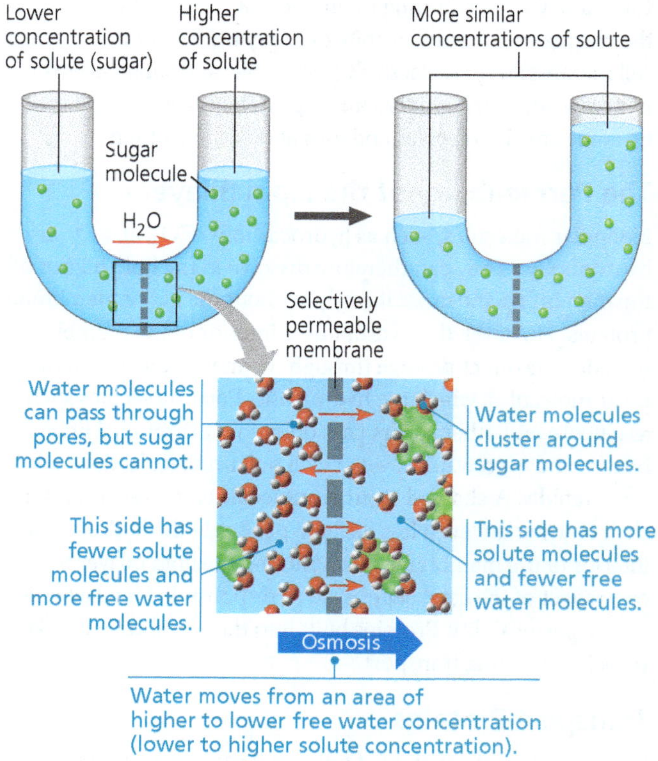

▲ Figure 10 Osmosis. Two sugar solutions of different concentrations are separated by a membrane that the solvent (water) can pass through but the solute (sugar) cannot. Water molecules move randomly and may cross in either direction, but overall, water diffuses from the solution with less concentrated solute to that with more concentrated solute. This passive transport of water, or osmosis, makes the sugar concentrations on both sides roughly equal.

WHAT IF? *If an orange dye capable of passing through the membrane was added to the left side of the tube above, how would it be distributed at the end of the experiment? (See Figure 9.) Would the final solution levels in the tube be affected?*

for sugar molecules to pass through but large enough for water molecules. However, tight clustering of water molecules around the hydrophilic solute molecules makes some of the water unavailable to cross the membrane. As a result, the solution with a higher solute concentration has a lower *free* water concentration. Water diffuses across the membrane from the region of higher free water concentration (lower solute concentration) to that of lower free water concentration (higher solute concentration) until the solute concentrations on both sides of the membrane are more nearly equal. The diffusion of free water across a selectively permeable membrane, whether artificial or cellular, is called **osmosis**. The movement of water across cell membranes and the balance of water between the cell and its environment are crucial to organisms. Let's now apply to living cells what we've learned about osmosis in this system to living cells.

(a) Animal cell. An animal cell fares best in an isotonic environment unless it has special adaptations that offset the osmotic uptake or loss of water.

(b) Plant cell. Plant cells are turgid (firm) and generally healthiest in a hypotonic environment, where the uptake of water is eventually balanced by the wall pushing back on the cell.

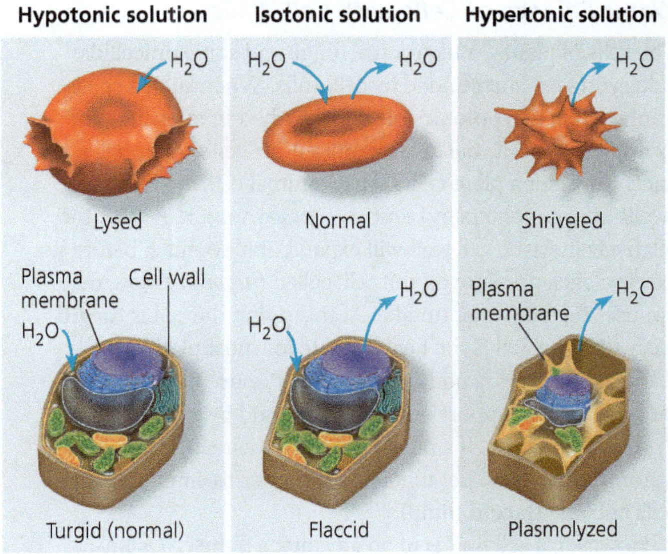

▲ **Figure 11 The water balance of living cells.** How living cells react to changes in the solute concentration of their environment depends on whether or not they have cell walls. **(a)** Animal cells, such as this red blood cell, do not have cell walls. **(b)** Plant cells do. (Arrows indicate net water movement after the cells were first placed in these solutions.)

Water Balance of Cells Without Cell Walls

To explain the behavior of a cell in a solution, we must consider both solute concentration and membrane permeability. Both factors are taken into account in the concept of **tonicity**, the ability of a surrounding solution to cause a cell to gain or lose water. The tonicity of a solution depends in part on its concentration of solutes that cannot cross the membrane (nonpenetrating solutes) relative to that inside the cell. If there is a higher concentration of nonpenetrating solutes in the surrounding solution, water will tend to leave the cell, and vice versa.

If a cell without a cell wall, such as an animal cell, is immersed in an environment that is **isotonic** to the cell (*iso* means "same"), there will be no *net* movement of water across the plasma membrane. Water diffuses across the membrane, but at the same rate in both directions. In an isotonic environment, the volume of an animal cell is stable, as shown in the middle of **Figure 11a**.

Let's transfer the cell to a solution that is **hypertonic** to the cell (*hyper* means "more," in this case referring to nonpenetrating solutes). The cell will lose water, shrivel, and probably die (see Figure 11a, right). This is why an increase in the salinity (saltiness) of a lake can kill the animals there; if the lake water becomes hypertonic to the animals' cells, they might shrivel and die. However, taking up too much water can be just as hazardous as losing water. If we place the cell in a solution that is **hypotonic** to the cell (*hypo* means "less"), water will enter the cell faster than it leaves, and the cell will swell and lyse (burst) like an overfilled water balloon (see Figure 11a, left).

A cell without rigid cell walls can tolerate neither excessive uptake nor excessive loss of water. This problem of water balance is automatically solved if such a cell lives in isotonic surroundings. Seawater is isotonic to many marine invertebrates. The cells of most terrestrial (land-dwelling) animals are bathed in an extracellular fluid that is isotonic to the cells. In hypertonic or hypotonic environments, however, organisms that lack rigid cell walls must have other adaptations for **osmoregulation**, the control of solute concentrations and water balance. For example, the unicellular protist *Paramecium caudatum* lives in pond water, which is hypotonic to the cell. Water continually enters the cell. The *P. caudatum* cell doesn't burst because it is equipped with a contractile vacuole, an organelle that functions as a bilge pump to force water out of the cell as fast as it enters by osmosis **(Figure 12)**.

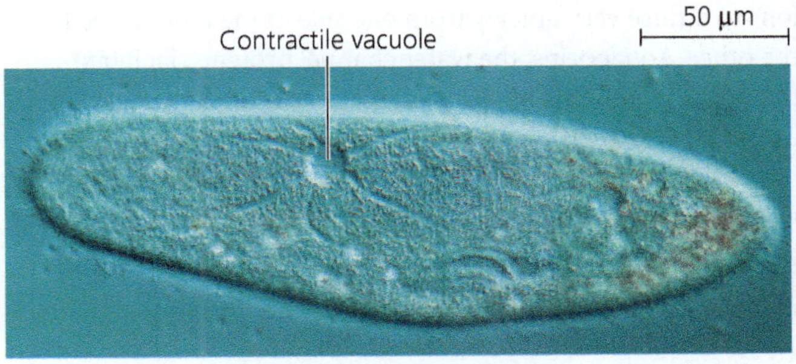

▲ **Figure 12 The contractile vacuole of *Paramecium caudatum*.** The vacuole collects fluid from a system of canals in the cytoplasm. When full, the vacuole and canals contract, expelling fluid from the cell (LM).

Water Balance of Cells with Cell Walls

The cells of plants, prokaryotes, fungi, and some unicellular eukaryotes are surrounded by cell walls. When such a cell is immersed in a hypotonic solution—bathed in rainwater, for example—the cell wall helps maintain the cell's water balance. Consider a plant cell. Like an animal cell, the plant cell swells as water enters by osmosis **(Figure 11b)**. However, the relatively inelastic cell wall will expand only so much before it exerts a back pressure on the cell, called *turgor pressure*, that opposes further water uptake. At this point, the cell is **turgid** (very firm), which is the healthy state for most plant cells. Plants that are not woody, such as most houseplants, depend for mechanical support on cells kept turgid by a surrounding hypotonic solution. If a plant's cells and their surroundings are isotonic, there is no net tendency for water to enter, and the cells become **flaccid** (limp).

However, a cell wall is of no advantage if the cell is immersed in a hypertonic environment. In this case, a plant cell, like an animal cell, will lose water to its surroundings and shrink. As the plant cell shrivels, its plasma membrane pulls away from the cell wall at multiple places. This phenomenon, called **plasmolysis**, causes the plant to wilt and can lead to plant death. The walled cells of bacteria and fungi also plasmolyze in hypertonic environments.

Facilitated Diffusion: Passive Transport Aided by Proteins

Let's look more closely at how water and certain hydrophilic solutes cross a membrane. As mentioned earlier, many polar molecules and ions impeded by the lipid bilayer of the membrane diffuse passively with the help of transport proteins that span the membrane. This phenomenon is called **facilitated diffusion**. Cell biologists are still trying to learn exactly how various transport proteins facilitate diffusion. Most transport proteins are very specific: They transport some substances but not others.

As mentioned earlier, the two types of transport proteins are channel proteins and carrier proteins. Channel proteins simply provide corridors that allow specific molecules or ions to cross the membrane **(Figure 13a)**. The hydrophilic passageways provided by these proteins can allow water molecules or small ions to diffuse very quickly from one side of the membrane to the other. Aquaporins, the water channel proteins, facilitate the massive amounts of diffusion that occur in plant cells and in animal cells such as red blood cells. Certain kidney cells also have many aquaporin molecules, allowing them to reclaim water from urine before it is excreted. If the kidneys did not perform this function, you would excrete about 180 L of urine per day—and have to drink an equal volume of water!

Channel proteins that transport ions are called **ion channels**. Many ion channels function as **gated channels**, which open or close in response to a stimulus. For some gated channels, the stimulus is electrical. In a nerve cell, for example, an ion channel opens in response to an electrical stimulus, allowing a stream of potassium ions to leave the cell. This restores the cell's ability to fire again. Other gated channels open or close when a specific substance other than the one to be transported binds to the channel. These gated channels are also important in the functioning of the nervous system.

Carrier proteins, such as the glucose transporter mentioned earlier, seem to undergo a subtle change in shape that somehow translocates the solute-binding site across the membrane **(Figure 13b)**. Such a change in shape may be triggered by the binding and release of the transported molecule. Like ion channels, carrier proteins involved in facilitated diffusion result in the net movement of a substance down its concentration gradient. No energy input is required: This is passive transport. The **Scientific Skills Exercise** gives you an opportunity to work with data from an experiment related to glucose transport.

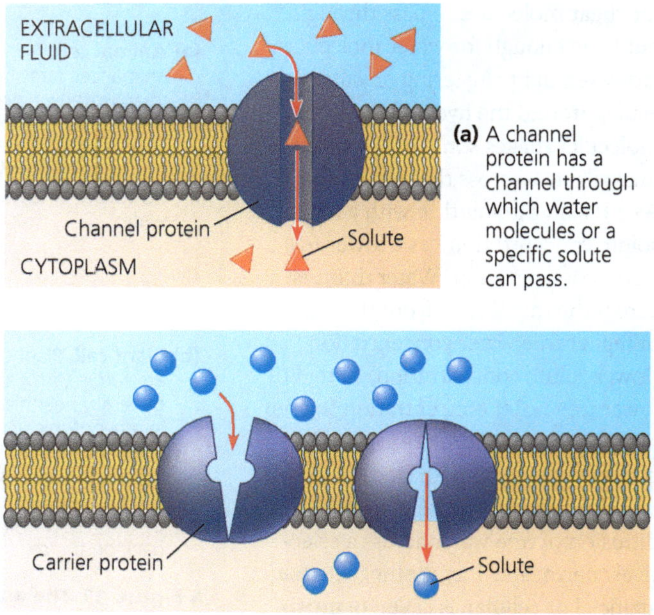

(a) A channel protein has a channel through which water molecules or a specific solute can pass.

(b) A carrier protein alternates between two shapes, moving a solute across the membrane during the shape change.

▲ **Figure 13 Two types of transport proteins that carry out facilitated diffusion.** In both cases, the protein can transport the solute in either direction, but the net movement is down the concentration gradient of the solute.

CONCEPT CHECK 3

1. How do you think a cell performing cellular respiration rids itself of the resulting CO_2?
2. **WHAT IF?** If a *Paramecium caudatum* cell swims from a hypotonic to an isotonic environment, will its contractile vacuole become more active or less? Why?

Scientific Skills Exercise

Interpreting a Scatter Plot with Two Sets of Data

Is Glucose Uptake into Cells Affected by Age? Glucose, an important energy source for animals, is transported into cells by facilitated diffusion using protein carriers. In this exercise, you will interpret a graph with two sets of data from an experiment that examined glucose uptake over time in red blood cells from guinea pigs of different ages. You will determine if the age of the guinea pigs affected their cells' rate of glucose uptake.

How the Experiment Was Done Researchers incubated guinea pig red blood cells in a 300 m*M* (millimolar) radioactive glucose solution at pH 7.4 at 25°C. Every 10 or 15 minutes, they removed a sample of cells from the solution and measured the concentration of radioactive glucose inside those cells. The cells came from either a 15-day-old or 1-month-old guinea pig.

Data from the Experiment When you have multiple sets of data, it can be useful to plot them on the same graph for comparison. In the graph here, each set of dots (dots of the same color) forms a *scatter plot,* in which every data point represents two numerical values, one for each variable. For each data set, a curve that best fits the points has been drawn to make it easier to see the trends. (For additional information about graphs, see the Scientific Skills Review in the Study Area in MasteringBiology.)

▶ 15-day-old and 1-month-old guinea pigs

Glucose Uptake over Time in Guinea Pig Red Blood Cells

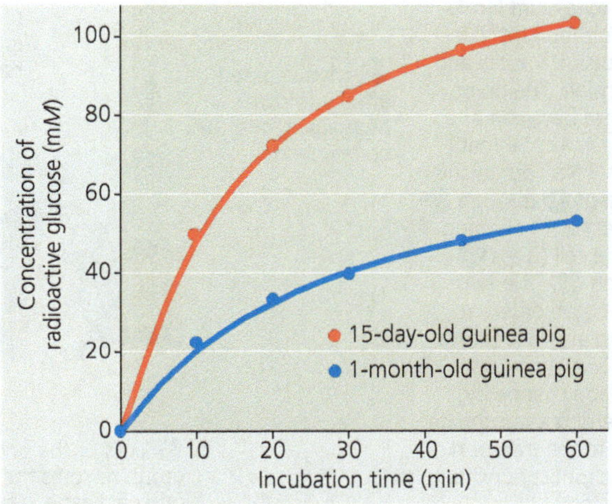

Data from T. Kondo and E. Beutler, Developmental changes in glucose transport of guinea pig erythrocytes, *Journal of Clinical Investigation* 65:1–4 (1980).

INTERPRET THE DATA

1. First make sure you understand the parts of the graph. (a) Which variable is the independent variable—the variable that was controlled by the researchers? (b) Which variable is the dependent variable—the variable that depended on the treatment and was measured by the researchers? (c) What do the red dots represent? (d) The blue dots?
2. From the data points on the graph, construct a table of the data. Put "Incubation Time (min)" in the left column of the table.
3. What does the graph show? Compare and contrast glucose uptake in red blood cells from a 15-day-old and a 1-month-old guinea pig.
4. Develop a hypothesis to explain the difference between glucose uptake in red blood cells from a 15-day-old and a 1-month-old guinea pig. (Think about how glucose gets into cells.)
5. Design an experiment to test your hypothesis.

(MB) A version of this Scientific Skills Exercise can be assigned in MasteringBiology.

CONCEPT 4

Active transport uses energy to move solutes against their gradients

Despite the help of transport proteins, facilitated diffusion is considered passive transport because the solute is moving down its concentration gradient, a process that requires no energy. Facilitated diffusion speeds transport of a solute by providing efficient passage through the membrane, but it does not alter the direction of transport. Some other transport proteins, however, can move solutes against their concentration gradients, across the plasma membrane from the side where they are less concentrated (whether inside or outside) to the side where they are more concentrated.

The Need for Energy in Active Transport

To pump a solute across a membrane against its gradient requires work; the cell must expend energy. Therefore, this type of membrane traffic is called **active transport**. The transport proteins that move solutes against their concentration gradients are all carrier proteins rather than channel proteins. This makes sense because when channel proteins are open, they merely allow solutes to diffuse down their concentration gradients rather than picking them up and transporting them against their gradients. Active transport enables a cell to maintain internal concentrations of small solutes that differ from concentrations in its environment. For example, compared with its surroundings, an animal cell has a much higher concentration of potassium ions (K^+) and a much lower concentration of sodium ions (Na^+). The plasma membrane helps maintain these steep gradients by pumping Na^+ out of the cell and K^+ into the cell.

As in other types of cellular work, ATP supplies the energy for most active transport. One way ATP can power active transport is by transferring its terminal phosphate group directly to the transport protein. This can induce the protein to change its shape in a manner that translocates a solute bound to the protein across the membrane. One transport system

▶ **Figure 14 The sodium-potassium pump: a specific case of active transport.** This transport system pumps ions against steep concentration gradients: Sodium ion concentration ($[Na^+]$) is high outside the cell and low inside, while potassium ion concentration ($[K^+]$) is low outside the cell and high inside. The pump oscillates between two shapes in a cycle that moves 3 Na^+ out of the cell (steps 1 through 3) for every 2 K^+ pumped into the cell (steps 4 through 6). The two shapes have different binding affinities for Na^+ and K^+. ATP powers the shape change by transferring a phosphate group to the transport protein (phosphorylating the protein).

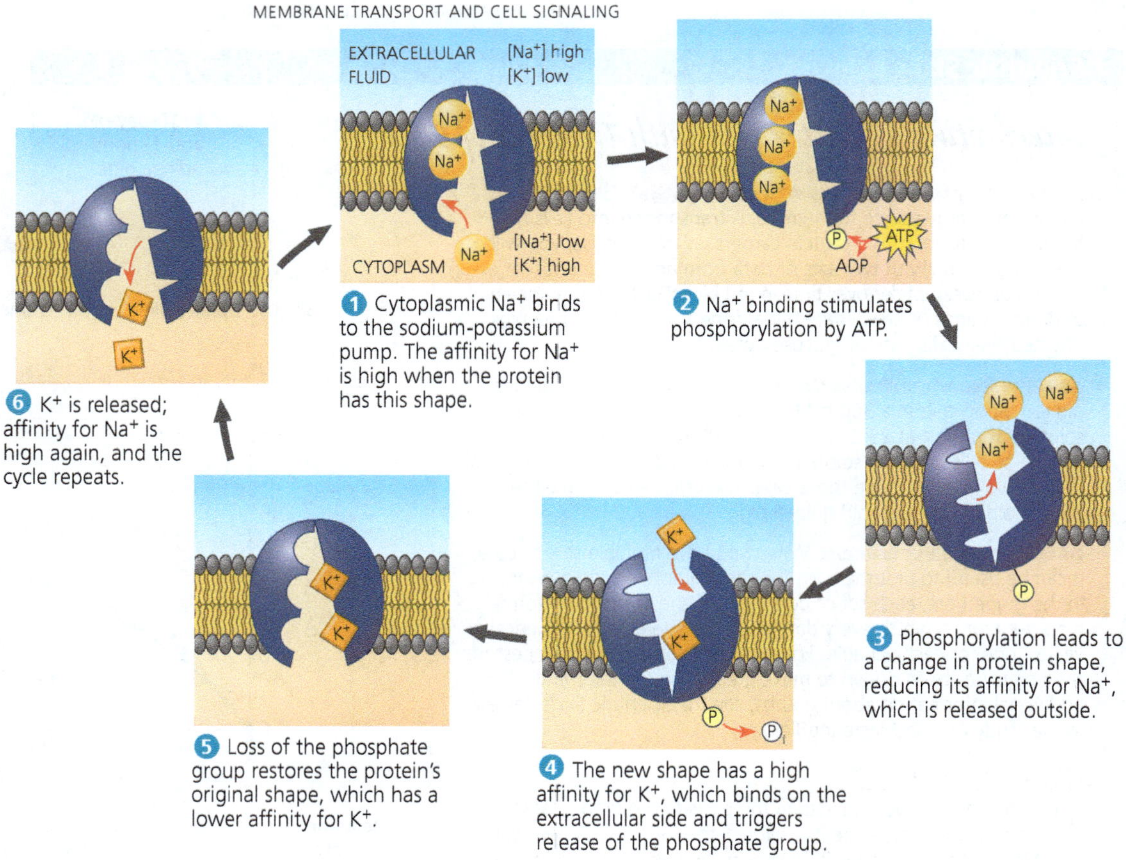

that works this way is the **sodium-potassium pump**, which exchanges Na^+ for K^+ across the plasma membrane of animal cells **(Figure 14)**. The distinction between passive transport and active transport is reviewed in **Figure 15**.

How Ion Pumps Maintain Membrane Potential

All cells have voltages across their plasma membranes. Voltage is electrical potential energy—a separation of opposite charges. The cytoplasmic side of the membrane is negative in charge relative to the extracellular side because of an unequal distribution of anions and cations on the two sides. The voltage across a membrane, called a **membrane potential**, ranges from about –50 to –200 millivolts (mV). (The minus sign indicates that the inside of the cell is negative relative to the outside.)

The membrane potential acts like a battery, an energy source that affects the traffic of all charged substances across the membrane. Because the inside of the cell is negative compared with the outside, the membrane potential favors the passive transport of cations into the cell and anions out of the cell. Thus, *two* forces drive the diffusion of ions across a membrane: a chemical force (the ion's concentration gradient) and an electrical force (the effect of the membrane potential on the ion's movement). This combination of forces acting on an ion is called the **electrochemical gradient**.

▼ **Figure 15 Review: passive and active transport.**

Passive transport. Substances diffuse spontaneously down their concentration gradients, crossing a membrane with no expenditure of energy by the cell. The rate of diffusion can be greatly increased by transport proteins in the membrane.

Active transport. Some transport proteins act as pumps, moving substances across a membrane against their concentration (or electrochemical) gradients. Energy for this work is usually supplied by ATP.

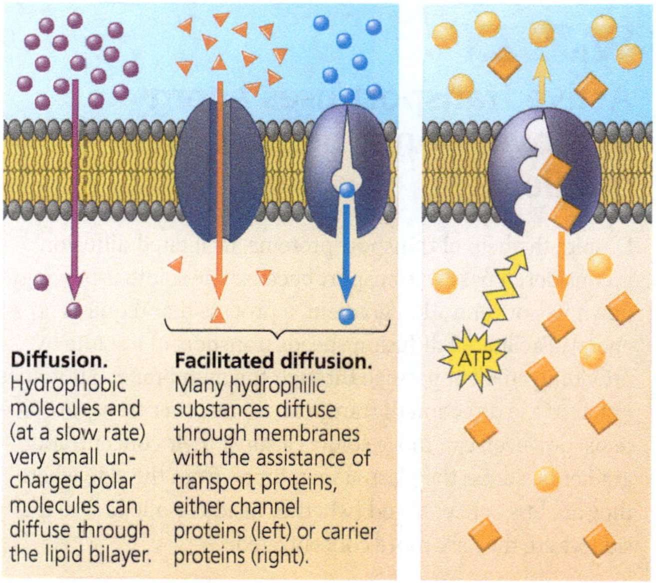

? *For each solute in the right panel, describe its direction of movement, and state whether it is moving with or against its concentration gradient.*

In the case of ions, then, we must refine our concept of passive transport: An ion diffuses not simply down its *concentration* gradient but, more exactly, down its *electrochemical* gradient. For example, the concentration of Na^+ inside a resting nerve cell is much lower than outside it. When the cell is stimulated, gated channels open that facilitate Na^+ diffusion. Sodium ions then "fall" down their electrochemical gradient, driven by the concentration gradient of Na^+ and by the attraction of these cations to the negative side (inside) of the membrane. In this example, both electrical and chemical contributions to the electrochemical gradient act in the same direction across the membrane, but this is not always so. In cases where electrical forces due to the membrane potential oppose the simple diffusion of an ion down its concentration gradient, active transport may be necessary.

Some membrane proteins that actively transport ions contribute to the membrane potential. An example is the sodium-potassium pump. Notice in Figure 14 that the pump does not translocate Na^+ and K^+ one for one, but pumps three sodium ions out of the cell for every two potassium ions it pumps into the cell. With each "crank" of the pump, there is a net transfer of one positive charge from the cytoplasm to the extracellular fluid, a process that stores energy as voltage. A transport protein that generates voltage across a membrane is called an **electrogenic pump**. The sodium-potassium pump appears to be the major electrogenic pump of animal cells. The main electrogenic pump of plants, fungi, and bacteria is a **proton pump**, which actively transports protons (hydrogen ions, H^+) out of the cell. The pumping of H^+ transfers positive charge from the cytoplasm to the extracellular solution **(Figure 16)**. By generating voltage across membranes, electrogenic pumps help store energy that can be tapped for cellular work. One important use of proton gradients in the cell is for ATP synthesis during cellular respiration). Another is a type of membrane traffic called cotransport.

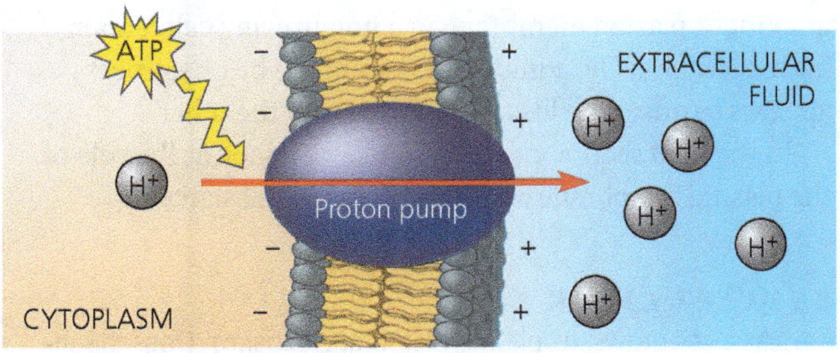

▲ **Figure 16 A proton pump.** Proton pumps are electrogenic pumps that store energy by generating voltage (charge separation) across membranes. A proton pump translocates positive charge in the form of hydrogen ions (that is, protons). The voltage and H^+ concentration gradient represent a dual energy source that can drive other processes, such as the uptake of nutrients. Most proton pumps are powered by ATP.

Cotransport: Coupled Transport by a Membrane Protein

A solute that exists in different concentrations across a membrane can do work as it moves across that membrane by diffusion down its concentration gradient. This is analogous to water that has been pumped uphill and performs work as it flows back down. In a mechanism called **cotransport**, a transport protein (a cotransporter) can couple the "downhill" diffusion of the solute to the "uphill" transport of a second substance against its own concentration (or electrochemical) gradient. For instance, a plant cell uses the gradient of H^+ generated by its ATP-powered proton pumps to drive the active transport of amino acids, sugars, and several other nutrients into the cell. In the example shown in **Figure 17**, a cotransporter couples the return of H^+ to the transport of sucrose into the cell. This protein can translocate sucrose into the cell against its concentration gradient, but only if the sucrose molecule travels in the company of an H^+. The H^+ uses the transport protein as an avenue to diffuse down its own electrochemical gradient, which is maintained by the proton pump. Plants use sucrose-H^+ cotransport to load sucrose produced by photosynthesis into cells in the veins of leaves. The vascular tissue of the plant can then distribute the sugar to nonphotosynthetic organs, such as roots.

What we know about cotransport proteins in animal cells has helped us find more effective treatments for diarrhea, a serious problem in developing countries. Normally, sodium in waste is reabsorbed in the colon, maintaining constant levels

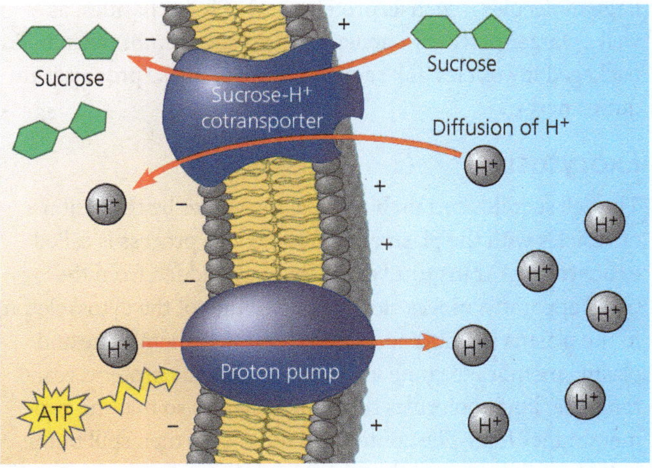

▲ **Figure 17 Cotransport: active transport driven by a concentration gradient.** A carrier protein, such as this sucrose-H^+ cotransporter in a plant cell (top), is able to use the diffusion of H^+ down its electrochemical gradient into the cell to drive the uptake of sucrose against its concentration gradient. (The cell wall is not shown.) Although not technically part of the cotransport process, an ATP-driven proton pump is shown here (bottom), which concentrates H^+ outside the cell. The resulting H^+ gradient represents potential energy that can be used for active transport—of sucrose, in this case. Thus, ATP indirectly provides the energy necessary for cotransport.

in the body, but diarrhea expels waste so rapidly that reabsorption is not possible, and sodium levels fall precipitously. To treat this life-threatening condition, patients are given a solution to drink containing high concentrations of salt (NaCl) and glucose. The solutes are taken up by sodium-glucose cotransporters on the surface of intestinal cells and passed through the cells into the blood. This simple treatment has lowered infant mortality worldwide.

CONCEPT CHECK 4

1. Sodium-potassium pumps help nerve cells establish a voltage across their plasma membranes. Do these pumps use ATP or produce ATP? Explain.
2. Explain why the sodium-potassium pump in Figure 14 would not be considered a cotransporter.
3. **MAKE CONNECTIONS** Review the characteristics of the lysosome. Given the internal environment of a lysosome, what transport protein might you expect to see in its membrane?

CONCEPT 5

Bulk transport across the plasma membrane occurs by exocytosis and endocytosis

Water and small solutes enter and leave the cell by diffusing through the lipid bilayer of the plasma membrane or by being moved across the membrane by transport proteins. However, large molecules—such as proteins and polysaccharides, as well as larger particles—generally cross the membrane in bulk, packaged in vesicles. Like active transport, these processes require energy.

Exocytosis

The cell secretes certain biological molecules by the fusion of vesicles with the plasma membrane; this process is called **exocytosis**. A transport vesicle that has budded from the Golgi apparatus moves along microtubules of the cytoskeleton to the plasma membrane. When the vesicle membrane and plasma membrane come into contact, specific proteins rearrange the lipid molecules of the two bilayers so that the two membranes fuse. The contents of the vesicle then spill to the outside of the cell, and the vesicle membrane becomes part of the plasma membrane (see Figure 8, step 4).

Many secretory cells use exocytosis to export products. For example, the cells in the pancreas that make insulin secrete it into the extracellular fluid by exocytosis. In another example, nerve cells use exocytosis to release neurotransmitters that signal other neurons or muscle cells. When plant cells are making cell walls, exocytosis delivers proteins and carbohydrates from Golgi vesicles to the outside of the cell.

Endocytosis

In **endocytosis**, the cell takes in molecules and particulate matter by forming new vesicles from the plasma membrane. Although the proteins involved in the two processes are different, the events of endocytosis look like the reverse of exocytosis. First, a small area of the plasma membrane sinks inward to form a pocket. Then, as the pocket deepens, it pinches in, forming a vesicle containing material that had been outside the cell. Study **Figure 18** carefully to understand the three types of endocytosis: phagocytosis ("cellular eating"), pinocytosis ("cellular drinking"), and receptor-mediated endocytosis.

Human cells use receptor-mediated endocytosis to take in cholesterol for membrane synthesis and the synthesis of other steroids. Cholesterol travels in the blood in particles called low-density lipoproteins (LDLs), each a complex of lipids and a protein. LDLs bind to LDL receptors on plasma membranes and then enter the cells by endocytosis. In the inherited disease familial hypercholesterolemia, characterized by a very high level of cholesterol in the blood, LDLs cannot enter cells because the LDL receptor proteins are defective or missing:

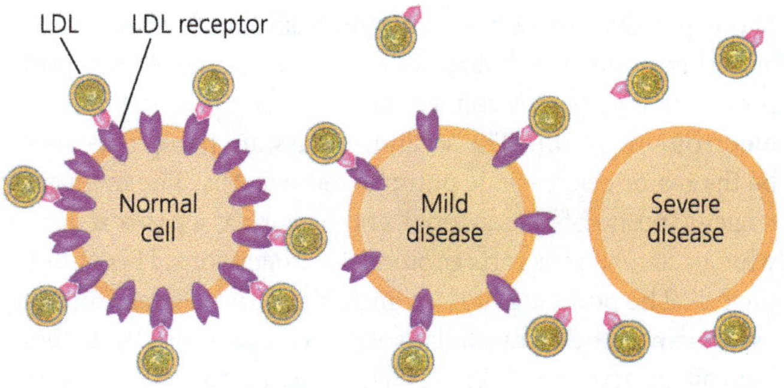

Consequently, cholesterol accumulates in the blood, where it contributes to early atherosclerosis, the buildup of lipid deposits within the walls of blood vessels. This buildup narrows the space in the vessels and impedes blood flow, potentially resulting in heart damage or stroke.

Endocytosis and exocytosis also provide mechanisms for rejuvenating or remodeling the plasma membrane. These processes occur continually in most eukaryotic cells, yet the amount of plasma membrane in a nongrowing cell remains fairly constant. The addition of membrane by one process appears to offset the loss of membrane by the other.

In the final section of this chapter, we'll look at the role of the plasma membrane and its proteins in cell signaling.

CONCEPT CHECK 5

1. As a cell grows, its plasma membrane expands. Does this involve endocytosis or exocytosis? Explain.
2. **DRAW IT** Return to Figure 8, and circle a patch of plasma membrane that is coming from a vesicle involved in exocytosis.
3. **MAKE CONNECTIONS** Animal cells make an extracellular matrix (ECM). Describe the cellular pathway of synthesis and deposition of an ECM glycoprotein.

▼ Figure 18 **Exploring Endocytosis in Animal Cells**

Phagocytosis

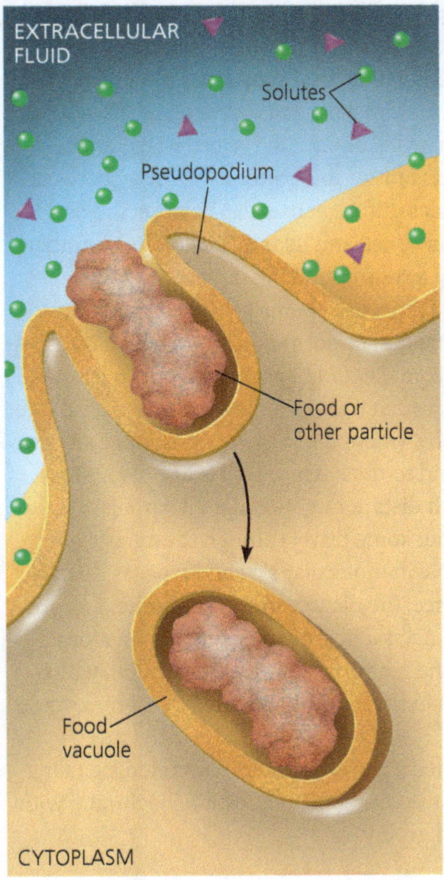

In **phagocytosis**, a cell engulfs a particle by extending pseudopodia (singular, *pseudopodium*) around it and packaging it within a membranous sac called a food vacuole. The particle will be digested after the food vacuole fuses with a lysosome containing hydrolytic enzymes

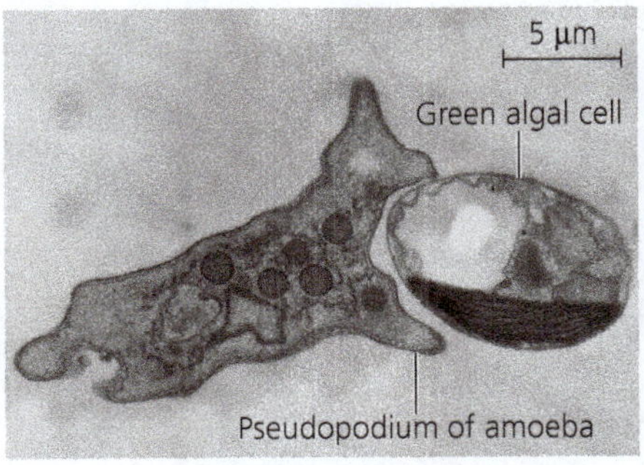

An amoeba engulfing a green algal cell via phagocytosis (TEM).

Pinocytosis

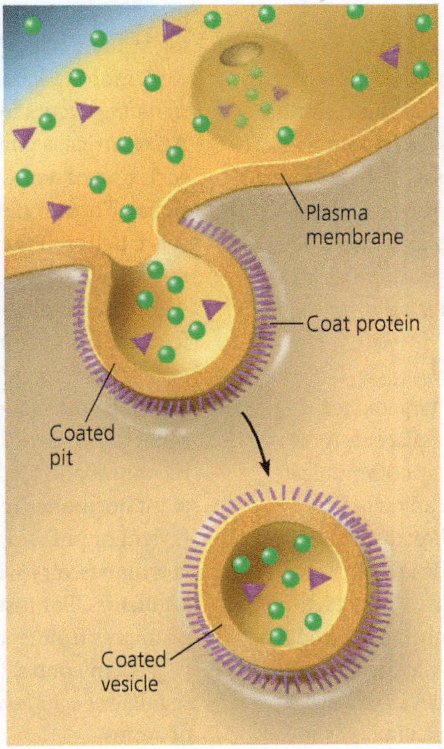

In **pinocytosis**, a cell continually "gulps" droplets of extracellular fluid into tiny vesicles, formed by infoldings of the plasma membrane. In this way, the cell obtains molecules dissolved in the droplets. Because any and all solutes are taken into the cell, pinocytosis as shown here is nonspecific for the substances it transports. In many cases, as above, the parts of the plasma membrane that form vesicles are lined on their cytoplasmic side by a fuzzy layer of coat protein; the "pits" and resulting vesicles are said to be "coated."

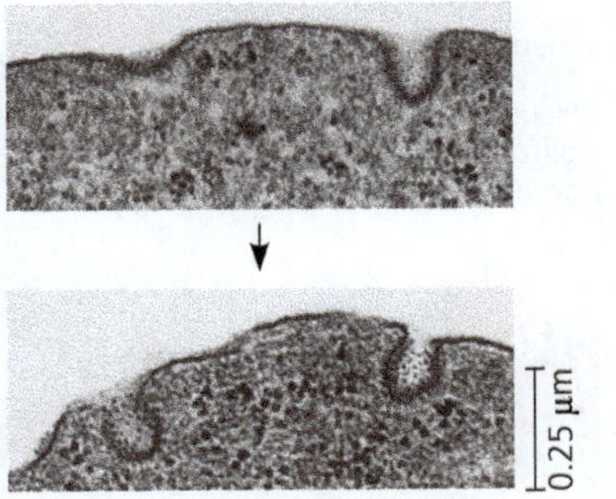

Pinocytotic vesicles forming (TEMs).

Receptor-Mediated Endocytosis

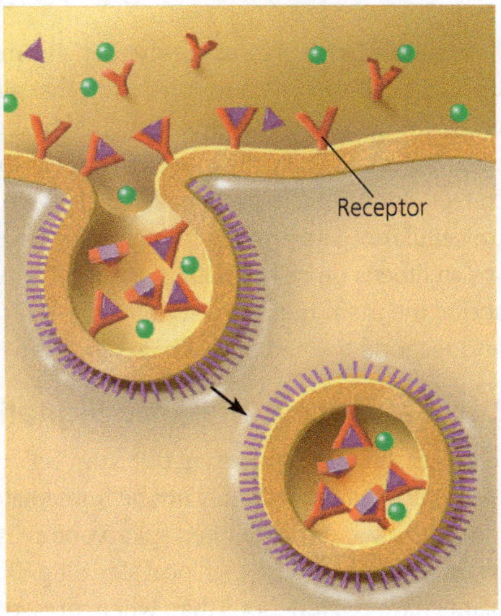

Receptor-mediated endocytosis is a specialized type of pinocytosis that enables the cell to acquire bulk quantities of specific substances, even though those substances may not be very concentrated in the extracellular fluid. Embedded in the plasma membrane are proteins with receptor sites exposed to the extracellular fluid. Specific solutes bind to the sites. The receptor proteins then cluster in coated pits, and each coated pit forms a vesicle containing the bound molecules. Notice that there are relatively more bound molecules (purple triangles) inside the vesicle, but other molecules (green balls) are also present. After the ingested material is liberated from the vesicle, the emptied receptors are recycled to the plasma membrane by the same vesicle (not shown).

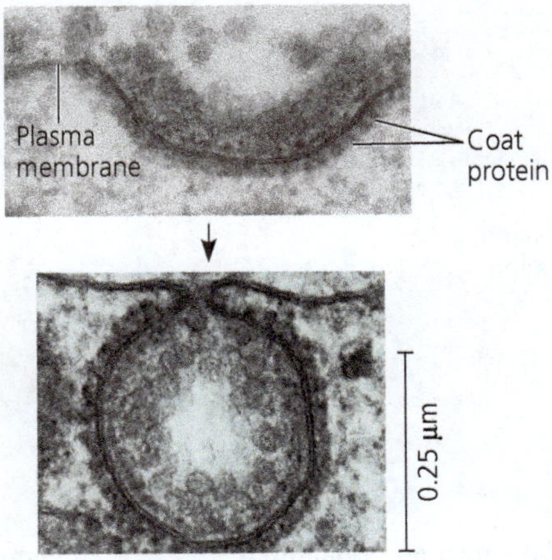

Top: A coated pit. *Bottom*: A coated vesicle forming during receptor-mediated endocytosis (TEMs).

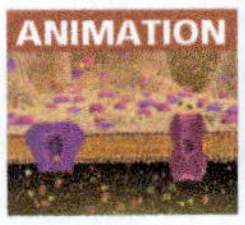

Visit the Study Area in **MasteringBiology** for the BioFlix® 3-D Animation on Membrane Transport.

CONCEPT 6

The plasma membrane plays a key role in most cell signaling

In a multicellular organism, whether a human being or an oak tree, it is cell-to-cell communication that allows the trillions of cells of the body to coordinate their activities, and the communication process usually involves the cells' plasma membranes. In fact, communication between cells is also essential for many unicellular organisms, including prokaryotes. However, here we will focus on cell signaling in animals and plants. We'll describe the main mechanisms by which cells receive, process, and respond to chemical signals sent from other cells.

Local and Long-Distance Signaling

The signaling molecules sent out from cells are targeted for other cells that may or may not be immediately adjacent. As discussed earlier in this chapter, eukaryotic cells may communicate by direct contact, a type of local signaling. Both animals and plants have cell junctions that, where present, directly connect the cytoplasms of adjacent cells; in animals, these are gap junctions, and in plants, plasmodesmata. In these cases, signaling substances dissolved in the cytosol can pass freely between adjacent cells. Also, animal cells may communicate via direct contact between membrane-bound cell-surface molecules in cell-cell recognition (see Figure 7d). This sort of local signaling is especially important in embryonic development and in the immune response.

In many other cases of local signaling, the signaling cell secretes messenger molecules. Some of these travel only short distances; such local regulators influence cells in the vicinity. One class of local regulators in animals, *growth factors*, are compounds that stimulate nearby target cells to grow and divide. Numerous cells can simultaneously receive and respond to the molecules of growth factor produced by a nearby cell. This type of local signaling in animals is called *paracrine signaling* **(Figure 19a)**.

A more specialized type of local signaling called *synaptic signaling* occurs in the animal nervous system **(Figure 19b)**. An electrical signal moving along a nerve cell triggers the secretion of neurotransmitter molecules carrying a chemical signal. These molecules diffuse across the synapse, the narrow space between the nerve cell and its target cell (often another nerve cell), triggering a response in the target cell.

Both animals and plants use chemicals called **hormones** for long-distance signaling. In hormonal signaling in animals, also known as *endocrine signaling*, specialized cells release hormone molecules, which travel via the circulatory system to other parts of the body, where they reach target cells that can recognize and respond to the hormones **(Figure 19c)**. Most plant hormones reach distant targets via plant vascular tissues (xylem or phloem), but some travel through the air as a gas. Hormones vary widely in molecular size and type, as do local regulators. For instance, the plant hormone ethylene, a gas that promotes fruit ripening, is a hydrocarbon of only six atoms (C_2H_4). In contrast, the mammalian hormone insulin, which regulates sugar levels in the blood, is a protein with thousands of atoms.

What happens when a cell encounters a secreted signaling molecule? We will now consider this question, beginning with a bit of historical background.

▼ Figure 19 Local and long-distance cell signaling by secreted molecules in animals. In both local and long-distance signaling, only specific target cells that can recognize a given signaling molecule will respond to it.

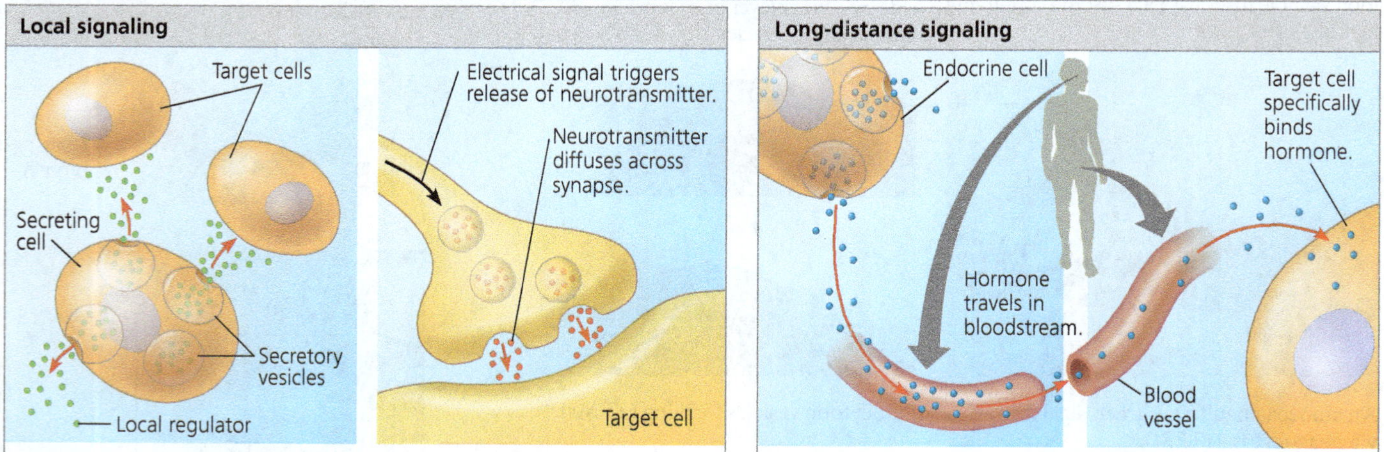

(a) Paracrine signaling. A secreting cell acts on nearby target cells by secreting molecules of a local regulator (a growth factor, for example).

(b) Synaptic signaling. A nerve cell releases neurotransmitter molecules into a synapse, stimulating the target cell, such as a muscle or nerve cell.

(c) Endocrine (hormonal) signaling. Specialized endocrine cells secrete hormones into body fluids, often blood. Hormones reach virtually all body cells, but are bound only by some cells.

The Three Stages of Cell Signaling: *A Preview*

Our current understanding of how chemical messengers act on cells had its origins in the pioneering work of the American Earl W. Sutherland about a half-century ago. He was investigating how the animal hormone epinephrine (also called adrenaline) triggers the "fight-or-flight" response in animals by stimulating the breakdown of the storage polysaccharide glycogen within liver cells and skeletal muscle cells. Glycogen breakdown releases the sugar glucose 1-phosphate, which the cell converts to glucose 6-phosphate. The liver or muscle cell can then use this compound, an early intermediate in glycolysis, for energy production. Alternatively, the compound can be stripped of phosphate and released from the cell into the blood as glucose, which can fuel cells throughout the body. Thus, one effect of epinephrine is the mobilization of fuel reserves, which can be used by an animal to either defend itself (fight) or escape whatever elicited a scare (flight), as this impala is doing.

Sutherland's research team discovered that epinephrine stimulates glycogen breakdown by activating a cytosolic enzyme (glycogen phosphorylase) while never actually entering the glycogen-containing cells. This discovery provided two insights. First, epinephrine does not interact directly with glycogen phosphorylase; an intermediate step or series of steps must be occurring in the cell. Second, the plasma membrane must somehow be involved in transmitting the signal. Sutherland's research suggested that the process going on at the receiving end of a cell-to-cell message can be divided into three stages: reception, transduction, and response (Figure 20). ❶ **Reception** is the target cell's detection of a signaling molecule coming from outside the cell. A chemical signal is "detected" when the signaling molecule binds to a receptor protein located at the cell's surface or, in some cases, inside the cell. ❷ **Transduction** is a step or series of steps that converts the signal to a form that can bring about a specific cellular response. Transduction usually requires a sequence of changes in a series of different molecules—a **signal transduction pathway**. The molecules in the pathway are often called relay molecules. ❸ In the third stage of cell signaling, the transduced signal finally triggers a specific cellular **response**. The response may be almost any imaginable cellular activity, such as catalysis by an enzyme (for example, glycogen phosphorylase), rearrangement of the cytoskeleton, or activation of specific genes in the nucleus. The cell-signaling process helps ensure that crucial activities like these occur in the right cells, at the right time, and in proper coordination with the activities of other cells of the organism. We'll now explore the mechanisms of cell signaling in more detail.

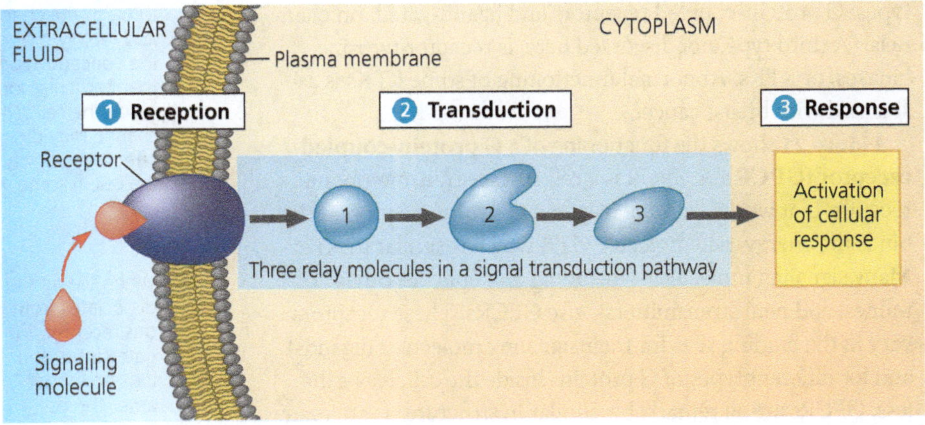

▲ **Figure 20 Overview of cell signaling.** From the perspective of the cell receiving the message, cell signaling can be divided into three stages: signal reception, signal transduction, and cellular response. When reception occurs at the plasma membrane, as shown here, the transduction stage is usually a pathway of several steps, with each specific relay molecule in the pathway bringing about a change in the next molecule. The final molecule in the pathway triggers the cell's response. The three stages are explained in more detail in the text.

Reception, the Binding of a Signaling Molecule to a Receptor Protein

A radio station broadcasts its signal indiscriminately, but it can be picked up only by radios tuned to the right frequency; reception of the signal depends on the receiver. Similarly, in the case of epinephrine, the hormone encounters many types of cells as it circulates in the blood, but only certain target cells detect and react to the epinephrine molecule. A receptor protein on or in the target cell allows the cell to detect the signal and respond to it. The signaling molecule is complementary in shape to a specific site on the receptor and attaches there, like a key in a lock. The signaling molecule acts as a **ligand**, a molecule that specifically binds to another molecule, often a larger one. (LDLs, mentioned in Concept 5, act as ligands when they bind to their receptors, as do the molecules that bind to enzymes.) Ligand binding generally causes a receptor protein to undergo a change in shape. For many receptors, this shape change directly activates the receptor, enabling it to interact with other cellular molecules.

Most signal receptors are plasma membrane proteins. Their ligands are water-soluble and generally too large to pass freely through the plasma membrane. Other signal receptors, however, are located inside the cell. We discuss both of these types next.

Receptors in the Plasma Membrane

Most water-soluble signaling molecules bind to specific sites on receptor proteins that span the cell's plasma membrane. Such a transmembrane receptor transmits information from the extracellular environment to the inside of the cell by

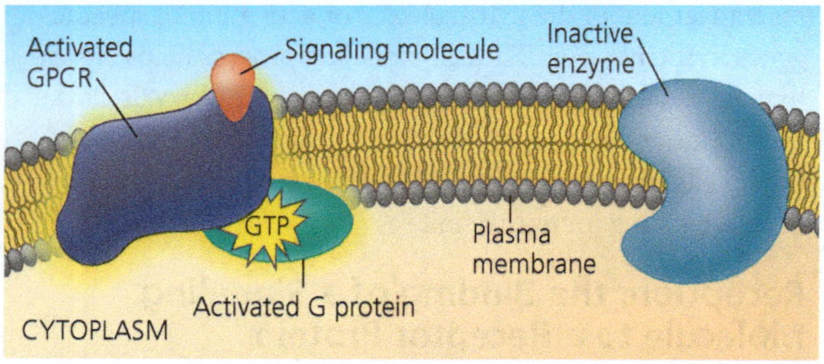

① When the appropriate signaling molecule binds to the extracellular side of the receptor, the receptor is activated and changes shape. Its cytoplasmic side then binds and activates a G protein. The activated G protein carries a GTP molecule.

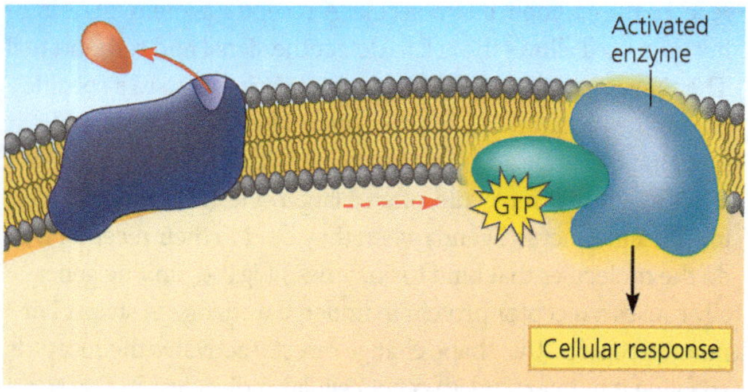

② The activated G protein leaves the receptor, diffuses along the membrane, and then binds to an enzyme, altering the enzyme's shape and activity. Once activated, the enzyme can trigger the next step leading to a cellular response. Binding of signaling molecules is reversible. The activating change in the GPCR, as well as the changes in the G protein and enzyme, are only temporary; these molecules soon become available for reuse.

▲ **Figure 21 A G protein-coupled receptor (GPCR) in action.**

changing shape when a specific ligand binds to it. We can see how transmembrane receptors work by looking at two major types: G protein-coupled receptors and ligand-gated ion channels. (A third type, not discussed here, is receptor tyrosine kinases, or RTKs. Abnormal functioning of some RTKs is associated with breast cancer.)

Figure 21 shows the functioning of a **G protein-coupled receptor (GPCR)**. A GPCR is a cell-surface transmembrane receptor that works with the help of a **G protein**, a protein that binds the energy-rich molecule GTP, which is similar to ATP. Many signaling molecules—including epinephrine, other hormones, and neurotransmitters—use GPCRs. These receptors vary in the binding sites for their signaling molecules (ligands) and for different types of G proteins inside the cell. Nevertheless, GPCRs are all remarkably similar in structure, as are many G proteins, suggesting that these signaling systems evolved very early in the history of life.

The nearly 1,000 GPCRs examined to date make up the largest family of cell-surface receptors in mammals. GPCR pathways are extremely diverse in their functions, which include roles in embryonic development and the senses of sight, smell, and taste. They are also involved in many human diseases. For example, cholera, pertussis (whooping cough), and botulism are caused by bacterial toxins that interfere with G protein function. Up to 60% of all medicines used today exert their effects by influencing G protein pathways.

A **ligand-gated ion channel** is a membrane receptor with a region that can act as a "gate" for ions when the receptor assumes a certain shape **(Figure 22)**. When a signaling molecule binds as a ligand to the receptor protein, the gate opens or closes, allowing or blocking the diffusion of specific ions, such as Na^+ or Ca^{2+}, through a channel in the protein. Like other membrane receptors, these proteins bind the ligand at a specific site on their extracellular side.

Ligand-gated ion channels are very important in the nervous system. For example, the neurotransmitter molecules released at a synapse between two nerve cells (see Figure 19b) bind as ligands to ion channels on the receiving cell, causing

① Here we show a ligand-gated ion channel receptor in which the gate remains closed until a ligand binds to the receptor.

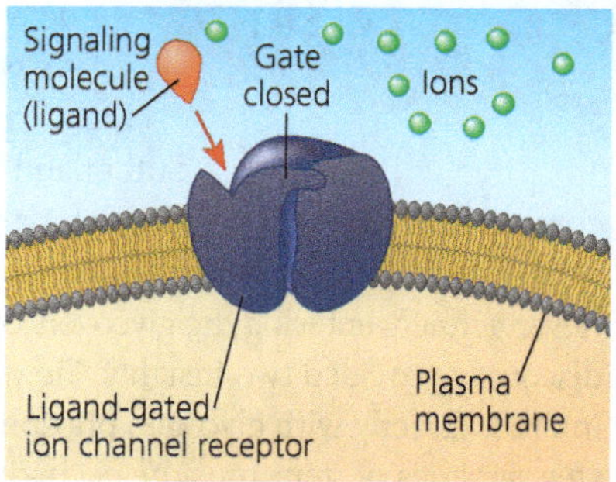

② When the ligand binds to the receptor and the gate opens, specific ions can flow through the channel and rapidly change the concentration of that particular ion inside the cell. This change may directly affect the activity of the cell in some way.

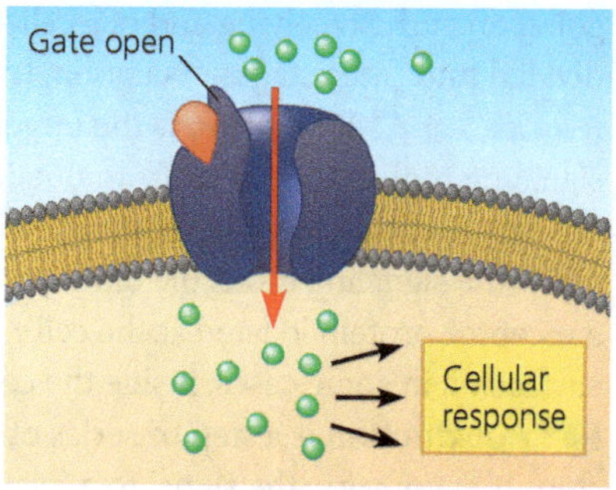

③ When the ligand dissociates from this receptor, the gate closes and ions no longer enter the cell.

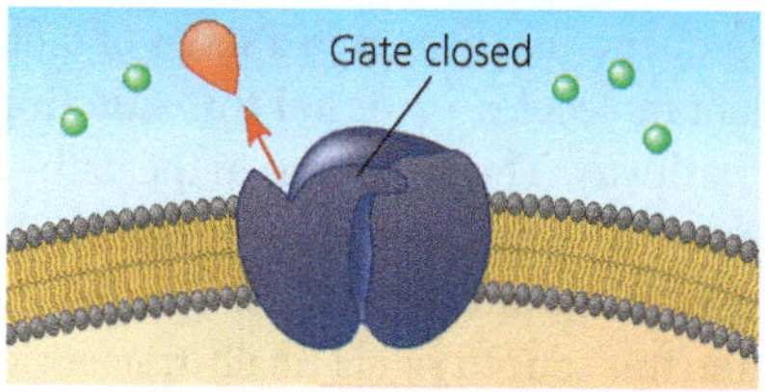

▲ **Figure 22 Ion channel receptor.** This is a ligand-gated ion channel, a type of receptor protein that regulates the passage of specific ions across the membrane. Whether the channel is open or closed depends on whether a specific ligand is bound to the protein.

the channels to open. The diffusion of ions through the open channels may trigger an electrical signal that propagates down the length of the receiving cell.

Intracellular Receptors

Intracellular receptor proteins are found in either the cytoplasm or nucleus of target cells. To reach such a receptor, a signaling molecule passes through the target cell's plasma membrane. A number of important signaling molecules can do this because they are hydrophobic enough to cross the hydrophobic interior of the membrane. These hydrophobic chemical messengers include the steroid hormones and thyroid hormones of animals. In both animals and plants, another chemical signaling molecule with an intracellular receptor is nitric oxide (NO), a gas; its very small, hydrophobic molecules can easily pass between the membrane phospholipids.

The behavior of aldosterone is representative of steroid hormones. This hormone is secreted by cells of the adrenal gland, a gland that lies over the kidney. It then travels through the blood and enters cells all over the body. However, a response occurs only in kidney cells, which contain receptor molecules for aldosterone. In these cells, the hormone binds to the receptor protein, activating it **(Figure 23)**. With the hormone attached, the active form of the receptor protein then enters the nucleus and turns on specific genes that control water and sodium flow in kidney cells, ultimately affecting blood volume.

How does the activated hormone-receptor complex turn on genes? Recall that the genes in a cell's DNA function by being transcribed and processed into messenger RNA (mRNA), which leaves the nucleus and is translated into a specific protein by ribosomes in the cytoplasm. Special proteins called *transcription factors* control which genes are turned on—that is, which genes are transcribed into mRNA—in a particular cell at a particular time. When the aldosterone receptor is activated, it acts as a transcription factor that turns on specific genes.

By acting as a transcription factor, the aldosterone receptor itself carries out the transduction part of the signaling pathway. Most other intracellular receptors function in the same way, although many of them, such as the thyroid hormone receptor, are already in the nucleus before the signaling molecule reaches them. Interestingly, many of these intracellular receptor proteins are structurally similar, suggesting an evolutionary kinship.

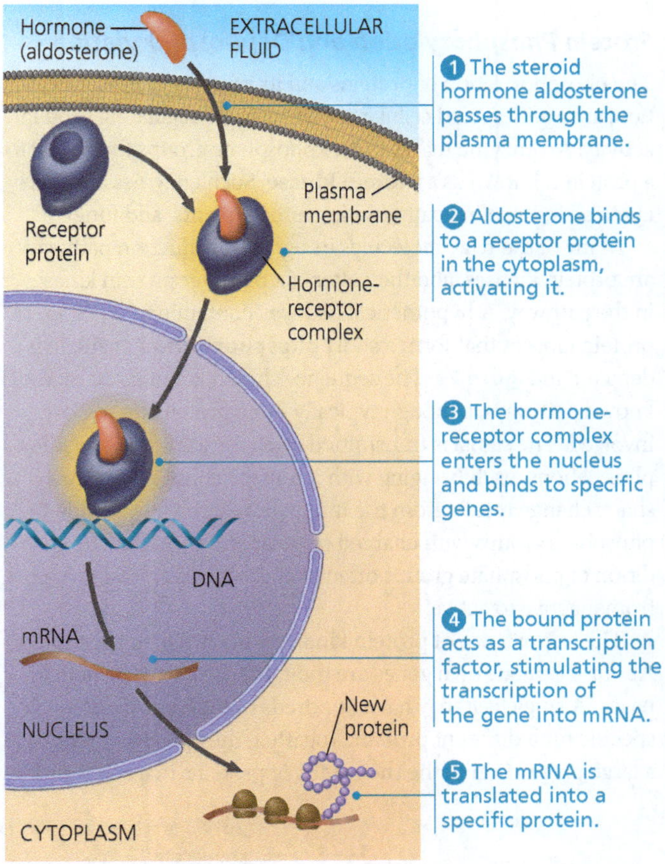

▲ **Figure 23** **Steroid hormone interacting with an intracellular receptor.**

? *Why is a cell-surface receptor protein not required for this steroid hormone to enter the cell?*

Transduction by Cascades of Molecular Interactions

When receptors for signaling molecules are plasma membrane proteins, like most of those we have discussed, the transduction stage of cell signaling is usually a multistep pathway involving many molecules. Steps often include activation of proteins by addition or removal of phosphate groups or release of other small molecules or ions that act as messengers. One benefit of multiple steps is the possibility of greatly amplifying a signal. If each molecule in a pathway transmits the signal to numerous molecules at the next step in the series, the result is a geometric increase in the number of activated molecules by the end of the pathway. Moreover, multistep pathways provide more opportunities for coordination and control than do simpler systems.

The binding of a specific signaling molecule to a receptor in the plasma membrane triggers the first step in the chain of molecular interactions—the signal transduction pathway—that leads to a particular response within the cell. Like falling dominoes, the signal-activated receptor activates another molecule, which activates yet another molecule, and so on, until the protein that produces the final cellular response is activated. The molecules that relay a signal from receptor to response, which we call relay molecules in this text, are often proteins. The interaction of proteins is a major theme of cell signaling.

Keep in mind that the original signaling molecule is not physically passed along a signaling pathway; in most cases, it never even enters the cell. When we say that the signal is relayed along a pathway, we mean that certain information is passed on. At each step, the signal is transduced into a different form, commonly via a shape change in a protein. Very often, the shape change is brought about by phosphorylation, the addition of phosphate groups to a protein.

Protein Phosphorylation and Dephosphorylation

The phosphorylation of proteins and its reverse, dephosphorylation, are a widespread cellular mechanism for regulating protein activity. An enzyme that transfers phosphate groups from ATP to a protein is known as a **protein kinase**. Such enzymes are widely involved in signaling pathways in animals, plants, and fungi.

Many of the relay molecules in signal transduction pathways are protein kinases, and they often act on other protein kinases in the pathway. A hypothetical pathway containing two different protein kinases that form a short **phosphorylation cascade** is depicted in **Figure 24**. The sequence shown is similar to many known pathways, although typically three protein kinases are involved. The signal is transmitted by a cascade of protein phosphorylations, each bringing with it a shape change. Each such shape change results from the interaction of the newly added phosphate groups with charged or polar amino acids. The addition of phosphate groups often changes the form of a protein from inactive to active.

The importance of protein kinases can hardly be overstated. About 2% of our own genes are thought to code for protein kinases. A single cell may have hundreds of different kinds, each specific for a different protein. Together, they probably regulate a large proportion of the thousands of proteins in a cell. Among these are most of the proteins that, in turn, regulate cell division. Abnormal activity of such a kinase can cause abnormal cell division and contribute to the development of cancer.

Equally important in the phosphorylation cascade are the **protein phosphatases** (see Figure 24), enzymes that can rapidly remove phosphate groups from proteins, a process called dephosphorylation. By dephosphorylating and thus inactivating protein kinases, phosphatases provide the mechanism for turning off the signal transduction pathway when the initial signal is no longer present. Phosphatases also make the protein kinases available for reuse, enabling the cell to respond again to an extracellular signal. A phosphorylation-dephosphorylation system acts as a molecular switch in the cell, turning an activity on or off, or up or down, as required. At any given moment, the activity of a protein regulated by phosphorylation depends on the balance in the cell between active kinase molecules and active phosphatase molecules.

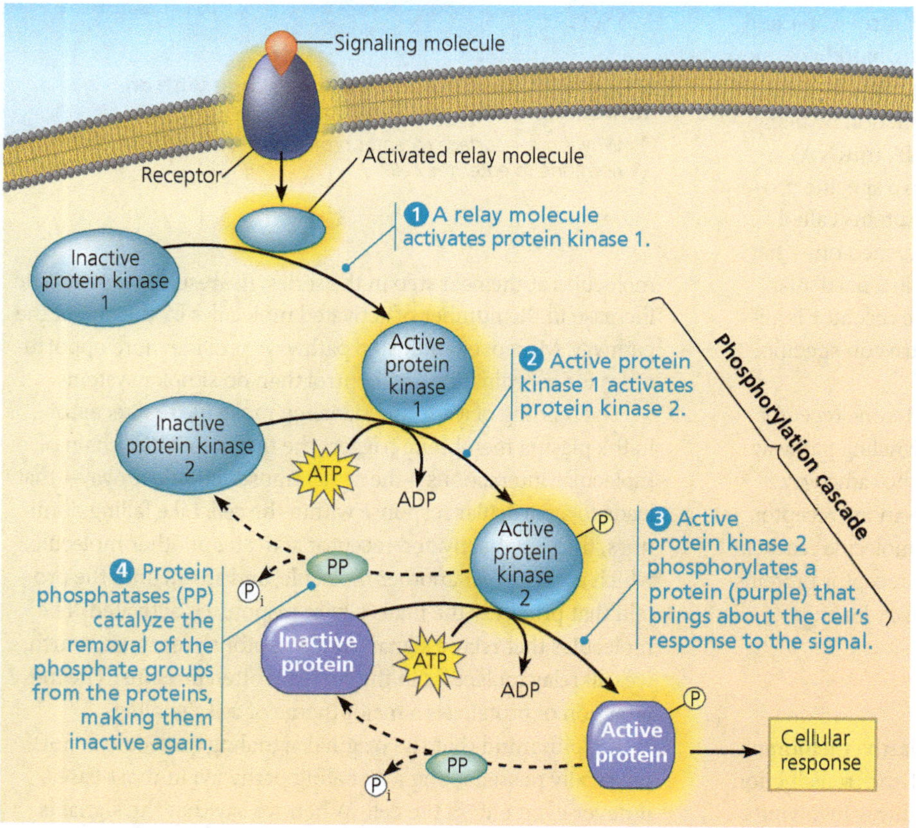

▲ **Figure 24 A phosphorylation cascade.** In a phosphorylation cascade, a series of different proteins in a pathway are phosphorylated in turn, each protein adding a phosphate group to the next one in line. Dephosphorylation by protein phosphatases (PP) can then return the protein to its inactive form.

? *Which protein is responsible for activation of protein kinase 2?*

Small Molecules and Ions as Second Messengers

Not all components of signal transduction pathways are proteins. Many signaling pathways also involve small, nonprotein, water-soluble molecules or ions called **second messengers**. (The pathway's "first messenger" is considered to be the extracellular signaling molecule that binds to the membrane receptor.) Because they are small, second messengers can readily spread throughout the cell by diffusion. The two most common second messengers are cyclic AMP and calcium ions, Ca^{2+}. Here we'll limit our discussion to cyclic AMP.

In his research on epinephrine, Earl Sutherland discovered that the binding of epinephrine to the plasma membrane of a liver cell elevates the cytosolic concentration of **cyclic AMP (cAMP**; cyclic adenosine monophosphate). The binding of epinephrine to a G protein-coupled receptor leads, via a G protein, to activation of adenylyl cyclase, an enzyme embedded in the plasma membrane that converts ATP to cAMP **(Figure 25)**. Each molecule of adenylyl cyclase can catalyze the synthesis of many molecules of cAMP. In this way, the normal cellular concentration of cAMP can be boosted 20-fold in a matter of seconds. The cAMP broadcasts the signal to the cytoplasm. It does not persist for long in the absence of the hormone because a different enzyme converts cAMP to AMP. Another surge of epinephrine is needed to boost the cytosolic concentration of cAMP again.

Subsequent research has revealed that epinephrine is only one of many

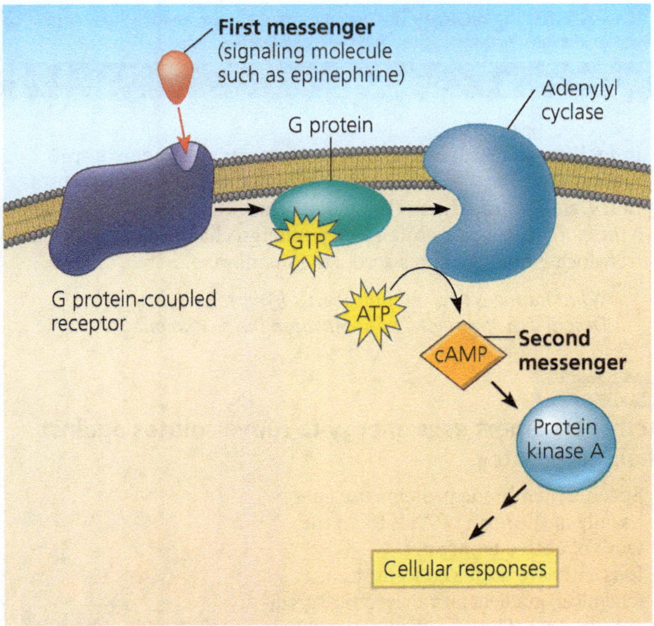

▲ **Figure 25** **cAMP as a second messenger in a G protein signaling pathway.** The first messenger activates a G protein-coupled receptor, which activates a specific G protein. In turn, the G protein activates adenylyl cyclase, which catalyzes the conversion of ATP to cAMP. The cAMP then acts as a second messenger and activates another protein, usually protein kinase A, leading to cellular responses.

hormones and other signaling molecules that trigger the formation of cAMP. The immediate effect of cAMP is usually the activation of a protein kinase called *protein kinase A*. The activated protein kinase A then phosphorylates various other proteins.

Response: Regulation of Transcription or Cytoplasmic Activities

What is the nature of the final step in a signaling pathway—the *response* to an external signal? Ultimately, a signal transduction pathway leads to the regulation of one or more cellular activities. The response may occur in the nucleus of the cell or in the cytoplasm.

Many signaling pathways ultimately regulate protein synthesis, usually by turning specific genes on or off in the nucleus. Like an activated steroid receptor (see Figure 23), the final activated molecule in a signaling pathway may function as a transcription factor. **Figure 26** shows an example in which a signaling pathway activates a transcription factor that turns a gene on: The response to this growth factor signal is transcription, the synthesis of one or more specific mRNAs, which will be translated in the cytoplasm into specific proteins. In other cases, the transcription factor might regulate a gene by turning it off. Often a transcription factor regulates several different genes.

Sometimes a signaling pathway may regulate the *activity* of proteins rather than causing their *synthesis* by activating gene expression. This directly affects proteins that function

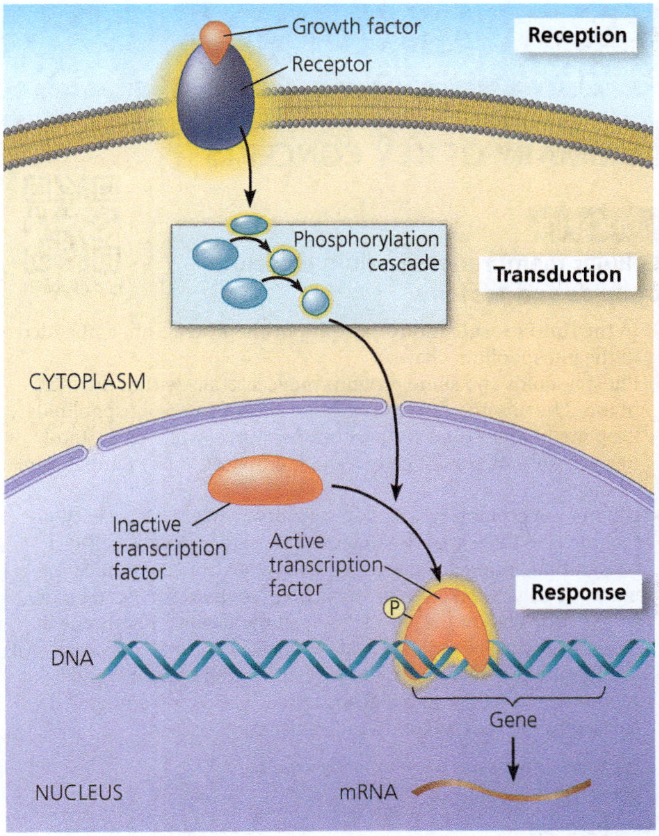

▲ **Figure 26** **Nuclear response to a signal: the activation of a specific gene by a growth factor.** This diagram shows a typical signaling pathway that leads to the regulation of gene activity in the cell nucleus. The initial signaling molecule, a local regulator called a growth factor, triggers a phosphorylation cascade. (The ATP molecules and phosphate groups are not shown.) Once phosphorylated, the last kinase in the sequence enters the nucleus and activates a transcription factor, which stimulates transcription of a specific gene. The resulting mRNAs then direct the synthesis of a particular protein in the cytoplasm.

outside the nucleus. For example, a signal may cause the opening or closing of an ion channel in the plasma membrane or a change in cell metabolism. As we have discussed, the response of cells to the hormone epinephrine helps regulate cellular energy metabolism by affecting the activity of an enzyme: The final step in the signaling pathway that begins with epinephrine binding activates the enzyme that catalyzes the breakdown of glycogen.

CONCEPT CHECK 6

1. During an epinephrine-initiated signal in liver cells, in which of the three stages of cell signaling does glycogen phosphorylase act?
2. When a signal transduction pathway involves a phosphorylation cascade, what turns off the cell's response?
3. **WHAT IF?** How can a target cell's response to a single hormone molecule result in a response that affects a million other molecules?

Chapter Review

Go to **MasteringBiology**® for Assignments, the eText, and the Study Area with Animations, Activities, Vocab Self-Quiz, and Practice Tests.

SUMMARY OF KEY CONCEPTS

VOCAB SELF-QUIZ
goo.gl/gbai8v

CONCEPT 1

Cellular membranes are fluid mosaics of lipids and proteins

- In the **fluid mosaic model**, **amphipathic** proteins are embedded in the phospholipid bilayer.
- Phospholipids and some proteins move laterally within the membrane. The unsaturated hydrocarbon tails of some phospholipids keep membranes fluid at lower temperatures, while cholesterol helps membranes resist changes in fluidity caused by temperature changes.
- Membrane proteins function in transport, enzymatic activity, attachment to the cytoskeleton and extracellular matrix, cell-cell recognition, intercellular joining, and signal transduction. Short chains of sugars linked to proteins (in **glycoproteins**) and lipids (in **glycolipids**) on the exterior side of the plasma membrane interact with surface molecules of other cells.
- Membrane proteins and lipids are synthesized in the ER and modified in the ER and Golgi apparatus. The inside and outside faces of membranes differ in molecular composition.

? *In what ways are membranes crucial to life?*

CONCEPT 2

Membrane structure results in selective permeability

- A cell must exchange molecules and ions with its surroundings, a process controlled by the **selective permeability** of the plasma membrane. Hydrophobic molecules are soluble in lipids and pass through membranes rapidly, whereas polar molecules and ions usually need specific **transport proteins**.

? *How do aquaporins affect the permeability of a membrane?*

CONCEPT 3

Passive transport is diffusion of a substance across a membrane with no energy investment

- **Diffusion** is the spontaneous movement of a substance down its **concentration gradient**. Water diffuses out through the permeable membrane of a cell (**osmosis**) if the solution outside has a higher solute concentration than the cytosol (is **hypertonic**); water enters the cell if the solution has a lower solute concentration (is **hypotonic**). If the concentrations are equal (**isotonic**), no net osmosis occurs. Cell survival depends on balancing water uptake and loss.

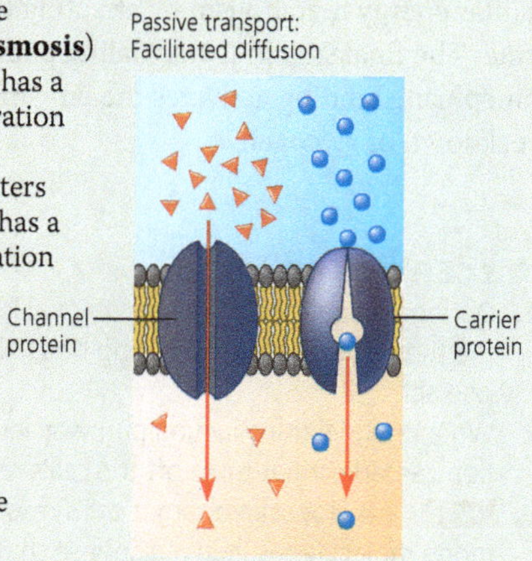

- In **facilitated diffusion**, a transport protein speeds the movement of water or a solute across a membrane down its concentration gradient. **Ion channels** facilitate the diffusion of ions across a membrane. Carrier proteins can undergo changes in shape that translocate bound solutes across the membrane.

? *What happens to a cell placed in a hypertonic solution? Describe the free water concentration inside and out.*

CONCEPT 4

Active transport uses energy to move solutes against their gradients

- Specific membrane proteins use energy, usually in the form of ATP, to do the work of **active transport**.
- Ions can have both a concentration (chemical) gradient and an electrical gradient (voltage). These gradients combine in the **electrochemical gradient**, which determines the net direction of ionic diffusion.
- **Cotransport** of two solutes occurs when a membrane protein enables the "downhill" diffusion of one solute to drive the "uphill" transport of the other.

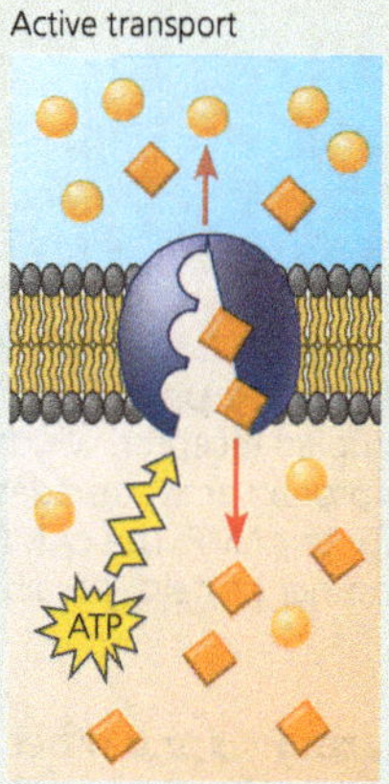

? *ATP is not directly involved in the functioning of a cotransporter. Why, then, is cotransport considered active transport?*

CONCEPT 5

Bulk transport across the plasma membrane occurs by exocytosis and endocytosis

- Three main types of **endocytosis** are **phagocytosis**, **pinocytosis**, and **receptor-mediated endocytosis**.

? *Which type of endocytosis involves the binding of specific substances in the extracellular fluid to membrane proteins? What does this type of transport enable a cell to do?*

CONCEPT 6

The plasma membrane plays a key role in most cell signaling

- Local signaling by animal cells involves direct contact or the secretion of growth factors and other signaling molecules. For long-distance signaling, animal and plant cells use **hormones**; animals also signal electrically.
- Signaling molecules that bind to membrane receptors trigger a three-stage cell-signaling pathway:

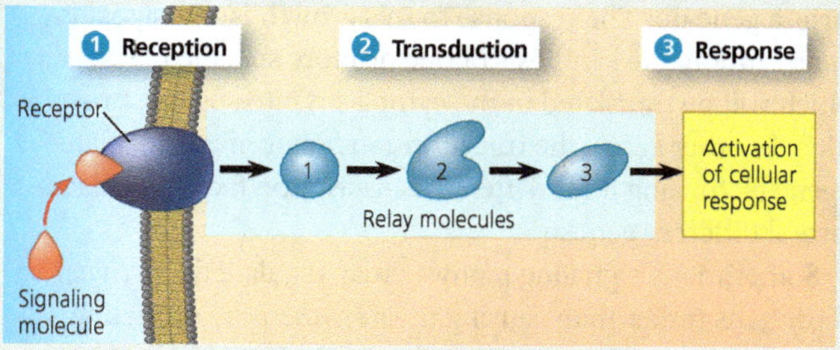

- In **reception**, a signaling molecule binds to a receptor protein, causing the protein to change shape. Two major types of membrane receptors are **G protein-coupled receptors (GPCRs)**, which work with the help of cytoplasmic **G proteins**, and **ligand-gated ion channels**, which open or close in response to binding by signaling molecules. Signaling molecules that are hydrophobic cross the plasma membrane and bind to receptors inside the cell.
- At each step in a **signal transduction pathway**, the signal is *transduced* into a different form, which commonly involves a change in a protein's shape. Many pathways include **phosphorylation cascades**, in which a series of **protein kinases** each add a phosphate group to the next one in line, activating it. The balance between phosphorylation and dephosphorylation, by **protein phosphatases**, regulates the activity of proteins in the pathway.
- **Second messengers**, such as the small molecule **cyclic AMP (cAMP)**, diffuse readily through the cytosol and thus help broadcast signals quickly. Many G proteins activate the enzyme that makes cAMP from ATP.
- The cell's **response** to a signal may be the regulation of transcription in the nucleus or of an activity in the cytoplasm.

? *What determines* whether *a cell responds to a hormone such as epinephrine? What determines* how *the cell responds?*

TEST YOUR UNDERSTANDING

PRACTICE TEST

goo.gl/CRZjvS

Level 1: Knowledge/Comprehension

1. In what way do the membranes of a eukaryotic cell vary?
   (A) Phospholipids are found only in certain membranes.
   (B) Certain proteins are unique to each kind of membrane.
   (C) Only certain membranes of the cell are selectively permeable.
   (D) Only certain membranes are constructed from amphipathic molecules.
2. Which of the following factors would tend to increase membrane fluidity?
   (A) a greater proportion of unsaturated phospholipids
   (B) a greater proportion of saturated phospholipids
   (C) a lower temperature
   (D) a relatively high protein content in the membrane
3. Phosphorylation cascades involving a series of protein kinases are useful for cellular signal transduction because
   (A) they are species specific.
   (B) they always lead to the same cellular response.
   (C) they amplify the original signal manyfold.
   (D) they counter the harmful effects of phosphatases.
4. Lipid-soluble signaling molecules, such as aldosterone, cross the membranes of all cells but affect only target cells because
   (A) only target cells retain the appropriate DNA segments.
   (B) intracellular receptors are present only in target cells.
   (C) only target cells have enzymes that break down aldosterone.
   (D) only in target cells is aldosterone able to initiate the phosphorylation cascade that turns genes on.

Level 2: Application/Analysis

5. Which of the following processes includes all the others?
   (A) osmosis
   (B) diffusion of a solute across a membrane
   (C) passive transport
   (D) transport of an ion down its electrochemical gradient
6. Based on Figure 17, which of these experimental treatments would increase the rate of sucrose transport into a plant cell?
   (A) decreasing extracellular sucrose concentration
   (B) decreasing extracellular pH
   (C) decreasing cytoplasmic pH
   (D) adding a substance that makes the membrane more permeable to hydrogen ions

Level 3: Synthesis/Evaluation

7. **SCIENTIFIC INQUIRY**
   An experiment is designed to study the mechanism of sucrose uptake by plant cells. Cells are immersed in a sucrose solution, and the pH of the solution is monitored. Samples of the cells are taken at intervals, and their sucrose concentration is measured. The pH is observed to decrease until it reaches a steady, slightly acidic level, and then sucrose uptake begins. (a) Evaluate these results and propose a hypothesis to explain them. (b) Predict what would happen if an inhibitor of ATP regeneration by the cell were added to the beaker once the pH was at a steady level? Explain your thinking.
8. **SCIENCE, TECHNOLOGY, AND SOCIETY**
   Extensive irrigation in arid regions causes salts to accumulate in the soil. (When water evaporates, salts that were dissolved in the water are left behind in the soil.) Based on what you have learned about water balance in plant cells, explain why increased soil salinity (saltiness) might be harmful to crops.
9. **FOCUS ON EVOLUTION**
   Paramecium and other unicellular eukaryotes that live in hypotonic environments have cell membranes that limit water uptake, while those living in isotonic environments have membranes that are more permeable to water. Describe what water regulation adaptations might have evolved in unicellular eukaryotes in hypertonic habitats such as the Great Salt Lake and in habitats with changing salt concentration.
10. **FOCUS ON INTERACTIONS**
    A human pancreatic cell obtains O_2—and necessary molecules such as glucose, amino acids, and cholesterol—from its environment, and it releases CO_2 as a waste product. In response to hormonal signals, the cell secretes digestive enzymes. It also regulates its ion concentrations by exchange with its environment. Based on what you have just learned about the structure and function of cellular membranes, write a short essay (100–150 words) to describe how such a cell accomplishes these interactions with its environment.
11. **SYNTHESIZE YOUR KNOWLEDGE**

In the supermarket, lettuce and other produce are often sprayed with water. Explain why this makes vegetables crisp.

Answers

Figure Questions

Figure 3

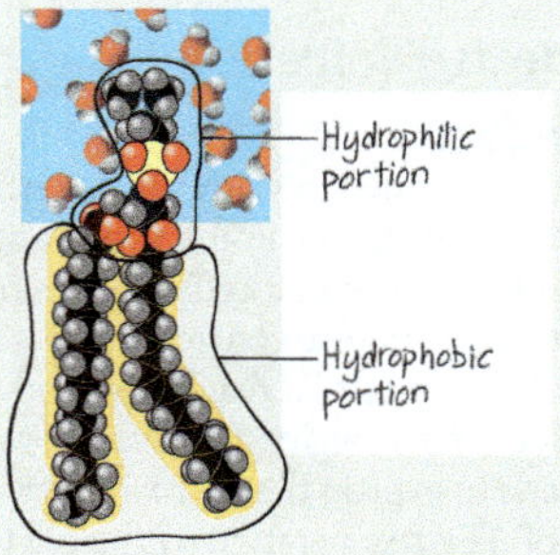

The hydrophilic portion is in contact with an aqueous environment (cytosol or extracellular fluid), and the hydrophobic portion is in contact with the hydrophobic portions of other phospholipids in the interior of the bilayer. **Figure 4** You couldn't rule out movement of proteins within membranes of the same species. You might propose that the membrane lipids and proteins from one species weren't able to mingle with those from the other species because of some incompatibility. **Figure 7** A transmembrane protein like the dimer in (f) might change its shape upon binding to a particular extracellular matrix (ECM) molecule. The new shape might enable the interior portion of the protein to bind to a second, cytoplasmic protein that would relay the message to the inside of the cell, as shown in (c).

Figure 8

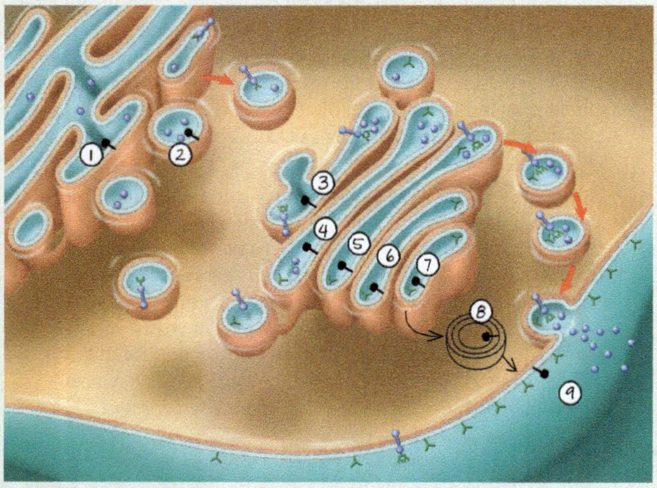

The protein would contact the extracellular fluid. **Figure 10** The orange dye would be evenly distributed throughout the solution on both sides of the membrane. The solution levels would not be affected because the orange dye can diffuse through the membrane and equalize its concentration. Thus, no additional osmosis would take place in either direction. **Figure 15** The diamond solutes are moving into the cell (down), and the round solutes are moving out of the cell (up); each is moving against its concentration gradient. **Figure 23** The aldosterone molecule is hydrophobic and can therefore pass directly through the lipid bilayer of the plasma membrane into the cell. (Hydrophilic molecules cannot do this.) **Figure 24** The active form of protein kinase 1

Concept Check 1

1. They are on the inner side of the transport vesicle membrane. **2.** The grasses living in the cooler region would be expected to have more unsaturated fatty acids in their membranes because those fatty acids remain fluid at lower temperatures. The grasses living immediately adjacent to the hot springs would be expected to have more saturated fatty acids, which would allow the fatty acids to "stack" more closely, making the membranes less fluid and therefore helping them to stay intact at higher temperatures. (Cholesterol could not moderate the effects of temperature on membrane fluidity in this case because it is not found within plant cell membranes.)

Concept Check 2

1. O_2 and CO_2 are both nonpolar molecules that can easily pass through the hydrophobic interior of a membrane. **2.** Water is a polar molecule, so it cannot pass very rapidly through the hydrophobic region in the middle of a phospholipid bilayer. **3.** The hydronium ion is charged, while glycerol is not. Charge is probably more significant than size as a basis for exclusion by the aquaporin channel.

Concept Check 3

1. CO_2 is a nonpolar molecule that can diffuse through the plasma membrane. As long as it diffuses away so that the concentration remains low outside the cell, it will continue to exit the cell in this way. (This is the opposite of the case for O_2, described in this section of the text.) **2.** The activity of *Paramecium caudatum*'s contractile vacuole will decrease. The vacuole pumps out excess water that accumulates in the cell; this accumulation occurs only in a hypotonic environment.

Concept Check 4

1. The pump uses ATP. To establish a voltage, ions have to be pumped against their gradients, which requires energy. **2.** Each ion is being transported against its electrochemical gradient. If either ion were transported down its electrochemical gradient, this process *would* be considered cotransport. **3.** The internal environment of a lysosome is acidic, so it has a higher concentration of H^+ than does the cytoplasm. Therefore, you might expect the membrane of the lysosome to have a proton pump such as that shown in Figure 16 to pump H^+ into the lysosome.

Concept Check 5

1. Exocytosis. When a transport vesicle fuses with the plasma membrane, the vesicle membrane becomes part of the plasma membrane.
2.

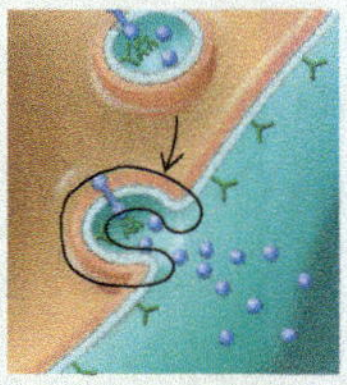

3. The glycoprotein would be synthesized in the ER lumen, move through the Golgi apparatus, and then travel in a vesicle to the plasma membrane, where it would undergo exocytosis and become part of the ECM.

Concept Check 6

1. Glycogen phosphorylase acts in the third stage, the response to epinephrine signaling. **2.** Protein phosphatases reverse the effects of the kinases, and unless the signaling molecule is at a high enough concentration that it is continuously rebinding the receptor, the kinase molecules will all be returned to their inactive states by phosphatases. **3.** At each step in a cascade of sequential activations, one molecule or ion may activate numerous molecules functioning in the next step.

Summary of Key Concepts Questions

1 Plasma membranes define the cell by separating the cellular components from the external environment. This allows conditions inside cells to be controlled by membrane proteins, which regulate entry and exit of molecules and even cell function (see Figure 7). The processes of life can be carried out inside the controlled environment of the cell, so membranes are crucial. In eukaryotes, membranes also function to subdivide the cytoplasm into different compartments where distinct processes can occur, even under differing conditions such as low or high pH. **2** Aquaporins are channel proteins that greatly increase the permeability of a membrane to water molecules, which are polar and therefore do not readily diffuse through the hydrophobic interior of the membrane. **3** There will be a net diffusion of water out of a cell into a hypertonic solution. The free water concentration is higher inside the cell than in the solution (where many water molecules are not free, but are clustered around the higher concentration of solute particles). **4** One of the solutes moved by the cotransporter is actively transported against its concentration gradient. The energy for this transport comes from the concentration gradient of the other solute, which was established by an electrogenic pump that used energy (usually provided by ATP) to transport the other solute across the membrane. **5** In receptor-mediated endocytosis, specific molecules bind to receptors on the plasma membrane in a region where a coated pit develops. The cell can acquire bulk quantities of those specific molecules when the coated pit forms a vesicle and carries the bound molecules into the cell. **6** A cell is able to respond to a hormone only if it has a receptor protein on the cell surface or inside the cell that can bind to the hormone. The response to a hormone depends on the specific signal transduction pathway within the cell, which will lead to the specific cellular response. The response can vary for different types of cells.

Test Your Understanding

1. B **2.** A **3.** C **4.** B **5.** C **6.** B

Credits

Credits are listed in the order of appearance.

Photo Credits

1 Used by permission of B. L. de Groot. Related to work done for: Water Permeation Across Biological Membranes: Mechanism and Dynamics of Aquaporin-1 and GlpF. B. L. de Groot and H. Grubmüller. *Science* 294:2353–2357 (2001); **12** Michael Abbey/Science Source; **5 SSE** Photo Fun/Shutterstock; **18 phagocytosis** Biophoto Associates/Science Source; **pinocytosis** Don W. Fawcett/Science Source; **endocytosis** From M. M. Perry and A. B. Gilbert, *Journal of Cell Science* 39: 257–272, Fig. 11 (1979). © 1979 The Company of Biologists Ltd; **Chapter Review cheetah** Federico Veronesi/Alamy; **SYK** Kristoffer Tripplaar/Alamy.

Illustration and Text Credits

4 Based on L. D. Frye and M. Edidin, The rapid intermixing of cell surface antigens after formation of mouse-human heterokaryons, *Journal of Cell Science* 7:319 (1970); **6** PDB ID 3HAO: N.H. Joh, A. Oberai, D. Yang, J.P. Whitelegge, and J.U. Bowie. Similar energetic contributions of packing in the core of membrane and water-soluble proteins. *J. Am. Chem. Soc.* 131:10846–10847 (2009); **21** Adapted from *The World of the Cell*, 3rd Edition, by W.M. Becker, J.B. Reece, and M.F. Poenie, 1996. Copyright © 1996 by Pearson Education, Inc. Reprinted and electronically reproduced by permission of Pearson Education, Inc. Upper Saddle River, NJ; **25** Adapted from *The World of the Cell*, 3rd Edition, by W.M. Becker, J.B. Reece, and M.F. Poenie, 1996. Copyright © 1996 by Pearson Education, Inc. Reprinted and electronically reproduced by permission of Pearson Education, Inc. Upper Saddle River, NJ; **5 SSE.1** Data from Figure 1 in T. Kondo and E. Beutler, Developmental changes in glucose transport of guinea pig erythrocytes, *Journal of Clinical Investigation* 65:1–4 (1980).

Glossary

active transport The movement of a substance across a cell membrane against its concentration or electrochemical gradient, mediated by specific transport proteins and requiring an expenditure of energy.

amphipathic (am′-fē-path′-ik) Having both a hydrophilic region and a hydrophobic region.

aquaporin A channel protein in the plasma membrane of a plant, animal, or microorganism cell that specifically facilitates osmosis, the diffusion of free water across the membrane.

concentration gradient A region along which the density of a chemical substance increases or decreases.

cotransport The coupling of the "downhill" diffusion of one substance to the "uphill" transport of another against its own concentration gradient.

electrochemical gradient The diffusion gradient of an ion, which is affected by both the concentration difference of an ion across a membrane (a chemical force) and the ion's tendency to move relative to the membrane potential (an electrical force).

electrogenic pump An active transport protein that generates voltage across a membrane while pumping ions.

endocytosis (en′-dō-sī-tō′-sis) Cellular uptake of biological molecules and particulate matter via formation of vesicles from the plasma membrane.

exocytosis (ek′-sō-sī-tō′-sis) The cellular secretion of biological molecules by the fusion of vesicles containing them with the plasma membrane.

facilitated diffusion The passage of molecules or ions down their electrochemical gradient across a biological membrane with the assistance of specific transmembrane transport proteins, requiring no energy expenditure.

flaccid (flas′-id) Limp. Lacking turgor (stiffness or firmness), as in a plant cell in surroundings where there is a tendency for water to leave the cell. (A walled cell becomes flaccid if it has a higher water potential than its surroundings, resulting in the loss of water.)

fluid mosaic model The currently accepted model of cell membrane structure, which envisions the membrane as a mosaic of protein molecules drifting laterally in a fluid bilayer of phospholipids.

G protein A GTP-binding protein that relays signals from a plasma membrane signal receptor, known as a G protein-coupled receptor, to other signal transduction proteins inside the cell.

gated channel A transmembrane protein channel that opens or closes in response to a particular stimulus.

gated ion channel A gated channel for a specific ion. The opening or closing of such channels may alter a cell's membrane potential.

glycolipid A lipid with one or more covalently attached carbohydrates.

hormone In multicellular organisms, one of many types of secreted chemicals that are formed in specialized cells, travel in body fluids, and act on specific target cells in other parts of the body, changing the target cells' functioning. Hormones are thus important in long-distance signaling.

hypertonic Referring to a solution that, when surrounding a cell, will cause the cell to lose water.

hypotonic Referring to a solution that, when surrounding a cell, will cause the cell to take up water.

integral protein A transmembrane protein with hydrophobic regions that extend into and often completely span the hydrophobic interior of the membrane and with hydrophilic regions in contact with the aqueous solution on one or both sides of the membrane (or lining the channel in the case of a channel protein).

ion channel A transmembrane protein channel that allows a specific ion to diffuse across the membrane down its concentration or electrochemical gradient.

isotonic (ī′-sō-ton′-ik) Referring to a solution that, when surrounding a cell, causes no net movement of water into or out of the cell.

ligand (lig′-und) A molecule that binds specifically to another molecule, usually a larger one.

passive transport The diffusion of a substance across a biological membrane with no expenditure of energy.

peripheral protein A protein loosely bound to the surface of a membrane or to part of an integral protein and not embedded in the lipid bilayer.

phosphorylation cascade A series of chemical reactions during cell signaling mediated by enzymes (kinases), in which each kinase in turn phosphorylates and activates another, ultimately leading to phosphorylation of many proteins.

protein kinase An enzyme that transfers phosphate groups from ATP to a protein, thus phosphorylating the protein.

proton pump An active transport protein in a cell membrane that uses ATP to transport hydrogen ions out of a cell against their concentration gradient, generating a membrane potential in the process.

reception The binding of a signaling molecule to a receptor protein, activating the receptor by causing it to change shape.

second messenger A small, nonprotein, water-soluble molecule or ion, such as a calcium ion (Ca^{2+}) or cyclic AMP, that relays a signal to a cell's interior in response to a signaling molecule bound by a signal receptor protein.

selective permeability A property of biological membranes that allows them to regulate the passage of substances across them.

signal transduction pathway A series of steps linking a mechanical, chemical, or electrical stimulus to a specific cellular response.

tonicity The ability of a solution surrounding a cell to cause that cell to gain or lose water.

transport protein A transmembrane protein that helps a certain substance or class of closely related substances to cross the membrane.

An Introduction to Metabolism

From Chapter 6 of *Campbell Biology in Focus*, Second Edition. Lisa A. Urry, Michael L. Cain, Steven A. Wasserman, Peter V. Minorsky, and Jane B. Reece.

An Introduction to Metabolism

KEY CONCEPTS

1. An organism's metabolism transforms matter and energy
2. The free-energy change of a reaction tells us whether or not the reaction occurs spontaneously
3. ATP powers cellular work by coupling exergonic reactions to endergonic reactions
4. Enzymes speed up metabolic reactions by lowering energy barriers
5. Regulation of enzyme activity helps control metabolism

▲ **Figure 1** What causes these breaking waves to glow?

The Energy of Life

The living cell is a chemical factory in miniature, where thousands of reactions occur within a microscopic space. Sugars can be converted to amino acids that are linked together into proteins when needed, and when food is digested, proteins are dismantled into amino acids that can be converted to sugars. The process called cellular respiration drives the cellular economy by extracting the energy stored in sugars and other fuels. Cells apply this energy to perform various types of work. In an exotic example, the ocean waves shown in **Figure 1** are brightly illuminated from within by free-floating single-celled marine organisms called dinoflagellates. These dinoflagellates convert the energy stored in certain organic molecules to light, a process called bioluminescence. Such metabolic activities are precisely coordinated and controlled in the cell. In its complexity, its efficiency, and its responsiveness to subtle changes, the cell is peerless as a chemical factory. The concepts of metabolism that you learn in this chapter will help you understand how matter and energy flow during life's processes and how that flow is regulated.

CONCEPT 1

An organism's metabolism transforms matter and energy

The totality of an organism's chemical reactions is called **metabolism** (from the Greek *metabole*, change). Metabolism is an emergent property of life that arises from orderly interactions between molecules.

Metabolic Pathways

We can picture a cell's metabolism as an elaborate road map of many chemical reactions, arranged as intersecting metabolic pathways. In a **metabolic pathway**, a specific molecule is altered in a series of defined steps, resulting in a product. Each step of the pathway is catalyzed by a specific enzyme:

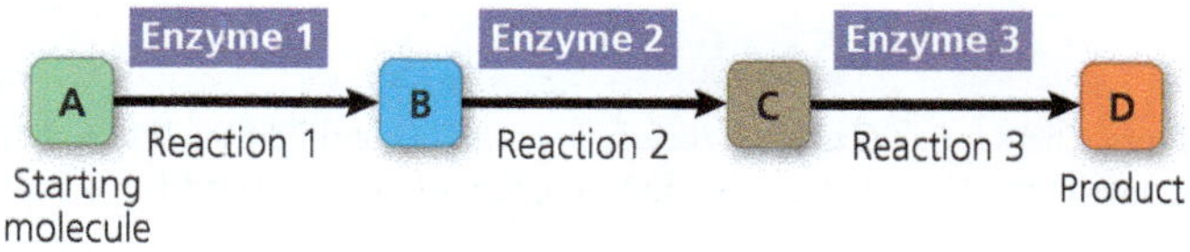

Analogous to the red, yellow, and green stoplights that control the flow of automobile traffic, mechanisms that regulate enzymes balance metabolic supply and demand.

Metabolism as a whole manages the material and energy resources of the cell. Some metabolic pathways release energy by breaking down complex molecules to simpler compounds. These degradative processes are called **catabolic pathways**, or breakdown pathways. A major pathway of catabolism is cellular respiration, in which the sugar glucose and other organic fuels are broken down in the presence of oxygen to carbon dioxide and water. Energy stored in the organic molecules becomes available to do the work of the cell, such as ciliary beating or membrane transport. **Anabolic pathways**, in contrast, consume energy to build complicated molecules from simpler ones; they are sometimes called biosynthetic pathways. Examples of anabolism are the synthesis of an amino acid from simpler molecules and the synthesis of a protein from amino acids. Catabolic and anabolic pathways are the "downhill" and "uphill" avenues of the metabolic landscape. Energy released from the downhill reactions of catabolic pathways can be stored and then used to drive the uphill reactions of anabolic pathways.

In this chapter, we will focus on mechanisms common to metabolic pathways. Because energy is fundamental to all metabolic processes, a basic knowledge of energy is necessary to understand how the living cell works. Although we will use some nonliving examples to study energy, the concepts demonstrated by these examples also apply to **bioenergetics**, the study of how energy flows through living organisms.

Forms of Energy

Energy is the capacity to cause change. In everyday life, energy is important because some forms of energy can be used to do work—that is, to move matter against opposing forces, such as gravity and friction. Put another way, energy is the ability to rearrange a collection of matter. For example, you expend energy to turn the pages of this text, and your cells expend energy in transporting certain substances across membranes. Energy exists in various forms, and the work of life depends on the ability of cells to transform energy from one form to another.

Energy can be associated with the relative motion of objects; this energy is called **kinetic energy**. Moving objects can perform work by imparting motion to other matter: Water gushing through a dam turns turbines, and the contraction of leg muscles pushes bicycle pedals. **Thermal energy** is kinetic energy associated with the random movement of atoms or molecules; thermal energy in transfer from one object to another is called **heat**. Light is also a type of energy that can be harnessed to perform work, such as powering photosynthesis in green plants.

An object not presently moving may still possess energy. Energy that is not kinetic is called **potential energy**; it is energy that matter possesses because of its location or structure. Water behind a dam, for instance, possesses energy because of its altitude above sea level. Molecules possess energy because of the arrangement of electrons in the bonds between their atoms. **Chemical energy** is a term used by biologists to refer to the potential energy available for release in a chemical reaction. Recall that catabolic pathways release energy by breaking down complex molecules. Biologists say that these complex molecules, such as glucose, are high in chemical energy. During a catabolic reaction, some bonds are broken and others are formed, releasing energy and resulting in lower-energy breakdown products. This transformation also occurs in the engine of a car when the hydrocarbons of gasoline react explosively with oxygen, releasing the energy that pushes the pistons and producing exhaust. Although less explosive, a similar reaction of food molecules with oxygen provides chemical energy in biological systems, producing carbon dioxide and water as waste products. Biochemical pathways, carried out in the context of cellular structures, enable cells to release chemical energy from food molecules and use the energy to power life processes.

How is energy converted from one form to another? Consider **Figure 2**. The young woman climbing the ladder to the diving platform is releasing chemical energy from the food she ate for lunch and using some of that energy to perform the work of climbing. The kinetic energy of muscle movement is thus being transformed into potential energy due to her increasing height above the water. The young man diving is converting his potential energy to kinetic energy, which is then transferred to the water as he enters it. A small amount of energy is lost as heat due to friction.

Now let's consider the original source of the organic food molecules that provided the necessary chemical energy for the diver to climb the steps. This chemical energy was itself derived from light energy absorbed by plants during photosynthesis. Organisms are energy transformers.

▲ **Figure 2 Transformations between potential and kinetic energy.**

The Laws of Energy Transformation

The study of the energy transformations that occur in a collection of matter is called **thermodynamics**. Scientists use the word *system* to denote the matter under study; they refer to the rest of the universe—everything outside the system—as the *surroundings*. An *isolated system*, such as that approximated by liquid in a thermos bottle, is unable to exchange either energy or matter with its surroundings outside the thermos. In an *open system*, energy and matter can be transferred between the system and its surroundings. Organisms are open systems. They absorb energy—for instance, light energy or chemical energy in the form of organic molecules—and release heat and metabolic waste products, such as carbon dioxide, to the surroundings. Two laws of thermodynamics govern energy transformations in organisms and all other collections of matter.

The First Law of Thermodynamics

According to the **first law of thermodynamics**, the energy of the universe is constant: *Energy can be transferred and or transformed, but it cannot be created or destroyed.* The first law is also known as the *principle of conservation of energy*. The electric company does not make energy, but merely converts it to a form that is convenient for us to use. By converting sunlight to chemical energy, a plant acts as an energy transformer, not an energy producer.

The brown bear in **Figure 3a** will convert the chemical energy of the organic molecules in its food to kinetic and other forms of energy as it carries out biological processes. What happens to this energy after it has performed work? The second law of thermodynamics helps to answer this question.

The Second Law of Thermodynamics

If energy cannot be destroyed, why can't organisms simply recycle their energy over and over again? It turns out that during every energy transfer or transformation, some energy is converted to thermal energy and released as heat, becoming unavailable to do work. Only a small fraction of the chemical energy from the food in Figure 3a is transformed into the motion of the brown bear shown in **Figure 3b**; most is lost as heat, which dissipates rapidly through the surroundings.

A system can put thermal energy to work only when there is a temperature difference that results in the thermal energy flowing as heat from a warmer location to a cooler one. If temperature is uniform, as it is in a living cell, then the heat generated during a chemical reaction will simply warm a body of matter, such as the organism. (This can make a room crowded with people uncomfortably warm, as each person is carrying out a multitude of chemical reactions!)

A logical consequence of the loss of usable energy as heat to the surroundings is that each energy transfer or transformation makes the universe more disordered. Scientists use a quantity called **entropy** as a measure of disorder, or randomness. The more randomly arranged a collection of matter is, the greater its entropy. We can now state the **second law of thermodynamics**: *Every energy transfer or transformation increases the entropy of the universe.* Although order can increase locally, there is an unstoppable trend toward randomization of the universe as a whole.

In many cases, increased entropy is evident in the physical disintegration of a system's organized structure. For example, you can observe increasing entropy in the gradual decay of an unmaintained building. Much of the increasing entropy of the universe is less obvious, however, because it takes the form of increasing amounts of heat and less ordered forms of matter. As the bear in Figure 3b converts chemical energy to kinetic energy, it is also increasing the disorder of its surroundings by producing heat and small molecules, such as the CO_2 it exhales, that are the breakdown products of food.

The concept of entropy helps us understand why certain processes are energetically favorable and occur on their own. It turns out that if a given process, by itself, leads to an increase in entropy, that process can proceed without requiring

(a) First law of thermodynamics: Energy can be transferred or transformed but neither created nor destroyed. For example, chemical reactions in this brown bear will convert the chemical (potential) energy in the fish into the kinetic energy of running.

(b) Second law of thermodynamics: Every energy transfer or transformation increases the disorder (entropy) of the universe. For example, as the bear runs, disorder is increased around it by the release of heat and small molecules that are the by-products of metabolism. A brown bear can run at speeds up to 35 miles per hour (56 km/hr) —as fast as a racehorse.

▲ **Figure 3 The two laws of thermodynamics.**

an input of energy. Such a process is called a **spontaneous process**. Note that as we're using it here, the word *spontaneous* does not imply that the process would occur quickly; rather, the word signifies that it is energetically favorable. (In fact, it may be helpful for you to think of the phrase "energetically favorable" when you read the formal term "spontaneous.") Some spontaneous processes, such as an explosion, may be virtually instantaneous, while others, such as the rusting of an old car over time, are much slower.

A process that, on its own, leads to a decrease in entropy is said to be nonspontaneous: It will happen only if energy is supplied. We know from experience that certain events occur spontaneously and others do not. For instance, we know that water flows downhill spontaneously but moves uphill only with an input of energy, such as when a machine pumps the water against gravity. This understanding gives us another way to state the second law: *For a process to occur spontaneously, it must increase the entropy of the universe.*

Biological Order and Disorder

Living systems increase the entropy of their surroundings, as predicted by thermodynamic law. It is true that cells create ordered structures from less organized starting materials. For example, simpler molecules are ordered into the more complex structure of an amino acid, and amino acids are ordered into polypeptide chains. At the organismal level as well, complex and beautifully ordered structures result from biological processes that use simpler starting materials (**Figure 4**). On the other hand, an organism also takes in organized forms of matter and energy from the surroundings and replaces them with less ordered forms. For example, an animal obtains starch, proteins, and other complex molecules from the food it eats. As catabolic pathways break these molecules down, the animal releases carbon dioxide and water—small molecules that possess less chemical energy than the food did. The depletion of chemical energy is accounted for by heat generated during metabolism. On a larger scale, energy flows into most ecosystems in the form of light and exits in the form of heat.

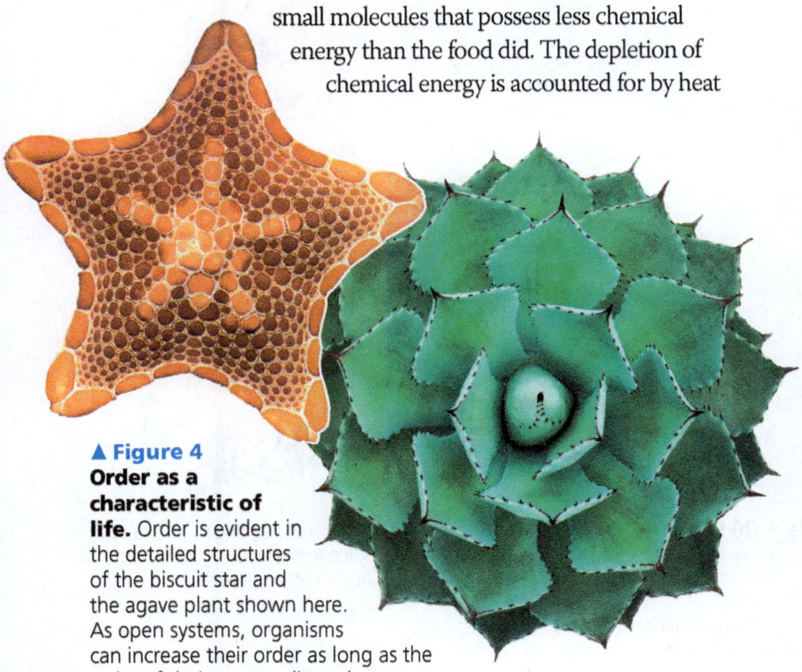

▲ **Figure 4** **Order as a characteristic of life.** Order is evident in the detailed structures of the biscuit star and the agave plant shown here. As open systems, organisms can increase their order as long as the order of their surroundings decreases.

During the early history of life, complex organisms evolved from simpler ancestors. For instance, we can trace the ancestry of the plant kingdom from much simpler organisms called green algae to more complex flowering plants. However, this increase in organization over time in no way violates the second law. The entropy of a particular system, such as an organism, may actually decrease as long as the total entropy of the *universe*—the system plus its surroundings—increases. Thus, organisms are islands of low entropy in an increasingly random universe. The evolution of biological order is perfectly consistent with the laws of thermodynamics.

CONCEPT CHECK 1

1. **MAKE CONNECTIONS** How does the second law of thermodynamics help explain the diffusion of a substance across a membrane?
2. Describe the forms of energy found in an apple as it grows on a tree, falls, and then is digested by someone who eats it.

CONCEPT 2

The free-energy change of a reaction tells us whether or not the reaction occurs spontaneously

The laws of thermodynamics that we've just discussed apply to the universe as a whole. As biologists, we want to understand the chemical reactions of life—for example, which reactions occur spontaneously and which ones require some input of energy from outside. But how can we know this without assessing the energy and entropy changes in the entire universe for each separate reaction?

Free-Energy Change (ΔG), Stability, and Equilibrium

Recall that the universe is really equivalent to "the system" plus "the surroundings." In 1878, J. Willard Gibbs, a professor at Yale, defined a very useful function called the Gibbs free energy of a system (without considering its surroundings), symbolized by the letter G. We'll refer to the Gibbs free energy simply as free energy. **Free energy** is the portion of a system's energy that can perform work when temperature and pressure are uniform throughout the system, as in a living cell. Biologists find it most informative to focus on the *change* in free energy (ΔG) during the chemical reactions of life. ΔG represents the difference between the free energy of the final state and the free energy of the initial state:

$$\Delta G = G_{\text{final state}} - G_{\text{initial state}}$$

Using chemical methods, we can measure ΔG for any reaction. More than a century of experiments has shown that only reactions with a negative ΔG can occur with no input of energy, so the value of ΔG tells us whether a particular reaction is a spontaneous one. This principle is very important in the study of metabolism, where a major goal is to determine which reactions occur spontaneously and can be harnessed to supply energy for cellular work.

For a reaction to have a negative ΔG, the system must lose free energy during the change from initial state to final state. Because it has less free energy, the system in its final state is less likely to change and is therefore more stable than it was previously. We can think of free energy as a measure of a system's instability—its tendency to change to a more stable state. Unstable systems (higher G) tend to change in such a way that they become more stable (lower G), as shown in **Figure 5**.

Another term that describes a state of maximum stability is chemical *equilibrium*. At equilibrium, the forward and reverse reactions occur at the same rate, and there is no further net change in the relative concentration of products and reactants. For a system at equilibrium, G is at its lowest possible value in that system. We can think of the equilibrium state as a free-energy valley. Any change from the equilibrium position will have a positive ΔG and will not be spontaneous. For this reason, systems never spontaneously move away from equilibrium. Because a system at equilibrium cannot spontaneously change, it can do no work. *A process is spontaneous and can perform work only when it is moving toward equilibrium.*

Free Energy and Metabolism

We can now apply the free-energy concept more specifically to the chemistry of life's processes.

Exergonic and Endergonic Reactions in Metabolism

Based on their free-energy changes, chemical reactions can be classified as either exergonic ("energy outward") or endergonic ("energy inward"). An **exergonic reaction** proceeds with a net release of free energy **(Figure 6a)**. Because the chemical mixture loses free energy (G decreases), ΔG is negative for an exergonic reaction. Using ΔG as a standard for spontaneity, exergonic reactions are those that occur spontaneously. (Remember, the word *spontaneous* implies that it is energetically favorable, not that it will occur rapidly.) The magnitude of ΔG for an exergonic reaction represents the maximum amount of work the reaction can perform (some of the free energy is released as heat and cannot do work). The greater the decrease in free energy, the greater the amount of work that can be done.

Consider the overall reaction for cellular respiration:

$$C_6H_{12}O_6 + 6\,O_2 \rightarrow 6\,CO_2 + 6\,H_2O$$
$$\Delta G = -686 \text{ kcal/mol } (-2{,}870 \text{ kJ/mol})$$

686 kcal (2,870 kJ) of energy are made available for work for each mole (180 g) of glucose broken down by respiration under "standard conditions" (1 *M* of each reactant and product, 25°C, pH 7). Because energy must be conserved, the products of respiration store 686 kcal less free energy per mole than the reactants. The products are the "exhaust" of a process that tapped the free energy stored in the bonds of the sugar molecules.

It is important to realize that the breaking of bonds does not release energy; on the contrary, as you will soon see, it requires energy. The phrase "energy stored in bonds" is shorthand for the potential energy that can be released when new bonds are

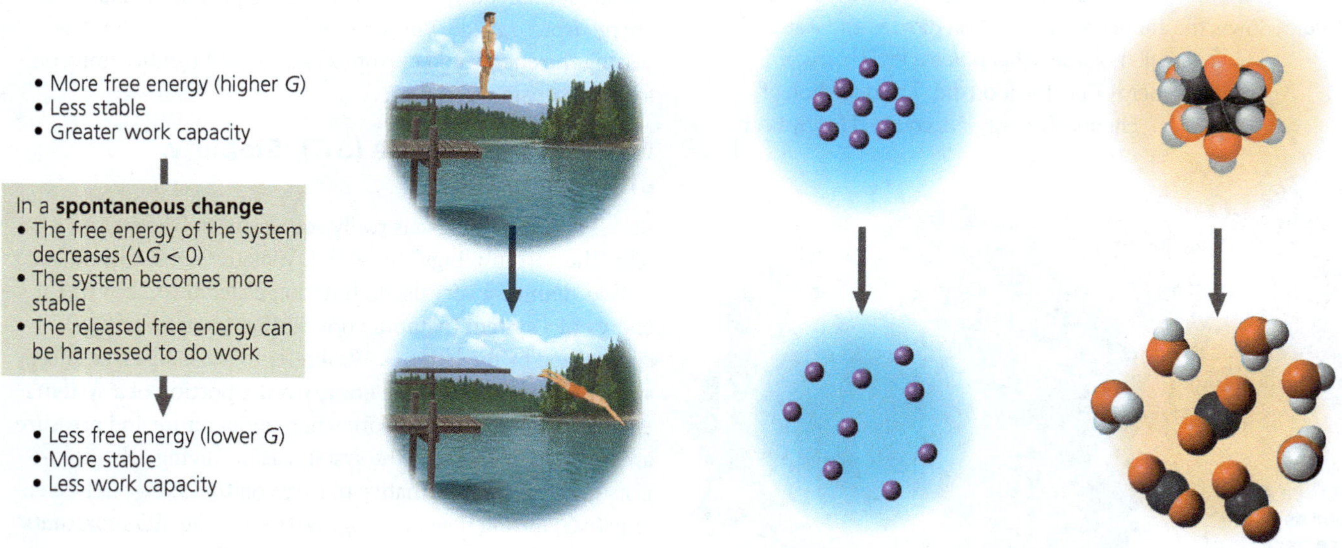

(a) Gravitational motion. Objects move spontaneously from a higher altitude to a lower one.

(b) Diffusion. Molecules in a drop of dye diffuse until they are randomly dispersed.

(c) Chemical reaction. In a cell, a glucose molecule is broken down into simpler molecules.

▲ **Figure 5 The relationship of free energy to stability, work capacity, and spontaneous change.** Unstable systems (top) are rich in free energy, G. They have a tendency to change spontaneously to a more stable state (bottom), and it is possible to harness this "downhill" change to perform work.

▼ **Figure 6 Free energy changes (Δ*G*) in exergonic and endergonic reactions.**

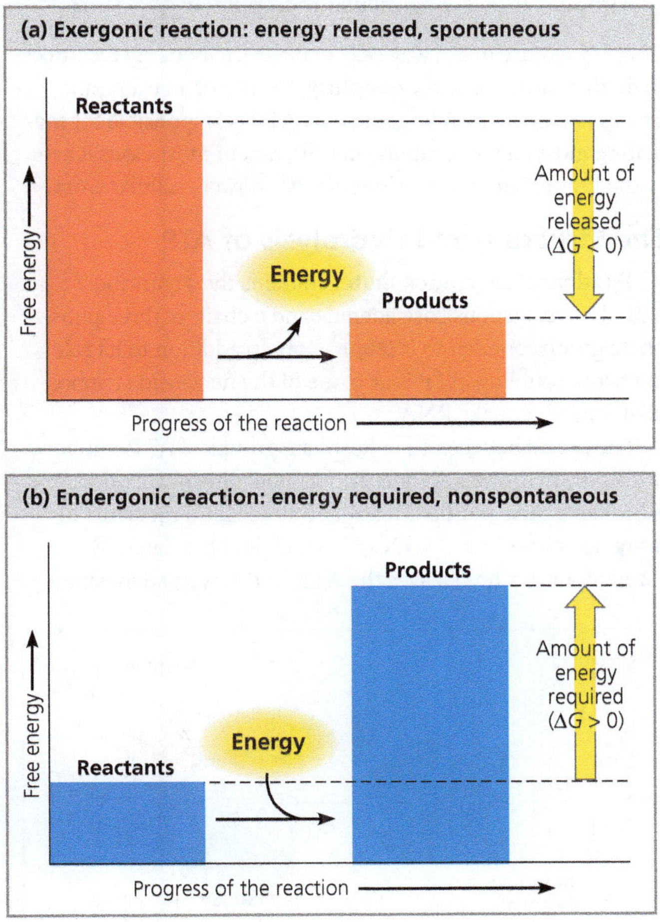

formed after the original bonds break, as long as the products are of lower free energy than the reactants.

An **endergonic reaction** is one that absorbs free energy from its surroundings **(Figure 6b)**. Because this kind of reaction essentially *stores* free energy in molecules (*G* increases), ΔG is positive. Such reactions are nonspontaneous, and the magnitude of ΔG is the quantity of energy required to drive the reaction. If a chemical process is exergonic (downhill), releasing energy in one direction, then the reverse process must be endergonic (uphill), using energy. A reversible process cannot be downhill in both directions. If $\Delta G = -686$ kcal/mol for respiration, which converts glucose and oxygen to carbon dioxide and water, then the reverse process—the conversion of carbon dioxide and water to glucose and oxygen—must be strongly endergonic, with $\Delta G = +686$ kcal/mol. Such a reaction would never happen by itself.

How, then, do plants make the sugar that organisms use for energy? Plants get the required energy—686 kcal to make a mole of glucose—from the environment by capturing light and converting its energy to chemical energy. Next, in a long series of exergonic steps, they gradually spend that chemical energy to assemble glucose molecules.

Equilibrium and Metabolism

Reactions in an isolated system eventually reach equilibrium and can then do no work, as illustrated by the isolated hydroelectric system in **Figure 7a**. The chemical reactions of metabolism are reversible, and they, too, would reach equilibrium if they occurred in the isolation of a test tube. Because systems at equilibrium are at a minimum of *G* and can do no work, a cell that has reached metabolic equilibrium is dead! The fact that metabolism as a whole is never at equilibrium is one of the defining features of life.

Like most systems, a living cell is not in equilibrium. The constant flow of materials in and out of the cell keeps the metabolic pathways from ever reaching equilibrium, and the cell continues to do work throughout its life. This principle is illustrated by the open (and more realistic) hydroelectric system in **Figure 7b**. However, unlike this simple single-step system,

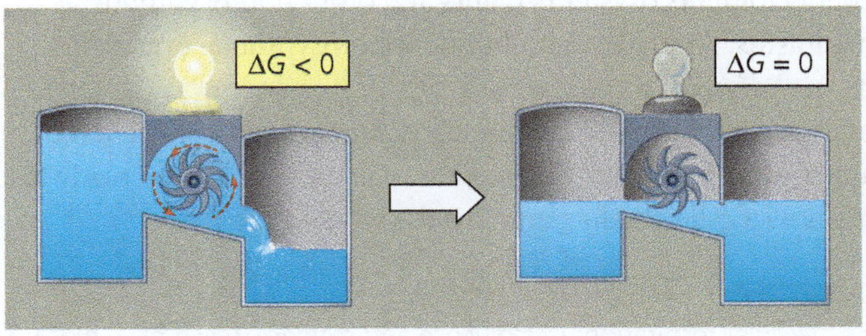

(a) An isolated hydroelectric system. Water flowing downhill turns a turbine that drives a generator providing electricity to a lightbulb, but only until the system reaches equilibrium.

(b) An open hydroelectric system. Flowing water keeps driving the generator because intake and outflow of water keep the system from reaching equilibrium.

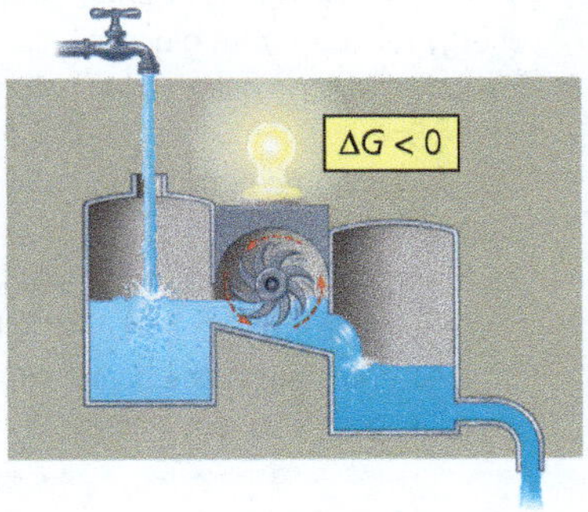

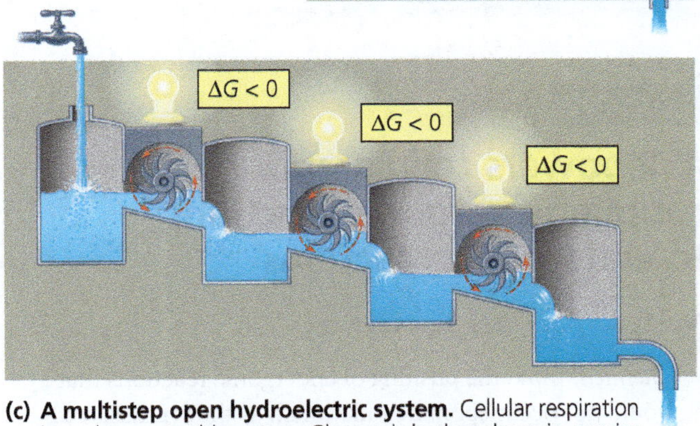

(c) A multistep open hydroelectric system. Cellular respiration is analogous to this system: Glucose is broken down in a series of exergonic reactions that power the work of the cell. The product of each reaction is used as the reactant for the next, so no reaction reaches equilibrium.

▲ **Figure 7 Equilibrium and work in isolated and open systems.**

a catabolic pathway in a cell releases free energy in a series of reactions. An example is cellular respiration, illustrated by the hydroelectric system analogy in **Figure 7c**. Some of the reversible reactions of respiration are constantly "pulled" in one direction—that is, they are kept out of equilibrium. The key to maintaining this lack of equilibrium is that the product of a reaction does not accumulate but instead becomes a reactant in the next step; finally, waste products are expelled from the cell. The overall sequence of reactions is kept going by the huge free-energy difference between glucose and oxygen at the top of the energy "hill" and carbon dioxide and water at the "downhill" end. As long as our cells have a steady supply of glucose or other fuels and oxygen and are able to expel waste products to the surroundings, their metabolic pathways never reach equilibrium and can continue to do the work of life.

We see once again how important it is to think of organisms as open systems. Sunlight provides a daily source of free energy for an ecosystem's plants and other photosynthetic organisms. Animals and other nonphotosynthetic organisms in an ecosystem must have a source of free energy in the form of the organic products of photosynthesis. Now that we have applied the free-energy concept to metabolism, we are ready to see how a cell actually performs the work of life.

CONCEPT CHECK 2

1. Cellular respiration uses glucose and oxygen, which have high levels of free energy, and releases CO_2 and water, which have low levels of free energy. Is cellular respiration spontaneous or not? Is it exergonic or endergonic? What happens to the energy released from glucose?
2. How do the processes of catabolism and anabolism relate to Figure 5c?
3. WHAT IF? Some nighttime partygoers wear glow-in-the-dark necklaces. The necklaces start glowing once they are "activated" by snapping the necklace in a way that allows two chemicals to react and emit light in the form of chemiluminescence. Is this chemical reaction exergonic or endergonic? Explain your answer.

CONCEPT 3

ATP powers cellular work by coupling exergonic reactions to endergonic reactions

A cell does three main kinds of work:

- *Chemical work*, the pushing of endergonic reactions that would not occur spontaneously, such as the synthesis of polymers from monomers (chemical work will be discussed further in this chapter)
- *Transport work*, the pumping of substances across membranes against the direction of spontaneous movement
- *Mechanical work*, such as the beating of cilia Concept6, the contraction of muscle cells, and the movement of chromosomes during cellular reproduction

A key feature in the way cells manage their energy resources to do this work is **energy coupling**, the use of an exergonic process to drive an endergonic one. ATP is responsible for mediating most energy coupling in cells, and in most cases it acts as the immediate source of energy that powers cellular work.

The Structure and Hydrolysis of ATP

ATP (adenosine triphosphate) contains the sugar ribose, with the nitrogenous base adenine and a chain of three phosphate groups bonded to it **(Figure 8a)**. In addition to its role in energy coupling, ATP is also one of the nucleoside triphosphates used to make RNA.

The bonds between the phosphate groups of ATP can be broken by hydrolysis. When the terminal phosphate bond is broken by the addition of a water molecule, a molecule of inorganic phosphate ($HOPO_3^{2-}$, which is abbreviated Ⓟ$_i$ throughout this text) leaves the ATP. In this way, adenosine

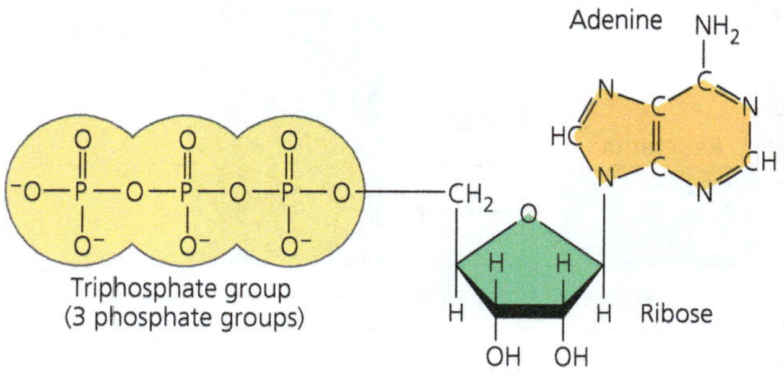

(a) The structure of ATP. In the cell, most hydroxyl groups of phosphates are ionized (—O⁻).

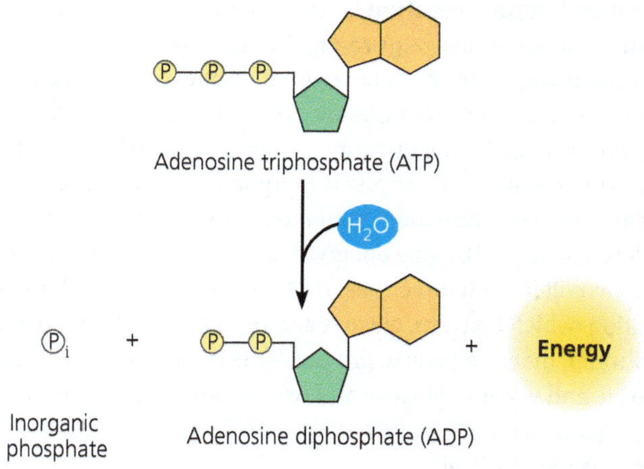

(b) The hydrolysis of ATP. The reaction of ATP and water yields inorganic phosphate (Ⓟ$_i$) and ADP and releases energy.

▲ **Figure 8 The structure and hydrolysis of adenosine triphosphate (ATP).** Throughout this text, the chemical structure of the triphosphate group seen in (a) will be represented by the three joined yellow circles shown in (b).

*tri*phosphate becomes adenosine *di*phosphate, or ADP **(Figure 8b)**. The reaction is exergonic and releases 7.3 kcal of energy per mole of ATP hydrolyzed:

$$ATP + H_2O \rightarrow ADP + \text{Ⓟ}_i$$
$$\Delta G = -7.3 \text{ kcal/mol } (-30.5 \text{ kJ/mol})$$

This is the free-energy change measured under standard conditions. In the cell, conditions do not conform to standard conditions, primarily because reactant and product concentrations differ from 1 *M*. For example, when ATP hydrolysis occurs under cellular conditions, the actual ΔG is about −13 kcal/mol, 78% greater than the energy released by ATP hydrolysis under standard conditions.

Because their hydrolysis releases energy, the phosphate bonds of ATP are sometimes referred to as high-energy phosphate bonds, but the term is misleading. The phosphate bonds of ATP are not unusually strong bonds, as "high-energy" may imply; rather, the reactants (ATP and water) themselves have high energy relative to the energy of the products (ADP and Ⓟ_i). The release of energy during the hydrolysis of ATP comes from the chemical change to a state of lower free energy, not from the phosphate bonds themselves.

ATP is useful to the cell because the energy it releases on losing a phosphate group is somewhat greater than the energy most other molecules could deliver. But why does this hydrolysis release so much energy? If we reexamine the ATP molecule in Figure 8a, we can see that all three phosphate groups are negatively charged. These like charges are crowded together, and their mutual repulsion contributes to the instability of this region of the ATP molecule. The triphosphate tail of ATP is the chemical equivalent of a compressed spring.

How the Hydrolysis of ATP Performs Work

When ATP is hydrolyzed in a test tube, the release of free energy merely heats the surrounding water. In an organism, this same generation of heat can sometimes be beneficial. For instance, the process of shivering uses ATP hydrolysis during muscle contraction to warm the body. In most cases in the cell, however, the generation of heat alone would be an inefficient (and potentially dangerous) use of a valuable energy resource. Instead, the cell's proteins harness the energy released during ATP hydrolysis in several ways to perform the three types of cellular work—chemical, transport, and mechanical.

For example, with the help of specific enzymes, the cell is able to use the energy released by ATP hydrolysis directly to drive chemical reactions that, by themselves, are endergonic **(Figure 9)**. If the ΔG of an endergonic reaction is less than the amount of energy released by ATP hydrolysis, then the two reactions can be coupled so that, overall, the coupled reactions are exergonic. This usually involves phosphorylation, the transfer of a phosphate group from ATP to some other

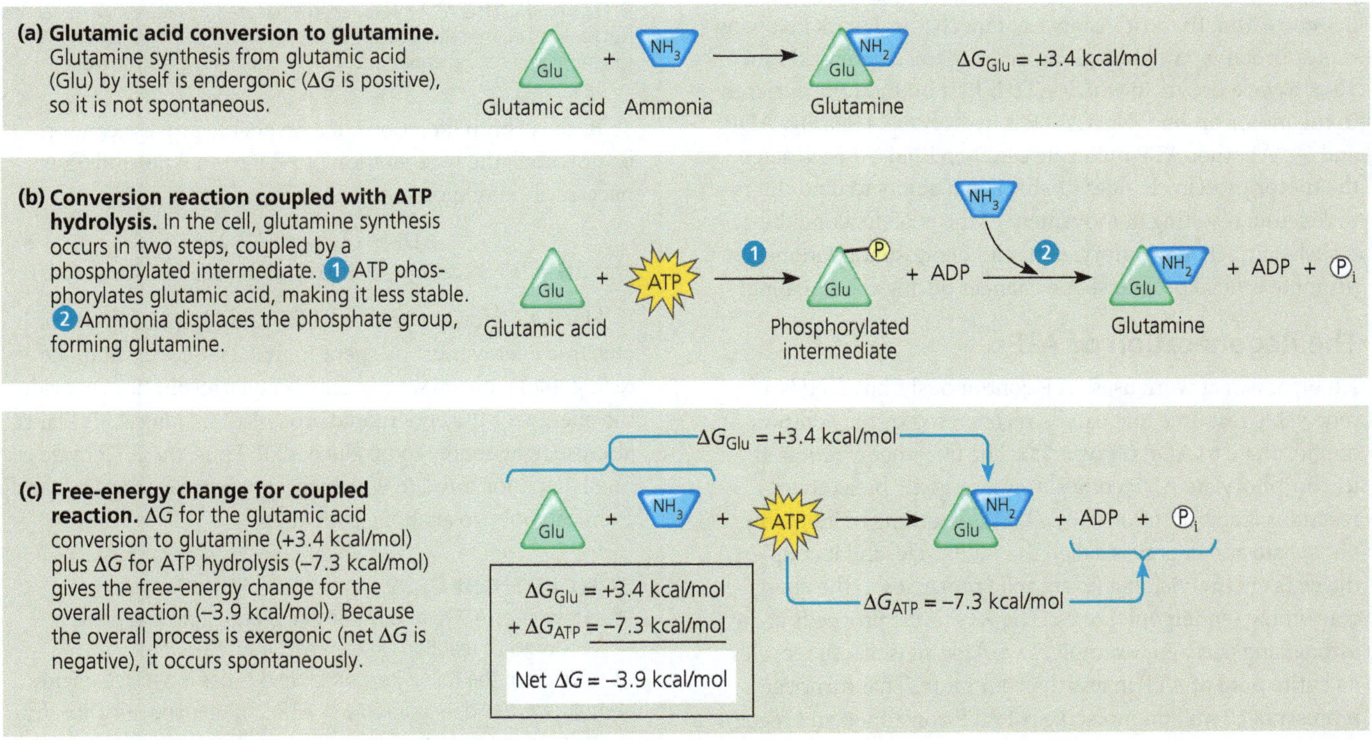

▲ **Figure 9 How ATP drives chemical work: energy coupling using ATP hydrolysis.** In this example, the exergonic process of ATP hydrolysis is used to drive an endergonic process—the cellular synthesis of the amino acid glutamine from glutamic acid and ammonia.

MAKE CONNECTIONS *Explain why glutamine is drawn as a glutamic acid (Glu) with an amino group attached.*

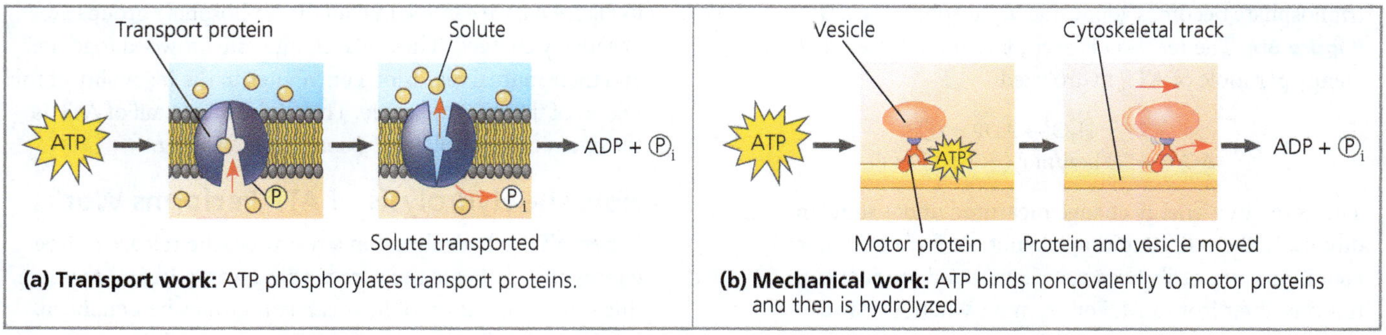

▲ **Figure 10 How ATP drives transport and mechanical work.** ATP hydrolysis causes changes in the shapes and binding affinities of proteins. This can occur either **(a)** directly, by phosphorylation, as shown for a membrane protein carrying out active transport of a solute, or **(b)** indirectly, via noncovalent binding of ATP and its hydrolytic products, as is the case for motor proteins that move vesicles (and other organelles) along cytoskeletal "tracks" in the cell.

molecule, such as the reactant (see Figure 9b). The recipient with the phosphate group covalently bonded to it is then called a **phosphorylated intermediate**. The key to coupling exergonic and endergonic reactions is the formation of this phosphorylated intermediate, which is more reactive (less stable) than the original unphosphorylated molecule.

Transport and mechanical work in the cell are also nearly always powered by the hydrolysis of ATP. In these cases, ATP hydrolysis leads to a change in a protein's shape and often its ability to bind another molecule. Sometimes this occurs via a phosphorylated intermediate, as seen for the transport protein in **Figure 10a**. In most instances of mechanical work involving motor proteins "walking" along cytoskeletal elements **(Figure 10b)**, a cycle occurs in which ATP is first bound noncovalently to the motor protein. Next, ATP is hydrolyzed, releasing ADP and Ⓟ$_i$. Another ATP molecule can then bind. At each stage, the motor protein changes its shape and ability to bind the cytoskeleton, resulting in movement of the protein along the cytoskeletal track. Phosphorylation and dephosphorylation also promote crucial protein shape changes during cell signaling.

The Regeneration of ATP

An organism at work uses ATP continuously, but ATP is a renewable resource that can be regenerated by the addition of phosphate to ADP **(Figure 11)**. The free energy required to phosphorylate ADP comes from exergonic breakdown reactions (catabolism) in the cell. This shuttling of inorganic phosphate and energy is called the ATP cycle, and it couples the cell's energy-yielding (exergonic) processes to the energy-consuming (endergonic) ones. The ATP cycle proceeds at an astonishing pace. For example, a working muscle cell recycles its entire pool of ATP in less than a minute. That turnover represents 10 million molecules of ATP consumed and regenerated per second per cell. If ATP could not be regenerated by the phosphorylation of ADP, humans would use up nearly their body weight in ATP each day.

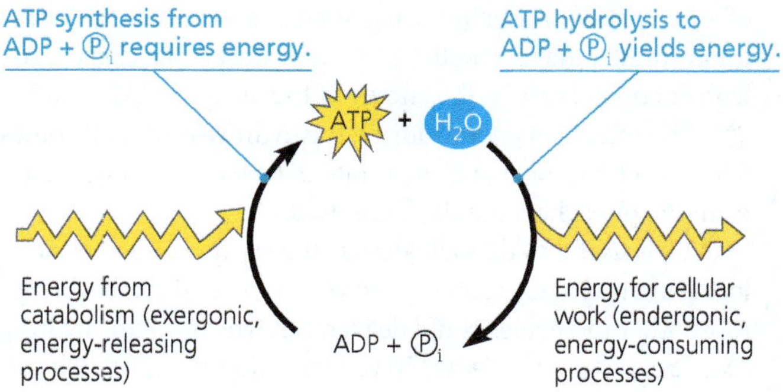

▲ **Figure 11 The ATP cycle.** Energy released by breakdown reactions (catabolism) in the cell is used to phosphorylate ADP, regenerating ATP. Chemical potential energy stored in ATP drives most cellular work.

Because both directions of a reversible process cannot be downhill, the regeneration of ATP from ADP and Ⓟ$_i$ is necessarily endergonic:

$$\text{ADP} + \text{Ⓟ}_i \rightarrow \text{ATP} + H_2O$$
$$\Delta G = +7.3 \text{ kcal/mol } (+30.5 \text{ kJ/mol}) \text{ (standard conditions)}$$

Since ATP formation from ADP and Ⓟ$_i$ is not spontaneous, free energy must be spent to make it occur. Catabolic (exergonic) pathways, especially cellular respiration, provide the energy for the endergonic process of making ATP. Plants also use light energy to produce ATP. Thus, the ATP cycle is a revolving door through which energy passes during its transfer from catabolic to anabolic pathways.

CONCEPT CHECK 3

1. How does ATP typically transfer energy from exergonic to endergonic reactions in the cell?
2. Which of the following combinations has more free energy: glutamic acid + ammonia + ATP or glutamine + ADP + Ⓟ$_i$? Explain your answer.
3. **MAKE CONNECTIONS** Does Figure 10a show passive or active transport? Explain.

CONCEPT 4

Enzymes speed up metabolic reactions by lowering energy barriers

The laws of thermodynamics tell us what will and will not happen under given conditions but say nothing about the *rate* of these processes. A spontaneous chemical reaction occurs without any requirement for outside energy, but it may occur so slowly that it is imperceptible. For example, even though the hydrolysis of sucrose (table sugar) to glucose and fructose is exergonic, occurring spontaneously with a release of free energy ($\Delta G = -7$ kcal/mol), a solution of sucrose dissolved in sterile water will sit for years at room temperature with no appreciable hydrolysis. However, if we add a small amount of the enzyme sucrase to the solution, then all the sucrose may be hydrolyzed within seconds, as shown here:

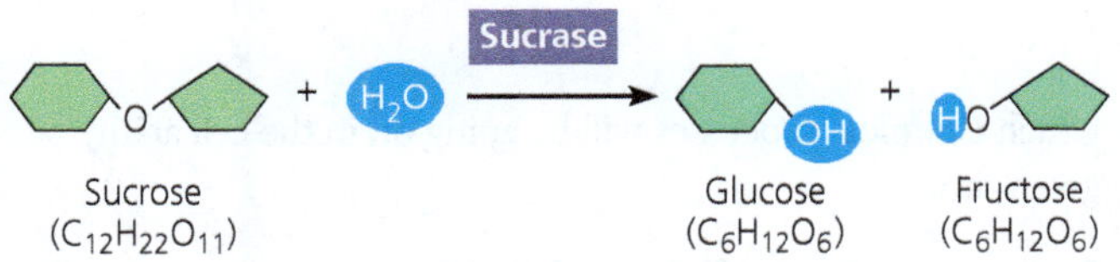

How does the enzyme do this?

An **enzyme** is a macromolecule that acts as a **catalyst**, a chemical agent that speeds up a reaction without being consumed by the reaction. In this chapter, we are focusing on enzymes that are proteins. (Some RNA molecules, called ribozymes, can function as enzymes.) Without regulation by enzymes, chemical traffic through the pathways of metabolism would become terribly congested because many chemical reactions would take such a long time. In the next two sections, we will see why spontaneous reactions can be slow and how an enzyme changes the situation.

The Activation Energy Barrier

Every chemical reaction between molecules involves both bond breaking and bond forming. For example, the hydrolysis of sucrose involves breaking the bond between glucose and fructose and one of the bonds of a water molecule and then forming two new bonds, as shown above. Changing one molecule into another generally involves contorting the starting molecule into a highly unstable state before the reaction can proceed. This contortion can be compared to the bending of a metal key ring when you pry it open to add a new key. The key ring is highly unstable in its opened form but returns to a stable state once the key is threaded all the way onto the ring. To reach the contorted state where bonds can change, reactant molecules must absorb energy from their surroundings. When the new bonds of the product molecules form, energy is released as heat, and the molecules return to stable shapes with lower energy than the contorted state.

The initial investment of energy for starting a reaction—the energy required to contort the reactant molecules so the bonds can break—is known as the *free energy of activation*, or **activation energy**, abbreviated E_A in this text. We can think of activation energy as the amount of energy needed to push the reactants to the top of an energy barrier, or uphill, so that the "downhill" part of the reaction can begin. Activation energy is often supplied by heat in the form of thermal energy that the reactant molecules absorb from the surroundings. The absorption of thermal energy accelerates the reactant molecules, so they collide more often and more forcefully. It also agitates the atoms within the molecules, making the breakage of bonds more likely. When the molecules have absorbed enough energy for the bonds to break, the reactants are in an unstable condition known as the *transition state*.

Figure 12 graphs the energy changes for a hypothetical exergonic reaction that swaps portions of two reactant molecules:

$$\underset{\text{Reactants}}{AB + CD} \rightarrow \underset{\text{Products}}{AC + BD}$$

The activation of the reactants is represented by the uphill portion of the graph, in which the free-energy content of the reactant molecules is increasing. At the summit, when energy

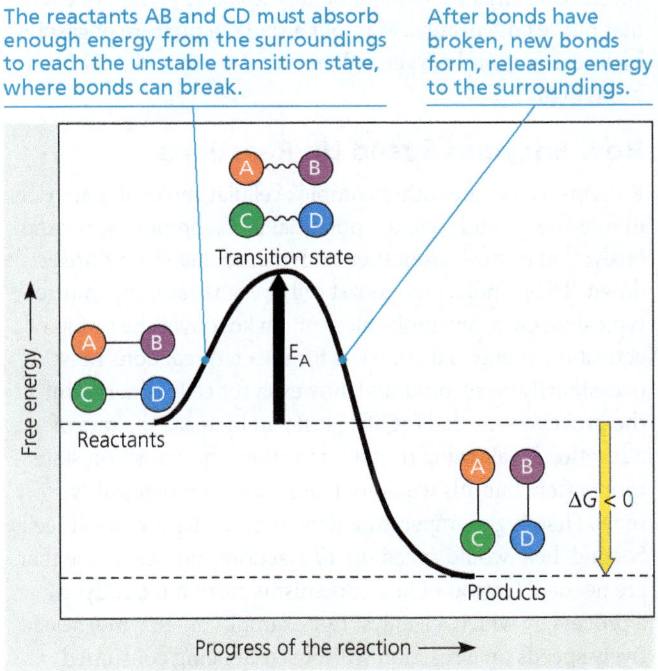

▲ **Figure 12 Energy profile of an exergonic reaction.** The "molecules" are hypothetical, with A, B, C, and D representing portions of the molecules. Thermodynamically, this is an exergonic reaction, with a negative ΔG, and the reaction occurs spontaneously. However, the activation energy (E_A) provides a barrier that determines the rate of the reaction.

DRAW IT *Graph the progress of an endergonic reaction in which EF and GH form products EG and FH, assuming that the reactants must pass through a transition state.*

equivalent to E_A has been absorbed, the reactants are in the transition state: They are activated, and their bonds can be broken. As the atoms then settle into their new, more stable bonding arrangements, energy is released to the surroundings. This corresponds to the downhill part of the curve, which shows the loss of free energy by the molecules. The overall decrease in free energy means that E_A is repaid with dividends, as the formation of new bonds releases more energy than was invested in the breaking of old bonds.

The reaction shown in Figure 12 is exergonic and occurs spontaneously ($\Delta G < 0$). However, the activation energy provides a barrier that determines the rate of the reaction. The reactants must absorb enough energy to reach the top of the activation energy barrier before the reaction can occur. For some reactions, E_A is modest enough that even at room temperature there is sufficient thermal energy for many of the reactant molecules to reach the transition state in a short time. In most cases, however, E_A is so high and the transition state is reached so rarely that the reaction will hardly proceed at all. In these cases, the reaction will occur at a noticeable rate only if energy is provided, usually as heat. For example, the reaction of gasoline and oxygen is exergonic and will occur spontaneously, but energy is required for the molecules to reach the transition state and react. Only when the spark plugs fire in an automobile engine can there be the explosive release of energy that pushes the pistons. Without a spark, a mixture of gasoline hydrocarbons and oxygen will not react because the E_A barrier is too high.

How Enzymes Speed Up Reactions

Proteins, DNA, and other complex cellular molecules are rich in free energy and have the potential to decompose spontaneously; that is, the laws of thermodynamics favor their breakdown. These molecules persist only because at temperatures typical for cells, few molecules can make it over the hump of activation energy. The barriers for selected reactions must occasionally be surmounted, however, for cells to carry out the processes needed for life. Heat can increase the rate of a reaction by allowing reactants to attain the transition state more often, but this would not work well in biological systems. First, high temperature denatures proteins and kills cells. Second, heat would speed up *all* reactions, not just those that are needed. Instead of heat, organisms carry out **catalysis**, a process by which a catalyst (for example, an enzyme) selectively speeds up a reaction without itself being consumed.

An enzyme catalyzes a reaction by lowering the E_A barrier **(Figure 13)**, enabling the reactant molecules to absorb enough energy to reach the transition state even at moderate temperatures. An enzyme cannot change the ΔG for a reaction; it cannot make an endergonic reaction exergonic. Enzymes can only hasten reactions that would eventually occur anyway, but this enables the cell to have a dynamic metabolism, routing chemicals smoothly through metabolic pathways. Also, enzymes are very specific for the reactions they catalyze, so they determine which chemical processes will be going on in the cell at any given time.

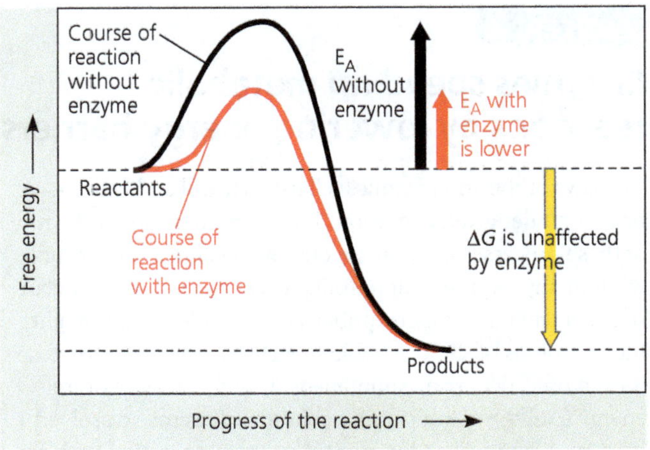

▲ **Figure 13 The effect of an enzyme on activation energy.** Without affecting the free-energy change (ΔG) for a reaction, an enzyme speeds the reaction by reducing its activation energy (E_A).

Substrate Specificity of Enzymes

The reactant an enzyme acts on is referred to as the enzyme's **substrate**. The enzyme binds to its substrate (or substrates, when there are two or more reactants), forming an **enzyme-substrate complex**. While enzyme and substrate are joined, the catalytic action of the enzyme converts the substrate to the product (or products) of the reaction. The overall process can be summarized as follows:

$$\text{Enzyme + Substrate(s)} \rightleftharpoons \text{Enzyme-substrate complex} \rightleftharpoons \text{Enzyme + Product(s)}$$

For example, the enzyme sucrase (most enzyme names end in *-ase*) catalyzes the hydrolysis of the disaccharide sucrose into its two monosaccharides, glucose and fructose (see the illustrated equation at the beginning of Concept 4):

$$\text{Sucrase + Sucrose + } H_2O \rightleftharpoons \text{Sucrase-sucrose-}H_2O\text{ complex} \rightleftharpoons \text{Sucrase + Glucose + Fructose}$$

The reaction catalyzed by each enzyme is very specific; an enzyme can recognize its specific substrate even among closely related compounds. For instance, sucrase will act only on sucrose and will not bind to other disaccharides, such as maltose. What accounts for this molecular recognition? Recall that most enzymes are proteins, and proteins are macromolecules with unique three-dimensional configurations. The specificity of an enzyme results from its shape, which is a consequence of its amino acid sequence.

Only a restricted region of the enzyme molecule actually binds to the substrate. This region, known as the **active site**, is typically a pocket or groove on the surface of the enzyme where catalysis occurs (**Figure 14a**). Usually, the active site is formed by only a few of the enzyme's amino acids, with the rest of the protein molecule providing a framework that determines the shape of the active site. The specificity of an enzyme is attributed to a complementary fit between the shape of its active site and the shape of the substrate, like that seen in the binding of a signaling molecule to a receptor protein.

An enzyme is not a stiff structure locked into a given shape. In fact, recent work by biochemists has shown that enzymes (and other proteins) seem to "dance" between subtly different shapes in a dynamic equilibrium, with slight differences in free energy for each "pose." The shape that best fits the substrate isn't necessarily the one with the lowest energy, but during the very short time the enzyme takes on this shape, its active site can bind to the substrate. The active site itself is also not a rigid receptacle for the substrate. As the substrate enters the active site, the enzyme changes shape slightly due to interactions between the substrate's chemical groups and chemical groups on the side chains of the amino acids that form the active site. This shape change makes the active site fit even more snugly around the substrate (**Figure 14b**). This tightening of the binding after initial contact—called **induced fit**—is like a clasping handshake. Induced fit brings chemical groups of the active site into positions that enhance their ability to catalyze the chemical reaction.

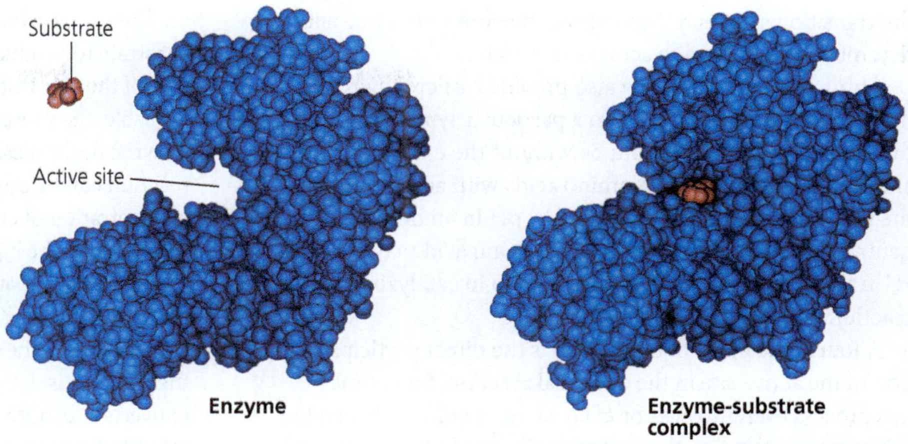

(a) In this computer graphic model, the active site of this enzyme (hexokinase, shown in blue) forms a groove on its surface. Its substrate is glucose (red).

(b) When the substrate enters the active site, it forms weak bonds with the enzyme, inducing a change in the shape of the protein. This change allows additional weak bonds to form, causing the active site to enfold the substrate and hold it in place.

▲ **Figure 14 Induced fit between an enzyme and its substrate.**

Catalysis in the Enzyme's Active Site

In most enzymatic reactions, the substrate is held in the active site by so-called weak interactions, such as hydrogen bonds and ionic bonds. The R groups of a few of the amino acids that make up the active site catalyze the conversion of substrate to product, and the product departs from the active site. The enzyme is then free to take another substrate molecule into its active site. The entire cycle happens so fast that a single enzyme molecule typically acts on about 1,000 substrate molecules per second, and some enzymes are even faster. Enzymes, like other catalysts, emerge from the reaction in their original form. Therefore, very small amounts of enzyme can have a huge metabolic impact by functioning over and over again in catalytic cycles. **Figure 15** shows a catalytic cycle involving two substrates and two products.

Most metabolic reactions are reversible, and an enzyme can catalyze either the forward or the reverse reaction, depending on which direction has a negative ΔG. This in turn depends mainly on the relative concentrations of reactants and products. The net effect is always in the direction of equilibrium.

Enzymes use a variety of mechanisms that lower activation energy and speed up a reaction. First, in reactions involving two or more reactants, the active site provides a template on which the substrates can come together in the proper orientation for a reaction to occur between them (see Figure 15, step 2). Second, as the active site of an enzyme clutches the bound substrates, the enzyme may stretch the substrate molecules toward their transition state form, stressing and bending critical chemical bonds that must be broken during the reaction. Because E_A is proportional to the difficulty of breaking the bonds, distorting the substrate helps it approach

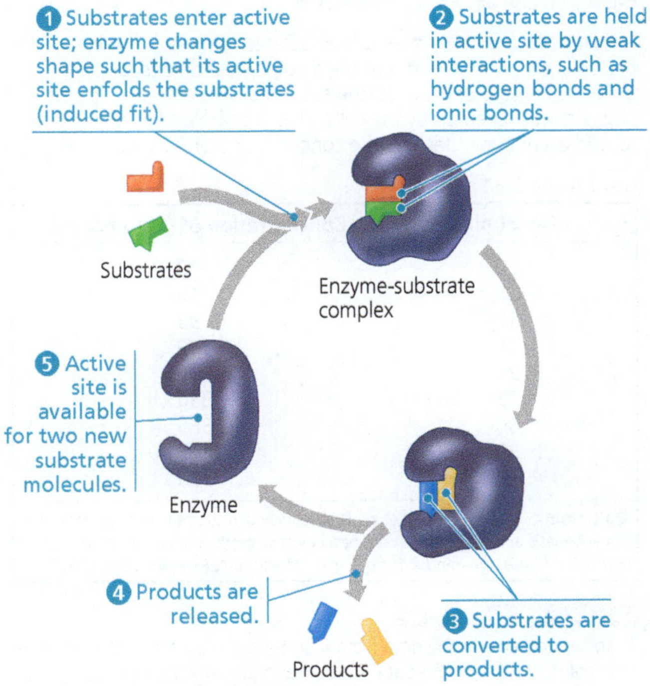

▲ **Figure 15 The active site and catalytic cycle of an enzyme.** An enzyme can convert one or more reactant molecules to one or more product molecules. The enzyme shown here converts two substrate molecules to two product molecules.

the transition state and thus reduces the amount of free energy that must be absorbed to achieve that state.

Third, the active site may also provide a microenvironment that is more conducive to a particular type of reaction than the solution itself would be without the enzyme. For example, if the active site has amino acids with acidic R groups, the active site may be a pocket of low pH in an otherwise neutral cell. In such cases, an acidic amino acid may facilitate H^+ transfer to the substrate as a key step in catalyzing the reaction.

A fourth mechanism of catalysis is the direct participation of the active site in the chemical reaction. Sometimes this process even involves brief covalent bonding between the substrate and the side chain of an amino acid of the enzyme. Subsequent steps of the reaction restore the side chains to their original states, so that the active site is the same after the reaction as it was before.

The rate at which a particular amount of enzyme converts substrate to product is partly a function of the initial concentration of the substrate: The more substrate molecules that are available, the more frequently they access the active sites of the enzyme molecules. However, there is a limit to how fast the reaction can be pushed by adding more substrate to a fixed concentration of enzyme. At some point, the concentration of substrate will be high enough that all enzyme molecules will have their active sites engaged. As soon as the product exits an active site, another substrate molecule enters. At this substrate concentration, the enzyme is said to be *saturated*, and the rate of the reaction is determined by the speed at which the active site converts substrate to product. When an enzyme population is saturated, the only way to increase the rate of product formation is to add more enzyme. Cells often increase the rate of a reaction by producing more enzyme molecules. You can graph the progress of an enzymatic reaction in the **Scientific Skills Exercise**.

Scientific Skills Exercise

Making a Line Graph and Calculating a Slope

Does the Rate of Glucose 6-Phosphatase Activity Change over Time in Isolated Liver Cells? Glucose 6-phosphatase, which is found in mammalian liver cells, is a key enzyme in control of blood glucose levels. The enzyme catalyzes the breakdown of glucose 6-phosphate into glucose and inorganic phosphate (Ⓟ$_i$). These products are transported out of liver cells into the blood, increasing blood glucose levels. In this exercise, you will graph data from a time-course experiment that measured Ⓟ$_i$ concentration in the buffer outside isolated liver cells, thus indirectly measuring glucose 6-phosphatase activity inside the cells.

How the Experiment Was Done Isolated rat liver cells were placed in a dish with buffer at physiological conditions (pH 7.4, 37°C). Glucose 6-phosphate (the substrate) was added to the dish, where it was taken up by the cells. Then a sample of buffer was removed every 5 minutes and the concentration of Ⓟ$_i$ determined.

Data from the Experiment

Time (min)	Concentration of Ⓟ$_i$ (μmol/mL)
0	0
5	10
10	90
15	180
20	270
25	330
30	355
35	355
40	355

Data from S. R. Commerford et al., Diets enriched in sucrose or fat increase gluconeogenesis and G-6-Pase but not basal glucose production in rats, *American Journal of Physiology—Endocrinology and Metabolism* 283:E545–E555 (2002).

INTERPRET THE DATA

1. To see patterns in the data from a time-course experiment like this, it is helpful to graph the data. First, determine which set of data goes on each axis. (a) What did the researchers intentionally vary in the experiment? This is the independent variable, which goes on the *x*-axis. (b) What are the units (abbreviated) for the independent variable? Explain in words what the abbreviation stands for. (c) What was measured by the researchers? This is the dependent variable, which goes on the *y*-axis. (d) What does the units abbreviation stand for? Label each axis, including the units.
2. Next, you'll want to mark off the axes with just enough evenly spaced tick marks to accommodate the full set of data. Determine the range of data values for each axis. (a) What is the largest value to go on the *x*-axis? What is a reasonable spacing for the tick marks, and what should be the highest one? (b) What is the largest value to go on the *y*-axis? What is a reasonable spacing for the tick marks, and what should be the highest one?
3. Plot the data points on your graph. Match each *x*-value with its partner *y*-value and place a point on the graph at that coordinate. Draw a line that connects the points. (For additional information about graphs, see the Scientific Skills Review in the Study Area in MasteringBiology.)
4. Examine your graph and look for patterns in the data. (a) Does the concentration of Ⓟ$_i$ increase evenly through the course of the experiment? To answer this question, describe the pattern you see in the graph. (b) What part of the graph shows the highest rate of enzyme activity? Consider that the rate of enzyme activity is related to the slope of the line, $\Delta y/\Delta x$ (the "rise" over the "run"), in μmol/(mL · min), with the steepest slope indicating the highest rate of enzyme activity. Calculate the rate of enzyme activity (slope) where the graph is steepest. (c) Can you think of a biological explanation for the pattern you see?
5. If your blood sugar level is low from skipping lunch, what reaction (discussed in this exercise) will occur in your liver cells? Write out the reaction and put the name of the enzyme over the reaction arrow. How will this reaction affect your blood sugar level?

MB A version of this Scientific Skills Exercise can be assigned in MasteringBiology.

Effects of Local Conditions on Enzyme Activity

The activity of an enzyme—how efficiently the enzyme functions—is affected by general environmental factors, such as temperature and pH. It can also be affected by chemicals that specifically influence that enzyme. In fact, researchers have learned much about enzyme function by employing such chemicals.

Effects of Temperature and pH

The three-dimensional structures of proteins are sensitive to their environment. As a consequence, each enzyme works better under some conditions than under other conditions, because these *optimal conditions* favor the most active shape for the enzyme.

Temperature and pH are environmental factors important in the activity of an enzyme. Up to a point, the rate of an enzymatic reaction increases with increasing temperature, partly because substrates collide with active sites more frequently when the molecules move rapidly. Above that temperature, however, the speed of the enzymatic reaction drops sharply. The thermal agitation of the enzyme molecule disrupts the hydrogen bonds, ionic bonds, and other weak interactions that stabilize the active shape of the enzyme, and the protein molecule eventually denatures. Each enzyme has an optimal temperature at which its reaction rate is greatest. Without denaturing the enzyme, this temperature allows the greatest number of molecular collisions and the fastest conversion of the reactants to product molecules. Most human enzymes have optimal temperatures of about 35–40°C (close to human body temperature). The thermophilic bacteria that live in hot springs contain enzymes with optimal temperatures of 70°C or higher **(Figure 16a)**.

Just as each enzyme has an optimal temperature, it also has a pH at which it is most active. The optimal pH values for most enzymes fall in the range of pH 6–8, but there are exceptions. For example, pepsin, a digestive enzyme in the human stomach, works best at pH 2. Such an acidic environment denatures most enzymes, but pepsin is adapted to maintain its functional three-dimensional structure in the acidic environment of the stomach. In contrast, trypsin, a digestive enzyme residing in the alkaline environment of the human intestine, has an optimal pH of 8 and would be denatured in the stomach **(Figure 16b)**.

Cofactors

Many enzymes require nonprotein helpers for catalytic activity. These adjuncts, called **cofactors**, may be bound tightly to the enzyme as permanent residents, or they may bind loosely and reversibly along with the substrate. The cofactors of some enzymes are inorganic, such as the metal atoms zinc, iron, and copper in ionic form. If the cofactor is an organic molecule, it is referred to, more specifically, as a **coenzyme**. Most vitamins are important in nutrition because they act as coenzymes or raw materials from which coenzymes are made. Cofactors function in various ways, but in all cases where they are used, they perform a crucial chemical function in catalysis. You'll encounter examples of cofactors later in the text.

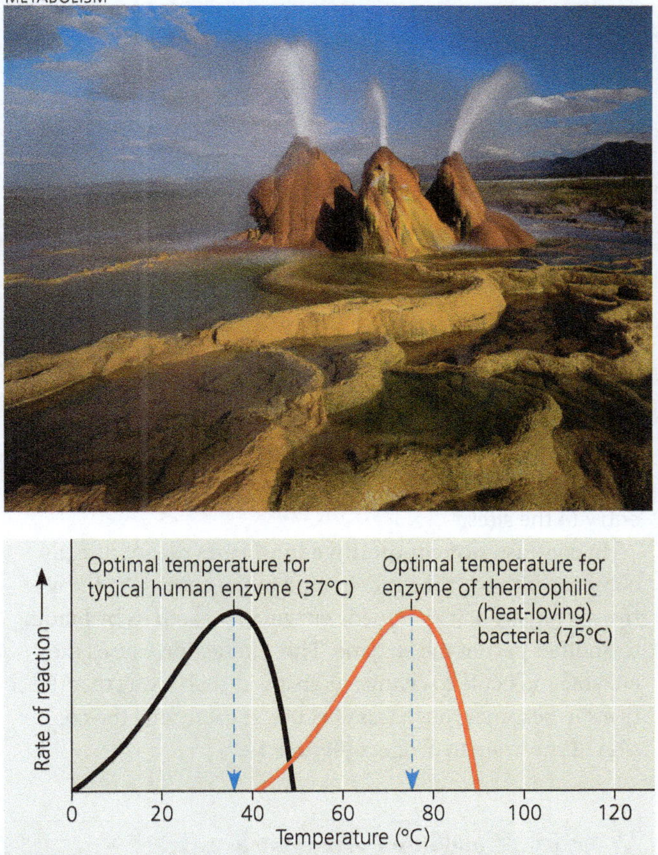

(a) The photo shows thermophilic cyanobacteria (green) thriving in the hot water of a Nevada geyser. The graph compares the optimal temperatures for an enzyme from the thermophilic bacterium *Thermus oshimai* (75°C) and human enzymes (body temperature, 37°C).

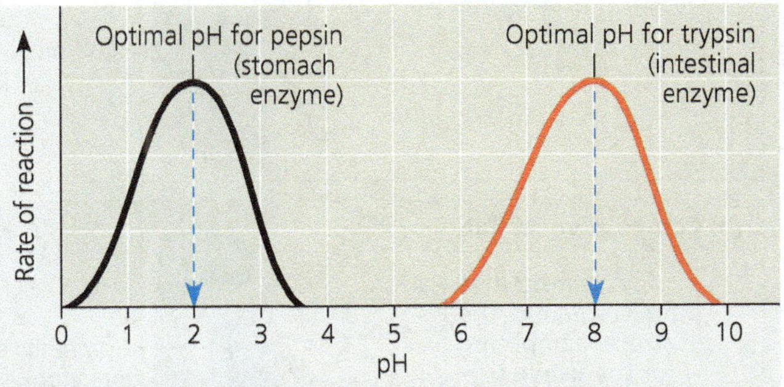

(b) Enzymes work best at different pHs, correlated with the conditions under which they function in the body.

▲ Figure 16 Environmental factors affecting enzyme activity. Each enzyme has an optimal **(a)** temperature and **(b)** pH that favor the most active shape of the protein molecule.

DRAW IT *Given that a mature lysosome has an internal pH of around 4.5, draw a curve in (b) showing what you would predict for a lysosomal enzyme, labeling its optimal pH.*

Enzyme Inhibitors

Certain chemicals selectively inhibit the action of specific enzymes. Sometimes the inhibitor attaches to the enzyme by covalent bonds, in which case the inhibition is usually irreversible. Many enzyme inhibitors, however, bind to the enzyme by weak interactions, and when this occurs, the inhibition is reversible. Some reversible inhibitors resemble the normal substrate molecule and compete for admission into the active site **(Figure 17a** and **b)**. These mimics, called **competitive inhibitors**, reduce the productivity of enzymes by blocking substrates from entering active sites. This kind of inhibition can be overcome by increasing the concentration of substrate so that as active sites become available, more substrate molecules than inhibitor molecules are around to gain entry to the sites.

In contrast, **noncompetitive inhibitors** do not directly compete with the substrate to bind to the enzyme at the active site. Instead, they impede enzymatic reactions by binding to another part of the enzyme. This interaction causes the enzyme molecule to change its shape in such a way that the active site becomes much less effective at catalyzing the conversion of substrate to product **(Figure 17c)**.

▼ **Figure 17 Inhibition of enzyme activity.**

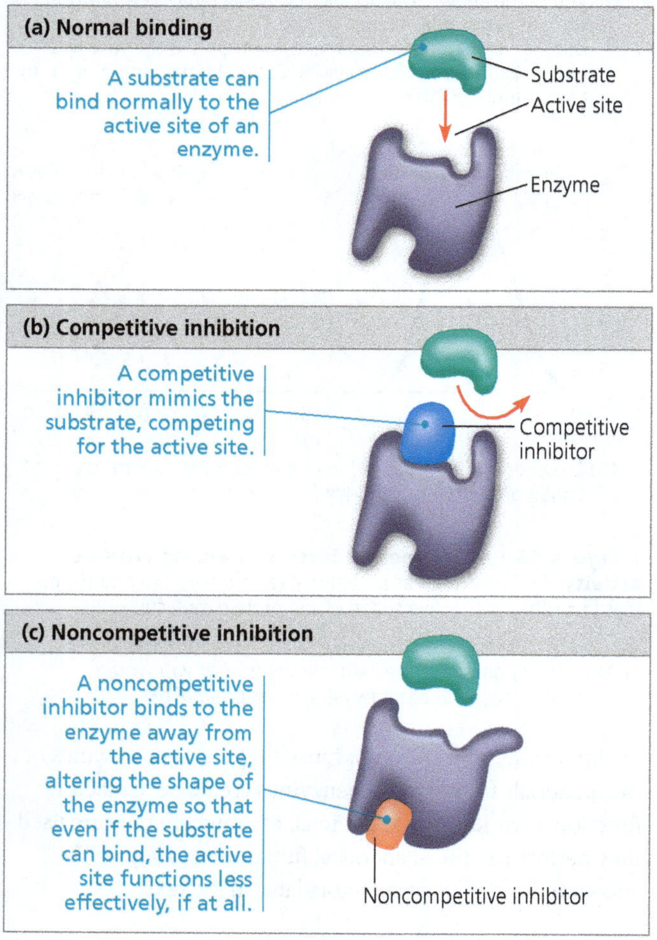

Toxins and poisons are often irreversible enzyme inhibitors. An example is sarin, a nerve gas. Sarin was released by terrorists in the Tokyo subway in 1995, killing several people and injuring many others. This small molecule binds covalently to the R group on the amino acid serine, which is found in the active site of acetylcholinesterase, an enzyme important in the nervous system. Other examples include the pesticides DDT and parathion, inhibitors of key enzymes in the nervous system. Finally, many antibiotics are inhibitors of specific enzymes in bacteria. For instance, penicillin blocks the active site of an enzyme that many bacteria use to make cell walls.

Citing enzyme inhibitors that are metabolic poisons may give the impression that enzyme inhibition is generally abnormal and harmful. In fact, molecules naturally present in the cell often regulate enzyme activity by acting as inhibitors. Such regulation—selective inhibition—is essential to the control of cellular metabolism, as we will discuss in Concept 5.

The Evolution of Enzymes

EVOLUTION Thus far, biochemists have identified more than 4,000 different enzymes in various species, most likely a very small fraction of all enzymes. How did this grand profusion of enzymes arise? Recall that most enzymes are proteins, and proteins are encoded by genes. A permanent change in a gene, known as a *mutation*, can result in a protein with one or more changed amino acids. In the case of an enzyme, if the changed amino acids are in the active site or some other crucial region, the altered enzyme might have a novel activity or might bind to a different substrate. Under environmental conditions where the new function benefits the organism, natural selection would tend to favor the mutated form of the gene, causing it to persist in the population. This simplified model is generally accepted as the main way in which the multitude of different enzymes arose over the past few billion years of life's history.

CONCEPT CHECK 4

1. Many spontaneous reactions occur very slowly. Why don't all spontaneous reactions occur instantly?
2. Why do enzymes act only on very specific substrates?
3. WHAT IF? Malonate is an inhibitor of the enzyme succinate dehydrogenase. How would you determine whether malonate is a competitive or noncompetitive inhibitor?

CONCEPT 5

Regulation of enzyme activity helps control metabolism

Chemical chaos would result if all of a cell's metabolic pathways were operating simultaneously. Intrinsic to life's processes is a cell's ability to tightly regulate its metabolic pathways by controlling when and where its various enzymes are active. It does this either by switching on and off the genes that encode specific enzymes or, as we discuss here, by regulating the activity of enzymes once they are made.

Allosteric Regulation of Enzymes

In many cases, the molecules that naturally regulate enzyme activity in a cell behave something like reversible noncompetitive inhibitors (see Figure 17c): These regulatory molecules change an enzyme's shape and the functioning of its active site by binding to a site elsewhere on the molecule, via noncovalent interactions. **Allosteric regulation** is the term used to describe any case in which a protein's function at one site is affected by the binding of a regulatory molecule to a separate site. It may result in either inhibition or stimulation of an enzyme's activity.

Allosteric Activation and Inhibition

Most enzymes known to be allosterically regulated are constructed from two or more subunits, each composed of a polypeptide chain with its own active site. The entire complex oscillates between two different shapes, one catalytically active and the other inactive **(Figure 18a)**. In the simplest kind of allosteric regulation, an activating or inhibiting regulatory molecule binds to a regulatory site (sometimes called an allosteric site), often located where subunits join. The binding of an *activator* to a regulatory site stabilizes the shape that has functional active sites, whereas the binding of an *inhibitor* stabilizes the inactive form of the enzyme. The subunits of an allosteric enzyme fit together in such a way that a shape change in one subunit is transmitted to all others. Through this interaction of subunits, a single activator or inhibitor molecule that binds to one regulatory site will affect the active sites of all subunits.

Fluctuating concentrations of regulators can cause a sophisticated pattern of response in the activity of cellular enzymes. The products of ATP hydrolysis (ADP and $Ⓟ_i$), for example, play a complex role in balancing the flow of traffic between anabolic and catabolic pathways by their effects on key enzymes. ATP binds to several catabolic enzymes allosterically, lowering their affinity for substrate and thus inhibiting their activity. ADP, however, functions as an activator of the same enzymes. This is logical because catabolism functions in regenerating ATP. If ATP production lags behind its use, ADP accumulates and activates the enzymes that speed up catabolism, producing more ATP. If the supply of ATP exceeds demand, then catabolism slows down as ATP molecules accumulate and bind to the same enzymes, inhibiting them. ATP, ADP, and other related molecules also affect key enzymes in anabolic pathways. In this way, allosteric enzymes control the rates of important reactions in both sorts of metabolic pathways.

In another kind of allosteric activation, a *substrate* molecule binding to one active site in a multisubunit enzyme triggers a shape change in all the subunits, thereby increasing catalytic activity at the other active sites **(Figure 18b)**.

▼ Figure 18 Allosteric regulation of enzyme activity.

(a) Allosteric activators and inhibitors

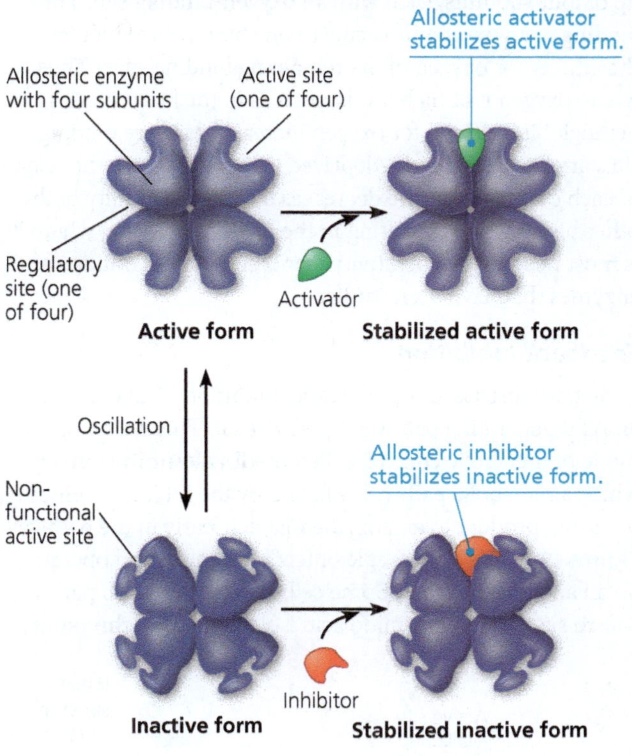

At low concentrations, activators and inhibitors dissociate from the enzyme. The enzyme can then oscillate again.

(b) Cooperativity: another type of allosteric activation

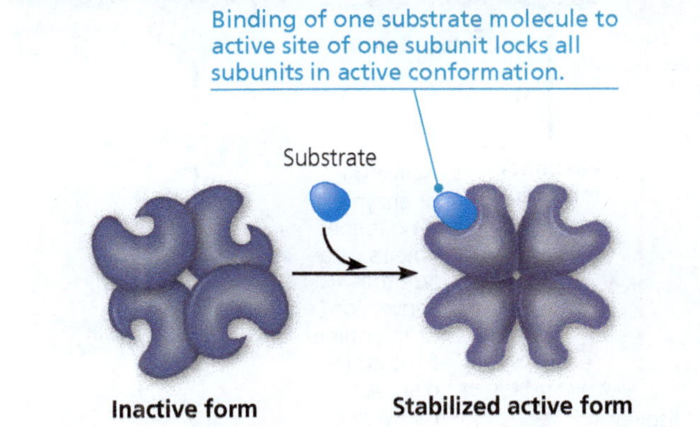

The inactive form shown on the left oscillates with the active form when the active form is not stabilized by substrate.

Called **cooperativity**, this mechanism amplifies the response of enzymes to substrates: One substrate molecule primes an enzyme to act on additional substrate molecules more readily. Cooperativity is considered allosteric regulation because binding of the substrate to one active site affects catalysis in another active site.

Although hemoglobin is not an enzyme (it carries O_2 rather than catalyzing a reaction), classic studies of hemoglobin have elucidated the principle of cooperativity. Hemoglobin is made up of four subunits, each with an oxygen-binding site. The binding of an oxygen molecule to one binding site increases the affinity for oxygen of the remaining binding sites. Thus, where oxygen is at high levels, such as in the lungs or gills, hemoglobin's affinity for oxygen increases as more binding sites are filled. In oxygen-deprived tissues, however, the release of each oxygen molecule decreases the oxygen affinity of the other binding sites, resulting in the release of oxygen where it is most needed. Cooperativity works similarly in multisubunit enzymes that have been studied.

Feedback Inhibition

Earlier, we discussed the allosteric inhibition of an enzyme in an ATP-generating pathway by ATP itself. This is a common mode of metabolic control, called **feedback inhibition**, in which a metabolic pathway is halted by the inhibitory binding of its end product to an enzyme that acts early in the pathway. **Figure 19** shows an example of feedback inhibition operating on an anabolic pathway. Some cells use this five-step pathway to synthesize the amino acid isoleucine from threonine, another amino acid. As isoleucine accumulates, it slows down its own synthesis by allosterically inhibiting the enzyme for the first step of the pathway. Feedback inhibition thereby prevents the cell from making more isoleucine than is necessary, and thus wasting chemical resources.

Organization of Enzymes Within the Cell

The cell is not just a bag of chemicals with thousands of different kinds of enzymes and substrates in a random mix. The cell is compartmentalized, and cellular structures help bring order to metabolic pathways. In some cases, a team of enzymes for several steps of a metabolic pathway is assembled into a multienzyme complex. The arrangement facilitates the sequence of reactions, with the product from the first enzyme becoming the substrate for an adjacent enzyme in the complex, and so on, until the end product is released. Some enzymes and enzyme complexes have fixed locations within the cell and act as structural components of particular membranes. Others are in solution within particular membrane-enclosed eukaryotic organelles, each with its own internal chemical environment. For example, in eukaryotic cells, the enzymes for cellular respiration reside in specific locations within mitochondria **(Figure 20)**.

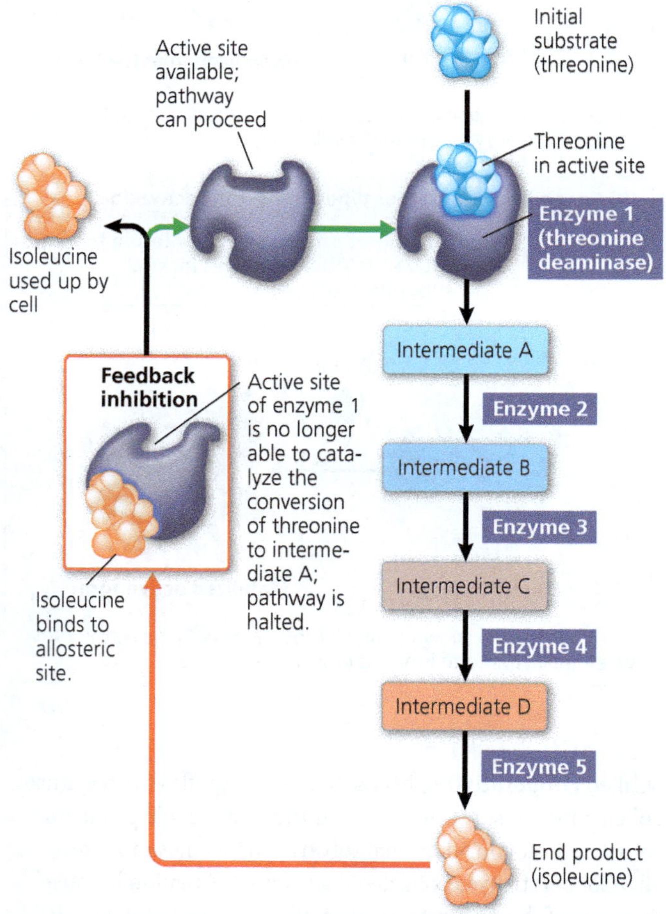

▲ **Figure 19 Feedback inhibition in isoleucine synthesis.**

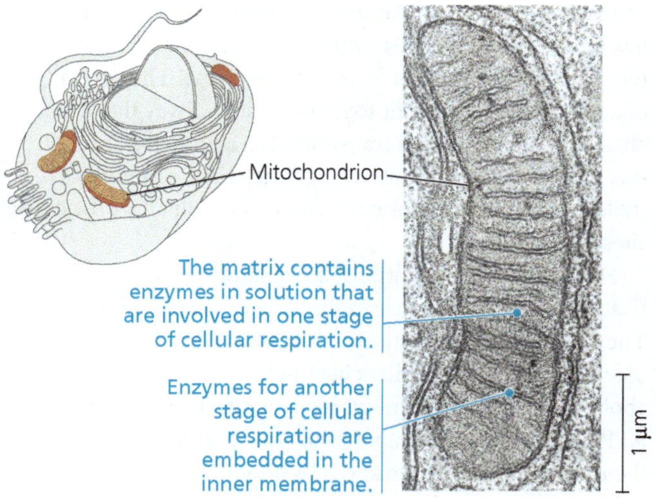

▲ **Figure 20 Organelles and structural order in metabolism.** Organelles such as the mitochondrion (TEM) contain enzymes that carry out specific functions, in this case cellular respiration.

In this chapter, you have learned that metabolism, the intersecting set of chemical pathways characteristic of life, is a choreographed interplay of thousands of different kinds of cellular molecules.

CONCEPT CHECK 5

1. How do an activator and an inhibitor have different effects on an allosterically regulated enzyme?

Go to **MasteringBiology®** for Assignments, the eText, and the Study Area with Animations, Activities, Vocab Self-Quiz, and Practice Tests.

Chapter Review

SUMMARY OF KEY CONCEPTS

VOCAB SELF-QUIZ
goo.gl/gbai8v

CONCEPT 1

An organism's metabolism transforms matter and energy

- **Metabolism** is the collection of chemical reactions that occur in an organism. Enzymes catalyze reactions in intersecting **metabolic pathways**, which may be **catabolic** (breaking down molecules, releasing energy) or **anabolic** (building molecules, consuming energy).
- **Energy** is the capacity to cause change; some forms of energy do work by moving matter. **Kinetic energy** is associated with motion and includes **thermal energy**, associated with the random motion of atoms or molecules. **Heat** is thermal energy in transfer from one object to another. **Potential energy** is related to the location or structure of matter and includes **chemical energy** possessed by a molecule due to its structure.
- The **first law of thermodynamics**, conservation of energy, states that energy cannot be created or destroyed, only transferred or transformed. The **second law of thermodynamics** states that **spontaneous processes**, those requiring no outside input of energy, increase the **entropy** (disorder) of the universe.

? *Explain how the highly ordered structure of a cell does not conflict with the second law of thermodynamics.*

CONCEPT 2

The free-energy change of a reaction tells us whether or not the reaction occurs spontaneously

- A living system's **free energy** is energy that can do work under cellular conditions. Organisms live at the expense of free energy. The change in free energy (ΔG) during a biological process tells us if the process is spontaneous. During a spontaneous process, free energy decreases and the stability of a system increases. At maximum stability, the system is at equilibrium and can do no work.
- In an **exergonic** (spontaneous) chemical reaction, the products have less free energy than the reactants ($-\Delta G$). **Endergonic** (nonspontaneous) reactions require an input of energy ($+\Delta G$). The addition of starting materials and the removal of end products prevent metabolism from reaching equilibrium.

? *Why are spontaneous reactions important in the metabolism of a cell?*

CONCEPT 3

ATP powers cellular work by coupling exergonic reactions to endergonic reactions

- **ATP** is the cell's energy shuttle. Hydrolysis of its terminal phosphate yields ADP and $\textcircled{P}_i$ and releases free energy.
- Through **energy coupling**, the exergonic process of ATP hydrolysis drives endergonic reactions by transfer of a phosphate group to specific reactants, forming a **phosphorylated intermediate** that is more reactive. ATP hydrolysis (sometimes with protein phosphorylation) also causes changes in the shape and binding affinities of transport and motor proteins.
- Catabolic pathways drive regeneration of ATP from ADP + $\textcircled{P}_i$.

? *Describe the ATP cycle: How is ATP used and regenerated in a cell?*

CONCEPT 4

Enzymes speed up metabolic reactions by lowering energy barriers

- In a chemical reaction, the energy necessary to break the bonds of the reactants is the **activation energy**, E_A.
- **Enzymes** lower the E_A barrier:

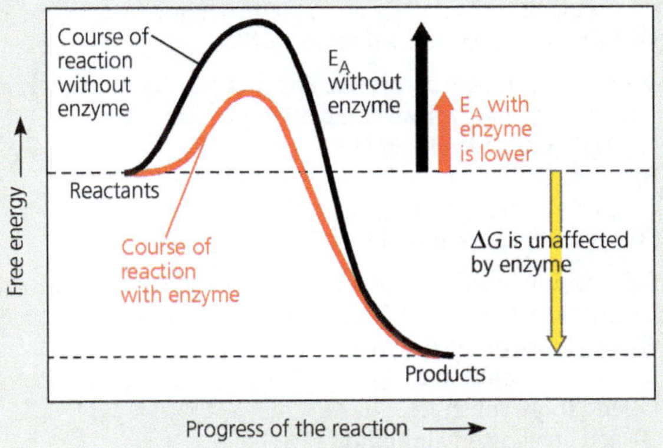

- Each type of enzyme has a unique **active site** that combines specifically with its **substrate(s)**, the reactant(s) on which it acts. It then changes shape, binding the substrate(s) more tightly (**induced fit**).
- The active site can lower an E_A barrier by orienting substrates correctly, straining their bonds, providing a favorable microenvironment, or even covalently bonding with the substrate.
- Each enzyme has an optimal temperature and pH. Inhibitors reduce enzyme function. A **competitive inhibitor** binds to the active site, whereas a **noncompetitive inhibitor** binds to a different site on the enzyme.
- Natural selection, acting on organisms with variant enzymes, is responsible for the diversity of enzymes found in organisms.

? *How do both activation energy barriers and enzymes help maintain the structural and metabolic order of life?*

CONCEPT 5

Regulation of enzyme activity helps control metabolism

- Many enzymes are subject to **allosteric regulation**: Regulatory molecules, either activators or inhibitors, bind to specific regulatory sites, affecting the shape and function of the enzyme. In **cooperativity**, binding of one substrate molecule can stimulate binding or activity at other active sites. In **feedback inhibition**, the end product of a metabolic pathway allosterically inhibits the enzyme for a previous step in the pathway.
- Some enzymes are grouped into complexes, some are incorporated into membranes, and some are contained inside organelles, increasing the efficiency of metabolic processes.

? *What roles do allosteric regulation and feedback inhibition play in the metabolism of a cell?*

TEST YOUR UNDERSTANDING

PRACTICE TEST
goo.gl/CRZjvS

Level 1: Knowledge/Comprehension

1. Choose the pair of terms that correctly completes this sentence: Catabolism is to anabolism as ____________ is to ____________.
   (A) exergonic; spontaneous
   (B) exergonic; endergonic
   (C) free energy; entropy
   (D) work; energy

2. Most cells cannot harness heat to perform work because
   (A) heat does not involve a transfer of energy.
   (B) cells do not have much heat; they are relatively cool.
   (C) temperature is usually uniform throughout a cell.
   (D) heat can never be used to do work.

3. Which of the following metabolic processes can occur without a net influx of energy from some other process?
   (A) ADP + Ⓟ$_i$ $\rightarrow$ ATP + H_2O
   (B) $C_6H_{12}O_6 + 6\,O_2 \rightarrow 6\,CO_2 + 6\,H_2O$
   (C) $6\,CO_2 + 6\,H_2O \rightarrow C_6H_{12}O_6 + 6\,O_2$
   (D) amino acids $\rightarrow$ protein

4. If an enzyme in solution is saturated with substrate, the most effective way to obtain a faster yield of products is to
   (A) add more of the enzyme.
   (B) heat the solution to 90°C.
   (C) add more substrate.
   (D) add an allosteric inhibitor.

5. Some bacteria are metabolically active in hot springs because
   (A) they are able to maintain a lower internal temperature.
   (B) high temperatures make catalysis unnecessary.
   (C) their enzymes have high optimal temperatures.
   (D) their enzymes are completely insensitive to temperature.

Level 2: Application/Analysis

6. If an enzyme is added to a solution where its substrate and product are in equilibrium, what will occur?
   (A) Additional product will be formed.
   (B) The reaction will change from endergonic to exergonic.
   (C) The free energy of the system will change.
   (D) Nothing; the reaction will stay at equilibrium.

Level 3: Synthesis/Evaluation

7. **DRAW IT** Using a series of arrows, draw the branched metabolic reaction pathway described by the following statements. Then answer the question at the end. Use red arrows and minus signs to indicate inhibition.

   L can form either M or N.
   M can form O.
   O can form either P or R.
   P can form Q.
   R can form S.
   O inhibits the reaction of L to form M.
   Q inhibits the reaction of O to form P.
   S inhibits the reaction of O to form R.

   Which reaction would prevail if both Q and S were present in the cell at high concentrations?

   (A) L $\rightarrow$ M
   (B) M $\rightarrow$ O
   (C) L $\rightarrow$ N
   (D) O $\rightarrow$ P

8. **SCIENTIFIC INQUIRY**
   DRAW IT A researcher has developed an assay to measure the activity of an important enzyme present in liver cells growing in culture. She adds the enzyme's substrate to a dish of cells and then measures the appearance of reaction products. The results are graphed as the amount of product on the *y*-axis versus time on the *x*-axis. The researcher notes four sections of the graph. For a short period of time, no products appear (section A). Then (section B) the reaction rate is quite high (the slope of the line is steep). Next, the reaction gradually slows down (section C). Finally, the graph line becomes flat (section D). Draw and label the graph, and propose a model to explain the molecular events occurring at each stage of this reaction profile.

9. **SCIENCE, TECHNOLOGY, AND SOCIETY**
   Organophosphates (organic compounds containing phosphate groups) are commonly used as insecticides to improve crop yield. Organophosphates typically interfere with nerve signal transmission by inhibiting the enzymes that degrade transmitter molecules. They affect humans and other vertebrates as well as insects. Thus, the use of organophosphate pesticides poses some health risks. On the other hand, these molecules break down rapidly upon exposure to air and sunlight. As a consumer, what level of risk are you willing to accept in exchange for an abundant and affordable food supply? Explain your thinking.

10. **FOCUS ON EVOLUTION**
    A recent revival of the antievolutionary "intelligent design" argument holds that biochemical pathways are too complex to have evolved, because all intermediate steps in a given pathway must be present to produce the final product. Critique this argument. How could you use the diversity of metabolic pathways that produce the same or similar products to support your case?

11. **FOCUS ON ENERGY AND MATTER**
    Life requires energy. In a short essay (100–150 words), describe the basic principles of bioenergetics in an animal cell. How is the flow and transformation of energy different in a photosynthesizing cell? Include the role of ATP and enzymes in your discussion.

12. **SYNTHESIZE YOUR KNOWLEDGE**

Explain what is happening in this photo in terms of kinetic energy and potential energy. Include the energy conversions that occur when the penguins eat fish and climb back up on the glacier. Describe the role of ATP and enzymes in the underlying molecular processes, including what happens to the free energy of some of the molecules involved.

Answers

Figure Questions

Figure 9 Glutamic acid (Glu) has a carboxyl group at the end of its R group. Glutamine (Gln) has exactly the same structure as glutamic acid, except that there is an amino group in place of the —OH on the R group. Thus, in this figure, Gln is drawn as a Glu with an attached NH_2.

Figure 12

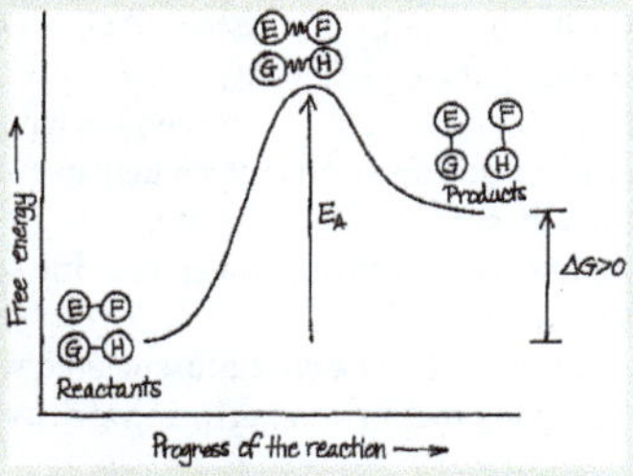

Figure 16

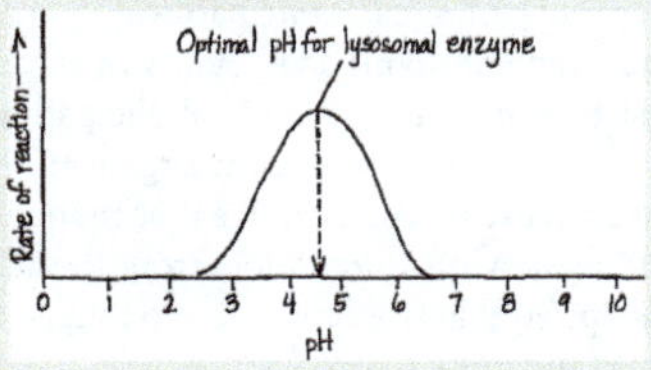

Concept Check 1

1. The second law is the trend toward randomization, or increasing entropy. When the concentrations of a substance on both sides of a membrane are equal, the distribution is more random than when they are unequal. Diffusion of a substance to a region where it is initially less concentrated increases entropy, making it an energetically favorable (spontaneous) process as described by the second law. (This explains the process seen in Figure 9.) **2.** The apple has potential energy in its position hanging on the tree, and the sugars and other nutrients it contains have chemical energy. The apple has kinetic energy as it falls from the tree to the ground. Finally, when the apple is digested and its molecules broken down, some of the chemical energy is used to do work, and the rest is lost as thermal energy.

Concept Check 2

1. Cellular respiration is a spontaneous and exergonic process. The energy released from glucose is used to do work in the cell or is lost as heat. **2.** Catabolism breaks down organic molecules, releasing their chemical energy and resulting in smaller products with more entropy, as when moving from the top to the bottom of part (c). Anabolism consumes energy to synthesize larger molecules from simpler ones, as when moving from the bottom to the top of part (c). **3.** The reaction is exergonic because it releases energy—in this case, in the form of light. (This is a nonbiological version of the bioluminescence seen in Figure 1.)

Concept Check 3

1. ATP usually transfers energy to endergonic processes by phosphorylating (adding phosphate groups to) other molecules. (Exergonic processes phosphorylate ADP to regenerate ATP.) **2.** A set of coupled reactions can transform the first combination into the second. Since this is an exergonic process overall, ΔG is negative and the first combination must have more free energy (see Figure 9). **3.** Active transport: The solute is being transported against its concentration gradient, which requires energy, provided by ATP hydrolysis.

Concept Check 4

1. A spontaneous reaction is a reaction that is exergonic. However, if it has a high activation energy that is rarely attained, the rate of the reaction may be low. **2.** Only the specific substrate(s) will fit properly into the active site of an enzyme, the part of the enzyme that carries out catalysis. **3.** In the presence of malonate, increase the concentration of the normal substrate (succinate) and see whether the rate of reaction increases. If it does, malonate is a competitive inhibitor.

Concept Check 5

1. The activator binds in such a way that it stabilizes the active form of an enzyme, whereas the inhibitor stabilizes the inactive form.

Summary of Key Concepts Questions

1 The process of "ordering" a cell's structure is accompanied by an increase in the entropy, or disorder, of the universe. For example, an animal cell takes in highly ordered organic molecules as the source of matter and energy used to build and maintain its structures. In the same process, however, the cell releases heat and the simple molecules of carbon dioxide and water to the surroundings. The increase in entropy of the latter process offsets the entropy decrease in the former. **2** Spontaneous reactions supply the energy to perform cellular work. **3** The free energy released from the hydrolysis of ATP may drive endergonic reactions through the transfer of a phosphate group to a reactant molecule, forming a more reactive phosphorylated intermediate. ATP hydrolysis also powers the mechanical and transport work of a cell, often by powering shape changes in the relevant motor proteins. Cellular respiration, the catabolic breakdown of glucose, provides the energy for the endergonic regeneration of ATP from ADP and Ⓟ$_i$. **4** Activation energy barriers prevent the complex molecules of the cell, which are rich in free energy, from spontaneously breaking down to less ordered, more stable molecules. Enzymes permit a regulated metabolism by binding to specific substrates and forming enzyme-substrate complexes that selectively lower the E_A for the chemical reactions in a cell. **5** A cell tightly regulates its metabolic pathways in response to fluctuating needs for energy and materials. The binding of activators or inhibitors to regulatory sites on allosteric enzymes stabilizes either the active or inactive form of the subunits. For example, the binding of ATP to a catabolic enzyme in a cell with excess ATP would inhibit that pathway. Such types of feedback inhibition preserve chemical resources within a cell. If ATP supplies are depleted, binding of ADP to the regulatory site of catabolic enzymes will activate that pathway, generating more ATP.

Test Your Understanding

1. B **2.** C **3.** B **4.** A **5.** C **6.** D **7.** C

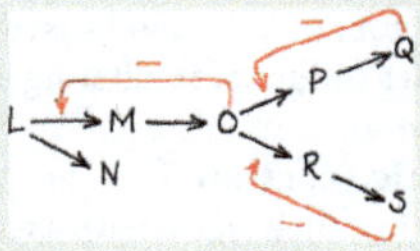

8.

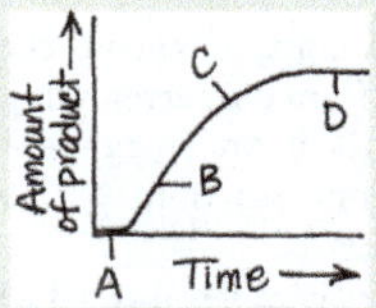

A. The substrate molecules are entering the cells, so no product is made yet.
B. There is sufficient substrate, so the reaction is proceeding at a maximum rate.
C. As the substrate is used up, the rate decreases (the slope is less steep).
D. The line is flat because no new substrate remains and thus no new product appears.

Credits

Credits are listed in the order of appearance.

Photo Credits

1 Doug Perrine/Nature Picture Library; **2** Stephen Simpson/Photolibrary/Getty Images; **3** Robert N. Johnson; **4a** Image Quest Marine; **4b** asharkyu/Shutterstock;
14 Thomas A. Steitz; **6 SSE** Fer Gregory/Shutterstock; **16** Jack Dykinga/Nature Picture Library;
20 Nicolae Simionescu; **Chapter Review SYK** PayPal/Flickr/Moment Select/Getty Images.

Illustration and Text Credits

SSE.1 Data from S. R. Commerford et al., Diets enriched in sucrose or fat increase gluconeogenesis and G-6-Pase but not basal glucose production in rats, *American Journal of Physiology–Endocrinology and Metabolism* 283:E545–E555 (2002).

Glossary

activation energy The amount of energy that reactants must absorb before a chemical reaction will start; also called free energy of activation.

active site The specific region of an enzyme that binds the substrate and that forms the pocket in which catalysis occurs.

allosteric regulation The binding of a regulatory molecule to a protein at one site that affects the function of the protein at a different site.

anabolic pathway (an′-uh-bol′-ik) A metabolic pathway that consumes energy to synthesize a complex molecule from simpler molecules.

ATP (adenosine triphosphate) (a-den′-ō-sēn trī-fos′-fāt) An adenine-containing nucleoside triphosphate that releases free energy when its phosphate bonds are hydrolyzed. This energy is used to drive endergonic reactions in cells.

bioenergetics (1) The overall flow and transformation of energy in an organism. (2) The study of how energy flows through organisms.

catabolic pathway (kat′-uh-bol′-ik) A metabolic pathway that releases energy by breaking down complex molecules to simpler molecules.

catalysis (kuh-ta′-luh-sis) A process by which a chemical agent called a catalyst selectively increases the rate of a reaction without being consumed by the reaction.

chemical energy Energy available in molecules for release in a chemical reaction; a form of potential energy.

coenzyme (kō-en′-zīm) An organic molecule serving as a cofactor. Most vitamins function as coenzymes in metabolic reactions.

cofactor Any nonprotein molecule or ion that is required for the proper functioning of an enzyme. Cofactors can be permanently bound to the active site or may bind loosely and reversibly, along with the substrate, during catalysis.

competitive inhibitor A substance that reduces the activity of an enzyme by entering the active site in place of the substrate, whose structure it mimics.

cooperativity A kind of allosteric regulation whereby a shape change in one subunit of a protein caused by substrate binding is transmitted to all the other subunits, facilitating binding of additional substrate molecules to those subunits.

endergonic reaction (en′-der-gon′-ik) A nonspontaneous chemical reaction, in which free energy is absorbed from the surroundings.

energy coupling In cellular metabolism, the use of energy released from an exergonic reaction to drive an endergonic reaction.

entropy A measure of disorder, or randomness.

enzyme-substrate complex A temporary complex formed when an enzyme binds to its substrate molecule(s).

exergonic reaction (ek′-ser-gon′-ik) A spontaneous chemical reaction, in which there is a net release of free energy.

feedback inhibition A method of metabolic control in which the end product of a metabolic pathway acts as an inhibitor of an enzyme within that pathway.

first law of thermodynamics The principle of conservation of energy: Energy can be transferred and transformed, but it cannot be created or destroyed.

free energy The portion of a biological system's energy that can perform work when temperature and pressure are uniform throughout the system. The change in free energy of a system (ΔG) is $G_{\text{final state}} - G_{\text{initial state.}}$ It can be calculated by the equation $\Delta G = \Delta H - T\Delta S$, where ΔH is the change in enthalpy (in biological systems, equivalent to total energy), T is the absolute temperature, and ΔS is the change in entropy.

induced fit Caused by entry of the substrate, the change in shape of the active site of an enzyme so that it binds more snugly to the substrate.

kinetic energy (kuh-net′-ik) The energy associated with the relative motion of objects. Moving matter can perform work by imparting motion to other matter.

metabolic pathway A series of chemical reactions that either builds a complex molecule (anabolic pathway) or breaks down a complex molecule to simpler molecules (catabolic pathway).

metabolism (muh-tab′-uh-lizm) The totality of an organism's chemical reactions, consisting of catabolic and anabolic pathways, which manage the material and energy resources of the organism.

noncompetitive inhibitor A substance that reduces the activity of an enzyme by binding to a location remote from the active site, changing the enzyme's shape so that the active site no longer effectively catalyzes the conversion of substrate to product.

phosphorylated intermediate A molecule (often a reactant) with a phosphate group covalently bound to it, making it more reactive (less stable) than the unphosphorylated molecule.

potential energy The energy that matter possesses as a result of its location or spatial arrangement (structure).

second law of thermodynamics The principle stating that every energy transfer or transformation increases the entropy of the universe. Usable forms of energy are at least partly converted to heat.

spontaneous process A process that occurs without an overall input of energy; a process that is energetically favorable.

substrate The reactant on which an enzyme works.

thermal energy Kinetic energy due to the random motion of atoms and molecules; energy in its most random form.

thermodynamics (ther′-mō-dī-nam′-iks) The study of energy transformations that occur in a collection of matter. *See also* first law of thermodynamics; second law of thermodynamics.

Cellular Respiration and Fermentation

KEY CONCEPTS

1. Catabolic pathways yield energy by oxidizing organic fuels
2. Glycolysis harvests chemical energy by oxidizing glucose to pyruvate
3. After pyruvate is oxidized, the citric acid cycle completes the energy-yielding oxidation of organic molecules
4. During oxidative phosphorylation, chemiosmosis couples electron transport to ATP synthesis
5. Fermentation and anaerobic respiration enable cells to produce ATP without the use of oxygen
6. Glycolysis and the citric acid cycle connect to many other metabolic pathways

▲ **Figure 1** How do these leaves power the work of life for this giraffe?

Life Is Work

Living cells require transfusions of energy from outside sources to perform their many tasks—for example, assembling polymers, pumping substances across membranes, moving, and reproducing. The giraffe in **Figure 1** obtains energy for its cells by eating plants; some animals feed on other organisms that eat plants. The energy stored in the organic molecules of food ultimately comes from the sun. Energy flows into an ecosystem as sunlight and exits as heat; in contrast, the chemical elements essential to life are recycled **(Figure 2)**. Photosynthesis generates oxygen and organic molecules that are used by the mitochondria of eukaryotes (including plants and algae) as fuel for cellular respiration. Respiration breaks down this fuel, generating ATP. The waste products of this type of respiration, carbon dioxide and water, are the raw materials for photosynthesis.

In this chapter, we'll consider how cells harvest the chemical energy stored in organic molecules and use it to generate ATP, the molecule that drives most cellular work. After presenting some basics about respiration, we'll focus on three key pathways of respiration: glycolysis, the citric acid cycle, and oxidative phosphorylation. We'll also consider fermentation, a somewhat simpler pathway coupled to glycolysis that has deep evolutionary roots.

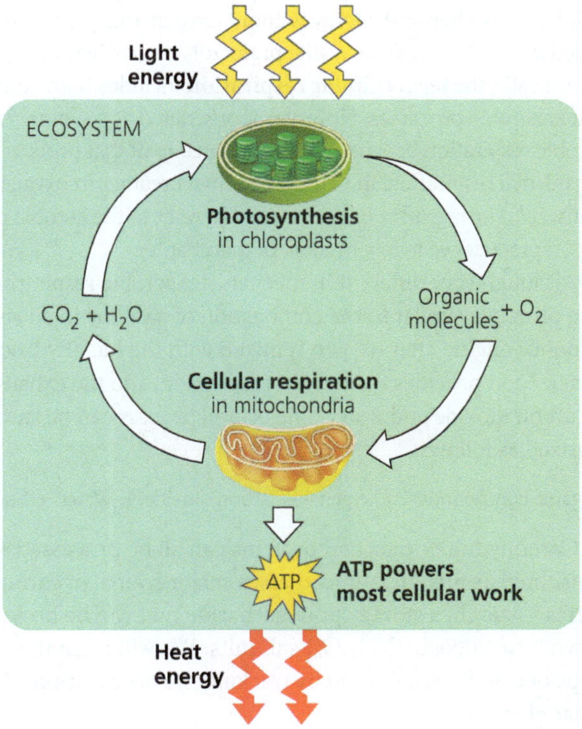

▲ **Figure 2 Energy flow and chemical recycling in ecosystems.** Energy flows into an ecosystem as sunlight and ultimately leaves as heat, while the chemical elements essential to life are recycled.

CONCEPT 1

Catabolic pathways yield energy by oxidizing organic fuels

Metabolic pathways that release stored energy by breaking down complex molecules are called catabolic pathways. Electron transfer plays a major role in these pathways. In this section, we'll consider these processes, which are central to cellular respiration.

Catabolic Pathways and Production of ATP

Organic compounds possess potential energy as a result of the arrangement of electrons in the bonds between their atoms. Compounds that can participate in exergonic reactions can act as fuels. Through the activity of enzymes, a cell systematically degrades complex organic molecules that are rich in potential energy to simpler waste products that have less energy. Some of the energy taken out of chemical storage can be used to do work; the rest is dissipated as heat.

One catabolic process, **fermentation**, is a partial degradation of sugars or other organic fuel that occurs without the use of oxygen. However, the most efficient catabolic pathway is **aerobic respiration**, in which oxygen is consumed as a reactant along with the organic fuel (*aerobic* is from the Greek *aer*, air, and *bios*, life). The cells of most eukaryotic and many prokaryotic organisms can carry out aerobic respiration. Some prokaryotes use substances other than oxygen as reactants in a similar process that harvests chemical energy without oxygen; this process is called *anaerobic respiration* (the prefix *an-* means "without"). Technically, the term **cellular respiration** includes both aerobic and anaerobic processes. However, it originated as a synonym for aerobic respiration because of the relationship of that process to organismal respiration, in which an animal breathes in oxygen. Thus, *cellular respiration* is often used to refer to the aerobic process, a practice we follow in most of this chapter.

Although very different in mechanism, aerobic respiration is in principle similar to the combustion of gasoline in an automobile engine after oxygen is mixed with the fuel (hydrocarbons). Food provides the fuel for respiration, and the exhaust is carbon dioxide and water. The overall process can be summarized as follows:

Organic compounds + Oxygen → Carbon dioxide + Water + Energy

Carbohydrates, fats, and proteins can all be processed and consumed as fuel. In animal diets, a major source of carbohydrates is starch, a storage polysaccharide that can be broken down into glucose ($C_6H_{12}O_6$) subunits. We will learn the steps of cellular respiration by tracking the degradation of the sugar glucose:

$$C_6H_{12}O_6 + 6\,O_2 \rightarrow 6\,CO_2 + 6\,H_2O + \text{Energy (ATP + heat)}$$

This breakdown of glucose is exergonic, having a free-energy change of −686 kcal (2,870 kJ) per mole of glucose decomposed ($\Delta G = -686$ kcal/mol). Recall that a negative ΔG indicates that the products of the chemical process store less energy than the reactants and that the reaction can happen spontaneously—in other words, without an input of energy.

Catabolic pathways do not directly move flagella, pump solutes across membranes, polymerize monomers, or perform other cellular work. Catabolism is linked to work by a chemical drive shaft—ATP. To keep working, the cell must regenerate its supply of ATP from ADP and Ⓟi. To understand how cellular respiration accomplishes this, let's examine the fundamental chemical processes known as oxidation and reduction.

Redox Reactions: Oxidation and Reduction

How do the catabolic pathways that decompose glucose and other organic fuels yield energy? The answer is based on the transfer of electrons during the chemical reactions. The relocation of electrons releases energy stored in organic molecules, and this energy ultimately is used to synthesize ATP.

The Principle of Redox

In many chemical reactions, there is a transfer of one or more electrons (e^-) from one reactant to another. These electron transfers are called oxidation-reduction reactions, or **redox reactions** for short. In a redox reaction, the loss of electrons from one substance is called **oxidation**, and the addition of electrons to another substance is known as **reduction**. (Note that *adding* electrons is called *reduction*; adding negatively charged electrons to an atom *reduces* the amount of positive charge of that atom.) To take a simple, nonbiological example, consider the reaction between the elements sodium (Na) and chlorine (Cl) that forms table salt:

becomes oxidized (loses electron)

$$Na + Cl \longrightarrow Na^+ + Cl^-$$

becomes reduced (gains electron)

We could generalize a redox reaction this way:

becomes oxidized

$$Xe^- + Y \longrightarrow X + Ye^-$$

becomes reduced

In the generalized reaction, substance Xe^-, the electron donor, is called the **reducing agent**; it reduces Y, which accepts the donated electron. Substance Y, the electron acceptor, is the **oxidizing agent**; it oxidizes Xe^- by removing its electron. Because an electron transfer requires both an electron donor and an acceptor, oxidation and reduction always go hand in hand.

Not all redox reactions involve the complete transfer of electrons from one substance to another; some change the degree of electron sharing in covalent bonds. Methane combustion,

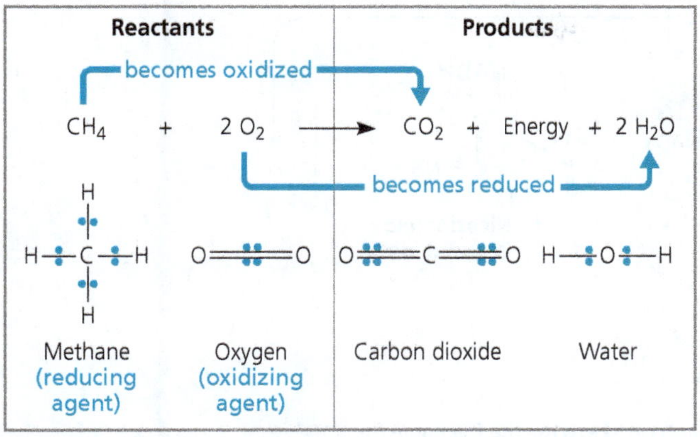

▲ **Figure 3 Methane combustion as an energy-yielding redox reaction.** The reaction releases energy to the surroundings because the electrons lose potential energy when they end up being shared unequally, spending more time near electronegative atoms such as oxygen.

shown in **Figure 3**, is an example. The covalent electrons in methane are shared nearly equally between the bonded atoms because carbon and hydrogen have about the same affinity for valence electrons; they are about equally electronegative. But when methane reacts with oxygen, forming carbon dioxide, electrons end up shared less equally between the carbon atom and its new covalent partners, the oxygen atoms, which are very electronegative. In effect, the carbon atom has partially "lost" its shared electrons; thus, methane has been oxidized.

Now let's examine the fate of the reactant O_2. The two atoms of the oxygen molecule (O_2) share their electrons equally. But when oxygen reacts with the hydrogen from methane, forming water, the electrons of the covalent bonds spend more time near the oxygen (see Figure 3). In effect, each oxygen atom has partially "gained" electrons, so the oxygen molecule has been reduced. Because oxygen is so electronegative, it is one of the most powerful of all oxidizing agents.

Energy must be added to pull an electron away from an atom, just as energy is required to push a ball uphill. The more electronegative the atom (the stronger its pull on electrons), the more energy is required to take an electron away from it. An electron loses potential energy when it shifts from a less electronegative atom toward a more electronegative one, just as a ball loses potential energy when it rolls downhill. A redox reaction that moves electrons closer to oxygen, such as the burning (oxidation) of methane, therefore releases chemical energy that can be put to work.

Oxidation of Organic Fuel Molecules During Cellular Respiration

The oxidation of methane by oxygen is the main combustion reaction that occurs at the burner of a gas stove. The combustion of gasoline in an automobile engine is also a redox reaction; the energy released pushes the pistons. But the energy-yielding redox process of greatest interest to biologists is respiration: the oxidation of glucose and other molecules in food. Examine again the summary equation for cellular respiration, but this time think of it as a redox process:

$$\underbrace{C_6H_{12}O_6}_{\text{becomes oxidized} \rightarrow 6\,CO_2} + \underbrace{6\,O_2}_{\text{becomes reduced} \rightarrow 6\,H_2O} \longrightarrow 6\,CO_2 + 6\,H_2O + \text{Energy}$$

As in the combustion of methane or gasoline, the fuel (glucose) is oxidized and oxygen is reduced. The electrons lose potential energy along the way, and energy is released.

In general, organic molecules that have an abundance of hydrogen are excellent fuels because their bonds are a source of "hilltop" electrons, whose energy may be released as these electrons "fall" down an energy gradient when they are transferred to oxygen. The summary equation for respiration indicates that hydrogen is transferred from glucose to oxygen. But the important point, not visible in the summary equation, is that the energy state of the electron changes as hydrogen (with its electron) is transferred to oxygen. In respiration, the oxidation of glucose transfers electrons to a lower energy state, liberating energy that becomes available for ATP synthesis.

The main energy-yielding foods, carbohydrates and fats, are reservoirs of electrons associated with hydrogen. Only the barrier of activation energy holds back the flood of electrons to a lower energy state. Without this barrier, a food substance like glucose would combine almost instantaneously with O_2. If we supply the activation energy by igniting glucose, it burns in air, releasing 686 kcal (2,870 kJ) of heat per mole of glucose (about 180 g). Body temperature is not high enough to initiate burning, of course. Instead, if you swallow some glucose, enzymes in your cells will lower the barrier of activation energy, allowing the sugar to be oxidized in a series of steps.

Stepwise Energy Harvest via NAD^+ and the Electron Transport Chain

If energy is released from a fuel all at once, it cannot be harnessed efficiently for constructive work. For example, if a gasoline tank explodes, it cannot drive a car very far. Cellular respiration does not oxidize glucose (or any other organic fuel) in a single explosive step, either. Rather, glucose is broken down in a series of steps, each one catalyzed by an enzyme. At key steps, electrons are stripped from the glucose. As is often the case in oxidation reactions, each electron travels with a proton—thus, as a hydrogen atom. The hydrogen atoms are not transferred directly to oxygen, but instead are usually passed first to an electron carrier, a coenzyme called **NAD^+** (nicotinamide adenine dinucleotide, a derivative of the vitamin niacin). NAD^+ is well suited as an electron carrier because it can cycle easily between oxidized (NAD^+) and reduced (NADH) states. As an electron acceptor, NAD^+ functions as an oxidizing agent during respiration.

How does NAD^+ trap electrons from glucose and other organic molecules in food? Enzymes called dehydrogenases remove a pair of hydrogen atoms (2 electrons and 2 protons) from the

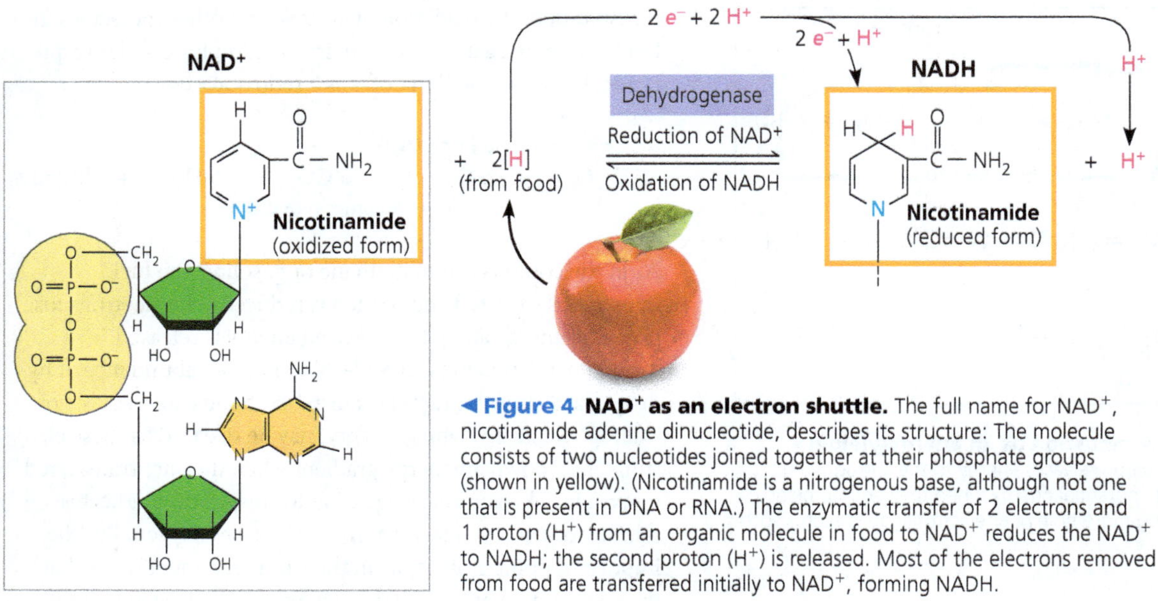

◀ **Figure 4 NAD$^+$ as an electron shuttle.** The full name for NAD$^+$, nicotinamide adenine dinucleotide, describes its structure: The molecule consists of two nucleotides joined together at their phosphate groups (shown in yellow). (Nicotinamide is a nitrogenous base, although not one that is present in DNA or RNA.) The enzymatic transfer of 2 electrons and 1 proton (H^+) from an organic molecule in food to NAD$^+$ reduces the NAD$^+$ to NADH; the second proton (H^+) is released. Most of the electrons removed from food are transferred initially to NAD$^+$, forming NADH.

substrate (glucose, in this example), thereby oxidizing it. The enzyme delivers the 2 electrons along with 1 proton to its coenzyme, NAD$^+$, forming NADH **(Figure 4)**. The other proton is released as a hydrogen ion (H^+) into the surrounding solution:

$$\text{H}-\overset{|}{\underset{|}{\text{C}}}-\text{OH} + \text{NAD}^+ \xrightarrow{\text{Dehydrogenase}} \overset{|}{\underset{|}{\text{C}}}=\text{O} + \text{NADH} + \text{H}^+$$

By receiving 2 negatively charged electrons but only 1 positively charged proton, the nicotinamide portion of NAD$^+$ has its charge neutralized when NAD$^+$ is reduced to NADH. The name NADH shows the hydrogen that has been received in the reaction. NAD$^+$ is the most versatile electron acceptor in cellular respiration and functions in several of the redox steps during the breakdown of glucose.

Electrons lose very little of their potential energy when they are transferred from glucose to NAD$^+$. Each NADH molecule formed during respiration represents stored energy that can be tapped to make ATP when the electrons complete their "fall" down an energy gradient from NADH to oxygen.

How do electrons that are extracted from glucose and stored as potential energy in NADH finally reach oxygen? It will help to compare the redox chemistry of cellular respiration to a much simpler reaction: the reaction between hydrogen and oxygen to form water **(Figure 5a)**. Mix H_2 and O_2, provide a spark for activation energy, and the gases combine explosively. In fact, combustion of liquid H_2 and O_2 is harnessed to help power the rocket engines that boost satellites into orbit and launch spacecraft. The explosion represents a release of energy as the electrons of hydrogen "fall" closer to the electronegative oxygen atoms. Cellular respiration also brings hydrogen and oxygen together to form water, but there are two important differences. First, in cellular respiration, the hydrogen that reacts with oxygen is derived from organic molecules rather than H_2. Second, instead of occurring in one explosive reaction, respiration uses an electron transport chain to break the fall of electrons

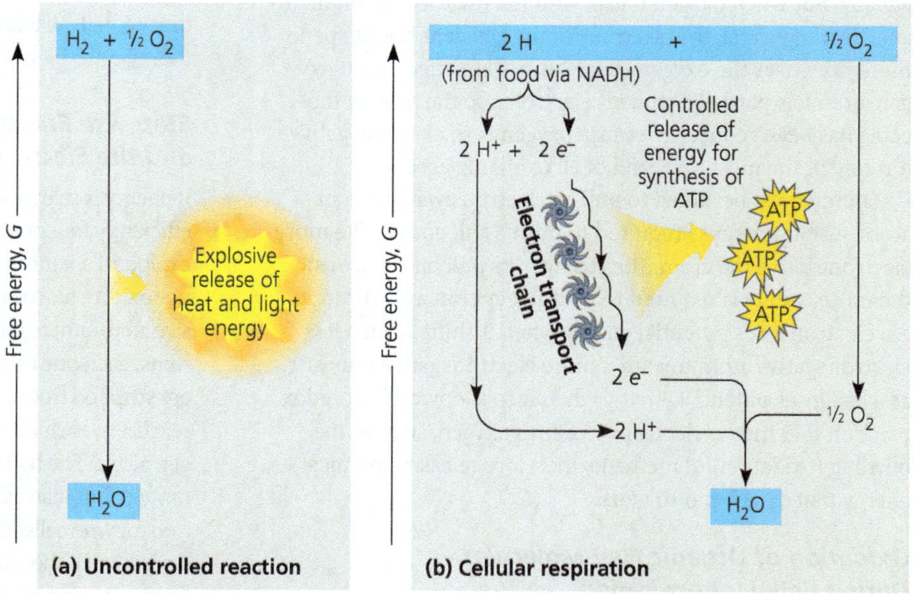

▲ **Figure 5 An introduction to electron transport chains. (a)** The one-step exergonic reaction of hydrogen with oxygen to form water releases a large amount of energy in the form of heat and light: an explosion. **(b)** In cellular respiration, the same reaction occurs in stages: An electron transport chain breaks the "fall" of electrons in this reaction into a series of smaller steps and stores some of the released energy in a form that can be used to make ATP. (The rest of the energy is released as heat.)

to oxygen into several energy-releasing steps **(Figure 5b)**. An **electron transport chain** consists of a number of molecules, mostly proteins, built into the inner membrane of the mitochondria of eukaryotic cells and the plasma membrane of aerobically respiring prokaryotes. Electrons removed from glucose are shuttled by NADH to the "top," higher-energy end of the chain. At the "bottom," lower-energy end, O_2 captures these electrons along with hydrogen nuclei (H^+), forming water.

Electron transfer from NADH to oxygen is an exergonic reaction with a free-energy change of −53 kcal/mol (−222 kJ/mol). Instead of this energy being released and wasted in a single explosive step, electrons cascade down the chain from one carrier molecule to the next in a series of redox reactions, losing a small amount of energy with each step until they finally reach oxygen, the terminal electron acceptor, which has a very great affinity for electrons. Each "downhill" carrier is more electronegative than, and thus capable of oxidizing, its "uphill" neighbor, with oxygen at the bottom of the chain. Therefore, the electrons transferred from glucose to NAD^+, forming NADH, will fall down an energy gradient in the electron transport chain to a far more stable location in the electronegative oxygen atom. Put another way, oxygen pulls electrons down the chain in an energy-yielding tumble analogous to gravity pulling objects downhill.

In summary, during cellular respiration, most electrons travel the following "downhill" route: glucose → NADH → electron transport chain → oxygen. Later in this chapter, you will learn more about how the cell uses the energy released from this exergonic electron fall to regenerate its supply of ATP. For now, having covered the basic redox mechanisms of cellular respiration, let's look at the entire process by which energy is harvested from organic fuels.

The Stages of Cellular Respiration: *A Preview*

The harvesting of energy from glucose by cellular respiration is a cumulative function of three metabolic stages. We list them here along with a color-coding scheme that we will use throughout the chapter to help you keep track of the big picture.

1. GLYCOLYSIS (color-coded blue throughout the chapter)
2. PYRUVATE OXIDATION and the CITRIC ACID CYCLE (color-coded orange)
3. OXIDATIVE PHOSPHORYLATION: Electron transport and chemiosmosis (color-coded purple)

Biochemists usually reserve the term *cellular respiration* for stages 2 and 3 together. In this text, however, we include glycolysis as a part of cellular respiration because most respiring cells deriving energy from glucose use glycolysis to produce the starting material for the citric acid cycle.

As diagrammed in **Figure 6**, glycolysis and pyruvate oxidation followed by the citric acid cycle are the catabolic pathways that break down glucose and other organic fuels. **Glycolysis,** which occurs in the cytosol, begins the degradation process by breaking glucose into two molecules of a compound called pyruvate. In eukaryotes, pyruvate enters the mitochondrion and is oxidized to a compound called acetyl CoA, which enters the **citric acid cycle** (also called the Krebs cycle). There, the breakdown of glucose to carbon dioxide is completed. (In prokaryotes, these processes take place in the cytosol.) Thus, the carbon dioxide produced by respiration represents fragments of oxidized organic molecules.

Some of the steps of glycolysis and the citric acid cycle are redox reactions in which dehydrogenases transfer electrons from substrates to NAD^+, forming NADH. In the third stage

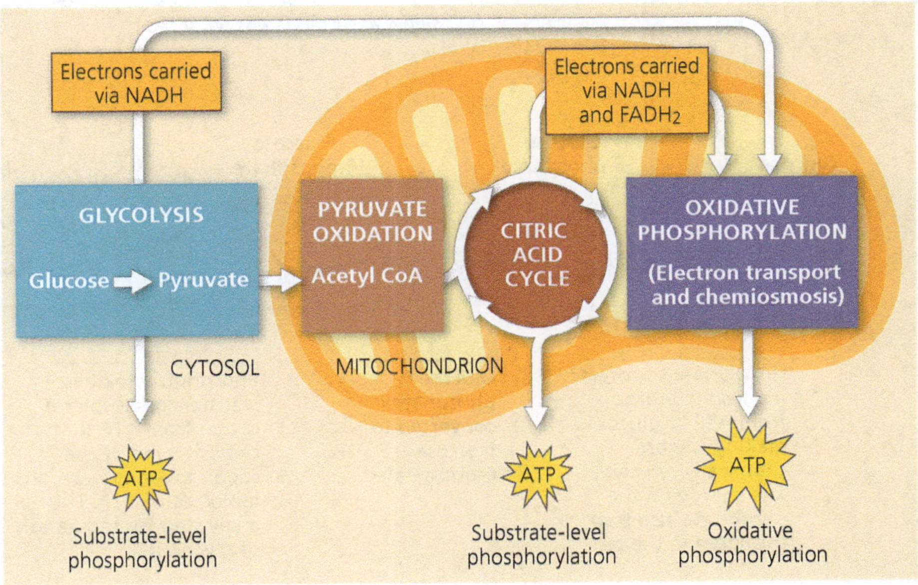

▶ **Figure 6 An overview of cellular respiration.** During glycolysis, each glucose molecule is broken down into two molecules of pyruvate. In eukaryotic cells, as shown here, the pyruvate enters the mitochondrion. There it is oxidized to acetyl CoA, which will be further oxidized to CO_2 in the citric acid cycle. The electron carriers NADH and $FADH_2$ transfer electrons derived from glucose to electron transport chains. During oxidative phosphorylation, electron transport chains convert the chemical energy to a form used for ATP synthesis in the process called chemiosmosis. (During earlier steps of cellular respiration, smaller amounts of ATP are synthesized in a process called substrate-level phosphorylation.)

Visit the Study Area in **MasteringBiology** for the BioFlix® 3-D Animation on Cellular Respiration.

of respiration, the electron transport chain accepts electrons (most often via NADH) from the breakdown products of the first two stages and passes these electrons from one molecule to another. At the end of the chain, the electrons are combined with molecular oxygen and hydrogen ions (H^+), forming water (see Figure 5b). The energy released at each step of the chain is stored in a form the mitochondrion (or prokaryotic cell) can use to make ATP from ADP. This mode of ATP synthesis is called **oxidative phosphorylation** because it is powered by the redox reactions of the electron transport chain.

In eukaryotic cells, the inner membrane of the mitochondrion is the site of electron transport and chemiosmosis, the processes that together constitute oxidative phosphorylation. (In prokaryotes, these processes take place in the plasma membrane.) Oxidative phosphorylation accounts for almost 90% of the ATP generated by respiration. A smaller amount of ATP is formed directly in a few reactions of glycolysis and the citric acid cycle by a mechanism called **substrate-level phosphorylation** (**Figure 7**). This mode of ATP synthesis occurs when an enzyme transfers a phosphate group from a substrate molecule to ADP, rather than adding an inorganic phosphate to ADP as in oxidative phosphorylation. "Substrate molecule" here refers to an organic molecule generated as an intermediate during the catabolism of glucose.

For each molecule of glucose degraded to carbon dioxide and water by respiration, the cell makes up to about 32 molecules of ATP, each with 7.3 kcal/mol of free energy. Respiration cashes in the large denomination of energy banked in a single molecule of glucose (686 kcal/mol) for the small change of many molecules of ATP, which is more practical for the cell to spend on its work.

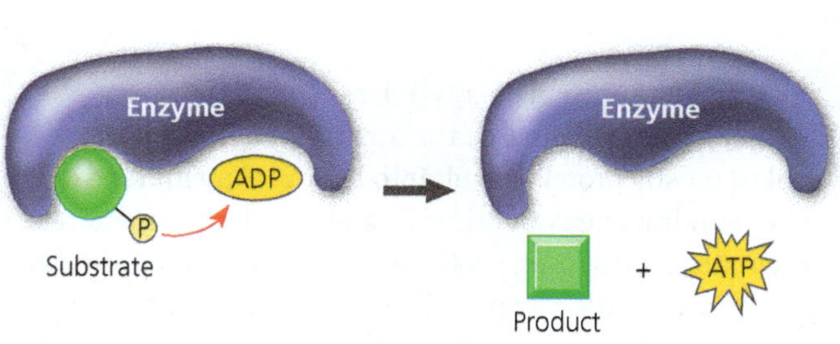

▲ **Figure 7 Substrate-level phosphorylation.** Some ATP is made by direct transfer of a phosphate group from an organic substrate to ADP by an enzyme. (For examples in glycolysis, see Figure 9, steps 7 and 10.)

MAKE CONNECTIONS *Do you think the potential energy is higher for the reactants or the products in the reaction shown above? Explain.*

This preview has introduced you to how glycolysis, the citric acid cycle, and oxidative phosphorylation fit into the process of cellular respiration. We are now ready to take a closer look at each of these three stages of respiration.

CONCEPT CHECK 1

1. Compare and contrast aerobic and anaerobic respiration.
2. Name and describe the two ways in which ATP is made during cellular respiration. During what stage(s) in the process does each type occur?
3. **WHAT IF?** If the following redox reaction occurred, which compound would be oxidized? Which reduced?

$$C_4H_6O_5 + NAD^+ \rightarrow C_4H_4O_5 + NADH + H^+$$

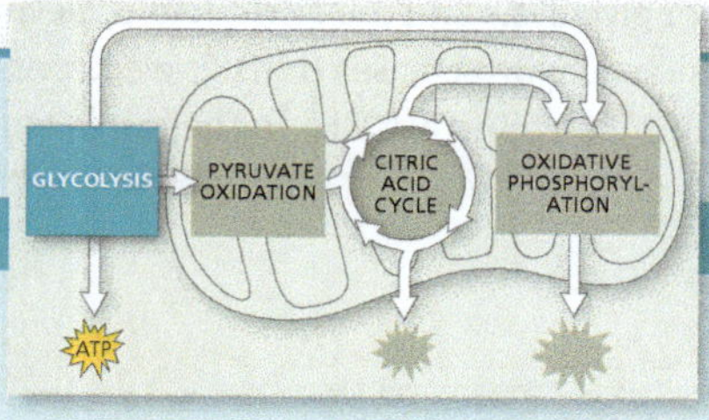

▼ **Figure 9 A closer look at glycolysis.** Note that glycolysis is a source of ATP and NADH.

GLYCOLYSIS: Energy Investment Phase

WHAT IF? *What would happen if you removed the dihydroxyacetone phosphate generated in step 4 as fast as it was produced?*

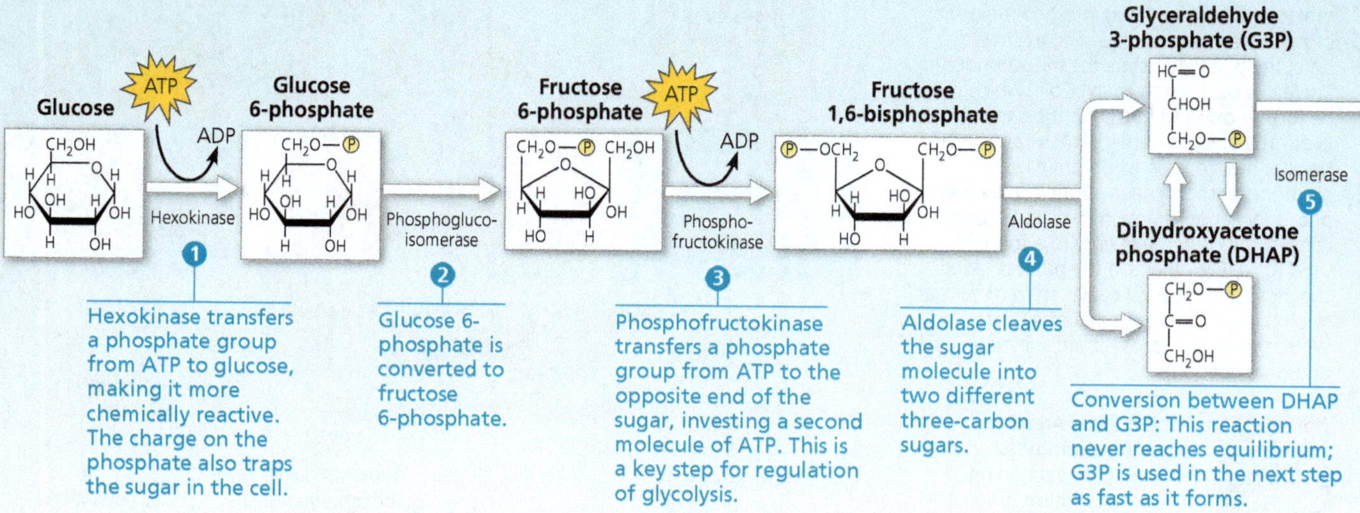

CONCEPT 2

Glycolysis harvests chemical energy by oxidizing glucose to pyruvate

The word *glycolysis* means "sugar splitting," and that is exactly what happens during this pathway. Glucose, a six-carbon sugar, is split into two three-carbon sugars. These smaller sugars are then oxidized and their remaining atoms rearranged to form two molecules of pyruvate. (Pyruvate is the ionized form of pyruvic acid.)

As summarized in **Figure 8**, glycolysis can be divided into two phases: energy investment and energy payoff. During the energy investment phase, the cell actually spends ATP. This investment is repaid with interest during the energy payoff phase, when ATP is produced by substrate-level phosphorylation and NAD^+ is reduced to NADH by electrons released from the oxidation of glucose. The net energy yield from glycolysis, per glucose molecule, is 2 ATP plus 2 NADH.

Because glycolysis is a fundamental core process shared by bacteria, archaea, and eukaryotes alike, we will use it as an example of a biochemical pathway. The ten steps of the glycolytic pathway are shown in **Figure 9**.

All of the carbon originally present in glucose is accounted for in the two molecules of pyruvate; no carbon is released as CO_2 during glycolysis. Glycolysis occurs whether or not O_2 is present. However, if O_2 *is* present, the chemical energy stored in pyruvate and NADH can be extracted by pyruvate oxidation, the citric acid cycle, and oxidative phosphorylation.

▼ Figure 8 The energy input and output of glycolysis.

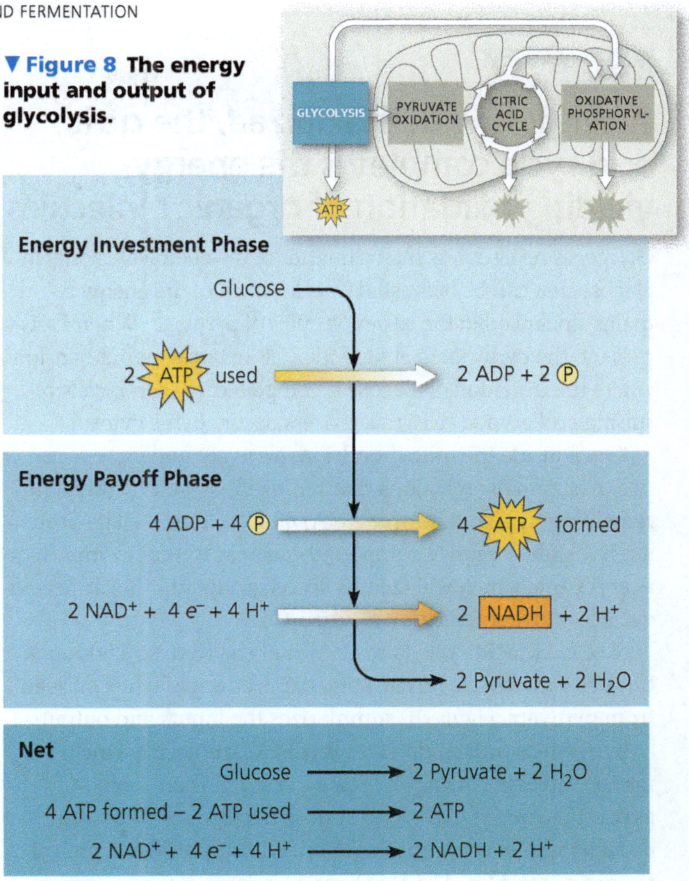

CONCEPT CHECK 2

1. During step 6 in Figure 9, which molecule acts as the oxidizing agent? The reducing agent?

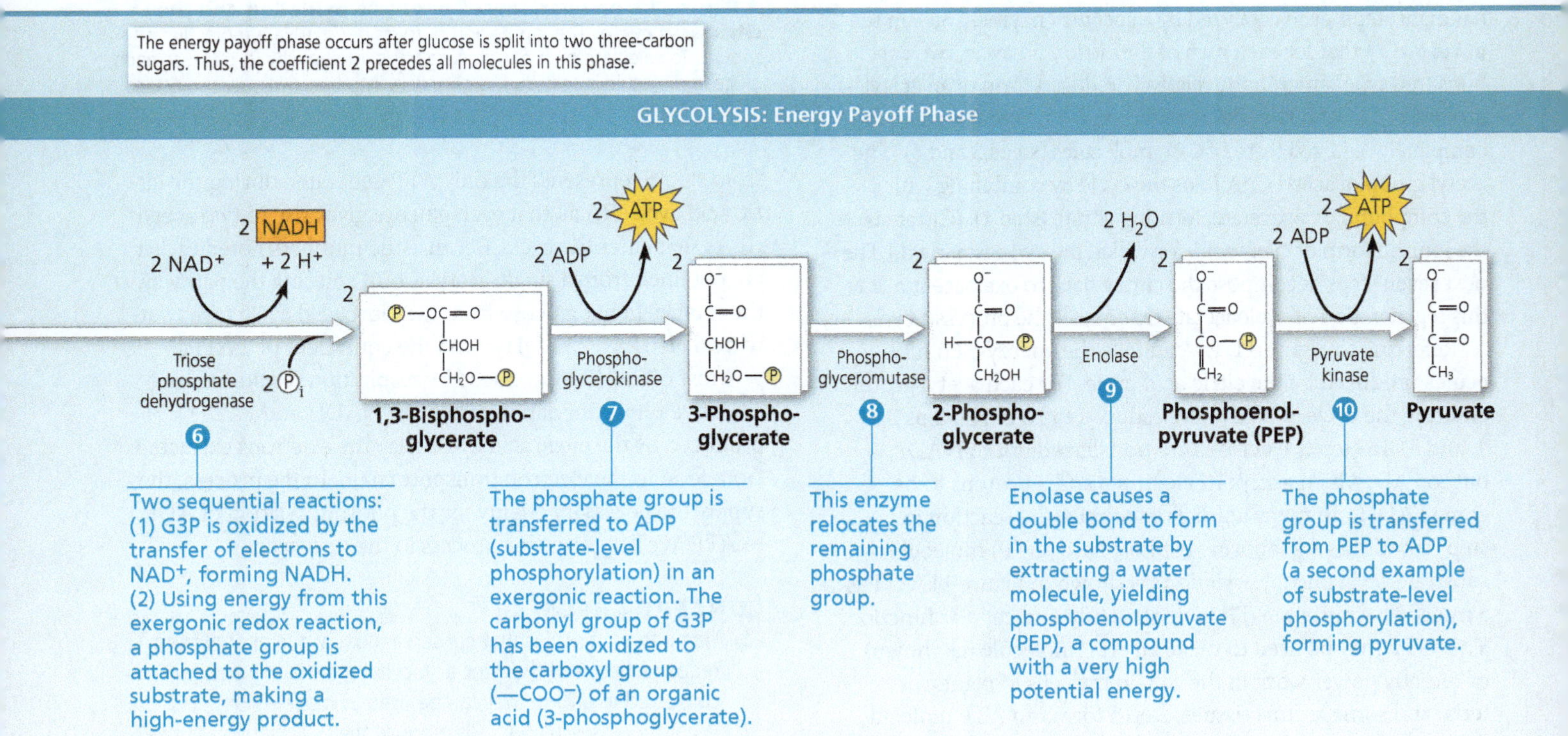

CONCEPT 3

After pyruvate is oxidized, the citric acid cycle completes the energy-yielding oxidation of organic molecules

Glycolysis releases less than a quarter of the chemical energy in glucose that can be harvested by cells; most of the energy remains stockpiled in the two molecules of pyruvate. When O_2 is present, the pyruvate in eukaryotic cells enters a mitochondrion, where the oxidation of glucose is completed. (In aerobically respiring prokaryotic cells, this process occurs in the cytosol.)

Once inside the mitochondrion, pyruvate undergoes a series of enzymatic reactions that remove CO_2 and oxidizes the remaining fragment, forming NADH from NAD^+. The product is a highly reactive compound called acetyl coenzyme A, or **acetyl CoA**, which will feed its acetyl group into the citric acid cycle for further oxidation **(Figure 10)**.

The citric acid cycle (also known as the Krebs cycle) functions as a metabolic furnace that oxidizes organic fuel derived from pyruvate. Figure 10 summarizes the inputs and outputs as pyruvate is broken down to three CO_2 molecules, including the molecule of CO_2 released during the conversion of pyruvate to acetyl CoA. The cycle generates 1 ATP per turn by substrate-level phosphorylation, but most of the chemical energy is transferred to NAD^+ and a related electron carrier, the coenzyme FAD (flavin adenine dinucleotide, derived from riboflavin, a B vitamin), during the redox reactions. The reduced coenzymes, NADH and $FADH_2$, shuttle their cargo of high-energy electrons into the electron transport chain.

Now let's look at the citric acid cycle in more detail. The cycle has eight steps, each catalyzed by a specific enzyme. You can see in **Figure 11** that for each turn of the citric acid cycle, two carbons (red type) enter in the relatively reduced form of an acetyl group (step 1), and two different carbons (blue type) leave in the completely oxidized form of CO_2 molecules (steps 3 and 4). The acetyl group of acetyl CoA joins the cycle by combining with the compound oxaloacetate, forming citrate (step 1). (Citrate is the ionized form of citric acid, for which the cycle is named.) The next seven steps decompose the citrate back to oxaloacetate. It is this regeneration of oxaloacetate that makes the process a *cycle*.

Referring to Figure 11, we can tally the energy-rich molecules produced by the citric acid cycle. For each acetyl group entering the cycle, 3 NAD^+ are reduced to NADH (steps 3, 4, and 8). In step 6, electrons are transferred not to NAD^+, but to FAD, which accepts 2 electrons and 2 protons to become $FADH_2$. In many animal tissue cells, the reaction in step 5 produces a guanosine triphosphate (GTP) molecule by substrate-level phosphorylation, as shown in Figure 11. GTP is a molecule similar to ATP in its structure and cellular function. This GTP may be used to make an ATP molecule (as shown) or directly power work in the cell. In the cells of plants, bacteria, and some animal tissues, step 5 forms an ATP molecule directly by substrate-level phosphorylation. The output

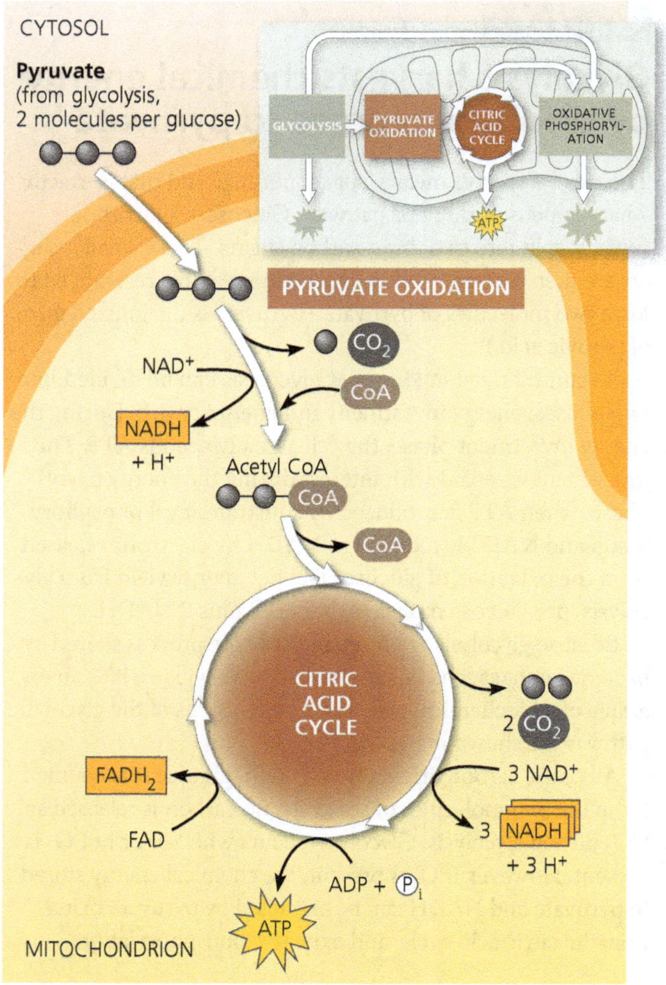

▲ **Figure 10 An overview of pyruvate oxidation and the citric acid cycle.** The inputs and outputs per pyruvate molecule are shown. To calculate on a per-glucose basis, multiply by 2, because each glucose molecule is split during glycolysis into two pyruvate molecules.

from step 5 represents the only ATP generated during the citric acid cycle. Recall that each glucose gives rise to two acetyl CoAs that enter the cycle. Because the numbers noted earlier are obtained from a single acetyl group entering the pathway, the total yield per glucose from the citric acid cycle turns out to be 6 NADHs, 2 $FADH_2$s, and the equivalent of 2 ATPs.

Most of the ATP produced by respiration results from oxidative phosphorylation, when the NADH and $FADH_2$ produced by the citric acid cycle relay the electrons extracted from food to the electron transport chain. In the process, they supply the necessary energy for the phosphorylation of ADP to ATP. We'll explore this process in the next section.

CONCEPT CHECK 3

1. Name the molecules that conserve most of the energy from redox reactions of the citric acid cycle. How is this energy converted to a form that can be used to make ATP?
2. What processes in your cells produce the CO_2 that you exhale?

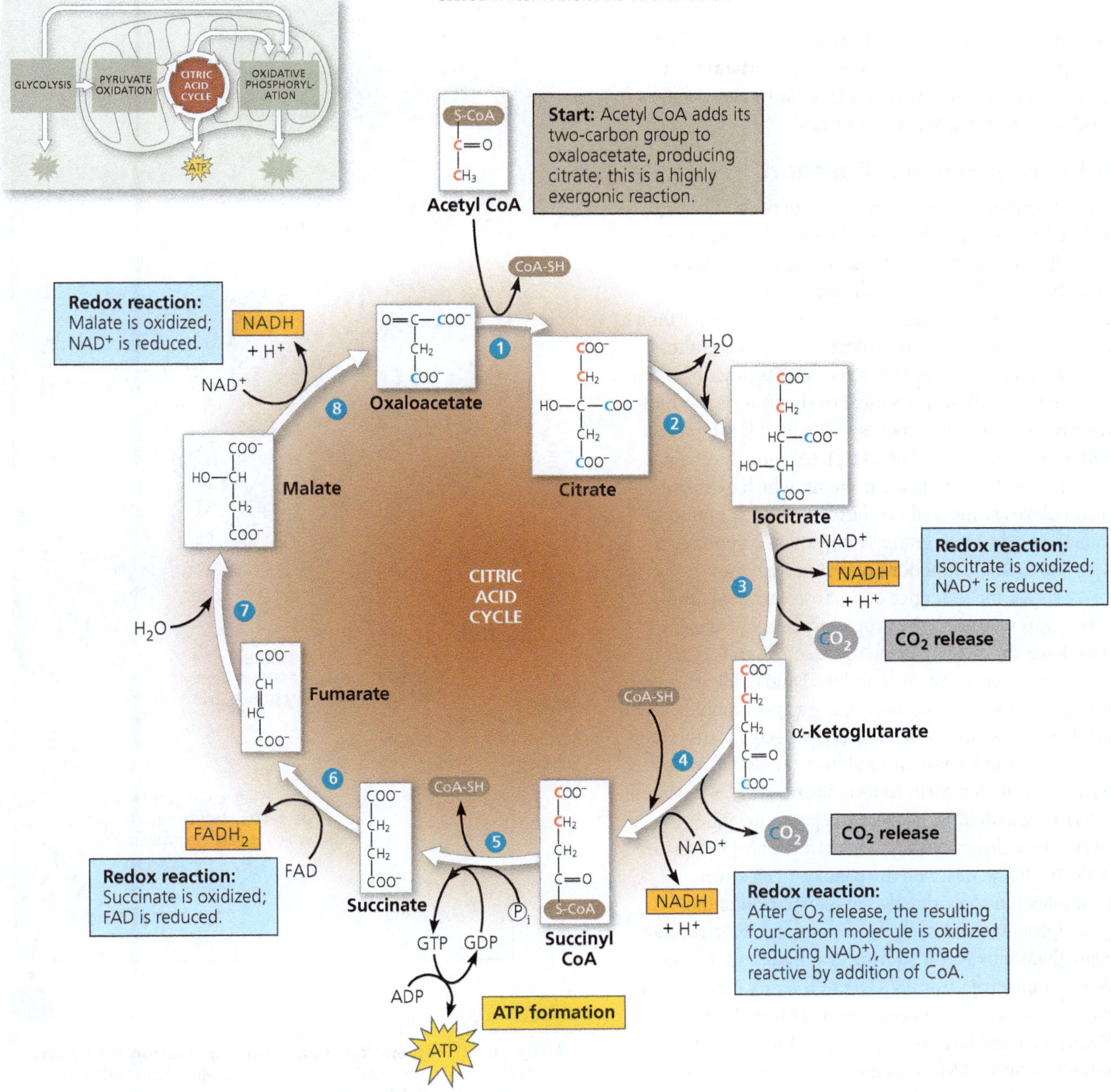

▲ **Figure 11 A closer look at the citric acid cycle.** Key steps (redox reactions, CO_2 release, and ATP formation) are labeled. In the chemical structures, red type traces the fate of the two carbon atoms that enter the cycle via acetyl CoA (step 1), and blue type indicates the two carbons that exit the cycle as CO_2 in steps 3 and 4. (The red type goes only through step 5 because the succinate molecule is symmetrical; the two ends cannot be distinguished from each other.) Notice that the carbon atoms that enter the cycle from acetyl CoA do not leave the cycle in the same turn. They remain in the cycle, occupying a different location in the molecules on their next turn, after another acetyl group is added. Therefore, the oxaloacetate regenerated at step 8 is made up of different carbon atoms each time around.

CONCEPT 4

During oxidative phosphorylation, chemiosmosis couples electron transport to ATP synthesis

Our main objective in this chapter is to learn how cells harvest the energy of glucose and other nutrients in food to make ATP. But the metabolic components of respiration we have dissected so far, glycolysis and the citric acid cycle, produce only 4 ATP molecules per glucose molecule, all by substrate-level phosphorylation: 2 net ATP from glycolysis and 2 ATP from the citric acid cycle. At this point, molecules of NADH (and $FADH_2$) account for most of the energy extracted from each glucose molecule. These electron escorts link glycolysis and the citric acid cycle to the machinery of oxidative phosphorylation, which uses energy released by the electron transport chain to power ATP synthesis. In this section, you will learn first how the electron transport chain works and then how electron flow down the chain is coupled to ATP synthesis.

The Pathway of Electron Transport

The electron transport chain is a collection of molecules embedded in the inner membrane of the mitochondrion in eukaryotic cells. (In prokaryotes, these molecules reside in the plasma membrane.) The folding of the inner membrane to form cristae increases its surface area, providing space for thousands of copies of the electron transport chain in each mitochondrion. Once again, we see that structure fits function—the infolded membrane with its placement of electron carrier molecules in a row, one after the other, is well-suited for the series of sequential redox reactions that take place along the chain. Most components of the chain are proteins, which exist in multiprotein complexes numbered I through IV. Tightly bound to these proteins are *prosthetic groups*, nonprotein components essential for the catalytic functions of certain enzymes.

Figure 12 shows the sequence of electron carriers in the electron transport chain and the drop in free energy as electrons travel down the chain. During this electron transport, electron carriers alternate between reduced and oxidized states as they accept and then donate electrons. Each component of the chain becomes reduced when it accepts electrons from its "uphill" neighbor, which has a lower affinity for electrons (is less electronegative). It then returns to its oxidized form as it passes electrons to its "downhill," more electronegative neighbor.

Now let's take a closer look at the electron transport chain in Figure 12. We'll first describe the passage of electrons through complex I in some detail as an illustration of the general principles involved in electron transport. Electrons acquired from glucose by NAD^+ during glycolysis and the citric acid cycle are transferred from NADH to the first molecule of the electron transport chain in complex I. This molecule is a flavoprotein, so named because it has a prosthetic group called flavin mononucleotide (FMN). In the next redox reaction, the flavoprotein returns to its oxidized form as it passes electrons to an iron-sulfur protein (Fe · S in complex I), one of a family of proteins with both iron and sulfur tightly bound. The iron-sulfur protein then passes the electrons to a compound called ubiquinone (Q in Figure 12). This electron carrier is a small hydrophobic molecule, the only member of the electron transport chain that is not a protein. Ubiquinone is individually mobile within the membrane rather than residing in a particular complex. (Another name for ubiquinone is coenzyme Q, or CoQ; you may have seen it sold as a nutritional supplement.)

Most of the remaining electron carriers between ubiquinone and oxygen are proteins called **cytochromes**. Their prosthetic group, called a heme group, has an iron atom that accepts and donates electrons. (The heme group in a cytochrome is similar to the heme group in hemoglobin, the protein of red

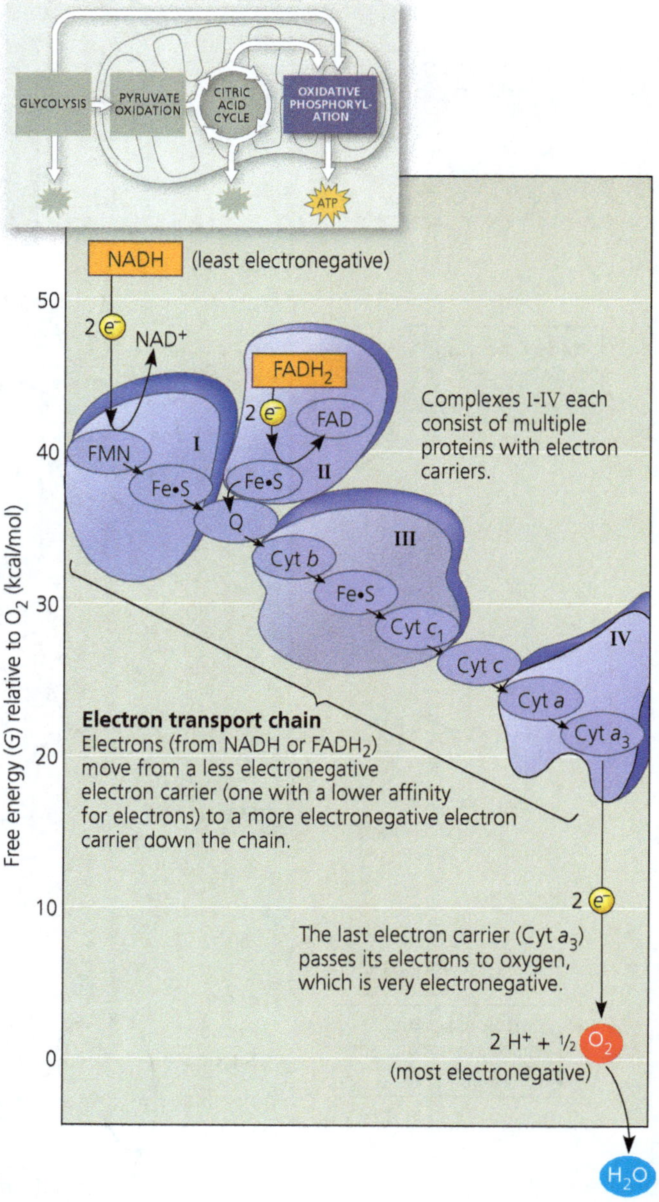

▲ **Figure 12 Free-energy change during electron transport.** The overall energy drop (ΔG) for electrons traveling from NADH to oxygen is 53 kcal/mol, but this "fall" is broken up into a series of smaller steps by the electron transport chain. (An oxygen atom is represented here as ½ O_2 to emphasize that the electron transport chain reduces molecular oxygen, O_2, not individual oxygen atoms.)

blood cells, except that the iron in hemoglobin carries oxygen, not electrons.) The electron transport chain has several types of cytochromes, each a different protein with a slightly different electron-carrying heme group. The last cytochrome of the chain, Cyt a_3, passes its electrons to oxygen, which is *very* electronegative. Each oxygen atom also picks up a pair of hydrogen ions (protons) from the aqueous solution, neutralizing the −2 charge of the added electrons and forming water.

Another source of electrons for the transport chain is $FADH_2$, the other reduced product of the citric acid cycle.

Notice in Figure 12 that $FADH_2$ adds its electrons to the electron transport chain from within complex II, at a lower energy level than NADH does. Consequently, although NADH and $FADH_2$ each donate an equivalent number of electrons (2) for oxygen reduction, the electron transport chain provides about one-third less energy for ATP synthesis when the electron donor is $FADH_2$ rather than NADH. We'll see why in the next section.

The electron transport chain makes no ATP directly. Instead, it eases the fall of electrons from food to oxygen, breaking a large free-energy drop into a series of smaller steps that release energy in manageable amounts. How does the mitochondrion (or the plasma membrane in prokaryotes) couple this electron transport and energy release to ATP synthesis? The answer is a mechanism called chemiosmosis.

Chemiosmosis: The Energy-Coupling Mechanism

Populating the inner membrane of the mitochondrion or the prokaryotic plasma membrane are many copies of a protein complex called **ATP synthase**, the enzyme that actually makes ATP from ADP and inorganic phosphate. ATP synthase works like an ion pump running in reverse. Ion pumps usually use ATP as an energy source to transport ions against their gradients. Enzymes can catalyze a reaction in either direction, depending on the ΔG for the reaction, which is affected by the local concentrations of reactants and products. Rather than hydrolyzing ATP to pump protons against their concentration gradient, under the conditions of cellular respiration ATP synthase uses the energy of an existing ion gradient to power ATP synthesis. The power source for ATP synthase is a difference in the concentration of H^+ on opposite sides of the inner mitochondrial membrane. (We can also think of this gradient as a difference in pH, since pH is a measure of H^+ concentration.) This process, in which energy stored in the form of a hydrogen ion gradient across a membrane is used to drive cellular work such as the synthesis of ATP, is called **chemiosmosis** (from the Greek *osmos*, push). We have previously used the word *osmosis* in discussing water transport, but here it refers to the flow of H^+ across a membrane.

From studying the structure of ATP synthase, scientists have learned how the flow of H^+ through this large enzyme powers ATP generation. ATP synthase is a multisubunit complex with four main parts, each made up of multiple polypeptides. Protons move one by one into binding sites on one of the parts (the rotor), causing it to spin in a way that catalyzes ATP production from ADP and inorganic phosphate **(Figure 13)**. The flow of protons thus behaves somewhat like a rushing stream that turns a waterwheel. ATP synthase is the smallest molecular rotary motor known in nature.

How does the inner mitochondrial membrane or the prokaryotic plasma membrane generate and maintain the H^+ gradient that drives ATP synthesis by the ATP synthase protein complex? Establishing the H^+ gradient across the inner

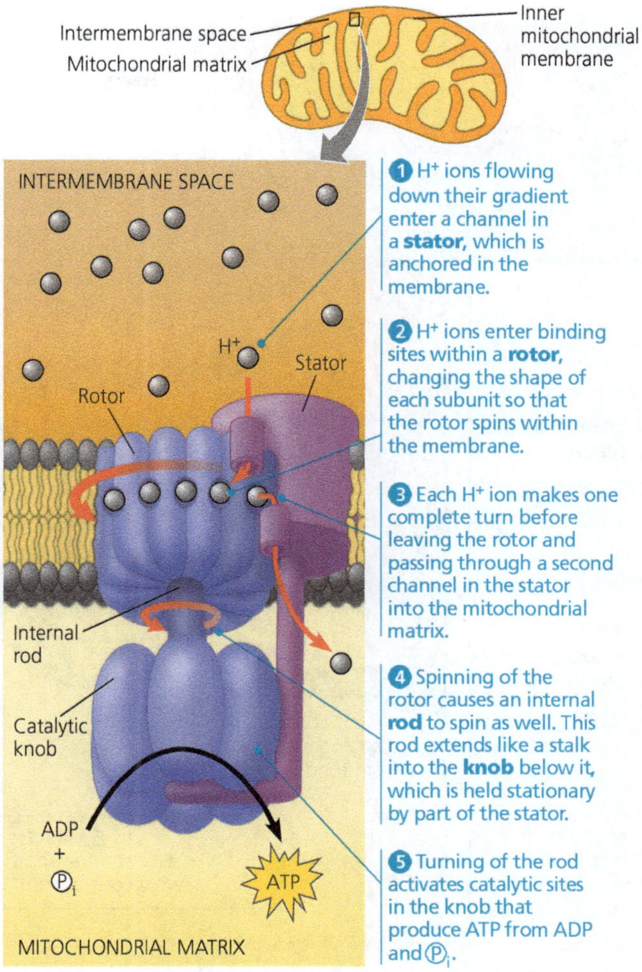

(a) The ATP synthase protein complex functions as a mill, powered by the flow of hydrogen ions.

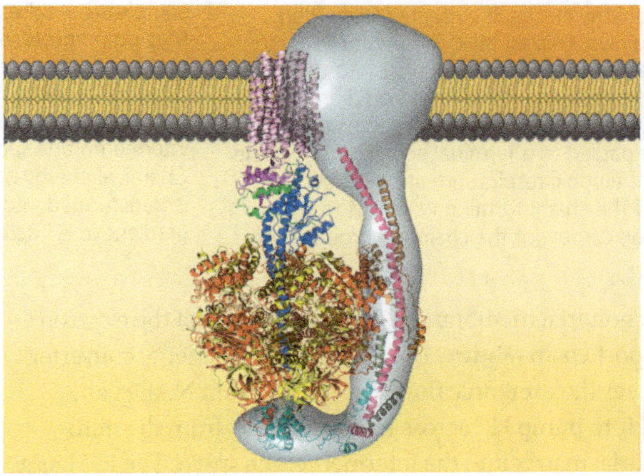

(b) This computer model shows the four parts of ATP synthase. Each part consists of a number of polypeptide subunits. The entire structure of the gray region has not yet been determined and is an area of active research.

▲ **Figure 13 ATP synthase, a molecular mill.** Multiple ATP synthases reside in eukaryotic mitochondrial and chloroplast membranes and in prokaryotic plasma membranes.

DRAW IT *Label the rotor, stator, internal rod, and catalytic knob in part (b), the computer model.*

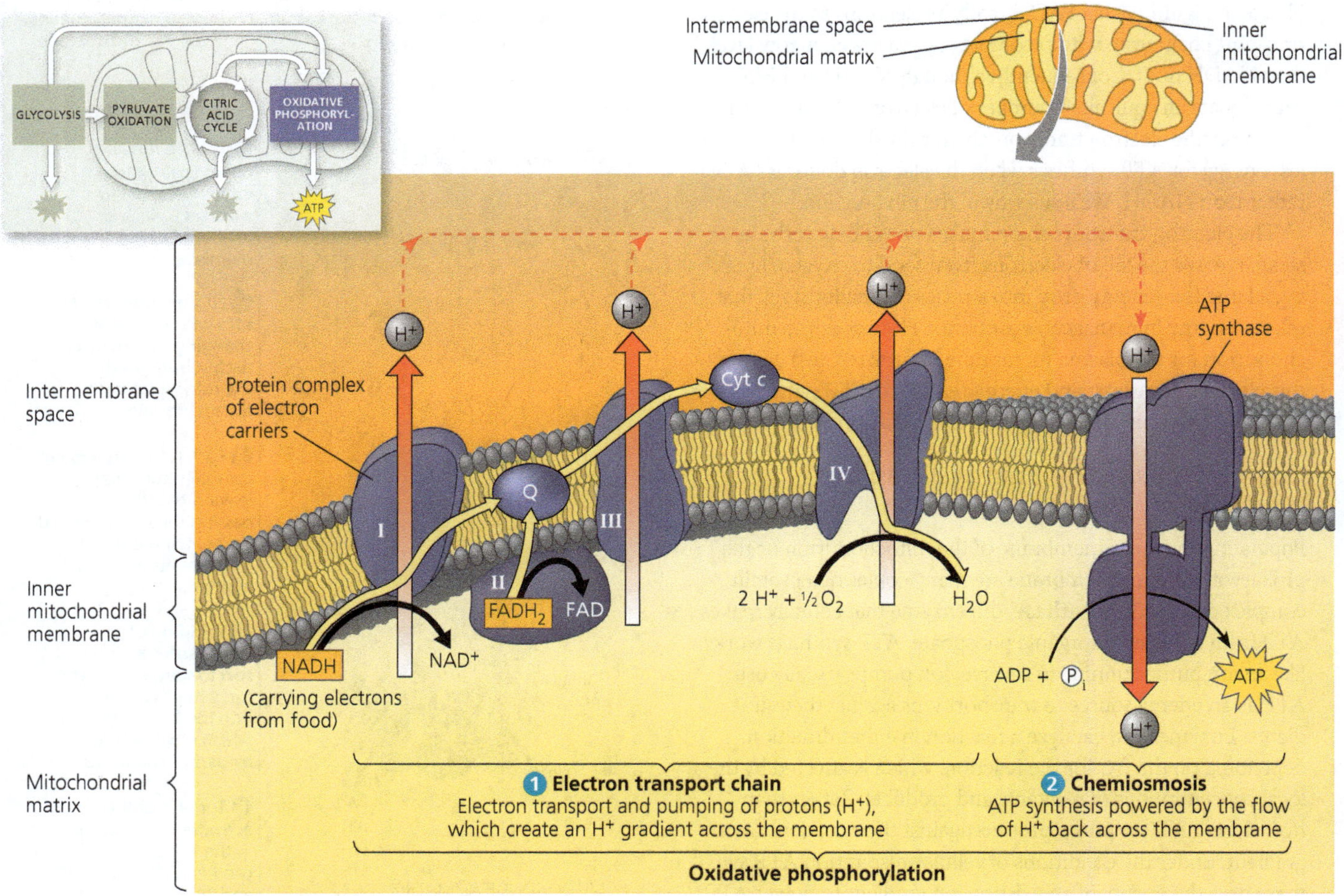

▲ **Figure 14 Chemiosmosis couples the electron transport chain to ATP synthesis.** 1 NADH and $FADH_2$ shuttle high-energy electrons extracted from food during glycolysis and the citric acid cycle into an electron transport chain built into the inner mitochondrial membrane. The gold arrows trace the transport of electrons, which are finally passed to a terminal acceptor (O_2, in the case of aerobic respiration) at the "downhill" end of the chain, forming water. Most of the electron carriers of the chain are grouped into four complexes (I–IV). Two mobile carriers, ubiquinone (Q) and cytochrome c (Cyt c), move rapidly, ferrying electrons between the large complexes. As the complexes shuttle electrons, they pump protons from the mitochondrial matrix into the intermembrane space. $FADH_2$ deposits its electrons via complex II and so results in fewer protons being pumped into the intermembrane space than occurs with NADH. Chemical energy originally harvested from food is transformed into a proton-motive force, a gradient of H^+ across the membrane.

2 During chemiosmosis, the protons flow back down their gradient via ATP synthase, which is built into the membrane nearby. The ATP synthase harnesses the proton-motive force to phosphorylate ADP, forming ATP. Together, electron transport and chemiosmosis make up oxidative phosphorylation.

WHAT IF? *If complex IV were nonfunctional, could chemiosmosis produce any ATP, and if so, how would the rate of synthesis differ?*

mitochondrial membrane is a major function of the electron transport chain **(Figure 14)**. The chain is an energy converter that uses the exergonic flow of electrons from NADH and $FADH_2$ to pump H^+ across the membrane, from the mitochondrial matrix into the intermembrane space. The H^+ has a tendency to move back across the membrane, diffusing down its gradient. And the ATP synthases are the only sites that provide a route through the membrane for H^+. As we described previously, the passage of H^+ through ATP synthase uses the exergonic flow of H^+ to drive the phosphorylation of ADP. Thus, the energy stored in an H^+ gradient across a membrane couples the redox reactions of the electron transport chain to ATP synthesis.

At this point, you may be wondering how the electron transport chain pumps hydrogen ions. Researchers have found that certain members of the electron transport chain accept and release protons (H^+) along with electrons. (The aqueous solutions inside and surrounding the cell are a ready source of H^+.) At certain steps along the chain, electron transfers cause H^+ to be taken up and released into the surrounding solution. In eukaryotic cells, the electron carriers are spatially arranged in the inner mitochondrial membrane in such a way that H^+ is accepted from the mitochondrial matrix and deposited in the intermembrane space (see Figure 14). The H^+ gradient that results is referred to as a **proton-motive force**, emphasizing the capacity of the gradient to perform work. The force drives H^+ back across the membrane through the H^+ channels provided by ATP synthases.

In general terms, *chemiosmosis is an energy-coupling mechanism that uses energy stored in the form of an H^+ gradient* *across a membrane to drive cellular work.* In mitochondria, the energy for gradient formation comes from exergonic redox reactions, and ATP synthesis is the work performed. But chemiosmosis also occurs elsewhere and in other variations. Chloroplasts use chemiosmosis to generate ATP during photosynthesis; in these organelles, light (rather than chemical energy) drives both electron flow down an electron transport chain and the resulting H^+ gradient formation. Prokaryotes, as already mentioned, generate H^+ gradients across their plasma membranes. They then tap the proton-motive force not only to make ATP inside the cell but also to rotate their flagella and to pump nutrients and waste products across the membrane. Because of its central importance to energy conversions in prokaryotes and eukaryotes, chemiosmosis has helped unify the study of bioenergetics. Peter Mitchell was awarded the Nobel Prize in 1978 for originally proposing the chemiosmotic model.

An Accounting of ATP Production by Cellular Respiration

In the last few sections, we have looked rather closely at the key processes of cellular respiration. Now let's take a step back and remind ourselves of its overall function: harvesting the energy of glucose for ATP synthesis.

During respiration, most energy flows in this sequence: glucose → NADH → electron transport chain → proton-motive force → ATP. We can do some bookkeeping to calculate the ATP profit when cellular respiration oxidizes a molecule of glucose to six molecules of carbon dioxide. The three main departments of this metabolic enterprise are glycolysis, the citric acid cycle, and the electron transport chain, which drives oxidative phosphorylation. **Figure 15** gives a detailed accounting of the ATP yield per glucose molecule oxidized. The tally adds the 4 ATP produced directly by substrate-level phosphorylation during glycolysis and the citric acid cycle to the many more molecules of ATP generated by oxidative phosphorylation. Each NADH that transfers a pair of electrons from glucose to the electron transport chain contributes enough to the proton-motive force to generate a maximum of about 3 ATP.

Why are the numbers in Figure 15 inexact? There are three reasons we cannot state an exact number of ATP molecules generated by the breakdown of one molecule of glucose. First, phosphorylation and the redox reactions are not directly coupled to each other, so the ratio of the number of NADH molecules to the number of ATP molecules is not a whole number. We know that 1 NADH results in 10 H^+ being transported out across the inner mitochondrial membrane, but the exact number of H^+ that must reenter the mitochondrial matrix via ATP synthase to generate 1 ATP has long been debated. Based on experimental data, however, most biochemists now agree that the most accurate number is 4 H^+. Therefore, a single molecule of NADH generates enough proton-motive

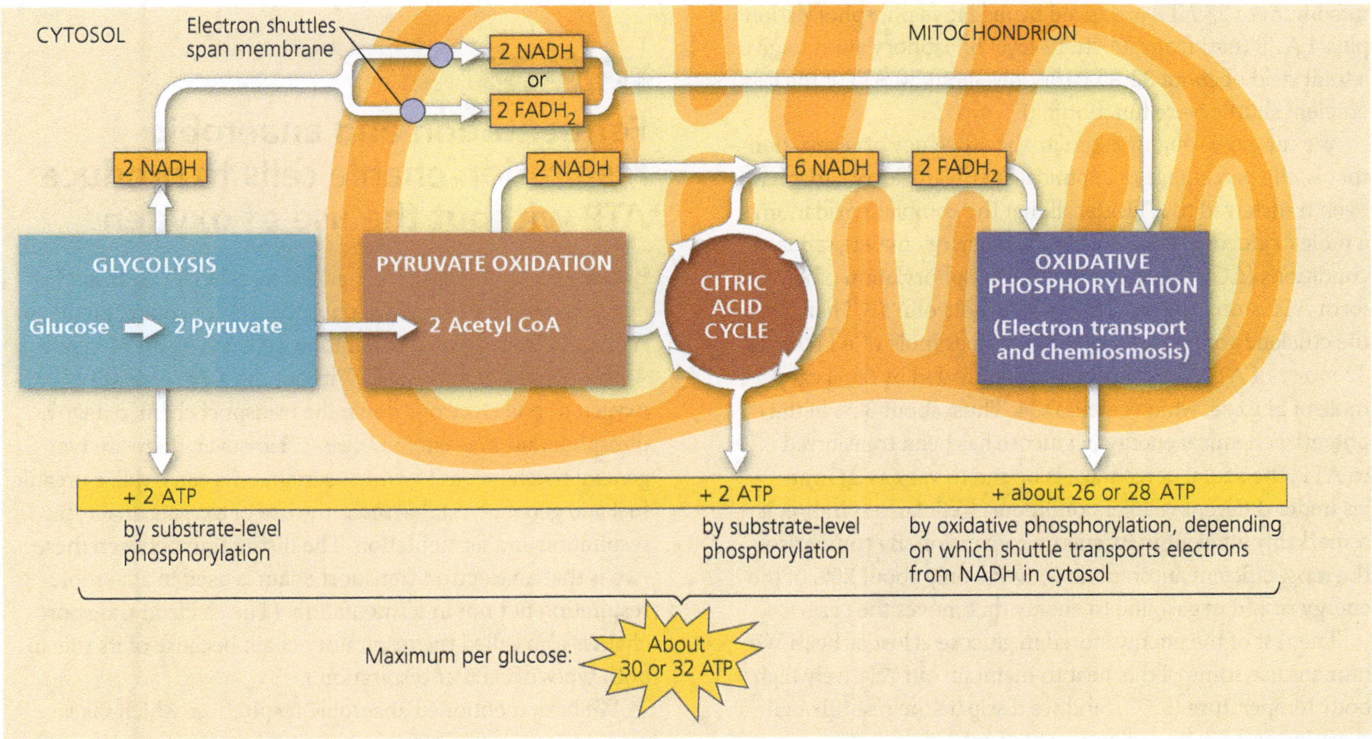

▲ **Figure 15 ATP yield per molecule of glucose at each stage of cellular respiration.**

? *Explain exactly how the total of 26 or 28 ATP (see the yellow bar) was calculated.*

force for the synthesis of 2.5 ATP. The citric acid cycle also supplies electrons to the electron transport chain via $FADH_2$, but since its electrons enter later in the chain, each molecule of this electron carrier is responsible for transport of only enough H^+ for the synthesis of 1.5 ATP. These numbers also take into account the slight energetic cost of moving the ATP formed in the mitochondrion out into the cytosol, where it will be used.

Second, the ATP yield varies slightly depending on the type of shuttle used to transport electrons from the cytosol into the mitochondrion. The mitochondrial inner membrane is impermeable to NADH, so NADH in the cytosol is segregated from the machinery of oxidative phosphorylation. The 2 electrons of NADH captured in glycolysis must be conveyed into the mitochondrion by one of several electron shuttle systems. Depending on the kind of shuttle in a particular cell type, the electrons are passed either to NAD^+ or to FAD in the mitochondrial matrix (see Figure 15). If the electrons are passed to FAD, as in brain cells, only about 1.5 ATP can result from each NADH that was originally generated in the cytosol. If the electrons are passed to mitochondrial NAD^+, as in liver cells and heart cells, the yield is about 2.5 ATP per NADH.

A third variable that reduces the yield of ATP is the use of the proton-motive force generated by the redox reactions of respiration to drive other kinds of work. For example, the proton-motive force powers the mitochondrion's uptake of pyruvate from the cytosol. However, if *all* the proton-motive force generated by the electron transport chain were used to drive ATP synthesis, one glucose molecule could generate a maximum of 28 ATP produced by oxidative phosphorylation plus 4 ATP (net) from substrate-level phosphorylation to give a total yield of about 32 ATP (or only about 30 ATP if the less efficient shuttle were functioning).

We can now roughly estimate the efficiency of respiration—that is, the percentage of chemical energy in glucose that has been transferred to ATP. Recall that the complete oxidation of a mole of glucose releases 686 kcal of energy under standard conditions ($\Delta G = -686$ kcal/mol). Phosphorylation of ADP to form ATP stores at least 7.3 kcal per mole of ATP. Therefore, the efficiency of respiration is 7.3 kcal per mole of ATP times 32 moles of ATP per mole of glucose divided by 686 kcal per mole of glucose, which equals 0.34. Thus, about 34% of the potential chemical energy in glucose has been transferred to ATP; the actual percentage is bound to vary as ΔG varies under different cellular conditions. Cellular respiration is remarkably efficient in its energy conversion. By comparison, the most efficient automobile converts only about 25% of the energy stored in gasoline to energy that moves the car.

The rest of the energy stored in glucose is lost as heat. We humans use some of this heat to maintain our relatively high body temperature (37°C), and we dissipate the rest through sweating and other cooling mechanisms.

Surprisingly, perhaps, it may be beneficial under certain conditions to reduce the efficiency of cellular respiration. A remarkable adaptation is shown by hibernating mammals, which overwinter in a state of inactivity and lowered metabolism. Although their internal body temperature is lower than normal, it still must be kept significantly higher than the external air temperature. One type of tissue, called brown fat, is made up of cells packed full of mitochondria. The inner mitochondrial membrane contains a channel protein called the uncoupling protein, which allows protons to flow back down their concentration gradient without generating ATP. Activation of these proteins in hibernating mammals results in ongoing oxidation of stored fuel stores (fats), generating heat without any ATP production. In the absence of such an adaptation, the buildup of ATP would eventually cause cellular respiration to be shut down by regulatory mechanisms in the cell. In the **Scientific Skills Exercise**, you can work with data in a related but different case where a decrease in metabolic efficiency in cells is used to generate heat.

CONCEPT CHECK 4

1. What effect would an absence of O_2 have on the process shown in Figure 14?
2. **WHAT IF?** In the absence of O_2, as in question 1, what do you think would happen if you decreased the pH of the intermembrane space of the mitochondrion? Explain your answer.
3. **MAKE CONNECTIONS** Membranes must be fluid to function properly. How does the operation of the electron transport chain support that assertion?

CONCEPT 5

Fermentation and anaerobic respiration enable cells to produce ATP without the use of oxygen

Because most of the ATP generated by cellular respiration is due to the work of oxidative phosphorylation, our estimate of ATP yield from aerobic respiration depends on an adequate supply of oxygen to the cell. Without the electronegative oxygen to pull electrons down the transport chain, oxidative phosphorylation eventually ceases. However, there are two general mechanisms by which certain cells can oxidize organic fuel and generate ATP *without* the use of oxygen: anaerobic respiration and fermentation. The distinction between these two is that an electron transport chain is used in anaerobic respiration but not in fermentation. (The electron transport chain is also called the respiratory chain because of its role in both types of cellular respiration.)

We have mentioned anaerobic respiration, which takes place in some prokaryotic organisms living in environments lacking oxygen. These organisms have an electron transport chain but do not use oxygen as a final electron acceptor at the

Scientific Skills Exercise

Making a Bar Graph and Evaluating a Hypothesis

Does Thyroid Hormone Level Affect Oxygen Consumption in Cells? Some animals, such as mammals and birds, maintain a relatively constant body temperature, above that of their environment, by using heat produced as a by-product of metabolism. When the core temperature of these animals drops below an internal set point, their cells are triggered to reduce the efficiency of ATP production by the electron transport chains in mitochondria. At lower efficiency, extra fuel must be consumed to produce the same number of ATPs, generating additional heat. Because the response is moderated by the endocrine system, researchers hypothesized that thyroid hormone might trigger this cellular response. In this exercise, you will use a bar graph to visualize data from an experiment that compared the metabolic rates (by measuring oxygen consumption) in mitochondria of cells from animals with different levels of thyroid hormone.

How the Experiment Was Done Liver cells were isolated from sibling rats that had low, normal, or elevated thyroid hormone levels. The oxygen consumption rate due to activity of the mitochondrial electron transport chains of each type of cell was measured under controlled conditions.

Data from the Experiment

Thyroid Hormone Level	Oxygen Consumption Rate [nmol O_2/(min · mg cells)]
Low	4.3
Normal	4.8
Elevated	8.7

Data from M. E. Harper and M. D. Brand, The quantitative contributions of mitochondrial proton leak and ATP turnover reactions to the changed respiration rates of hepatocytes from rats of different thyroid status, *Journal of Biological Chemistry* 268:14850–14860 (1993).

INTERPRET THE DATA

1. To visualize any differences in oxygen consumption between cell types, it will be useful to graph the data in a bar graph. First, set up the axes. (a) What is the independent variable (intentionally varied by the researchers), which goes on the *x*-axis? List the categories along the *x*-axis; because they are discrete rather than continuous, you can list them in any order. (b) What is the dependent variable (measured by the researchers), which goes on the *y*-axis? (c) What units (abbreviated) should go on the *y*-axis? Label the *y*-axis, including the units specified in the data table. Determine the range of values of the data that will need to go on the *y*-axis. What is the largest value? Draw evenly spaced tick marks and label them, starting with 0 at the bottom.
2. Graph the data for each sample. Match each *x*-value with its *y*-value and place a mark on the graph at that coordinate, then draw a bar from the *x*-axis up to the correct height for each sample. Why is a bar graph more appropriate than a scatter plot or line graph? (For additional information about graphs, see the Scientific Skills Review in the Study Area in MasteringBiology.)
3. Examine your graph and look for a pattern in the data. (a) Which cell type had the highest rate of oxygen consumption, and which had the lowest? (b) Does this support the researchers' hypothesis? Explain. (c) Based on what you know about mitochondrial electron transport and heat production, predict which rats had the highest, and which had the lowest, body temperature.

(MB) A version of this Scientific Skills Exercise can be assigned in MasteringBiology.

end of the chain. Oxygen performs this function very well because it is extremely electronegative, but other, less electronegative substances can also serve as final electron acceptors. Some "sulfate-reducing" marine bacteria, for instance, use the sulfate ion (SO_4^{2-}) at the end of their respiratory chain. Operation of the chain builds up a proton-motive force used to produce ATP, but H_2S (hydrogen sulfide) is made as a by-product rather than water. The rotten-egg odor you may have smelled while walking through a salt marsh or a mudflat signals the presence of sulfate-reducing bacteria.

Fermentation is a way of harvesting chemical energy without using either oxygen or any electron transport chain—in other words, without cellular respiration. How can food be oxidized without cellular respiration? Remember, oxidation simply refers to the loss of electrons to an electron acceptor, so it does not need to involve oxygen. Glycolysis oxidizes glucose to two molecules of pyruvate. The oxidizing agent of glycolysis is NAD^+, and neither oxygen nor any electron transfer chain is involved. Overall, glycolysis is exergonic, and some of the energy made available is used to produce 2 ATP (net) by substrate-level phosphorylation. If oxygen *is* present, then additional ATP is made by oxidative phosphorylation when NADH passes electrons removed from glucose to the electron transport chain. But glycolysis generates 2 ATP whether oxygen is present or not—that is, whether conditions are aerobic or anaerobic.

As an alternative to respiratory oxidation of organic nutrients, fermentation is an extension of glycolysis that allows continuous generation of ATP by the substrate-level phosphorylation of glycolysis. For this to occur, there must be a sufficient supply of NAD^+ to accept electrons during the oxidation step of glycolysis. Without some mechanism to recycle NAD^+ from NADH, glycolysis would soon deplete the cell's pool of NAD^+ by reducing it all to NADH and would shut itself down for lack of an oxidizing agent. Under aerobic conditions, NAD^+ is recycled from NADH by the transfer of electrons to the electron transport chain. An anaerobic alternative is to transfer electrons from NADH to pyruvate, the end product of glycolysis.

Types of Fermentation

Fermentation consists of glycolysis plus reactions that regenerate NAD^+ by transferring electrons from NADH to pyruvate or derivatives of pyruvate. The NAD^+ can then be reused to oxidize sugar by glycolysis, which nets two molecules of ATP by substrate-level phosphorylation. There are many types of fermentation, differing in the end products formed from pyruvate. Two common types are alcohol fermentation and lactic acid fermentation, and both are harnessed by humans in food production.

In **alcohol fermentation** **(Figure 16a)**, pyruvate is converted to ethanol (ethyl alcohol) in two steps. The first step releases carbon dioxide from the pyruvate, which is converted to the two-carbon compound acetaldehyde. In the second step, acetaldehyde is reduced by NADH to ethanol. This regenerates the supply of NAD^+ needed for the continuation of glycolysis. Many bacteria carry out alcohol fermentation under anaerobic conditions. Yeast (a fungus) also carries out alcohol fermentation. For thousands of years, humans have used yeast in brewing, winemaking, and baking. The CO_2 bubbles generated by baker's yeast during alcohol fermentation allow bread to rise.

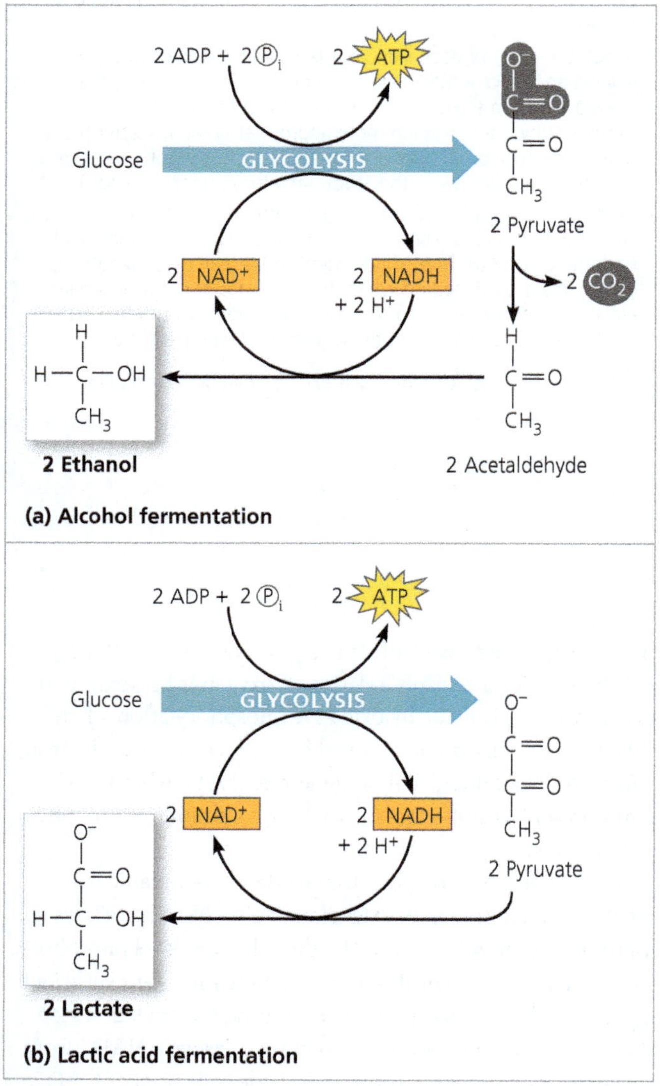

▲ **Figure 16 Fermentation.** In the absence of oxygen, many cells use fermentation to produce ATP by substrate-level phosphorylation. Pyruvate, the end product of glycolysis, serves as an electron acceptor for oxidizing NADH back to NAD^+, which can then be reused in glycolysis. Two of the common end products formed from fermentation are **(a)** ethanol and **(b)** lactate, the ionized form of lactic acid.

During **lactic acid fermentation** **(Figure 16b)**, pyruvate is reduced directly by NADH to form lactate as an end product, with no release of CO_2. (Lactate is the ionized form of lactic acid.) Lactic acid fermentation by certain fungi and bacteria is used in the dairy industry to make cheese and yogurt.

Human muscle cells make ATP by lactic acid fermentation when oxygen is scarce. This occurs during strenuous exercise, when sugar catabolism for ATP production outpaces the muscle's supply of oxygen from the blood. Under these conditions, the cells switch from aerobic respiration to fermentation. The lactate that accumulates was previously thought to cause the muscle fatigue and pain that occurs a day or so after intense exercise. However, evidence shows that within an hour, blood carries the excess lactate from the muscles to the liver, where it is converted back to pyruvate by liver cells. Because oxygen is available, this pyruvate can then enter the mitochondria in liver cells and complete cellular respiration. Next-day muscle soreness is more likely caused by trauma to small muscle fibers, which leads to inflammation and pain.

Comparing Fermentation with Anaerobic and Aerobic Respiration

Fermentation, anaerobic respiration, and aerobic respiration are three alternative cellular pathways for producing ATP by harvesting the chemical energy of food. All three use glycolysis to oxidize glucose and other organic fuels to pyruvate, with a net production of 2 ATP by substrate-level phosphorylation. And in all three pathways, NAD^+ is the oxidizing agent that accepts electrons from food during glycolysis.

A key difference is the contrasting mechanisms for oxidizing NADH back to NAD^+, which is required to sustain glycolysis. In fermentation, the final electron acceptor is an organic molecule such as pyruvate (lactic acid fermentation) or acetaldehyde (alcohol fermentation). In cellular respiration, by contrast, electrons carried by NADH are transferred to an electron transport chain, which regenerates the NAD^+ required for glycolysis.

Another major difference is the amount of ATP produced. Fermentation yields two molecules of ATP, produced by substrate-level phosphorylation. In the absence of an electron transport chain, the energy stored in pyruvate is unavailable. In cellular respiration, however, pyruvate is completely oxidized in the mitochondrion. Most of the chemical energy from this process is shuttled by NADH and $FADH_2$ in the form of electrons to the electron transport chain. There, the electrons move stepwise down a series of redox reactions to a final electron acceptor. (In aerobic respiration, the final electron acceptor is oxygen; in anaerobic respiration, the final acceptor is another molecule that is electronegative, although less so than oxygen.) Stepwise electron transport drives oxidative phosphorylation, yielding ATPs. Thus, cellular respiration harvests much more energy from each sugar molecule than fermentation can. In fact, aerobic respiration yields up to 32 molecules of ATP per glucose molecule—up to 16 times as much as does fermentation.

Some organisms, called **obligate anaerobes**, carry out only fermentation or anaerobic respiration. In fact, these organisms cannot survive in the presence of oxygen. A few cell types can carry out only aerobic oxidation of pyruvate, not fermentation. Other organisms, including yeasts and many bacteria, can make enough ATP to survive using either fermentation or respiration. Such species are called **facultative anaerobes**. On the cellular level, our muscle cells behave as facultative anaerobes. In such cells, pyruvate is a fork in the metabolic road that leads to two alternative catabolic routes **(Figure 17)**. Under aerobic conditions, pyruvate can be converted to acetyl CoA, which enters the citric acid cycle. Under anaerobic conditions, lactic acid fermentation occurs: Pyruvate is diverted from the citric acid cycle, serving instead as an electron acceptor to

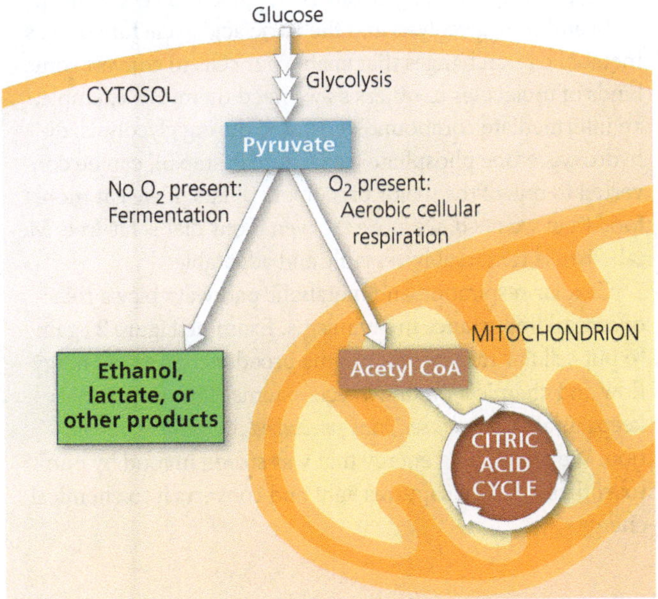

▲ **Figure 17 Pyruvate as a key juncture in catabolism.** Glycolysis is common to fermentation and cellular respiration. The end product of glycolysis, pyruvate, represents a fork in the catabolic pathways of glucose oxidation. In a facultative anaerobe or a muscle cell, which are capable of both aerobic cellular respiration and fermentation, pyruvate is committed to one of those two pathways, usually depending on whether or not oxygen is present.

recycle NAD^+. To make the same amount of ATP, a facultative anaerobe has to consume sugar at a much faster rate when fermenting than when respiring.

The Evolutionary Significance of Glycolysis

EVOLUTION The role of glycolysis in both fermentation and respiration has an evolutionary basis. Ancient prokaryotes are thought to have used glycolysis to make ATP long before oxygen was present in Earth's atmosphere. The oldest known fossils of bacteria date back 3.5 billion years, but appreciable quantities of oxygen probably did not begin to accumulate in the atmosphere until about 2.7 billion years ago, produced by photosynthesizing cyanobacteria. Therefore, early prokaryotes may have generated ATP exclusively from glycolysis. The fact that glycolysis is today the most widespread metabolic pathway among Earth's organisms suggests that it evolved very early in the history of life. The cytosolic location of glycolysis also implies great antiquity; the pathway does not require any of the membrane-enclosed organelles of the eukaryotic cell, which evolved approximately 1 billion years after the prokaryotic cell. Glycolysis is a metabolic heirloom from early cells that continues to function in fermentation and as the first stage in the breakdown of organic molecules by respiration.

CONCEPT CHECK 5

1. Consider the NADH formed during glycolysis. What is the final acceptor for its electrons during fermentation? What is the final acceptor for its electrons during aerobic respiration?
2. WHAT IF? A glucose-fed yeast cell is moved from an aerobic environment to an anaerobic one. How would its rate of glucose consumption change if ATP were to be generated at the same rate?

CONCEPT 6

Glycolysis and the citric acid cycle connect to many other metabolic pathways

So far, we have treated the oxidative breakdown of glucose in isolation from the cell's overall metabolic economy. In this section, you will learn that glycolysis and the citric acid cycle are major intersections of the cell's catabolic (breakdown) and anabolic (biosynthetic) pathways.

The Versatility of Catabolism

Throughout this chapter, we have used glucose as an example of a fuel for cellular respiration. But free glucose molecules are not common in the diets of humans and other animals. We obtain most of our calories in the form of fats, proteins, and carbohydrates such as sucrose and other disaccharides, and

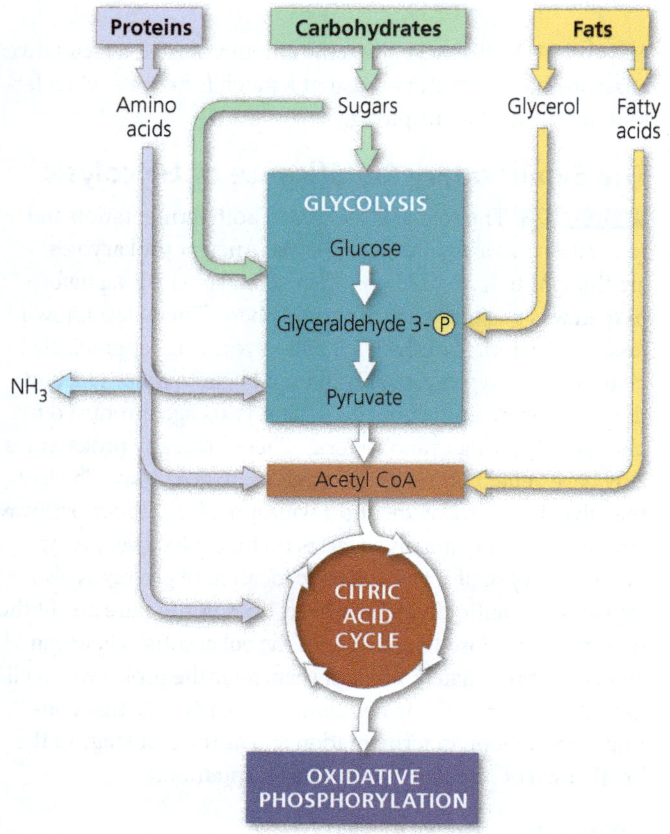

▲ **Figure 18 The catabolism of various molecules from food.** Carbohydrates, fats, and proteins can all be used as fuel for cellular respiration. Monomers of these molecules enter glycolysis or the citric acid cycle at various points. Glycolysis and the citric acid cycle are catabolic funnels through which electrons from all kinds of organic molecules flow on their exergonic fall to oxygen.

starch, a polysaccharide. All these organic molecules in food can be used by cellular respiration to make ATP **(Figure 18)**.

Glycolysis can accept a wide range of carbohydrates for catabolism. In the digestive tract, starch is hydrolyzed to glucose, which is broken down in cells by glycolysis and the citric acid cycle. Similarly, glycogen, the polysaccharide that humans and many other animals store in their liver and muscle cells, can be hydrolyzed to glucose between meals as fuel for respiration. The digestion of disaccharides, including sucrose, provides glucose and other monosaccharides as fuel for respiration.

Proteins can also be used for fuel, but first they must be digested to their constituent amino acids. Many of the amino acids are used by the organism to build new proteins. Amino acids present in excess are converted by enzymes to intermediates of glycolysis and the citric acid cycle. Before amino acids can feed into glycolysis or the citric acid cycle, their amino groups must be removed, a process called *deamination*. The nitrogenous waste is excreted from the animal in the form of ammonia (NH_3), urea, or other waste products.

Catabolism can also harvest energy stored in fats obtained either from food or from fat cells. After fats are digested to glycerol and fatty acids, the glycerol is converted to glyceraldehyde 3-phosphate, an intermediate of glycolysis. Most of the energy of a fat is stored in the fatty acids. A metabolic sequence called **beta oxidation** breaks the fatty acids down to two-carbon fragments, which enter the citric acid cycle as acetyl CoA. NADH and $FADH_2$ are also generated during beta oxidation, resulting in further ATP production. Fats make excellent fuels, largely due to their chemical structure and the high energy level of their electrons compared with those of carbohydrates. A gram of fat oxidized by respiration produces more than twice as much ATP as a gram of carbohydrate.

Biosynthesis (Anabolic Pathways)

Cells need substance as well as energy. Not all the organic molecules of food are destined to be oxidized as fuel to make ATP. In addition to calories, food must also provide the carbon skeletons that cells require to make their own molecules. Some organic monomers obtained from digestion can be used directly. For example, as previously mentioned, amino acids from the hydrolysis of proteins in food can be incorporated into the organism's own proteins. Often, however, the body needs specific molecules that are not present as such in food. Compounds formed as intermediates of glycolysis and the citric acid cycle can be diverted into anabolic pathways as precursors from which the cell can synthesize the molecules it requires. For example, humans can make about half of the 20 amino acids in proteins by modifying compounds siphoned away from the citric acid cycle; the rest are "essential amino acids" that must be obtained in the diet. Also, glucose can be made from pyruvate, and fatty acids can be synthesized from acetyl CoA. Of course, these anabolic, or biosynthetic, pathways do not generate ATP, but instead consume it.

In addition, glycolysis and the citric acid cycle function as metabolic interchanges that enable our cells to convert some kinds of molecules to others as we need them. For example, an intermediate compound generated during glycolysis, dihydroxyacetone phosphate (see Figure 9, step 5), can be converted to one of the major precursors of fats. If we eat more food than we need, we store fat even if our diet is fat-free. Metabolism is remarkably versatile and adaptable.

Cellular respiration and metabolic pathways play a role of central importance in organisms. Examine Figure 2 again to put cellular respiration into the broader context of energy flow and chemical cycling in ecosystems. The energy that keeps us alive is *released*, not *produced*, by cellular respiration. We are tapping energy that was stored in food by photosynthesis, which captures light and converts it to chemical energy.

CONCEPT CHECK 6

1. **MAKE CONNECTIONS** Compare the structure of a fat with that of a carbohydrate. What features of their structures make fat a much better fuel?
2. When might your body synthesize fat molecules?
3. **WHAT IF?** During intense exercise, can a muscle cell use fat as a concentrated source of chemical energy? Explain.

Chapter Review

SUMMARY OF KEY CONCEPTS

VOCAB SELF-QUIZ
goo.gl/gbai8v

CONCEPT 1

Catabolic pathways yield energy by oxidizing organic fuels

- Cells break down glucose and other organic fuels to yield chemical energy in the form of ATP. **Fermentation** is a process that results in the partial degradation of glucose without the use of oxygen. **Cellular respiration** is a more complete breakdown of glucose; in **aerobic respiration**, oxygen is used as a reactant. The cell taps the energy stored in food molecules through **redox reactions**, in which one substance partially or totally shifts electrons to another. **Oxidation** is the loss of electrons from one substance, while **reduction** is the addition of electrons to the other.
- During aerobic respiration, glucose ($C_6H_{12}O_6$) is oxidized to CO_2, and O_2 is reduced to H_2O. Electrons lose potential energy during their transfer from glucose or other organic compounds to oxygen. Electrons are usually passed first to **NAD^+**, reducing it to NADH, and then from NADH to an **electron transport chain**, which conducts them to O_2 in energy-releasing steps. The energy is used to make ATP.
- Aerobic respiration occurs in three stages: (1) **glycolysis**, (2) pyruvate oxidation and the **citric acid cycle**, and (3) **oxidative phosphorylation** (electron transport and chemiosmosis).

? *Describe the difference between the two processes in cellular respiration that produce ATP: oxidative phosphorylation and substrate-level phosphorylation.*

CONCEPT 2

Glycolysis harvests chemical energy by oxidizing glucose to pyruvate

- Glycolysis ("splitting of sugar") is a series of reactions that breaks down glucose into two pyruvate molecules, which may go on to enter the citric acid cycle, and nets 2 ATP and 2 NADH per glucose molecule.

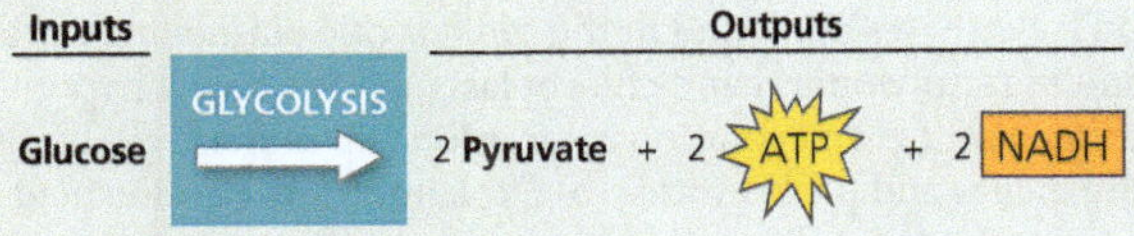

? *Which reactions are the source of energy for the formation of ATP and NADH in glycolysis?*

CONCEPT 3

After pyruvate is oxidized, the citric acid cycle completes the energy-yielding oxidation of organic molecules

- In eukaryotic cells, pyruvate enters the mitochondrion and is oxidized to **acetyl CoA**, which is further oxidized in the citric acid cycle.

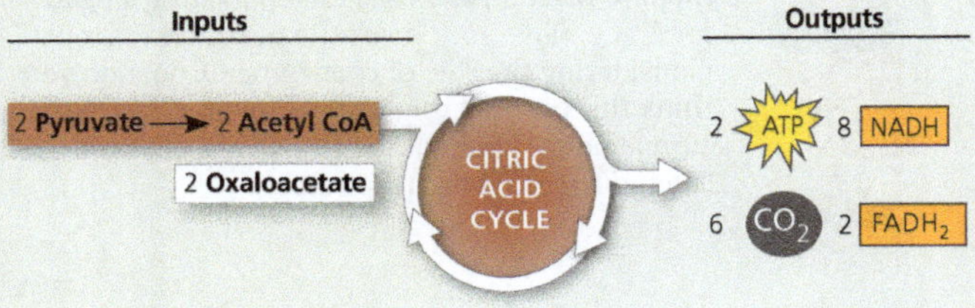

? *What molecular products indicate the complete oxidation of glucose during cellular respiration?*

CONCEPT 4

During oxidative phosphorylation, chemiosmosis couples electron transport to ATP synthesis

- NADH and $FADH_2$ transfer electrons to the electron transport chain. Electrons move down the chain, losing energy in several energy-releasing steps. Finally, electrons are passed to O_2, reducing it to H_2O.

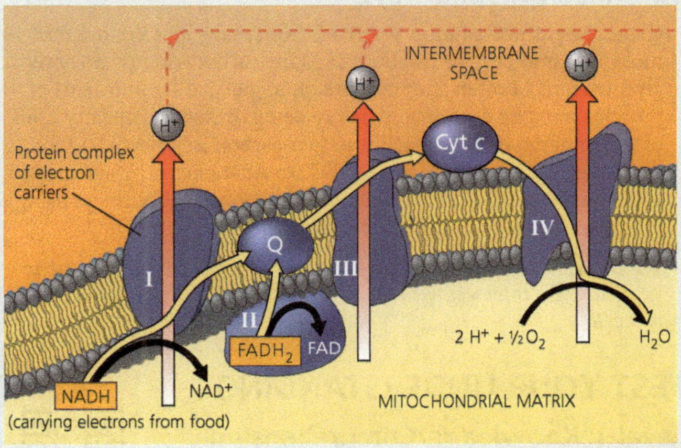

- Along the electron transport chain, electron transfer causes protein complexes to move H^+ from the mitochondrial matrix (in eukaryotes) to the intermembrane space, storing energy as a **proton-motive force** (H^+ gradient). As H^+ diffuses back into the matrix through **ATP synthase**, its passage drives the phosphorylation of ADP, an energy-coupling mechanism called **chemiosmosis**.
- About 34% of the energy stored in a glucose molecule is transferred to ATP during cellular respiration, producing a maximum of about 32 ATP.

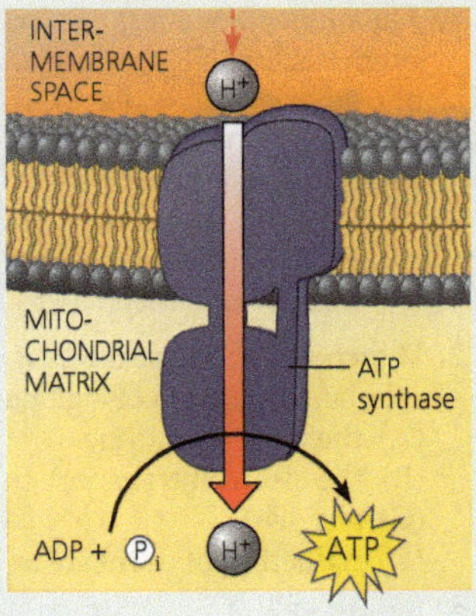

? *Briefly explain the mechanism by which ATP synthase produces ATP. List three locations in which ATP synthases are found.*

CONCEPT 5

Fermentation and anaerobic respiration enable cells to produce ATP without the use of oxygen

- Glycolysis nets 2 ATP by substrate-level phosphorylation, whether oxygen is present or not. Under anaerobic conditions, either anaerobic respiration or fermentation can take place. In anaerobic respiration, an electron transport chain is present with a final electron acceptor other than oxygen. In fermentation, the electrons from NADH are passed to pyruvate or a derivative of pyruvate, regenerating the NAD^+ required to oxidize more glucose. Two common types of fermentation are **alcohol fermentation** and **lactic acid fermentation**.

- Fermentation, anaerobic respiration, and aerobic respiration all use glycolysis to oxidize glucose, but they differ in their final electron acceptor and whether an electron transport chain is used (respiration) or not (fermentation). Respiration yields more ATP; aerobic respiration, with O_2 as the final electron acceptor, yields about 16 times as much ATP as does fermentation.
- Glycolysis occurs in nearly all organisms and is thought to have evolved in ancient prokaryotes before there was O_2 in the atmosphere.

? *Which process yields more ATP, fermentation or anaerobic respiration? Explain.*

CONCEPT 6

Glycolysis and the citric acid cycle connect to many other metabolic pathways

- Catabolic pathways funnel electrons from many kinds of organic molecules into cellular respiration. Many carbohydrates can enter glycolysis, most often after conversion to glucose. Amino acids of proteins must be deaminated before being oxidized. The fatty acids of fats undergo **beta oxidation** to two-carbon fragments and then enter the citric acid cycle as acetyl CoA. Anabolic pathways can use small molecules from food directly or build other substances using intermediates of glycolysis or the citric acid cycle.

? *Describe how the catabolic pathways of glycolysis and the citric acid cycle intersect with anabolic pathways in the metabolism of a cell.*

TEST YOUR UNDERSTANDING

PRACTICE TEST

goo.gl/CRZjvS

Level 1: Knowledge/Comprehension

1. The *immediate* energy source that drives ATP synthesis by ATP synthase during oxidative phosphorylation is the
(A) oxidation of glucose and other organic compounds.
(B) flow of electrons down the electron transport chain.
(C) H^+ concentration gradient across the membrane holding ATP synthase.
(D) transfer of phosphate to ADP.

2. Which metabolic pathway is common to both fermentation and cellular respiration of a glucose molecule?
(A) the citric acid cycle
(B) the electron transport chain
(C) glycolysis
(D) reduction of pyruvate to lactate

3. In mitochondria, exergonic redox reactions
(A) are the source of energy driving prokaryotic ATP synthesis.
(B) provide the energy that establishes the proton gradient.
(C) reduce carbon atoms to carbon dioxide.
(D) are coupled via phosphorylated intermediates to endergonic processes.

4. The final electron acceptor of the electron transport chain that functions in aerobic oxidative phosphorylation is
(A) oxygen.
(B) water.
(C) NAD^+.
(D) pyruvate.

Level 2: Application/Analysis

5. What is the oxidizing agent in the following reaction?

Pyruvate + NADH + H^+ → Lactate + NAD^+

(A) oxygen
(B) NADH
(C) lactate
(D) pyruvate

6. When electrons flow along the electron transport chains of mitochondria, which of the following changes occurs?
(A) The pH of the matrix increases.
(B) ATP synthase pumps protons by active transport.
(C) The electrons gain free energy.
(D) NAD^+ is oxidized.

7. Most CO_2 from catabolism is released during
(A) glycolysis.
(B) the citric acid cycle.
(C) lactate fermentation.
(D) electron transport.

Level 3: Synthesis/Evaluation

8. DRAW IT The graph here shows the pH difference across the inner mitochondrial membrane over time in an actively respiring cell. At the time indicated by the vertical arrow, a metabolic poison is added that specifically and completely inhibits all function of mitochondrial ATP synthase. Draw what you would expect to see for the rest of the graphed line, and explain your graph.

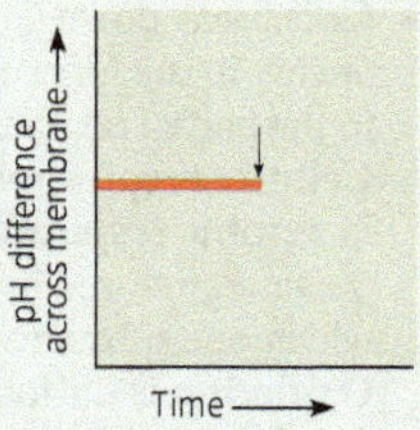

9. INTERPRET THE DATA Phosphofructokinase is an enzyme that acts on fructose 6-phosphate at an early step in glucose breakdown (step 3 in Figure 9). Negative regulation of this enzyme by ATP and positive regulation by AMP control whether the sugar will continue on in the glycolytic pathway. Considering this graph, under which condition is phosphofructokinase more active? Given this enzyme's role in glycolysis, explain why it makes sense that ATP and AMP have these effects.

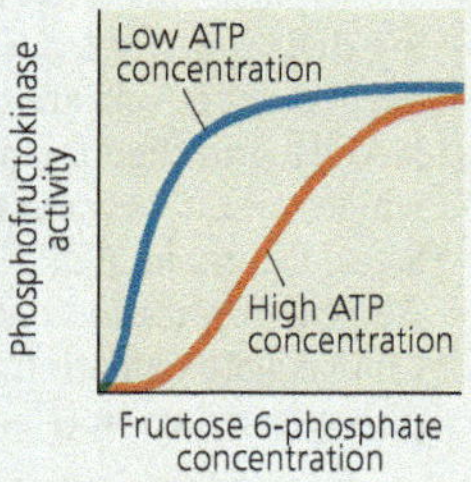

10. SCIENTIFIC INQUIRY
In the 1930s, some physicians prescribed low doses of a compound called dinitrophenol (DNP) to help patients lose weight. This unsafe method was abandoned after some patients died. DNP uncouples the chemiosmotic machinery by making the lipid bilayer of the inner mitochondrial membrane leaky to H^+. Explain how this could cause weight loss and death.

11. FOCUS ON EVOLUTION
ATP synthases are found in the prokaryotic plasma membrane and in mitochondria and chloroplasts. (a) Propose a hypothesis to account for an evolutionary relationship of these eukaryotic organelles and prokaryotes. (b) Explain how the amino acid sequences of the ATP synthases from the different sources might either support or fail to support your hypothesis.

12. FOCUS ON ORGANIZATION
In a short essay (100–150 words), explain how oxidative phosphorylation—the production of ATP using energy derived from the redox reactions of a spatially organized electron transport chain followed by chemiosmosis—is an example of how new properties emerge at each level of the biological hierarchy.

13. SYNTHESIZE YOUR KNOWLEDGE

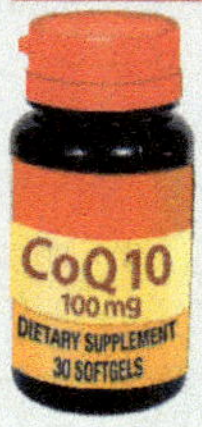

Coenzyme Q (CoQ) is sold as a nutritional supplement. One company uses this marketing slogan for CoQ: "Give your heart the fuel it craves most." Considering the role of coenzyme Q, how do you think this product might function as a nutritional supplement to benefit the heart? Is CoQ used as a "fuel" during cellular respiration?

Answers

Figure Questions

Figure 7 Because there is no external source of energy for the reaction, it must be exergonic, and the reactants must be at a higher energy level than the products.
Figure 9 The removal would probably stop glycolysis, or at least slow it down, since it would push the equilibrium for step 5 toward the bottom (toward DHAP). If less (or no) glyceraldehyde 3-phosphate were available, step 6 would slow down (or be unable to occur).
Figure 13

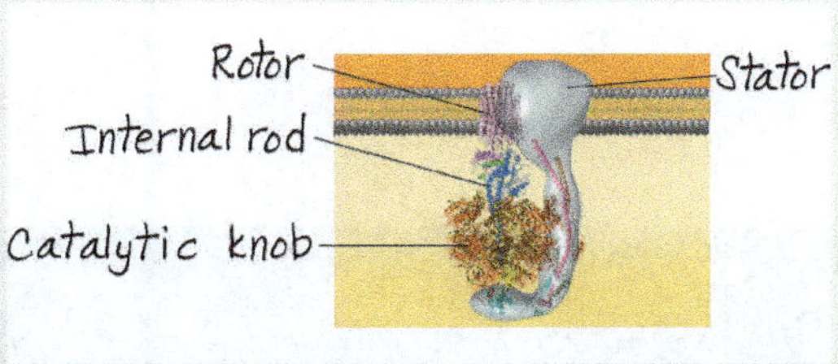

Figure 14 At first, some ATP could be made, since electron transport could proceed as far as complex III, and a small H^+ gradient could be built up. Soon, however, no more electrons could be passed to complex III because it could not be reoxidized by passing its electrons to complex IV. **Figure 15** First, there are 2 NADH from the oxidation of pyruvate plus 6 NADH from the citric acid cycle (CAC); 8 NADH $\times$ 2.5 ATP/NADH = 20 ATP. Second, there are 2 $FADH_2$ from the CAC; 2 $FADH_2 \times 1.5$ ATP/$FADH_2$ = 3 ATP. Third, the 2 NADH from glycolysis enter the mitochondrion through one of two types of shuttle. They pass their electrons either to 2 FAD, which become $FADH_2$ and result in 3 ATP, or to 2 NAD^+, which become NADH and result in 5 ATP. Thus, 20 + 3 + 3 = 26 ATP or 20 + 3 + 5 = 28 ATP from all NADH and $FADH_2$.

Concept Check 1

1. Both processes include glycolysis, the citric acid cycle, and oxidative phosphorylation. In aerobic respiration, the final electron acceptor is molecular oxygen (O_2); in anaerobic respiration, the final electron acceptor is a different substance. **2.** Substrate-level phosphorylation, which occurs during glycolysis and the citric acid cycle, involves the direct transfer of a phosphate group from an organic substrate to ADP by an enzyme. The process of oxidative phosphorylation occurs during the third stage of cellular respiration, which is called oxidative phosphorylation. In this process, the synthesis of ATP from ADP and inorganic phosphate (Ⓟi) is powered by the redox reactions of the electron transport chain. **3.** $C_4H_6O_5$ would be oxidized, and NAD^+ would be reduced.

Concept Check 2

1. NAD^+ acts as the oxidizing agent in step 6, accepting electrons from glyceraldehyde 3-phosphate (G3P), which thus acts as the reducing agent.

Concept Check 3

1. NADH and $FADH_2$; they will donate electrons to the electron transport chain.
2. CO_2 is released from the pyruvate that is the end product of glycolysis, and CO_2 is also released during the citric acid cycle.

Concept Check 4

1. Oxidative phosphorylation would eventually stop entirely, resulting in no ATP production by this process. Without oxygen to "pull" electrons down the electron transport chain, H^+ would not be pumped into the mitochondrion's intermembrane space and chemiosmosis would not occur. **2.** Decreasing the pH means the addition of H^+. This would establish a proton gradient even without the function of the electron transport chain, and we would expect ATP synthase to function and synthesize ATP. (In fact, it was experiments like this that provided support for chemiosmosis as an energy-coupling mechanism.) **3.** One of the components of the electron transport chain, ubiquinone (Q), must be able to diffuse within the membrane. It could not do so if the membrane components were locked rigidly into place.

Concept Check 5

1. A derivative of pyruvate, such as acetaldehyde during alcohol fermentation, or pyruvate itself during lactic acid fermentation; oxygen during aerobic respiration
2. The cell would need to consume glucose at a rate about 16 times the consumption rate in the aerobic environment (2 ATP are generated by fermentation versus up to 32 ATP by cellular respiration).

Concept Check 6

1. The fat is much more reduced; it has many —CH_2— units, and in all these bonds the electrons are equally shared. The electrons present in a carbohydrate molecule are already somewhat oxidized (shared unequally in bonds), as quite a few of them are bound to oxygen. Electrons that are equally shared, as in fat, have a higher energy level than electrons that are unequally shared, as in carbohydrates. Thus, fats are much better fuels than carbohydrates. **2.** When you consume more food than necessary for metabolic processes, your body synthesizes fat as a way of storing energy for later use. **3.** When oxygen is present, the fatty acid chains containing most of the energy of a fat are oxidized and fed into the citric acid cycle and the electron transport chain. During intense exercise, however, oxygen is scarce in muscle cells, so ATP must be generated by glycolysis alone. A very small part of the fat molecule, the glycerol backbone, can be oxidized via glycolysis, but the amount of energy released by this portion is insignificant compared with that released by the fatty acid chains. (This is why moderate exercise, staying below 70% maximum heart rate, is better for burning fat—because enough oxygen remains available to the muscles.)

Summary of Key Concepts Questions

1 Most of the ATP produced in cellular respiration comes from oxidative phosphorylation, in which the energy released from redox reactions in an electron transport chain is used to produce ATP. In substrate-level phosphorylation, an enzyme directly transfers a phosphate group to ADP from an intermediate substrate. All ATP production in glycolysis occurs by substrate-level phosphorylation; this form of ATP production also occurs at one step in the citric acid cycle. **2** The oxidation of the three-carbon sugar glyceraldehyde 3-phosphate yields energy. In this oxidation, electrons and H^+ are transferred to NAD^+, forming NADH, and a phosphate group is attached to the oxidized substrate. ATP is then formed by substrate-level phosphorylation when this phosphate group is transferred to ADP. **3** The release of six molecules of CO_2 represents the complete oxidation of glucose. During the processing of two pyruvates to acetyl CoA, the fully oxidized carboxyl groups (—COO^-) are given off as 2 CO_2. The remaining four carbons are released as CO_2 in the citric acid cycle as citrate is oxidized back to oxaloacetate. **4** The flow of H^+ through the ATP synthase complex causes the rotor and attached rod to rotate, exposing catalytic sites in the knob portion that produce ATP from ADP and Ⓟi. ATP synthases are found in the inner mitochondrial membrane, the plasma membrane of prokaryotes, and membranes within chloroplasts. **5** Anaerobic respiration yields more ATP. The 2 ATP produced by substrate-level phosphorylation in glycolysis represent the total energy yield of fermentation. NADH passes its "high-energy" electrons to pyruvate or a derivative of pyruvate, recycling NAD^+ and allowing glycolysis to continue. In anaerobic respiration, the NADH produced during glycolysis, as well as additional molecules of NADH produced as pyruvate is oxidized, are used to generate ATP molecules. An electron transport chain captures the energy of the electrons in NADH via a series of redox reactions; ultimately, the electrons are transferred to an electronegative molecule other than oxygen. Also, additional molecules of NADH are produced in anaerobic respiration as pyruvate is oxidized. **6** The ATP produced by catabolic pathways is used to drive anabolic pathways. Also, many of the intermediates of glycolysis and the citric acid cycle are used in the biosynthesis of a cell's molecules.

Test Your Understanding

1. C **2.** C **3.** B **4.** A **5.** D **6.** A **7.** B
8.

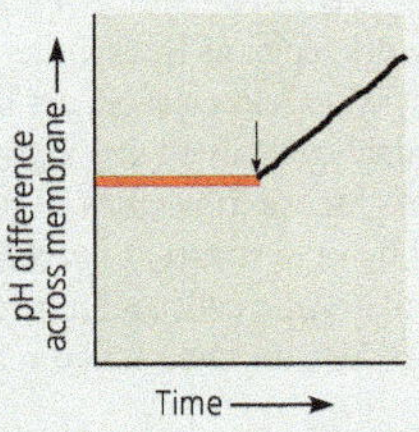

H^+ would continue to be pumped across the membrane into the intermembrane space, increasing the difference between the matrix pH and the intermembrane space pH. H^+ would not be able to flow back through ATP synthase, since the enzyme is inhibited by the poison, so rather than maintaining a constant difference across the membrane, the difference would continue to increase. (Ultimately, the H^+ concentration in the intermembrane space would be so high that no more H^+ would be able to be pumped against the gradient, but this isn't shown in the graph.)

Credits

Credits are listed in the order of appearance.

Photo Credits

1 Sergey Uryadnikov/Shutterstock; **4** Dionisvera/fotolia; **13b** The image of the ATP synthase is reproduced with the kind permission of Professor Sir John Walker and Dr. Martin Montgomery of the Medical Research Council's Mitochondrial Biology Unit; **SSE** Kitchin and Hurst/All Canada Photos/age fotostock; **Chapter Review SYK** Stephen Rees/Shutterstock.

Illustration and Text Credits

5 Adapted from *Molecular Biology of the Cell*, 4th ed., by B. Alberts, et al. Copyright © 2002 by Garland Science/Taylor & Francis Books, LLC. Reprinted with permission; **9a** Adapted from *Biochemistry*, 2nd ed., by C.K. Matthews and K.E. van Holde, 1996. Copyright © 1996 by Pearson Education, Inc. Reprinted and electronically reproduced by permission of Pearson Education, Inc. Upper Saddle River, NJ; **9b** Adapted from *Biochemistry*, 2nd ed., by C.K. Matthews and K.E. van Holde, 1996. Copyright © 1996 by Pearson Education, Inc. Reprinted and electronically reproduced by permission of Pearson Education, Inc. Upper Saddle River, NJ; **SSE.1** Data from M. E. Harper and M. D. Brand, The quantitative contributions of mitochondrial proton leak and ATP turnover reactions to the changed respiration rates of hepatocytes from rats of different thyroid status, *Journal of Biological Chemistry* 268:14850–14860 (1993).

Glossary

acetyl CoA Acetyl coenzyme A; the entry compound for the citric acid cycle in cellular respiration, formed from a two-carbon fragment of pyruvate attached to a coenzyme.

aerobic respiration A catabolic pathway for organic molecules, using oxygen (O_2) as the final electron acceptor in an electron transport chain and ultimately producing ATP. This is the most efficient catabolic pathway and is carried out in most eukaryotic cells and many prokaryotic organisms.

alcohol fermentation Glycolysis followed by the reduction of pyruvate to ethyl alcohol, regenerating NAD^+ and releasing carbon dioxide.

ATP synthase A complex of several membrane proteins that functions in chemiosmosis with adjacent electron transport chains, using the energy of a hydrogen ion (proton) concentration gradient to make ATP. ATP synthases are found in the inner mitochondrial membranes of eukaryotic cells and in the plasma membranes of prokaryotes.

beta oxidation A metabolic sequence that breaks fatty acids down to two-carbon fragments that enter the citric acid cycle as acetyl CoA.

cellular respiration The catabolic pathways of aerobic and anaerobic respiration, which break down organic molecules and use an electron transport chain for the production of ATP.

chemiosmosis (kem'-ē-oz-mō'-sis) An energy-coupling mechanism that uses energy stored in the form of a hydrogen ion gradient across a membrane to drive cellular work, such as the synthesis of ATP. Under aerobic conditions, most ATP synthesis in cells occurs by chemiosmosis.

citric acid cycle A chemical cycle involving eight steps that completes the metabolic breakdown of glucose molecules begun in glycolysis by oxidizing acetyl CoA (derived from pyruvate) to carbon dioxide; occurs within the mitochondrion in eukaryotic cells and in the cytosol of prokaryotes; together with pyruvate oxidation, the second major stage in cellular respiration.

cytochrome (sī'-tō-krōm) An iron-containing protein that is a component of electron transport chains in the mitochondria and chloroplasts of eukaryotic cells and the plasma membranes of prokaryotic cells.

electron transport chain A sequence of electron carrier molecules (membrane proteins) that shuttle electrons down a series of redox reactions that release energy used to make ATP.

facultative anaerobe (fak'-ul-tā'-tiv an'-uh-rōb) An organism that makes ATP by aerobic respiration if oxygen is present but that switches to anaerobic respiration or fermentation if oxygen is not present.

fermentation A catabolic process that makes a limited amount of ATP from glucose (or other organic molecules) without an electron transport chain and that produces a characteristic end product, such as ethyl alcohol or lactic acid.

glycolysis (glī-kol'-uh-sis) A series of reactions that ultimately splits glucose into pyruvate. Glycolysis occurs in almost all living cells, serving as the starting point for fermentation or cellular respiration.

lactic acid fermentation Glycolysis followed by the reduction of pyruvate to lactate, regenerating NAD^+ with no release of carbon dioxide.

NAD^+ Nicotinamide adenine dinucleotide, a coenzyme that cycles easily between oxidized (NAD^+) and reduced (NADH) states, thus acting as an electron carrier.

obligate anaerobe (ob'-lig-et an'-uh-rōb) An organism that only carries out fermentation or anaerobic respiration. Such organisms cannot use oxygen and in fact may be poisoned by it.

oxidation The complete or partial loss of electrons from a substance involved in a redox reaction.

oxidative phosphorylation (fos'-fōr-uh-lā'-shun) The production of ATP using energy derived from the redox reactions of an electron transport chain; the third major stage of cellular respiration.

oxidizing agent The electron acceptor in a redox reaction.

proton-motive force The potential energy stored in the form of a proton electrochemical gradient, generated by the pumping of hydrogen ions (H^+) across a biological membrane during chemiosmosis.

redox reaction (rē'-doks) A chemical reaction involving the complete or partial transfer of one or more electrons from one reactant to another; short for **red**uction-**ox**idation reaction.

reducing agent The electron donor in a redox reaction.

reduction The complete or partial addition of electrons to a substance involved in a redox reaction.

substrate-level phosphorylation The enzyme-catalyzed formation of ATP by direct transfer of a phosphate group to ADP from an intermediate substrate in catabolism.

The Cell Cycle

The Cell Cycle

KEY CONCEPTS

1. Most cell division results in genetically identical daughter cells
2. The mitotic phase alternates with interphase in the cell cycle
3. The eukaryotic cell cycle is regulated by a molecular control system

▲ **Figure 1** How do dividing cells distribute chromosomes to daughter cells?

The Key Roles of Cell Division

The ability of organisms to produce more of their own kind is the one characteristic that best distinguishes living things from nonliving matter. This unique capacity to procreate, like all biological functions, has a cellular basis. In 1855, Rudolf Virchow, a German physician, put it this way: "Where a cell exists, there must have been a preexisting cell, just as the animal arises only from an animal and the plant only from a plant." He summarized this concept with the Latin axiom "*Omnis cellula e cellula*," meaning "Every cell from a cell." The continuity of life is based on the reproduction of cells, or **cell division**. The series of confocal fluorescence micrographs in **Figure 1**, starting at the upper left and reading across both rows left to right, follows the events of cell division as the cells of a two-celled embryo become four.

Cell division plays several important roles in life. The division of one prokaryotic cell reproduces an entire organism. The same is true of a unicellular eukaryote **(Figure 2a)**. Cell division also enables multicellular eukaryotes to develop from a single cell, like the fertilized egg that gave rise to the two-celled embryo in **Figure 2b**. And after such an organism is fully grown, cell division continues to function in renewal and repair, replacing cells that die from accidents or normal wear and tear. For example, dividing cells in your bone marrow continuously make new blood cells **(Figure 2c)**.

The cell division process is an integral part of the **cell cycle**, the life of a cell from the time it is first formed during division of a parent cell until its own division into two daughter cells. (Biologists use the words *daughter* or *sister* in relation to cells, but this is not meant to imply gender.) Passing identical genetic material to cellular offspring is a crucial function of cell division. In this chapter, you'll learn how this process occurs. After studying the mechanics of cell division in eukaryotes and bacteria, you'll learn about the molecular control system that regulates progress through the eukaryotic cell cycle and what happens when the control system malfunctions. Because a breakdown in cell cycle control plays a major role in cancer development, this aspect of cell biology is an active area of research.

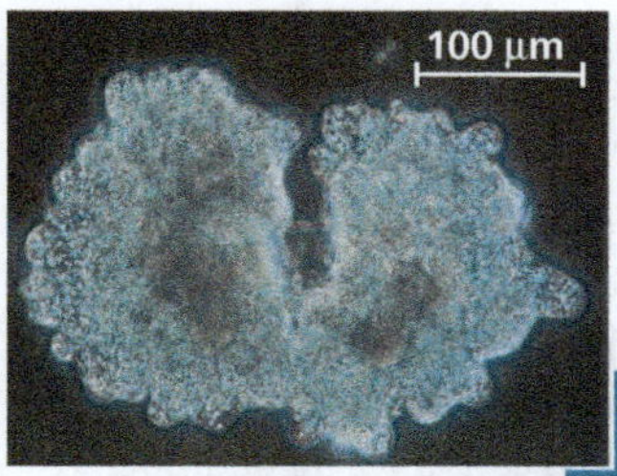

◀ **(a) Reproduction.** An amoeba, a single-celled eukaryote, is dividing into two cells. Each new cell will be an individual organism (LM).

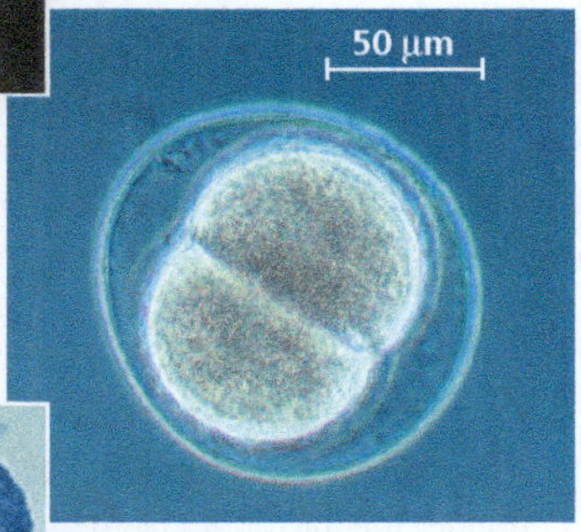

▶ **(b) Growth and development.** This micrograph shows a sand dollar embryo shortly after the fertilized egg divided, forming two cells (LM).

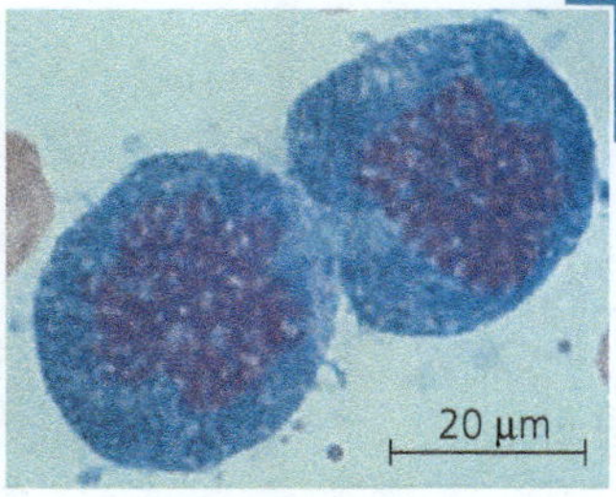

◀ **(c) Tissue renewal.** These dividing bone marrow cells will give rise to new blood cells (LM).

▲ **Figure 2 The functions of cell division.**

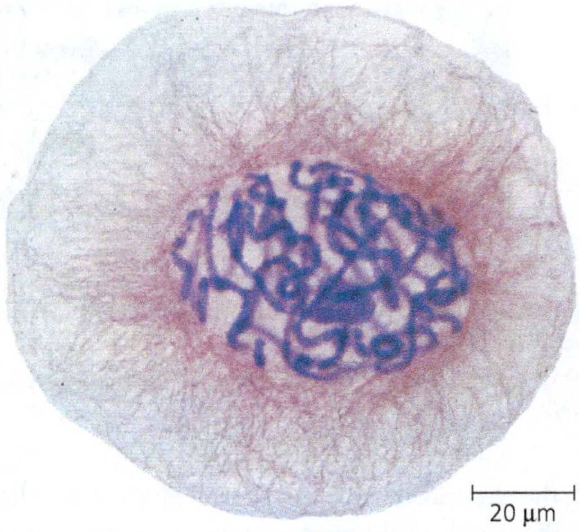

▲ **Figure 3 Eukaryotic chromosomes.** Chromosomes (stained purple) are visible within the nucleus of this cell from an African blood lily. The thinner red threads in the surrounding cytoplasm are the cytoskeleton. The cell is preparing to divide (LM).

CONCEPT 1

Most cell division results in genetically identical daughter cells

The reproduction of a cell, with all its complexity, cannot occur by a mere pinching in half; a cell is not like a soap bubble that simply enlarges and splits in two. In both prokaryotes and eukaryotes, most cell division involves the distribution of identical genetic material—DNA—to two daughter cells. (The exception is meiosis, the special type of eukaryotic cell division that can produce sperm and eggs.) What is most remarkable about cell division is the fidelity with which the DNA is passed from one generation of cells to the next. A dividing cell replicates its DNA, allocates the two copies to opposite ends of the cell, and only then splits into daughter cells.

Cellular Organization of the Genetic Material

A cell's endowment of DNA, its genetic information, is called its **genome**. Although a prokaryotic genome is often a single DNA molecule, eukaryotic genomes usually consist of a number of DNA molecules. The overall length of DNA in a eukaryotic cell is enormous. A typical human cell, for example, has about 2 m of DNA—a length about 250,000 times greater than the cell's diameter. Before the cell can divide to form genetically identical daughter cells, all of this DNA must be copied, or replicated, and then the two copies must be separated so that each daughter cell ends up with a complete genome.

The replication and distribution of so much DNA is manageable because the DNA molecules are packaged into structures called **chromosomes** (from the Greek *chroma*, color, and *soma*, body), so named because they take up certain dyes used in microscopy **(Figure 3)**. Each eukaryotic chromosome consists of one very long, linear DNA molecule associated with many proteins. The DNA molecule carries several hundred to a few thousand genes, the units of information that specify an organism's inherited traits. The associated proteins maintain the structure of the chromosome and help control the activity of the genes. Together, the entire complex of DNA and proteins that is the building material of chromosomes is referred to as **chromatin**. As you will soon see, the chromatin of a chromosome varies in its degree of condensation during the process of cell division.

Every eukaryotic species has a characteristic number of chromosomes in each cell's nucleus. For example, the nuclei of human **somatic cells** (all body cells except the reproductive cells) each contain 46 chromosomes, made up of two sets of 23, one set inherited from each parent. Reproductive cells, or **gametes**—sperm and eggs—have half as many chromosomes as somatic cells, or one set of 23 chromosomes in humans. The number of chromosomes in somatic cells varies widely among species: 18 in cabbage plants, 48 in chimpanzees, 56 in elephants, 90 in hedgehogs, and 148 in one species of alga. We'll now consider how these chromosomes behave during cell division.

Distribution of Chromosomes During Eukaryotic Cell Division

When a cell is not dividing, and even as it replicates its DNA in preparation for cell division, each chromosome is in the form of a long, thin chromatin fiber. After DNA replication, however, the chromosomes condense as a part of cell division: Each chromatin fiber becomes densely coiled and folded,

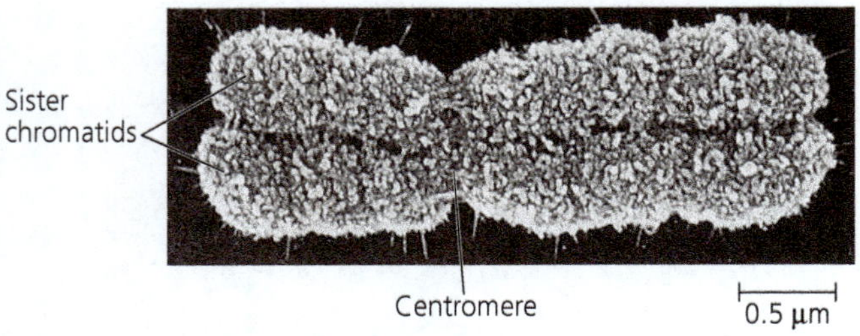

▲ **Figure 4 A highly condensed, duplicated human chromosome (SEM).**

DRAW IT *Circle one sister chromatid of the chromosome in this micrograph.*

making the chromosomes much shorter and so thick that we can see them with a light microscope.

Each duplicated chromosome has two **sister chromatids**, which are joined copies of the original chromosome **(Figure 4)**. The two chromatids, each containing an identical DNA molecule, are initially attached all along their lengths by protein complexes called *cohesins*; this attachment is known as *sister chromatid cohesion*. Each sister chromatid has a **centromere**, a region of the chromosomal DNA where the chromatid is attached most closely to its sister chromatid. This attachment is mediated by proteins bound to the centromeric DNA sequences and gives the condensed, duplicated chromosome a narrow "waist." The part of a chromatid to either side of the centromere is referred to as an *arm* of the chromatid. (An uncondensed, unduplicated chromosome has a single centromere and two arms.)

Later in the cell division process, the two sister chromatids of each duplicated chromosome separate and move into two new nuclei, one forming at each end of the cell. Once the sister chromatids separate, they are no longer called sister chromatids but are considered individual chromosomes; this step essentially doubles the number of chromosomes in the cell. Thus, each new nucleus receives a collection of chromosomes identical to that of the parent cell **(Figure 5)**. **Mitosis**, the division of the genetic material in the nucleus, is usually followed immediately by **cytokinesis**, the division of the cytoplasm. One cell has become two, each the genetic equivalent of the parent cell.

From a fertilized egg, mitosis and cytokinesis produced the 200 trillion somatic cells that now make up your body, and the same processes continue to generate new cells to replace dead and damaged ones. In contrast, you produce gametes—eggs or sperm—by a variation of cell division called *meiosis*, which yields daughter cells with only one set of chromosomes, half as many chromosomes as the parent cell. Meiosis in humans occurs only in special cells in the ovaries or testes (the gonads). Generating gametes, meiosis reduces the chromosome number from 46 (two sets) to 23 (one set). Fertilization fuses two gametes together and returns the chromosome number to 46 (two sets). Mitosis then conserves that number in every somatic cell nucleus of the new human individual. In the remainder of this chapter, we focus on mitosis and the rest of the cell cycle in eukaryotes.

CONCEPT CHECK 1

1. How many chromosomes are drawn in each part of Figure 5? (Ignore the micrograph in part 2.)
2. WHAT IF? A chicken has 78 chromosomes in its somatic cells. How many chromosomes did the chicken inherit from each parent? How many chromosomes are in each of the chicken's gametes? How many chromosomes will be in each somatic cell of the chicken's offspring?

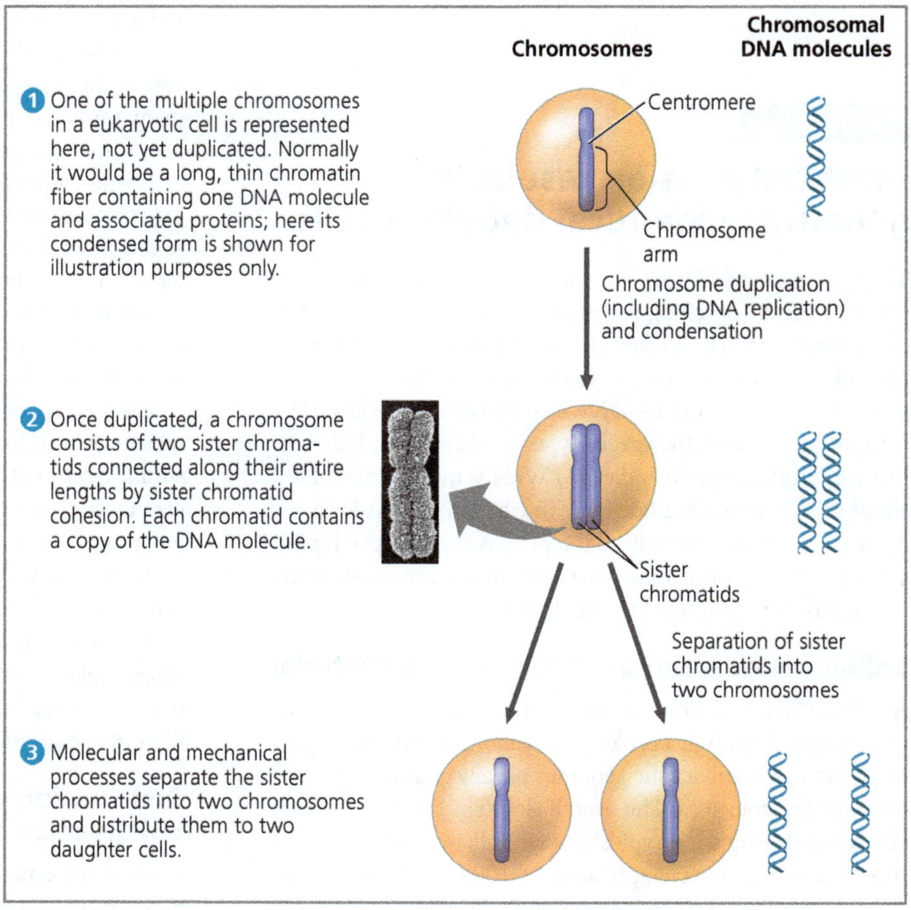

▲ **Figure 5 Chromosome duplication and distribution during cell division.**

? *How many chromatid arms does the chromosome in step 2 have? Identify the point in the figure where one chromosome becomes two.*

CONCEPT 2

The mitotic phase alternates with interphase in the cell cycle

In 1882, a German anatomist named Walther Flemming developed dyes that allowed him to observe, for the first time, the behavior of chromosomes during mitosis and cytokinesis. (In fact, Flemming coined the terms *mitosis* and *chromatin*.) It appeared to Flemming that during the period between one cell division and the next, the cell was simply growing larger. But we now know that many critical events occur during this stage in the life of a cell.

Phases of the Cell Cycle

Mitosis is just one part of the cell cycle **(Figure 6)**. In fact, the **mitotic (M) phase**, which includes both mitosis and cytokinesis, is usually the shortest part of the cell cycle. The mitotic phase alternates with a much longer stage called **interphase**, which often accounts for about 90% of the cycle. Interphase can be divided into subphases: the **G_1 phase** ("first gap"), the **S phase** ("synthesis"), and the **G_2 phase** ("second gap"). The G phases were misnamed as "gaps" when they were first observed because the cells appeared inactive, but we now know that intense metabolic activity and growth occur throughout interphase. During all three subphases of interphase, in fact, a cell grows by producing proteins and cytoplasmic organelles such as mitochondria and endoplasmic reticulum. Duplication of the chromosomes, crucial for eventual division of the cell, occurs entirely during the S phase. Thus, a cell grows (G_1), continues to grow as it copies its chromosomes (S), grows more as it completes preparations for cell division (G_2), and divides (M). The daughter cells may then repeat the cycle.

A particular human cell might undergo one division in 24 hours. Of this time, the M phase would occupy less than 1 hour, while the S phase might occupy about 10–12 hours, or about half the cycle. The rest of the time would be apportioned between the G_1 and G_2 phases. The G_2 phase usually takes 4–6 hours; in our example, G_1 would occupy about 5–6 hours. G_1 is the most variable in length in different types of cells. Some cells in a multicellular organism divide very infrequently or not at all. These cells spend their time in G_1 (or a related phase called G_0) doing their job in the organism—a nerve cell carries impulses, for example.

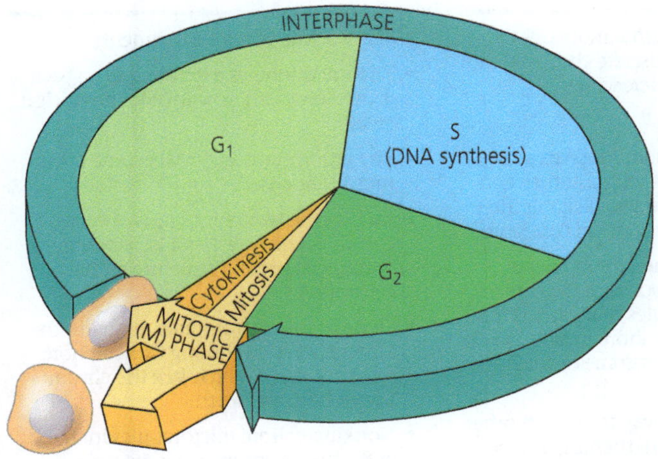

▲ **Figure 6 The cell cycle.** In a dividing cell, the mitotic (M) phase alternates with interphase, a growth period. The first part of interphase (G_1) is followed by the S phase, when the chromosomes duplicate; G_2 is the last part of interphase. In the M phase, mitosis distributes the daughter chromosomes to daughter nuclei, and cytokinesis divides the cytoplasm, producing two daughter cells.

Mitosis is conventionally broken down into five stages: **prophase**, **prometaphase**, **metaphase**, **anaphase**, and **telophase**. Overlapping with the latter stages of mitosis, cytokinesis completes the mitotic phase. **Figure 7** describes these stages in an animal cell. Study this figure thoroughly before progressing to the next two sections, which examine mitosis and cytokinesis more closely.

The Mitotic Spindle: *A Closer Look*

Many of the events of mitosis depend on the **mitotic spindle**, which begins to form in the cytoplasm during prophase. This structure consists of fibers made of microtubules and associated proteins. While the mitotic spindle assembles, the other microtubules of the cytoskeleton partially disassemble, providing the material used to construct the spindle. The spindle microtubules elongate (polymerize) by incorporating more subunits of the protein tubulin and shorten (depolymerize) by losing subunits.

In animal cells, the assembly of spindle microtubules starts at the **centrosome**, a subcellular region containing material that functions throughout the cell cycle to organize the cell's microtubules. (It is also a type of *microtubule-organizing center*.) A pair of centrioles is located at the center of the centrosome, but they are not essential for cell division: If the centrioles are destroyed with a laser microbeam, a spindle nevertheless forms during mitosis. In fact, centrioles are not even present in plant cells, which do form mitotic spindles.

During interphase in animal cells, the single centrosome duplicates, forming two centrosomes, which remain together near the nucleus (see Figure 7). The two centrosomes move apart during prophase and prometaphase of mitosis as spindle microtubules grow out from them. By the end of prometaphase, the two centrosomes, one at each pole of the spindle, are at opposite ends of the cell. An **aster**, a radial array of short microtubules, extends from each centrosome. The spindle includes the centrosomes, the spindle microtubules, and the asters.

Each of the two sister chromatids of a duplicated chromosome has a **kinetochore**, a structure made up of proteins that have assembled on specific sections of chromosomal DNA at each centromere. The chromosome's two kinetochores face in opposite directions. During prometaphase, some of the spindle microtubules attach to the kinetochores; these are called kinetochore microtubules. (The number of microtubules attached to a kinetochore varies among species, from one microtubule in yeast cells to 40 or so in some mammalian cells.)

▼ Figure 7 Exploring Mitosis in an Animal Cell

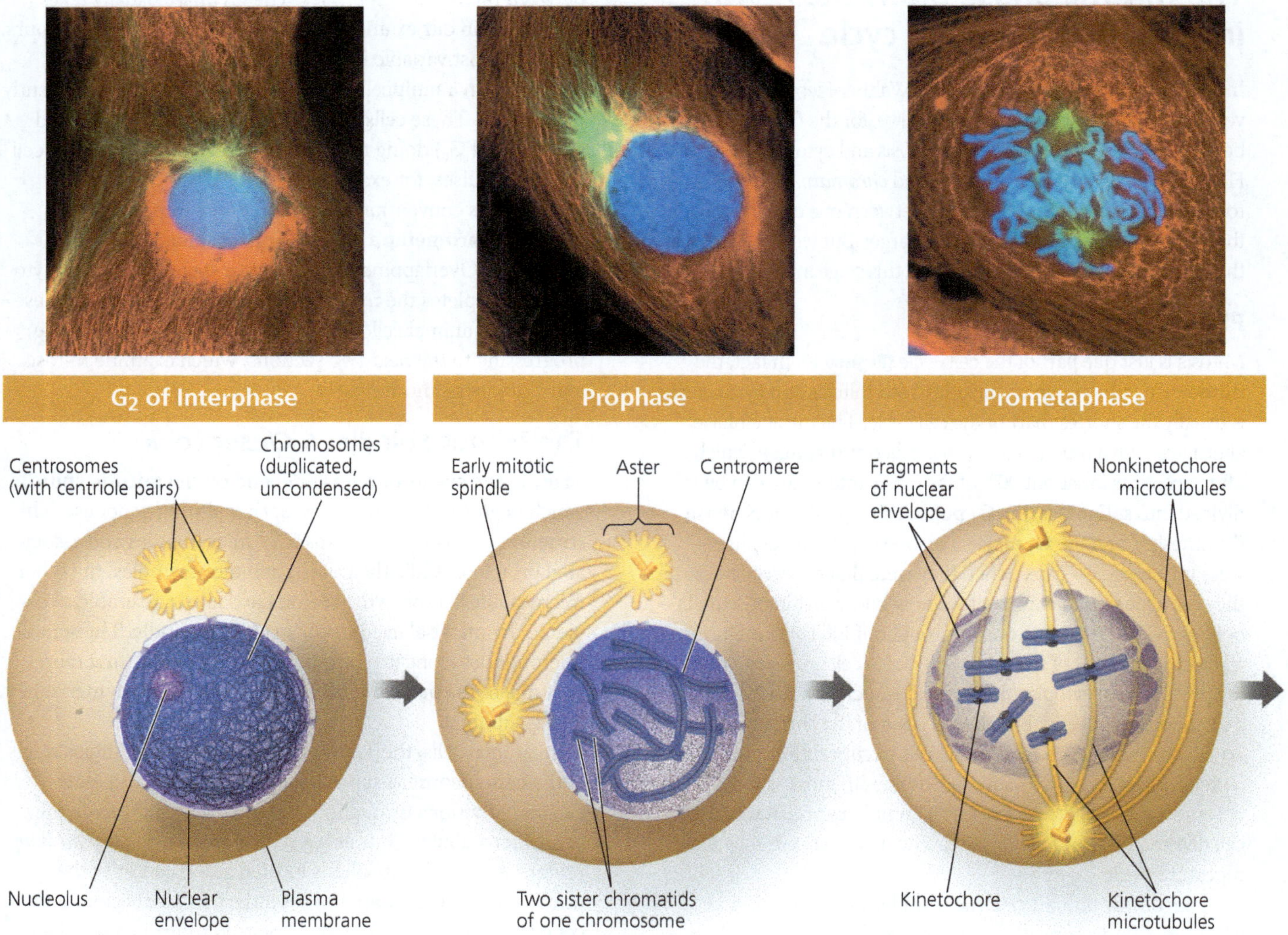

G_2 of Interphase

- A nuclear envelope encloses the nucleus.
- The nucleus contains one or more nucleoli (singular, *nucleolus*).
- Two centrosomes have formed by duplication of a single centrosome. Centrosomes are regions in animal cells that organize the microtubules of the spindle. Each centrosome contains two centrioles.
- Chromosomes, duplicated during S phase, cannot be seen individually because they have not yet condensed.

The fluorescence micrographs show dividing lung cells from a newt; this species has 22 chromosomes. Chromosomes appear blue, microtubules green, and intermediate filaments red. For simplicity, the drawings show only 6 chromosomes.

Prophase

- The chromatin fibers become more tightly coiled, condensing into discrete chromosomes observable with a light microscope.
- The nucleoli disappear.
- Each duplicated chromosome appears as two identical sister chromatids joined at their centromeres and, in some species, all along their arms by cohesins (sister chromatid cohesion).
- The mitotic spindle (named for its shape) begins to form. It is composed of the centrosomes and the microtubules that extend from them. The radial arrays of shorter microtubules that extend from the centrosomes are called asters ("stars").
- The centrosomes move away from each other, propelled partly by the lengthening microtubules between them.

Prometaphase

- The nuclear envelope fragments.
- The microtubules extending from each centrosome can now invade the nuclear area.
- The chromosomes have become even more condensed.
- Each of the two chromatids of each chromosome now has a kinetochore, a specialized protein structure at the centromere.
- Some of the microtubules attach to the kinetochores, becoming "kinetochore microtubules," which jerk the chromosomes back and forth.
- Nonkinetochore microtubules interact with those from the opposite pole of the spindle.

? *How many molecules of DNA are in the prometaphase drawing? How many molecules per chromosome? How many double helices are there per chromosome? Per chromatid?*

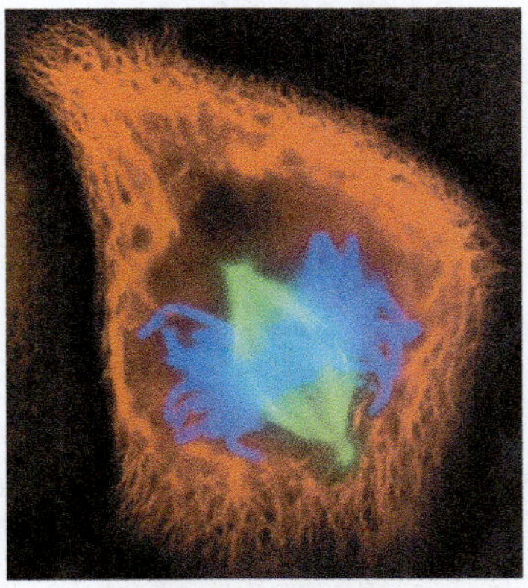

Metaphase

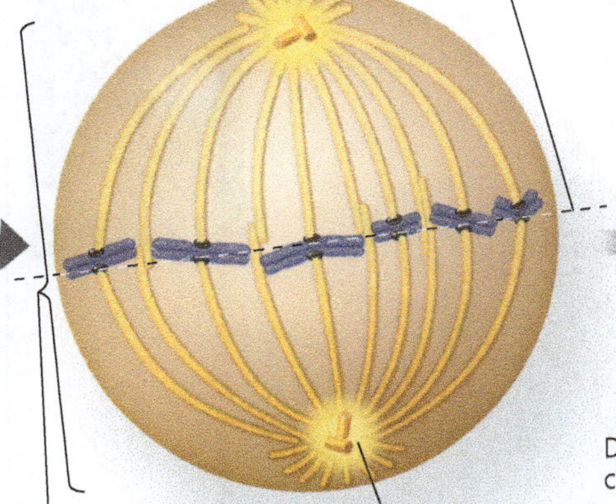

Metaphase

- The centrosomes are now at opposite poles of the cell.
- The chromosomes have all arrived at the *metaphase plate*, a plane that is equidistant between the spindle's two poles. The chromosomes' centromeres lie at the metaphase plate.
- For each chromosome, the kinetochores of the sister chromatids are attached to kinetochore microtubules coming from opposite poles.

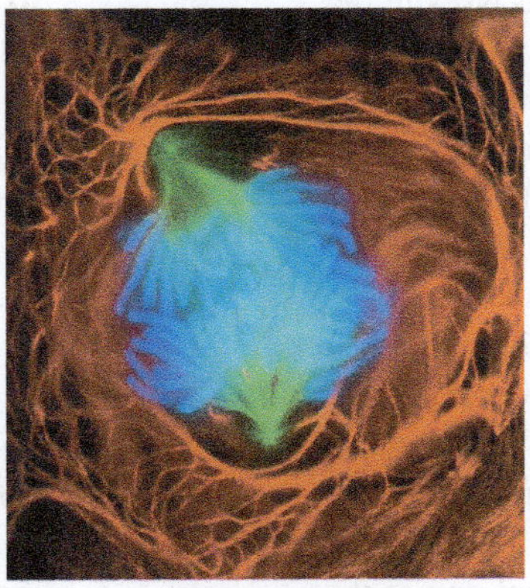

Anaphase

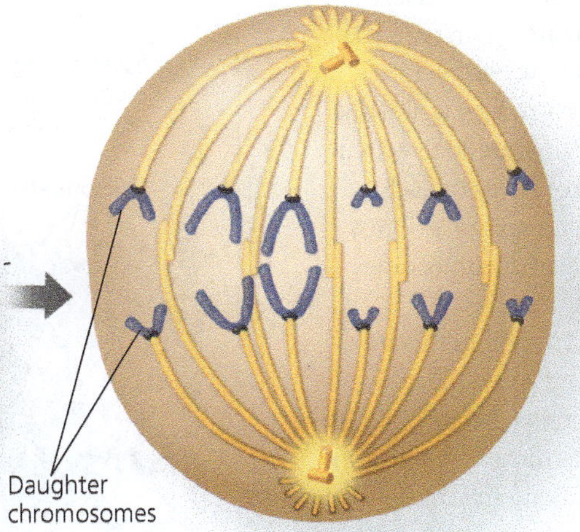

Anaphase

- Anaphase is the shortest stage of mitosis, often lasting only a few minutes.
- Anaphase begins when the cohesin proteins are cleaved. This allows the two sister chromatids of each pair to part suddenly. Each chromatid thus becomes a full-fledged chromosome.
- The two liberated daughter chromosomes begin moving toward opposite ends of the cell as their kinetochore microtubules shorten. Because these microtubules are attached at the centromere region, the chromosomes move centromere first (at about 1 μm/min).
- The cell elongates as the nonkinetochore microtubules lengthen.
- By the end of anaphase, the two ends of the cell have equivalent—and complete—collections of chromosomes.

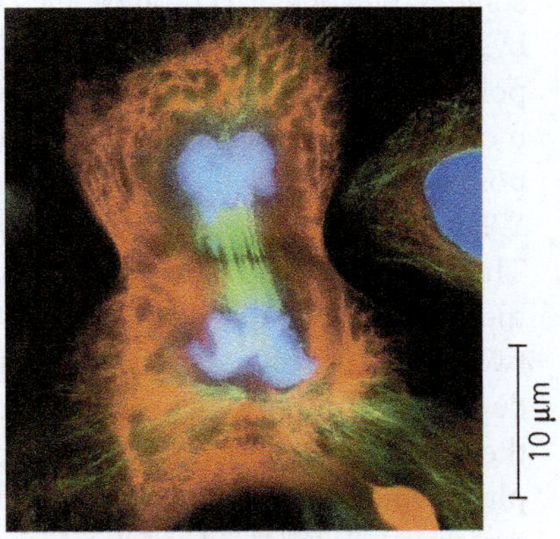

Telophase and Cytokinesis

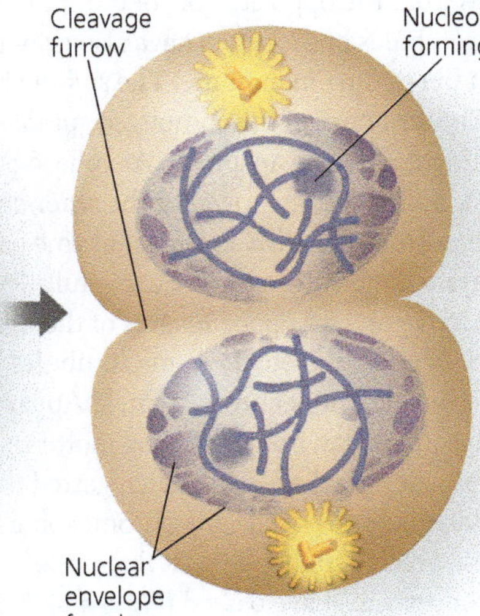

Telophase

- Two daughter nuclei form in the cell. Nuclear envelopes arise from the fragments of the parent cell's nuclear envelope and other portions of the endomembrane system.
- Nucleoli reappear.
- The chromosomes become less condensed.
- Any remaining spindle microtubules are depolymerized.
- Mitosis, the division of one nucleus into two genetically identical nuclei, is now complete.

Cytokinesis

- The division of the cytoplasm is usually well under way by late telophase, so the two daughter cells appear shortly after the end of mitosis.
- In animal cells, cytokinesis involves the formation of a cleavage furrow, which pinches the cell in two.

ANIMATION

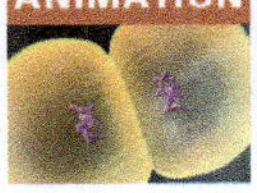

Visit the Study Area in **MasteringBiology** for the BioFlix® 3-D Animation on Mitosis.

The kinetochore acts as a coupling device that attaches the motor of the spindle to the cargo that it moves—the chromosome. When one of a chromosome's kinetochores is "captured" by microtubules, the chromosome begins to move toward the pole from which those microtubules extend. However, this movement is checked as soon as microtubules from the opposite pole attach to the kinetochore on the other chromatid. What happens next is like a tug-of-war that ends in a draw. The chromosome moves first in one direction and then in the other, back and forth, finally settling midway between the two ends of the cell. At metaphase, the centromeres of all the duplicated chromosomes are on a plane midway between the spindle's two poles. This plane is called the **metaphase plate**, which is an imaginary plate rather than an actual cellular structure **(Figure 8)**. Meanwhile, microtubules that do not attach to kinetochores have been elongating, and by metaphase they overlap and interact with other nonkinetochore microtubules from the opposite pole of the spindle. By metaphase, the microtubules of the asters have also grown and are in contact with the plasma membrane. The spindle is now complete.

The structure of the completed spindle correlates well with its function during anaphase. Anaphase begins suddenly when the cohesins holding together the sister chromatids of each chromosome are cleaved by an enzyme called *separase*. Once separated, the chromatids become full-fledged chromosomes that move toward opposite ends of the cell.

How do the kinetochore microtubules function in this poleward movement of chromosomes? Apparently, two mechanisms are in play, both involving motor proteins. Results of a cleverly designed experiment suggested that motor proteins on the kinetochores "walk" the chromosomes along the microtubules, which depolymerize at their kinetochore ends after the motor proteins have passed **(Figure 9)**. (This is referred to as the "Pac-man" mechanism because of its resemblance to the arcade game character that moves by eating all the dots in its path.) However, other researchers, working with different cell types or cells from other species, have shown that chromosomes are "reeled in" by motor proteins at the spindle poles and that the microtubules depolymerize after they pass by these motor proteins. The general consensus now is that both mechanisms are used and that their relative contributions vary among cell types.

In a dividing animal cell, the nonkinetochore microtubules are responsible for elongating the whole cell during anaphase. Nonkinetochore microtubules from opposite poles overlap each other extensively during metaphase (see Figure 8). During anaphase, the region of overlap is reduced as motor proteins attached to the microtubules walk them away from one another, using energy from ATP. As the microtubules push apart from each other, their spindle poles are pushed apart, elongating the cell. At the same time, the microtubules lengthen somewhat by the addition of tubulin subunits to their overlapping ends. As a result, the microtubules continue to overlap.

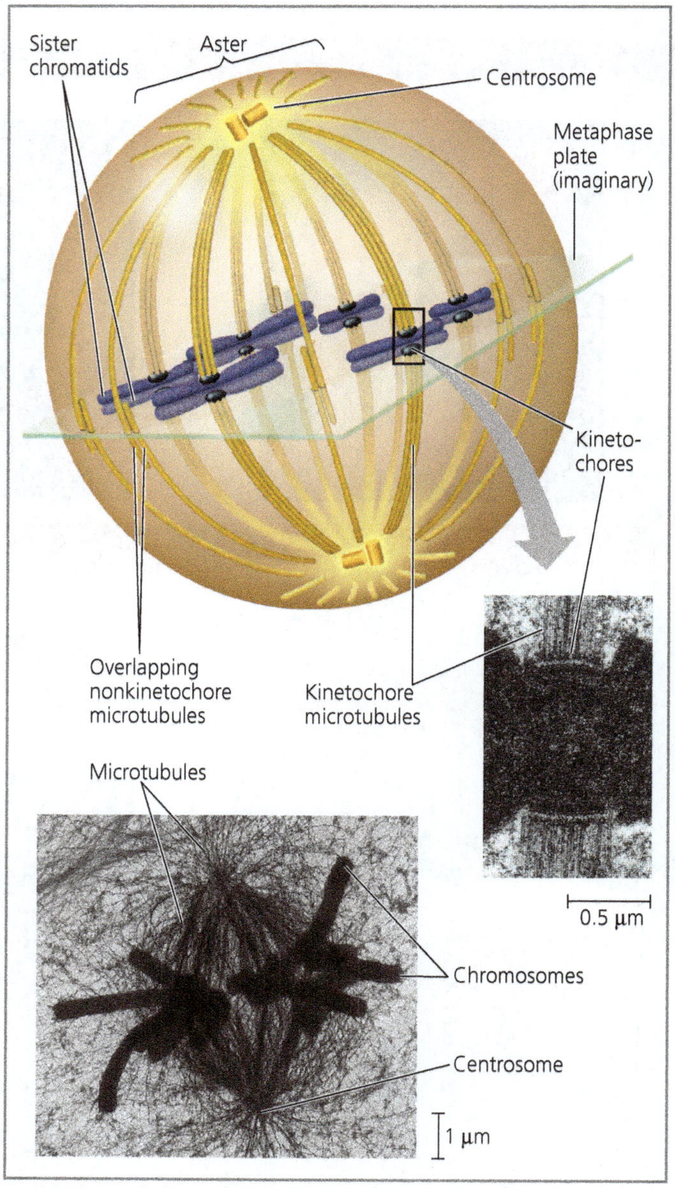

▲ **Figure 8 The mitotic spindle at metaphase.** The kinetochores of each chromosome's two sister chromatids face in opposite directions. Here, each kinetochore is attached to a cluster of kinetochore microtubules extending from the nearest centrosome. Nonkinetochore microtubules overlap at the metaphase plate (TEMs).

DRAW IT *On the lower micrograph, draw a line indicating the position of the metaphase plate. Circle an aster. Draw arrows indicating the directions of chromosome movement once anaphase begins.*

At the end of anaphase, duplicate groups of chromosomes have arrived at opposite ends of the elongated parent cell. Nuclei re-form during telophase. Cytokinesis generally begins during anaphase or telophase, and the spindle eventually disassembles by depolymerization of microtubules.

Cytokinesis: *A Closer Look*

In animal cells, cytokinesis occurs by a process known as **cleavage**. The first sign of cleavage is the appearance of a **cleavage furrow**, a shallow groove in the cell surface near the old metaphase plate **(Figure 10a)**. On the cytoplasmic side

▼ Figure 9 Inquiry

At which end do kinetochore microtubules shorten during anaphase?

Experiment Gary Borisy and colleagues at the University of Wisconsin wanted to determine whether kinetochore microtubules depolymerize at the kinetochore end or the pole end as chromosomes move toward the poles during mitosis. First they labeled the microtubules of a pig kidney cell in early anaphase with a yellow fluorescent dye. (Nonkinetochore microtubules are not shown.)

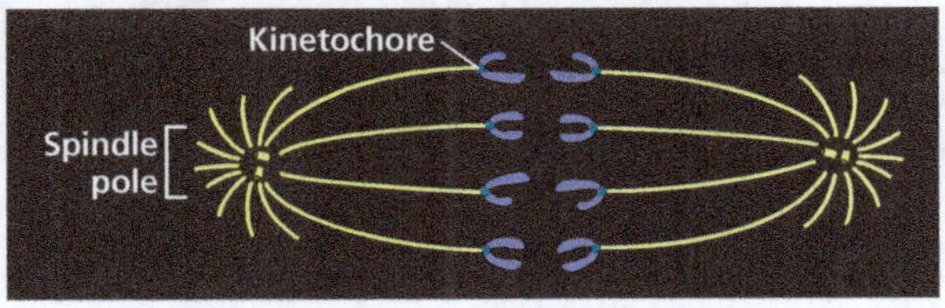

Then they marked a region of the kinetochore microtubules between one spindle pole and the chromosomes by using a laser to eliminate the fluorescence from that region, while leaving the microtubules intact (see below). As anaphase proceeded, they monitored the changes in microtubule length on either side of the mark.

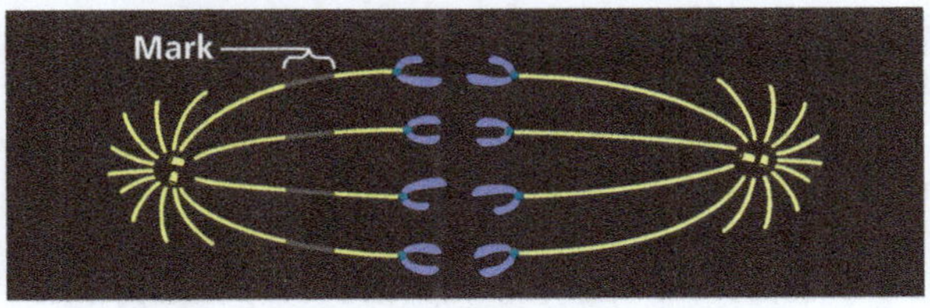

Results As the chromosomes moved poleward, the microtubule segments on the kinetochore side of the mark shortened, while those on the spindle pole side stayed the same length.

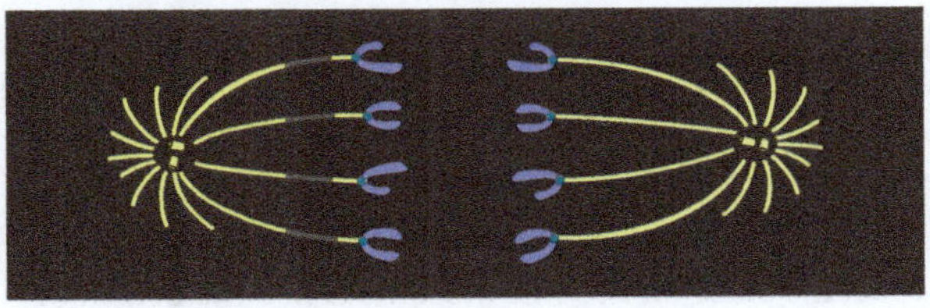

Conclusion During anaphase in this cell type, chromosome movement is correlated with kinetochore microtubules shortening at their kinetochore ends and not at their spindle pole ends. This experiment supports the hypothesis that during anaphase, a chromosome is walked along a microtubule as the microtubule depolymerizes at its kinetochore end, releasing tubulin subunits.

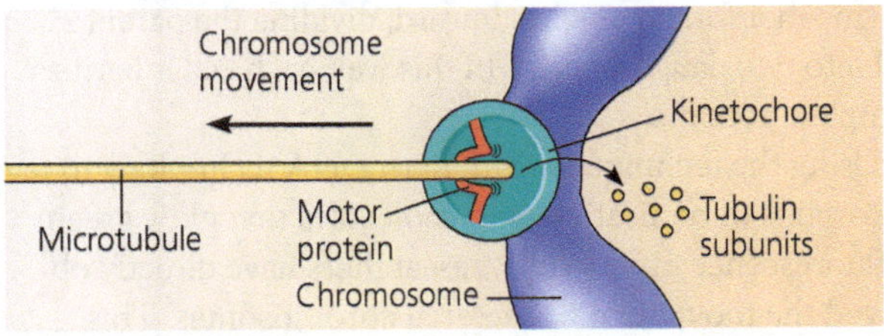

Data from G. J. Gorbsky, P. J. Sammak, and G. G. Borisy, Chromosomes move poleward in anaphase along stationary microtubules that coordinately disassemble from their kinetochore ends, *Journal of Cell Biology* 104:9–18 (1987).

WHAT IF? If this experiment had been done on a cell type in which "reeling in" at the poles was the main cause of chromosome movement, how would the mark have moved relative to the poles? How would the microtubule segment lengths have changed?

▼ Figure 10 Cytokinesis in animal and plant cells.

(a) Cleavage of an animal cell (SEM)

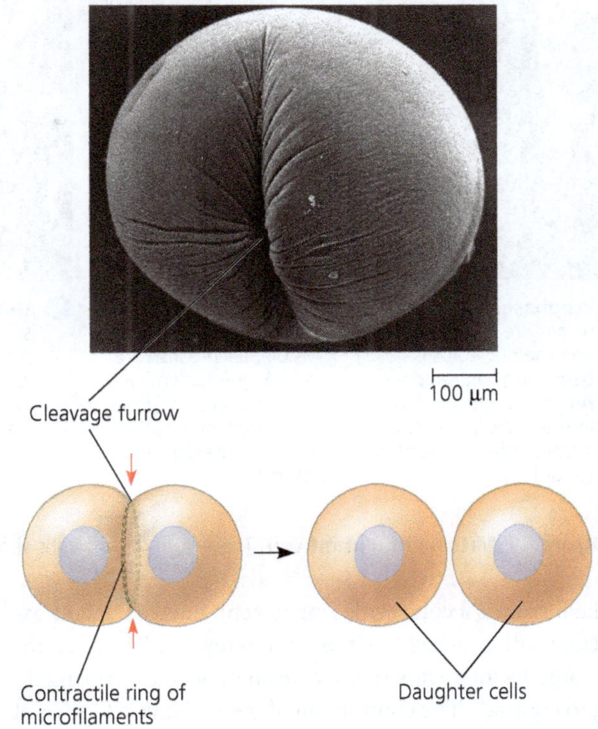

(b) Cell plate formation in a plant cell (TEM)

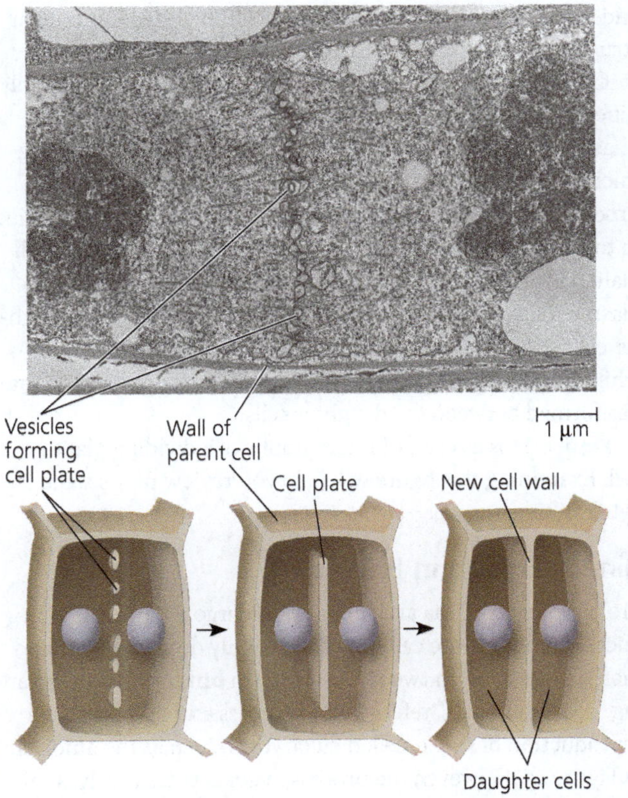

1 **Prophase.** The chromosomes are condensing and the nucleolus is beginning to disappear. Although not yet visible in the micrograph, the mitotic spindle is starting to form.

2 **Prometaphase.** Discrete chromosomes are now visible; each consists of two aligned, identical sister chromatids. Later in prometaphase, the nuclear envelope will fragment.

3 **Metaphase.** The spindle is complete, and the chromosomes, attached to microtubules at their kinetochores, are all at the metaphase plate.

4 **Anaphase.** The chromatids of each chromosome have separated, and the daughter chromosomes are moving to the ends of the cell as their kinetochore microtubules shorten.

5 **Telophase.** Daughter nuclei are forming. Meanwhile, cytokinesis has started: The cell plate, which will divide the cytoplasm in two, is growing toward the perimeter of the parent cell.

▲ **Figure 11 Mitosis in a plant cell.** These light micrographs show mitosis in cells of an onion root.

of the furrow is a contractile ring of actin microfilaments associated with molecules of the protein myosin. The actin microfilaments interact with the myosin molecules, causing the ring to contract. The contraction of the dividing cell's ring of microfilaments is like the pulling of a drawstring. The cleavage furrow deepens until the parent cell is pinched in two, producing two completely separated cells, each with its own nucleus and its own share of cytosol, organelles, and other subcellular structures.

Cytokinesis in plant cells, which have cell walls, is markedly different. There is no cleavage furrow. Instead, during telophase, vesicles derived from the Golgi apparatus move along microtubules to the middle of the cell, where they coalesce, producing a **cell plate** (**Figure 10b**). Cell wall materials carried in the vesicles collect inside the cell plate as it grows. The cell plate enlarges until its surrounding membrane fuses with the plasma membrane along the perimeter of the cell. Two daughter cells result, each with its own plasma membrane. Meanwhile, a new cell wall arising from the contents of the cell plate has formed between the daughter cells.

Figure 11 is a series of micrographs of a dividing plant cell. Examining this figure will help you review mitosis and cytokinesis.

Binary Fission in Bacteria

Prokaryotes (bacteria and archaea) undergo a type of reproduction in which the cell grows to roughly double its size and then divides to form two cells. The term **binary fission**, meaning "division in half," refers to this process and to the asexual reproduction of single-celled eukaryotes, such as the amoeba in Figure 2a. However, the process in eukaryotes involves mitosis, while that in prokaryotes does not.

In bacteria, most genes are carried on a single bacterial chromosome that consists of a circular DNA molecule and associated proteins. Although bacteria are smaller and simpler than eukaryotic cells, the challenge of replicating their genomes in an orderly fashion and distributing the copies equally to two daughter cells is still formidable. For example, when it is fully stretched out, the chromosome of the bacterium *Escherichia coli* is about 500 times as long as the cell. For such a long chromosome to fit within the cell, it must be highly coiled and folded.

In *E. coli*, the process of cell division is initiated when the DNA of the bacterial chromosome begins to replicate at a specific place on the chromosome called the **origin of replication**, producing two origins. As the chromosome continues to replicate, one origin moves rapidly toward the opposite end of the cell (**Figure 12**). While the chromosome is replicating, the cell elongates. When replication is complete and the bacterium has reached about twice its initial size, its plasma membrane pinches inward, dividing the parent *E. coli* cell into two daughter cells. In this way, each cell inherits a complete genome.

Using the techniques of modern DNA technology to tag the origins of replication with molecules that glow green in fluorescence microscopy, researchers have directly observed the movement of bacterial chromosomes. This movement is reminiscent of the poleward movements of the centromere regions of eukaryotic chromosomes during anaphase of mitosis, but bacteria don't have visible mitotic spindles or even microtubules. In most bacterial species studied, the two origins of replication end up at opposite ends of the cell or in some other very specific location, possibly anchored there by one or more proteins. How bacterial

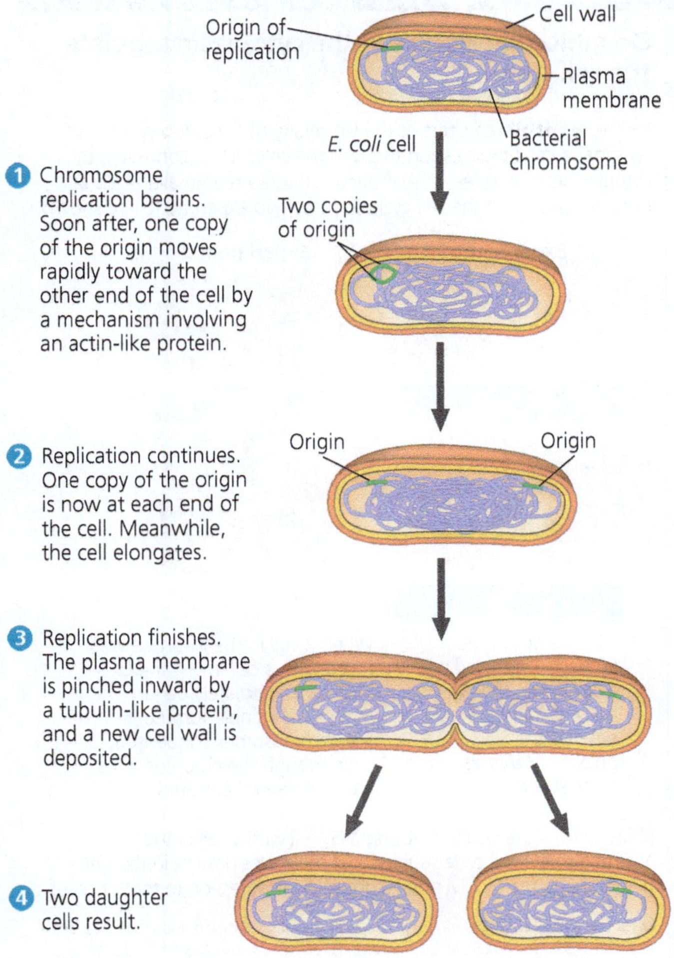

▲ **Figure 12 Bacterial cell division by binary fission.** The bacterium *E. coli*, shown here, has a single, circular chromosome.

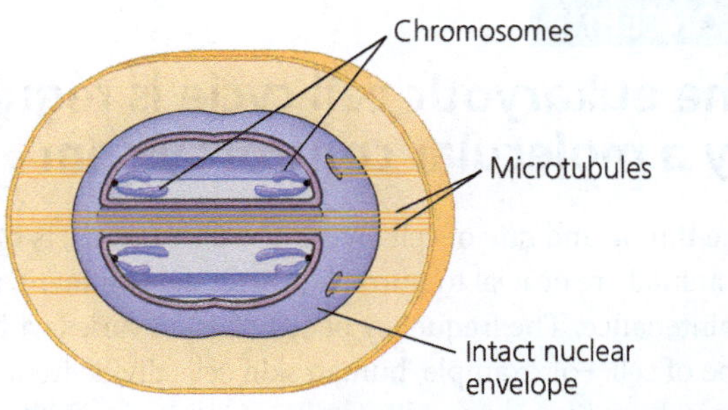

(a) Dinoflagellates. In unicellular eukaryotes called dinoflagellates, the chromosomes attach to the nuclear envelope, which remains intact during cell division. Microtubules pass through the nucleus inside cytoplasmic tunnels, reinforcing the spatial orientation of the nucleus, which then divides in a process reminiscent of bacterial binary fission.

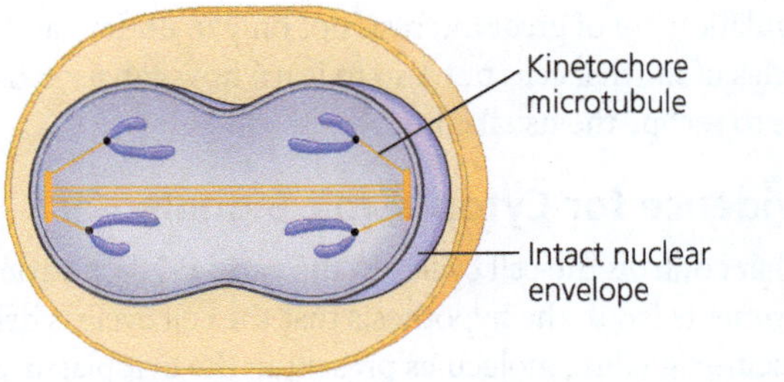

(b) Diatoms and some yeasts. In two other groups of unicellular eukaryotes, diatoms and some yeasts, the nuclear envelope also remains intact during cell division. In these organisms, the microtubules form a spindle *within* the nucleus. Microtubules separate the chromosomes, and the nucleus splits into two daughter nuclei.

▲ **Figure 13 Mechanisms of cell division.** Some unicellular eukaryotes existing today have mechanisms of cell division that may resemble intermediate steps in the evolution of mitosis.

chromosomes move and how their specific location is established and maintained are active areas of research. Several proteins have been identified that play important roles. Polymerization of one protein resembling eukaryotic actin apparently functions in bacterial chromosome movement during cell division, and another protein that is related to tubulin helps pinch the plasma membrane inward, separating the two bacterial daughter cells.

The Evolution of Mitosis

EVOLUTION Given that prokaryotes preceded eukaryotes on Earth by more than a billion years, we might hypothesize that mitosis evolved from simpler prokaryotic mechanisms of cell reproduction. The fact that some of the proteins involved in bacterial binary fission are related to eukaryotic proteins that function in mitosis supports that hypothesis.

As eukaryotes with nuclear envelopes and larger genomes evolved, the ancestral process of binary fission, seen today in bacteria, somehow gave rise to mitosis. Variations on cell division exist in different groups. These variant processes may be similar to mechanisms used by ancestral species and thus may resemble steps in the evolution of mitosis from a binary fission-like process carried out by very early bacteria. Possible intermediate stages are suggested by two unusual types of nuclear division found today in certain unicellular eukaryotes—dinoflagellates, diatoms, and some yeasts (**Figure 13**). These are thought to be cases where ancestral mechanisms have remained relatively unchanged over evolutionary time. In both types, the nuclear envelope remains intact, in contrast to what happens in most eukaryotic cells.

CONCEPT CHECK 2

1. How many chromosomes are shown in the drawing in Figure 8? Are they duplicated? How many chromatids are shown?
2. Compare cytokinesis in animal cells and plant cells.
3. During which stages of the cell cycle does a chromosome consist of two identical chromatids?
4. Compare the roles of tubulin and actin during eukaryotic cell division with the roles of tubulin-like and actin-like proteins during bacterial binary fission.

CONCEPT 3

The eukaryotic cell cycle is regulated by a molecular control system

The timing and rate of cell division in different parts of a plant or animal are crucial to normal growth, development, and maintenance. The frequency of cell division varies with the type of cell. For example, human skin cells divide frequently throughout life, whereas liver cells maintain the ability to divide but keep it in reserve until an appropriate need arises—say, to repair a wound. Some of the most specialized cells, such as fully formed nerve cells and muscle cells, do not divide at all in a mature human. These cell cycle differences result from regulation at the molecular level. The mechanisms of this regulation are of great interest, not only to understand the life cycles of normal cells but also to learn how cancer cells manage to escape the usual controls.

Evidence for Cytoplasmic Signals

What controls the cell cycle? In the early 1970s, a variety of experiments led to the hypothesis that the cell cycle is driven by specific signaling molecules present in the cytoplasm. Some of the first strong evidence for this hypothesis came from experiments with mammalian cells grown in culture **(Figure 14)**. In these experiments, two cells in different phases of the cell cycle were fused to form a single cell with two nuclei. If one of the original cells was in the S phase and the other was in G_1, the G_1 nucleus immediately entered the S phase, as though stimulated by signaling molecules present in the cytoplasm of the first cell. Similarly, if a cell undergoing mitosis (M phase) was fused with another cell in any stage of its cell cycle, even G_1, the second nucleus immediately entered mitosis, with condensation of the chromatin and formation of a mitotic spindle.

Checkpoints of the Cell Cycle Control System

The experiment shown in Figure 14 and other experiments on animal cells and yeasts demonstrated that the sequential events of the cell cycle are directed by a distinct **cell cycle control system**, a cyclically operating set of molecules in the cell that both triggers and coordinates key events in the cell cycle. The cell cycle control system has been compared to the control device of a washing machine **(Figure 15)**. Like the washer's timing device, the cell cycle control system proceeds on its own, according to a built-in clock. However, just as a washer's cycle is subject to both internal control (such as the sensor that detects when the tub is filled with water) and external adjustment (such as starting the machine), the cell cycle is regulated at certain checkpoints by both internal and external signals.

A **checkpoint** in the cell cycle is a control point where stop and go-ahead signals can regulate the cycle. Animal cells generally have built-in stop signals that halt the cell cycle at checkpoints

▼ **Figure 14 Inquiry**

Do molecular signals in the cytoplasm regulate the cell cycle?

Experiment Researchers at the University of Colorado wondered whether a cell's progression through the cell cycle is controlled by cytoplasmic molecules. They induced cultured mammalian cells at different phases of the cell cycle to fuse. Two experiments are shown.

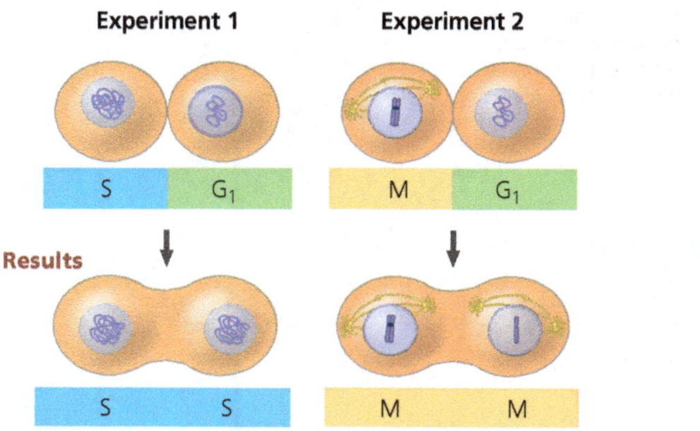

When a cell in the S phase was fused with a cell in G_1, the G_1 nucleus immediately entered the S phase—DNA was synthesized.

When a cell in the M phase was fused with a cell in G_1, the G_1 nucleus immediately began mitosis—a spindle formed and the chromosomes condensed, even though the chromosomes had not been duplicated.

Conclusion The results of fusing a G_1 cell with a cell in the S or M phase of the cell cycle suggest that molecules present in the cytoplasm during the S or M phase control the progression to those phases.

Data from R. T. Johnson and P. N. Rao, Mammalian cell fusion: Induction of premature chromosome condensation in interphase nuclei, *Nature* 226:717–722 (1970).

WHAT IF? If the progression of phases did not depend on cytoplasmic molecules and each phase began when the previous one was complete, how would the results have differed?

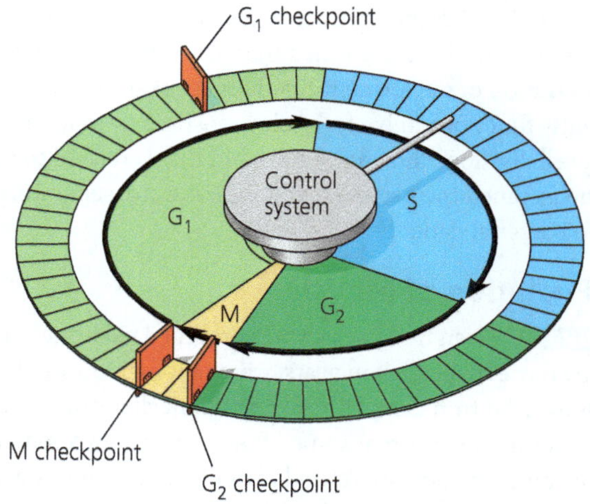

▲ **Figure 15 Mechanical analogy for the cell cycle control system.** In this diagram of the cell cycle, the flat "stepping stones" around the perimeter represent sequential events. Like the control device of an automatic washer, the cell cycle control system proceeds on its own, driven by a built-in clock. However, the system is subject to internal and external regulation at various checkpoints; three important checkpoints are shown (red).

until overridden by go-ahead signals. Many signals registered at checkpoints come from cellular surveillance mechanisms inside the cell. These signals report whether crucial cellular processes that should have occurred by that point have in fact been completed correctly and thus whether or not the cell cycle should proceed. Checkpoints also register signals from outside the cell, as we'll discuss later. Three important checkpoints are those in the G_1, G_2, and M phases (red gates in Figure 15).

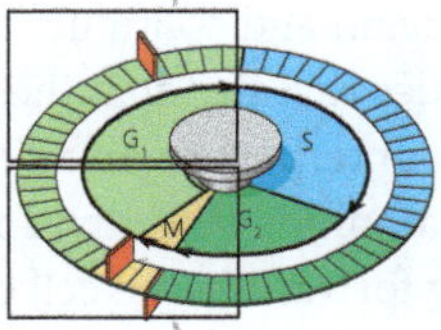

For many cells, the G_1 checkpoint—dubbed the "restriction point" in mammalian cells—seems to be the most important. If a cell receives a go-ahead signal at the G_1 checkpoint, it will usually complete the G_1, S, G_2, and M phases and divide. If it does not receive a go-ahead signal at that point, it may exit the cycle, switching into a nondividing state called the **G_0 phase** **(Figure 16a)**. Most cells of the human body are actually in the G_0 phase. As mentioned earlier, mature nerve cells and muscle cells never divide. Other cells, such as liver cells, can be "called back" from the G_0 phase to the cell cycle by external cues, such as growth factors released during injury.

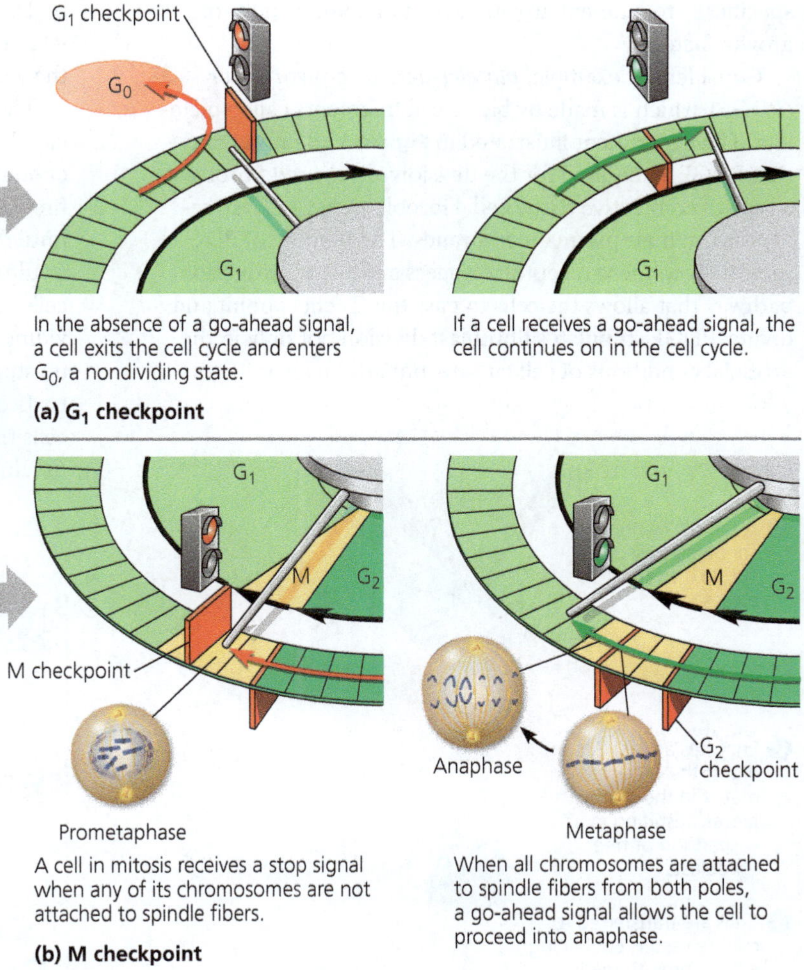

▲ **Figure 16 Two important checkpoints.** At certain checkpoints in the cell cycle (red gates), cells do different things depending on the signals they receive. Events of the **(a)** G_1 and **(b)** M checkpoints are shown. In part (b), the G_2 checkpoint has already been passed by the cell.

WHAT IF? *What might be the result if the cell ignored either checkpoint and progressed through the cell cycle?*

The cell cycle is regulated at the molecular level by a set of regulatory proteins and protein complexes, including proteins called *cyclins*, and other proteins interacting with cyclins that are kinases (enzymes that activate or inactivate other proteins by phosphorylating them). To understand how a cell progresses through the cycle, let's consider the checkpoint signals that can make the cell cycle clock pause or continue.

Biologists are currently working out the pathways that link signals originating inside and outside the cell with the responses by cyclins, kinases, and other proteins. An example of an internal signal occurs at the third important checkpoint, the M phase checkpoint **(Figure 16b)**. Anaphase, the separation of sister chromatids, does not begin until all the chromosomes are properly attached to the spindle at the metaphase plate. Researchers have learned that as long as some kinetochores are unattached to spindle microtubules, the sister chromatids remain together, delaying anaphase. Only when the kinetochores of all the chromosomes are properly attached to the spindle does the appropriate regulatory protein complex become activated. Once activated, the complex sets off a chain of molecular events that activates the enzyme separase, which cleaves the cohesins, allowing the sister chromatids to separate. This mechanism ensures that daughter cells do not end up with missing or extra chromosomes.

Studies using animal cells in culture have led to the identification of many external factors, both chemical and physical, that can influence cell division. For example, cells fail to divide if an essential nutrient is lacking in the culture medium. (This is analogous to trying to run a washing machine without the water supply hooked up; an internal sensor won't allow the machine to continue past the point where water is needed.) And even if all other conditions are favorable, most types of mammalian cells divide in culture only if the growth medium includes specific growth factors. A **growth factor** is a protein released by certain cells that stimulates other cells to divide. Different cell types respond specifically to different growth factors or combinations of growth factors.

Consider, for example, *platelet-derived growth factor (PDGF)*, which is made by blood cell fragments called platelets. The experiment illustrated in **Figure 17** demonstrates that PDGF is required for the division of cultured fibroblasts, a type of connective tissue cell. Fibroblasts have PDGF receptors on their plasma membranes. The binding of PDGF molecules to these receptors triggers a signal transduction pathway that allows the cells to pass the G_1 checkpoint and divide. PDGF stimulates fibroblast division not only in the artificial conditions of cell culture, but also in an animal's body. When an injury occurs, platelets release PDGF in the vicinity. The resulting proliferation of fibroblasts helps heal the wound.

The effect of an external physical factor on cell division is clearly seen in **density-dependent inhibition**, a phenomenon in which crowded cells stop dividing **(Figure 18a)**. As first observed many years ago, cultured cells normally divide until they form a single layer of cells on the inner surface of a culture flask, at which point the cells stop dividing. If some cells are removed, those bordering the open space begin dividing again and continue until the vacancy is filled. Follow-up studies revealed that the binding of a cell-surface protein to its counterpart on an adjoining cell sends a signal to both cells that inhibits cell division, preventing the cells from moving forward in the cell cycle. Growth factors also have a

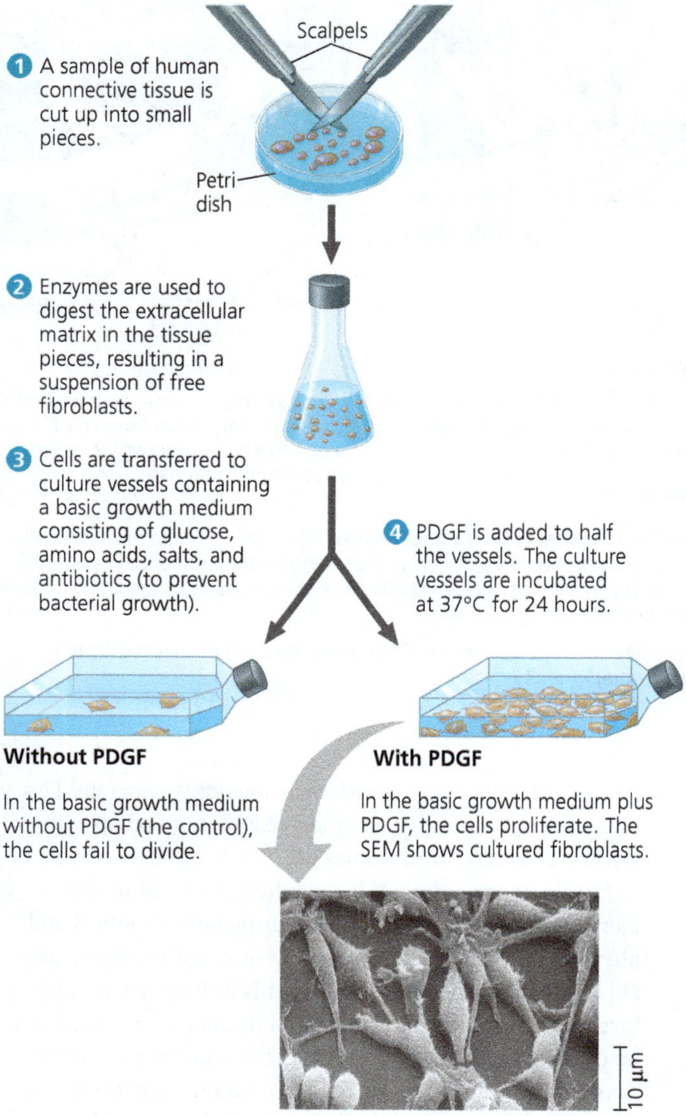

▲ **Figure 17 The effect of platelet-derived growth factor (PDGF) on cell division.**

MAKE CONNECTIONS *PDGF signals cells by binding to a cell-surface receptor that then becomes phosphorylated, activating it so that it sends a signal. If you added a chemical that blocked phosphorylation, how would the results differ?*

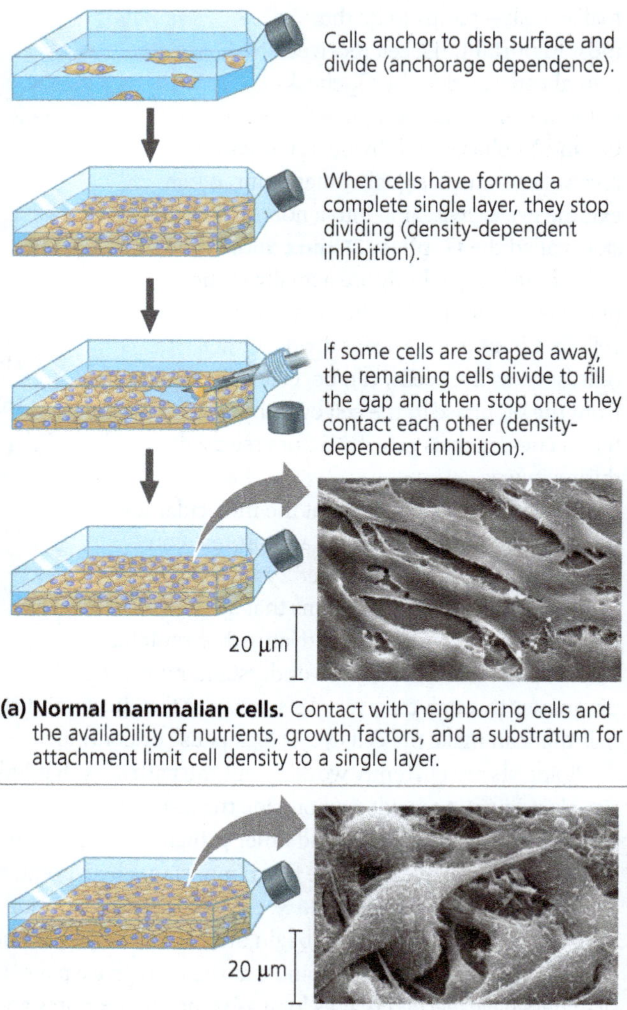

(a) Normal mammalian cells. Contact with neighboring cells and the availability of nutrients, growth factors, and a substratum for attachment limit cell density to a single layer.

(b) Cancer cells. Cancer cells usually continue to divide well beyond a single layer, forming a clump of overlapping cells. They do not exhibit anchorage dependence or density-dependent inhibition.

▲ **Figure 18 Density-dependent inhibition and anchorage dependence of cell division.** Individual cells are shown disproportionately large in the drawings.

role in determining the density that cells attain before ceasing division.

Most animal cells also exhibit **anchorage dependence** (see Figure 18a). To divide, they must be attached to a substratum, such as the inside of a culture flask or the extracellular matrix of a tissue. Experiments suggest that like cell density, anchorage is signaled to the cell cycle control system via pathways involving plasma membrane proteins and elements of the cytoskeleton linked to them.

Density-dependent inhibition and anchorage dependence appear to function not only in cell culture but also in the body's tissues, checking the growth of cells at some optimal density and location during embryonic development and throughout an organism's life. Cancer cells, which we discuss next, exhibit neither density-dependent inhibition nor anchorage dependence **(Figure 18b)**.

Loss of Cell Cycle Controls in Cancer Cells

Cancer cells do not heed the normal signals that regulate the cell cycle. In culture, they do not stop dividing when growth factors are depleted. A logical hypothesis is that cancer cells do not need growth factors in their culture medium to grow and divide. They may make a required growth factor themselves, or they may have an abnormality in the signaling pathway that conveys the growth factor's signal to the cell cycle control system even in the absence of that factor. Another possibility is an abnormal cell cycle control system. In these scenarios, the underlying basis of the abnormality is almost always a change in one or more genes (for example, a mutation) that alters the function of their protein products, resulting in faulty cell cycle control.

There are other important differences between normal cells and cancer cells that reflect derangements of the cell cycle. If and when they stop dividing, cancer cells do so at random points in the cycle, rather than at the normal checkpoints. Moreover, cancer cells can go on dividing indefinitely in culture if they are given a continual supply of nutrients; in essence, they are "immortal." A striking example is a cell line that has been reproducing in culture since 1951. Cells of this line are called HeLa cells because their original source was a tumor removed from a woman named *He*nrietta *La*cks. Cells in culture that acquire the ability to divide indefinitely are said to have undergone a process called **transformation**, causing them to behave (in cell division, at least) like cancer cells. By contrast, nearly all normal, nontransformed mammalian cells growing in culture divide only about 20 to 50 times before they stop dividing, age, and die. Finally, cancer cells evade the normal controls that trigger a cell to undergo a type of programmed cell death called *apoptosis* when something is wrong—for example, when an irreparable mistake has occurred during DNA replication preceding mitosis.

The abnormal behavior of cancer cells can be catastrophic when it occurs in the body. The problem begins when a single cell in a tissue undergoes the first changes of the multistep process that converts a normal cell to a cancer cell. Such a cell often has altered proteins on its surface, and the body's immune system normally recognizes the cell as an insurgent—and destroys it. However, if the cell evades destruction, it may proliferate and form a tumor, a mass of abnormal cells within otherwise normal tissue. The abnormal cells may remain at the original site if they have too few genetic and cellular changes to survive at another site. In that case, the tumor is called a **benign tumor**. Most benign tumors do not cause serious problems and can be removed by surgery. In contrast, a **malignant tumor** includes cells whose genetic and cellular changes enable them to spread to new tissues and impair the functions of one or more organs; based on their ability to divide indefinitely in culture, these cells are also considered *transformed* cells. An individual with a malignant tumor is said to have cancer; **Figure 19** shows the development of breast cancer.

The changes that have occurred in cells of malignant tumors show up in many ways besides excessive proliferation. These

▼ **Figure 19 The growth and metastasis of a malignant breast tumor.** A series of genetic and cellular changes contribute to a tumor becoming malignant (cancerous). The cells of malignant tumors grow in an uncontrolled way and can spread to neighboring tissues and, via lymph and blood vessels, to other parts of the body (metastasis).

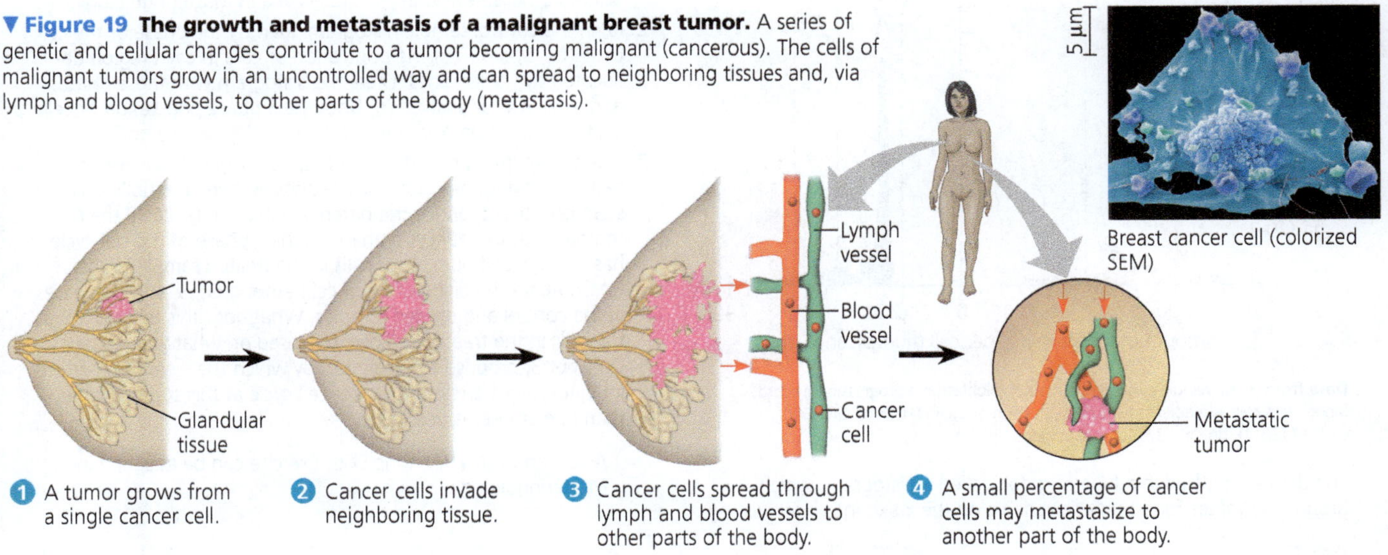

cells may have unusual numbers of chromosomes, though whether this is a cause or an effect of transformation is an ongoing topic of debate. Their metabolism may be altered, and they may cease to function in any constructive way. Abnormal changes on the cell surface cause cancer cells to lose attachments to neighboring cells and the extracellular matrix, allowing them to spread into nearby tissues. Cancer cells may also secrete signaling molecules that cause blood vessels to grow toward the tumor. A few tumor cells may separate from the original tumor, enter blood vessels and lymph vessels, and travel to other parts of the body. There, they may proliferate and form a new tumor. This spread of cancer cells to locations distant from their original site is called **metastasis** (see Figure 19).

A tumor that appears to be localized may be treated with high-energy radiation, which damages DNA in cancer cells much more than DNA in normal cells, apparently because the majority of cancer cells have lost the ability to repair such damage. To treat known or suspected metastatic tumors, chemotherapy is used, in which drugs that are toxic to actively dividing cells are administered through the circulatory system. As you might expect, chemotherapeutic drugs interfere with specific steps in the cell cycle. For example, the drug Taxol freezes the mitotic spindle by preventing microtubule depolymerization, which stops actively dividing cells from proceeding past metaphase and leads to their destruction. The side effects of chemotherapy are due to the drugs' effects on normal cells that divide often. For example, nausea results from chemotherapy's effects on intestinal cells, hair loss from effects on hair follicle cells, and susceptibility to infection from effects on immune system cells. You'll work with data from an experiment involving a potential chemotherapeutic agent in the **Scientific Skills Exercise**.

Scientific Skills Exercise

Interpreting Histograms

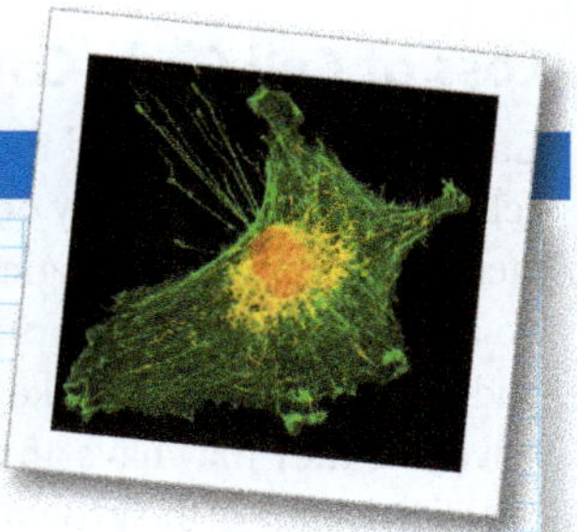

▶ Human glioblastoma cell

At What Phase Is the Cell Cycle Arrested by an Inhibitor? Many medical treatments are aimed at stopping cancer cell proliferation by blocking the cell cycle of cancerous tumor cells. One potential treatment is a cell cycle inhibitor derived from human umbilical cord stem cells. In this exercise, you will compare two histograms to determine where in the cell cycle the inhibitor blocks the division of cancer cells.

How the Experiment Was Done In the treated sample, human glioblastoma (brain cancer) cells were grown in tissue culture in the presence of the inhibitor, while control sample cells were grown in its absence. After 72 hours of growth, the two cell samples were harvested. To get a "snapshot" of the phase of the cell cycle each cell was in at that time, the samples were treated with a fluorescent chemical that binds to DNA and then run through a flow cytometer, an instrument that records the fluorescence level of each cell. Computer software then graphed the number of cells in each sample with a particular fluorescence level, as shown below.

Data from the Experiment

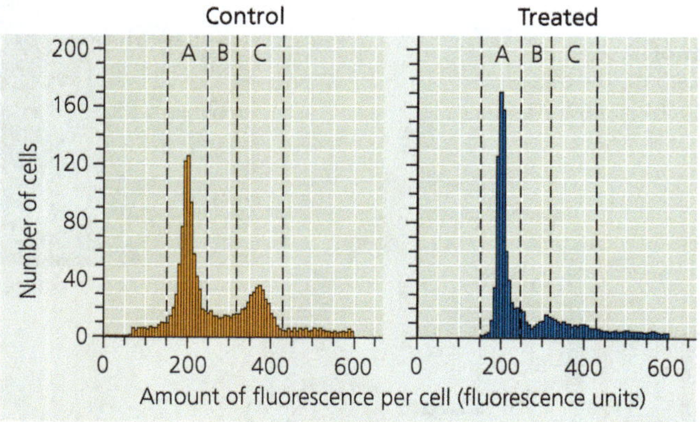

Data from K. K. Velpula et al., Regulation of glioblastoma progression by cord blood stem cells is mediated by downregulation of cyclin D1, *PLoS ONE* 6(3): e18017 (2011). doi:10.1371/journal.pone.0018017

The data are plotted in a type of graph called a histogram, which groups the values for a numeric variable on the *x*-axis into intervals. A histogram allows you to see how all the experimental subjects (cells, in this case) are distributed along a continuous variable (amount of fluorescence). In these histograms, the bars are so narrow that the data appear to follow a curve for which you can detect peaks and dips. Each narrow bar represents the number of cells observed to have a level of fluorescence in the range of that interval. This in turn indicates the relative amount of DNA in those cells. Overall, comparing the two histograms allows you to see how the DNA content of this cell population is altered by the treatment.

INTERPRET THE DATA

1. Study the data in the histograms. (a) Which axis indirectly shows the relative amount of DNA per cell? Explain your answer. (b) In the control sample, compare the first peak in the histogram (in region A) to the second peak (in region C). Which peak shows the population of cells with the higher amount of DNA per cell? Explain. (For additional information about graphs, see the Scientific Skills Review in the Study Area in MasteringBiology.)
2. (a) In the control sample histogram, identify the phase of the cell cycle (G_1, S, or G_2) of the population of cells in each region delineated by vertical lines. Label the histogram with these phases and explain your answer. (b) Does the S phase population of cells show a distinct peak in the histogram? Why or why not?
3. The histogram representing the treated sample shows the effect of growing the cancer cells alongside human umbilical cord stem cells that produce the potential inhibitor. (a) Label the histogram with the cell cycle phases. Which phase of the cell cycle has the greatest number of cells in the treated sample? Explain. (b) Compare the distribution of cells among G_1, S, and G_2 phases in the control and treated samples. What does this tell you about the cells in the treated sample? (c) Based on what you learned in Concept 3, propose a mechanism by which the stem cell–derived inhibitor might arrest the cancer cell cycle at this stage. (More than one answer is possible.)

(MB) A version of this Scientific Skills Exercise can be assigned in MasteringBiology.

Over the past several decades, researchers have produced a flood of valuable information about cell-signaling pathways and how their malfunction contributes to the development of cancer through effects on the cell cycle. Coupled with new molecular techniques, such as the ability to rapidly sequence the DNA of cells in a particular tumor, medical treatments for cancer are beginning to become more "personalized" to a particular patient's tumor.

For example, the cells of roughly 20% of breast cancer tumors show abnormally high amounts of a cell-surface receptor tyrosine kinase called HER2, and many show an increase in the number of estrogen receptor (ER) molecules, intracellular receptors that can trigger cell division. Based on lab findings, a physician can prescribe chemotherapy with a molecule that blocks the function of the specific protein (Herceptin for HER2 and tamoxifen for ERs). Treatment using these agents, when appropriate, has led to increased survival rates and fewer cancer recurrences.

CONCEPT CHECK 3

1. In Figure 14, why do the nuclei resulting from experiment 2 contain different amounts of DNA?
2. What phase are most of your body cells in?
3. Compare and contrast a benign tumor and a malignant tumor.
4. **WHAT IF?** What would happen if you performed the experiment in Figure 17 with cancer cells?

Chapter Review

SUMMARY OF KEY CONCEPTS

VOCAB SELF-QUIZ
goo.gl/gbai8v

- Unicellular organisms reproduce by **cell division**; multicellular organisms depend on cell division for their development from a fertilized egg and for growth and repair. Cell division is part of the **cell cycle**, an ordered sequence of events in the life of a cell.

CONCEPT 1

Most cell division results in genetically identical daughter cells

- The genetic material (DNA) of a cell—its **genome**—is partitioned among **chromosomes**. Each eukaryotic chromosome consists of one DNA molecule associated with many proteins. Together, the complex of DNA and associated proteins is called **chromatin**. The chromatin of a chromosome exists in different states of condensation at different times. In animals, **gametes** have one set of chromosomes, and **somatic cells** have two sets.
- Cells replicate their genetic material before they divide, each daughter cell receiving a copy of the DNA. Prior to cell division, chromosomes are duplicated. Each one then consists of two identical **sister chromatids** joined along their lengths by sister chromatid cohesion and held most tightly together at a constricted region at the **centromeres**. When this cohesion is broken, the chromatids separate during cell division, becoming the chromosomes of the daughter cells. Eukaryotic cell division consists of **mitosis** (division of the nucleus) and **cytokinesis** (division of the cytoplasm).

? *Differentiate between these terms: chromosome, chromatin, and chromatid.*

CONCEPT 2

The mitotic phase alternates with interphase in the cell cycle

- Between divisions, a cell is in **interphase**: the **G_1, S,** and **G_2 phases**. The cell grows throughout interphase, with DNA being replicated only during the synthesis (S) phase. Mitosis and cytokinesis make up the **mitotic (M) phase** of the cell cycle.

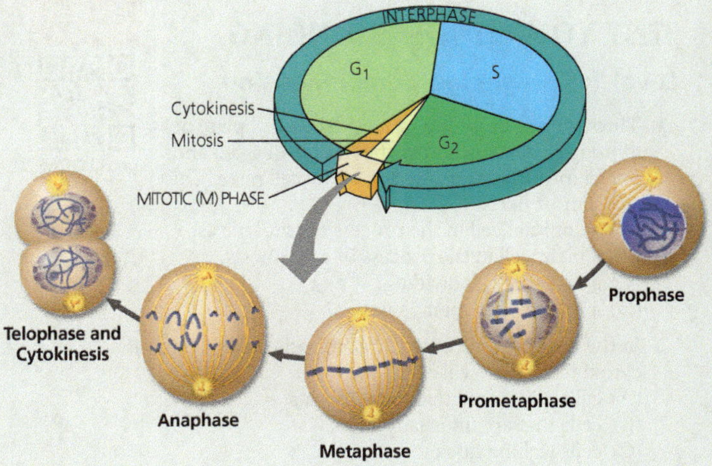

- The **mitotic spindle**, made up of microtubules, controls chromosome movement during mitosis. In animal cells, the mitotic spindle arises from the **centrosomes** and includes spindle microtubules and **asters**. Some spindle microtubules attach to the **kinetochores** of chromosomes and move the chromosomes to the **metaphase plate**. After sister chromatids separate, motor proteins move them along kinetochore microtubules toward opposite ends of the cell. The cell elongates when motor proteins push nonkinetochore microtubules from opposite poles away from each other.
- Mitosis is usually followed by cytokinesis. Animal cells carry out cytokinesis by **cleavage**, and plant cells form a **cell plate**.
- During **binary fission** in bacteria, the chromosome replicates and the daughter chromosomes actively move apart. Some of the proteins involved in bacterial binary fission are related to eukaryotic actin and tubulin. Since prokaryotes preceded eukaryotes by more than a billion years, it is likely that mitosis evolved from prokaryotic cell division.

? *In which of the three subphases of interphase and the stages of mitosis do chromosomes exist as single DNA molecules?*

CONCEPT 3

The eukaryotic cell cycle is regulated by a molecular control system

- Signaling molecules present in the cytoplasm regulate progress through the cell cycle.
- The **cell cycle control system** is molecularly based; key regulatory proteins are cyclins and kinases. The cell cycle clock has specific **checkpoints** where the cell cycle stops until a go-ahead signal is received; important checkpoints occur in the G_1, G_2, and M phases. Cell culture has enabled researchers to study the molecular details of cell division. Both internal signals and external signals control the cell cycle checkpoints via signal transduction pathways. Most cells exhibit **density-dependent inhibition** of cell division as well as **anchorage dependence**.
- Cancer cells elude normal cell cycle regulation and divide unchecked, forming tumors. **Malignant tumors** invade nearby tissues and can undergo **metastasis**, exporting cancer cells to other sites, where they may form secondary tumors. Recent cell cycle and cell-signaling research, and new techniques for sequencing DNA, have led to improved cancer treatments.

? *Explain the significance of the G_1 and M checkpoints and the go-ahead signals involved in the cell cycle control system.*

TEST YOUR UNDERSTANDING

PRACTICE TEST
goo.gl/CRZjvS

Level 1: Knowledge/Comprehension

1. Through a microscope, you can see a cell plate beginning to develop across the middle of a cell and nuclei forming on either side of the cell plate. This cell is most likely
(A) an animal cell in the process of cytokinesis.
(B) a plant cell in the process of cytokinesis.
(C) a bacterial cell dividing.
(D) a plant cell in metaphase.

2. In the cells of some organisms, mitosis occurs without cytokinesis. This will result in
(A) cells with more than one nucleus.
(B) cells that are unusually small.
(C) cells lacking nuclei.
(D) cell cycles lacking an S phase.

3. Which of the following does *not* occur during mitosis?
(A) condensation of the chromosomes
(B) replication of the DNA
(C) separation of sister chromatids
(D) spindle formation

Level 2: Application/Analysis

4. A particular cell has half as much DNA as some other cells in a mitotically active tissue. The cell in question is most likely in
(A) G_1.
(B) G_2.
(C) prophase.
(D) metaphase.

5. The drug cytochalasin B blocks the function of actin. Which of the following aspects of the animal cell cycle would be most disrupted by cytochalasin B?
(A) spindle formation
(B) spindle attachment to kinetochores
(C) cell elongation during anaphase
(D) cleavage furrow formation and cytokinesis

6. DRAW IT Draw one eukaryotic chromosome as it would appear during interphase, during each of the stages of mitosis, and during cytokinesis. Also draw and label the nuclear envelope and any microtubules attached to the chromosome(s).

7. The light micrograph shows dividing cells near the tip of an onion root. Identify a cell in each of the following stages: prophase, prometaphase, metaphase, anaphase, and telophase. Describe the major events occurring at each stage.

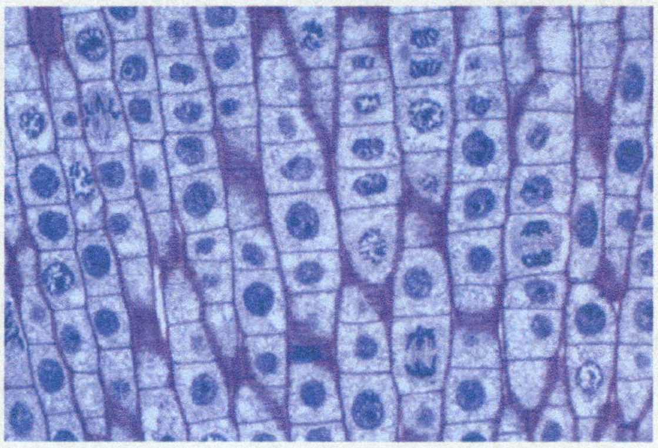

Level 3: Synthesis/Evaluation

8. SCIENTIFIC INQUIRY
Although both ends of a microtubule can gain or lose subunits, one end (called the plus end) polymerizes and depolymerizes at a higher rate than the other end (the minus end). For spindle microtubules, the plus ends are in the center of the spindle, and the minus ends are at the poles. Motor proteins that move along microtubules specialize in walking either toward the plus end or toward the minus end; the two types are called plus end directed and minus end directed motor proteins, respectively. Given what you know about chromosome movement and spindle changes during anaphase, predict which type of motor proteins would be present on (a) kinetochore microtubules and (b) nonkinetochore microtubules.

9. FOCUS ON EVOLUTION
The result of mitosis is that the daughter cells end up with the same number of chromosomes that the parent cell had. Another way to maintain the number of chromosomes would be to carry out cell division first and then duplicate the chromosomes in each daughter cell. Assess whether this would be an equally good way of organizing the cell cycle. Explain why evolution has not led to this alternative.

10. FOCUS ON INFORMATION
The continuity of life is based on heritable information in the form of DNA. In a short essay (100–150 words), explain how the process of mitosis faithfully parcels out exact copies of this heritable information in the production of genetically identical daughter cells.

11. SYNTHESIZE YOUR KNOWLEDGE

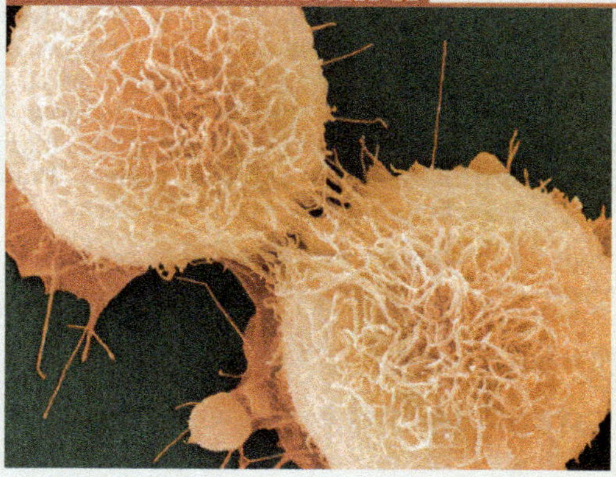

Shown here are two HeLa cancer cells that are just completing cytokinesis. Explain how the cell division of cancer cells like these is misregulated. What genetic and other changes might have caused these cells to escape normal cell cycle regulation?

Answers

Figure Questions

Figure 4

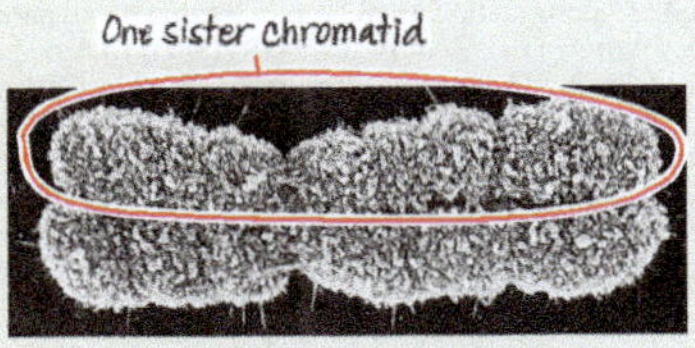

Circling the other chromatid instead would also be correct. **Figure 5** The chromosome has four chromatid arms. The single (duplicated) chromosome in step 2 becomes two (unduplicated) chromosomes in step 3. The duplicated chromosome in step 2 is considered one single chromosome. **Figure 7** 12; 2; 2; 1

Figure 8

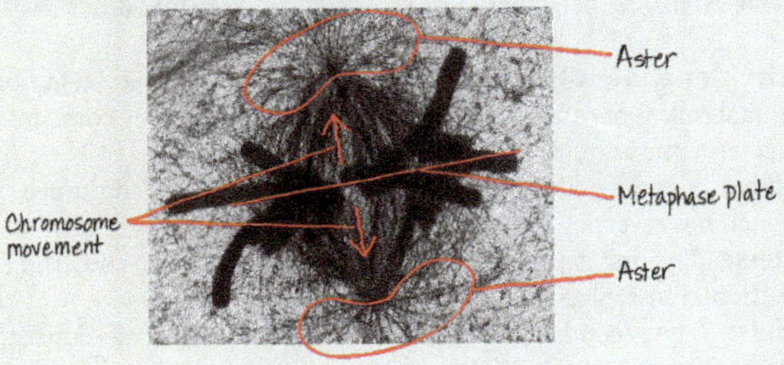

Figure 9 The mark would have moved toward the nearer pole. The lengths of fluorescent microtubules between that pole and the mark would have decreased, while the lengths between the chromosomes and the mark would have remained the same. **Figure 14** In both cases, the G_1 nucleus would have remained in G_1 until the time it normally would have entered the S phase. Chromosome condensation and spindle formation would not have occurred until the S and G_2 phases had been completed. **Figure 16** The cell would divide under conditions where it was inappropriate to do so. If the daughter cells and their descendants also ignored either of the checkpoints and divided, there would soon be an abnormal mass of cells. (This type of inappropriate cell division can contribute to the development of cancer.) **Figure 17** The cells in the vessel with PDGF would not be able to respond to the growth factor signal and thus would not divide. The culture would resemble that without the added PDGF.

Concept Check 1

1. 1; 1; 2 **2.** 39; 39; 78

Concept Check 2

1. 6 chromosomes; they are duplicated; 12 chromatids **2.** Following mitosis, cytokinesis results in two genetically identical daughter cells in both plant cells and animal cells. However, the mechanism of dividing the cytoplasm is different in animals and plants. In an animal cell, cytokinesis occurs by cleavage, which divides the parent cell in two with a contractile ring of actin filaments. In a plant cell, a cell plate forms in the middle of the cell and grows until its membrane fuses with the plasma membrane of the parent cell. A new cell wall grows inside the cell plate. **3.** From the end of S phase in interphase through the end of metaphase in mitosis **4.** During eukaryotic cell division, tubulin is involved in spindle formation and chromosome movement, while actin functions during cytokinesis. In bacterial binary fission, it's the opposite: Tubulin-like molecules are thought to act in daughter cell separation, and actin-like molecules are thought to move the daughter bacterial chromosomes to opposite ends of the cell.

Concept Check 3

1. The nucleus on the right was originally in the G_1 phase; therefore, it had not yet duplicated its chromosomes. The nucleus on the left was in the M phase, so it had already duplicated its chromosomes. **2.** Most body cells are in a nondividing state called G_0. **3.** Both types of tumors consist of abnormal cells, but their characteristics are different. A benign tumor stays at the original site and can usually be surgically removed; the cells have some genetic and cellular changes from normal, non-tumor cells. Cancer cells from a malignant tumor have more significant genetic and cellular changes, can spread from the original site by metastasis, and may impair the functions of one or more organs. **4.** The cells might divide even in the absence of PDGF. In addition, they would not stop when the surface of the culture vessel was covered; they would continue to divide, piling on top of one another.

Summary of Key Concepts Questions

1 The DNA of a eukaryotic cell is packaged into structures called *chromosomes*. Each chromosome is a long molecule of DNA, which carries hundreds to thousands of genes, with associated proteins that maintain chromosome structure and help control gene activity. This DNA-protein complex is called *chromatin*. The chromatin of each chromosome is long and thin when the cell is not dividing. Prior to cell division, each chromosome is duplicated, and the resulting sister *chromatids* are attached to each other by proteins at the centromeres and, for many species, all along their lengths (a phenomenon called sister chromatid cohesion). **2** Chromosomes exist as single DNA molecules in G_1 of interphase and in anaphase and telophase of mitosis. During S phase, DNA replication produces sister chromatids, which persist during G_2 of interphase and through prophase, prometaphase, and metaphase of mitosis. **3** Checkpoints allow cellular surveillance mechanisms to determine whether the cell is prepared to go to the next stage. Internal and external signals move a cell past these checkpoints. The G_1 checkpoint, called the "restriction point" in mammalian cells, determines whether a cell will complete the cell cycle and divide or switch into the G_0 phase. The signals to pass this checkpoint often are external—such as growth factors. Regulation of the cell cycle is carried out by a molecular system, including proteins called cyclins and other proteins that are kinases. The signal to pass the M phase checkpoint is not activated until all chromosomes are attached to kinetochore fibers and are aligned at the metaphase plate. Only then will sister chromatid separation occur.

Test Your Understanding

1. B **2.** A **3.** B **4.** A **5.** D

6.

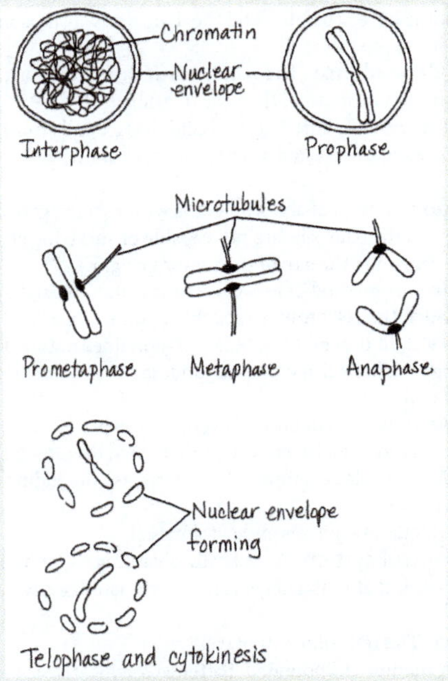

7. See Figure 7 for a description of major events.

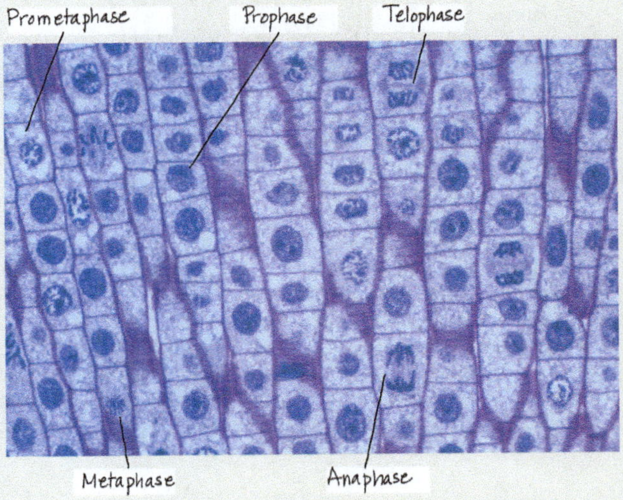

Only one cell is indicated for each stage, but other correct answers are also present in this micrograph.

Credits

Credits are listed in the order of appearance.

Photo Credits

1 George von Dassow; **2a** Biophoto Associates/Science Source; **2b** Biology Pics/Science Source; **2c** Biophoto/Science Source; **3** J. M. Murray, Univ. of Pennsylvania Medical School; **4, 5b** Biophoto/Science Source; **7** Conly L. Rieder; **8.1 top** J. Richard McIntosh; **8.2** From: The kinetochore fiber structure in the acentric spindles of the green alga *Oedogonium*. M. J. Schibler and J. D. Pickettheaps. *Protoplasma*, 1987; 137(1): 29–44. Fig. 17. Springer-Verlag GmbH & Co. KG; **10a** Don W. Fawcett/Science Source; **10b** Eldon H. Newcomb; **11** Elizabeth Pierson, Pearson Education; **17** Guenter Albrecht-Buehler, Northwestern University; **18** Lan Bo Chen; **19** Anne Weston/Wellcome Institute Library; **9 SSE** Mike Davidson; **Chapter Review mitosis** J.L. Carson/Custom Medical Stock Photo; **SYK** Steve Gschmeissner/Science Source.

Illustration and Text Credits

p.182 Quote ("Where a cell . . .") Rudolf Virchow, 1855, quoted in *Biology*, Helena Curtis, second edition, Worth, p. 90; **9** Data from G. J. Gorbsky, P. J. Sammak, and G. G. Borisy, Chromosomes move poleward in anaphase along stationary microtubules that coordinately disassemble from their kinetochore ends, *Journal of Cell Biology* 104:9–18 (1987); **13** Adapted from *Molecular Biology of the Cell*, 4th ed., by B. Alberts, et al. Copyright © 2002 by Garland Science/Taylor & Francis Books, LLC. Reprinted with permission; **14** Data from R. T. Johnson and P. N. Rao, Mammalian cell fusion: Induction of premature chromosome condensation in interphase nuclei, *Nature* 226:717–722 (1970); **SSE.1** Data from K. K. Velpula et al., Regulation of glioblastoma progression by cord blood stem cells is mediated by downregulation of cyclin D1, *PLoS ONE* 6(3): e18017 (2011). doi:10.1371/journal.pone.0018017.

Glossary

anaphase The fourth stage of mitosis, in which the chromatids of each chromosome have separated and the daughter chromosomes are moving to the poles of the cell.

anchorage dependence The requirement that a cell must be attached to a substratum in order to initiate cell division.

aster A radial array of short microtubules that extends from each centrosome toward the plasma membrane in an animal cell undergoing mitosis.

benign tumor A mass of abnormal cells with specific genetic and cellular changes such that the cells are not capable of surviving at a new site and generally remain at the site of the tumor's origin.

binary fission A method of asexual reproduction in single-celled organisms in which the cell grows to roughly double its size and then divides into two cells. In prokaryotes, binary fission does not involve mitosis, but in single-celled eukaryotes that undergo binary fission, mitosis is part of the process.

cell cycle An ordered sequence of events in the life of a cell, from its origin in the division of a parent cell until its own division into two. The eukaryotic cell cycle is composed of interphase (including G_1, S, and G_2 subphases) and M phase (including mitosis and cytokinesis).

cell cycle control system A cyclically operating set of molecules in the eukaryotic cell that both triggers and coordinates key events in the cell cycle.

cell division The reproduction of cells.

cell plate A membrane-bounded, flattened sac located at the midline of a dividing plant cell, inside which the new cell wall forms during cytokinesis.

centromere (sen′-trō-mēr) In a duplicated chromosome, the region on each sister chromatid where it is most closely attached to its sister chromatid by proteins that bind to specific DNA sequences; this close attachment causes a constriction in the condensed chromosome. (An uncondensed, unduplicated chromosome has a single centromere, identified by its DNA sequence.)

checkpoint A control point in the cell cycle where stop and go-ahead signals can regulate the cycle.

cleavage (1) The process of cytokinesis in animal cells, characterized by pinching of the plasma membrane. (2) The succession of rapid cell divisions without significant growth during early embryonic development that converts the zygote to a ball of cells.

cleavage furrow The first sign of cleavage in an animal cell; a shallow groove around the cell in the cell surface near the old metaphase plate.

cytokinesis (sī′-tō-kuh-nē′-sis) The division of the cytoplasm to form two separate daughter cells immediately after mitosis, meiosis I, or meiosis II.

density-dependent inhibition The phenomenon observed in normal animal cells that causes them to stop dividing when they come into contact with one another.

G_0 phase A nondividing state occupied by cells that have left the cell cycle, sometimes reversibly.

G_1 phase The first gap, or growth phase, of the cell cycle, consisting of the portion of interphase before DNA synthesis begins.

G_2 phase The second gap, or growth phase, of the cell cycle, consisting of the portion of interphase after DNA synthesis occurs.

gamete (gam′-ēt) A haploid reproductive cell, such as an egg or sperm. Gametes unite during sexual reproduction to produce a diploid zygote.

growth factor (1) A protein that must be present in the extracellular environment (culture medium or animal body) for the growth and normal development of certain types of cells. (2) A local regulator that acts on nearby cells to stimulate cell proliferation and differentiation.

interphase The period in the cell cycle when the cell is not dividing. During interphase, cellular metabolic activity is high, chromosomes and organelles are duplicated, and cell size may increase. Interphase often accounts for about 90% of the cell cycle.

kinetochore (kuh-net′-uh-kōr) A structure of proteins attached to the centromere that links each sister chromatid to the mitotic spindle.

malignant tumor A cancerous tumor containing cells that have significant genetic and cellular changes and are capable of invading and surviving in new sites. Malignant tumors can impair the functions of one or more organs.

metaphase The third stage of mitosis, in which the spindle is complete and the chromosomes, attached to microtubules at their kinetochores, are all aligned at the metaphase plate.

metaphase plate An imaginary structure located at a plane midway between the two poles of a cell in metaphase on which the centromeres of all the duplicated chromosomes are located.

metastasis (muh-tas′-tuh-sis) The spread of cancer cells to locations distant from their original site.

mitosis (mī-tō′-sis) A process of nuclear division in eukaryotic cells conventionally divided into five stages: prophase, prometaphase, metaphase, anaphase, and telophase. Mitosis conserves chromosome number by allocating replicated chromosomes equally to each of the daughter nuclei.

mitotic (M) phase The phase of the cell cycle that includes mitosis and cytokinesis.

mitotic spindle An assemblage of microtubules and associated proteins that is involved in the movement of chromosomes during mitosis.

origin of replication Site where the replication of a DNA molecule begins, consisting of a specific sequence of nucleotides.

prometaphase The second stage of mitosis, in which the nuclear envelope fragments and the spindle microtubules attach to the kinetochores of the chromosomes.

prophase The first stage of mitosis, in which the chromatin condenses into discrete chromosomes visible with a light microscope, the mitotic spindle begins to form, and the nucleolus disappears but the nucleus remains intact.

S phase The synthesis phase of the cell cycle; the portion of interphase during which DNA is replicated.

sister chromatids Two copies of a duplicated chromosome attached to each other by proteins at the centromere and, sometimes, along the arms. While joined, two sister chromatids make up one chromosome. Chromatids are eventually separated during mitosis or meiosis II.

somatic cell (sō-mat′-ik) Any cell in a multicellular organism except a sperm or egg or their precursors.

telophase The fifth and final stage of mitosis, in which daughter nuclei are forming and cytokinesis has typically begun.

Meiosis and Sexual Life Cycles

Meiosis and Sexual Life Cycles

KEY CONCEPTS

1 Offspring acquire genes from parents by inheriting chromosomes

2 Fertilization and meiosis alternate in sexual life cycles

3 Meiosis reduces the number of chromosome sets from diploid to haploid

4 Genetic variation produced in sexual life cycles contributes to evolution

▲ **Figure 1** What accounts for family resemblance?

Variations on a Theme

Most people who send out birth announcements mention the sex of the baby, but they don't feel the need to specify that their offspring is a human being! One of the characteristics of life is the ability of organisms to reproduce their own kind—elephants produce little elephants, and oak trees generate oak saplings. Exceptions to this rule show up only as sensational but highly suspect stories in tabloid newspapers.

Another rule often taken for granted is that offspring resemble their parents more than they do unrelated individuals. If you examine the family members shown in **Figure 1**, you can pick out some similar features among them. The transmission of traits from one generation to the next is called inheritance, or **heredity** (from the Latin *heres*, heir). However, sons and daughters are not identical copies of either parent or of their siblings. Along with inherited similarity, there is also **variation**. Farmers have exploited the principles of heredity and variation for thousands of years, breeding plants and animals for desired traits. But what are the biological mechanisms leading to the hereditary similarity and variation that we call a "family resemblance"? A detailed answer to this question eluded biologists until the advance of genetics in the 20th century.

Genetics is the scientific study of heredity and inherited variation. Rapid acquisition and analysis of the genome sequences of many species, including our own, have taught us a great deal about evolution on the molecular level—in other words, evolution of the genome itself. In fact, genetic methods and discoveries are catalyzing progress in all areas of biology, from cell biology to physiology, developmental biology, behavior, and even ecology.

We begin our study of genetics in this chapter by examining how chromosomes pass from parents to offspring in sexually reproducing organisms. The processes of meiosis (a special type of cell division) and fertilization (the fusion of sperm and egg) maintain a species' chromosome count during the sexual life cycle. We'll describe the cellular mechanics of meiosis and explain how this process differs from mitosis. Finally, we'll consider how both meiosis and fertilization contribute to genetic variation, such as the variation obvious in the family shown in Figure 1.

CONCEPT 1

Offspring acquire genes from parents by inheriting chromosomes

Family friends may tell you that you have your mother's nose or your father's eyes. Of course, parents do not, in any literal sense, give their children a nose, eyes, hair, or any other traits. What, then, *is* actually inherited?

Inheritance of Genes

Parents endow their offspring with coded information in the form of hereditary units called **genes**. The genes we inherit from our mothers and fathers are our genetic link to our parents, and they account for family resemblances such as shared eye color or freckles. Our genes program specific traits that emerge as we develop from fertilized eggs into adults.

The genetic program is written in the language of DNA, the polymer of four different nucleotides. Inherited information is passed on in the form of each gene's specific sequence of DNA nucleotides, much as printed information is communicated in the form of meaningful sequences of letters. In both cases, the language is symbolic. Just as your brain translates the word *apple* into a mental image of the fruit, cells translate genes into freckles and other features. Most genes program cells to synthesize specific enzymes and other proteins, whose cumulative action produces an organism's inherited traits. The programming of these traits in the form of DNA is one of the unifying themes of biology.

The transmission of hereditary traits has its molecular basis in the replication of DNA, which produces copies of genes that can be passed from parents to offspring. In animals and plants, reproductive cells called **gametes** are the vehicles that transmit genes from one generation to the next. During fertilization, male and female gametes (sperm and eggs) unite, passing on genes of both parents to their offspring.

Except for small amounts of DNA in mitochondria and chloroplasts, the DNA of a eukaryotic cell is packaged into chromosomes within the nucleus. Every species has a characteristic number of chromosomes. For example, humans have 46 chromosomes in their **somatic cells**—all the cells of the body except the gametes and their precursors. Each chromosome consists of a single long DNA molecule elaborately coiled in association with various proteins. One chromosome includes several hundred to a few thousand genes, each of which is a specific sequence of nucleotides within the DNA molecule. A gene's specific location along the length of a chromosome is called the gene's **locus** (plural, *loci*; from the Latin, meaning "place"). Our genetic endowment (our genome) consists of the genes and other DNA that make up the chromosomes we inherited from our parents.

Comparison of Asexual and Sexual Reproduction

Only organisms that reproduce asexually have offspring that are exact genetic copies of themselves. In **asexual reproduction**, a single individual is the sole parent and passes copies of all its genes to its offspring without the fusion of gametes. For example, single-celled eukaryotic organisms can reproduce asexually by mitotic cell division, in which DNA is copied and allocated equally to two daughter cells. The genomes of the offspring are virtually exact copies of the parent's genome. Some multicellular organisms are also capable of reproducing asexually **(Figure 2)**. Because the cells of the offspring arise via mitosis in the parent, the offspring is usually genetically identical to its parent. An individual that reproduces asexually gives rise to a **clone**, a group of genetically identical individuals. Genetic differences occasionally arise in asexually reproducing organisms as a result of changes in the DNA called mutations.

In **sexual reproduction**, two parents give rise to offspring that have unique combinations of genes inherited from the two parents. In contrast to a clone, offspring of sexual reproduction vary genetically from their siblings and both parents: They are variations on a common theme of family resemblance, not exact replicas. Genetic variation like that shown in Figure 1 is an important consequence of sexual reproduction. What mechanisms generate this genetic variation? The key is the behavior of chromosomes during the sexual life cycle.

(a) Hydra **(b) Redwoods**

▲ **Figure 2 Asexual reproduction in two multicellular organisms. (a)** This relatively simple animal, a hydra, reproduces by budding. The bud, a localized mass of mitotically dividing cells, develops into a small hydra, which detaches from the parent (LM). **(b)** All the trees in this circle of redwoods arose asexually from a single parent tree, whose stump is in the center of the circle.

CONCEPT CHECK 1

1. **MAKE CONNECTIONS** Using what you know of gene expression in a cell, explain what causes traits of parents (such as hair color) to show up in their offspring.
2. How does an asexually reproducing eukaryotic organism produce offspring that are genetically identical to each other and to their parent?
3. **WHAT IF?** A horticulturalist breeds orchids, trying to obtain a plant with a unique combination of desirable traits. After many years, she finally succeeds. To produce more plants like this one, should she crossbreed it with another plant or clone it? Why?

CONCEPT 2

Fertilization and meiosis alternate in sexual life cycles

A **life cycle** is the generation-to-generation sequence of stages in the reproductive history of an organism, from conception to production of its own offspring. In this section, we use humans as an example to track the behavior of chromosomes through the sexual life cycle. We begin by considering the chromosome count in human somatic cells and gametes. We will then explore how the behavior of chromosomes relates to the human life cycle and other types of sexual life cycles.

Sets of Chromosomes in Human Cells

In humans, each somatic cell has 46 chromosomes. During mitosis, the chromosomes become condensed enough to be visible under a light microscope. At this point, they can be distinguished from one another by their size, the position of their centromeres, and the pattern of colored bands produced by certain chromatin-binding stains.

Careful examination of a micrograph of the 46 human chromosomes from a single cell in mitosis reveals that there are two chromosomes of each of 23 types. This becomes clear when images of the chromosomes are arranged in pairs, starting with the longest chromosomes. The resulting ordered display is called a **karyotype** (Figure 3). The two chromosomes of a pair have the same length, centromere position, and staining pattern: These are called **homologous chromosomes** (or **homologs**, for short) or a **homologous pair**. Both chromosomes of each pair carry genes controlling the same inherited characters. For example, if a gene for eye color is situated at a particular locus on a certain chromosome, its homologous chromosome will also have a version of the eye color gene at the equivalent locus.

The two chromosomes referred to as X and Y are an important exception to the general pattern of homologous chromosomes in human somatic cells. Human females have a homologous pair of X chromosomes (XX), but males have one X and one Y chromosome (XY; see Figure 3). Only

▼ Figure 3 Research Method

Preparing a Karyotype

Application A karyotype is a display of condensed chromosomes arranged in pairs. Karyotyping can be used to screen for defective chromosomes or abnormal numbers of chromosomes associated with certain congenital disorders, such as Down syndrome.

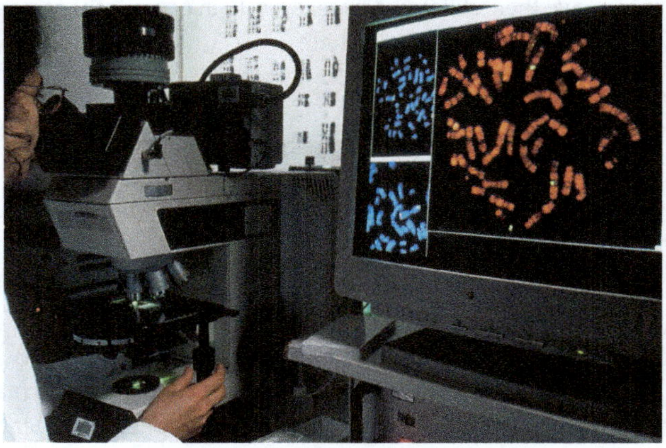

Technique Karyotypes are prepared from isolated somatic cells, which are treated with a drug to stimulate mitosis and then grown in culture for several days. Cells arrested when chromosomes are most highly condensed—in metaphase—are stained and then viewed with a microscope equipped with a digital camera. An image of the chromosomes is displayed on a computer monitor, and digital software is used to arrange them in pairs according to their appearance.

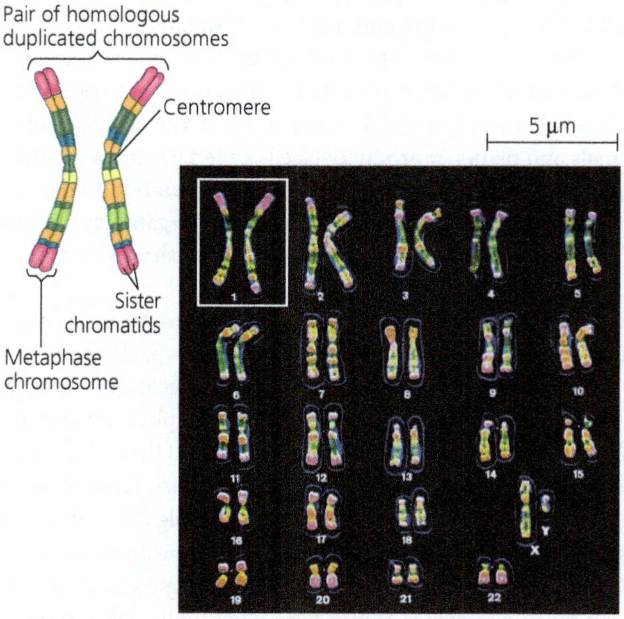

Results This karyotype shows the chromosomes from a human male, digitally colored to emphasize the chromosome banding patterns. The size of the chromosome, position of the centromere, and pattern of stained bands help identify specific chromosomes. Although difficult to discern in the karyotype, each metaphase chromosome consists of two closely attached sister chromatids (see the diagram of the boxed pair of homologous duplicated chromosomes).

small parts of the X and Y are homologous. Most of the genes carried on the X chromosome do not have counterparts on the tiny Y, and the Y chromosome has genes lacking on the X. Because they determine an individual's sex, the X and Y chromosomes are called **sex chromosomes**. The other chromosomes are called **autosomes**.

The occurrence of pairs of homologous chromosomes in each human somatic cell is a consequence of our sexual origins. We inherit one chromosome of each pair from each parent. Thus, the 46 chromosomes in our somatic cells are actually two sets of 23 chromosomes—a maternal set (from our mother) and a paternal set (from our father). The number of chromosomes in a single set is represented by n. Any cell with two chromosome sets is called a **diploid cell** and has a diploid number of chromosomes, abbreviated $2n$. For humans, the diploid number is 46 ($2n = 46$), the number of chromosomes in our somatic cells. In a cell in which DNA synthesis has occurred, all the chromosomes are duplicated, and therefore each consists of two identical sister chromatids, associated closely at the centromere and along the arms. (Even though the chromosomes are duplicated, we still say the cell is diploid, or $2n$, because it has only two sets of information.) **Figure 4** helps clarify the various terms that we use to describe duplicated chromosomes in a diploid cell.

Unlike somatic cells, gametes contain a single set of chromosomes. Such cells are called **haploid cells**, and each has a haploid number of chromosomes (n). For humans, the haploid number is 23 ($n = 23$). The set of 23 consists of the 22 autosomes plus a single sex chromosome. An unfertilized egg contains an X chromosome, but a sperm may contain an X or a Y chromosome.

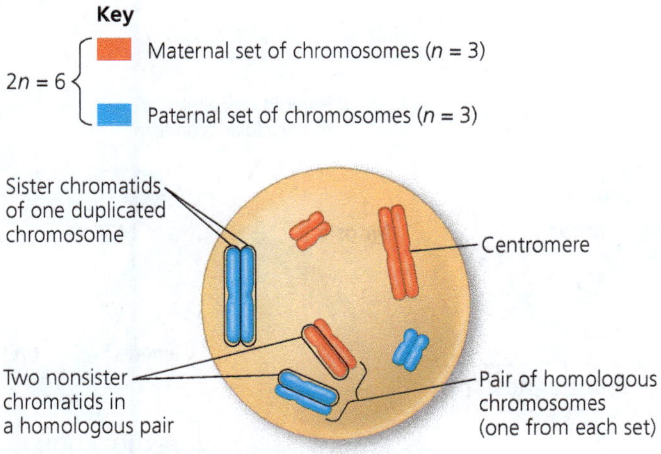

▲ **Figure 4 Describing chromosomes.** A cell from an organism with a diploid number of 6 ($2n = 6$) is depicted here following chromosome duplication and condensation. Each of the six duplicated chromosomes consists of two sister chromatids associated closely along their lengths. Each homologous pair is composed of one chromosome from the maternal set (red) and one from the paternal set (blue). Each set is made up of three chromosomes in this example (long, medium, and short). Together, one maternal and one paternal chromatid in a pair of homologous chromosomes are called nonsister chromatids.

? *How many sets of chromosomes are present in this diagram? How many pairs of homologs are present?*

Each sexually reproducing species has a characteristic diploid and haploid number. For example, the fruit fly *Drosophila melanogaster* has a diploid number ($2n$) of 8 and a haploid number (n) of 4, while for dogs, $2n$ is 78 and n is 39.

Now let's consider chromosome behavior during sexual life cycles. We'll use the human life cycle as an example.

Behavior of Chromosome Sets in the Human Life Cycle

The human life cycle begins when a haploid sperm from the father fuses with a haploid egg from the mother **(Figure 5)**. This union of gametes, culminating in fusion of their nuclei, is called **fertilization**. The resulting fertilized egg, or **zygote**, is diploid because it contains two haploid sets of chromosomes bearing genes representing the maternal and paternal family lines. As a human develops into a sexually mature adult, mitosis of the zygote and its descendant cells generates all the somatic cells of the body. Both chromosome sets in the zygote and all the genes they carry are passed with precision to the somatic cells.

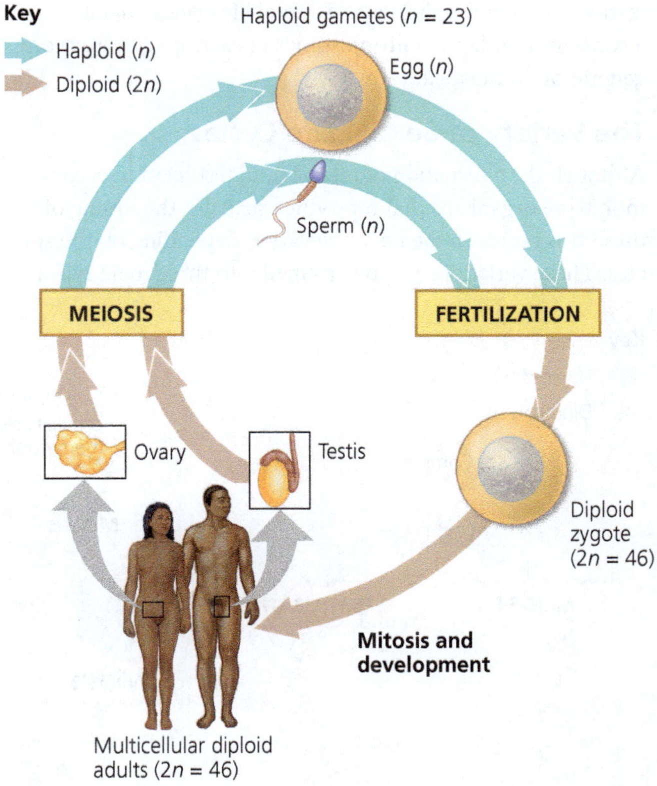

▲ **Figure 5 The human life cycle.** In each generation, the number of chromosome sets doubles at fertilization but is halved during meiosis. For humans, the number of chromosomes in a haploid cell is 23, consisting of one set ($n = 23$); the number of chromosomes in the diploid zygote and all somatic cells arising from it is 46, consisting of two sets ($2n = 46$).

This figure introduces a color code that will be used for other life cycles later in this text. The aqua arrows identify haploid stages of a life cycle, and the tan arrows identify diploid stages.

The only cells of the human body not produced by mitosis are the gametes, which develop from specialized cells called *germ cells* in the gonads—ovaries in females and testes in males (see Figure 5). Imagine what would happen if human gametes were made by mitosis: They would be diploid like the somatic cells. At the next round of fertilization, when two gametes fused, the normal chromosome number of 46 would double to 92, and each subsequent generation would double the number of chromosomes yet again. This does not happen, however, because in sexually reproducing organisms, gamete formation involves a type of cell division called **meiosis**. This type of cell division reduces the number of sets of chromosomes from two to one in the gametes, counterbalancing the doubling that occurs at fertilization. As a result of meiosis, each human sperm and egg is haploid ($n = 23$). Fertilization restores the diploid condition by combining two haploid sets of chromosomes, and the human life cycle is repeated, generation after generation (see Figure 5).

In general, the steps of the human life cycle are typical of many sexually reproducing animals. Indeed, the processes of fertilization and meiosis are the hallmarks of sexual reproduction in plants, fungi, and protists as well as in animals. Fertilization and meiosis alternate in sexual life cycles, maintaining a constant number of chromosomes in each species from one generation to the next.

The Variety of Sexual Life Cycles

Although the alternation of meiosis and fertilization is common to all organisms that reproduce sexually, the timing of these two events in the life cycle varies, depending on the species. These variations can be grouped into three main types of life cycles. In the type that occurs in humans and most other animals, gametes are the only haploid cells **(Figure 6a)**. Meiosis occurs in germ cells during the production of gametes, which undergo no further cell division prior to fertilization. After fertilization, the diploid zygote divides by mitosis, producing a multicellular organism that is diploid.

Plants and some species of algae exhibit a second type of life cycle called **alternation of generations** **(Figure 6b)**. This type includes both diploid and haploid stages that are multicellular. The multicellular diploid stage is called the *sporophyte*. Meiosis in the sporophyte produces haploid cells called *spores*. Unlike a gamete, a haploid spore doesn't fuse with another cell but divides mitotically, generating a multicellular haploid stage called the *gametophyte*. Cells of the gametophyte give rise to gametes by mitosis. Fusion of two haploid gametes at fertilization results in a diploid zygote, which develops into the next sporophyte generation. Therefore, in this type of life cycle, the sporophyte generation produces a gametophyte as its offspring, and the gametophyte generation produces the next sporophyte generation. The term *alternation of generations* fits well as a name for this type of life cycle.

A third type of life cycle occurs in most fungi and some protists, including some algae **(Figure 6c)**. After gametes fuse and form a diploid zygote, meiosis occurs without a multicellular diploid offspring developing. Meiosis produces not gametes but haploid cells that then divide by mitosis and give rise to either unicellular descendants or a haploid multicellular adult organism. Subsequently, the haploid organism carries out further mitoses, producing the cells that develop into gametes. The only diploid stage found in these species is the single-celled zygote.

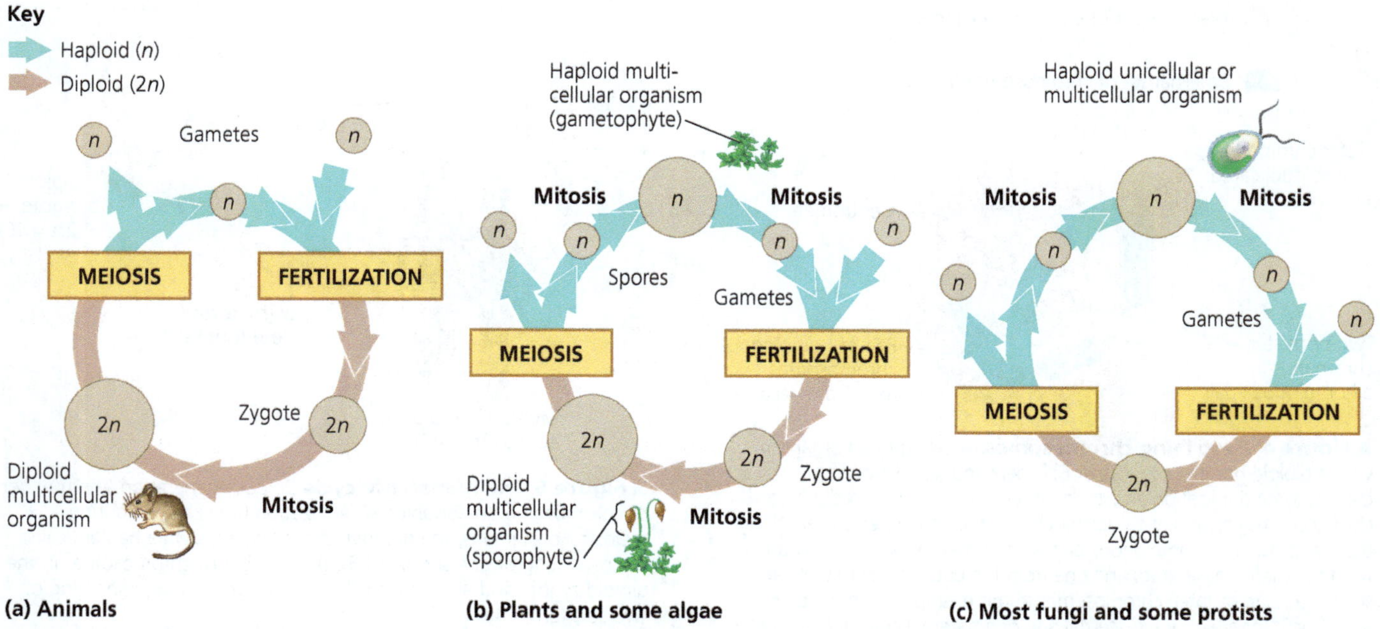

▲ **Figure 6 Three types of sexual life cycles.** The common feature of all three cycles is the alternation of meiosis and fertilization, key events that contribute to genetic variation among offspring. The cycles differ in the timing of these two key events. (Small circles are cells; large circles are organisms.)

Note that *either* haploid or diploid cells can divide by mitosis, depending on the type of life cycle. Only diploid cells, however, can undergo meiosis because haploid cells have a single set of chromosomes that cannot be further reduced. Though the three types of sexual life cycles differ in the timing of meiosis and fertilization, they share a fundamental result: genetic variation among offspring. A closer look at meiosis will reveal the sources of this variation.

CONCEPT CHECK 2

1. MAKE CONNECTIONS In Figure 4, how many DNA molecules (double helices) are present? What is the haploid number of this cell? Is a set of chromosomes haploid or diploid?
2. In the karyotype shown in Figure 3, how many pairs of chromosomes are present? How many sets?
3. Each sperm of a pea plant contains seven chromosomes. What are the haploid and diploid numbers for this species?
4. WHAT IF? A certain eukaryote lives as a unicellular organism, but during environmental stress, it produces gametes. The gametes fuse, and the resulting zygote undergoes meiosis, generating new single cells. What type of organism could this be?

CONCEPT 3

Meiosis reduces the number of chromosome sets from diploid to haploid

Several steps of meiosis closely resemble corresponding steps in mitosis. Meiosis, like mitosis, is preceded by the duplication of chromosomes. However, this single duplication is followed by not one but two consecutive cell divisions, called **meiosis I** and **meiosis II**. These two divisions result in four daughter cells (rather than the two daughter cells of mitosis), each with only half as many chromosomes as the parent cell—one set, rather than two.

The Stages of Meiosis

The overview of meiosis in **Figure 7** shows, for a single pair of homologous chromosomes in a diploid cell, that both members of the pair are duplicated and the copies sorted into four haploid daughter cells. Recall that sister chromatids are two copies of *one* chromosome, closely associated all along their lengths; this association is called *sister chromatid cohesion*. Together, the sister chromatids make up one duplicated chromosome (see Figure 4). In contrast, the two chromosomes of a homologous pair are individual chromosomes that were inherited from different parents. Homologs appear alike in the microscope, but they may have different versions of genes at corresponding loci; each version is called an *allele* of that gene. For example, one chromosome may have an allele for freckles, but the homologous chromosome may have an allele for the absence of freckles at the same locus. Homologs are not associated with each other in any obvious way except during meiosis, as you will soon see.

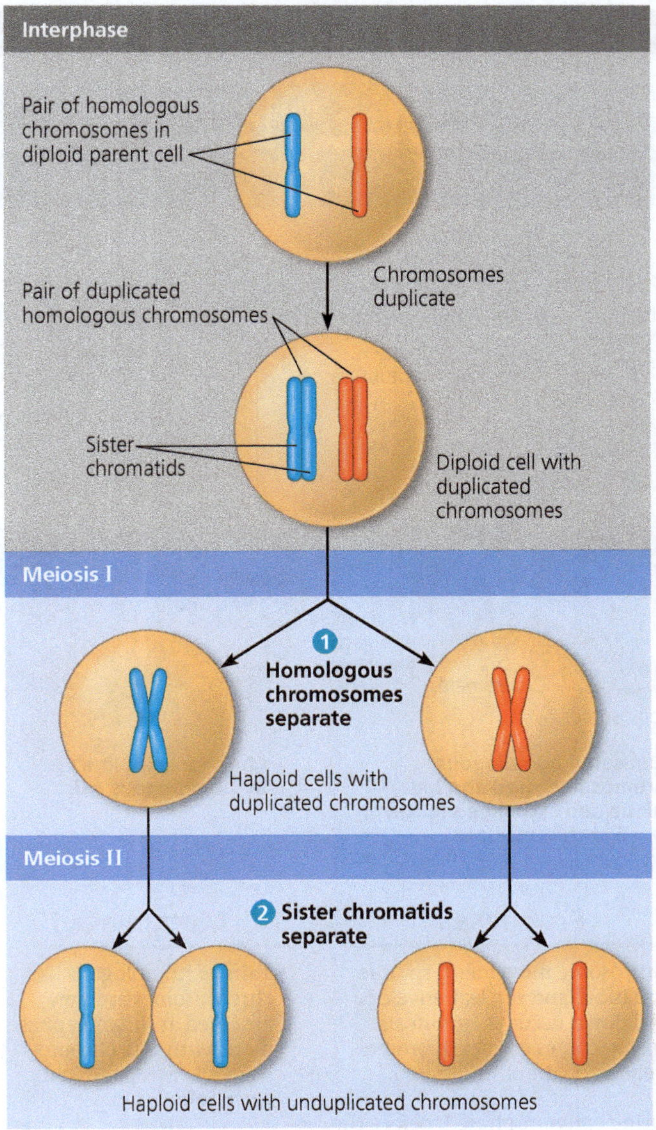

▲ **Figure 7 Overview of meiosis: how meiosis reduces chromosome number.** After the chromosomes duplicate in interphase, the diploid cell divides *twice*, yielding four haploid daughter cells. This overview tracks just one pair of homologous chromosomes, which for the sake of simplicity are drawn in the condensed state throughout. (They would not normally be condensed during interphase.) The red chromosome was inherited from the female parent, the blue chromosome from the male parent.

DRAW IT *Redraw the cells in this figure using a simple double helix to represent each DNA molecule.*

Figure 8 describes in detail the stages of the two divisions of meiosis for an animal cell whose diploid number is 6. Meiosis halves the total number of chromosomes in a very specific way, reducing the number of sets from two to one, with each daughter cell receiving one set of chromosomes.Study Figure 8 thoroughly before going on.

▼ **Figure 8** **Exploring Meiosis in an Animal Cell**

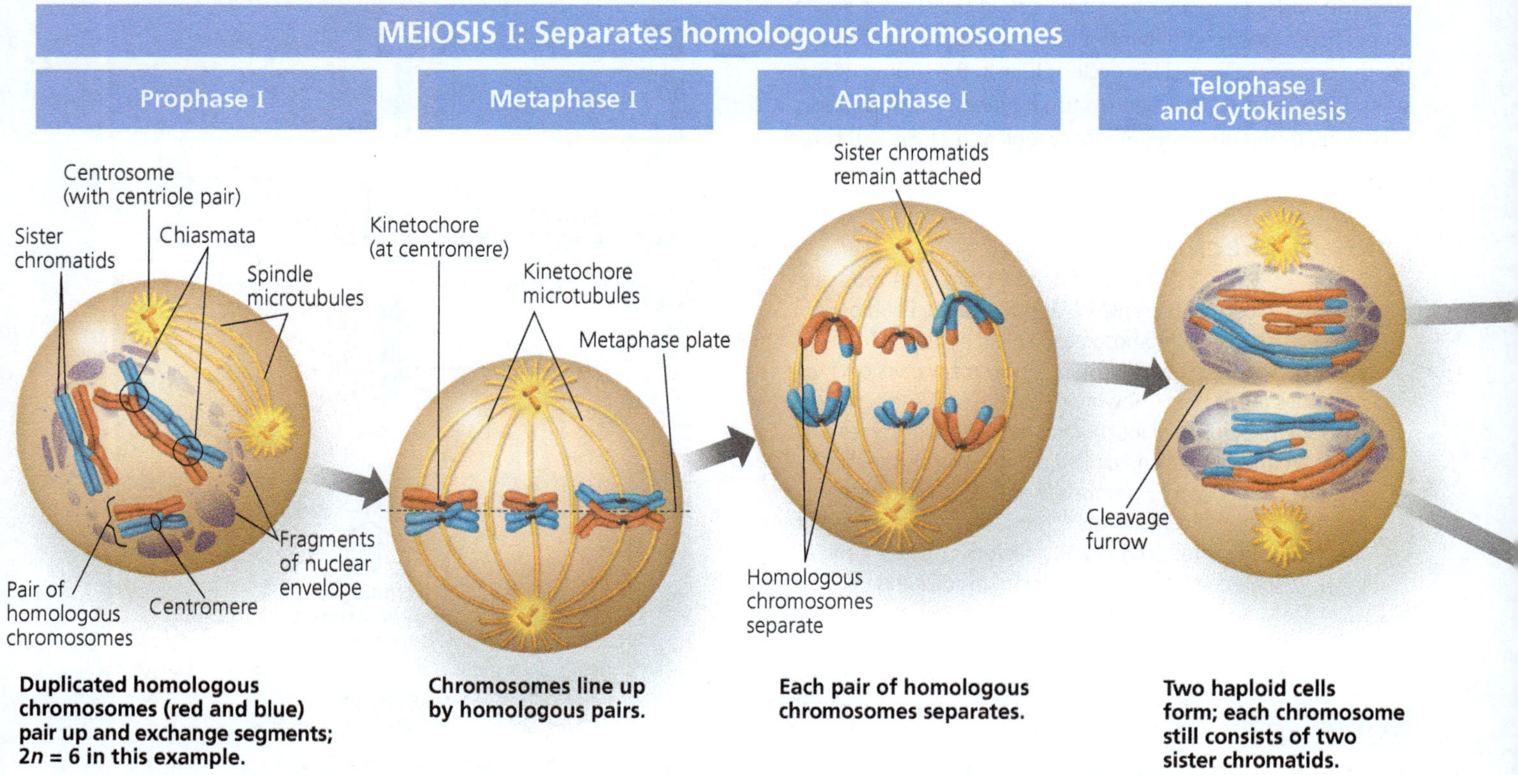

Prophase I

- Centrosome movement, spindle formation, and nuclear envelope breakdown occur as in mitosis. Chromosomes condense progressively throughout prophase I.
- During early prophase I, before the stage shown above, each chromosome pairs with its homolog, aligned gene by gene, and **crossing over** occurs: The DNA molecules of nonsister chromatids are broken (by proteins) and are rejoined to each other.
- At the stage shown above, each homologous pair has one or more X-shaped regions called **chiasmata** (singular, *chiasma*), where crossovers have occurred.
- Later in prophase I, after the stage shown above, microtubules from one pole or the other attach to the kinetochores, one at the centromere of each homolog. (Each homolog acts as if it has a single kinetochore.) Microtubules move the homologous pairs toward the metaphase plate (see the metaphase I diagram).

Metaphase I

- Pairs of homologous chromosomes are now arranged at the metaphase plate, with one chromosome of each pair facing each pole.
- Both chromatids of one homolog are attached to kinetochore microtubules from one pole; the chromatids of the other homolog are attached to microtubules from the opposite pole.

Anaphase I

- Breakdown of proteins that are responsible for sister chromatid cohesion along chromatid arms allows homologs to separate.
- The homologs move toward opposite poles, guided by the spindle apparatus.
- Sister chromatid cohesion persists at the centromere, causing chromatids to move as a unit toward the same pole.

Telophase I and Cytokinesis

- When telophase I begins, each half of the cell has a complete haploid set of duplicated chromosomes. Each chromosome is composed of two sister chromatids; one or both chromatids include regions of nonsister chromatid DNA.
- Cytokinesis (division of the cytoplasm) usually occurs simultaneously with telophase I, forming two haploid daughter cells.
- In animal cells like these, a cleavage furrow forms. (In plant cells, a cell plate forms.)
- In some species, chromosomes decondense and nuclear envelopes form.
- No chromosome duplication occurs between meiosis I and meiosis II.

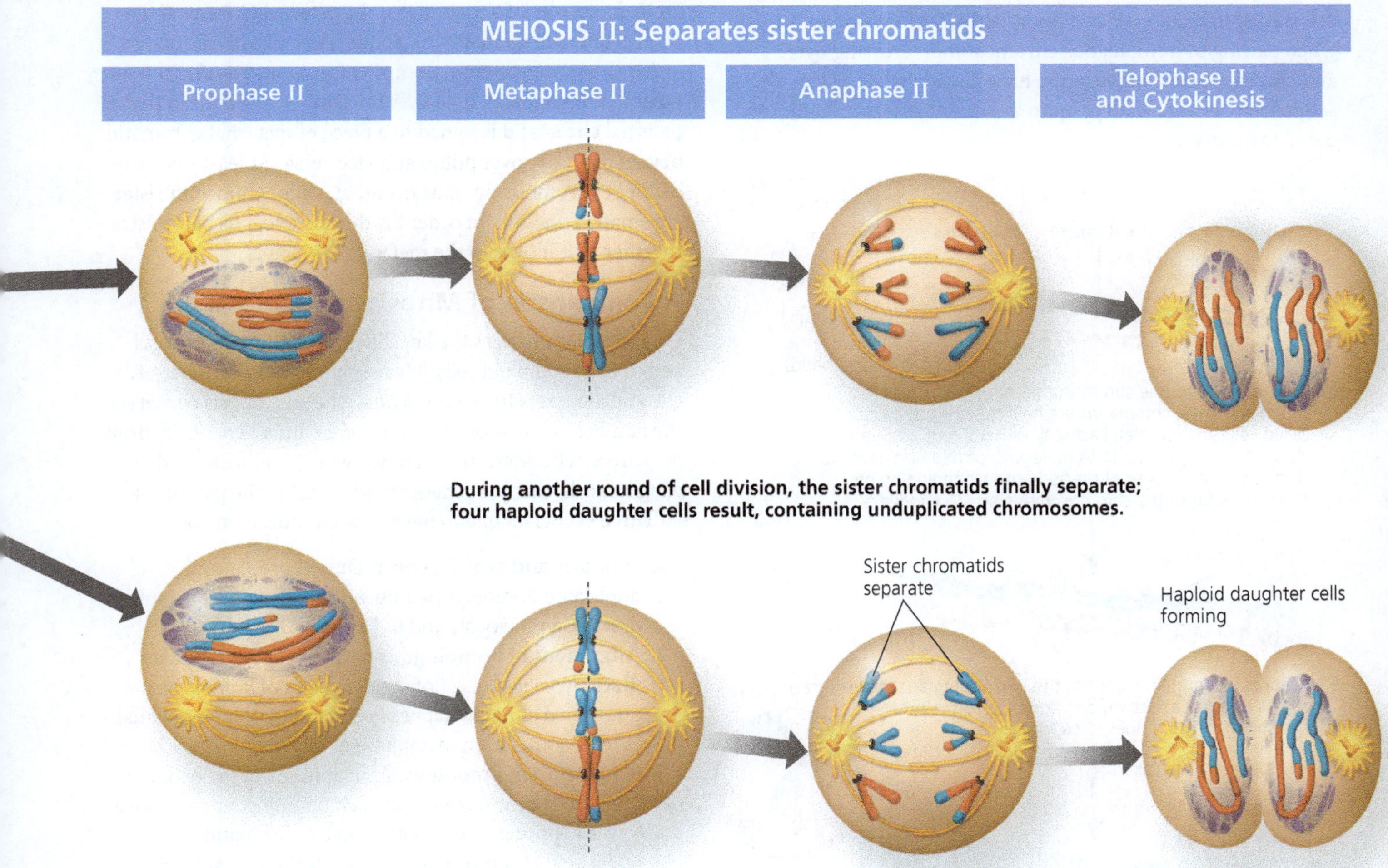

Prophase II

- A spindle apparatus forms.
- In late prophase II (not shown here), chromosomes, each still composed of two chromatids associated at the centromere, are moved by microtubules toward the metaphase II plate.

Metaphase II

- The chromosomes are positioned at the metaphase plate as in mitosis.
- Because of crossing over in meiosis I, the two sister chromatids of each chromosome are not genetically identical.
- The kinetochores of sister chromatids are attached to microtubules extending from opposite poles.

Anaphase II

- Breakdown of proteins holding the sister chromatids together at the centromere allows the chromatids to separate. The chromatids move toward opposite poles as individual chromosomes.

Telophase II and Cytokinesis

- Nuclei form, the chromosomes begin decondensing, and cytokinesis occurs.
- The meiotic division of one parent cell produces four daughter cells, each with a haploid set of (unduplicated) chromosomes.
- The four daughter cells are genetically distinct from one another and from the parent cell.

MAKE CONNECTIONS *Look at Figure 7 and imagine the two daughter cells undergoing another round of mitosis, yielding four cells. Compare the number of chromosomes in each of those four cells, after mitosis, with the number in each cell in Figure 8, after meiosis. Explain how the process of meiosis results in this difference, even though meiosis also includes two cell divisions.*

ANIMATION Visit the Study Area in **MasteringBiology** for the BioFlix® 3-D Animation on Meiosis.

Crossing Over and Synapsis During Prophase I

Prophase I of meiosis is a very busy time. The prophase I cell shown in Figure 8 is at a point fairly late in prophase I, when pairing of homologous chromosomes, crossing over, and chromosome condensation have already taken place. The sequence of events leading up to that point is shown in more detail in **Figure 9**.

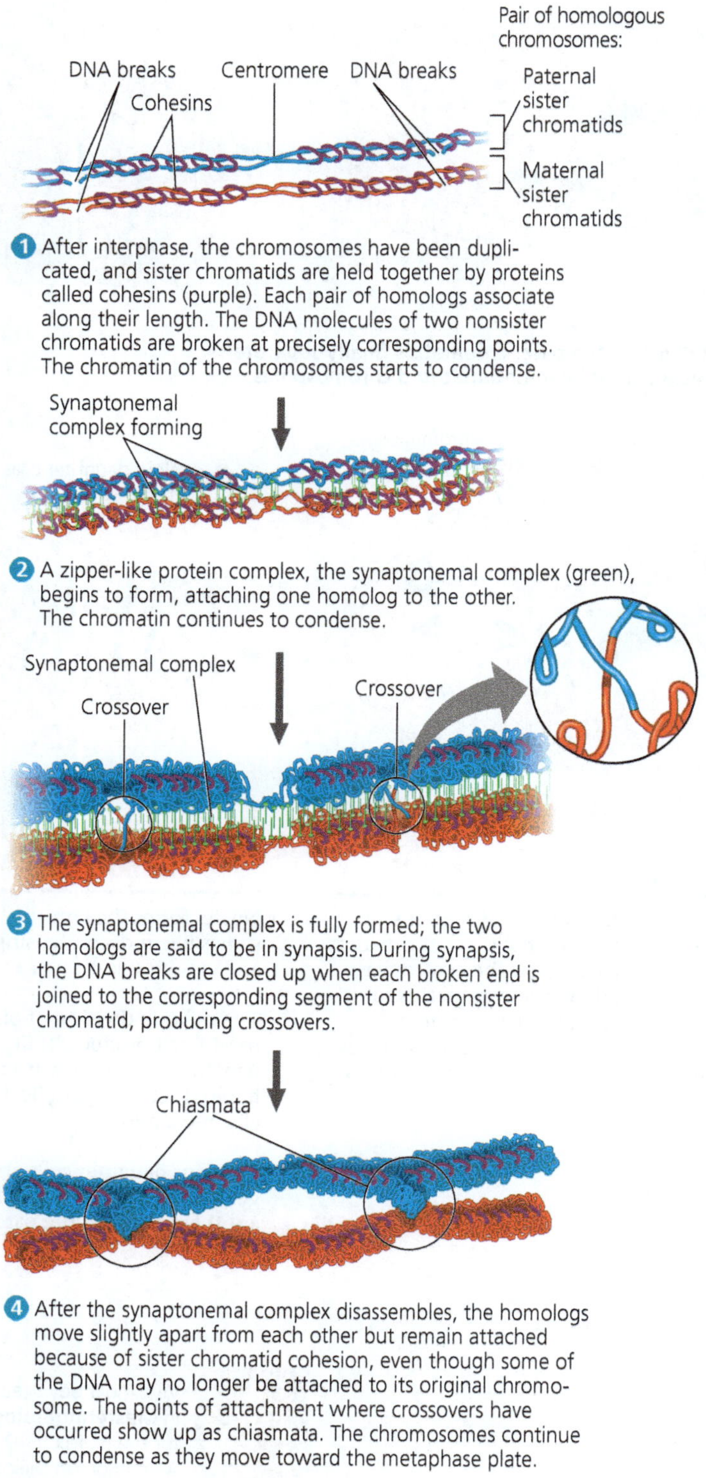

▲ **Figure 9 Crossing over and synapsis in prophase I.**

As shown in this figure, the two members of a homologous pair associate along their length, aligned allele by allele. The DNA molecules of a maternal and a paternal chromatid are broken at precisely matching points. A zipper-like structure called the **synaptonemal complex** forms, and during this attachment (**synapsis**), the DNA breaks are closed up so that a paternal chromatid is joined to a piece of maternal chromatid beyond the crossover point, and vice versa. At least one crossover per chromosome must occur, in conjunction with sister chromatid cohesion, in order for the homologous pair to stay together as it moves to the metaphase I plate.

A Comparison of Mitosis and Meiosis

Figure 10 summarizes the key differences between meiosis and mitosis in diploid cells. Meiosis reduces the number of chromosome sets from two to one, whereas mitosis conserves the number. Meiosis produces cells that differ genetically from the parent cell and from each other, whereas mitosis produces daughter cells that are genetically identical to the parent cell.

Three events unique to meiosis occur during meiosis I:

1. **Synapsis and crossing over.** During prophase I, duplicated homologs pair up, and crossing over occurs as described previously and in Figure 9.
2. **Alignment of homologous pairs at the metaphase plate.** At metaphase I of meiosis, pairs of homologs are positioned at the metaphase plate, rather than individual chromosomes as in metaphase of mitosis.
3. **Separation of homologs.** At anaphase I of meiosis, the duplicated chromosomes of each homologous pair move toward opposite poles, but the sister chromatids of each duplicated chromosome remain attached. In anaphase of mitosis, by contrast, sister chromatids separate.

Sister chromatids stay together due to sister chromatid cohesion, mediated by cohesion proteins. In mitosis, this attachment lasts until the end of metaphase, when enzymes cleave the cohesins, freeing the sister chromatids to move to opposite poles of the cell. In meiosis, sister chromatid cohesion is released in two steps, one at the start of anaphase I and one at anaphase II. In metaphase I, the two homologs of each pair are held together by cohesion between sister chromatid arms in regions beyond points of crossing over, where stretches of sister chromatids now belong to different chromosomes. The combination of crossing over and sister chromatid cohesion along the arms results in the formation of a chiasma. Chiasmata hold homologs together as the spindle forms for the first meiotic division. At the onset of anaphase I, the release of cohesion along sister chromatid arms allows homologs to separate. At anaphase II, the release of sister chromatid cohesion at the centromeres allows the sister chromatids to separate. Thus, sister chromatid cohesion and crossing over, acting together, play an essential role in the lining up of chromosomes by homologous pairs at metaphase I.

Meiosis I reduces the number of chromosome sets per cell from two (diploid) to one set (haploid). During the second

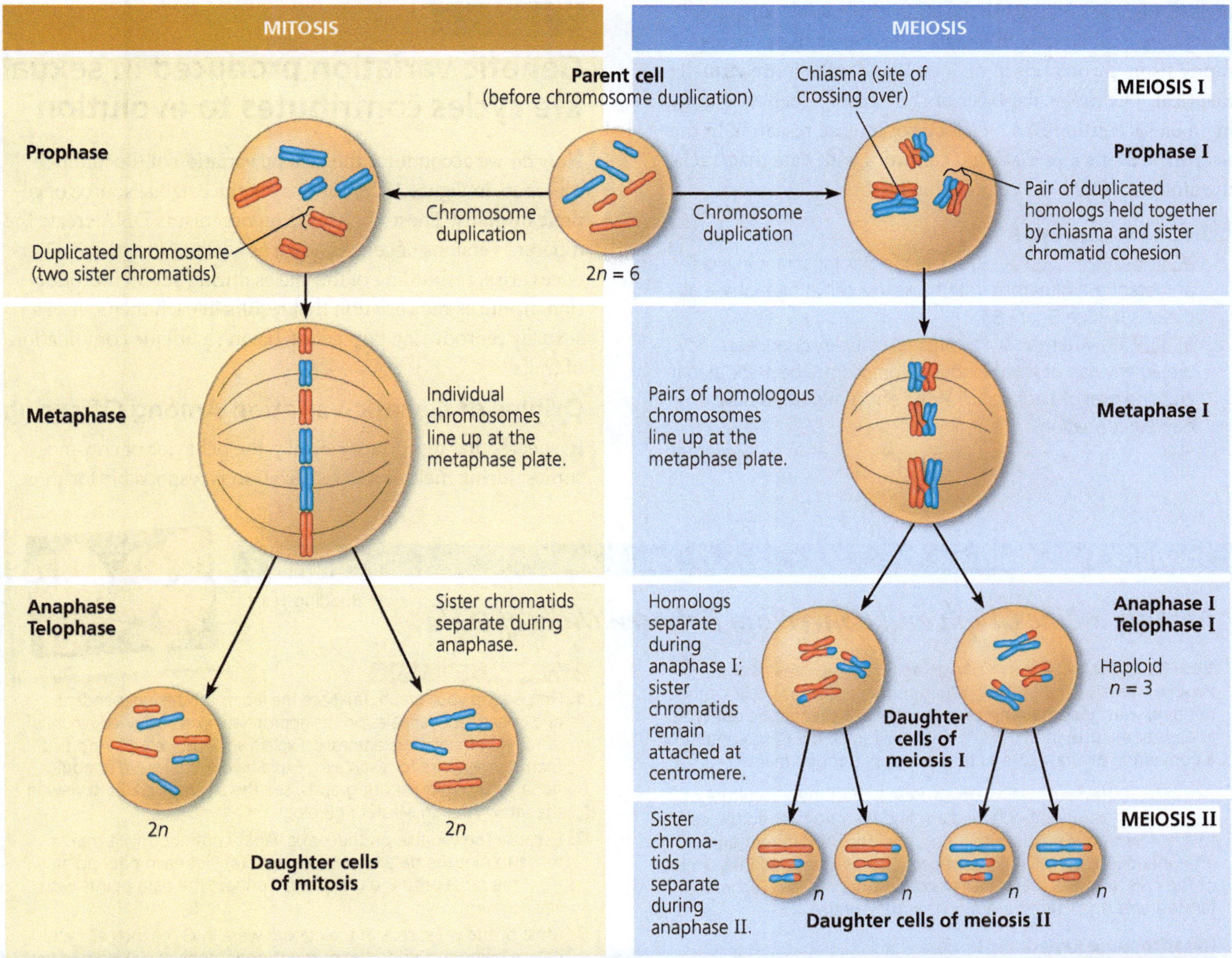

SUMMARY

Property	**Mitosis** (occurs in both diploid and haploid cells)	**Meiosis** (can only occur in diploid cells)
DNA replication	Occurs during interphase before mitosis begins	Occurs during interphase before meiosis I begins
Number of divisions	One, including prophase, prometaphase, metaphase, anaphase, and telophase	Two, each including prophase, metaphase, anaphase, and telophase
Synapsis of homologous chromosomes	Does not occur	Occurs during prophase I along with crossing over between nonsister chromatids; resulting chiasmata hold pairs together due to sister chromatid cohesion
Number of daughter cells and genetic composition	Two, each genetically identical to the parent cell, with the same number of chromosomes	Four, each haploid (n); genetically different from the parent cell and from each other
Role in the animal or plant body	Enables multicellular animal or plant (gametophyte or sporophyte) to arise from a single cell; produces cells for growth, repair, and, in some species, asexual reproduction; produces gametes in the gametophyte plant	Produces gametes (in animals) or spores (in the sporophyte plant); reduces number of chromosome sets by half and introduces genetic variability among the gametes or spores

▲ **Figure 10 A comparison of mitosis and meiosis.**

DRAW IT *Could any other combinations of chromosomes be generated during meiosis II from the specific cells shown in telophase I? Explain. (Hint: Draw the cells as they would appear in metaphase II.)*

meiotic division, meiosis II, the sister chromatids separate, producing haploid daughter cells. The mechanisms for separating sister chromatids in meiosis II and mitosis are virtually identical. The molecular basis of chromosome behavior during meiosis continues to be a focus of intense research. In the **Scientific Skills Exercise**, you can work with data that tracks the amount of DNA in cells during meiosis.

CONCEPT CHECK 3

1. **MAKE CONNECTIONS** Compare the chromosomes in a cell at metaphase of mitosis with those in a cell at metaphase of meiosis II. (See Figure 8.)
2. **WHAT IF?** After the synaptonemal complex disappears, how would any pair of homologous chromosomes be associated if crossing over did not occur? What effect might this have on gamete formation?

CONCEPT 4

Genetic variation produced in sexual life cycles contributes to evolution

How do we account for the genetic variation of the family members in Figure 1? Mutations are the original source of genetic diversity. These changes in an organism's DNA create the different versions of genes known as alleles. Once these differences arise, reshuffling of the alleles during sexual reproduction produces the variation that results in each member of a sexually reproducing population having a unique combination of traits.

Origins of Genetic Variation Among Offspring

In species that reproduce sexually, the behavior of chromosomes during meiosis and fertilization is responsible for most

Scientific Skills Exercise

Making a Line Graph and Converting Between Units of Data

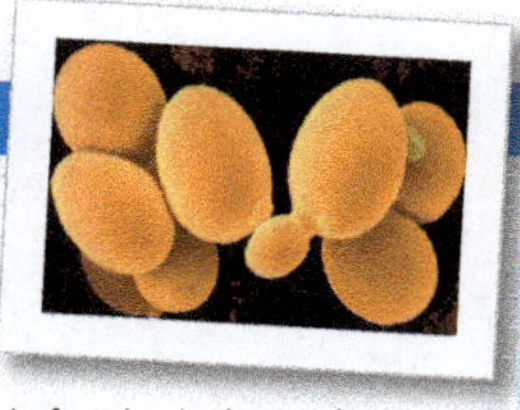

▶ Budding yeast

How Does DNA Content Change as Budding Yeast Cells Proceed Through Meiosis? When nutrients are low, cells of the budding yeast (*Saccharomyces cerevisiae*) exit the mitotic cell cycle and enter meiosis. In this exercise you will track the DNA content of a population of yeast cells as they progress through meiosis.

How the Experiment Was Done Researchers grew a culture of yeast cells in a nutrient-rich medium and then transferred the cells to a nutrient-poor medium to induce meiosis. At different times after induction, the DNA content per cell was measured in a sample of the cells, and the average DNA content per cell was recorded in femtograms (fg; 1 femtogram = 1×10^{-15} gram).

Data from the Experiment

Time After Induction (hours)	Average Amount of DNA per Cell (fg)
0.0	24.0
1.0	24.0
2.0	40.0
3.0	47.0
4.0	47.5
5.0	48.0
6.0	48.0
7.0	47.5
7.5	25.0
8.0	24.0
9.0	23.5
9.5	14.0
10.0	13.0
11.0	12.5
12.0	12.0
13.0	12.5
14.0	12.0

INTERPRET THE DATA

1. First, set up your graph. (a) Place the labels for the independent and dependent variables on the appropriate axes, followed by units of measurement in parentheses. Explain your choices. (b) Add tick marks and values for each axis. Explain your choices. (For additional information about graphs, see the Scientific Skills Review in the Study Area in MasteringBiology.)
2. Because the variable on the *x*-axis varies continuously, it makes sense to plot the data on a line graph. (a) Plot each data point from the table onto the graph. (b) Connect the data points with line segments.
3. Most of the yeast cells in the culture were in G_1 of the cell cycle before being moved to the nutrient-poor medium. (a) How many femtograms of DNA are there in each yeast cell in G_1? Estimate this value from the data in your graph. (b) How many femtograms of DNA should be present in each cell in G_2? At the end of meiosis I (MI)? At the end of meiosis II (MII)? (See Figure 7.) (c) Using these values as a guideline, distinguish the different phases by inserting vertical dashed lines in the graph between phases and label each phase (G_1, S, G_2, MI, MII). You can figure out where to put the dividing lines based on what you know about the DNA content of each phase (see Figure 7). (d) Think carefully about the point where the line at the highest value begins to slope downward. What specific point of meiosis does this "corner" represent? What stage(s) correspond to the downward sloping line?
4. Given the fact that 1 fg of DNA = 9.78×10^5 base pairs (on average), you can convert the amount of DNA per cell to the length of DNA in numbers of base pairs. (a) Calculate the number of base pairs of DNA in the haploid yeast genome. Express your answer in millions of base pairs (Mb), a standard unit for expressing genome size. Show your work. (b) How many base pairs per minute were synthesized during the S phase of these yeast cells?

MB A version of this Scientific Skills Exercise can be assigned in MasteringBiology.

Further Reading G. Simchen, Commitment to meiosis: What determines the mode of division in budding yeast? *BioEssays* 31:169–177 (2009). doi:10.1002/bies.200800124 of the variation that arises in each generation. Three mechanisms contribute to the genetic variation arising from sexual reproduction: independent assortment of chromosomes, crossing over, and random fertilization.

Independent Assortment of Chromosomes

One aspect of sexual reproduction that generates genetic variation is the random orientation of pairs of homologous chromosomes at metaphase of meiosis I. At metaphase I, the homologous pairs, each consisting of one maternal and one paternal chromosome, are situated at the metaphase plate. (Note that the terms *maternal* and *paternal* refer, respectively, to whether the chromosome in question was contributed by the mother or the father of the individual whose cells are undergoing meiosis.) Each pair may orient with either its maternal or paternal homolog closer to a given pole—its orientation is as random as the flip of a coin. Thus, there is a 50% chance that a given daughter cell of meiosis I will get the maternal chromosome of a certain homologous pair and a 50% chance that it will get the paternal chromosome.

Because each pair of homologous chromosomes is positioned independently of the other pairs at metaphase I, the first meiotic division results in each pair sorting its maternal and paternal homologs into daughter cells independently of every other pair. This is called *independent assortment.* Each daughter cell represents one outcome of all possible combinations of maternal and paternal chromosomes. As shown in **Figure 11**, the number of combinations possible for daughter cells formed by meiosis of a diploid cell with two pairs of homologous chromosomes ($n = 2$) is four: two possible arrangements for the first pair times two possible arrangements for the second pair. Note that only two of the four combinations of daughter cells shown in the figure would result from meiosis of a *single* diploid cell, because a single parent cell would have one or the other possible chromosomal arrangement at metaphase I, but not both. However, the population of daughter cells resulting from meiosis of a large number of diploid cells contains all four types in approximately equal numbers. In the case of $n = 3$, eight combinations of chromosomes are possible for daughter cells. More generally, the number of possible combinations when chromosomes sort independently during meiosis is 2^n, where n is the haploid number of the organism.

In the case of humans ($n = 23$), the number of possible combinations of maternal and paternal chromosomes in the resulting gametes is 2^{23}, or about 8.4 million. Each gamete that you produce in your lifetime contains one of roughly 8.4 million possible combinations of chromosomes.

Crossing Over

As a consequence of the independent assortment of chromosomes during meiosis, each of us produces a collection of gametes differing greatly in their combinations of the chromosomes we inherited from our two parents. Figure 11 suggests that each chromosome in a gamete is exclusively maternal or paternal in origin. In fact, this is *not* the case because crossing over produces **recombinant chromosomes,** individual chromosomes that carry genes (DNA) from two different parents. In meiosis in humans, an average of one to three crossover events occur per chromosome pair, depending on the size of the chromosomes and the position of their centromeres. As you learned in Figure 9, crossing over produces chromosomes with new combinations of maternal and paternal alleles. At metaphase II, chromosomes that contain one or more recombinant chromatids can be oriented in two nonequivalent ways with respect to other chromosomes because their sister chromatids are no longer identical. The possible arrangements of nonidentical sister chromatids during meiosis II increase the number of genetic types of daughter cells that can result from meiosis.

▶ Figure 11 The independent assortment of homologous chromosomes in meiosis.

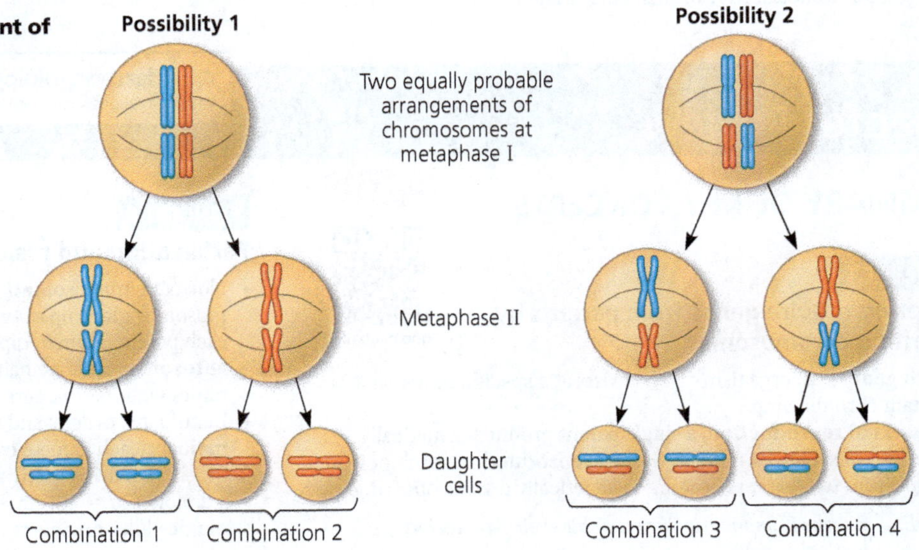

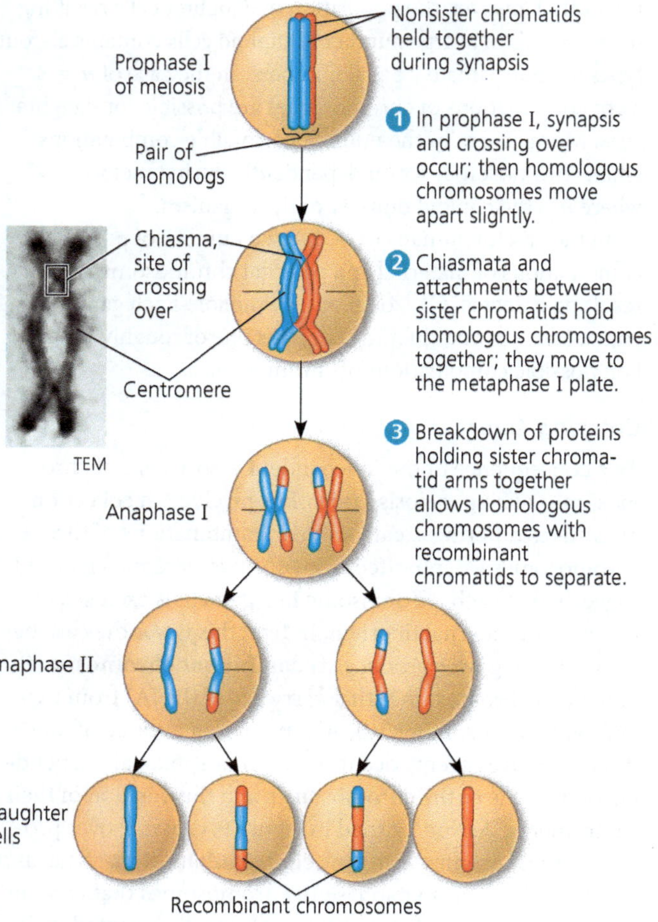

▲ **Figure 12 The results of crossing over during meiosis.**

Figure 12 provides an overview of how crossing over increases genetic variation by producing recombinant chromosomes that include DNA from two parents.

Random Fertilization

The random nature of fertilization adds to the genetic variation arising from meiosis. In humans, each male and female gamete represents one of about 8.4 million (2^{23}) possible chromosome combinations due to independent assortment. The fusion of a male gamete with a female gamete during fertilization will produce a zygote with any of about 70 trillion ($2^{23} \times 2^{23}$) diploid combinations. With the variation from crossing over, the number of possibilities is astronomical. You really *are* unique.

The Evolutionary Significance of Genetic Variation Within Populations

EVOLUTION How does the genetic variation in a population relate to evolution? A population evolves through differential reproductive success of its members. On average, those best suited to the environment leave more offspring, thereby transmitting their genes. Thus, natural selection leads to an increase in genetic variations favored by the environment. As the environment changes, the population may survive if some members can cope with the new conditions. Mutations, the original source of different alleles, are mixed and matched during meiosis. New combinations of alleles may work better than those that previously prevailed.

We have seen how sexual reproduction greatly increases the genetic variation present in a population. Although Darwin realized that heritable variation makes evolution possible, he could not explain why offspring resemble—but are not identical to—their parents. Ironically, Gregor Mendel, a contemporary of Darwin, published a theory of inheritance that helps explain genetic variation, but his discoveries had no impact on biologists until 1900, more than 15 years after Darwin (1809–1882) and Mendel (1822–1884) had died.

CONCEPT CHECK 4

1. What is the source of variation among alleles of a gene?
2. WHAT IF? If maternal and paternal chromatids have the identical two alleles for every gene, will crossing over lead to genetic variation?

Chapter Review

Go to MasteringBiology® for Assignments, the eText, and the Study Area with Animations, Activities, Vocab Self-Quiz, and Practice Tests.

SUMMARY OF KEY CONCEPTS

VOCAB SELF-QUIZ
goo.gl/gbai8v

CONCEPT 1

Offspring acquire genes from parents by inheriting chromosomes

- Each **gene** in an organism's DNA exists at a specific **locus** on a certain chromosome.
- In **asexual reproduction**, a single parent produces genetically identical offspring by mitosis. **Sexual reproduction** combines genes from two parents, leading to genetically diverse offspring.

? *Explain why human offspring resemble their parents but are not identical to them.*

CONCEPT 2

Fertilization and meiosis alternate in sexual life cycles

- Normal human **somatic cells** are **diploid**. They have 46 chromosomes made up of two sets of 23 chromosomes, one set from each parent. Human diploid cells have 22 homologous pairs of **autosomes** and one pair of **sex chromosomes**; the latter determines whether the person is female (XX) or male (XY).
- In humans, ovaries and testes produce haploid gametes by **meiosis**, each gamete containing a single set of 23 chromosomes ($n = 23$). During **fertilization**, an egg and sperm unite, forming a diploid ($2n = 46$) single-celled **zygote**, which develops into a multicellular organism by mitosis.

- Sexual **life cycles** differ in the timing of meiosis relative to fertilization and in the point(s) of the cycle at which a multicellular organism is produced by mitosis.

? *Compare the life cycles of animals and plants, mentioning their similarities and differences.*

CONCEPT 3

Meiosis reduces the number of chromosome sets from diploid to haploid

- **Meiosis I** and **meiosis II** produce four haploid daughter cells. The number of chromosome sets is reduced from two (diploid) to one (haploid) during meiosis I.
- Meiosis is distinguished from mitosis by three events of meiosis I:

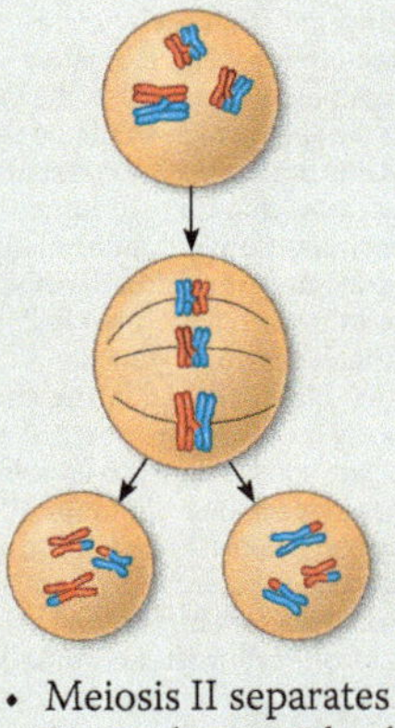

- Meiosis II separates the sister chromatids.
- Sister chromatid cohesion and crossing over allow chiasmata to hold homologs together until anaphase I. Cohesins are cleaved along the arms at anaphase I, allowing homologs to separate, and at the centromeres in anaphase II, releasing sister chromatids.

? *In prophase I, homologous chromosomes pair up and undergo crossing over. Can this also occur during prophase II? Explain.*

CONCEPT 4

Genetic variation produced in sexual life cycles contributes to evolution

- Three events in sexual reproduction contribute to genetic variation in a population: independent assortment of chromosomes during meiosis, crossing over during meiosis I, and random fertilization of egg cells by sperm. During crossing over, DNA of nonsister chromatids in a homologous pair is broken and rejoined.
- Genetic variation is the raw material for evolution by natural selection. Mutations are the original source of this variation; recombination of variant genes generates additional diversity.

? *Explain how three processes unique to sexual reproduction generate a great deal of genetic variation.*

TEST YOUR UNDERSTANDING

PRACTICE TEST

goo.gl/CRZjvS

Level 1: Knowledge/Comprehension

1. A human cell containing 22 autosomes and a Y chromosome is
(A) a sperm.
(B) an egg.
(C) a zygote.
(D) a somatic cell of a male.

2. Homologous chromosomes move toward opposite poles of a dividing cell during
(A) mitosis.
(B) meiosis I.
(C) meiosis II.
(D) fertilization.

Level 2: Application/Analysis

3. If the DNA content of a diploid cell in the G_1 phase of the cell cycle is represented by x, then the DNA content of the same cell at metaphase of meiosis I would be
(A) $0.25x$. (B) $0.5x$. (C) x. (D) $2x$.

4. If we continued to follow the cell lineage from question 3, the DNA content of a cell at metaphase of meiosis II would be
(A) $0.25x$. (B) $0.5x$. (C) x. (D) $2x$.

5. DRAW IT The diagram shows a cell in meiosis. (a) Label the appropriate structures with these terms: chromosome (label as duplicated or unduplicated), centromere, kinetochore, sister chromatids, nonsister chromatids, homologous pair (use a bracket when labeling), homolog (label each one), chiasma, sister chromatid cohesion, gene loci, alleles of the F gene, alleles of the H gene. (b) Identify the stage of meiosis shown. (c) Describe the makeup of a haploid set and a diploid set.

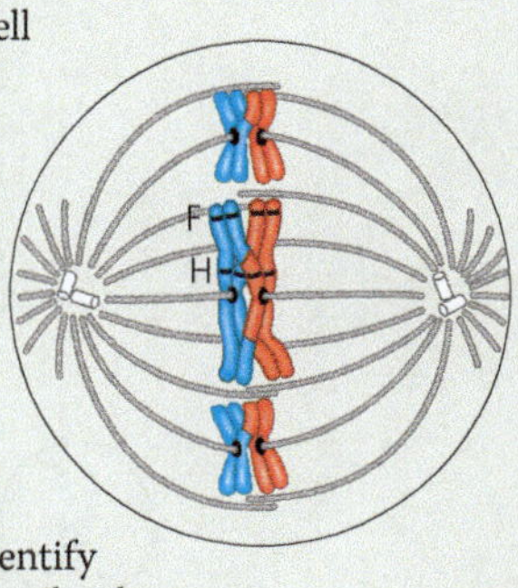

Level 3: Synthesis/Evaluation

6. Explain how you can tell that the cell in question 5 is undergoing meiosis, not mitosis.

7. SCIENTIFIC INQUIRY
The diagram in question 5 represents a meiotic cell. Assume the freckles gene is at the locus marked F, and the hair-color gene is located at the locus marked H, both on the long chromosome. The person from whom this cell was taken has inherited different alleles for each gene ("freckles" and "black hair" from one parent and "no freckles" and "blond hair" from the other). Predict allele combinations in the gametes resulting from this meiotic event. List other possible combinations of these alleles in this person's gametes.

8. FOCUS ON EVOLUTION
Many species can reproduce either asexually or sexually. What might be the evolutionary significance of the switch from asexual to sexual reproduction that occurs in some organisms when the environment becomes unfavorable?

9. FOCUS ON INFORMATION
The continuity of life is based on heritable information in the form of DNA. In a short essay (100–150 words), explain how chromosome behavior during sexual reproduction in animals ensures perpetuation of parental traits in offspring and, at the same time, genetic variation among offspring.

10. SYNTHESIZE YOUR KNOWLEDGE

The Cavendish banana, the world's most popular fruit, is threatened by extinction due to a fungus. This banana variety is "triploid" ($3n$, with three sets of chromosomes) and can only reproduce through cloning by cultivators. Explain how the triploid number accounts for its inability to form normal gametes. Discuss how the absence of sexual reproduction might make this species vulnerable to infection.

Answers

Figure Questions

Figure 4 Two sets of chromosomes are present. Three pairs of homologous chromosomes are present.

Figure 7

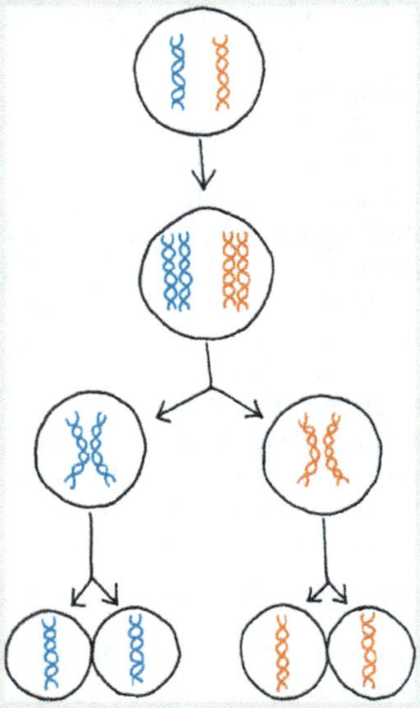

(A short strand of DNA is shown here for simplicity, but each chromosome or chromatid contains a very long coiled and folded DNA molecule.) **Figure 8** If a cell with six chromosomes undergoes two rounds of mitosis, each of the four resulting cells will have six chromosomes, while the four cells resulting from meiosis in Figure 8 each have three chromosomes. In mitosis, DNA replication (and thus chromosome duplication) precedes each prophase, ensuring that daughter cells have the same number of chromosomes as the parent cell. In meiosis, in contrast, DNA replication occurs only before prophase I (not before prophase II). Thus, in two rounds of mitosis, the chromosomes duplicate twice and divide twice, while in meiosis, the chromosomes duplicate once and divide twice. **Figure 10** Yes. Each of the six chromosomes (three per cell) shown in telophase I has one nonrecombinant chromatid and one recombinant chromatid. Therefore, eight possible sets of chromosomes can be generated for the cell on the left and eight for the cell on the right.

Concept Check 1

1. Parents pass genes to their offspring; by dictating the production of messenger RNAs (mRNAs), the genes program cells to make specific enzymes and other proteins, whose cumulative action produces an individual's inherited traits. **2.** Such organisms reproduce by mitosis, which generates offspring whose genomes are exact copies of the parent's genome (in the absence of mutations). **3.** She should clone it. Crossbreeding it with another plant would generate offspring that have additional variation, which she no longer desires now that she has obtained her ideal orchid.

Concept Check 2

1. Each of the six chromosomes is duplicated, so each contains two DNA double helices. Therefore, there are 12 DNA molecules in the cell. The haploid number, n, is 3. One set is always haploid. **2.** There are 23 pairs of chromosomes and two sets. **3.** The haploid number (n) is 7; the diploid number ($2n$) is 14. **4.** This organism has the life cycle shown in Figure 6c. Therefore, it must be a fungus or a protist, perhaps an alga.

Concept Check 3

1. The chromosomes are similar in that each is composed of two sister chromatids, and the individual chromosomes are positioned similarly at the metaphase plate. The chromosomes differ in that in a mitotically dividing cell, sister chromatids of each chromosome are genetically identical, but in a meiotically dividing cell, sister chromatids are genetically distinct because of crossing over in meiosis I. Moreover, the chromosomes in metaphase of mitosis can be a diploid set or a haploid set, but the chromosomes in metaphase of meiosis II always consist of a haploid set. **2.** If crossing over did not occur, the two homologs would not be associated in any way; each sister chromatid would be either all maternal or all paternal, and would only be attached to its sister, not to a nonsister chromatid. This might result in incorrect arrangement of homologs during metaphase I and ultimately in formation of gametes with an abnormal number of chromosomes.

Concept Check 4

1. Mutations in a gene lead to the different versions (alleles) of that gene. **2.** If the segments of the maternal and paternal chromatids that undergo crossing over are genetically identical and thus have the same two alleles for every gene, then the recombinant chromosomes will be genetically equivalent to the parental chromosomes. Crossing over contributes to genetic variation only when it involves the rearrangement of different alleles.

Summary of Key Concepts Questions

1 Genes program specific traits, and offspring inherit their genes from each parent, accounting for similarities in their appearance to one or the other parent. Humans reproduce sexually, which ensures new combinations of genes (and thus traits) in the offspring. Consequently, the offspring are not clones of their parents (which would be the case if humans reproduced asexually). **2** Animals and plants both reproduce sexually, alternating meiosis with fertilization. Both have haploid gametes that unite to form a diploid zygote, which then goes on to divide mitotically, forming a diploid multicellular organism. In animals, haploid cells become gametes and don't undergo mitosis, while in plants, the haploid cells resulting from meiosis undergo mitosis to form a haploid multicellular organism, the gametophyte. This organism then goes on to generate haploid gametes. (In plants such as trees, the gametophyte is quite reduced in size and not obvious to the casual observer.) **3** At the end of meiosis I, the two members of a homologous pair end up in different cells, so they cannot pair up and undergo crossing over. **4** First, during independent assortment in metaphase I, each pair of homologous chromosomes lines up independent of every other pair at the metaphase plate, so a daughter cell of meiosis I randomly inherits either a maternal or paternal chromosome. Second, due to crossing over, each chromosome is not exclusively maternal or paternal, but includes regions at the ends of the chromatid from a nonsister chromatid (a chromatid of the other homolog). (The nonsister segment can also be in an internal region of the chromatid if a second crossover occurs beyond the first one before the end of the chromatid.) This provides much additional diversity in the form of new combinations of alleles. Third, random fertilization ensures even more variation, since any sperm of a large number containing many possible genetic combinations can fertilize any egg of a similarly large number of possible combinations.

Test Your Understanding

1. A **2.** B **3.** D **4.** C

5. (a)

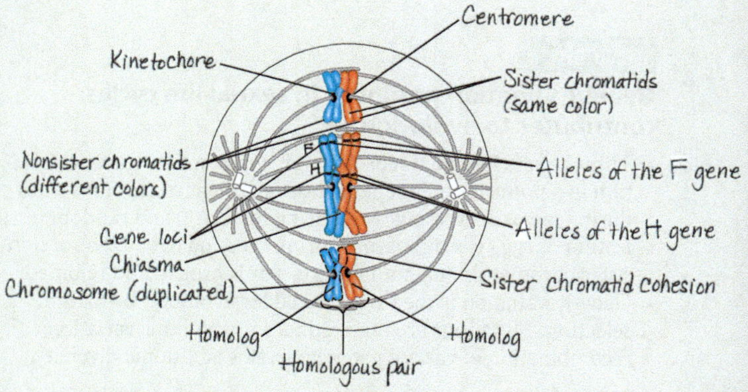

(b) Metaphase I (c) A haploid set is made up of one long, one medium, and one short chromosome, no matter what combination of colors. For example, one red long, one blue medium, and one red short chromosome make up a haploid set. (In cases where crossovers have occurred, a haploid set of one color may include segments of chromatids of the other color.) All red and blue chromosomes together make up a diploid set. **6.** This cell must be undergoing meiosis because homologous chromosomes are associated with each other at the metaphase plate; this does not occur in mitosis.

Credits

Credits are listed in the order of appearance.

Photo Credits

1 Mango Productions/Comet/Corbis; **2a** Roland Birke/OKAPIA/Science Source; **2b** George Ostertag/SuperStock; **3a** Ermakoff/Science Source; **3c** CNRI/Science Source; **SSE** SciMAT/Science Source; **12** Mark Petronczki and Maria Siomos; **Chapter Review SYK** Randy Ploetz.

Illustration and Text Credits

SSE.1 Data from G. Simchen, Commitment to meiosis: What determines the mode of division in budding yeast? *BioEssays* 31:169–177 (2009). doi:10.1002/bies.200800124.

Glossary

alternation of generations A life cycle in which there is both a multicellular diploid form, the sporophyte, and a multicellular haploid form, the gametophyte; characteristic of plants and some algae.

autosome (ot′-ō-sōm) A chromosome that is not directly involved in determining sex; not a sex chromosome.

clone (1) A lineage of genetically identical individuals or cells. (2) In popular usage, an individual that is genetically identical to another individual. (3) As a verb, to make one or more genetic replicas of an individual or cell.

diploid cell (dip′-loyd) A cell containing two sets of chromosomes ($2n$), one set inherited from each parent.

genetics The scientific study of heredity and hereditary variation.

haploid cell (hap′-loyd) A cell containing only one set of chromosomes (n).

heredity The transmission of traits from one generation to the next.

homologous chromosomes (hō-mol′-uh-gus) A pair of chromosomes of the same length, centromere position, and staining pattern that possess genes for the same characters at corresponding loci. One homologous chromosome is inherited from the organism's father, the other from the mother. Also called homologs, or a homologous pair.

homologs *See* homologous chromosomes.

karyotype (kār′-ē-ō-tīp) A display of the chromosome pairs of a cell arranged by size and shape.

life cycle The generation-to-generation sequence of stages in the reproductive history of an organism.

locus (plural, **loci**) (lō′-kus, lō′-sī) A specific place along the length of a chromosome where a given gene is located.

meiosis (mī-ō′-sis) A modified type of cell division in sexually reproducing organisms consisting of two rounds of cell division but only one round of DNA replication. It results in cells with half the number of chromosome sets as the original cell.

meiosis I The first division of a two-stage process of cell division in sexually reproducing organisms that results in cells with half the number of chromosome sets as the original cell.

meiosis II The second division of a two-stage process of cell division in sexually reproducing organisms that results in cells with half the number of chromosome sets as the original cell.

recombinant chromosome A chromosome created when crossing over combines DNA from two parents into a single chromosome.

sex chromosome A chromosome responsible for determining the sex of an individual.

synapsis (si-nap′-sis) The pairing and physical connection of duplicated homologous chromosomes during prophase I of meiosis.

synaptonemal complex A zipper-like structure composed of proteins, which connects two homologous chromosomes tightly along their lengths during part of prophase I of meiosis.

variation Differences between members of the same species.

Mendel and the Gene Idea

Mendel and the Gene Idea

KEY CONCEPTS

1 Mendel used the scientific approach to identify two laws of inheritance
2 Probability laws govern Mendelian inheritance
3 Inheritance patterns are often more complex than predicted by simple Mendelian genetics
4 Many human traits follow Mendelian patterns of inheritance

▲ **Figure 1** What principles of inheritance did Gregor Mendel discover by breeding pea plants?

Drawing from the Deck of Genes

The crowd at a soccer match displays the marvelous variety and diversity of humankind. Brown, blue, or gray eyes; black, brown, or blond hair—these are just a few examples of heritable variations among individuals in a population. What principles account for the transmission of such traits from parents to offspring?

The explanation of heredity most widely in favor during the 1800s was the "blending" hypothesis: the idea that genetic material contributed by the two parents mixes, just as blue and yellow paints blend to make green. This hypothesis predicts that over many generations a freely mating population will give rise to a uniform population of individuals—something we don't see. The blending hypothesis also fails to explain how traits can reappear after skipping a generation.

An alternative to the blending model is a "particulate" hypothesis of inheritance: the gene idea. In this model, parents pass on discrete heritable units—genes—that retain their separate identities in offspring. An organism's collection of genes is more like a deck of cards than a bucket of paint. Like cards, genes can be shuffled and passed along, generation after generation, in undiluted form.

Modern genetics had its genesis in an abbey garden, where a monk named Gregor Mendel documented a particulate mechanism for inheritance using pea plants **(Figure 1)**. Mendel developed his theory of inheritance several decades before chromosomes were observed under the microscope and the significance of their behavior during mitosis or meiosis was understood. In this chapter, we'll step into Mendel's garden to re-create his experiments and explain how he arrived at his theory of inheritance. We'll also explore inheritance patterns more complex than those observed by Mendel in garden peas. Finally, we'll see how the Mendelian model applies to the inheritance of human variations, including hereditary disorders such as sickle-cell disease.

CONCEPT 1

Mendel used the scientific approach to identify two laws of inheritance

Mendel discovered the basic principles of heredity by breeding garden peas in carefully planned experiments. As we retrace his work, you'll recognize the key elements of the scientific process.

Mendel's Experimental, Quantitative Approach

One reason Mendel probably chose to work with peas is that they are available in many varieties. For example, one variety has purple flowers, while another variety has white flowers. A heritable feature that varies among individuals, such as flower color, is called a **character**. Each variant for a character, such as purple or white color for flowers, is called a **trait**.

Mendel could strictly control mating between plants. Each pea flower has both pollen-producing organs (stamens) and an egg-bearing organ (carpel). In nature, pea plants usually self-fertilize: Pollen grains from the stamens land on the carpel of the same flower, and sperm released from the pollen grains fertilize eggs present in the carpel. To achieve cross-pollination of two plants, Mendel removed the immature stamens of a plant before they produced pollen and then dusted pollen from another plant onto the altered flowers **(Figure 2)**. Each resulting zygote then developed into a plant embryo encased in a seed (pea). His method allowed Mendel to always be sure of the parentage of new seeds.

▼ Figure 2 Research Method

Crossing Pea Plants

Application By crossing (mating) two true-breeding varieties of an organism, scientists can study patterns of inheritance. In this example, Mendel crossed pea plants that varied in flower color.

Technique

Results When pollen from a white flower was transferred to a purple flower, the first-generation hybrids all had purple flowers. The result was the same for the reciprocal cross, which involved the transfer of pollen from purple flowers to white flowers.

Mendel chose to track only those characters that occurred in two distinct, alternative forms, such as purple or white flower color. He also made sure that he started his experiments with varieties that, over many generations of self-pollination, had produced only the same variety as the parent plant. Such plants are said to be **true-breeding**. For example, a plant with purple flowers is true-breeding if the seeds produced by self-pollination in successive generations all give rise to plants that also have purple flowers.

In a typical breeding experiment, Mendel cross-pollinated two contrasting, true-breeding pea varieties—for example, purple-flowered plants and white-flowered plants (see Figure 2). This mating, or *crossing*, of two true-breeding varieties is called **hybridization**. The true-breeding parents are referred to as the **P generation** (parental generation), and their hybrid offspring are the **F_1 generation** (first filial generation, the word *filial* from the Latin word for "son"). Allowing these F_1 hybrids to self-pollinate (or to cross-pollinate with other F_1 hybrids) produces an **F_2 generation** (second filial generation). Mendel usually followed traits for at least the P, F_1, and F_2 generations. Had Mendel stopped his experiments with the F_1 generation, the basic patterns of inheritance would have eluded him. Mendel's quantitative analysis of the F_2 plants from thousands of genetic crosses like these allowed him to deduce two fundamental principles of heredity, now called the law of segregation and the law of independent assortment.

The Law of Segregation

If the blending model of inheritance were correct, the F_1 hybrids from a cross between purple-flowered and white-flowered pea plants would have pale purple flowers, a trait intermediate between those of the P generation. Notice in Figure 2 that the experiment produced a very different result: All the F_1 offspring had flowers of the same color as the purple-flowered parents. What happened to the white-flowered plants' genetic contribution to the hybrids? If it were lost, then the F_1 plants could produce only purple-flowered offspring in the F_2 generation. But when Mendel allowed the F_1 plants to self-pollinate and planted their seeds, the white-flower trait reappeared in the F_2 generation.

Mendel used very large sample sizes and kept accurate records of his results: 705 of the F_2 plants had purple flowers, and 224 had white flowers. These data fit a ratio of approximately three purple to one white **(Figure 3)**. Mendel reasoned that the heritable factor for white flowers did not disappear in the F_1 plants, but was somehow hidden, or masked, when the purple-flower factor was present. In Mendel's terminology, purple flower color is a *dominant* trait, and white flower color is a *recessive* trait. The reappearance of white-flowered plants in the F_2 generation was evidence that the heritable factor causing white flowers had not been diluted or destroyed by coexisting with the purple-flower factor in the F_1 hybrids.

Mendel observed the same pattern of inheritance in six other characters, each represented by two distinctly different traits **(Table 1)**. For example, when Mendel crossed a true-breeding variety that produced smooth, round pea seeds with one that produced wrinkled seeds, all the F_1 hybrids produced round seeds; this is the dominant trait for seed shape. In the F_2 generation, approximately 75% of the seeds were round and 25% were wrinkled—a 3:1 ratio, as in Figure 3. Now let's see how Mendel deduced the law of segregation from his experimental results. In the discussion that follows, we will use modern terms instead of some of the terms used by Mendel. (For example, we'll use "gene" instead of Mendel's "heritable factor.")

▼ Figure 3 Inquiry

When F_1 hybrid pea plants self- or cross-pollinate, which traits appear in the F_2 generation?

Experiment Mendel crossed true-breeding purple-flowered plants and white-flowered plants (crosses are symbolized by ×). The resulting F_1 hybrids were allowed to self-pollinate or were cross-pollinated with other F_1 hybrids. The F_2 generation plants were then observed for flower color.

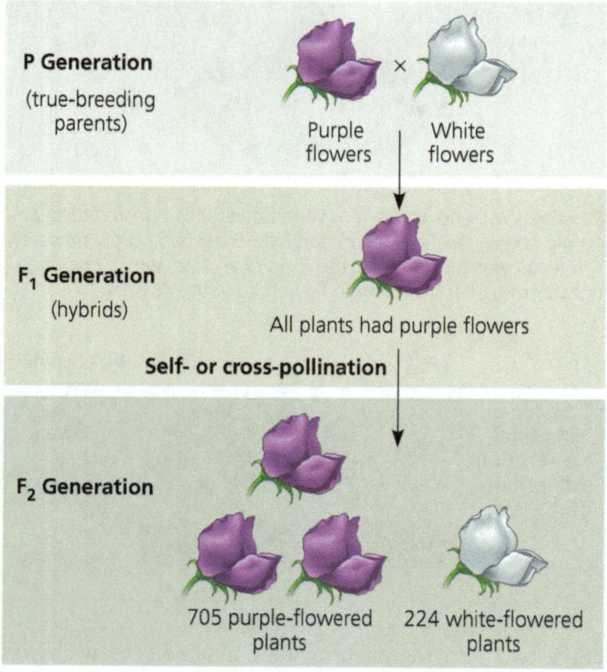

Results Both purple-flowered and white-flowered plants appeared in the F_2 generation, in a ratio of approximately 3:1.

Conclusion The "heritable factor" for the recessive trait (white flowers) had not been destroyed, deleted, or "blended" in the F_1 generation but was merely masked by the presence of the factor for purple flowers, which is the dominant trait.

Data from G. Mendel, Experiments in plant hybridization, *Proceedings of the Natural History Society of Brünn* 4:3–47 (1866).

WHAT IF? If you mated two purple-flowered plants from the P generation, what ratio of traits would you expect to observe in the offspring? Explain.

Table 1 The Results of Mendel's F_1 Crosses for Seven Characters in Pea Plants

Character	Dominant Trait	×	Recessive Trait	F_2 Generation Dominant: Recessive	Ratio
Flower color	Purple 	×	White 	705:224	3.15:1
Seed color	Yellow 	×	Green 	6,022:2,001	3.01:1
Seed shape	Round 	×	Wrinkled 	5,474:1,850	2.96:1
Pod shape	Inflated 	×	Constricted 	882:299	2.95:1
Pod color	Green 	×	Yellow 	428:152	2.82:1
Flower position	Axial 	×	Terminal 	651:207	3.14:1
Stem length	Tall 	×	Dwarf 	787:277	2.84:1

Mendel's Model

Mendel developed a model to explain the 3:1 inheritance pattern that he consistently observed among the F_2 offspring in his pea experiments. We describe four related concepts making up this model, the fourth of which is the law of segregation.

First, *alternative versions of genes account for variations in inherited characters.* The gene for flower color in pea plants, for example, exists in two versions, one for purple flowers and the other for white flowers. These alternative versions of a gene are called **alleles**. Today, we can relate this concept to chromosomes and DNA. As shown in **Figure 4**, each gene is a sequence of nucleotides at a specific place, or locus, along a particular chromosome. The DNA at that locus, however, can vary slightly in its nucleotide sequence. This variation in information content can affect the function of the encoded protein and thus an inherited character of the organism. The purple-flower allele and the white-flower allele are two DNA sequence variations possible at the flower-color locus on one of a pea plant's chromosomes. The purple-flower allele sequence allows synthesis of purple pigment, and the white-flower allele sequence does not.

Second, *for each character, an organism inherits two copies (that is, two alleles) of a gene, one from each parent.* Remarkably, Mendel made this deduction without knowing about the role, or even the existence, of chromosomes. Each somatic cell in a diploid organism has two sets of chromosomes, one set inherited from each parent. Thus, a genetic locus is actually represented twice in a diploid cell, once on each homolog of a specific pair of chromosomes. The two alleles at a particular locus may be identical, as in the true-breeding plants of Mendel's P generation, or the alleles may differ, as in the F_1 hybrids (see Figure 4).

Third, *if the two alleles at a locus differ, then one, the **dominant allele**, determines the organism's appearance; the other, the **recessive allele**, has no noticeable effect on the organism's appearance.* Accordingly, Mendel's F_1 plants had purple flowers because the allele for that trait is dominant and the allele for white flowers is recessive.

The fourth and final part of Mendel's model, the **law of segregation**, states that *the two alleles for a heritable character segregate (separate from each other) during gamete formation and end up in different gametes.* Thus, an egg or a sperm gets only one of the two alleles that are present in the somatic cells of the organism making the gamete. In terms of chromosomes, this segregation corresponds to the distribution of copies of the two members of a pair of homologous chromosomes to different gametes in meiosis. Note that if an organism has identical alleles for a particular character—that is, the organism is true-breeding for that character—then that allele is present in all gametes. But if different alleles are present, as in the F_1 hybrids, then 50% of the gametes receive the dominant allele and 50% receive the recessive allele.

Does Mendel's segregation model account for the 3:1 ratio he observed in the F_2 generation of his numerous crosses? For the flower-color character, the model predicts that the two different alleles present in an F_1 individual will segregate into gametes such that half the gametes will have the purple-flower allele and half will have the white-flower allele. During self-pollination, gametes of each class unite randomly. An egg with a purple-flower allele has an equal chance of being fertilized by a sperm with a purple-flower allele or one with a white-flower allele. Since the same is true for an egg with a white-flower allele, there are four equally likely combinations of sperm

▶ **Figure 4 Alleles, alternative versions of a gene.** This diagram shows a pair of homologous chromosomes in an F_1 hybrid pea plant, with the actual DNA sequence from the flower color allele of each chromosome. The paternally inherited chromosome (blue) has an allele for purple flowers, which codes for a protein that indirectly controls synthesis of purple pigment. The maternally inherited chromosome (red) has an allele for white flowers, which results in no functional protein being made.

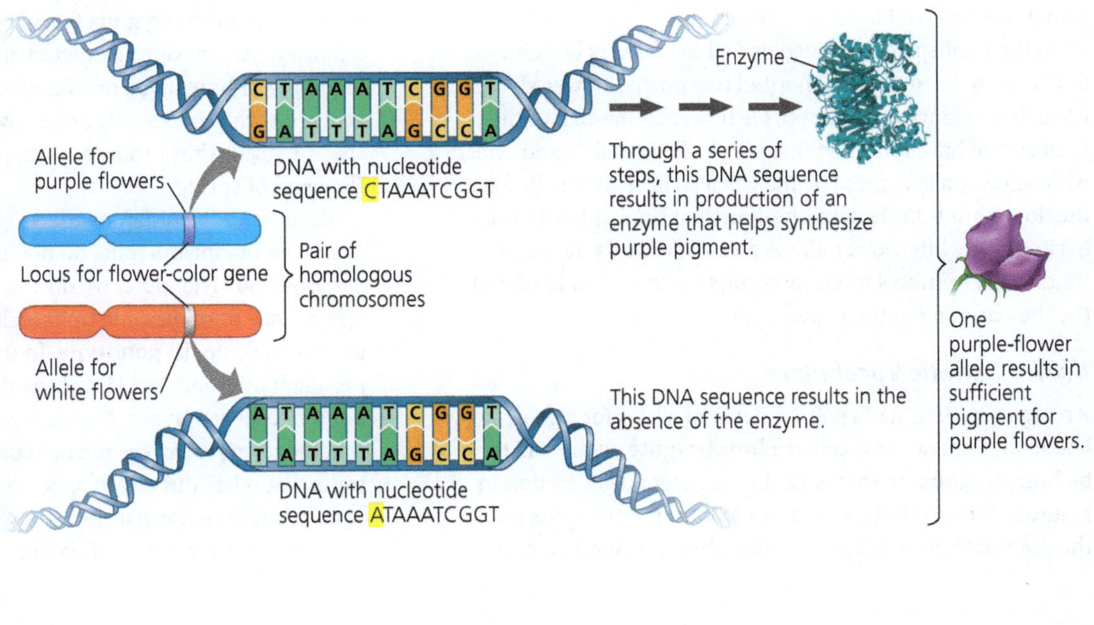

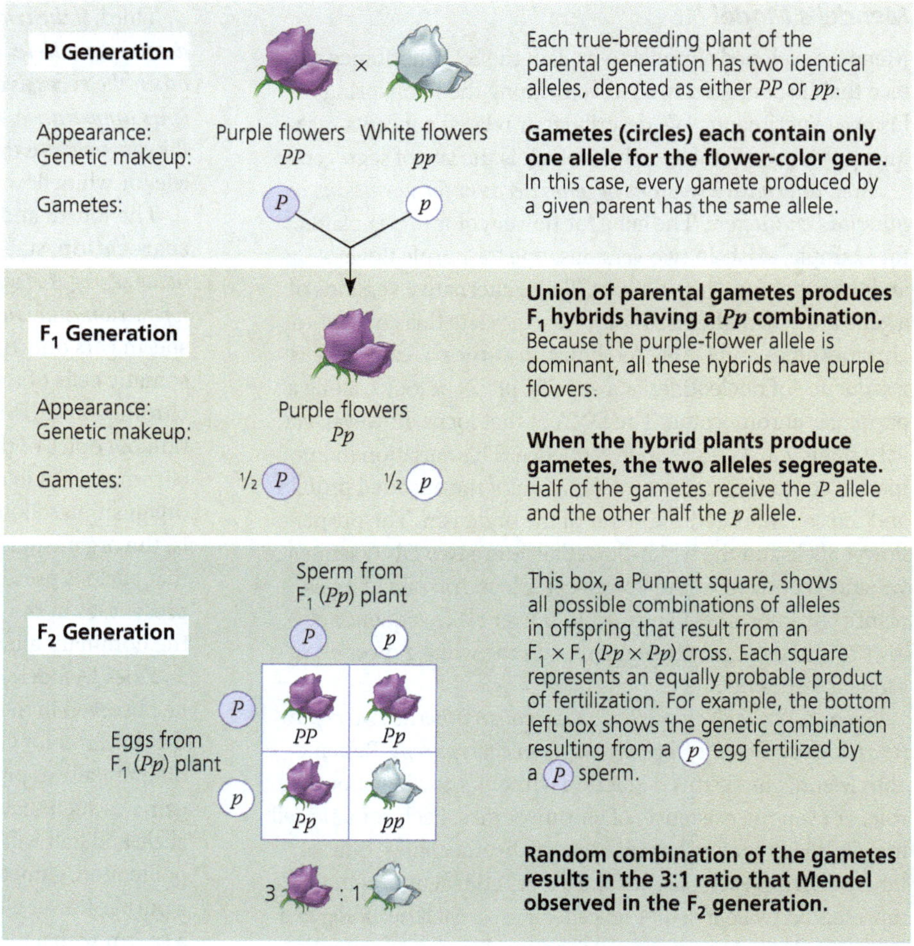

▶ **Figure 5 Mendel's law of segregation.** This diagram shows the genetic makeup of the generations in Figure 3. It illustrates Mendel's model for inheritance of the alleles of a single gene. Each plant has two alleles for the gene controlling flower color, one allele inherited from each of the plant's parents. To construct a Punnett square that predicts the F_2 generation offspring, we list all the possible gametes from one parent (here, the F_1 female) along the left side of the square and all the possible gametes from the other parent (here, the F_1 male) along the top. The boxes represent the offspring resulting from all the possible unions of male and female gametes.

and egg. **Figure 5** illustrates these combinations using a **Punnett square**, a handy diagrammatic device for predicting the allele composition of all offspring resulting from a cross between individuals of known genetic makeup. Notice that we use a capital letter to symbolize a dominant allele and a lowercase letter for a recessive allele. In our example, *P* is the purple-flower allele, and *p* is the white-flower allele; the gene itself is sometimes referred to as the *P*/*p* gene.

In the F_2 offspring, what color will the flowers be? One-fourth of the plants have inherited two purple-flower alleles; these plants will have purple flowers. One-half of the F_2 offspring have inherited one purple-flower allele and one white-flower allele; these plants will also have purple flowers, the dominant trait. Finally, one-fourth of the F_2 plants have inherited two white-flower alleles and will express the recessive trait. Thus, Mendel's model accounts for the 3:1 ratio of traits that he observed in the F_2 generation.

Useful Genetic Vocabulary

An organism that has a pair of identical alleles for a gene encoding a character is called a **homozygote** and is said to be **homozygous** for that gene. In the parental generation in Figure 5, the purple-flowered pea plant is homozygous for the dominant allele (*PP*), while the white-flowered plant is homozygous for the recessive allele (*pp*). Homozygous plants "breed true" because all of their gametes contain the same allele—either *P* or *p* in this example. If we cross dominant homozygotes with recessive homozygotes, every offspring will have two different alleles—*Pp* in the case of the F_1 hybrids of our flower-color experiment (see Figure 5). An organism that has two different alleles for a gene is called a **heterozygote** and is said to be **heterozygous** for that gene. Unlike homozygotes, heterozygotes produce gametes with different alleles, so they are not true-breeding. For example, *P*- and *p*-containing gametes are both produced by our F_1 hybrids. Self-pollination of the F_1 hybrids thus produces both purple-flowered and white-flowered offspring.

Because of the different effects of dominant and recessive alleles, an organism's traits do not always reveal its genetic composition. Therefore, we distinguish between an organism's appearance or observable traits, called its **phenotype**, and its genetic makeup, its **genotype**. In the case of flower color in pea plants, *PP* and *Pp* plants have the same phenotype (purple) but different genotypes. **Figure 6** reviews these terms. Note that the term *phenotype* refers to physiological traits as well as traits that relate directly to appearance. For example, one pea variety lacks the normal ability to self-pollinate, which is a phenotypic trait (non-self-pollination).

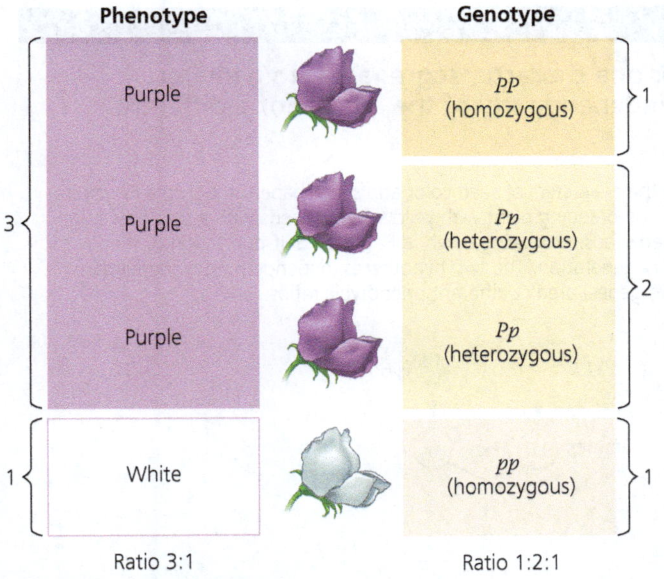

▲ **Figure 6 Phenotype versus genotype.** Grouping F_2 offspring from a cross for flower color according to phenotype results in the typical 3:1 phenotypic ratio. In terms of genotype, however, there are actually two categories of purple-flowered plants, *PP* (homozygous) and *Pp* (heterozygous), giving a 1:2:1 genotypic ratio.

The Testcross

Given a purple-flowered pea plant, we cannot tell if it is homozygous (*PP*) or heterozygous (*Pp*) because both genotypes result in the same purple phenotype. To determine the genotype, we can cross this plant with a white-flowered plant (*pp*), which will make only gametes with the recessive allele (*p*). The allele in the gamete contributed by the mystery plant will therefore determine the appearance of the offspring (**Figure 7**). If all the offspring of the cross have purple flowers, then the purple-flowered mystery plant must be homozygous for the dominant allele, because a *PP* × *pp* cross produces all *Pp* offspring. But if both the purple and the white phenotypes appear among the offspring, then the purple-flowered parent must be heterozygous. The offspring of a *Pp* × *pp* cross will be expected to have a 1:1 phenotypic ratio. Breeding an organism of unknown genotype with a recessive homozygote is called a **testcross** because it can reveal the genotype of that organism. The testcross was devised by Mendel and continues to be used by geneticists.

▼ **Figure 7 Research Method**

The Testcross

Application An organism that exhibits a dominant trait, such as purple flowers in pea plants, can be either homozygous for the dominant allele or heterozygous. To determine the organism's genotype, geneticists can perform a testcross.

Technique In a testcross, the individual with the unknown genotype is crossed with a homozygous individual expressing the recessive trait (white flowers in this example), and Punnett squares are used to predict the possible outcomes.

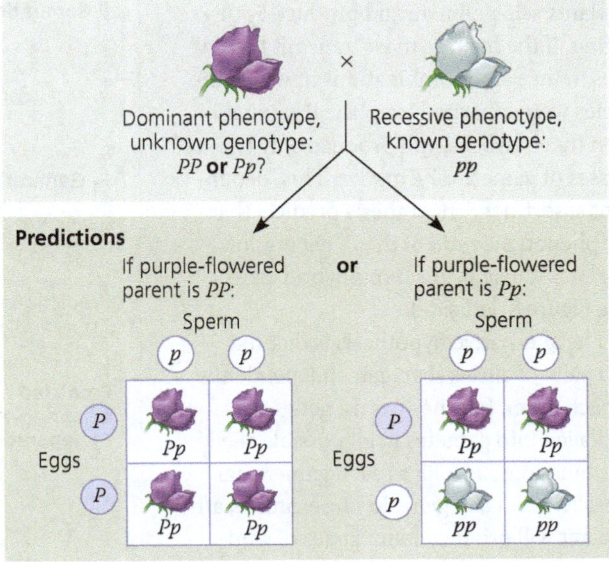

Results Matching the results to either prediction identifies the unknown parental genotype (either *PP* or *Pp* in this example). In this testcross, we transferred pollen from a white-flowered plant to the carpels of a purple-flowered plant; the opposite (reciprocal) cross would have led to the same results.

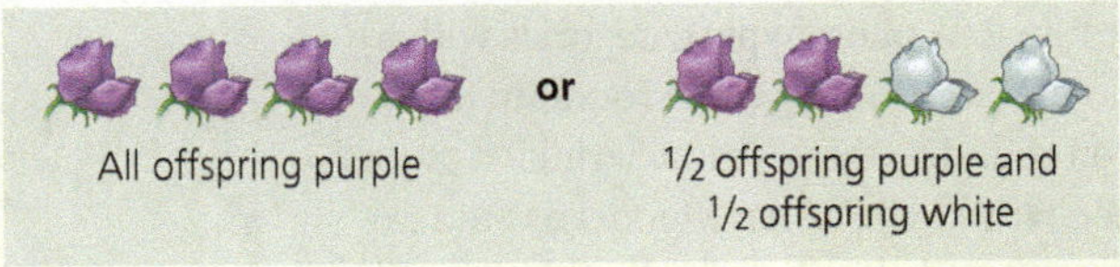

The Law of Independent Assortment

Mendel derived the law of segregation from experiments in which he followed only a *single* character, such as flower color. All the F_1 progeny produced in his crosses of true-breeding parents were **monohybrids**, meaning that they were heterozygous for the one particular character being followed in the cross. We refer to a cross between such heterozygotes as a **monohybrid cross**.

Mendel identified his second law of inheritance by following *two* characters at the same time, such as seed color and seed shape. Seeds (peas) may be either yellow or green. They also may be either round (smooth) or wrinkled. From single-character crosses, Mendel knew that the allele for yellow seeds (*Y*) is dominant and the allele for green seeds (*y*) is recessive. For the seed-shape character, the allele for round (*R*) is dominant, and the allele for wrinkled (*r*) is recessive.

Imagine crossing two true-breeding pea varieties that differ in *both* of these characters—a cross between a plant with yellow round seeds (*YYRR*) and a plant with green wrinkled seeds (*yyrr*). The F_1 plants will be **dihybrids**, individuals heterozygous for the two characters being followed in the cross (*YyRr*). But are these two characters transmitted from parents to offspring as a package? That is, will the *Y* and *R* alleles always stay together, generation after generation? Or are seed color and seed shape inherited independently? **Figure 8** shows how a **dihybrid cross**, a cross between F_1 dihybrids, can determine which of these two hypotheses is correct.

The F_1 plants, of genotype *YyRr*, exhibit both dominant phenotypes, yellow seeds with round shapes, no matter which hypothesis is correct. The key step in the experiment is to see what happens when F_1 plants self-pollinate and produce F_2 offspring. If the hybrids must transmit their alleles in the same combinations in which the alleles were inherited from the P generation, then the F_1 hybrids will produce only two classes of gametes: *YR* and *yr*. This "dependent assortment" hypothesis predicts that the phenotypic ratio of the F_2 generation will be 3:1, just as in a monohybrid cross (see Figure 8, left side).

The alternative hypothesis is that the two pairs of alleles segregate independently of each other. In other words, genes are packaged into gametes in all possible allelic combinations, as long as each gamete has one allele for each gene. In our example, an F_1 plant will produce four classes of gametes in equal quantities: *YR*, *Yr*, *yR*, and *yr*. If sperm of the four classes fertilize eggs of the four classes, there will be 16 (4×4) equally probable ways in which the alleles can combine in the F_2 generation, as shown in Figure 8, right side. These combinations result in four phenotypic categories with a ratio of 9:3:3:1 (nine yellow round to three green round to three yellow wrinkled to one green wrinkled). When Mendel did the experiment and classified the F_2 offspring, his results were close to the predicted 9:3:3:1 phenotypic ratio, supporting the hypothesis that the alleles for one gene—controlling seed color, for example—segregate into gametes independently of the alleles of any other gene, such as seed shape.

Mendel tested his seven pea characters in various dihybrid combinations and always observed a 9:3:3:1 phenotypic ratio in the F_2 generation. Is this consistent with the 3:1 phenotypic ratio seen for the monohybrid cross shown in Figure 5? If you calculate the ratio of yellow and green peas, ignoring shape, you will see that the color

▼ Figure 8 Inquiry

Do the alleles for one character segregate into gametes dependently or independently of the alleles for a different character?

Experiment To follow the characters of seed color and seed shape through the F_2 generation, Mendel crossed a true-breeding plant with yellow round seeds with a true-breeding plant with green wrinkled seeds, producing dihybrid F_1 plants. Self-pollination of the F_1 dihybrids produced the F_2 generation. The two hypotheses (dependent and independent "assortment" of the two genes) predict different phenotypic ratios.

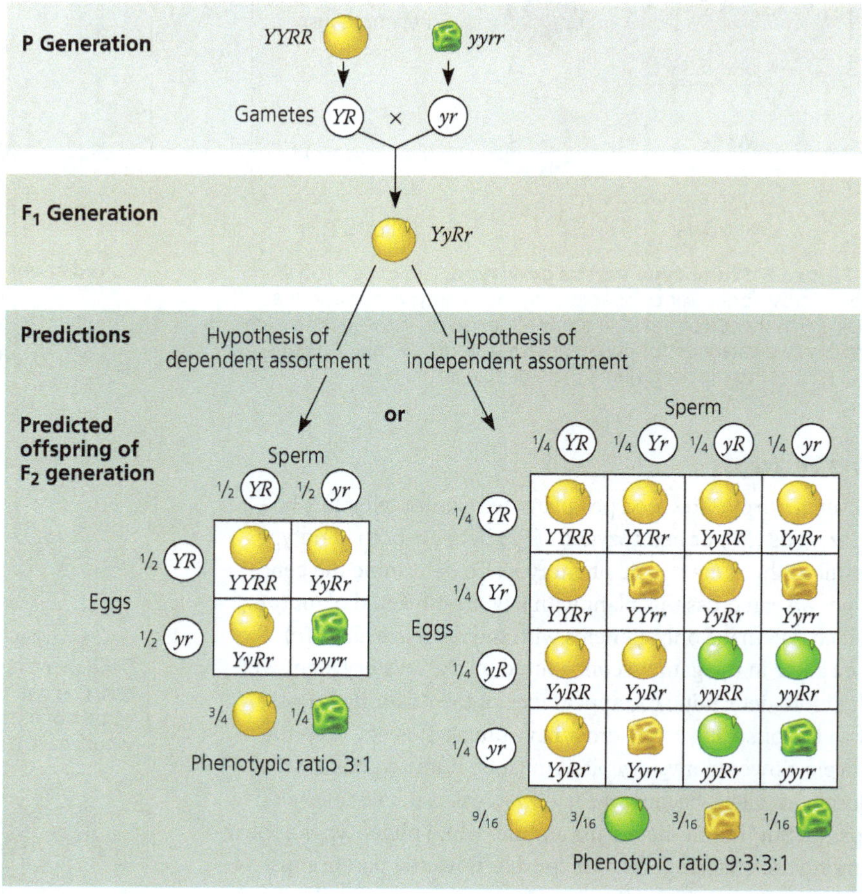

Results

Conclusion Only the hypothesis of independent assortment predicts two of the observed phenotypes: green round seeds and yellow wrinkled seeds (see the right-hand Punnett square). The alleles for each gene segregate independently of those of the other, and the two genes are said to assort independently.

Data from G. Mendel, Experiments in plant hybridization, *Proceedings of the Natural History Society of Brünn* 4:3–47 (1866).

WHAT IF? Suppose Mendel had transferred pollen from an F_1 plant to the carpel of a plant that was homozygous recessive for both genes. Set up the cross and draw Punnett squares that predict the offspring for both hypotheses. Would this cross have supported the hypothesis of independent assortment equally well?

alleles segregate as if this were a monohybrid cross (3:1). The results of Mendel's dihybrid experiments are the basis for what we now call the **law of independent assortment**, which states that *two or more genes assort independently—that is, each pair of alleles segregates independently of any other pair during gamete formation.*

This law applies only to genes (allele pairs) located on different chromosomes—that is, on chromosomes that are not homologous—or very far apart on the same chromosome. All the pea characters Mendel chose for analysis were controlled by genes on different chromosomes or far apart on the same chromosome; this greatly simplified the interpretation of his multicharacter pea crosses. All the examples we consider in the rest of this chapter involve genes located on different chromosomes.

CONCEPT CHECK 1

1. DRAW IT Pea plants heterozygous for flower position and stem length (*AaTt*) are allowed to self-pollinate, and 400 of the resulting seeds are planted. Draw a Punnett square for this cross. How many offspring would be predicted to have terminal flowers and be dwarf? (See Table 1.)
2. List all gametes that could be made by a pea plant heterozygous for seed color, seed shape, and pod shape (*YyRrIi*; see Table 1). How large a Punnett square is needed to predict the offspring of a self-pollination of this "trihybrid"?
3. MAKE CONNECTIONS In some pea plant crosses, the plants are self-pollinated. Explain whether self-pollination is considered asexual or sexual reproduction.

CONCEPT 2

Probability laws govern Mendelian inheritance

Mendel's laws of segregation and independent assortment reflect the same rules of probability that apply to tossing coins, rolling dice, and drawing cards from a deck. The probability scale ranges from 0 to 1. An event that is certain to occur has a probability of 1, while an event that is certain *not* to occur has a probability of 0. With a coin that has heads on both sides, the probability of tossing heads is 1, and the probability of tossing tails is 0. With a normal coin, the chance of tossing heads is ½, and the chance of tossing tails is ½. The probability of drawing the ace of spades from a 52-card deck is $1/52$. The probabilities of all possible outcomes for an event must add up to 1. With a deck of cards, the chance of picking a card other than the ace of spades is $51/52$.

Tossing a coin illustrates an important lesson about probability. For every toss, the probability of heads is ½. The outcome of any particular toss is unaffected by what has happened on previous trials. We refer to phenomena such as coin tosses as independent events. Each toss of a coin, whether done sequentially with one coin or simultaneously with many, is independent of every other toss. And like two separate coin tosses, the alleles of one gene segregate into gametes independently of another gene's alleles (the law of independent assortment). We'll now look at two basic rules of probability that help us predict the outcome of the fusion of such gametes in simple monohybrid crosses and more complicated crosses as well.

The Multiplication and Addition Rules Applied to Monohybrid Crosses

How do we determine the probability that two or more independent events will occur together in some specific combination? For example, what is the chance that two coins tossed simultaneously will both land heads up? The **multiplication rule** states that to determine this probability, we multiply the probability of one event (one coin coming up heads) by the probability of the other event (the other coin coming up heads). By the multiplication rule, then, the probability that both coins will land heads up is ½ × ½ = ¼.

We can apply the same reasoning to an F_1 monohybrid cross **(Figure 9)**. With seed shape in pea plants as the heritable character, the genotype of F_1 plants is *Rr*. Segregation in a heterozygous plant is like flipping a coin in terms of calculating

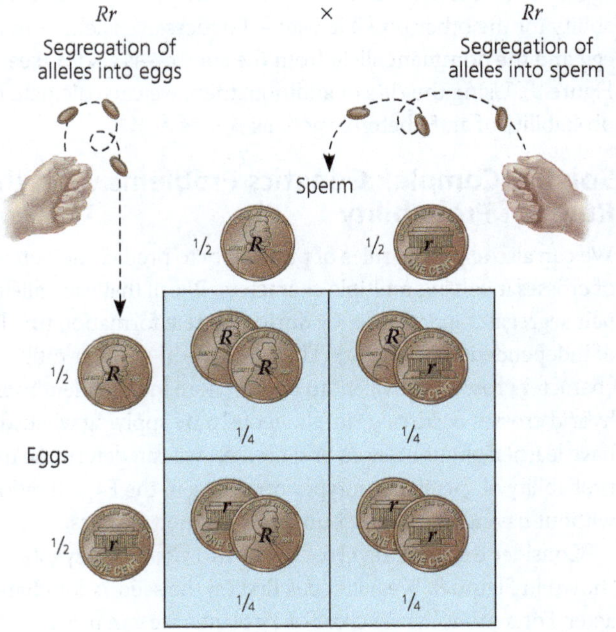

▲ Figure 9 Segregation of alleles and fertilization as chance events. When a heterozygote (*Rr*) forms gametes, whether a particular gamete ends up with an *R* or an *r* is like the toss of a coin. We can determine the probability for any genotype among the offspring of two heterozygotes by multiplying together the individual probabilities of an egg and sperm having a particular allele (*R* or *r* in this example).

the probability of each outcome: Each egg produced has a ½ chance of carrying the dominant allele (*R*) and a ½ chance of carrying the recessive allele (*r*). The same odds apply to each sperm cell produced. For a particular F_2 plant to have wrinkled seeds, the recessive trait, both the egg and the sperm that come together must carry the *r* allele. The probability that an *r* allele will be present in both gametes at fertilization is found by multiplying ½ (the probability that the egg will have an *r*) × ½ (the probability that the sperm will have an *r*). Thus, the multiplication rule tells us that the probability of an F_2 plant having wrinkled seeds (*rr*) is ¼ (see Figure 9). Likewise, the probability of an F_2 plant carrying both dominant alleles for seed shape (*RR*) is ¼.

To figure out the probability that an F_2 plant from a monohybrid cross will be heterozygous rather than homozygous, we need to invoke a second rule. Notice in Figure 9 that the dominant allele can come from the egg and the recessive allele from the sperm, or vice versa. That is, F_1 gametes can combine to produce *Rr* offspring in two *mutually exclusive* ways: For any particular heterozygous F_2 plant, the dominant allele can come from the egg *or* the sperm, but not from both. According to the **addition rule**, the probability that any one of two or more mutually exclusive events will occur is calculated by adding their individual probabilities. As we have just seen, the multiplication rule gives us the individual probabilities that we will now add together. The probability for one possible way of obtaining an F_2 heterozygote—the dominant allele from the egg and the recessive allele from the sperm—is ¼. The probability for the other possible way—the recessive allele from the egg and the dominant allele from the sperm—is also ¼ (see Figure 9). Using the rule of addition, then, we can calculate the probability of an F_2 heterozygote as ¼ + ¼ = ½.

Solving Complex Genetics Problems with the Rules of Probability

We can also apply the rules of probability to predict the outcome of crosses involving multiple characters. Recall that each allelic pair segregates independently during gamete formation (the law of independent assortment). Thus, a dihybrid or other multicharacter cross is equivalent to two or more independent monohybrid crosses occurring simultaneously. By applying what we have learned about monohybrid crosses, we can determine the probability of specific genotypes occurring in the F_2 generation without having to construct unwieldy Punnett squares.

Consider the dihybrid cross between *YyRr* heterozygotes shown in Figure 8. We will focus first on the seed-color character. For a monohybrid cross of *Yy* plants, we can use a simple Punnett square to determine that the probabilities of the offspring genotypes are ¼ for *YY*, ½ for *Yy*, and ¼ for *yy*. We can draw a second Punnett square to determine that the same probabilities apply to the offspring genotypes for seed shape: ¼ *RR*, ½ *Rr*, and ¼ *rr*. Knowing these probabilities, we can simply use the multiplication rule to determine the probability of each of the genotypes in the F_2 generation. To give two examples, the calculations for finding the probabilities of two of the possible F_2 genotypes (*YYRR* and *YyRR*) are shown below:

Probability of *YYRR* = ¼ (probability of *YY*) × ¼ (*RR*) = 1/16

Probability of *YyRR* = ½ (*Yy*) × ¼ (*RR*) = ⅛

The *YYRR* genotype corresponds to the upper left box in the larger Punnett square in Figure 8 (one box = 1/16). Looking closely at the larger Punnett square in Figure 8, you will see that 2 of the 16 boxes (⅛) correspond to the *YyRR* genotype.

Now let's see how we can combine the multiplication and addition rules to solve even more complex problems in Mendelian genetics. Imagine a cross of two pea varieties in which we track the inheritance of three characters. Let's cross a trihybrid with purple flowers and yellow round seeds (heterozygous for all three genes) with a plant with purple flowers and green wrinkled seeds (heterozygous for flower color but homozygous recessive for the other two characters). Using Mendelian symbols, our cross is *PpYyRr* × *Ppyyrr*. What fraction of offspring from this cross is predicted to exhibit the recessive phenotypes for *at least two* of the three characters?

To answer this question, we can start by listing all genotypes we could get that fulfill this condition: *ppyyRr*, *ppYyrr*, *Ppyyrr*, *PPyyrr*, and *ppyyrr*. (Because the condition is *at least two* recessive traits, it includes the last genotype, which shows all three recessive traits.) Next, we calculate the probability for each of these genotypes resulting from our *PpYyRr* × *Ppyyrr* cross by multiplying together the individual probabilities for the allele pairs, just as we did in our dihybrid example. Note that in a cross involving heterozygous and homozygous allele pairs (for example, *Yy* × *yy*), the probability of heterozygous offspring is ½ and the probability of homozygous offspring is ½. Finally, we use the addition rule to add the probabilities for all the different genotypes that fulfill the condition of at least two recessive traits, as shown below:

ppyyRr	¼ (probability of *pp*) × ½ (*yy*) × ½ (*Rr*)	= 1/16
ppYyrr	¼ × ½ × ½	= 1/16
Ppyyrr	½ × ½ × ½	= 2/16
PPyyrr	¼ × ½ × ½	= 1/16
ppyyrr	¼ × ½ × ½	= 1/16
Chance of *at least two* recessive traits		= 6/16 or ⅜

In time, you'll be able to solve genetics problems faster by using the rules of probability than by filling in Punnett squares.

We cannot predict with certainty the exact numbers of progeny of different genotypes resulting from a genetic cross. But the rules of probability give us the *likelihood* of various outcomes. Usually, the larger the sample size, the closer the results will conform to our predictions. The reason Mendel counted so many offspring from his crosses is that he understood this statistical feature of inheritance and had a keen sense of the rules of chance.

CONCEPT CHECK 2

1. For any gene with a dominant allele *A* and recessive allele *a*, what proportions of the offspring from an $AA \times Aa$ cross are expected to be homozygous dominant, homozygous recessive, and heterozygous?
2. Two organisms, with genotypes *BbDD* and *BBDd*, are mated. Assuming independent assortment of the *B/b* and *D/d* genes, write the genotypes of all possible offspring from this cross and use the rules of probability to calculate the chance of each genotype occurring.
3. WHAT IF? Three characters (flower color, seed color, and pod shape) are considered in a cross between two pea plants: $PpYyIi \times ppYyii$. What fraction of offspring is predicted to be homozygous recessive for at least two of the three characters?

CONCEPT 3

Inheritance patterns are often more complex than predicted by simple Mendelian genetics

In the 20th century, geneticists extended Mendelian principles not only to diverse organisms but also to patterns of inheritance more complex than those described by Mendel. For the work that led to his two laws of inheritance, Mendel chose pea plant characters that turn out to have a relatively simple genetic basis: Each character is determined by one gene, for which there are only two alleles, one completely dominant and the other completely recessive. (There is one exception: Mendel's pod-shape character is actually determined by two genes.) Few heritable characters are determined so simply, and the relationship between genotype and phenotype is rarely so straightforward. Mendel himself realized that he could not explain the more complicated patterns he observed in crosses involving other pea characters or other plant species. This does not diminish the utility of Mendelian genetics, however, because the basic principles of segregation and independent assortment apply even to more complex patterns of inheritance. In this section, we'll extend Mendelian genetics to hereditary patterns that were not reported by Mendel.

Extending Mendelian Genetics for a Single Gene

The inheritance of characters determined by a single gene deviates from simple Mendelian patterns when alleles are not completely dominant or recessive, when a particular gene has more than two alleles, or when a single gene produces multiple phenotypes. We'll describe examples of each of these situations in this section.

Degrees of Dominance

Alleles can show different degrees of dominance and recessiveness in relation to each other. In Mendel's classic pea crosses, the F_1 offspring always looked like one of the two parental varieties because one allele in a pair showed **complete dominance** over the other. In such situations, the phenotypes of the heterozygote and the dominant homozygote are indistinguishable.

For some genes, however, neither allele is completely dominant, and the F_1 hybrids have a phenotype somewhere between those of the two parental varieties. This phenomenon, called **incomplete dominance**, is seen when red snapdragons are crossed with white snapdragons: All the F_1 hybrids have pink flowers **(Figure 10)**. This third, intermediate phenotype results from flowers of the heterozygotes having less red pigment than the red homozygotes. (This is unlike the case of Mendel's pea plants, where the *Pp* heterozygotes make enough pigment for the flowers to be purple, indistinguishable from those of *PP* plants.)

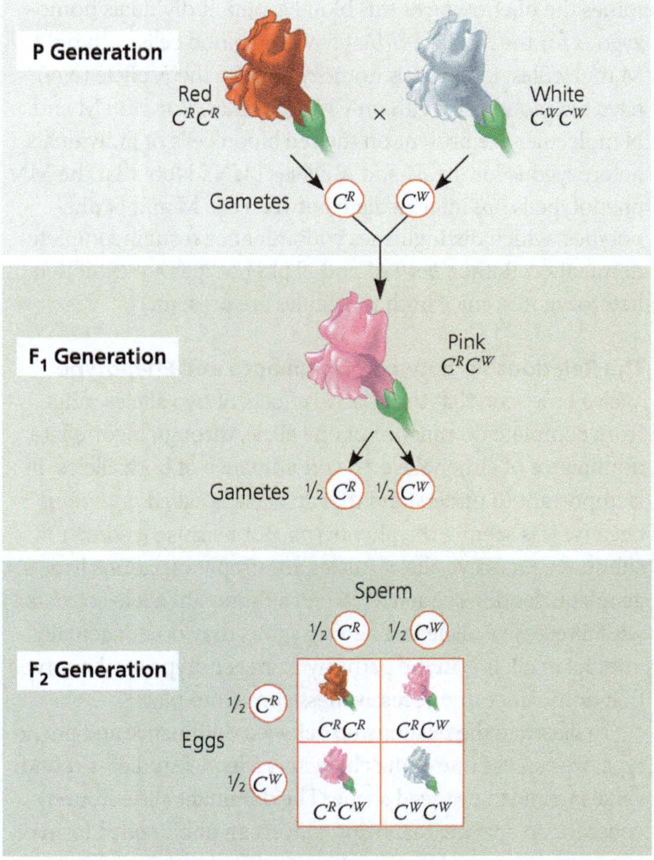

▲ **Figure 10 Incomplete dominance in snapdragon color.** When red snapdragons are crossed with white ones, the F_1 hybrids have pink flowers. Segregation of alleles into gametes of the F_1 plants results in an F_2 generation with a 1:2:1 ratio for both genotype and phenotype. Neither allele is dominant, so rather than using upper- and lowercase letters, we use the letter *C* with a superscript to indicate an allele for flower color: C^R for red and C^W for white.

? *Suppose a classmate argues that this figure supports the blending hypothesis for inheritance. What might your classmate say, and how would you respond?*

At first glance, incomplete dominance of either allele seems to provide evidence for the blending hypothesis of inheritance, which would predict that the red or white trait could never reappear among offspring of the pink hybrids. In fact, interbreeding F_1 hybrids produces F_2 offspring with a phenotypic ratio of one red to two pink to one white. (Because heterozygotes have a separate phenotype, the genotypic and phenotypic ratios for the F_2 generation are the same, 1:2:1.) The segregation of the red-flower and white-flower alleles in the gametes produced by the pink-flowered plants confirms that the alleles for flower color are heritable factors that maintain their identity in the hybrids; that is, inheritance is particulate.

Another variation on dominance relationships between alleles is called **codominance**; in this variation, the two alleles each affect the phenotype in separate, distinguishable ways. For example, the human MN blood group is determined by codominant alleles for two specific molecules located on the surface of red blood cells, the M and N molecules. A single gene locus, at which two allelic variations are possible, determines the phenotype of this blood group. Individuals homozygous for the *M* allele (*MM*) have red blood cells with only M molecules; individuals homozygous for the *N* allele (*NN*) have red blood cells with only N molecules. But *both* M and N molecules are present on the red blood cells of individuals heterozygous for the *M* and *N* alleles (*MN*). Note that the MN phenotype is *not* intermediate between the M and N phenotypes, which distinguishes codominance from incomplete dominance. Rather, *both* M and N phenotypes are exhibited by heterozygotes, since both molecules are present.

The Relationship Between Dominance and Phenotype

We've now seen that the relative effects of two alleles range from complete dominance of one allele, through incomplete dominance of either allele, to codominance of both alleles. It is important to understand that an allele is called *dominant* because it is seen in the phenotype, not because it somehow subdues a recessive allele. Alleles are simply variations in a gene's nucleotide sequence. When a dominant allele coexists with a recessive allele in a heterozygote, they do not actually interact at all. It is in the pathway from genotype to phenotype that dominance and recessiveness come into play.

To illustrate the relationship between dominance and phenotype, we can use one of the characters Mendel studied—round versus wrinkled pea seed shape. The dominant allele (round) codes for an enzyme that helps convert an unbranched form of starch to a branched form in the seed. The recessive allele (wrinkled) codes for a defective form of this enzyme, leading to an accumulation of unbranched starch, which causes excess water to enter the seed by osmosis. Later, when the seed dries, it wrinkles. If a dominant allele is present, no excess water enters the seed and it does not wrinkle when it dries. One dominant allele results in enough of the enzyme to synthesize adequate amounts of branched starch, which means that dominant homozygotes and heterozygotes have the same phenotype: round seeds.

A closer look at the relationship between dominance and phenotype reveals an intriguing fact: For any character, the observed dominant/recessive relationship of alleles depends on the level at which we examine the phenotype. **Tay-Sachs disease**, an inherited disorder in humans, provides an example. The brain cells of a child with Tay-Sachs disease cannot metabolize certain lipids because a crucial enzyme does not work properly. As these lipids accumulate in brain cells, the child begins to suffer seizures, blindness, and degeneration of motor and mental performance and dies within a few years.

Only children who inherit two copies of the Tay-Sachs allele (homozygotes) have the disease. Thus, at the *organismal* level, the Tay-Sachs allele qualifies as recessive. However, the activity level of the lipid-metabolizing enzyme in heterozygotes is intermediate between the activity level in individuals homozygous for the normal allele and the activity level in individuals with Tay-Sachs disease. (The term *normal* is used in the genetic sense to refer to the allele coding for the enzyme that functions properly.) The intermediate phenotype observed at the *biochemical* level is characteristic of incomplete dominance of either allele. Fortunately, the heterozygote condition does not lead to disease symptoms, apparently because half the normal enzyme activity is sufficient to prevent lipid accumulation in the brain. Extending our analysis to yet another level, we find that heterozygous individuals produce equal numbers of normal and dysfunctional enzyme molecules. Thus, at the *molecular* level, the normal allele and the Tay-Sachs allele are codominant. As you can see, whether alleles appear to be completely dominant, incompletely dominant, or codominant depends on the level at which the phenotype is analyzed.

Frequency of Dominant Alleles While you might assume that the dominant allele for a particular character would be more common in a population than the recessive one, this is not always so. For example, about one baby out of 400 in the United States is born with extra digits (fingers or toes), a condition known as polydactyly. Some cases are caused by the presence of a dominant allele. The low frequency of polydactyly indicates that the recessive allele, which results in five digits per appendage when homozygous, is far more prevalent than the dominant allele in the population.

Multiple Alleles

Only two alleles exist for each of the seven pea characters that Mendel studied, but most genes exist in more than two allelic forms. The ABO blood groups in humans, for instance, are determined by that person's two alleles of the blood group gene; there are three possible alleles: I^A, I^B, and *i*. A person's blood group may be one of four types: A, B, AB, or O. These letters refer to two carbohydrates—A and B—that are found on the surface of red blood cells. An individual's blood cells may have carbohydrate A (type A blood), carbohydrate B (type B), both (type AB), or neither (type O), as shown schematically in **Figure 11**. Matching compatible blood groups is critical for safe blood transfusions.

(a) The three alleles for the ABO blood groups and their carbohydrates. Each allele codes for an enzyme that may add a specific carbohydrate (designated by the superscript on the allele and shown as a triangle or circle) to red blood cells.			
Allele	I^A	I^B	i
Carbohydrate	A △	B ○	none

(b) Blood group genotypes and phenotypes. There are six possible genotypes, resulting in four different phenotypes.				
Genotype	I^AI^A or I^Ai	I^BI^B or I^Bi	I^AI^B	ii
Red blood cell appearance	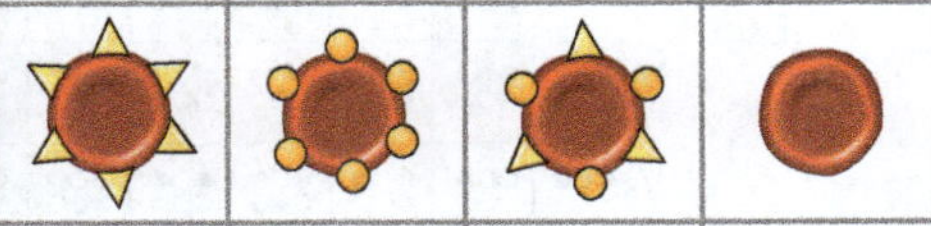			
Phenotype (blood group)	A	B	AB	O

▲ **Figure 11 Multiple alleles for the ABO blood groups.** The four blood groups result from different combinations of three alleles.

? *Based on the surface carbohydrate phenotype in (b), what are the dominance relationships among the alleles?*

Pleiotropy

So far, we have treated Mendelian inheritance as though each gene affects only one phenotypic character. Most genes, however, have multiple phenotypic effects, a property called **pleiotropy** (from the Greek *pleion*, more). In humans, for example, pleiotropic alleles are responsible for the multiple symptoms associated with certain hereditary diseases, such as cystic fibrosis and sickle-cell disease, discussed later in this chapter. In the garden pea, the gene that determines flower color also affects the color of the coating on the outer surface of the seed, which can be gray or white. Given the intricate molecular and cellular interactions responsible for an organism's development and physiology, it isn't surprising that a single gene can affect a number of characters.

Extending Mendelian Genetics for Two or More Genes

Dominance relationships, multiple alleles, and pleiotropy all have to do with the effects of the alleles of a single gene. We now consider two situations in which two or more genes are involved in determining a particular phenotype. In the first case, one gene affects the phenotype of another because the two gene products interact, whereas in the second case, multiple genes independently affect a single trait.

Epistasis

In **epistasis** (from the Greek for "standing upon"), the phenotypic expression of a gene at one locus alters that of a gene at a second locus. An example will help clarify this concept. In Labrador retrievers (commonly called Labs), black coat color is dominant to brown. Let's designate *B* and *b* as the two alleles for this character. For a Lab to have brown fur, its genotype must be *bb*; these dogs are called chocolate Labs. But there is more to the story. A second gene determines whether or not pigment will be deposited in the hair. The dominant allele, symbolized by *E*, results in the deposition of either black or brown pigment, depending on the genotype at the first locus. But if the Lab is homozygous recessive for the second locus (*ee*), then the coat is yellow, regardless of the genotype at the black/brown locus. In this case, the gene for pigment deposition (*E/e*) is said to be epistatic to the gene that codes for black or brown pigment (*B/b*).

What happens if we mate black Labs that are heterozygous for both genes (*BbEe*)? Although the two genes affect the same phenotypic character (coat color), they follow the law of independent assortment. Thus, our breeding experiment represents an F_1 dihybrid cross, like those that produced a 9:3:3:1 ratio in Mendel's experiments. We can use a Punnett square to represent the genotypes of the F_2 offspring **(Figure 12)**. As a result of epistasis, the phenotypic ratio among the F_2 offspring is 9 black to 3 chocolate to 4 yellow Labs. Other types of epistatic interactions produce different ratios, but all are modified versions of 9:3:3:1.

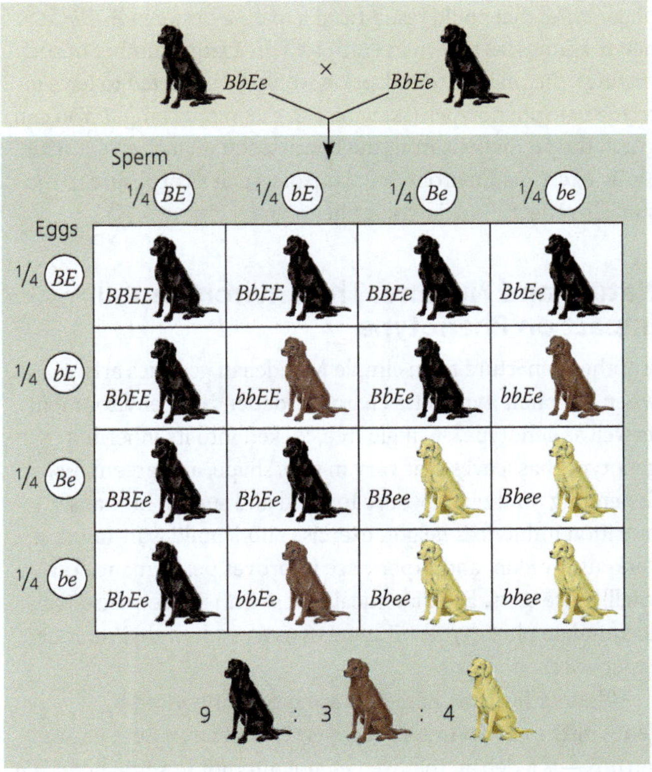

▲ **Figure 12 An example of epistasis.** This Punnett square illustrates the genotypes and phenotypes predicted for offspring of matings between two black Labrador retrievers of genotype *BbEe*. The *E/e* gene, which is epistatic to the *B/b* gene coding for hair pigment, controls whether or not pigment of any color will be deposited in the hair.

Polygenic Inheritance

Mendel studied characters that could be classified on an either-or basis, such as purple versus white flower color. But many characters, such as human skin color and height, are not one of two discrete characters, but instead vary in the population in gradations along a continuum. These are called **quantitative characters**. Quantitative variation usually indicates **polygenic inheritance**, an additive effect of two or more genes on a single phenotypic character. (In a way, this is the converse of pleiotropy, where a single gene affects several phenotypic characters.) Height is a good example of polygenic inheritance: Genomic studies have identified at least 180 gene variations that affect height.

Skin pigmentation in humans is also controlled by many separately inherited genes. Here, we'll simplify the story in order to understand the concept of polygenic inheritance. Let's consider three genes, with a dark-skin allele for each gene (*A*, *B*, or *C*) contributing one "unit" of darkness (also a simplification) to the phenotype and being incompletely dominant to the other, light-skin allele (*a*, *b*, or *c*). In our model, an *AABBCC* person would be very dark, whereas an *aabbcc* individual would be very light. An *AaBbCc* person would have skin of an intermediate shade. Because the alleles have a cumulative effect, the genotypes *AaBbCc* and *AABbcc* would make the same genetic contribution (three units) to skin darkness. There are seven skin-color phenotypes that could result from a mating between *AaBbCc* heterozygotes, as shown in **Figure 13**. In a large number of such matings, the majority of offspring would be expected to have intermediate phenotypes (skin color in the middle range). You can graph the predictions from the Punnett square in the **Scientific Skills Exercise**. Environmental factors, such as exposure to the sun, also affect the skin-color phenotype.

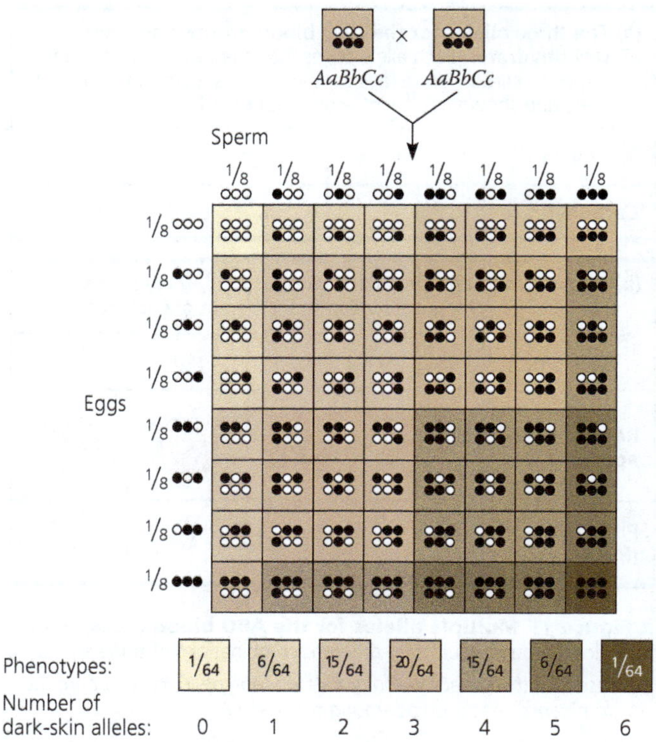

▲ **Figure 13 A simplified model for polygenic inheritance of skin color.** In this model, three separately inherited genes affect skin color. The heterozygous individuals (*AaBbCc*) represented by the two rectangles at the top of this figure each carry three dark-skin alleles (black circles, representing *A*, *B*, or *C*) and three light-skin alleles (white circles, representing *a*, *b*, or *c*). The Punnett square shows all the possible genetic combinations in gametes and offspring of many hypothetical matings between these heterozygotes. The results are summarized by the phenotypic frequencies (fractions) under the Punnett square. (The phenotypic ratio of the skin colors shown in the boxes is 1:6:15:20:15:6:1.)

Nature and Nurture: The Environmental Impact on Phenotype

Another departure from simple Mendelian genetics arises when the phenotype for a character depends on environment as well as genotype. A single tree, locked into its inherited genotype, has leaves that vary in size, shape, and greenness, depending on their exposure to wind and sun. In humans, nutrition influences height, exercise alters build, sun-tanning darkens the skin, and experience improves performance on intelligence tests. Even identical twins, who are genetic equals, accumulate phenotypic differences as a result of their unique experiences.

Whether human characters are more influenced by genes or the environment—in everyday terms, nature versus nurture—is a debate that we will not attempt to settle here. We can say, however, that a genotype generally is not associated with a rigidly defined phenotype, but rather with a range of phenotypic possibilities due to environmental influences. For some characters, such as the ABO blood group system, the range is extremely narrow; that is, a given genotype mandates a very specific phenotype. Other characters, such as a person's blood count of red and white cells, vary quite a bit, depending on such factors as the altitude, the customary level of physical activity, and the presence of infectious agents.

Generally, the phenotypic range is broadest for polygenic characters. Environment contributes to the quantitative nature of these characters, as we have seen in the continuous variation of skin color. Geneticists refer to such characters as **multifactorial**, meaning that many factors, both genetic and environmental, collectively influence phenotype.

A Mendelian View of Heredity and Variation

We have now broadened our view of Mendelian inheritance by exploring degrees of dominance as well as multiple alleles, pleiotropy, epistasis, polygenic inheritance, and the phenotypic impact of the environment. How can we integrate these refinements into a comprehensive theory of Mendelian genetics? The key is to make the transition from the reductionist emphasis on single genes and phenotypic characters to the emergent properties of the organism as a whole, one of the themes of this text.

Scientific Skills Exercise

Making a Histogram and Analyzing a Distribution Pattern

What Is the Distribution of Phenotypes Among Offspring of Two Parents Who Are Both Heterozygous for Three Additive Genes? Human skin color is a polygenic trait that is determined by the additive effects of many different genes. In this exercise, you will work with a simplified model of skin-color genetics where only three genes are assumed to affect the darkness of skin color and where each gene has two alleles—dark or light (see Figure 13). In this model, each dark allele contributes equally to the darkness of skin color, and each pair of alleles segregates independently of each other pair. Using a type of graph called a histogram, you will determine the distribution of phenotypes of offspring with different numbers of dark-skin alleles. (For additional information about graphs, see the Scientific Skills Review in the Study Area in MasteringBiology.)

How This Model Is Analyzed To predict the phenotypes of the offspring of parents heterozygous for the three genes in our simplified model, we can use the Punnett square in Figure 13. The heterozygous individuals (*AaBbCc*) represented by the two rectangles at the top of that figure each carry three dark-skin alleles (black circles, which represent *A*, *B*, or *C*) and three light-skin alleles (white circles, which represent *a*, *b*, or *c*). The Punnett square shows all the possible genetic combinations in gametes and in offspring of a large number of hypothetical matings between these heterozygotes.

Predictions from the Punnett Square If we assume that each square in the Punnett square represents one offspring of the heterozygous *AaBbCc* parents, then the squares below show the possible phenotypes. Below the squares are the predicted phenotypic frequencies of individuals with the same number of dark-skin alleles.

Phenotypes:	1/64	6/64	15/64	20/64	15/64	6/64	1/64
Number of dark-skin alleles:	0	1	2	3	4	5	6

INTERPRET THE DATA

1. A histogram is a bar graph that shows the distribution of numeric data (here, the number of dark skin alleles). To make a histogram of the allele distribution, put skin color (as the number of dark-skin alleles) along the *x*-axis and number of offspring (out of 64) with each phenotype on the *y*-axis. There are no gaps in our allele data, so draw the bars with no space between them.
2. You can see that the skin-color phenotypes are not distributed uniformly. (a) Which phenotype has the highest frequency? Draw a vertical dashed line through that bar. (b) Distributions of values like this one tend to show one of several common patterns. Sketch a rough curve that approximates the values and look at its shape. Is it symmetrically distributed around a central peak value (a "normal distribution," sometimes called a bell curve); is it skewed to one end of the *x*-axis or the other (a "skewed distribution"); or does it show two apparent groups of frequencies (a "bimodal distribution")? Explain the reason for the curve's shape. (It will help to read the text description that supports Figure 13.)

MB A version of this Scientific Skills Exercise can be assigned in MasteringBiology.

Further Reading R. A. Sturm, A golden age of human pigmentation genetics, *Trends in Genetics* 22:464–468 (2006).

The term *phenotype* can refer not only to specific characters, such as flower color and blood group, but also to an organism in its entirety—*all* aspects of its physical appearance, internal anatomy, physiology, and behavior. Similarly, the term *genotype* can refer to an organism's entire genetic makeup, not just its alleles for a single genetic locus. In most cases, a gene's impact on phenotype is affected by other genes and by the environment. In this integrated view of heredity and variation, an organism's phenotype reflects its overall genotype and unique environmental history.

Considering all that can occur in the pathway from genotype to phenotype, it is indeed impressive that Mendel could uncover the fundamental principles governing the transmission of individual genes from parents to offspring. Mendel's laws of segregation and independent assortment explain heritable variations in terms of alternative forms of genes (hereditary "particles," now known as the alleles of genes) that are passed along, generation after generation, according to simple rules of probability. This theory of inheritance is equally valid for peas, flies, fishes, birds, and human beings—indeed, for any organism with a sexual life cycle. Furthermore, by extending the principles of segregation and independent assortment to help explain such hereditary patterns as epistasis and quantitative characters, we begin to see how broadly Mendelian genetics applies. From Mendel's abbey garden came a theory of particulate inheritance that anchors modern genetics. In the last section of this chapter, we'll apply Mendelian genetics to human inheritance, with emphasis on the transmission of hereditary diseases.

CONCEPT CHECK 3

1. *Incomplete dominance* and *epistasis* are both terms that define genetic relationships. What is the most basic distinction between these terms?
2. If a man with type AB blood marries a woman with type O, what blood types would you expect in their children? What fraction would you expect of each type?
3. **WHAT IF?** A rooster with gray feathers and a hen of the same phenotype produce 15 gray, 6 black, and 8 white chicks. What is the simplest explanation for the inheritance of these colors in chickens? What phenotypes would you expect in the offspring of a cross between a gray rooster and a black hen?

CONCEPT 4

Many human traits follow Mendelian patterns of inheritance

Peas are convenient subjects for genetic research, but humans are not. The human generation span is long—about 20 years—and human parents produce many fewer offspring than peas and most other species. Even more important, it wouldn't be ethical to ask pairs of humans to breed so that the phenotypes of their offspring could be analyzed! In spite of these constraints, the study of human genetics continues, spurred on by our desire to understand our own inheritance. New molecular biological techniques have led to many breakthrough discoveries, but basic Mendelian genetics endures as the foundation of human genetics.

Pedigree Analysis

Unable to manipulate the matings of people, geneticists instead analyze the results of matings that have already occurred. They do so by collecting information about a family's history for a particular trait and assembling this information into a family tree describing the traits of parents and children across the generations—a family **pedigree**.

Figure 14a shows a three-generation pedigree that traces the occurrence of a pointed contour of the hairline on the forehead. This trait, called a widow's peak, is due to a dominant allele, *W*. Because the widow's-peak allele is dominant, all individuals who lack a widow's peak must be homozygous recessive (*ww*). The two grandparents with widow's peaks must have the *Ww* genotype, since some of their offspring are homozygous recessive. The offspring in the second generation who *do* have widow's peaks must also be heterozygous, because they are the products of *Ww* × *ww* matings. The third generation in this pedigree consists of two sisters. The one who has a widow's peak could be either homozygous (*WW*) or heterozygous (*Ww*), given what we know about the genotypes of her parents (both *Ww*).

Figure 14b is a pedigree of the same family, but this time we focus on a recessive trait, the inability of individuals to taste a chemical called PTC (phenylthiocarbamide). Compounds similar to PTC are found in broccoli, brussels sprouts, and related vegetables and account for the bitter taste some people report when eating these foods. We'll use *t* for the recessive allele and *T* for the dominant allele, which results in the ability to taste PTC. As you work your way through the pedigree, notice once again that you can apply what you have learned about Mendelian inheritance to understand the genotypes shown for the family members.

An important application of a pedigree is to help us calculate the probability that a future child will have a particular genotype and phenotype. Suppose that the couple represented in the second generation of Figure 14 decides to have one more

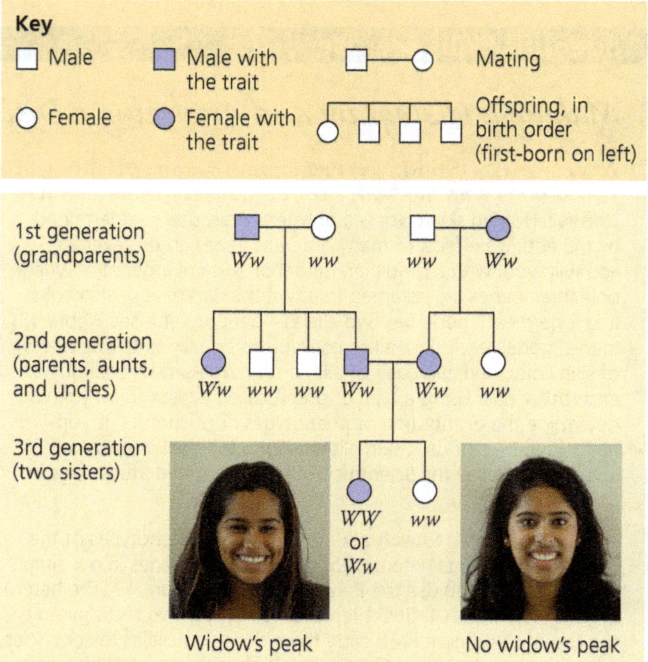

(a) Is a widow's peak a dominant or recessive trait?
Tips for pedigree analysis: Notice in the third generation that the second-born daughter lacks a widow's peak, although both of her parents had the trait. Such a pattern indicates that the trait is due to a dominant allele. If it were due to a *recessive* allele, and both parents had the recessive phenotype (straight hairline), *all* of their offspring would also have the recessive phenotype.

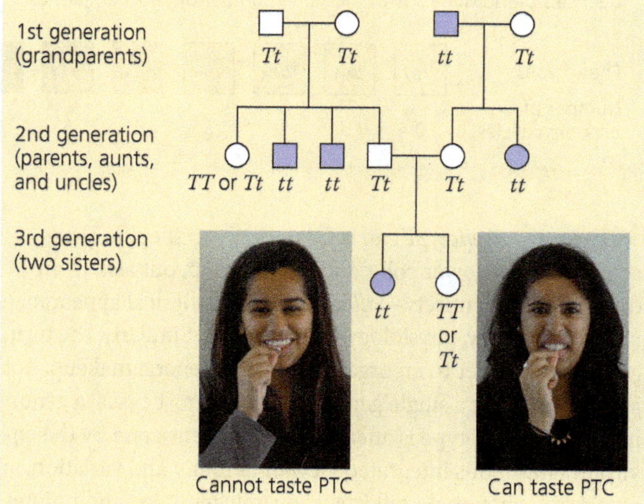

(b) Is the inability to taste a chemical called PTC a dominant or recessive trait?
Tips for pedigree analysis: Notice that the first-born daughter in the third generation has the trait (is unable to taste PTC), although both parents lack that trait (they *can* taste PTC). Such a pattern is explained if the non-taster phenotype is due to a recessive allele. (If it were due to a *dominant* allele, then at least one parent would also have had the trait.)

▲ **Figure 14 Pedigree analysis.** Each pedigree traces a trait through three generations of the same family. The two traits have different inheritance patterns. For the sake of understanding genetic principles, each trait is shown as being determined by two alleles of a single gene. This is a simplification because other genes may also affect each of these characters.

child. What is the probability that the child will have a widow's peak? This is equivalent to a Mendelian F_1 monohybrid cross ($Ww \times Ww$), and therefore the probability that a child will inherit a dominant allele and have a widow's peak is ¾ (¼ WW + ½ Ww). What is the probability that the child will be unable to taste PTC? We can also treat this as a monohybrid cross ($Tt \times Tt$), but this time we want to know the chance that the offspring will be homozygous recessive (tt). The probability is ¼. Finally, what is the chance the child will have a widow's peak *and* be unable to taste PTC? Assuming that the genes for these two characters are on different chromosomes, the two pairs of alleles will assort independently in this dihybrid cross ($WwTt \times WwTt$). Thus, we can use the multiplication rule: ¾ (chance of widow's peak) × ¼ (chance of inability to taste PTC) = 3/16 (chance of widow's peak and inability to taste PTC).

Pedigrees are a more serious matter when the alleles in question cause disabling or deadly diseases instead of an innocuous human variation, such as hairline configuration or ability to taste an innocuous chemical. However, for disorders inherited as simple Mendelian traits, the same techniques of pedigree analysis apply.

Recessively Inherited Disorders

Thousands of genetic disorders are known to be inherited as simple recessive traits. These disorders range in severity from relatively mild, such as albinism (lack of pigmentation, which results in susceptibility to skin cancers and vision problems), to life-threatening, such as cystic fibrosis.

The Behavior of Recessive Alleles

How can we account for the behavior of alleles that cause recessively inherited disorders? Recall that genes code for proteins of specific function. An allele that causes a genetic disorder (let's call it allele *a*) codes for either a malfunctioning protein or no protein at all. In the case of disorders classified as recessive, heterozygotes (*Aa*) typically have the normal phenotype because one copy of the normal allele (*A*) produces a sufficient amount of the specific protein. Thus, a recessively inherited disorder shows up only in the homozygous individuals (*aa*) who inherit a recessive allele from each parent. Although phenotypically normal with regard to the disorder, heterozygotes may transmit the recessive allele to their offspring and thus are called **carriers**. **Figure 15** illustrates these ideas using albinism as an example.

Most people who have recessive disorders are born to parents who are carriers of the disorder but have a normal phenotype, as is the case shown in the Punnett square in Figure 15. A mating between two carriers corresponds to a Mendelian F_1 monohybrid cross, so the predicted genotypic ratio for offspring is 1 *AA* : 2 *Aa* : 1 *aa*. Thus, each child has a ¼ chance of inheriting a double dose of the recessive allele; in the case of albinism, such a child will have albinism. From the genotypic ratio, we also can see that out of three offspring with the

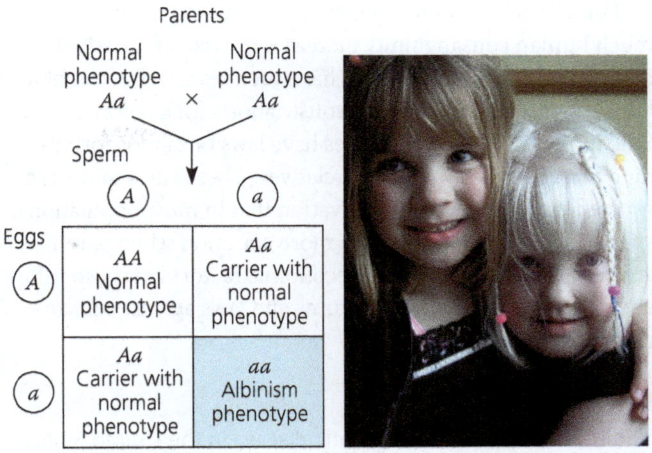

▲ **Figure 15 Albinism: a recessive trait.** One of the two sisters shown here has normal coloration; the other has albinism. Most recessive homozygotes are born to parents who are carriers of the disorder but themselves have a normal phenotype, the case shown in the Punnett square.

? *What is the probability that the sister with normal coloration is a carrier of the albinism allele?*

normal phenotype (one *AA* plus two *Aa*), two are predicted to be heterozygous carriers, a ⅔ chance. Recessive homozygotes could also result from *Aa* × *aa* and *aa* × *aa* matings, but if the disorder is lethal before reproductive age or results in sterility (neither of which is true for albinism), no *aa* individuals will reproduce. Even if recessive homozygotes are able to reproduce, this will occur relatively rarely because such individuals account for a much smaller percentage of the population than heterozygous carriers.

In general, genetic disorders are not evenly distributed among all groups of people. For example, the incidence of Tay-Sachs disease, which we described earlier in this chapter, is disproportionately high among Ashkenazic Jews, Jewish people whose ancestors lived in central Europe. In that population, Tay-Sachs disease occurs in one out of 3,600 births, an incidence about 100 times greater than that among non-Jews or Mediterranean (Sephardic) Jews. This uneven distribution results from the different genetic histories of the world's peoples during less technological times, when populations were more geographically (and hence genetically) isolated.

When a disease-causing recessive allele is rare, it is relatively unlikely that two carriers of the same harmful allele will meet and mate. The probability of passing on recessive traits increases greatly, however, if the man and woman are close relatives (for example, siblings or first cousins). This is because people with recent common ancestors are more likely to carry the same recessive alleles than are unrelated people. Thus, these consanguineous ("same blood") matings, indicated in pedigrees by double lines, are more likely to produce offspring homozygous for recessive traits—including harmful ones. Such effects can be observed in many types of domesticated and zoo animals that have become inbred.

There is debate among geneticists about exactly how much human consanguinity increases the risk of inherited diseases. Many harmful alleles have such severe effects that a homozygous embryo spontaneously aborts long before birth. Still, most societies and cultures have laws or taboos forbidding marriages between close relatives. These rules may have evolved out of empirical observation that in most populations, stillbirths and birth defects are more common when parents are closely related. Social and economic factors have also influenced the development of customs and laws against consanguineous marriages.

Cystic Fibrosis

The most common lethal genetic disease in the United States is **cystic fibrosis**, which strikes one out of every 2,500 people of European descent but is much rarer in other groups. Among people of European descent, one out of 25 (4%) are carriers of the cystic fibrosis allele. The normal allele for this gene codes for a membrane protein that functions in the transport of chloride ions between certain cells and the extracellular fluid. These chloride transport channels are defective or absent in the plasma membranes of children who inherit two recessive alleles for cystic fibrosis. The result is an abnormally high concentration of extracellular chloride, which causes the mucus that coats certain cells to become thicker and stickier than normal. The mucus builds up in the pancreas, lungs, digestive tract, and other organs, leading to multiple (pleiotropic) effects, including poor absorption of nutrients from the intestines, chronic bronchitis, and recurrent bacterial infections.

Untreated, cystic fibrosis can cause death by the age of 5. Daily doses of antibiotics to stop infection, gentle pounding on the chest to clear mucus from clogged airways, and other therapies can prolong life. In the United States, more than half of those with cystic fibrosis now survive into their 30s and beyond.

Sickle-Cell Disease: A Genetic Disorder with Evolutionary Implications

EVOLUTION The most common inherited disorder among people of African descent is **sickle-cell disease**, which affects one out of 400 African-Americans. Sickle-cell disease is caused by the substitution of a single amino acid in the hemoglobin protein of red blood cells; in homozygous individuals, all hemoglobin is of the sickle-cell (abnormal) variety. When the oxygen content of an affected individual's blood is low (at high altitudes or under physical stress, for instance), the sickle-cell hemoglobin molecules aggregate into long rods that deform the red cells into a sickle shape. Sickled cells may clump and clog small blood vessels, often leading to other symptoms throughout the body, including physical weakness, pain, organ damage, and even stroke and paralysis. Regular blood transfusions can ward off brain damage in children with sickle-cell disease, and new drugs can help prevent or treat other problems. There is currently no widely available cure, but the disease is the target of ongoing gene therapy research.

▼ Figure 16 Sickle-cell disease and sickle-cell trait.

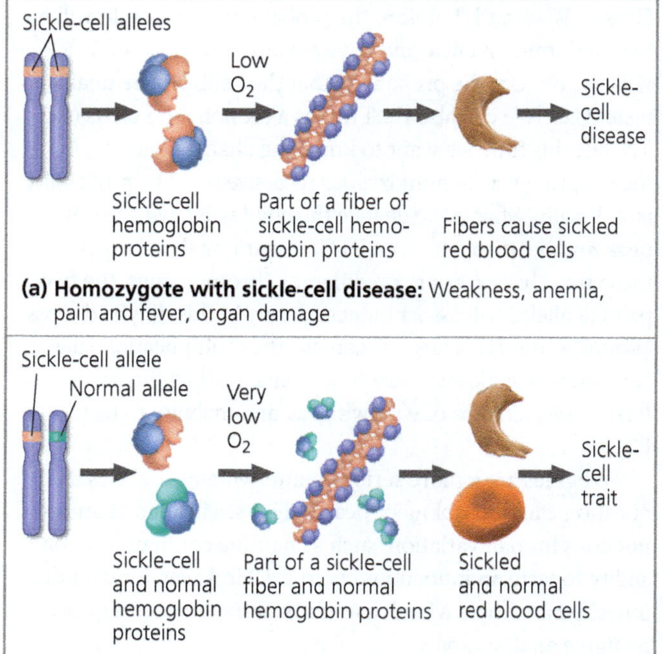

(a) Homozygote with sickle-cell disease: Weakness, anemia, pain and fever, organ damage

(b) Heterozygote with sickle-cell trait: Some symptoms when blood oxygen is very low; reduction of malaria symptoms

Although two sickle-cell alleles are necessary for an individual to manifest full-blown sickle-cell disease, the presence of one sickle-cell allele can affect the phenotype. Thus, at the organismal level, the normal allele is incompletely dominant to the sickle-cell allele **(Figure 16)**. At the molecular level, the two alleles are codominant; both normal and abnormal (sickle-cell) hemoglobins are made in heterozygotes (carriers), who are said to have *sickle-cell trait*. Heterozygotes are usually healthy but may suffer some symptoms during long periods of reduced blood oxygen.

About one out of ten African-Americans have sickle-cell trait, an unusually high frequency of heterozygotes for an allele with severe detrimental effects in homozygotes. Why haven't evolutionary processes resulted in the disappearance of the allele among this population? One explanation is that having a single copy of the sickle-cell allele reduces the frequency and severity of malaria attacks, especially among young children. The malaria parasite spends part of its life cycle in red blood cells, and the presence of even heterozygous amounts of sickle-cell hemoglobin results in lower parasite densities and hence reduced malaria symptoms. Thus, in tropical Africa, where infection with the malaria parasite is common, the sickle-cell allele confers an advantage to heterozygotes even though it is harmful in the homozygous state. The relatively high frequency of African-Americans with sickle-cell trait is a vestige of their African ancestry.

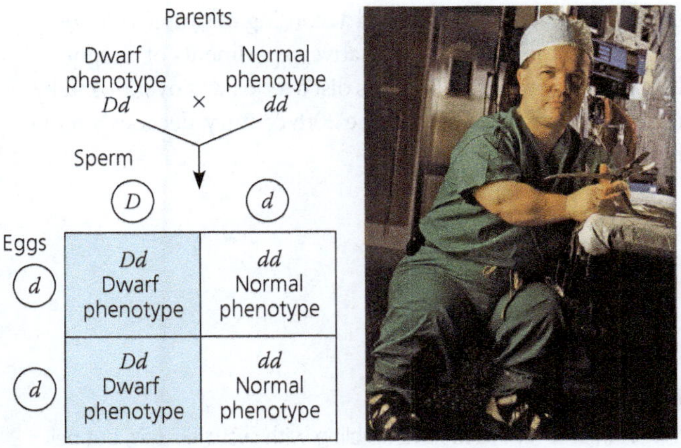

▲ **Figure 17 Achondroplasia: a dominant trait.** Dr. Michael C. Ain has achondroplasia, a form of dwarfism caused by a dominant allele. This has inspired his work: He is a specialist in the repair of bone defects caused by achondroplasia and other disorders. The dominant allele (*D*) might have arisen as a mutation in the egg or sperm of a parent or could have been inherited from an affected parent, as shown for an affected father in the Punnett square.

Dominantly Inherited Disorders

Although many harmful alleles are recessive, a number of human disorders are due to dominant alleles. One example is *achondroplasia*, a form of dwarfism that occurs in one of every 25,000 people. Heterozygous individuals have the dwarf phenotype **(Figure 17)**. Therefore, all people who do not have achondroplasia—99.99% of the population—are homozygous for the recessive allele. Like the presence of extra fingers or toes mentioned earlier, achondroplasia is a trait for which the recessive allele is much more prevalent than the corresponding dominant allele.

Unlike achondroplasia, which is relatively harmless, some dominant alleles cause lethal diseases. Those that do are much less common than recessive alleles that have lethal effects. A lethal recessive allele is only lethal when homozygous; it can be passed from one generation to the next by heterozygous carriers because the carriers themselves have normal phenotypes. A lethal dominant allele, however, often causes the death of afflicted individuals before they can mature and reproduce, and in this case the allele is not passed on to future generations.

A lethal dominant allele may be passed on, though, if the lethal disease symptoms first appear after reproductive age. In these cases, the individual may already have transmitted the allele to his or her children. For example, a degenerative disease of the nervous system called **Huntington's disease** is caused by a lethal dominant allele that has no obvious phenotypic effect until the individual is about 35 to 45 years old. Once the deterioration of the nervous system begins, it is irreversible and inevitably fatal. As with other dominant traits, a child born to a parent with the Huntington's disease allele has a 50% chance of inheriting the allele and the disorder (see the Punnett square in Figure 17). In the United States, this devastating disease afflicts about one in 10,000 people.

At one time, the onset of symptoms was the only way to know if a person had inherited the Huntington's allele, but this is no longer the case. By analyzing DNA samples from a large family with a high incidence of the disorder, geneticists tracked the Huntington's allele to a locus near the tip of chromosome 4, and the gene was sequenced in 1993. This information led to the development of a test that could detect the presence of the Huntington's allele in an individual's genome. The availability of this test poses an agonizing dilemma for those with a family history of Huntington's disease. Some individuals may want to be tested for this disease, whereas others may decide it would be too stressful to find out.

Multifactorial Disorders

The hereditary diseases we have discussed so far are sometimes described as simple Mendelian disorders because they result from abnormality of one or both alleles at a single genetic locus. Many more people are susceptible to diseases that have a multifactorial basis—a genetic component plus a significant environmental influence. Heart disease, diabetes, cancer, alcoholism, certain mental illnesses such as schizophrenia and bipolar disorder, and many other diseases are multifactorial. In these cases, the hereditary component is polygenic. For example, many genes affect cardiovascular health, making some of us more prone than others to heart attacks and strokes. No matter what our genotype, however, our lifestyle has a tremendous effect on phenotype for cardiovascular health and other multifactorial characters. Exercise, a healthful diet, abstinence from smoking, and an ability to handle stressful situations all reduce our risk of heart disease and some types of cancer.

Genetic Counseling Based on Mendelian Genetics

Avoiding simple Mendelian disorders is possible when the risk of a particular genetic disorder can be assessed before a child is conceived or during the early stages of the pregnancy. Many hospitals have genetic counselors who can provide information to prospective parents concerned about a family history for a specific disease.

Consider the case of a hypothetical couple, John and Carol. Each had a brother who died from the same recessively inherited lethal disease. Before conceiving their first child, John and Carol seek genetic counseling to determine the risk of having a child with the disease. From the information about their brothers, we know that both parents of John and both parents of Carol must have been carriers of the recessive allele. Thus, John and Carol are both products of $Aa \times Aa$ crosses, where a symbolizes the allele that causes this particular disease. We also know that John and Carol are not homozygous recessive (aa), because they do not have the disease. Therefore, their genotypes are either AA or Aa.

Given a genotypic ratio of 1 AA : 2 Aa : 1 aa for offspring of an $Aa \times Aa$ cross, John and Carol each have a ⅔ chance of being carriers (Aa). According to the rule of multiplication, the overall probability of their firstborn having the disorder is ⅔ (the chance that John is a carrier) times ⅔ (the chance that Carol is a carrier) times ¼ (the chance of two carriers having a child with the disease), which equals ⅑. Suppose that Carol and John decide to have a child—after all, there is an ⁸⁄₉ chance that their baby will not have the disorder. If, despite these odds, their child is born with the disease, then we would know that *both* John and Carol are, in fact, carriers (*Aa* genotype). If both John and Carol are carriers, there is a ¼ chance that any subsequent child this couple has will have the disease. The probability is higher for subsequent children because the diagnosis of the disease in the first child established that both parents are carriers, not because the genotype of the first child affects in any way that of future children.

When we use Mendel's laws to predict possible outcomes of matings, it is important to remember that each child represents an independent event in the sense that its genotype is unaffected by the genotypes of older siblings. Suppose that John and Carol have three more children, and *all three* have the hypothetical hereditary disease. There is only one chance in 64 (¼ × ¼ × ¼) that such an outcome will occur. Despite this run of misfortune, the chance that still another child of this couple will have the disease remains ¼.

Genetic counseling like this relies on the Mendelian model of inheritance. We owe the "gene idea"—the concept of heritable factors transmitted according to simple rules of chance—to the elegant quantitative experiments of Gregor Mendel. The importance of his discoveries was overlooked by most biologists until early in the 20th century, decades after he reported his findings.

CONCEPT CHECK 4

1. Beth and Tom each have a sibling with cystic fibrosis, but neither Beth nor Tom nor any of their parents have the disease. Calculate the probability that if this couple has a child, the child will have cystic fibrosis. What would be the probability if a test revealed that Tom is a carrier but Beth is not? Explain your answers.
2. **MAKE CONNECTIONS** In Table 1, note the phenotypic ratio of the dominant to recessive trait in the F_2 generation for the monohybrid cross involving flower color. Then determine the phenotypic ratio for the offspring of the second-generation couple in Figure 14b. What accounts for the difference in the two ratios?

11 Chapter Review

Go to **MasteringBiology®** for Assignments, the eText, and the Study Area with Animations, Activities, Vocab Self-Quiz, and Practice Tests.

SUMMARY OF KEY CONCEPTS

VOCAB SELF-QUIZ
goo.gl/gbai8v

CONCEPT 1

Mendel used the scientific approach to identify two laws of inheritance

- Gregor Mendel formulated a theory of inheritance based on experiments with garden peas, proposing that parents pass on to their offspring discrete genes that retain their identity through generations. This theory includes two "laws."
- The **law of segregation** states that genes have alternative forms, or **alleles**. In a diploid organism, the two alleles of a gene segregate (separate) during meiosis and gamete formation; each sperm or egg carries only one allele of each pair. This law explains the 3:1 ratio of F_2 phenotypes observed when **monohybrids** self-pollinate. Each organism inherits one allele for each gene from each parent. In **heterozygotes**, the two alleles are different: expression of the **dominant allele** masks the phenotypic effect of the **recessive allele**. **Homozygotes** have identical alleles of a given gene and are **true-breeding**.

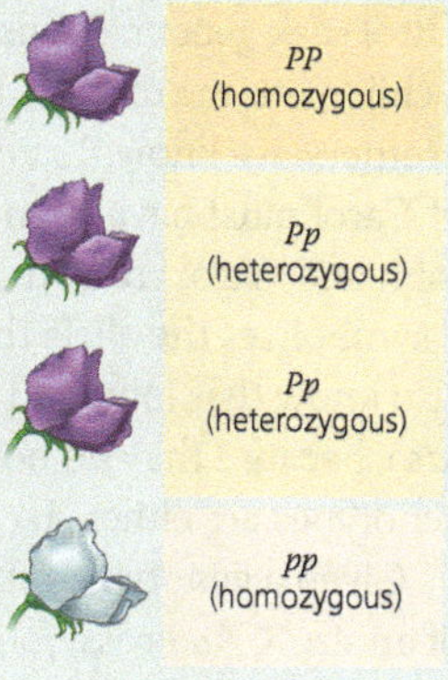

- The **law of independent assortment** states that the pair of alleles for a given gene segregates into gametes independently of the pair of alleles for any other gene. In a cross between **dihybrids** (individuals heterozygous for two genes), the offspring have four phenotypes in a 9:3:3:1 ratio.

? *When Mendel did crosses of true-breeding purple- and white-flowered pea plants, the white-flowered trait disappeared from the F_1 generation but reappeared in the F_2 generation. Use genetic terms to explain why that happened.*

CONCEPT 2

Probability laws govern Mendelian inheritance

- The **multiplication rule** states that the probability of two or more events occurring together is equal to the product of the individual probabilities of the independent single events. The **addition rule** states that the probability of an event that can occur in two or more independent, mutually exclusive ways is the sum of the individual probabilities.
- The rules of probability can be used to solve complex genetics problems. A dihybrid or other multicharacter cross is equivalent to two or more independent monohybrid crosses occurring simultaneously. In calculating the chances of the various offspring

genotypes from such crosses, each character is first considered separately and then the individual probabilities are multiplied.

DRAW IT *Redraw the Punnett square on the right side of Figure 8 as two smaller monohybrid Punnett squares, one for each gene. Below each square, list the fractions of each phenotype produced. Use the rule of multiplication to compute the overall fraction of each possible dihybrid phenotype. What is the phenotypic ratio?*

CONCEPT 3

Inheritance patterns are often more complex than predicted by simple Mendelian genetics

- Extensions of Mendelian genetics for a single gene:

Relationship among alleles of a single gene	Description	Example
Complete dominance of one allele	Heterozygous phenotype same as that of homozygous dominant	*PP* *Pp* 
Incomplete dominance of either allele	Heterozygous phenotype intermediate between the two homozygous phenotypes	 C^RC^R C^RC^W C^WC^W
Codominance	Both phenotypes expressed in heterozygotes	I^AI^B 
Multiple alleles	In the population, some genes have more than two alleles	ABO blood group alleles I^A, I^B, i
Pleiotropy	One gene affects multiple phenotypic characters	Sickle-cell disease

- Extensions of Mendelian genetics for two or more genes:

Relationship among two or more genes	Description	Example
Epistasis	The phenotypic expression of one gene affects the expression of another gene	*BbEe* × *BbEe* 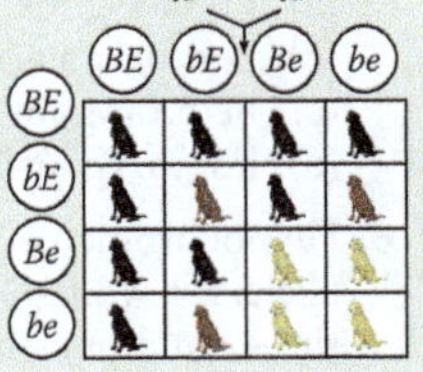 9 : 3 : 4
Polygenic inheritance	A single phenotypic character is affected by two or more genes	*AaBbCc* × *AaBbCc* 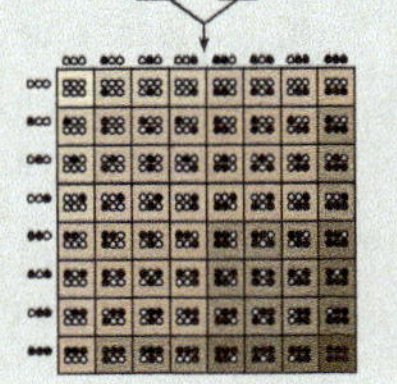

- The expression of a genotype can be affected by environmental influences. Polygenic characters that are also influenced by the environment are called **multifactorial** characters.

- An organism's overall phenotype reflects its overall genotype and unique environmental history. Even in more complex inheritance patterns, Mendel's fundamental laws still apply.

? *Which genetic relationships listed in the first column of the two tables are demonstrated by the inheritance pattern of the ABO blood group alleles? For each genetic relationship, explain why this inheritance pattern is or is not an example.*

CONCEPT 4

Many human traits follow Mendelian patterns of inheritance

- Analysis of family **pedigrees** can be used to deduce the possible genotypes of individuals and make predictions about future offspring. Such predictions are statistical probabilities rather than certainties.

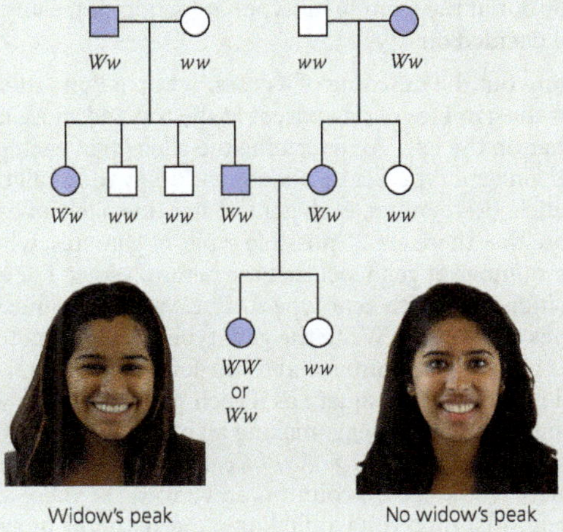

- Many genetic disorders are inherited as simple recessive traits. Most affected (homozygous recessive) individuals are children of phenotypically normal, heterozygous **carriers**.

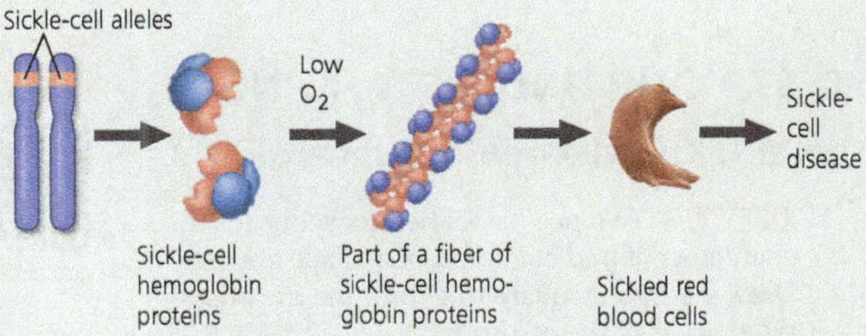

- The sickle-cell allele has probably persisted for evolutionary reasons: Heterozygotes have an advantage because one copy of the sickle-cell allele reduces both the frequency and severity of malaria attacks.
- Lethal dominant alleles are eliminated from the population if affected people die before reproducing. Nonlethal dominant alleles and lethal alleles that are expressed relatively late in life are inherited in a Mendelian way.
- Many human diseases are multifactorial—that is, they have both genetic and environmental components and do not follow simple Mendelian inheritance patterns.
- Using family histories, genetic counselors help couples determine the probability of their children having genetic disorders.

? *Both members of a couple know that they are carriers of the cystic fibrosis allele. None of their three children has cystic fibrosis, but any one of them might be a carrier. The couple would like to have a fourth child but are worried that he or she would very likely have the disease, since the first three do not. What would you tell the couple? Would it remove some more uncertainty in their prediction if they could find out whether the three children are carriers?*

TIPS FOR GENETICS PROBLEMS

1. Write down symbols for the alleles. (These may be given in the problem.) When represented by single letters, the dominant allele is uppercase and the recessive allele is lowercase.
2. Write down the possible genotypes, as determined by the phenotype.
   a. If the phenotype is that of the dominant trait (for example, purple flowers), then the genotype is either homozygous dominant or heterozygous (*PP* or *Pp*, in this example).
   b. If the phenotype is that of the recessive trait, the genotype must be homozygous recessive (for example, *pp*).
   c. If the problem says "true-breeding," the genotype is homozygous.
3. Determine what the problem is asking. If asked to do a cross, write it out in the form [genotype] × [genotype], using the alleles you've decided on.
4. To figure out the outcome of a cross, set up a Punnett square.
   a. Put the gametes of one parent at the top and those of the other on the left. To determine the allele(s) in each gamete for a given genotype, set up a systematic way to list all the possibilities. (Remember, each gamete has one allele of each gene.) Note that there are 2^n possible types of gametes, where n is the number of gene loci that are heterozygous. For example, an individual with genotype *AaBbCc* would produce $2^3 = 8$ types of gametes. Write the genotypes of the gametes in circles above the columns and to the left of the rows.
   b. Fill in the Punnett square as if each possible sperm were fertilizing each possible egg, making all of the possible offspring. In a cross of *AaBbCc* × *AaBbCc*, for example, the Punnett square would have 8 columns and 8 rows, so there are 64 different offspring; you would know the genotype of each and thus the phenotype. Count genotypes and phenotypes to obtain the genotypic and phenotypic ratios. Because the Punnett square is so large, this method is not the most efficient. Instead, see tip 5.
5. You can use the rules of probability if the Punnett square would be too big. (For example, see the question at the end of the summary for Concept 2 and question 7.) You can consider each gene separately (see the section Solving Complex Genetics Problems with the Rules of Probability in Concept 2).
6. If the problem gives you the phenotypic ratios of offspring but not the genotypes of the parents in a given cross, the phenotypes can help you deduce the parents' unknown genotypes.
   a. For example, if ½ of the offspring have the recessive phenotype and ½ the dominant, you know that the cross was between a heterozygote and a homozygous recessive.
   b. If the ratio is 3:1, the cross was between two heterozygotes.
   c. If two genes are involved and you see a 9:3:3:1 ratio in the offspring, you know that each parent is heterozygous for both genes. *Caution*: Don't assume that the reported numbers will exactly equal the predicted ratios. For example, if there are 13 offspring with the dominant trait and 11 with the recessive, assume that the ratio is one dominant to one recessive.
7. For pedigree problems, use the tips in Figure 14 and below to determine what kind of trait is involved.
   a. If parents without the trait have offspring with the trait, the trait must be recessive and both of the parents must be carriers.
   b. If the trait is seen in every generation, it is most likely dominant (see the next possibility, though).
   c. If both parents have the trait, then in order for it to be recessive, all offspring must show the trait.
   d. To determine the likely genotype of a certain individual in a pedigree, first label the genotypes of all the family members you can. Even if some of the genotypes are incomplete, label what you do know. For example, if an individual has the dominant phenotype, the genotype must be *AA* or *Aa*; you can write this as *A*–. Try different possibilities to see which fits the results. Use the rules of probability to calculate the probability of each possible genotype being the correct one.

TEST YOUR UNDERSTANDING

PRACTICE TEST
goo.gl/CRZjvS

Level 1: Knowledge/Comprehension

1. **DRAW IT** Two pea plants heterozygous for the characters of pod color and pod shape are crossed. Draw a Punnett square to determine the phenotypic ratios of the offspring.
2. A man with type A blood marries a woman with type B blood. Their child has type O blood. What are the genotypes of these three individuals? What genotypes, and in what frequencies, would you expect in future offspring from this marriage?
3. A man has six fingers on each hand and six toes on each foot. His wife and their daughter have the normal number of digits. Remember that extra digits is a dominant trait. What fraction of this couple's children would be expected to have extra digits?
4. **DRAW IT** A pea plant heterozygous for inflated pods (*Ii*) is crossed with a plant homozygous for constricted pods (*ii*). Draw a Punnett square for this cross to predict genotypic and phenotypic ratios. Assume that pollen comes from the *ii* plant.

Level 2: Application/Analysis

5. Flower position, stem length, and seed shape are three characters that Mendel studied. Each is controlled by an independently assorting gene and has dominant and recessive expression as indicated in Table 1. If a plant that is heterozygous for all three characters is allowed to self-fertilize, what proportion of the offspring would you expect to be as follows? (*Note*: Use the rules of probability instead of a huge Punnett square.)
   (a) homozygous for the three dominant traits
   (b) homozygous for the three recessive traits
   (c) heterozygous for all three characters
   (d) homozygous for axial and tall, heterozygous for seed shape
6. Hemochromatosis is an inherited disease caused by a recessive allele. If a woman and her husband, who are both carriers, have three children, what is the probability of each of the following?
   (a) All three children are of normal phenotype.
   (b) One or more of the three children have the disease.
   (c) All three children have the disease.
   (d) At least one child is phenotypically normal.

   (*Note*: It will help to remember that the probabilities of all possible outcomes always add up to 1.)
7. The genotype of F_1 individuals in a tetrahybrid cross is *AaBbCcDd*. Assuming independent assortment of these four genes, what are the probabilities that F_2 offspring will have the following genotypes?
   (a) *aabbccdd*
   (b) *AaBbCcDd*
   (c) *AABBCCDD*
   (d) *AaBBccDd*
   (e) *AaBBCCdd*

8. What is the probability that each of the following pairs of parents will produce the indicated offspring? (Assume independent assortment of all gene pairs.)
(a) $AABBCC \times aabbcc \rightarrow AaBbCc$
(b) $AABbCc \times AaBbCc \rightarrow AAbbCC$
(c) $AaBbCc \times AaBbCc \rightarrow AaBbCc$
(d) $aaBbCC \times AABbcc \rightarrow AaBbCc$

9. Karen and Steve each have a sibling with sickle-cell disease. Neither Karen nor Steve nor any of their parents have the disease, and none of them have been tested to see if they have the sickle-cell trait. Based on this incomplete information, calculate the probability that if this couple has a child, the child will have sickle-cell disease.

10. In 1981, a stray black cat with unusual rounded, curled-back ears was adopted by a family in California. Hundreds of descendants of the cat have since been born, and cat fanciers hope to develop the curl cat into a show breed. Suppose you owned the first curl cat and wanted to develop a true-breeding variety. How would you determine whether the curl allele is dominant or recessive? How would you obtain true-breeding curl cats? How could you be sure they are true-breeding?

11. In tigers, a recessive allele that is pleiotropic causes an absence of fur pigmentation (a white tiger) and a cross-eyed condition. If two phenotypically normal tigers that are heterozygous at this locus are mated, what percentage of their offspring will be cross-eyed? What percentage of cross-eyed tigers will be white?

12. In maize (corn) plants, a dominant allele *I* inhibits kernel color, while the recessive allele *i* permits color when homozygous. At a different locus, the dominant allele *P* causes purple kernel color, while the homozygous recessive genotype *pp* causes red kernels. If plants heterozygous at both loci are crossed, what will be the phenotypic ratio of the offspring?

13. The pedigree below traces the inheritance of alkaptonuria, a biochemical disorder. Affected individuals, indicated here by the colored circles and squares, are unable to metabolize a substance called alkapton, which colors the urine and stains body tissues. Does alkaptonuria appear to be caused by a dominant allele or by a recessive allele? Fill in the genotypes of the individuals whose genotypes can be deduced. What genotypes are possible for each of the other individuals?

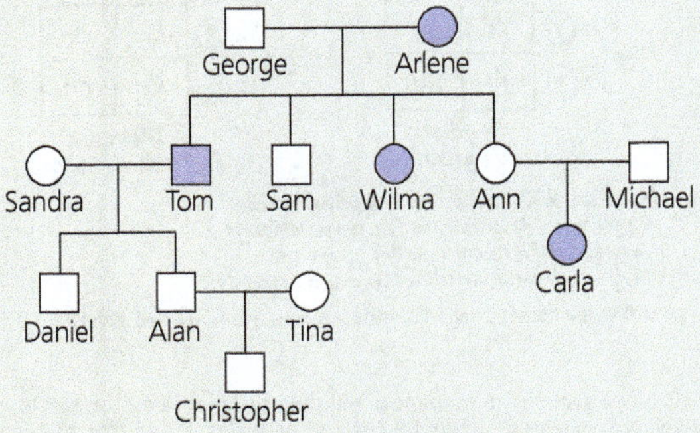

14. Imagine that you are a genetic counselor, and a couple planning to start a family comes to you for information. Charles was married once before, and he and his first wife had a child with cystic fibrosis. The brother of his current wife, Elaine, died of cystic fibrosis. What is the probability that Charles and Elaine will have a baby with cystic fibrosis? (Neither Charles, nor Elaine, nor their parents have cystic fibrosis.)

Level 3: Synthesis/Evaluation

15. FOCUS ON EVOLUTION
Over the past half century, there has been a trend in the United States and other developed countries for people to marry and start families later in life than did their parents and grandparents. Describe what effects this trend might have on the incidence (frequency) of late-acting dominant lethal alleles in the population.

16. SCIENTIFIC INQUIRY
You are handed a mystery pea plant with tall stems and axial flowers and asked to determine its genotype as quickly as possible. You know that the allele for tall stems (*T*) is dominant to that for dwarf stems (*t*) and that the allele for axial flowers (*A*) is dominant to that for terminal flowers (*a*).
(a) Identify ALL the possible genotypes for your mystery plant.
(b) Describe the ONE cross you would do, out in your garden, to determine the exact genotype of your mystery plant.
(c) While waiting for the results of your cross, you predict the results for each possible genotype listed in part a. Explain how you do this and why this is not called "performing a cross."
(d) Explain how the results of your cross and your predictions will help you learn the genotype of your mystery plant.

17. FOCUS ON INFORMATION
The continuity of life is based on heritable information in the form of DNA. In a short essay (100–150 words), explain how the passage of genes from parents to offspring, in the form of particular alleles, ensures perpetuation of parental traits in offspring and, at the same time, genetic variation among offspring. Use genetic terms in your explanation.

18. SYNTHESIZE YOUR KNOWLEDGE

Just for fun, imagine that "shirt-striping" is a phenotypic character caused by a single gene. Make up a genetic explanation for the appearance of the family in the above photograph, consistent with their "shirt phenotypes." Include in your answer the presumed allele combinations for "shirt-striping" in each family member. What is the inheritance pattern shown by the child?

Answers

Figure Questions

Figure 3 All offspring would have purple flowers. (The ratio would be 1 purple : 0 white.) The P generation plants are true-breeding, so mating two purple-flowered plants produces the same result as self-pollination: All the offspring have the same trait.
Figure 8

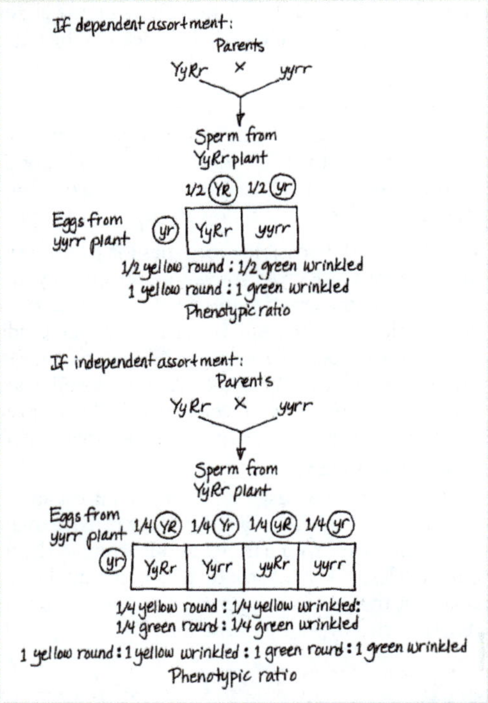

Yes, this cross would also have allowed Mendel to make different predictions for the two hypotheses, thereby allowing him to distinguish the correct one. **Figure 10** Your classmate would probably point out that the F_1 generation hybrids show an intermediate phenotype between those of the homozygous parents, which supports the blending hypothesis. You could respond that crossing the F_1 hybrids results in the reappearance of the white phenotype, rather than identical pink offspring, which fails to support the idea of traits blending during inheritance, in which case the white trait would have been lost after the F_1 generation. **Figure 11** Both the I^A and I^B alleles are dominant to the *i* allele, which is recessive and results in no attached carbohydrate. The I^A and I^B alleles are codominant; both are expressed in the phenotype of I^AI^B heterozygotes, who have type AB blood. **Figure 15** In the Punnett square, two of the three individuals with normal coloration are carriers, so the probability is ⅔. (Note that you must take into account everything you know when you calculate probability: You know she is not *aa*, so there are only three possible genotypes to consider.)

Concept Check 1

1. According to the law of independent assortment, 25 plants (1/16 of the offspring) are predicted to be *aatt*, or recessive for both characters. The actual result is likely to differ slightly from this value.

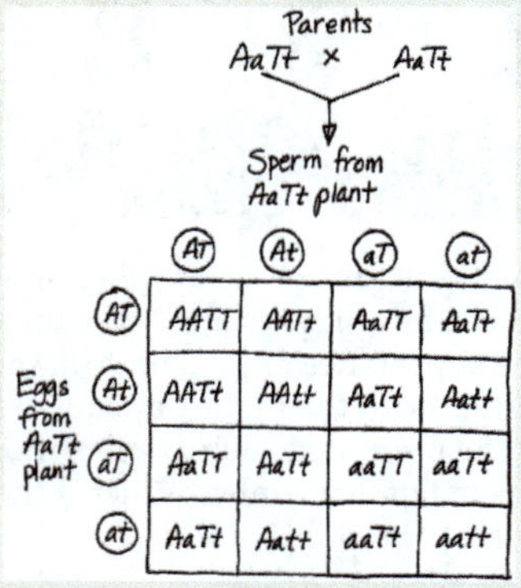

2. The plant could make eight different gametes (*YRI, YRi, YrI, Yri, yRI, yRi, yrI,* and *yri*). To fit all the possible gametes in a self-pollination, a Punnett square would need 8 rows and 8 columns. It would have spaces for the 64 possible unions of gametes in the offspring. **3.** Self-pollination is sexual reproduction because meiosis is involved in forming gametes, which unite during fertilization. As a result, the offspring in self-pollination are genetically different from the parent.

Concept Check 2

1. ½ homozygous dominant (*AA*), 0 homozygous recessive (*aa*), and ½ heterozygous (*Aa*) **2.** ¼ *BBDD*; ¼ *BbDD*; ¼ *BBDd*; ¼ *BbDd* **3.** The genotypes that fulfill this condition are *ppyyIi, ppYyii, Ppyyii, ppYYii,* and *ppyyii.* Use the multiplication rule to find the probability of getting each genotype and then use the addition rule to find the overall probability of meeting the conditions of this problem:

ppyyIi	½ (probability of pp) × ¼ (yy) × ½ (ii)	= 1/16
ppYyii	½ (pp) × ½ (Yy) × ½ (ii)	= 2/16
Ppyyii	½ (Pp) × ¼ (yy) × ½ (ii)	= 1/16
ppYYii	½ (pp) × ¼ (YY) × ½ (ii)	= 1/16
ppyyii	½ (pp) × ¼ (yy) × ½ (ii)	= 1/16
Fraction predicted to have at least two recessive traits		= 6/16 or 3/8

Concept Check 3

1. Incomplete dominance describes the relationship between two alleles of a single gene, whereas epistasis relates to the genetic relationship between two genes (and the respective alleles of each). **2.** Half of the children would be expected to have type A blood and half type B blood. **3.** The black and white alleles are incompletely dominant, with heterozygotes being gray in color. A cross between a gray rooster and a black hen should yield approximately equal numbers of gray and black offspring.

Concept Check 4

1. ⅑ (Since cystic fibrosis is caused by a recessive allele, Beth and Tom's siblings who have CF must be homozygous recessive. Therefore, each parent must be a carrier of the recessive allele. Since neither Beth nor Tom has CF, this means they each have a ⅔ chance of being a carrier. If they are both carriers, there is a ¼ chance that they will have a child with CF. ⅔ × ⅔ × ¼ = ⅑); 0 (Both Beth and Tom would have to be carriers to produce a child with the disease.) **2.** In the monohybrid cross involving flower color, the ratio is 3.15 purple : 1 white, while in the human family in the pedigree, the ratio in the third generation is 1 can taste PTC : 1 cannot taste PTC. The difference is due to the small sample size (two offspring) in the human family. If the second-generation couple in this pedigree were able to have 929 offspring as in the pea plant cross, the ratio would likely be closer to 3:1. (Note that none of the pea plant crosses in Table 1 yielded *exactly* a 3:1 ratio.)

Summary of Key Concepts Questions

1 Alternative versions of genes, called alleles, are passed from parent to offspring during sexual reproduction. In a cross between purple- and white-flowered homozygous parents, the F_1 offspring are all heterozygous, each inheriting a purple allele from one parent and a white allele from the other. Because the purple allele is dominant, it determines the phenotype of the F_1 offspring to be purple, and the expression of the recessive white allele is masked. Only in the F_2 generation is it possible for a white allele to exist in the homozygous state, which causes the white trait to be expressed.
2

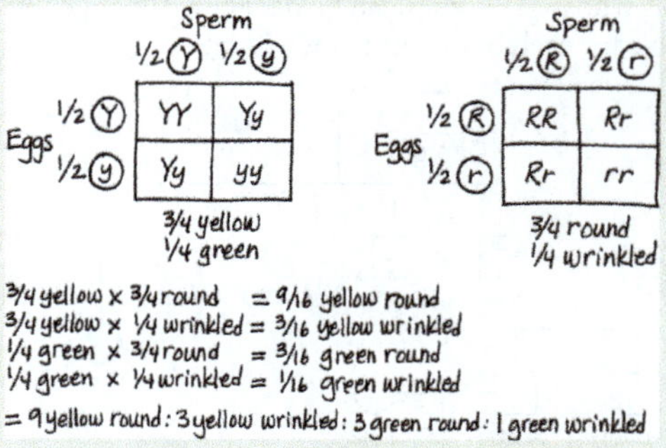

3 The ABO blood group is an example of multiple alleles because this single gene has more than two alleles (I^A, I^B, and *i*). Two of the alleles, I^A and I^B, exhibit codominance, since both carbohydrates (A and B) are present when these two alleles exist together in a genotype. I^A and I^B each exhibit complete dominance over the *i* allele. This situation is not an example of incomplete dominance because each allele affects the phenotype in a distinguishable way, so the result is not intermediate between the two phenotypes. Because this situation involves a single gene, it is not an example of epistasis or polygenic inheritance. **4** The chance of the fourth child having cystic fibrosis is ¼, as it was for each of the other children, because each birth is an independent event. We already know that both parents are carriers, so whether their first three children are carriers or not has no bearing on the probability that their next child will have the disease. The parents' genotypes provide the only relevant information.

Test Your Understanding

1.

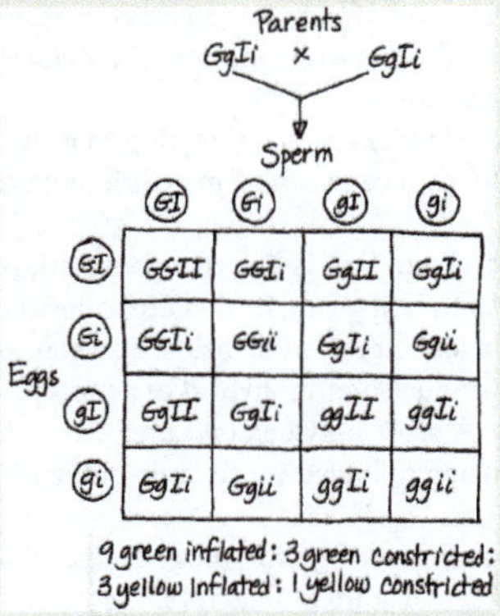

2. Man I^Ai; woman I^Bi; child ii. Genotypes for future children are predicted to be ¼ I^AI^B, ¼ I^Ai, ¼ I^Bi, ¼ ii. **3.** ½ **4.** A cross of $Ii \times ii$ would yield offspring with a genotypic ratio of 1 Ii : 1 ii (2:2 is an equivalent answer) and a phenotypic ratio of 1 inflated : 1 constricted (2:2 is equivalent).

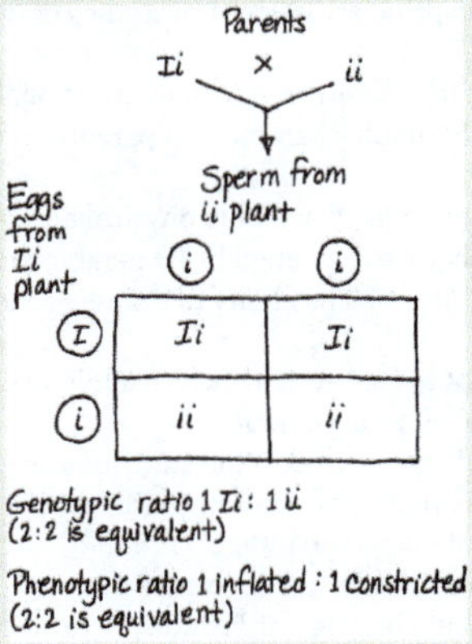

5. (a) 1⁄64; (b) 1⁄64; (c) ⅛; (d) 1⁄32 **6.** (a) ¾ × ¾ × ¾ = 27⁄64; (b) 1 − 27⁄64 = 37⁄64; (c) ¼ × ¼ × ¼ = 1⁄64; (d) 1 − 1⁄64 = 63⁄64 **7.** (a) 1⁄256; (b) 1⁄16; (c) 1⁄256; (d) 1⁄64; (e) 1⁄128 **8.** (a) 1; (b) 1⁄32; (c) ⅛; (d) ½ **9.** 1⁄9 **10.** Matings of the original mutant cat with true-breeding noncurl cats will produce both curl and noncurl F_1 offspring if the curl allele is dominant, but only noncurl offspring if the curl allele is recessive. You would obtain some true-breeding offspring homozygous for the curl allele from matings between the F_1 cats resulting from the original curl × noncurl crosses. If dominant, you wouldn't be able to tell true-breeding, homozygous offspring from heterozygotes without further crosses whether the curl trait is dominant or recessive. You know that cats are true-breeding when curl × curl matings produce only curl offspring for several generations. As it turns out, the allele that causes curled ears is dominant. 25%, or ¼, will be cross-eyed; all (100%) of the cross-eyed offspring will also be white. **12.** The dominant allele I is epistatic to the P/p locus, and thus the genotypic ratio for the F_1 generation will be 9 I–P– (colorless) : 3 I–pp (colorless) : 3 iiP– (purple) : 1 $iipp$ (red). Overall, the phenotypic ratio is 12 colorless : 3 purple : 1 red. **13.** Recessive. All affected individuals (Arlene, Tom, Wilma, and Carla) are homozygous recessive aa. George is Aa, since some of his children with Arlene are affected. Sam, Ann, Daniel, and Alan are each Aa, since they are all unaffected children with one affected parent. Michael also is Aa, since he has an affected child (Carla) with his heterozygous wife Ann. Sandra, Tina, and Christopher can each have either the AA or Aa genotype. **14.** 1⁄9

Credits

Credits are listed in the order of appearance.

Photo Credits

1 John Swithinbank/age fotostock; **11 SSE** Alberto Pomares/Getty Images; **14** Barbara Bowman, Pearson Education; **15** Rick Guidotti, Positive Exposure; **17** Michael Ciesielski Photography; **Chapter Review hairlines** Barbara Bowman, Pearson Education; **cat** Breeder/owner Patricia Speciale, photographer Norma JubinVille; **SYK** Rene Maltete/RAPHO/Gamma-Rapho via Getty Images.

Illustration and Text Credits

SSE.1 Data from R.A. Sturm, A golden age of human pigmentation genetics, *Trends in Genetics* 22:464–468 (2006); **3** Data from G. Mendel, Experiments in plant hybridization, *Proceedings of the Natural History Society of Brünn* 4:3–47 (1866). **8** Data from G. Mendel, Experiments in plant hybridization, *Proceedings of the Natural History Society of Brünn* 4:3–47 (1866).

Glossary

addition rule A rule of probability stating that the probability of any one of two or more mutually exclusive events occurring can be determined by adding their individual probabilities.

allele (uh-lē'-ul) Any of the alternative versions of a gene that may produce distinguishable phenotypic effects.

carrier In genetics, an individual who is heterozygous at a given genetic locus for a recessively inherited disorder. The heterozygote is generally phenotypically normal for the disorder but can pass on the recessive allele to offspring.

character An observable heritable feature that may vary among individuals.

codominance The situation in which the phenotypes of both alleles are exhibited in the heterozygote because both alleles affect the phenotype in separate, distinguishable ways.

complete dominance The situation in which the phenotypes of the heterozygote and dominant homozygote are indistinguishable.

cystic fibrosis (sis'-tik fī-brō'-sis) A human genetic disorder caused by a recessive allele for a chloride channel protein; characterized by an excessive secretion of mucus and consequent vulnerability to infection; fatal if untreated.

dihybrid (dī'-hī'-brid) An organism that is heterozygous with respect to two genes of interest. All the offspring from a cross between parents doubly homozygous for different alleles are dihybrids. For example, parents of genotypes *AABB* and *aabb* produce a dihybrid of genotype *AaBb*.

dihybrid cross A cross between two organisms that are each heterozygous for both of the characters being followed (or the self-pollination of a plant that is heterozygous for both characters).

dominant allele An allele that is fully expressed in the phenotype of a heterozygote.

epistasis (ep'-i-stā'-sis) A type of gene interaction in which the phenotypic expression of one gene alters that of another independently inherited gene.

F_1 generation The first filial, hybrid (heterozygous) offspring arising from a parental (P generation) cross.

F_2 generation The offspring resulting from interbreeding (or self-pollination) of the hybrid F_1 generation.

genotype (jē'-nō-tīp) The genetic makeup, or set of alleles, of an organism.

heterozygous (het'-er-ō-zī'-gus) Having two different alleles for a given gene.

homozygous (hō'-mō-zī'-gus) Having two identical alleles for a given gene.

Huntington's disease A human genetic disease caused by a dominant allele; characterized by uncontrollable body movements and degeneration of the nervous system; usually fatal 10 to 20 years after the onset of symptoms.

hybridization In genetics, the mating, or crossing, of two true-breeding varieties.

incomplete dominance The situation in which the phenotype of heterozygotes is intermediate between the phenotypes of individuals homozygous for either allele.

law of independent assortment Mendel's second law, stating that each pair of alleles segregates, or assorts, independently of each other pair during gamete formation; applies when genes for two characters are located on different pairs of homologous chromosomes or when they are far enough apart on the same chromosome to behave as though they are on different chromosomes.

law of segregation Mendel's first law, stating that the two alleles in a pair segregate (separate from each other) into different gametes during gamete formation.

monohybrid An organism that is heterozygous with respect to a single gene of interest. All the offspring from a cross between parents homozygous for different alleles are monohybrids. For example, parents of genotypes *AA* and *aa* produce a monohybrid of genotype *Aa*.

monohybrid cross A cross between two organisms that are heterozygous for the character being followed (or the self-pollination of a heterozygous plant).

multifactorial Referring to a phenotypic character that is influenced by multiple genes and environmental factors.

multiplication rule A rule of probability stating that the probability of two or more independent events occurring together can be determined by multiplying their individual probabilities.

P generation The true-breeding (homozygous) parent individuals from which F_1 hybrid offspring are derived in studies of inheritance; P stands for "parental."

pedigree A diagram of a family tree with conventional symbols, showing the occurrence of heritable characters in parents and offspring over multiple generations.

phenotype (fē'-nō-tīp) The observable physical and physiological traits of an organism, which are determined by its genetic makeup.

pleiotropy (plī'-o-truh-pē) The ability of a single gene to have multiple effects.

polygenic inheritance (pol'-ē-jen'-ik) An additive effect of two or more genes on a single phenotypic character.

Punnett square A diagram used in the study of inheritance to show the predicted genotypic results of random fertilization in genetic crosses between individuals of known genotype.

quantitative character A heritable feature that varies continuously over a range rather than in an either-or fashion.

Tay-Sachs disease A human genetic disease caused by a recessive allele for a dysfunctional enzyme, leading to accumulation of certain lipids in the brain. Seizures, blindness, and degeneration of motor and mental performance usually become manifest a few months after birth, followed by death within a few years.

testcross Breeding an organism of unknown genotype with a homozygous recessive individual to determine the unknown genotype. The ratio of phenotypes in the offspring reveals the unknown genotype.

trait One of two or more detectable variants in a genetic character.

true-breeding Referring to organisms that produce offspring of the same variety over many generations of self-pollination.

The Chromosomal Basis of Inheritance

The Chromosomal Basis of Inheritance

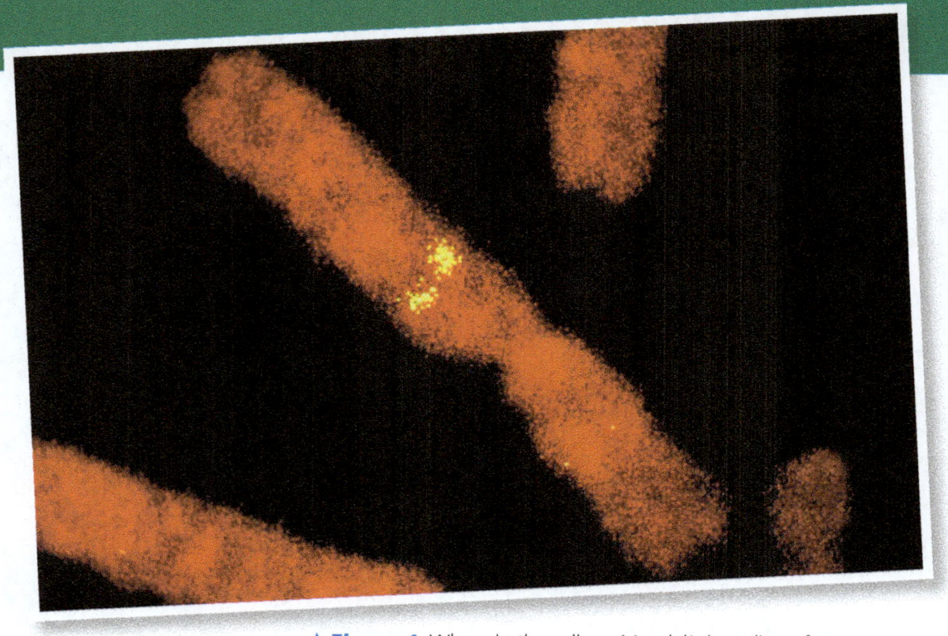

▲ **Figure 1** Where in the cell are Mendel's hereditary factors located?

KEY CONCEPTS

1. Morgan showed that Mendelian inheritance has its physical basis in the behavior of chromosomes: *scientific inquiry*
2. Sex-linked genes exhibit unique patterns of inheritance
3. Linked genes tend to be inherited together because they are located near each other on the same chromosome
4. Alterations of chromosome number or structure cause some genetic disorders

Locating Genes Along Chromosomes

Today, we know that genes—Mendel's "factors"—are segments of DNA located along chromosomes. We can see the location of a particular gene by tagging chromosomes with a fluorescent dye that highlights that gene. For example, the two yellow regions in **Figure 1** mark a specific gene on human chromosome 6. (The chromosome has duplicated, so the allele on that chromosome is present as two copies, one per sister chromatid.) However, Gregor Mendel's "hereditary factors" were purely an abstract concept when he proposed their existence in 1860. At that time, no cellular structures had been identified that could house these imaginary units, and most biologists were skeptical about Mendel's proposed laws of inheritance.

Using improved techniques of microscopy, cytologists worked out the process of mitosis in 1875 and meiosis in the 1890s. Cytology and genetics converged as biologists began to see parallels between the behavior of Mendel's proposed hereditary factors during sexual life cycles and the behavior of chromosomes. As shown in **Figure 2**, chromosomes and genes are both present in pairs in diploid cells; homologous chromosomes separate and alleles segregate during the process of meiosis. After meiosis, fertilization restores the paired condition for both chromosomes and genes. Around 1902, Walter S. Sutton, Theodor Boveri, and others noted these parallels and began to develop the **chromosome theory of inheritance**. According to this theory, Mendelian genes have specific loci (positions) along chromosomes, and it is the chromosomes that undergo segregation and independent assortment.

As you can see in Figure 2, the separation of homologs during anaphase I accounts for the segregation of the two alleles of a gene into separate gametes, and the random arrangement of chromosome pairs at metaphase I accounts for independent assortment of the alleles for two or more genes located on different homologous pairs. By carefully studying Figure 2, you can see how the behavior of chromosomes during meiosis in the F_1 generation and subsequent random fertilization give rise to the F_2 phenotypic ratio observed by Mendel.

In this chapter we will correlate the behavior of chromosomes with that of genes. After describing evidence from the fruit fly that strongly supported the chromosome theory, we'll explore the chromosomal basis for the transmission of genes from parents to offspring, including what happens when two genes are linked on the same chromosome. Finally, we'll discuss some important exceptions to the standard mode of inheritance.

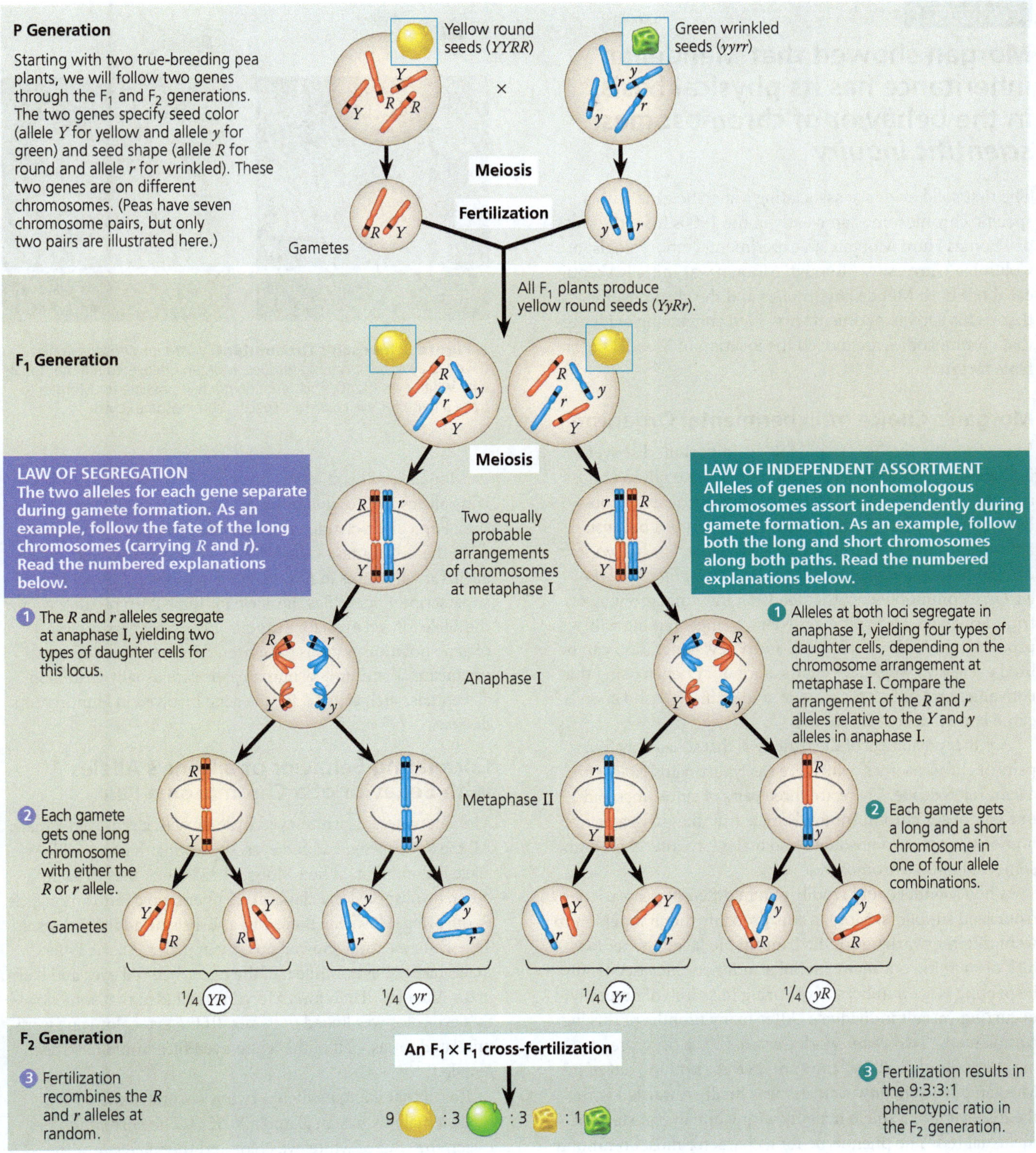

▲ **Figure 2 The chromosomal basis of Mendel's laws.** Here we correlate the results of one of Mendel's dihybrid crosses with the behavior of chromosomes during meiosis. The arrangement of chromosomes at metaphase I of meiosis and their movement during anaphase I account for, respectively, the independent assortment and segregation of the alleles for seed color and shape. Each cell that undergoes meiosis in an F_1 plant produces two kinds of gametes. If we count the results for all cells, however, each F_1 plant produces equal numbers of all four kinds of gametes because the alternative chromosome arrangements at metaphase I are equally likely.

? *If you crossed an F_1 plant with a plant that was homozygous recessive for both genes* (yyrr), *how would the phenotypic ratio of the offspring compare with the 9:3:3:1 ratio seen here?*

CONCEPT 1

Morgan showed that Mendelian inheritance has its physical basis in the behavior of chromosomes: *scientific inquiry*

The first solid evidence associating a specific gene with a specific chromosome came early in the 1900s from the work of Thomas Hunt Morgan, an experimental embryologist at Columbia University. Although Morgan was initially skeptical about both Mendelian genetics and the chromosome theory, his early experiments provided convincing evidence that chromosomes are indeed the location of Mendel's heritable factors.

Morgan's Choice of Experimental Organism

Many times in the history of biology, important discoveries have come to those insightful or lucky enough to choose an experimental organism suitable for the research problem being tackled. Mendel chose the garden pea because a number of distinct varieties were available. For his work, Morgan selected a species of fruit fly, *Drosophila melanogaster*, a common insect that feeds on the fungi growing on fruit. Fruit flies are prolific breeders; a single mating will produce hundreds of offspring, and a new generation can be bred every two weeks. Morgan's laboratory began using this convenient organism for genetic studies in 1907 and soon became known as "the fly room."

Another advantage of the fruit fly is that it has only four pairs of chromosomes, which are easily distinguishable with a light microscope. There are three pairs of autosomes and one pair of sex chromosomes. Female fruit flies have a pair of homologous X chromosomes, and males have one X chromosome and one Y chromosome.

While Mendel could readily obtain different pea varieties from seed suppliers, Morgan was probably the first person to want different varieties of the fruit fly. He faced the tedious task of carrying out many matings and then microscopically inspecting large numbers of offspring in search of naturally occurring variant individuals. After many months of this, he complained, "Two years' work wasted. I have been breeding those flies for all that time and I've got nothing out of it." Morgan persisted, however, and was finally rewarded with the discovery of a single male fly with white eyes instead of the usual red. The phenotype for a character most commonly observed in natural populations, such as red eyes in *Drosophila*, is called the **wild type** (**Figure 3**). Traits that are alternatives to the wild type, such as white eyes in *Drosophila*, are called *mutant phenotypes* because they are due to alleles assumed to have originated as changes, or mutations, in the wild-type allele.

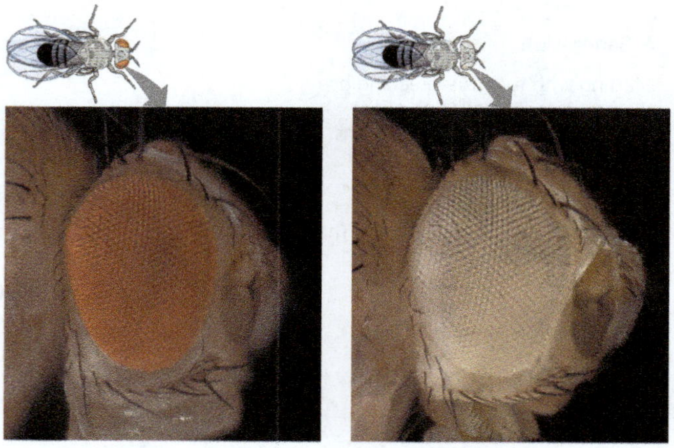

▲ **Figure 3 Morgan's first mutant.** Wild-type *Drosophila* flies have red eyes (left). Among his flies, Morgan discovered a mutant male with white eyes (right). This variation made it possible for Morgan to trace a gene for eye color to a specific chromosome (LMs).

Morgan and his students invented a notation for symbolizing alleles in *Drosophila* that is still widely used for fruit flies. For a given character in flies, the gene takes its symbol from the first mutant (non-wild type) discovered. Thus, the allele for white eyes in *Drosophila* is symbolized by *w*. The superscript + identifies the allele for the wild-type trait: w^+ for the allele for red eyes, for example. Over the years, a variety of gene notation systems have been developed for different organisms. For example, human genes are usually written in all capitals, such as *HTT* for the gene involved in Huntington's disease.

Correlating Behavior of a Gene's Alleles with Behavior of a Chromosome Pair

Morgan mated his white-eyed male fly with a red-eyed female. All the F_1 offspring had red eyes, suggesting that the wild-type allele is dominant. When Morgan bred the F_1 flies to each other, he observed the classical 3:1 phenotypic ratio among the F_2 offspring. However, there was a surprising additional result: The white-eye trait showed up only in males. All the F_2 females had red eyes, while half the males had red eyes and half had white eyes. Therefore, Morgan concluded that somehow a fly's eye color was linked to its sex. (If the eye-color gene were unrelated to sex, half of the white-eyed flies would have been female.)

Recall that a female fly has two X chromosomes (XX), while a male fly has an X and a Y (XY). The correlation between the trait of white eye color and the male sex of the affected F_2 flies suggested to Morgan that the gene involved in his white-eyed mutant was located exclusively on the X chromosome, with no corresponding allele present on the Y chromosome. His reasoning can be followed in **Figure 4**. For a male, a single copy of the mutant allele would confer white eyes; since a male has only one

▼ Figure 4 Inquiry

In a cross between a wild-type female fruit fly and a mutant white-eyed male, what color eyes will the F_1 and F_2 offspring have?

Experiment Thomas Hunt Morgan wanted to analyze the behavior of two alleles of a fruit fly eye-color gene. In crosses similar to those done by Mendel with pea plants, Morgan and his colleagues mated a wild-type (red-eyed) female with a mutant white-eyed male.

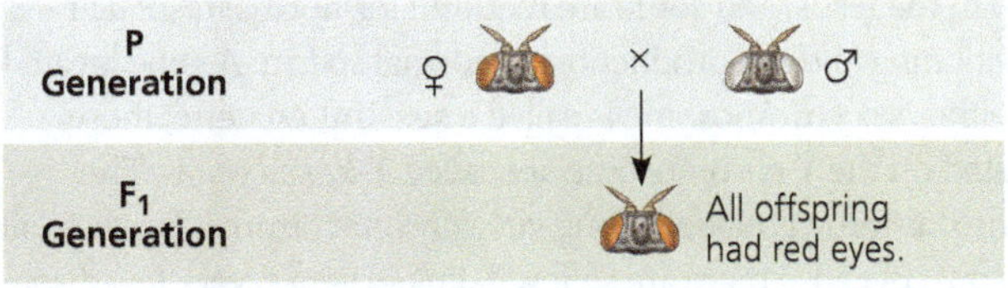

Morgan then bred an F_1 red-eyed female to an F_1 red-eyed male to produce the F_2 generation.

Results The F_2 generation showed a typical Mendelian ratio of 3 red-eyed flies to 1 white-eyed fly. However, all white-eyed flies were males; no females displayed the white-eye trait.

Conclusion All F_1 offspring had red eyes, so the mutant white-eye trait (w) must be recessive to the wild-type red-eye trait (w^+). Since the recessive trait—white eyes—was expressed only in males in the F_2 generation, Morgan deduced that this eye-color gene is located on the X chromosome and that there is no corresponding locus on the Y chromosome.

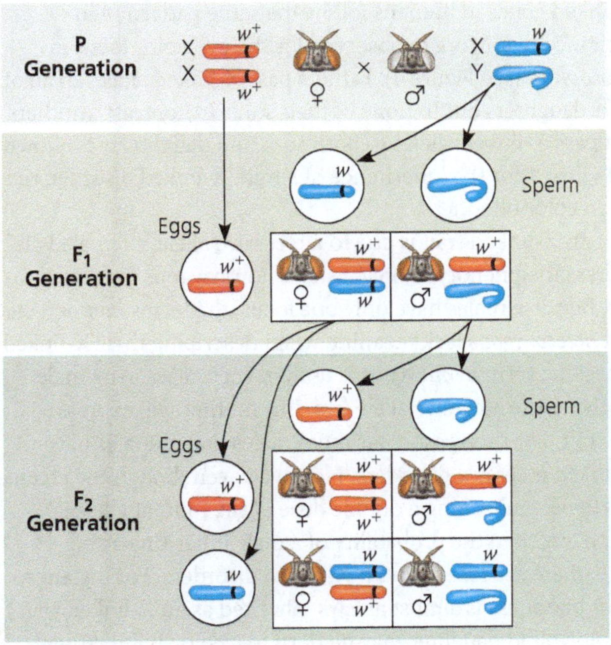

Data from T. H. Morgan, Sex-limited inheritance in *Drosophila, Science* 32:120–122 (1910).

MB A related Experimental Inquiry Tutorial can be assigned in MasteringBiology.

WHAT IF? Suppose this eye-color gene were located on an autosome. Predict the phenotypes (including gender) of the F_2 flies in this hypothetical cross. (*Hint*: Draw a Punnett square.)

X chromosome, there can be no wild-type allele (w^+) present to mask the recessive allele. However, a female could have white eyes only if both her X chromosomes carried the recessive mutant allele (w). This was impossible for the F_2 females in Morgan's experiment because all the F_1 fathers had red eyes, so each F_2 female received a w^+ allele on the X chromosome inherited from her father.

Morgan's finding of the correlation between a particular trait and an individual's sex provided support for the chromosome theory of inheritance, namely, that a specific gene is carried on a specific chromosome (in this case, an eye-color gene on the X chromosome). In addition, Morgan's work indicated that genes located on a sex chromosome exhibit unique inheritance patterns, which we'll discuss in the next section. Recognizing the importance of Morgan's early work, many bright students were attracted to his fly room.

CONCEPT CHECK 1

1. Which one of Mendel's laws relates to the inheritance of alleles for a single character? Which law relates to the inheritance of alleles for two characters in a dihybrid cross?
2. **MAKE CONNECTIONS** Review the description of meiosis and Mendel's laws of segregation and independent assortment. What is the physical basis for each of Mendel's laws?
3. **WHAT IF?** Propose a possible reason that the first naturally occurring mutant fruit fly Morgan saw involved a gene on a sex chromosome and was found in a male.

CONCEPT 2

Sex-linked genes exhibit unique patterns of inheritance

As you just learned, Morgan's discovery of a trait (white eyes) that correlated with the sex of flies was a key episode in the development of the chromosome theory of inheritance. Because the identity of the sex chromosomes in an individual could be inferred by observing the sex of the fly, the behavior of the two members of the pair of sex chromosomes could be correlated with the behavior of the two alleles of the eye-color gene. In this section, we'll take a closer look at the role of sex chromosomes in inheritance.

The Chromosomal Basis of Sex

Whether we are anatomically male or female may be one of our more obvious phenotypic characters. Although the anatomical and physiological differences between women and men are numerous, the chromosomal basis for determining sex is rather simple. Humans and other mammals have two types of sex chromosomes, designated X and Y. The Y chromosome is much smaller than the X chromosome

(**Figure 5**). A person who inherits two X chromosomes, one from each parent, usually develops as a female; a male inherits one X chromosome and one Y chromosome (**Figure 6**). Short segments at either end of the Y chromosome are the only regions that are homologous with regions on the X. These homologous regions allow the X and Y chromosomes in males to pair and behave like homologous chromosomes during meiosis in the testes.

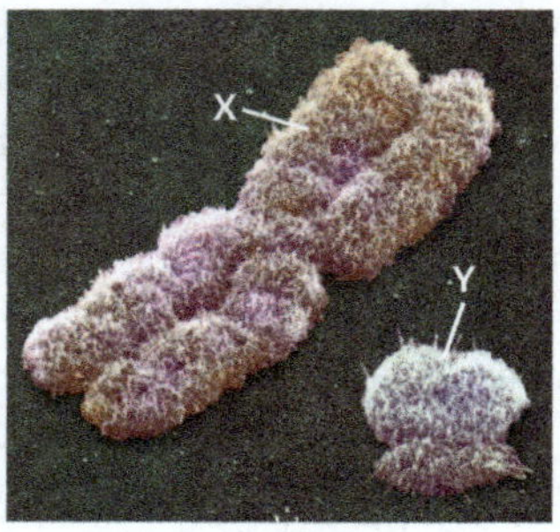

▲ **Figure 5 Human sex chromosomes.**

In mammalian testes and ovaries, the two sex chromosomes segregate during meiosis. Each egg receives one X chromosome. In contrast, sperm fall into two categories: Half the sperm cells a male produces receive an X chromosome, and half receive a Y chromosome. We can trace the sex of each offspring to the events of conception: If a sperm cell bearing an X chromosome fertilizes an egg, the zygote is XX, a female; if a sperm cell containing a Y chromosome fertilizes an egg, the zygote is XY, a male (see Figure 6). Thus, sex determination is a matter of chance—a fifty-fifty chance. In *Drosophila*, while males are also XY, sex depends not on the presence of the Y, but on the ratio between the number of X chromosomes and the number of autosome sets. There are other chromosomal systems as well, besides the X-Y system, for determining sex.

In humans, the anatomical signs of sex begin to emerge when the embryo is about 2 months old. Before then, the rudiments of the gonads are generic—they can develop into either testes or ovaries, depending on whether or not a Y chromosome is present. In 1990, a British research team identified a gene on the Y chromosome required for the development of testes. They named the gene *SRY*, for sex-determining region of Y. In the absence of *SRY*, the gonads develop into ovaries. The biochemical, physiological, and anatomical features that distinguish males and females are complex, and many genes are involved in their development. In fact, *SRY* codes for a protein that regulates other genes.

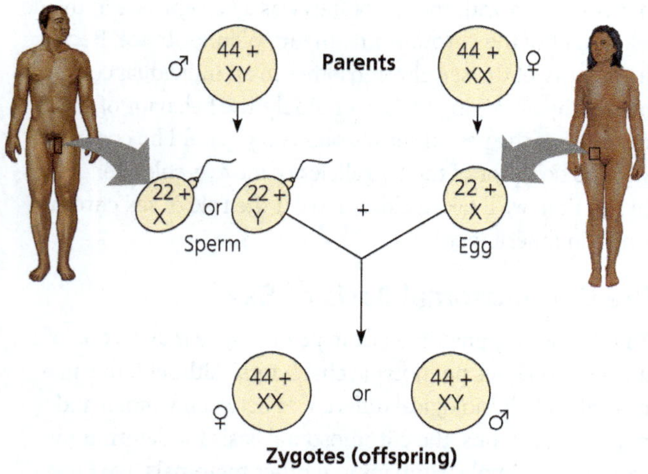

▲ **Figure 6 The mammalian X-Y chromosomal system of sex determination.** In mammals, the sex of an offspring depends on whether the sperm cell contains an X chromosome or a Y. (All eggs have an X.) Numerals indicate the number of autosomes.

Researchers have sequenced the human Y chromosome and have identified 78 genes that code for about 25 proteins (some genes are duplicates). About half of these genes are expressed only in the testis, and some are required for normal testicular functioning and the production of normal sperm. A gene located on either sex chromosome is called a **sex-linked gene**; those located on the Y chromosome are called *Y-linked genes*. The Y chromosome is passed along virtually intact from a father to all his sons. Because there are so few Y-linked genes, very few disorders are transferred from father to son on the Y chromosome. A rare example is that in the absence of certain Y-linked genes, an XY individual is male but does not produce normal sperm.

The human X chromosome contains approximately 1,100 genes, which are called **X-linked genes**. The fact that males and females inherit a different number of X chromosomes leads to a pattern of inheritance different from that produced by genes located on autosomes.

Inheritance of X-Linked Genes

While most Y-linked genes help determine sex, the X chromosomes have genes for many characters unrelated to sex. X-linked genes in humans follow the same pattern of inheritance that Morgan observed for the eye-color locus in *Drosophila* (see Figure 4). Fathers pass X-linked alleles to all of their daughters but to none of their sons. In contrast, mothers can pass X-linked alleles to both sons and daughters, as shown in **Figure 7** for the inheritance of a mild X-linked disorder, red-green color blindness.

If an X-linked trait is due to a recessive allele, a female will express the phenotype only if she is homozygous for that allele. Because males have only one locus, the terms *homozygous* and *heterozygous* lack meaning when describing their X-linked genes; the term *hemizygous* is used in such cases. Any male receiving the recessive allele from his mother will express the trait. For this reason, far more males than females have X-linked recessive disorders. However, even though the chance of a female inheriting a double dose of the mutant allele is much less than the probability of a male inheriting a single dose, there *are* females with X-linked disorders. For instance, color blindness is almost always inherited as an X-linked trait. A color-blind daughter may be born to a color-blind father whose mate is a carrier (see Figure 7c). Because the X-linked allele for color blindness is relatively rare, however, the probability that such a man and woman will mate is low.

A number of human X-linked disorders are much more serious than color blindness, such as **Duchenne muscular dystrophy**, which affects about one out of 3,500 males born in the United States. The disease is characterized by a progressive weakening of the muscles and loss of coordination. Affected

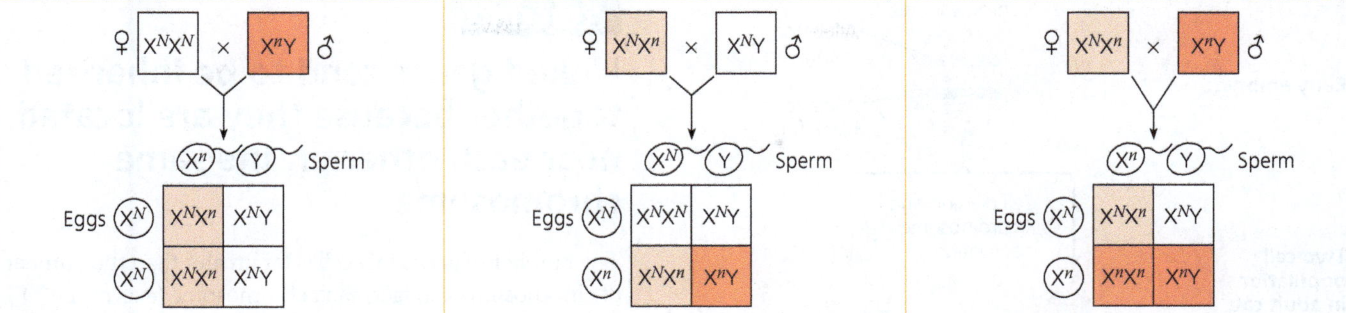

(a) A color-blind father will transmit the mutant allele to all daughters but to no sons. When the mother is a dominant homozygote, the daughters will have the normal phenotype but will be carriers of the mutation.

(b) If a carrier mates with a male who has normal color vision, there is a 50% chance that each daughter will be a carrier like her mother and a 50% chance that each son will have the disorder.

(c) If a carrier mates with a color-blind male, there is a 50% chance that each child born to them will have the disorder, regardless of sex. Daughters who have normal color vision will be carriers, whereas males who have normal color vision will be free of the recessive allele.

▲ **Figure 7 The transmission of X-linked recessive traits.** In this diagram, red-green color blindness is used as an example. The superscript *N* represents the dominant allele for normal color vision carried on the X chromosome, while *n* represents the recessive allele, which has a mutation causing color blindness. White boxes indicate unaffected individuals, light orange boxes indicate carriers, and dark orange boxes indicate color-blind individuals.

? *If a color-blind woman married a man who had normal color vision, what would be the probable phenotypes of their children?*

individuals rarely live past their early 20s. Researchers have traced the disorder to the absence of a key muscle protein called dystrophin and have mapped the gene for this protein to a specific locus on the X chromosome.

Hemophilia is an X-linked recessive disorder defined by the absence of one or more of the proteins required for blood clotting. When a person with hemophilia is injured, bleeding is prolonged because a firm clot is slow to form. Small cuts in the skin are usually not a problem, but bleeding in the muscles or joints can be painful and can lead to serious damage. In the 1800s, hemophilia was widespread among the royal families of Europe. Queen Victoria of England is known to have passed the allele to several of her descendants. Subsequent intermarriage with royal family members of other nations, such as Spain and Russia, further spread this X-linked trait, and its incidence is well documented in royal pedigrees. A few years ago, new genomic techniques allowed sequencing of DNA from tiny amounts isolated from the buried remains of royal family members. The genetic basis of the mutation, and how it resulted in a nonfunctional blood-clotting factor, is now understood. Today, people with hemophilia are treated as needed with intravenous injections of the protein that is missing.

X Inactivation in Female Mammals

Female mammals, including human females, inherit two X chromosomes—twice the number inherited by males—so you may wonder if females make twice as much of the proteins encoded by X-linked genes. In fact, almost all of one X chromosome in each cell in female mammals becomes inactivated during early embryonic development. As a result, the cells of females and males have the same effective dose (one active copy) of most X-linked genes. The inactive X in each cell of a female condenses into a compact object called a **Barr body** (discovered by Canadian anatomist Murray Barr), which lies along the inside of the nuclear envelope. Most of the genes of the X chromosome that forms the Barr body are not expressed. In the ovaries, however, Barr-body chromosomes are reactivated in the cells that give rise to eggs, resulting in every female gamete (egg) having an active X after meiosis.

British geneticist Mary Lyon demonstrated that the selection of which X chromosome will form the Barr body occurs randomly and independently in each embryonic cell present at the time of X inactivation. As a consequence, females consist of a *mosaic* of two types of cells: those with the active X derived from the father and those with the active X derived from the mother. After an X chromosome is inactivated in a particular cell, all mitotic descendants of that cell have the same inactive X. Thus, if a female is heterozygous for a sex-linked trait, about half of her cells will express one allele, while the others will express the alternate allele. **Figure 8** shows how this mosaicism results in the patchy coloration of a tortoiseshell cat. In humans, mosaicism can be observed in a recessive X-linked mutation that prevents the development of sweat glands. A woman who is heterozygous for this trait has patches of normal skin and patches of skin lacking sweat glands.

Inactivation of an X chromosome involves modification of the DNA and proteins bound to it called histones, including attachment of methyl groups ($—CH_3$) to DNA nucleotides. A particular region of each X chromosome contains several genes involved in the inactivation process. The two regions, one on each X chromosome, associate briefly with

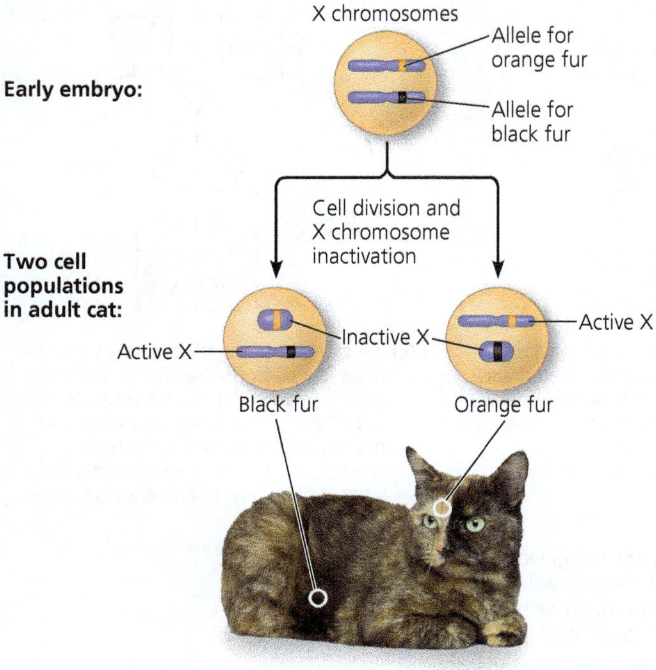

▲ **Figure 8 X inactivation and the tortoiseshell cat.** The tortoiseshell gene is on the X chromosome, and the tortoiseshell phenotype requires the presence of two different alleles, one for orange fur and one for black fur. Normally, only females can have both alleles, because only they have two X chromosomes. If a female cat is heterozygous for the tortoiseshell gene, she is tortoiseshell. Orange patches are formed by populations of cells in which the X chromosome with the orange allele is active; black patches have cells in which the X chromosome with the black allele is active. ("Calico" cats also have white areas, which are determined by yet another gene.)

each other in each cell at an early stage of embryonic development. Then one of the genes, called *XIST* (for X-inactive specific transcript), becomes active *only* on the chromosome that will become the Barr body. Multiple copies of the RNA product of this gene apparently attach to the X chromosome on which they are made, eventually almost covering it. Interaction of this RNA with the chromosome initiates X inactivation, and the RNA products of nearby genes help to regulate the process.

CONCEPT CHECK 2

1. A white-eyed female *Drosophila* is mated with a red-eyed (wild-type) male, the reciprocal cross of the one shown in Figure 4. What phenotypes and genotypes do you predict for the offspring?
2. Neither Tim nor Rhoda has Duchenne muscular dystrophy, but their firstborn son does have it. What is the probability that a second child of this couple will have the disease? What is the probability if the second child is a boy? A girl?
3. **MAKE CONNECTIONS** If a disorder were caused by a dominant X-linked allele, how would the inheritance pattern differ from what we see for recessive X-linked disorders?

CONCEPT 3

Linked genes tend to be inherited together because they are located near each other on the same chromosome

The number of genes in a cell is far greater than the number of chromosomes; in fact, each chromosome (except the Y) has hundreds or thousands of genes. Genes located near each other on the same chromosome tend to be inherited together in genetic crosses; such genes are said to be genetically linked and are called **linked genes**. (Note the distinction between the terms *sex-linked gene*, referring to a single gene on a sex chromosome, and *linked genes*, referring to two or more genes on the same chromosome that tend to be inherited together.) When geneticists follow linked genes in breeding experiments, the results deviate from those expected from Mendel's law of independent assortment.

How Linkage Affects Inheritance

To see how linkage between genes affects the inheritance of two different characters, let's examine another of Morgan's *Drosophila* experiments. In this case, the characters are body color and wing size, each with two different phenotypes. Wild-type flies have gray bodies and normal-sized wings. In addition to these flies, Morgan had managed to obtain, through breeding, doubly mutant flies with black bodies and wings much smaller than normal, called vestigial wings. The mutant alleles are recessive to the wild-type alleles, and neither gene is on a sex chromosome. In his investigation of these two genes, Morgan carried out the crosses shown in **Figure 9**. The first was a P generation cross to generate F_1 dihybrid flies, and the second was a testcross.

The resulting flies had a much higher proportion of the combinations of traits seen in the P generation flies (called parental phenotypes) than would be expected if the two genes assorted independently. Morgan thus concluded that body color and wing size are usually inherited together in specific (parental) combinations because the genes for these characters are near each other on the same chromosome:

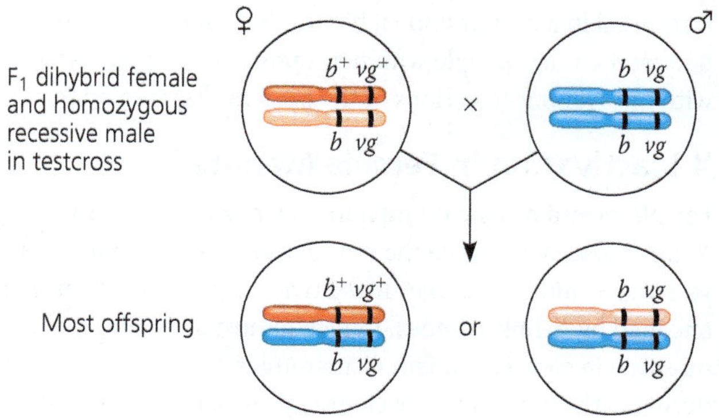

▼ Figure 9 Inquiry

How does linkage between two genes affect inheritance of characters?

Experiment Morgan wanted to know whether the genes for body color and wing size are genetically linked and, if so, how this affects their inheritance. The alleles for body color are b^+ (gray) and b (black), and those for wing size are vg^+ (normal) and vg (vestigial).

Morgan mated true-breeding P (parental) generation flies—wild-type flies with black, vestigial-winged flies—to produce heterozygous F_1 dihybrids ($b^+ b \;\; vg^+ vg$), all of which are wild-type in appearance.

He then mated wild-type F_1 dihybrid females with homozygous recessive males. This testcross will reveal the genotype of the eggs made by the dihybrid female.

The male's sperm contributes only recessive alleles, so the phenotype of the offspring reflects the genotype of the female's eggs.

Note: Although only females (with pointed abdomens) are shown, half the offspring in each class would be males (with rounded abdomens).

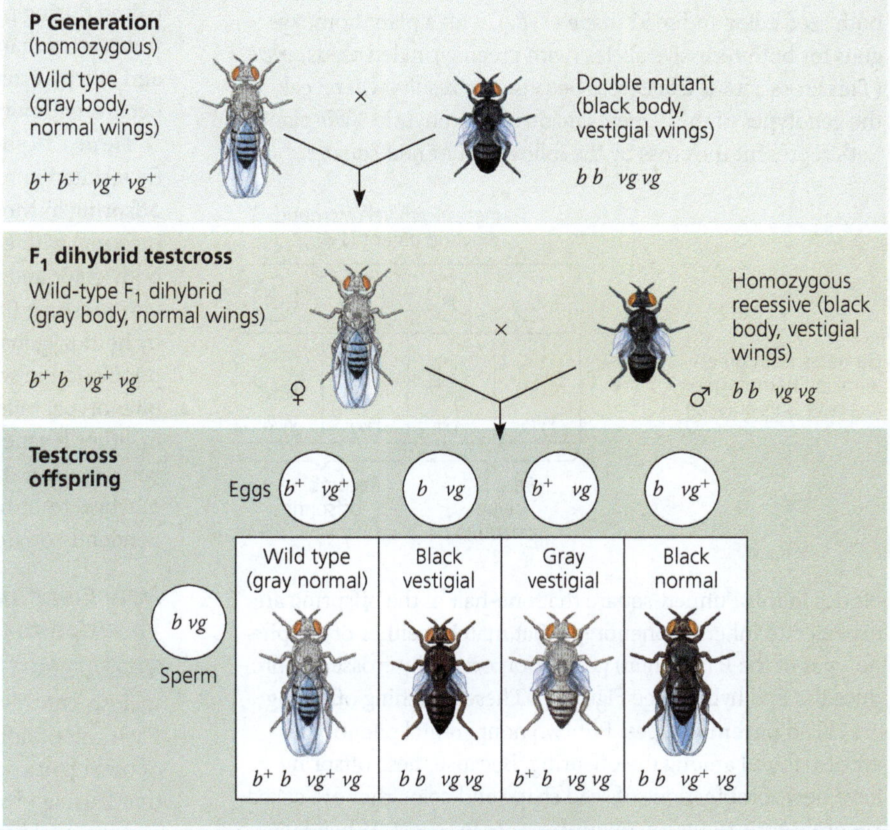

PREDICTED RATIOS

Predicted ratio if genes are located on different chromosomes:	1	:	1	:	1	:	1
Predicted ratio if genes are located on the same chromosome *and* parental alleles are always inherited together:	1	:	1	:	0	:	0
Results Data from Morgan's experiment:	965	:	944	:	206	:	185

Conclusion Since most offspring had a parental (P generation) phenotype, Morgan concluded that the genes for body color and wing size are genetically linked on the same chromosome. However, the production of a relatively small number of offspring with nonparental phenotypes indicated that some mechanism occasionally breaks the linkage between specific alleles of genes on the same chromosome.

Data from T. H. Morgan and C. J. Lynch, The linkage of two factors in *Drosophila* that are not sex-linked, *Biological Bulletin* 23:174–182 (1912).

WHAT IF? If the parental (P generation) flies had been true-breeding for gray body with vestigial wings and black body with normal wings, which phenotypic class(es) would be largest among the testcross offspring?

However, as Figure 9 shows, both of the combinations of traits not seen in the P generation (called nonparental phenotypes) were also produced in Morgan's experiments, suggesting that the body-color and wing-size alleles are not always linked genetically. To understand this conclusion, we need to further explore **genetic recombination**, the production of offspring with combinations of traits that differ from those found in either P generation parent.

Genetic Recombination and Linkage

Meiosis and random fertilization generate genetic variation among offspring of sexually reproducing organisms due to independent assortment of chromosomes and crossing over in meiosis I and the possibility of any sperm fertilizing any egg. Here we'll examine the chromosomal basis of recombination of alleles in relation to the genetic findings of Mendel and Morgan.

Recombination of Unlinked Genes: Independent Assortment of Chromosomes

Mendel learned from crosses in which he followed two characters that some offspring have combinations of traits that do not match those of either parent. For example, consider a cross of a dihybrid pea plant with yellow round seeds, heterozygous for both seed color and seed shape (*YyRr*), with a plant homozygous for both recessive alleles (with green wrinkled seeds, *yyrr*). (This cross acts as a testcross because the results will reveal the genotypes of the gametes made in the dihybrid *YyRr* plant.) Let's represent the cross by the following Punnett square:

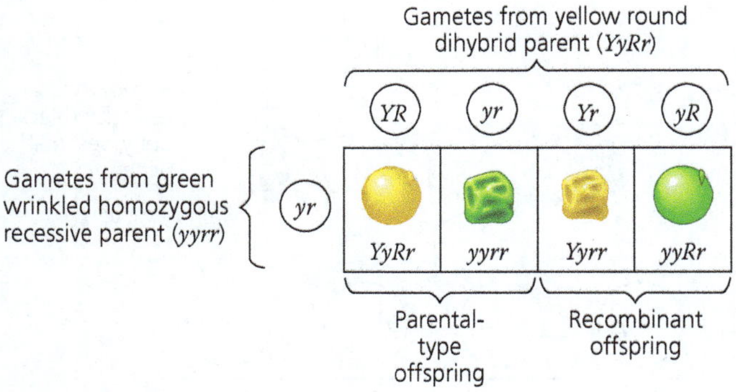

Notice in this Punnett square that one-half of the offspring are expected to inherit a phenotype that matches either of the phenotypes of the P (parental) generation originally crossed to produce the F_1 dihybrid (see Figure 2). These matching offspring are called **parental types**. But two nonparental phenotypes are also found among the offspring. Because these offspring have new combinations of seed shape and color, they are called **recombinant types**, or **recombinants** for short. When 50% of all offspring are recombinants, as in this example, geneticists say that there is a 50% frequency of recombination. The predicted phenotypic ratios among the offspring are similar to what Mendel actually found in his *YyRr* × *yyrr* crosses.

A 50% frequency of recombination in such testcrosses is observed for any two genes that are located on different chromosomes and thus cannot be linked. The physical basis of recombination between unlinked genes is the random orientation of homologous chromosomes at metaphase I of meiosis, which leads to the independent assortment of the two unlinked genes (see the question in the Figure 2 legend).

Recombination of Linked Genes: Crossing Over

Now let's explain the results of the *Drosophila* testcross in Figure 9. Recall that most of the offspring from the testcross for body color and wing size had parental phenotypes. That suggested that the two genes were on the same chromosome, since the occurrence of parental types with a frequency greater than 50% indicates that the genes are linked. About 17% of offspring, however, were recombinants.

Seeing these results, Morgan proposed that some process must occasionally break the physical connection between specific alleles of genes on the same chromosome. Later experiments showed that this process, now called **crossing over**, accounts for the recombination of linked genes. In crossing over, which occurs while replicated homologous chromosomes are paired during prophase of meiosis I, a set of proteins orchestrates an exchange of corresponding segments of one maternal and one paternal chromatid. In effect, when a single crossover occurs, end portions of two nonsister chromatids trade places.

Figure 10 shows how crossing over in a dihybrid female fly resulted in recombinant eggs and ultimately recombinant offspring in Morgan's testcross. Most of the eggs had a chromosome with either the b^+ vg^+ or *b vg* parental genotype for body color and wing size, but some eggs had a recombinant chromosome (b^+ *vg* or *b* vg^+). Fertilization of all classes of eggs by homozygous recessive sperm (*b vg*) produced an offspring population in which 17% exhibited a nonparental, recombinant phenotype, reflecting combinations of alleles not seen before in either P generation parent. In the **Scientific Skills Exercise**, you can use a statistical test to analyze the results from an F_1 dihybrid testcross and see whether the two genes assort independently or are linked.

New Combinations of Alleles: Variation for Natural Selection

EVOLUTION The physical behavior of chromosomes during meiosis contributes to the generation of variation in offspring. Each pair of homologous chromosomes lines up independently of other pairs during metaphase I, and crossing over prior to that, during prophase I, can mix and match parts of maternal and paternal homologs. Mendel's elegant experiments show that the behavior of the abstract entities known as genes—or, more concretely, alleles of genes—also leads to variation in offspring. Now, putting these different ideas together, you can see that the recombinant chromosomes resulting from crossing over may bring alleles together in new combinations, and the subsequent events of meiosis distribute to gametes the recombinant chromosomes in a multitude of combinations, such as the new variants seen in Figures 9 and 10. Random fertilization then increases even further the number of variant allele combinations that can be created.

This abundance of genetic variation provides the raw material on which natural selection works. If the traits conferred by particular combinations of alleles are better suited for a given environment, organisms possessing those genotypes will be expected to thrive and leave more offspring, ensuring the continuation of their genetic complement. In the next generation, of course, the alleles will be shuffled anew. Ultimately, the interplay between environment and genotype will determine which genetic combinations persist over time.

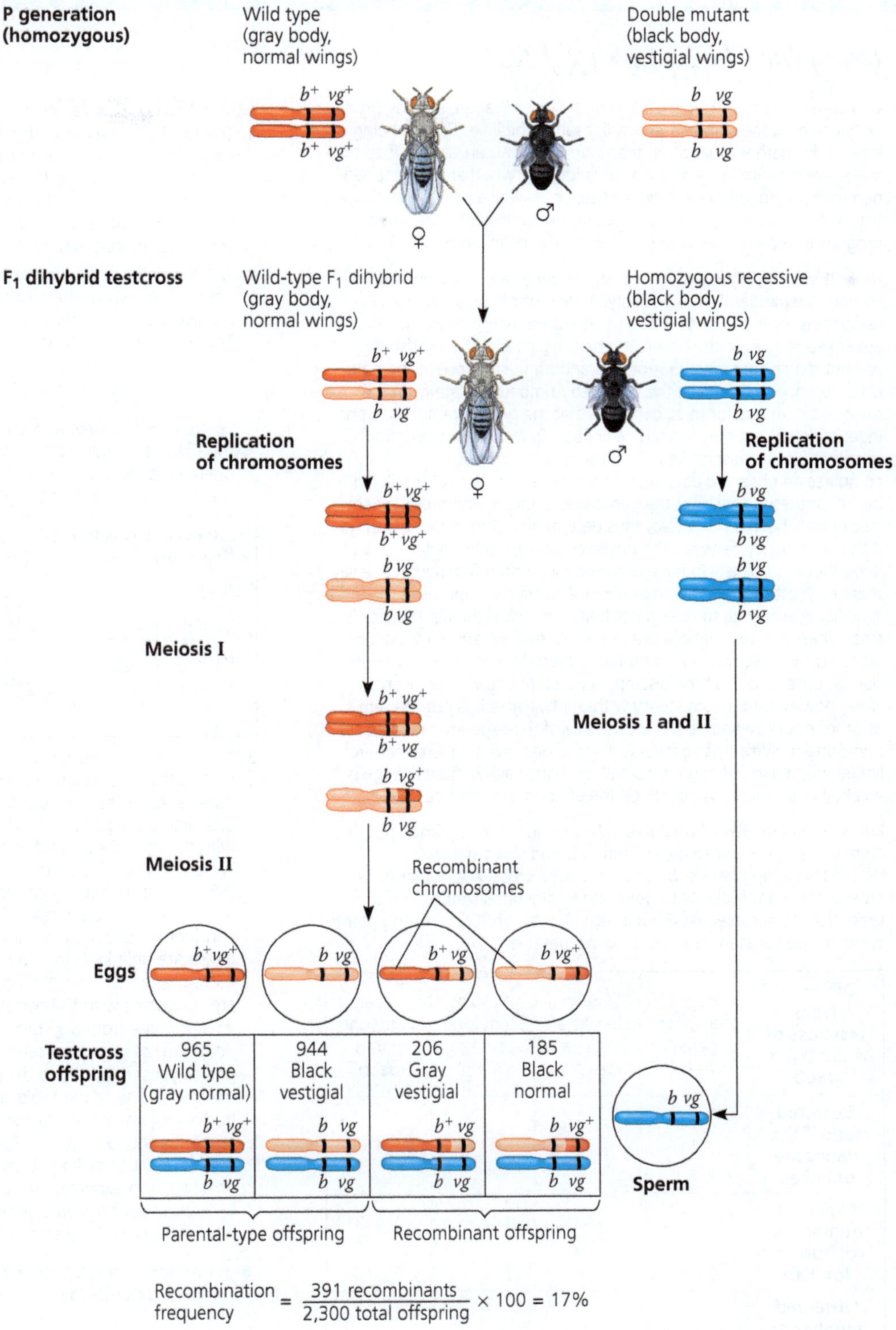

▶ Figure 10 Chromosomal basis for recombination of linked genes. In these diagrams recreating the testcross in Figure 9, we track chromosomes as well as genes. The maternal chromosomes (present in the wild-type F_1 dihybrid) are color-coded red and pink to distinguish one homolog from the other before any meiotic crossing over has occurred. Because crossing over between the b^+/b and vg^+/vg loci occurs in some, but not all, egg-producing cells, more eggs with parental-type chromosomes than with recombinant ones are produced in the mating females. Fertilization of the eggs by sperm of genotype $b\ vg$ gives rise to some recombinant offspring. The recombination frequency is the percentage of recombinant flies in the total pool of offspring.

DRAW IT *Suppose, as in the question at the bottom of Figure 9, the parental (P generation) flies were true-breeding for gray body with vestigial wings and black body with normal wings. Draw the chromosomes in each of the four possible kinds of eggs from an F_1 female, and label each chromosome as "parental" or "recombinant."*

Mapping the Distance Between Genes Using Recombination Data: *Scientific Inquiry*

The discovery of linked genes and recombination due to crossing over motivated one of Morgan's students, Alfred H. Sturtevant, to work out a method for constructing a **genetic map**, an ordered list of the genetic loci along a particular chromosome.

Sturtevant hypothesized that the percentage of recombinant offspring, the *recombination frequency*, calculated from experiments like the one in Figures 9 and 10, depends on the distance between genes on a chromosome. He assumed that crossing over is a random event, with the chance of crossing over approximately equal at all points along a chromosome.

Scientific Skills Exercise

Using the Chi-Square (χ^2) Test

Are Two Genes Linked or Unlinked? Genes that are in close proximity on the same chromosome will result in the linked alleles being inherited together more often than not. But how can you tell if certain alleles are inherited together due to linkage or whether they just happen to assort together? In this exercise, you will use a simple statistical test, the chi-square (χ^2) test, to analyze phenotypes of F_1 testcross progeny in order to see whether two genes are linked or unlinked.

How These Experiments Are Done If genes are unlinked and assorting independently, the phenotypic ratio of offspring from an F_1 testcross is expected to be 1:1:1:1 (see Figure 9). If the two genes are linked, however, the observed phenotypic ratio of the offspring will not match that ratio. Given that random fluctuations in the data do occur, how much must the observed numbers deviate from the expected numbers for us to conclude that the genes are not assorting independently but may instead be linked? To answer this question, scientists use a statistical test. This test, called a chi-square (χ^2) test, compares an observed data set with an expected data set predicted by a hypothesis (here, that the genes are unlinked) and measures the discrepancy between the two, thus determining the "goodness of fit." If the discrepancy between the observed and expected data sets is so large that it is unlikely to have occurred by random fluctuation, we say there is statistically significant evidence against the hypothesis (or, more specifically, evidence for the genes being linked). If the discrepancy is small, then our observations are well explained by random variation alone. In this case, we say that the observed data are consistent with our hypothesis, or that the discrepancy is statistically insignificant. Note, however, that consistency with our hypothesis is not the same as proof of our hypothesis. Also, the size of the experimental data set is important: With small data sets like this one, even if the genes are linked, discrepancies might be small by chance alone if the linkage is weak. (For simplicity, we overlook the effect of sample size here.)

Data from the Simulated Experiment In cosmos plants, purple stem (*A*) is dominant to green stem (*a*), and short petals (*B*) is dominant to long petals (*b*). In a simulated cross, *AABB* plants were crossed with *aabb* plants to generate F_1 dihybrids (*AaBb*), which were then testcrossed (*AaBb* × *aabb*). A total of 900 offspring plants were scored for stem color and flower petal length.

Offspring from testcross of *AaBb* (F_1) × *aabb*	Purple stem/short petals (*A–B–*)	Green stem/short petals (*aaB–*)	Purple stem/long petals (*A–bb*)	Green stem/long petals (*aabb*)
Expected ratio if the genes are unlinked	1	1	1	1
Expected number of offspring (of 900)				
Observed number of offspring (of 900)	220	210	231	239

INTERPRET THE DATA

1. The results in the data table are from a simulated F_1 dihybrid testcross. The hypothesis that the two genes are unlinked predicts that the offspring phenotypic ratio will be 1:1:1:1. Using this ratio, calculate the expected number of each phenotype out of the 900 total offspring, and enter the values in that data table.
2. The goodness of fit is measured by χ^2. This statistic measures the amounts by which the observed values differ from their respective predictions to indicate how closely the two sets of values match. The formula for calculating this value is

$$\chi^2 = \sum \frac{(o - e)^2}{e}$$

where o = observed and e = expected. Calculate the χ^2 value for the data using the table below. Fill out that table, carrying out the operations indicated in the top row. Then add up the entries in the last column to find the χ^2 value.

Testcross offspring	**Expected (*e*)**	**Observed (*o*)**	**Deviation ($o - e$)**	**$(o - e)^2$**	**$(o - e)^2/e$**
(*A–B–*)		220			
(*aaB–*)		210			
(*A–bb*)		231			
(*aabb*)		239			
				χ^2 = Sum	

3. The χ^2 value means nothing on its own—it is used to find the probability that, assuming the hypothesis is true, the observed data set could have resulted from random fluctuations. A low probability suggests that the observed data are not consistent with the hypothesis, and thus the hypothesis should be rejected. A standard cutoff point used by biologists is a probability of 0.05 (5%). If the probability corresponding to the χ^2 value is 0.05 or less, the differences between observed and expected values are considered statistically significant and the hypothesis (that the genes are unlinked) should be rejected. If the probability is above 0.05, the results are not statistically significant; the observed data are consistent with the hypothesis. To find the probability, locate your χ^2 value from a χ^2 distribution table. The "degrees of freedom" (df) of your data set is the number of categories (here, 4 phenotypes) minus 1, so df = 3. (a) Determine which values on the df = 3 line of the table your calculated χ^2 value lies between. (b) The column headings for these values show the probability range for your χ^2 number. Based on whether there are nonsignificant ($p > 0.05$) or significant ($p \leq 0.05$) differences between the observed and expected values, are the data consistent with the hypothesis that the two genes are unlinked and assorting independently, or is there enough evidence to reject this hypothesis?

(MB) A version of this Scientific Skills Exercise can be assigned in MasteringBiology.

Based on these assumptions, Sturtevant predicted that *the farther apart two genes are, the higher the probability that a crossover will occur between them and therefore the higher the recombination frequency.* His reasoning was simple: The greater the distance between two genes, the more points there are between them where crossing over can occur. Using recombination data from various fruit fly crosses, Sturtevant proceeded to assign relative

▼ Figure 11 Research Method

Constructing a Linkage Map

Application A linkage map shows the relative locations of genes along a chromosome.

Technique A linkage map is based on the assumption that the probability of a crossover between two genetic loci is proportional to the distance separating the loci. The recombination frequencies used to construct a linkage map for a particular chromosome are obtained from experimental crosses, such as the cross depicted in Figures 9 and 10. The distances between genes are expressed as map units, with one map unit equivalent to a 1% recombination frequency. Genes are arranged on the chromosome in the order that best fits the data.

Results In this example, the observed recombination frequencies between three *Drosophila* gene pairs (*b* and *cn*, 9%; *cn* and *vg*, 9.5%; and *b* and *vg*, 17%) best fit a linear order in which *cn* is positioned about halfway between the other two genes:

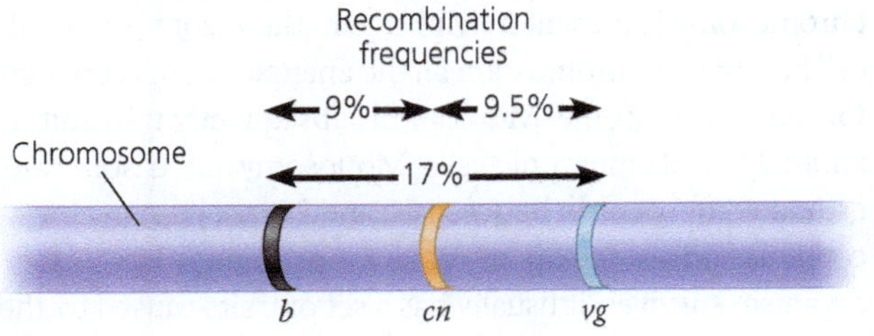

The *b-vg* recombination frequency (17%) is slightly less than the sum of the *b-cn* and *cn-vg* frequencies (9 + 9.5 = 18.5%) because of the few times that one crossover occurs between *b* and *cn* and another crossover occurs between *cn* and *vg*. The second crossover would "cancel out" the first, reducing the observed *b-vg* recombination frequency while contributing to the frequency between each of the closer pairs of genes. The value of 18.5% (18.5 map units) is closer to the actual distance between the genes. In practice, a geneticist would add the smaller distances in constructing a map.

positions to genes on the same chromosomes—that is, to *map* genes.

A genetic map based on recombination frequencies is called a **linkage map**. **Figure 11** shows Sturtevant's linkage map of three genes: the body-color (*b*) and wing-size (*vg*) genes depicted in Figure 10 and a third gene, called cinnabar (*cn*). Cinnabar is one of many *Drosophila* genes affecting eye color. Cinnabar eyes, a mutant phenotype, are a brighter red than the wild-type color. The recombination frequency between *cn* and *b* is 9%; that between *cn* and *vg*, 9.5%; and that between *b* and *vg*, 17%. In other words, crossovers between *cn* and *b* and between *cn* and *vg* are about half as frequent as crossovers between *b* and *vg*. Only a map that locates *cn* about midway between *b* and *vg* is consistent with these data, as you can prove to yourself by drawing alternative maps. Sturtevant expressed the distances between genes in **map units**, defining one map unit as equivalent to a 1% recombination frequency.

In practice, the interpretation of recombination data is more complicated than this example suggests. Some genes on a chromosome are so far from each other that a crossover between them is virtually certain. The observed frequency of recombination in crosses involving two such genes can have a maximum value of 50%, a result indistinguishable from that for genes on different chromosomes. In this case, the physical connection between genes on the same chromosome is not reflected in the results of genetic crosses. Despite being on the same chromosome and thus being *physically connected*, the genes are *genetically unlinked*; alleles of such genes assort independently, as if they were on different chromosomes. In fact, at least two of the genes for pea characters that Mendel studied are now known to be on the same chromosome, but the distance between them is so great that linkage is not observed in genetic crosses. Consequently, the two genes behaved as if they were on different chromosomes in Mendel's experiments. Genes located far apart on a chromosome are mapped by adding the recombination frequencies from crosses involving closer pairs of genes lying between the two distant genes.

Using recombination data, Sturtevant and his colleagues were able to map numerous *Drosophila* genes in linear arrays. They found that the genes clustered into four groups of linked genes (*linkage groups*). Light microscopy had revealed four pairs of chromosomes in *Drosophila*, so the linkage map provided additional evidence that genes are located on chromosomes. Each chromosome has a linear array of specific genes, each gene with its own locus **(Figure 12)**.

Because a linkage map is based strictly on recombination frequencies, it gives only an approximate picture of a chromosome. The frequency of crossing over is not actually uniform

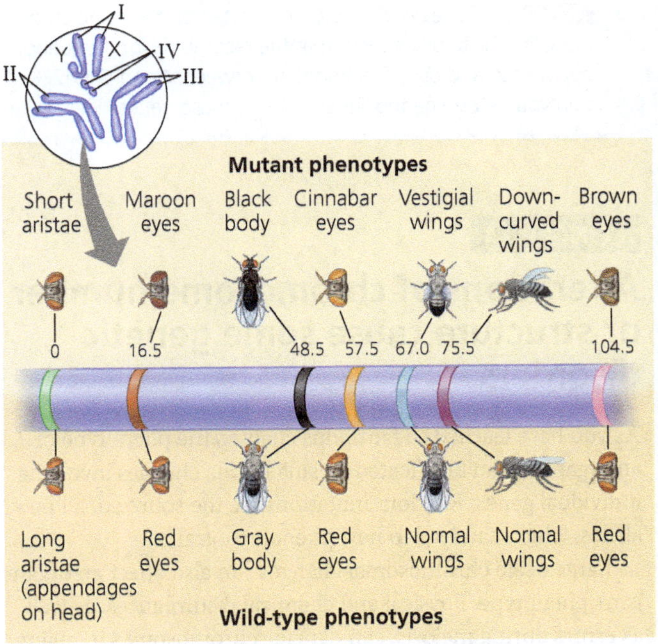

▲ Figure 12 A partial genetic (linkage) map of a *Drosophila* chromosome. This simplified map shows just seven of the genes that have been mapped on *Drosophila* chromosome 2. (DNA sequencing has revealed over 9,000 genes on that chromosome.) The number at each gene locus indicates the number of map units between that locus and the locus for arista length (left). Notice that more than one gene can affect a given phenotypic character, such as eye color.

over the length of a chromosome, as Sturtevant assumed, and therefore map units do not correspond to actual physical distances (in nanometers, for instance). A linkage map does portray the order of genes along a chromosome, but it does not accurately portray the precise locations of those genes. Other methods enable geneticists to construct **cytogenetic maps** of chromosomes, which locate genes with respect to chromosomal features, such as stained bands, that can be seen in the microscope. Technical advances over the last two decades have enormously increased the rate and affordability of DNA sequencing. Today, most researchers sequence whole genomes to map the locations of genes of a given species. The entire nucleotide sequence is the ultimate physical map of a chromosome, revealing the physical distances between gene loci in DNA nucleotides. Comparing a linkage map with such a physical map or with a cytogenetic map of the same chromosome, we find that the linear order of genes is identical in all the maps, but the spacing between genes is not.

CONCEPT CHECK 3

1. When two genes are located on the same chromosome, what is the physical basis for the production of recombinant offspring in a testcross between a dihybrid parent and a double-mutant (recessive) parent?
2. For each type of offspring of the testcross in Figure 9, explain the relationship between its phenotype and the alleles contributed by the female parent. (It will be useful to draw out the chromosomes of each fly and follow the alleles throughout the cross.)
3. WHAT IF? Genes *A*, *B*, and *C* are located on the same chromosome. Testcrosses show that the recombination frequency between *A* and *B* is 28% and that between *A* and *C* is 12%. Can you determine the linear order of these genes? Explain.

CONCEPT 4

Alterations of chromosome number or structure cause some genetic disorders

As you have learned so far in this chapter, the phenotype of an organism can be affected by small-scale changes involving individual genes. Random mutations are the source of all new alleles, which can lead to new phenotypic traits.

Large-scale chromosomal changes can also affect an organism's phenotype. Physical and chemical disturbances, as well as errors during meiosis, can damage chromosomes in major ways or alter their number in a cell. Large-scale chromosomal alterations in humans and other mammals often lead to spontaneous abortion (miscarriage) of a fetus, and individuals born with these types of genetic defects commonly exhibit various developmental disorders. Plants appear to tolerate such genetic defects better than animals do.

Abnormal Chromosome Number

Ideally, the meiotic spindle distributes chromosomes to daughter cells without error. But there is an occasional mishap, called a **nondisjunction**, in which the members of a pair of homologous chromosomes do not move apart properly during meiosis I or sister chromatids fail to separate during meiosis II (**Figure 13**). In nondisjunction, one gamete receives two of the same type of chromosome and another gamete receives no copy. The other chromosomes are usually distributed normally.

If either of the aberrant gametes unites with a normal one at fertilization, the zygote will also have an abnormal number of a particular chromosome, a condition known as **aneuploidy**. Fertilization involving a gamete that has no copy of a particular chromosome will lead to a missing chromosome in the zygote (so that the cell has $2n - 1$ chromosomes); the aneuploid zygote is said to be **monosomic** for that chromosome. If a chromosome is present in triplicate in the zygote (so that the cell has $2n + 1$ chromosomes), the aneuploid cell is **trisomic** for that chromosome. Mitosis will subsequently transmit the anomaly to all embryonic cells. Monosomy and trisomy are estimated to occur in between 10 and 25% of human conceptions and are the main reason for pregnancy loss. If the organism survives, it usually has a set of traits caused by the abnormal dose of the genes associated with the extra or missing chromosome. Down syndrome is an example of trisomy in humans that will be discussed later. Nondisjunction can also occur during mitosis. If such an error takes place early

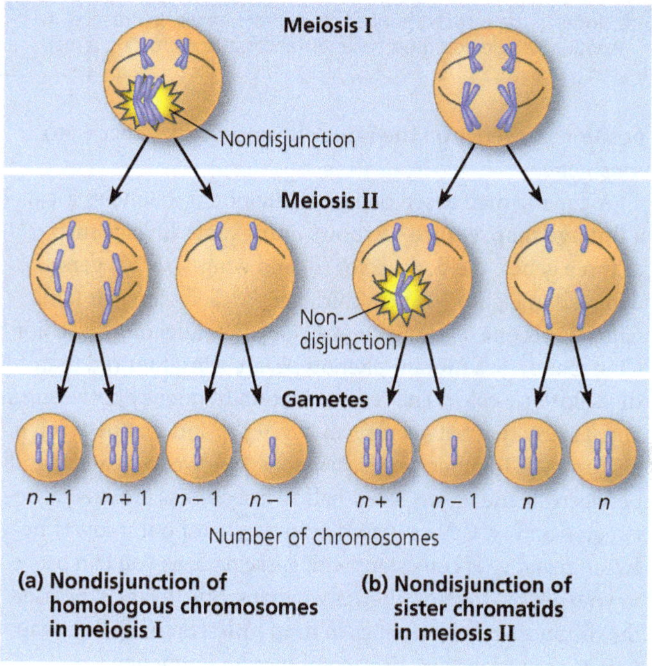

▲ **Figure 13 Meiotic nondisjunction.** Gametes with an abnormal chromosome number can arise by nondisjunction in either meiosis I or meiosis II. For simplicity, the figure does not show the spores formed by meiosis in plants. Ultimately, spores form gametes that have the defects shown.

in embryonic development, then the aneuploid condition is passed along by mitosis to a large number of cells and is likely to have a substantial effect on the organism.

Some organisms have more than two complete chromosome sets in all somatic cells. The general term for this chromosomal alteration is **polyploidy**; the specific terms *triploidy* ($3n$) and *tetraploidy* ($4n$) indicate three and four chromosomal sets, respectively. One way a triploid cell may arise is by the fertilization of an abnormal diploid egg produced by nondisjunction of all its chromosomes. Tetraploidy could result from the failure of a $2n$ zygote to divide after replicating its chromosomes. Subsequent normal mitotic divisions would then produce a $4n$ embryo.

Polyploidy is fairly common in plants; the spontaneous origin of polyploid individuals plays an important role in the evolution of plants. Many of the plant species we eat are polyploid; for example, bananas are triploid, wheat is hexaploid ($6n$), and strawberries are octoploid ($8n$).

Alterations of Chromosome Structure

Errors in meiosis or damaging agents such as radiation can cause breakage of a chromosome, which can lead to four types of changes in chromosome structure **(Figure 14)**. A **deletion** occurs when a chromosomal fragment is lost. The affected chromosome is then missing certain genes. The "deleted" fragment may become attached as an extra segment to a sister chromatid, producing a **duplication** of a portion of that chromosome. Alternatively, a detached fragment could attach to a nonsister chromatid of a homologous chromosome. In that case, though, the "duplicated" segments might not be identical because the homologs could carry different alleles of certain genes. A chromosomal fragment may also reattach to the original chromosome but in the reverse orientation, producing an **inversion**. A fourth possible result of chromosomal breakage is for the fragment to join a nonhomologous chromosome, a rearrangement called a **translocation**.

Deletions and duplications are especially likely to occur during meiosis. In crossing over, nonsister chromatids sometimes exchange unequal-sized segments of DNA, so that one partner gives up more genes than it receives. The products of such an unequal crossover are one chromosome with a deletion and one chromosome with a duplication.

A diploid embryo that is homozygous for a large deletion (or has a single X chromosome with a large deletion, in a male) is usually missing a number of essential genes, a condition that is typically lethal. Duplications and translocations also tend to be harmful. In reciprocal translocations, in which segments are exchanged between nonhomologous chromosomes, and in inversions, the balance of genes is not abnormal—all genes are present in their normal doses. Nevertheless, translocations and inversions can alter phenotype because a gene's expression can be influenced by its location among neighboring genes, which can have devastating effects.

▼ **Figure 14 Alterations of chromosome structure.** Red arrows indicate breakage points. Dark purple highlights the chromosomal parts affected by the rearrangements.

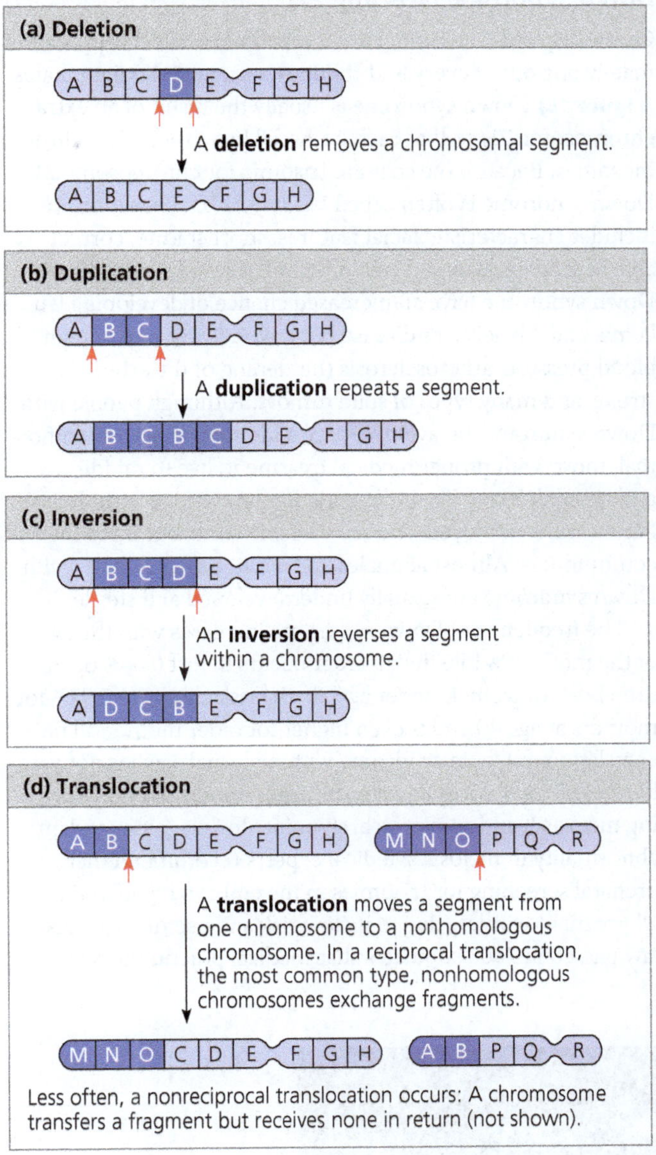

Human Disorders Due to Chromosomal Alterations

Alterations of chromosome number and structure are associated with a number of serious human disorders. As described earlier, nondisjunction in meiosis results in aneuploidy in gametes and any resulting zygotes. Although the frequency of aneuploid zygotes may be quite high in humans, most of these chromosomal alterations are so disastrous to development that the affected embryos are spontaneously aborted long before birth. However, some types of aneuploidy appear to upset the genetic balance less than others, where individuals with certain aneuploid conditions can survive to birth and beyond. These individuals have a set of traits—a *syndrome*—characteristic of the type of aneuploidy. Genetic disorders caused by aneuploidy can be diagnosed before birth by genetic testing of the fetus.

Down Syndrome (Trisomy 21)

One aneuploid condition, **Down syndrome**, affects approximately one out of every 830 children born in the United States **(Figure 15)**. Down syndrome is usually the result of an extra chromosome 21, so that each body cell has a total of 47 chromosomes. Because the cells are trisomic for chromosome 21, Down syndrome is often called *trisomy 21*. Down syndrome includes characteristic facial features, short stature, correctable heart defects, and developmental delays. Individuals with Down syndrome have an increased chance of developing leukemia and Alzheimer's disease but have a lower rate of high blood pressure, atherosclerosis (hardening of the arteries), stroke, and many types of solid tumors. Although people with Down syndrome, on average, have a life span shorter than normal, most, with proper medical treatment, live to middle age and beyond. Many live independently or at home with their families, are employed, and are valuable contributors to their communities. Almost all males and about half of females with Down syndrome are sexually underdeveloped and sterile.

The frequency of Down syndrome increases with the age of the mother. While the disorder occurs in just 0.04% of children born to women under age 30, the risk climbs to 0.92% for mothers at age 40 and is even higher for older mothers. The correlation of Down syndrome with maternal age has not yet been explained. Most cases result from nondisjunction during meiosis I, and some research points to an age-dependent abnormality in meiosis. Medical experts recommend that prenatal screening for trisomies in the embryo be offered to all pregnant women, due to its low risk and useful results. A law passed in 2008 stipulates that medical practitioners give accurate, up-to-date information about any prenatal or postnatal diagnosis received by parents and that they connect parents with appropriate support services.

▲ **Figure 15 Down syndrome.** The karyotype shows trisomy 21, the most common cause of Down syndrome. The child exhibits the facial features characteristic of this disorder.

Aneuploidy of Sex Chromosomes

Aneuploid conditions involving sex chromosomes appear to upset the genetic balance less than those involving autosomes. This may be because the Y chromosome carries relatively few genes. Also, extra copies of the X chromosome simply become inactivated as Barr bodies in somatic cells.

An extra X chromosome in a male, producing an XXY genotype, occurs approximately once in every 500 to 1,000 live male births. People with this disorder, called *Klinefelter syndrome*, have male sex organs, but the testes are abnormally small and the man is sterile. Even though the extra X is inactivated, some breast enlargement and other female body characteristics are common. Affected individuals may have subnormal intelligence. About one of every 1,000 males is born with an extra Y chromosome (XYY). These males undergo normal sexual development and do not exhibit any well-defined syndrome.

Females with trisomy X (XXX), which occurs once in approximately 1,000 live female births, are healthy and have no unusual physical features other than being slightly taller than average. Triple-X females are at risk for learning disabilities but are fertile. Monosomy X, called *Turner syndrome*, occurs about once in every 2,500 female births and is the only known viable monosomy in humans. Although these X0 individuals are phenotypically female, they are sterile because their sex organs do not mature. When provided with estrogen replacement therapy, girls with Turner syndrome do develop secondary sex characteristics. Most have normal intelligence.

Disorders Caused by Structurally Altered Chromosomes

Many deletions in human chromosomes, even in a heterozygous state, cause severe problems. One such syndrome, known as *cri du chat* ("cry of the cat"), results from a specific deletion in chromosome 5. A child born with this deletion is severely intellectually disabled, has a small head with unusual facial features, and has a cry that sounds like the mewing of a distressed cat. Such individuals usually die in infancy or early childhood.

Chromosomal translocations can also occur during mitosis; some have been implicated in certain cancers, including *chronic myelogenous leukemia* (*CML*). This disease occurs when a reciprocal translocation happens during mitosis of pre-white blood cells. The exchange of a large portion of chromosome 22 with a small fragment from a tip of chromosome 9 produces a much shortened, easily recognized chromosome 22, called the *Philadelphia chromosome* **(Figure 16)**. Such an exchange causes cancer by creating a new "fused" gene that leads to uncontrolled cell cycle progression.

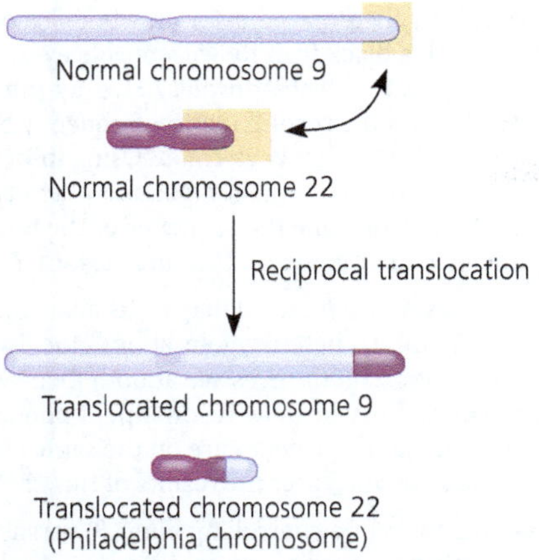

▲ **Figure 16 Translocation associated with chronic myelogenous leukemia (CML).** The cancerous cells in nearly all CML patients contain an abnormally short chromosome 22, the so-called Philadelphia chromosome, and an abnormally long chromosome 9. These altered chromosomes result from the reciprocal translocation shown here, which presumably occurred in a single white blood cell precursor undergoing mitosis and was then passed along to all descendant cells.

CONCEPT CHECK 4

1. About 5% of individuals with Down syndrome have a chromosomal translocation in which a third copy of chromosome 21 is attached to chromosome 14. If this translocation occurred in a parent's gonad, how could it lead to Down syndrome in a child?
2. **WHAT IF?** The ABO blood type locus has been mapped on chromosome 9. A father who has type AB blood and a mother who has type O blood have a child with trisomy 9 and type A blood. Using this information, can you tell in which parent the nondisjunction occurred? Explain your answer.
3. **MAKE CONNECTIONS** The gene that is activated on the Philadelphia chromosome codes for an intracellular kinase. Explain how the activation of this gene could contribute to the development of cancer.
4. Women born with an extra X chromosome (XXX) are generally healthy and indistinguishable in appearance from XX women. What is a likely explanation for this finding? How could you test this explanation?

Chapter Review

Go to **MasteringBiology**® for Assignments, the eText, and the Study Area with Animations, Activities, Vocab Self-Quiz, and Practice Tests.

SUMMARY OF KEY CONCEPTS

VOCAB SELF-QUIZ
goo.gl/gbai8v

CONCEPT 1

Morgan showed that Mendelian inheritance has its physical basis in the behavior of chromosomes: *scientific inquiry*

- Morgan's work with an eye-color gene in *Drosophila* led to the **chromosome theory of inheritance**, which states that genes are located on chromosomes and that the behavior of chromosomes during meiosis accounts for Mendel's laws.
- Morgan's discovery that transmission of the X chromosome in *Drosophila* correlates with inheritance of an eye-color trait was the first solid evidence indicating that a specific gene is associated with a specific chromosome.

? *What characteristic of the sex chromosomes allowed Morgan to correlate their behavior with that of the alleles of the eye-color gene?*

CONCEPT 2

Sex-linked genes exhibit unique patterns of inheritance

- Sex is often chromosomally based. Humans and other mammals have an X-Y system in which sex is determined by whether a Y chromosome is present.
- The sex chromosomes carry **sex-linked genes**, virtually all of which are on the X chromosome (X-linked). Any male who inherits a recessive X-linked allele (from his mother) will express the trait, such as color blindness.
- In mammalian females, one of the two X chromosomes in each cell is randomly inactivated during early embryonic development, becoming highly condensed into a **Barr body**.

? *Why are males affected by X-linked disorders much more often than females?*

CONCEPT 3

Linked genes tend to be inherited together because they are located near each other on the same chromosome

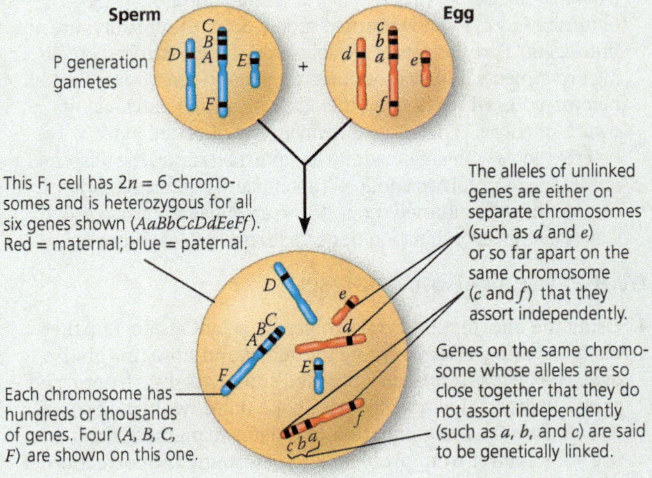

- An F_1 testcross yields **parental types** with the same combination of traits as those in the P generation parents and **recombinant types** with new combinations of traits. Unlinked genes exhibit a 50% frequency of recombination in the gametes. For genetically **linked genes**, **crossing over** accounts for the observed recombinants, always less than 50%.
- Recombination frequencies observed in genetic crosses allow construction of a **linkage map** (a type of **genetic map**).

? *Why are specific alleles of two distant genes more likely to show recombination than those of two closer genes?*

CONCEPT 4

Alterations of chromosome number or structure cause some genetic disorders

- **Aneuploidy**, an abnormal chromosome number, results from **nondisjunction** during meiosis. When a normal gamete unites with one containing two copies or no copies of a particular chromosome, the resulting zygote and its descendant cells either have one extra copy of that chromosome (**trisomy**, $2n + 1$) or are missing a copy (**monosomy**, $2n - 1$). **Polyploidy** (extra sets of chromosomes) can result from complete nondisjunction.
- Chromosome breakage can result in alterations of chromosome structure: **deletions**, **duplications**, **inversions**, and **translocations**.

? *Why are inversions and reciprocal translocations less likely to be lethal than are aneuploidy, duplications, and deletions?*

TEST YOUR UNDERSTANDING

PRACTICE TEST

goo.gl/CRZjvS

Level 1: Knowledge/Comprehension

1. A man with hemophilia (a recessive, sex-linked condition) has a daughter without the condition, who marries a man who does not have hemophilia. What is the probability that their daughter will have the condition? Their son? If they have four sons, that all will be affected?

2. Pseudohypertrophic muscular dystrophy is an inherited disorder that causes gradual deterioration of the muscles. It is seen almost exclusively in boys born to apparently unaffected parents and usually results in death in the early teens. Is this disorder caused by a dominant or a recessive allele? Is its inheritance sex-linked or autosomal? How do you know? Explain why this disorder is almost never seen in girls.

3. A space probe discovers a planet inhabited by creatures that reproduce with the same hereditary patterns seen in humans. Three phenotypic characters are height (T = tall, t = dwarf), head appendages (A = antennae, a = no antennae), and nose morphology (S = upturned snout, s = downturned snout). Since the creatures are not "intelligent," Earth scientists are able to do some controlled breeding experiments using various heterozygotes in testcrosses. For tall heterozygotes with antennae, the offspring are tall-antennae, 46; dwarf-antennae, 7; dwarf-no antennae, 42; tall-no antennae, 5. For heterozygotes with antennae and an upturned snout, the offspring are antennae-upturned snout, 47; antennae-downturned snout, 2; no antennae-downturned snout, 48; no antennae-upturned snout, 3. Calculate the recombination frequencies for both experiments.

Level 2: Application/Analysis

4. Using the information from problem 3, scientists do a further testcross using a heterozygote for height and nose morphology. The offspring are tall-upturned snout, 40; dwarf-upturned snout, 9; dwarf-downturned snout, 42; tall-downturned snout, 9. Calculate the recombination frequency from these data; then use your answer from problem 3 to determine the correct order of the three linked genes.

5. A man with red-green color blindness (a recessive, sex-linked condition) marries a woman with normal vision whose father was color-blind. What is the probability that they will have a color-blind daughter? That their first son will be color-blind? (Note the different wording in the two questions.)

6. You design *Drosophila* crosses to provide recombination data for gene *a*, which is located on the chromosome shown in Figure 12. Gene *a* has recombination frequencies of 14% with the vestigial-wing locus and 26% with the brown-eye locus. Approximately where is gene *a* located along the chromosome?

7. A wild-type fruit fly (heterozygous for gray body color and red eyes) is mated with a black fruit fly with purple eyes. The offspring are wild-type, 721; black-purple, 751; gray-purple, 49; black-red, 45. What is the recombination frequency between these genes for body color and eye color? Using information from Figure 9, what fruit flies (genotypes and phenotypes) would you mate to determine the sequence of the body-color, wing-size, and eye-color genes on the chromosome?

8. Assume that genes A and B are 50 map units apart on the same chromosome. An animal heterozygous at both loci is crossed with one that is homozygous recessive at both loci. What percentage of the offspring will show recombinant phenotypes? Without knowing that these genes are on the same chromosome, how would you interpret the results of this cross?

9. Two genes of a flower, one controlling blue (B) versus white (b) petals and the other controlling round (R) versus oval (r) stamens, are linked and are 10 map units apart. You cross a homozygous blue-oval plant with a homozygous white-round plant. The resulting F_1 progeny are crossed with homozygous white-oval plants, and 1,000 F_2 progeny are obtained. How many F_2 plants of each of the four phenotypes do you expect?

Level 3: Synthesis/Evaluation

10. **SCIENTIFIC INQUIRY**

DRAW IT Assume you are mapping genes A, B, C, and D in *Drosophila*. You know that these genes are linked on the same chromosome, and you determine the recombination frequencies between each pair of genes to be as follows: A and B, 8%; A and C, 28%; A and D, 25%; B and C, 20%, B and D, 33%.

(a) Describe how you determined the recombination frequency for each pair of genes.

(b) Draw a chromosome map based on your data.

11. **FOCUS ON EVOLUTION**

Crossing over is thought to be evolutionarily advantageous because it continually shuffles genetic alleles into novel combinations. Until recently, it was thought that Y-linked genes might degenerate because they lack homologous genes on the X chromosome with which to recombine. However, when the Y chromosome was sequenced, eight large regions were found to be internally homologous to each other, and quite a few of the 78 genes represent duplicates. Explain how this might be beneficial.

12. **FOCUS ON INFORMATION**

The continuity of life is based on heritable information in the form of DNA. In a short essay (100–150 words), relate the structure and behavior of chromosomes to inheritance in both asexually and sexually reproducing species.

13. **SYNTHESIZE YOUR KNOWLEDGE**

Butterflies have an X-Y sex determination system that is different from that of flies or humans. Female butterflies may be either XY or XO, while butterflies with two or more X chromosomes are males. This photograph shows a tiger swallowtail *gynandromorph*, an individual that is half male (left side) and half female (right side). Given that the first division of the zygote divides the embryo into the future right and left halves of the butterfly, propose a hypothesis that explains how nondisjunction during the first mitosis might have produced this unusual-looking butterfly.

Answers

Figure Questions

Figure 2 The ratio would be 1 yellow round : 1 green round : 1 yellow wrinkled : 1 green wrinkled. **Figure 4** About ¾ of the F_2 offspring would have red eyes and about ¼ would have white eyes. About half of the white-eyed flies would be female and half would be male; similarly, about half of the red-eyed flies would be female and half would be male. (Note that the homologs with the eye-color alleles would be the same shape in the Punnett square, and each offspring would inherit two alleles. The sex of the flies would be determined separately by inheritance of the sex chromosomes. Thus your Punnett square would have four possible combinations in sperm and four in eggs; it would have 16 squares altogether.) **Figure 7** All the males would be color-blind, and all the females would be carriers. (Another way to say this is that ½ the offspring would be color-blind males, and ½ the offspring would be carrier females.) **Figure 9** The two largest classes would still be the offspring with the phenotypes of the true-breeding P generation flies, but now they would be gray vestigial and black normal because those were the specific allele combinations in the P generation. **Figure 10** The two chromosomes on the left side of the sketch are like the two chromosomes inherited by the F_1 female, one from each P generation fly. They are passed by the F_1 female intact to the offspring and thus could be called "parental" chromosomes. The other two chromosomes result from crossing over during meiosis in the F_1 female. Because they have combinations of alleles not seen in either of the F_1 female's chromosomes, they can be called "recombinant" chromosomes. (Note that in this example, the alleles on the recombinant chromosomes, b^+ vg^+ and b vg, are the allele combinations that were on the parental chromosomes in the cross shown in Figures 9 and 10. The basis for calling them parental chromosomes is that they have the combination of alleles that was present on the P generation chromosomes.)

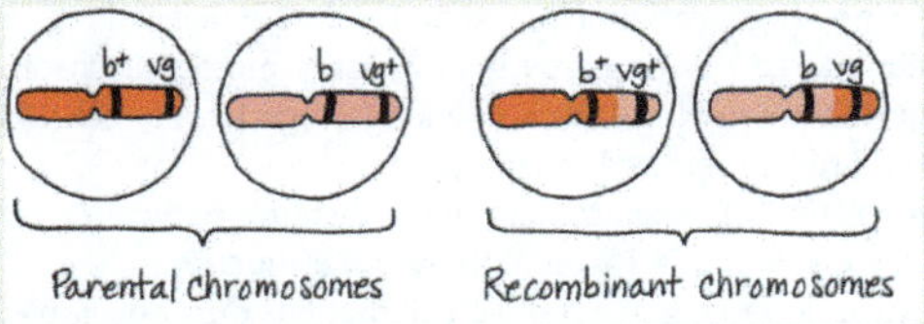

Concept Check 1

1. The law of segregation relates to the inheritance of alleles for a single character. The law of independent assortment of alleles relates to the inheritance of alleles for two characters. **2.** The physical basis for the law of segregation is the separation of homologs in anaphase I. The physical basis for the law of independent assortment is the alternative arrangements of all the different homologous chromosome pairs in metaphase I. **3.** To show the mutant phenotype, a male needs to possess only one mutant allele. If this gene had been on a pair of autosomes, the two alleles would both have had to be mutant for an individual to show the recessive mutant phenotype, a much less probable situation.

Concept Check 2

1. Because the gene for this eye-color character is located on the X chromosome, all female offspring will be red-eyed and heterozygous ($X^{w^+}X^w$); all male offspring will inherit a Y chromosome from the father and be white-eyed (X^wY). (Another way to say this is that ½ the offspring will be red-eyed, heterozygous [carrier] females, and ½ will be white-eyed males.) **2.** ¼ (½ chance that the child will inherit a Y chromosome from the father and be male × ½ chance that he will inherit the X carrying the disease allele from his mother); if the child is a boy, there is a ½ chance he will have the disease; a female would have zero chance (but ½ chance of being a carrier). **3.** With a disorder caused by a dominant allele, there is no such thing as a "carrier," since those with the allele have the disorder. Because the allele is dominant, the females lose any "advantage" in having two X chromosomes, since one disorder-associated allele is sufficient to result in the disorder. All fathers who have the dominant allele will pass it along to *all* their daughters, who will also have the disorder. A mother who has the allele (and thus the disorder) will pass it to half of her sons and half of her daughters.

Concept Check 3

1. Crossing over during meiosis I in the heterozygous parent produces some gametes with recombinant genotypes for the two genes. Offspring with a recombinant phenotype arise from fertilization of the recombinant gametes by homozygous recessive gametes from the double-mutant parent. **2.** In each case, the alleles contributed by the female parent (in the egg) determine the phenotype of the offspring because the male in this cross contributes only recessive alleles. Thus, identifying the phenotype of the offspring tells you what alleles were in its mother's (the dihybrid female's) egg. **3.** No. The order could be *A-C-B* or *C-A-B*. To determine which possibility is correct, you need to know the recombination frequency between *B* and *C*.

Concept Check 4

1. In meiosis, a combined 14-21 chromosome will behave as one chromosome. If a gamete receives the combined 14-21 chromosome and a normal copy of chromosome 21, trisomy 21 will result when this gamete combines with a normal gamete during fertilization. **2.** No. The child can be either I^AI^Ai or I^Aii. A sperm of genotype I^AI^A could result from nondisjunction in the father during meiosis II, while an egg with the genotype *ii* could result from nondisjunction in the mother during either meiosis I or meiosis II. **3.** Activation of this gene could lead to the production of too much of this kinase. If the kinase is involved in a signaling pathway that triggers cell division, too much of it could trigger unrestricted cell division, which in turn could contribute to the development of a cancer (in this case, a cancer of one type of white blood cell). **4.** The inactivation of two X chromosomes in XXX women would leave them with one genetically active X, as in women with the normal number of chromosomes. Microscopy should reveal two Barr bodies in XXX women.

Summary of Key Concepts Questions

1 Because the sex chromosomes are different from each other and because they determine the sex of the offspring, Morgan could use the sex of the offspring as a phenotypic character to follow the parental chromosomes. (He could also have followed them under a microscope, as the X and Y chromosomes look different.) At the same time, he could record eye color to follow the eye-color alleles. **2** Males have only one X chromosome, along with a Y chromosome, while females have two X chromosomes. The Y chromosome has very few genes on it, while the X has about 1,000. When a recessive X-linked allele that causes a disorder is inherited by a male on the X from his mother, there isn't a second allele present on the Y (males are hemizygous), so the male has the disorder. Because females have two X chromosomes, they must inherit two recessive alleles in order to have the disorder, a rarer occurrence. **3** Crossing over results in new combinations of alleles. Crossing over is a random occurrence, and the more distance there is between two genes, the more chances there are for crossing over to occur, leading to a new allele combination. **4** In inversions and reciprocal translocations, the same genetic material is present in the same relative amount but just organized differently. In aneuploidy, duplications, deletions, and nonreciprocal translocations, the balance of genetic material is upset, as large segments are either missing or present in more than one copy. Apparently, this type of imbalance is very damaging to the organism. (Although it isn't lethal in the developing embryo, the reciprocal translocation that produces the Philadelphia chromosome can lead to a serious condition, cancer, by altering the expression of important genes.)

Test Your Understanding

1. 0; ½; 1/16 **2.** Recessive; if the disorder were dominant, it would affect at least one parent of a child born with the disorder. The disorder's inheritance is sex-linked because it is seen only in boys. For a girl to have the disorder, she would have to inherit recessive alleles from *both* parents. This would be very rare, since males with the recessive allele on their X chromosome die in their early teens. **3.** Between *T* and *A*, 12%; between *A* and *S*, 5% **4.** Between *T* and *S*, 18%; sequence of genes is *T–A–S*. **5.** ¼ for each daughter (½ chance that the child will be female × ½ chance of a homozygous recessive genotype); ½ for first son **6.** About one-third of the distance from the vestigial-wing locus to the brown-eye locus **7.** 6%; wild-type heterozygous for normal wings and red eyes × recessive homozygous for vestigial wings and purple eyes **8.** Fifty percent of the offspring will show phenotypes resulting from crossovers. These results would be the same as those from a cross where *A* and *B* were *not* on the same chromosome, and you would interpret the results to mean that the genes are unlinked. (Further crosses involving other genes on the same chromosome would reveal the genetic linkage and map distances.) **9.** 450 each of blue-oval and white-round (parentals) and 50 each of blue-round and white-oval (recombinants) **10.** (a) For each pair of genes, you had to generate an F_1 dihybrid fly; let's use the *A* and *B* genes as an example. You obtained homozygous parental flies, either the first with dominant alleles of the two genes (*AABB*) and the second with recessive alleles (*aabb*), or the first with dominant alleles of gene *A* and recessive alleles of gene *B* (*AAbb*) and the second with recessive alleles of gene *A* and dominant alleles of gene *B* (*aaBB*). Breeding either of these pairs of P generation flies gave you an F_1 dihybrid, which you then testcrossed with a doubly homozygous recessive fly (*aabb*). You classed the offspring as parental or recombinant, based on the genotypes of the P generation parents (either of the two pairs described above). You added up the number of recombinant types and then divided by the total number of offspring. This gave you the recombination percentage (in this case, 8%), which you can translate into map units (8 map units) to construct your map.
(b)

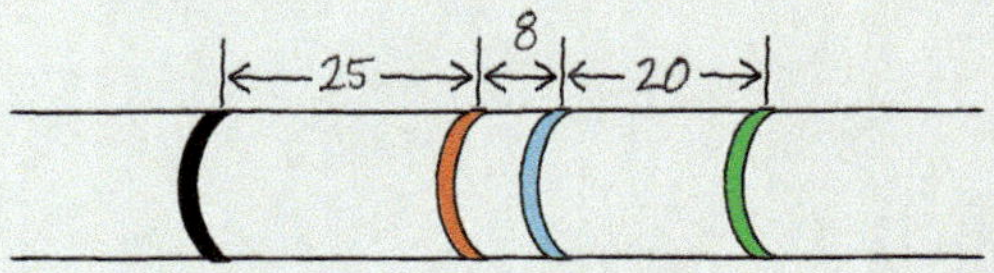

Credits

Credits are listed in the order of appearance.

Photo Credits

1 Peter Menzel/Science Source; **3** From: Learning to Fly: Phenotypic Markers in *Drosophila*. A poster of common phenotypic markers used in *Drosophila* genetics. Jennifer Childress, Richard Behringer, and Georg Halder. 2005. *Genesis* 43(1). Cover illustration; **5** Andrew Syred/Science Source; **8** Jagodka/Shutterstock; **12 SSE** Oliver911119/Shutterstock; **15a** SPL/Science Source; **15b** denys_kuvaiev/fotolia; **Chapter Review SYK** James K. Adams, Biology, Dalton State College, Dalton, Georgia.

Illustration and Text Credits

Quote ("Two years' work . . .") Thomas Hunt Morgan quoted in R.G. Harrison, Embryology and its relations, *Science* 85 (1937): 369–74. p. 370; **4** Data from T. H. Morgan, Sex-limited inheritance in *Drosophila*, *Science* 32:120–122 (1910); **9** Data from T. H. Morgan and C. J. Lynch, The linkage of two factors in *Drosophila* that are not sex-linked, *Biological Bulletin* 23:174–182 (1912).

Glossary

aneuploidy (an′-yū-ploy′-dē) A chromosomal aberration in which one or more chromosomes are present in extra copies or are deficient in number.

Barr body A dense object lying along the inside of the nuclear envelope in cells of female mammals, representing a highly condensed, inactivated X chromosome.

chromosome theory of inheritance A basic principle in biology stating that genes are located at specific positions (loci) on chromosomes and that the behavior of chromosomes during meiosis accounts for inheritance patterns.

crossing over The reciprocal exchange of genetic material between nonsister chromatids during prophase I of meiosis.

cytogenetic map A map of a chromosome that locates genes with respect to chromosomal features distinguishable in a microscope.

deletion (1) A deficiency in a chromosome resulting from the loss of a fragment through breakage. (2) A mutational loss of one or more nucleotide pairs from a gene.

Down syndrome A human genetic disease usually caused by the presence of an extra chromosome 21; characterized by developmental delays and heart and other defects that are generally treatable or non-life-threatening.

Duchenne muscular dystrophy (duh-shen′) A human genetic disease caused by a sex-linked recessive allele; characterized by progressive weakening and a loss of muscle tissue.

duplication An aberration in chromosome structure due to fusion with a fragment from a homologous chromosome, such that a portion of a chromosome is duplicated.

genetic map An ordered list of genetic loci (genes or other genetic markers) along a chromosome.

genetic recombination General term for the production of offspring with combinations of traits that differ from those found in either parent.

hemophilia (hē′-muh-fil′-ē-uh) A human genetic disease caused by a sex-linked recessive allele, resulting in the absence of one or more blood-clotting proteins; characterized by excessive bleeding following injury.

inversion An aberration in chromosome structure resulting from reattachment of a chromosomal fragment in a reverse orientation to the chromosome from which it originated.

linkage map A genetic map based on the frequencies of recombination between markers during crossing over of homologous chromosomes.

linked genes Genes located close enough together on a chromosome that they tend to be inherited together.

map unit A unit of measurement of the distance between genes. One map unit is equivalent to a 1% recombination frequency.

monosomic Referring to a diploid cell that has only one copy of a particular chromosome instead of the normal two.

nondisjunction An error in meiosis or mitosis in which members of a pair of homologous chromosomes or sister chromatids fail to separate properly from each other.

parental type An offspring with a phenotype that matches one of the true-breeding parental (P generation) phenotypes; also refers to the phenotype itself.

recombinant type (recombinant) An offspring whose phenotype differs from that of the true-breeding P generation parents; also refers to the phenotype itself.

sex-linked gene A gene located on either sex chromosome. Most sex-linked genes are on the X chromosome and show distinctive patterns of inheritance; there are very few genes on the Y chromosome.

trisomic Referring to a diploid cell that has three copies of a particular chromosome instead of the normal two.

wild type The phenotype most commonly observed in natural populations; also refers to the individual with that phenotype.

The Molecular Basis of Inheritance

KEY CONCEPTS

1 DNA is the genetic material
2 Many proteins work together in DNA replication and repair
3 A chromosome consists of a DNA molecule packed together with proteins
4 Understanding DNA structure and replication makes genetic engineering possible

▲ **Figure 1** What is the structure of DNA?

Life's Operating Instructions

The elegant double-helical structure of deoxyribonucleic acid, or DNA, has become an icon of modern biology **(Figure 1)**. James Watson and Francis Crick shook the scientific world in April 1953 with their DNA model, which they constructed from sheet metal and wire **(Figure 2)**. Gregor Mendel's heritable factors and Thomas Hunt Morgan's genes on chromosomes are, in fact, composed of DNA. Chemically speaking, your genetic endowment is the DNA you inherited from your parents. DNA, the substance of inheritance, is the most celebrated molecule of our time.

Of all nature's molecules, nucleic acids are unique in their ability to direct their own replication from monomers. Indeed, the resemblance of offspring to their parents has its basis in the accurate replication of DNA and its transmission from one generation to the next. Hereditary information in DNA directs the development of your biochemical, anatomical, physiological, and, to some extent, behavioral traits. In this chapter, you'll discover how biologists deduced that DNA is the genetic material and how Watson and Crick worked out its structure. You'll also learn how a molecule of DNA is copied during **DNA replication** and how cells repair their DNA. Next, you'll see how DNA is packaged together with proteins in a chromosome. Finally, you'll explore how an understanding of DNA-related processes has allowed scientists to directly manipulate genes for practical purposes.

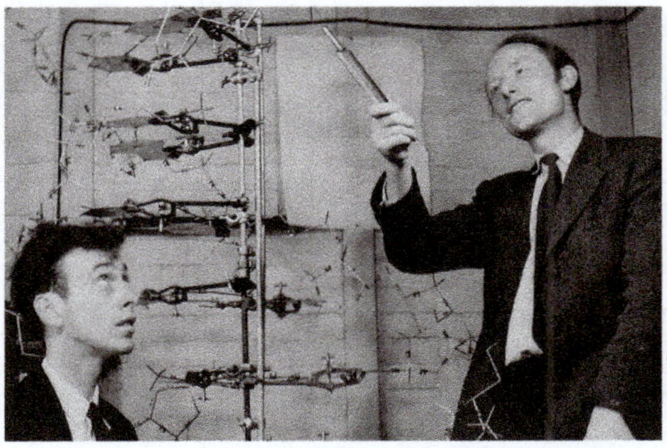

◀ **Figure 2** James Watson (left) and Francis Crick with their DNA model.

CONCEPT 1

DNA is the genetic material

Today, even schoolchildren have heard of DNA, and scientists routinely manipulate DNA in the laboratory, often to change the heritable traits of cells in their experiments. Early in the 20th century, however, identifying the molecules of inheritance loomed as a major challenge to biologists.

The Search for the Genetic Material: *Scientific Inquiry*

Once T. H. Morgan's group showed that genes exist as parts of chromosomes, the two chemical components of chromosomes—DNA and protein—emerged as the leading candidates for the genetic material. Until the 1940s, the case for proteins seemed stronger: Biochemists had identified proteins as a class of macromolecules with great heterogeneity and specificity of function, essential requirements for the hereditary material. Moreover, little was known about nucleic acids, whose physical and chemical properties seemed far too uniform to account for the multitude of specific inherited traits exhibited by every organism. This view gradually changed as the role of DNA in heredity was worked out in studies of bacteria and the viruses that infect them, systems far simpler than fruit flies or humans. Let's trace the search for the genetic material in some detail as a case study in scientific inquiry.

Evidence That DNA Can Transform Bacteria

In 1928, a British medical officer named Frederick Griffith was trying to develop a vaccine against pneumonia. He was studying *Streptococcus pneumoniae*, a bacterium that causes pneumonia in mammals. Griffith had two strains (varieties) of the bacterium, one pathogenic (disease-causing) and one nonpathogenic (harmless). He was surprised to find that when he killed the pathogenic bacteria with heat and then mixed the cell remains with living bacteria of the nonpathogenic strain, some of the living cells became pathogenic **(Figure 3)**. Furthermore, this newly acquired trait of pathogenicity was inherited by all the descendants of the transformed bacteria. Apparently, some chemical component of the dead pathogenic cells caused this heritable change, although the identity of the substance was not known. Griffith called the phenomenon **transformation**, now defined as a change in genotype and phenotype due to the assimilation of external DNA by a cell. Later work by Oswald Avery and others identified the transforming substance as DNA.

Scientists remained skeptical, however, since many still viewed proteins as better candidates for the genetic material. Also, many biologists were not convinced that bacterial genes would be similar in composition and function to those of more complex organisms. But the major reason for the continued doubt was that so little was known about DNA.

▼ Figure 3 Inquiry

Can a genetic trait be transferred between different bacterial strains?

Experiment Frederick Griffith studied two strains of the bacterium *Streptococcus pneumoniae*. The S (smooth) strain can cause pneumonia in mice; it is pathogenic because the cells have an outer capsule that protects them from an animal's immune system. Cells of the R (rough) strain lack a capsule and are nonpathogenic. To test for the trait of pathogenicity, Griffith injected mice with the two strains:

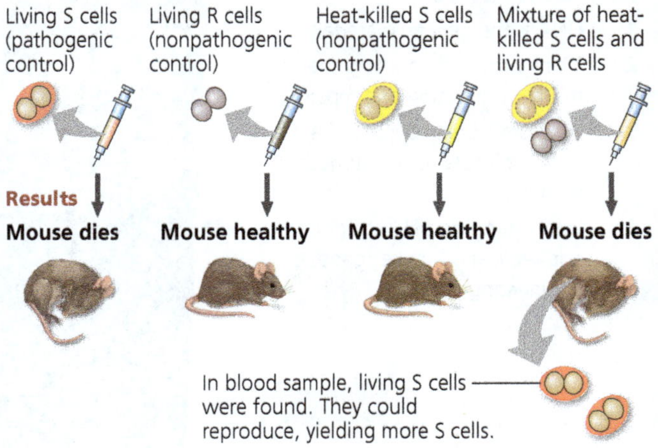

Conclusion The living R bacteria had been transformed into pathogenic S bacteria by an unknown, heritable substance from the dead S cells that enabled the R cells to make capsules.

Data from F. Griffith, The significance of pneumococcal types, *Journal of Hygiene* 27:113–159 (1928).

WHAT IF? How did this experiment rule out the possibility that the R cells simply used the dead S cells' capsules to become pathogenic?

Evidence That Viral DNA Can Program Cells

Additional evidence that DNA was the genetic material came from studies of viruses that infect bacteria. These viruses are called **bacteriophages** (meaning "bacteria-eaters"), or **phages** for short. Viruses are much simpler than cells. A **virus** is little more than DNA (or sometimes RNA) enclosed by a protective coat, which is often simply protein **(Figure 4)**. To produce

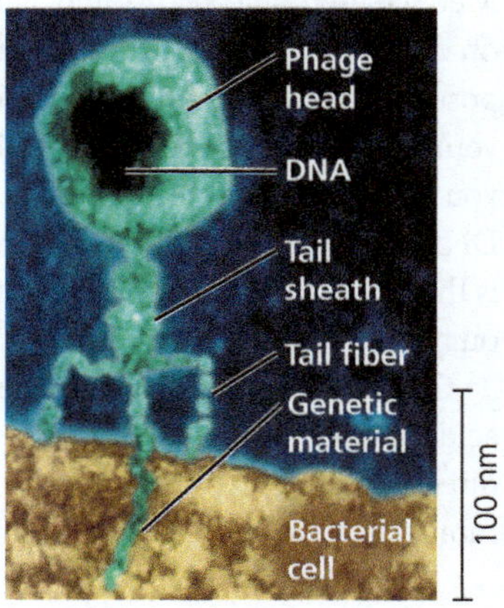

► Figure 4 A virus infecting a bacterial cell. A phage called T2 attaches to a host cell and injects its genetic material through the plasma membrane while the head and tail parts remain on the outer bacterial surface (colorized TEM).

more viruses, a virus must infect a cell and take over the cell's metabolic machinery.

Phages have been widely used as tools by researchers in molecular genetics. In 1952, Alfred Hershey and Martha Chase performed experiments showing that DNA is the genetic material of a phage known as T2. This is one of many phages that infect *Escherichia coli* (*E. coli*), a bacterium that normally lives in the intestines of mammals and is a model organism for molecular biologists. At that time, biologists already knew that T2, like many other phages, was composed almost entirely of DNA and protein. They also knew that the T2 phage could quickly turn an *E. coli* cell into a T2-producing factory that released many copies of new phages when the cell ruptured. Somehow, T2 could reprogram its host cell to produce viruses. But which viral component—protein or DNA—was responsible?

Hershey and Chase answered this question by devising an experiment showing that only one of the two components of T2 actually enters the *E. coli* cell during infection **(Figure 5)**. In their experiment, they used a radioactive isotope of sulfur to tag protein in one batch of T2 and a radioactive isotope of phosphorus to tag DNA in a second batch. Because protein, but not DNA, contains sulfur, radioactive sulfur atoms were incorporated only into the protein of the phage. In a similar way, the atoms of radioactive phosphorus labeled only the DNA, not the protein, because nearly all the phage's phosphorus is in its DNA. In the experiment, separate samples

▼ Figure 5 Inquiry

Is protein or DNA the genetic material of phage T2?

Experiment Alfred Hershey and Martha Chase used radioactive sulfur and phosphorus to trace the fates of protein and DNA, respectively, of T2 phages that infected bacterial cells. They wanted to see which of these molecules entered the cells and could reprogram them to make more phages.

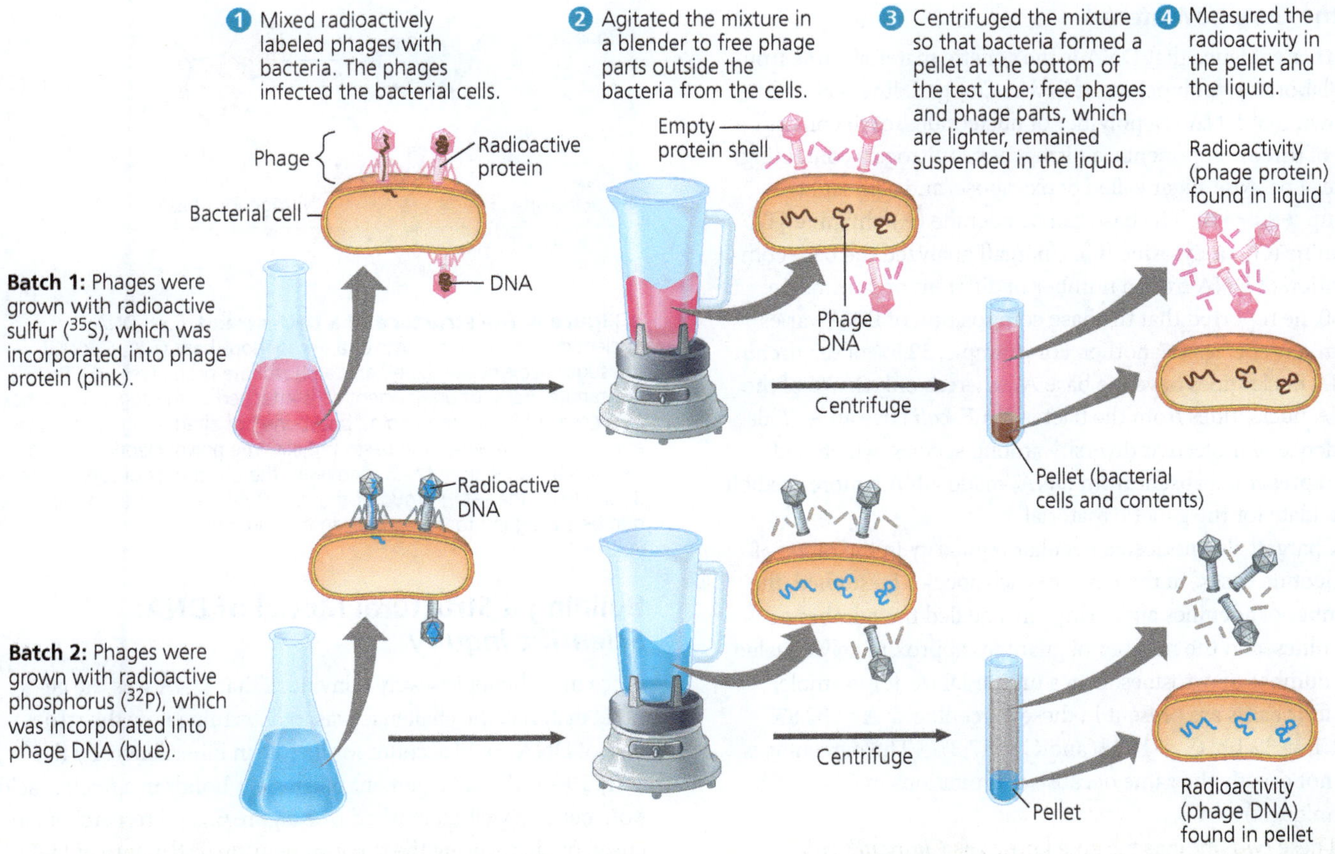

Results When proteins were labeled (batch 1), radioactivity remained outside the cells, but when DNA was labeled (batch 2), radioactivity was found inside the cells. Cells containing radioactive phage DNA released new phages with some radioactive phosphorus.

Conclusion Phage DNA entered bacterial cells, but phage proteins did not. Hershey and Chase concluded that DNA, not protein, functions as the genetic material of phage T2.

Data from A. D. Hershey and M. Chase, Independent functions of viral protein and nucleic acid in growth of bacteriophage, *Journal of General Physiology* 36:39–56 (1952).

WHAT IF? How would the results have differed if proteins carried the genetic information?

of nonradioactive *E. coli* cells were infected with the protein-labeled and DNA-labeled batches of T2. The researchers then tested the two samples shortly after the onset of infection to see which type of molecule—protein or DNA—had entered the bacterial cells and would therefore have been capable of reprogramming them.

Hershey and Chase found that the phage DNA entered the host cells, but the phage protein did not. Moreover, when these bacteria were returned to a culture medium and the infection ran its course, the *E. coli* released phages that contained some radioactive phosphorus. This result further showed that the DNA inside the cell played an ongoing role during the infection process.

Hershey and Chase concluded that the DNA injected by the phage must be the molecule carrying the genetic information that makes the cells produce new viral DNA and proteins. The Hershey-Chase experiment was a landmark study because it provided powerful evidence that nucleic acids, rather than proteins, are the hereditary material, at least for certain viruses.

Additional Evidence That DNA Is the Genetic Material

Further evidence that DNA is the genetic material came from the laboratory of biochemist Erwin Chargaff. It was already known that DNA is a polymer of nucleotides, each consisting of three components: a nitrogenous (nitrogen-containing) base, a pentose sugar called deoxyribose, and a phosphate group **(Figure 6)**. The base can be adenine (A), thymine (T), guanine (G), or cytosine (C). Chargaff analyzed the base composition of DNA from a number of different organisms. In 1950, he reported that the base composition of DNA varies from one species to another. For example, 32.8% of sea urchin DNA nucleotides have the base A, whereas only 24.7% of the DNA nucleotides from the bacterium *E. coli* have an A. This evidence of molecular diversity among species, which had been presumed absent from DNA, made DNA a more credible candidate for the genetic material.

Chargaff also noticed a peculiar regularity in the ratios of nucleotide bases. In the DNA of each species he studied, the number of adenines approximately equaled the number of thymines, and the number of guanines approximately equaled the number of cytosines. In sea urchin DNA, for example, the four bases are present in these percentages: A = 32.8% and T = 32.1%; G = 17.7% and C = 17.3%. (The percentages are not exactly the same because of limitations in Chargaff's techniques.)

These two findings became known as *Chargaff's rules*: (1) DNA base composition varies between species, and (2) for each species, the percentages of A and T bases are roughly equal, as are those of G and C bases. In the **Scientific Skills Exercise**, you can use Chargaff's rules to predict unknown percentages of nucleotide bases. The basis for these rules remained unexplained until the discovery of the double helix.

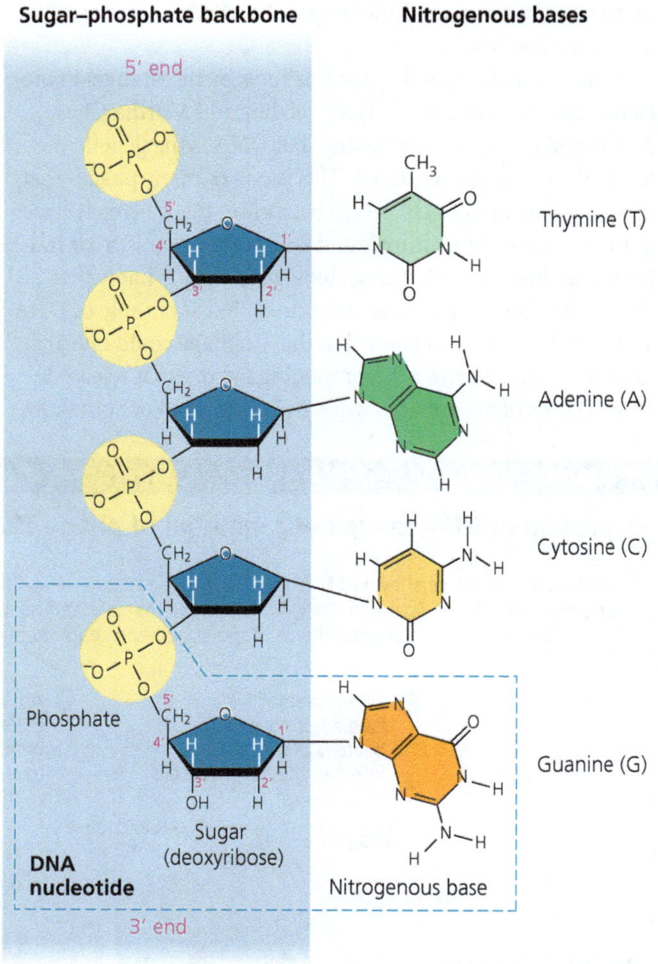

▲ **Figure 6 The structure of a DNA strand.** Each DNA nucleotide monomer consists of a nitrogenous base (T, A, C, or G), the sugar deoxyribose (blue), and a phosphate group (yellow). The phosphate group of one nucleotide is attached to the sugar of the next by a covalent bond, forming a "backbone" of alternating phosphates and sugars from which the bases project. The polynucleotide strand has directionality, from the 5′ end (with the phosphate group) to the 3′ end (with the —OH group of the sugar). 5′ and 3′ refer to the numbers assigned to the carbons in the sugar ring.

Building a Structural Model of DNA: *Scientific Inquiry*

Once most biologists were convinced that DNA was the genetic material, the challenge was to determine how the structure of DNA could account for its role in inheritance. By the early 1950s, the arrangement of covalent bonds in a nucleic acid polymer was well established (see Figure 6), and researchers focused on discovering the three-dimensional structure of DNA. Among the scientists working on the problem were Linus Pauling, at the California Institute of Technology, and Maurice Wilkins and Rosalind Franklin, at King's College in London. First to come up with the complete answer, however, were two scientists who were relatively unknown at the time—the American James Watson and the Englishman Francis Crick.

Scientific Skills Exercise

Working with Data in a Table

Given the Percentage Composition of One Nucleotide in a Genome, Can We Predict the Percentages of the Other Three Nucleotides? Even before the structure of DNA was elucidated, Erwin Chargaff and his coworkers noticed a pattern in the base composition of nucleotides from different species: The percentage of adenine (A) bases roughly equaled that of thymine (T) bases, and the percentage of cytosine (C) bases roughly equaled that of guanine (G) bases. Further, the percentage of each pair (A/T or C/G) varied from species to species. We now know that the 1:1 A/T and C/G ratios are due to complementary base pairing between A and T and between C and G in the DNA double helix, and interspecies differences are due to the unique sequences of bases along a DNA strand. In this exercise, you will apply Chargaff's rules to predict the composition of bases in a genome.

How the Experiments Were Done In Chargaff's experiments, DNA was extracted from the given organism, hydrolyzed to break apart the individual nucleotides, and then analyzed chemically. These experiments provided approximate values for each type of nucleotide. (Today, whole-genome sequencing allows base composition analysis to be done more precisely directly from the sequence data.)

Data from the Experiments Tables are useful for organizing sets of data representing a common set of values (here, percentages of A, G, C, and T) for a number of different samples (in this case, from different species). You can apply the patterns that you see in the known data to predict unknown values. In the table, complete base distribution data are given for sea urchin DNA and salmon DNA; you will use Chargaff's rules to fill in the rest of the table with predicted values.

▶ Sea urchin

	Base Percentage			
Source of DNA	**Adenine**	**Guanine**	**Cytosine**	**Thymine**
Sea urchin	32.8	17.7	17.3	32.1
Salmon	29.7	20.8	20.4	29.1
Wheat	28.1	21.8	22.7	
E. coli	24.7	26.0		
Human	30.4			30.1
Ox	29.0			
Average %				

Data from several papers by Chargaff, for example, E. Chargaff et al., Composition of the desoxypentose nucleic acids of four genera of sea-urchin, *Journal of Biological Chemistry* 195:155–160 (1952).

INTERPRET THE DATA

1. Explain how the sea urchin and salmon data demonstrate both of Chargaff's rules.
2. Using Chargaff's rules, fill in the table with your predictions of the missing percentages of bases, starting with the wheat genome and proceeding through *E. coli*, human, and ox. Show how you arrived at your answers.
3. If Chargaff's rule—that the amount of A equals the amount of T and the amount of C equals the amount of G—is valid, then hypothetically we could extrapolate this to the combined DNA of all species on Earth (like one huge Earth genome). To see whether the data in the table support this hypothesis, calculate the average percentage for each base in your completed table by averaging the values in each column. Does Chargaff's equivalence rule still hold true?

MB A version of this Scientific Skills Exercise can be assigned in MasteringBiology.

The brief but celebrated partnership that solved the puzzle of DNA structure began soon after Watson journeyed to Cambridge University, where Crick was studying protein structure with a technique called X-ray crystallography. While visiting the laboratory of Maurice Wilkins, Watson saw an X-ray diffraction image of DNA produced by Wilkins's accomplished colleague Rosalind Franklin **(Figure 7a)**. Images produced by X-ray crystallography are not actually pictures of molecules. The spots and smudges in **Figure 7b** were produced by X-rays that were diffracted (deflected) as they passed through aligned fibers of purified DNA. Watson was familiar with the type of X-ray diffraction pattern that helical molecules produce, and an examination of the photo that Wilkins showed him confirmed that DNA was helical in shape. The photo also augmented earlier data obtained by Franklin and others suggesting the width of the helix and the spacing of the nitrogenous bases along it. The pattern in this photo implied that the helix was made up of two strands, contrary to a three-stranded model that Linus Pauling had

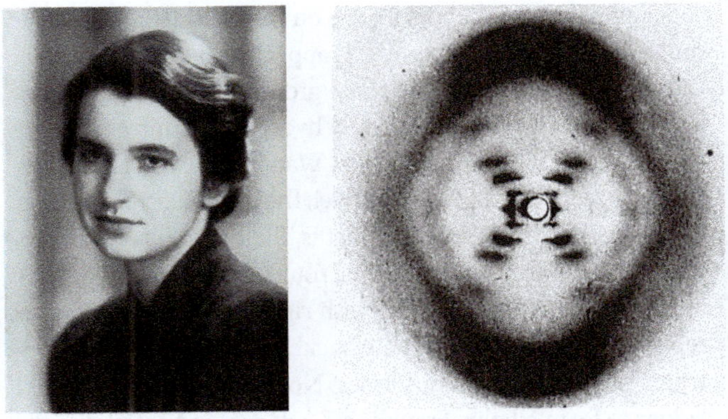

(a) Rosalind Franklin **(b) Franklin's X-ray diffraction photograph of DNA**

▲ **Figure 7 Rosalind Franklin and her X-ray diffraction photo of DNA.** Franklin, a very accomplished X-ray crystallographer, conducted critical experiments resulting in the photo that allowed Watson and Crick to deduce the double-helical structure of DNA.

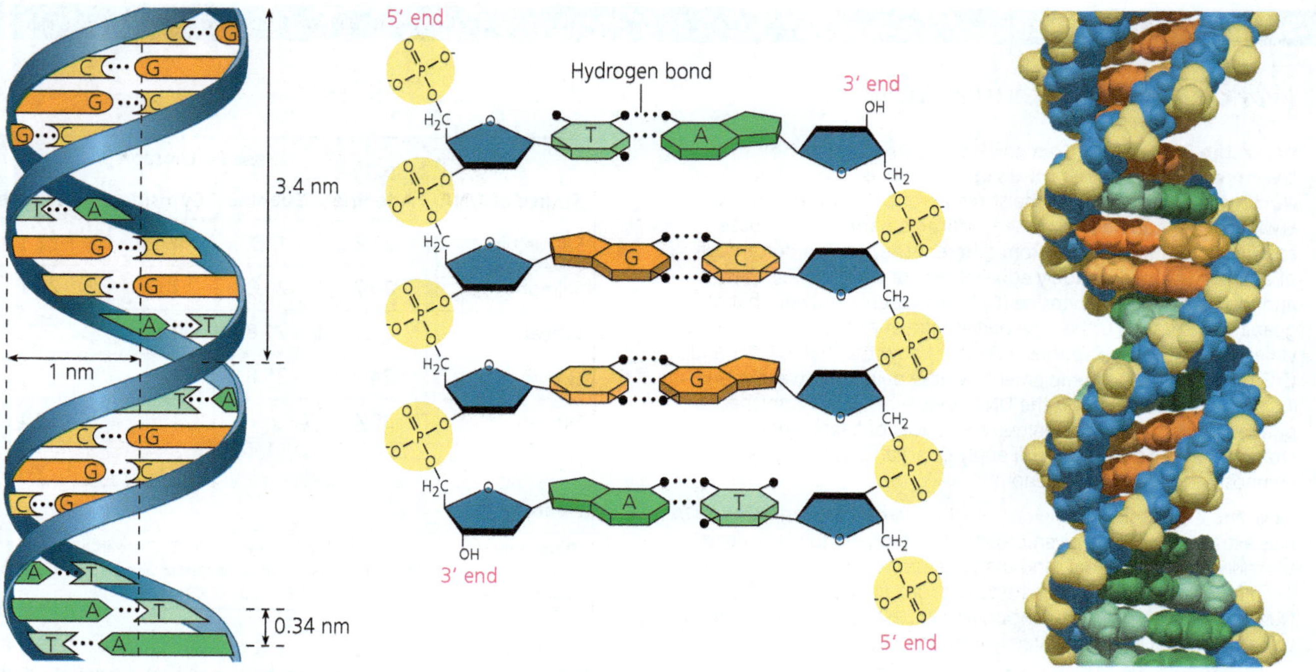

(a) Key features of DNA structure. The "ribbons" in this diagram represent the sugar-phosphate backbones of the two DNA strands. The helix is "right-handed," curving up to the right in the front. The two strands are held together by hydrogen bonds (dotted lines) between the nitrogenous bases, which are paired in the interior of the double helix.

(b) Partial chemical structure. For clarity, the two DNA strands are shown untwisted in this partial chemical structure. Strong covalent sugar-phosphate bonds link the nucleotides of each strand, while weaker hydrogen bonds between the bases hold one strand to the other. Notice that the strands are antiparallel, meaning that they are oriented in opposite directions, like the lanes of a divided street. From top to bottom, the left strand is oriented in the 5′ to 3′ direction, and the right strand in the 3′ to 5′ direction (see also Figure 6).

(c) Space-filling model. This space-filling model generated by a computer shows that the base pairs are tightly stacked. Van der Waals interactions between the stacked pairs play a major role in holding the molecule together.

▲ **Figure 8 The structure of the double helix.**

proposed a short time earlier. The presence of two strands accounts for the now-familiar term **double helix** (**Figure 8**).

Watson and Crick began building models of a double helix that would conform to the X-ray measurements and what was then known about the chemistry of DNA, including Chargaff's rules. They knew that Franklin had concluded that the sugar-phosphate backbones were on the outside of the DNA molecule. This arrangement was appealing because it put the negatively charged phosphate groups facing the aqueous surroundings, while the relatively hydrophobic nitrogenous bases were hidden in the interior. Watson constructed such a model (see Figure 2). In this model, the two sugar-phosphate backbones are **antiparallel**—that is, their subunits run in opposite directions (see Figure 8b). You can imagine the overall arrangement as a rope ladder with rigid rungs. The side ropes represent the sugar-phosphate backbones, and the rungs represent pairs of nitrogenous bases. Now imagine twisting the ladder to form a helix. Franklin's X-ray data indicated that the helix makes one full turn every 3.4 nm along its length. With the bases stacked just 0.34 nm apart, there are ten "rungs" of base pairs in each full turn of the helix.

The nitrogenous bases of the double helix are paired in specific combinations: adenine (A) with thymine (T), and guanine (G) with cytosine (C). It was mainly by trial and error that Watson and Crick arrived at this key feature of DNA. At first, Watson imagined that the bases paired like with like—for example, A with A and C with C. But this model did not fit the X-ray data, which suggested that the double helix had a uniform diameter. Why is this requirement inconsistent with like-with-like pairing of bases? Adenine and guanine are purines, nitrogenous bases with two organic rings, while cytosine and thymine are nitrogenous bases called pyrimidines, which have a single ring. Pairing a purine with a pyrimidine is the only combination that results in a uniform diameter for the double helix (**Figure 9**).

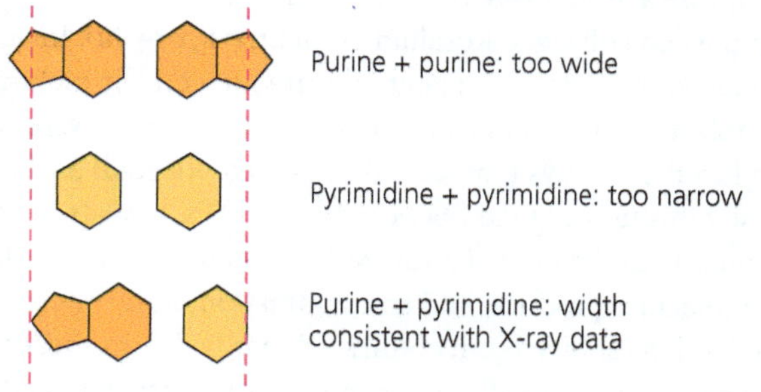

▲ **Figure 9 Possible base pairings in the DNA double helix.** Purines (A and G) are about twice as wide as pyrimidines (C and T). A purine-purine pair is too wide and a pyrimidine-pyrimidine pair is too narrow to account for the 2-nm diameter of the double helix, while a purine-pyrimidine pair fits the data well.

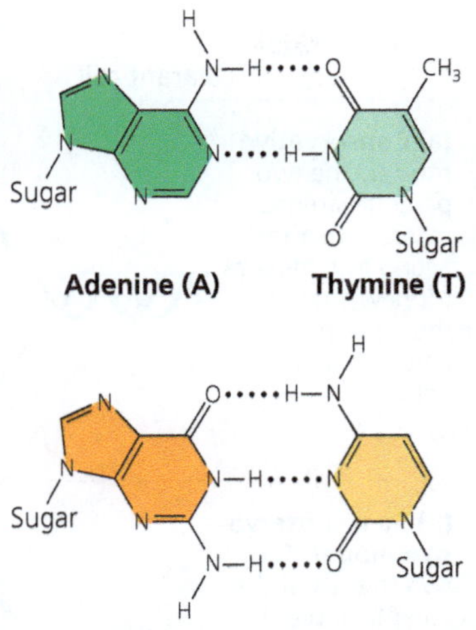

▶ **Figure 10 Base pairing in DNA.** The pairs of nitrogenous bases in a DNA double helix are held together by hydrogen bonds, shown here as black dotted lines.

Watson and Crick reasoned that there must be additional specificity of pairing dictated by the structure of the bases. Each base has chemical side groups that can form hydrogen bonds with its appropriate partner: Adenine can form two hydrogen bonds with thymine and only thymine; guanine forms three hydrogen bonds with cytosine and only cytosine. In shorthand, A pairs with T, and G pairs with C **(Figure 10)**.

The Watson-Crick model took into account Chargaff's ratios and ultimately explained them. Wherever one strand of a DNA molecule has an A, the partner strand has a T. Similarly, a G in one strand is always paired with a C in the complementary strand. Therefore, in the DNA of any organism, the amount of adenine equals the amount of thymine, and the amount of guanine equals the amount of cytosine. (Modern DNA sequencing techniques have confirmed that the amounts are exactly equal.) Although the base-pairing rules dictate the combinations of nitrogenous bases that form the "rungs" of the double helix, they do not restrict the sequence of nucleotides *along* each DNA strand. The linear sequence of the four bases can be varied in countless ways, and each gene has a unique order, or base sequence.

In April 1953, Watson and Crick surprised the scientific world with a succinct, one-page paper that reported their molecular model for DNA: the double helix, which has since become the symbol of molecular biology. Watson and Crick, along with Maurice Wilkins, were awarded the Nobel Prize in 1962 for this work. (Sadly, Rosalind Franklin had died at the age of 38 in 1958 and was thus ineligible for the prize.) The beauty of the double helix model was that the structure of DNA suggested the basic mechanism of its replication.

CONCEPT CHECK 1

1. Given a polynucleotide sequence such as GAATTC, explain what further information you would need in order to identify which end is the 5′ end. (See Figure 6.)
2. WHAT IF? Griffith did not expect transformation to occur in his experiment. What results was he expecting? Explain.

CONCEPT 2

Many proteins work together in DNA replication and repair

The relationship between structure and function is manifest in the double helix. The idea that there is specific pairing of nitrogenous bases in DNA was the flash of inspiration that led Watson and Crick to the double helix. At the same time, they saw the functional significance of the base-pairing rules. They ended their classic paper with this wry statement: "It has not escaped our notice that the specific pairing we have postulated immediately suggests a possible copying mechanism for the genetic material." In this section, you'll learn about the basic principle of DNA replication **(Figure 11)**, as well as some important details of the process.

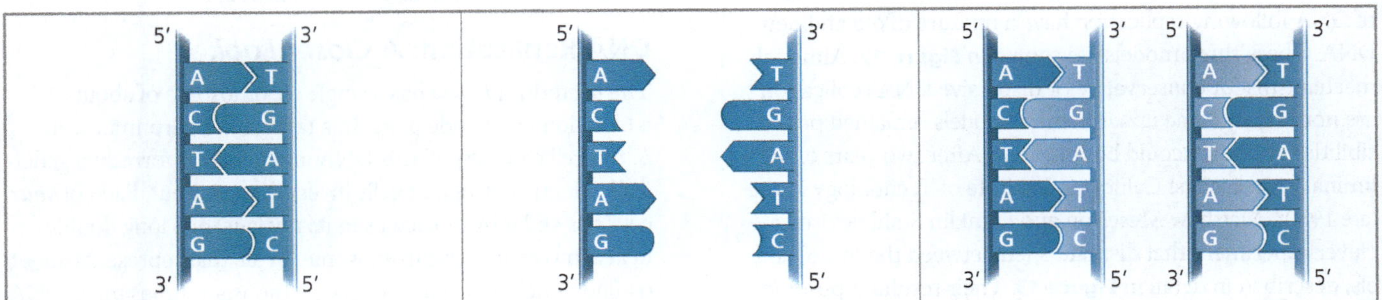

(a) The parental molecule has two complementary strands of DNA. Each base is paired by hydrogen bonding with its specific partner, A with T and G with C.

(b) First, the two DNA strands are separated. Each parental strand can now serve as a template for a new, complementary strand.

(c) Nucleotides complementary to the parental (dark blue) strand are connected to form the sugar-phosphate backbones of the new "daughter" (light blue) strands.

▲ **Figure 11 A model for DNA replication: the basic concept.** In this simplified illustration, a short segment of DNA has been untwisted. Simple shapes symbolize the four kinds of bases. Dark blue represents DNA strands present in the parental molecule; light blue represents newly synthesized DNA.

The Basic Principle: Base Pairing to a Template Strand

In a second paper, Watson and Crick stated their hypothesis for how DNA replicates:

> Now our model for deoxyribonucleic acid is, in effect, a pair of templates, each of which is complementary to the other. We imagine that prior to duplication the hydrogen bonds are broken, and the two chains unwind and separate. Each chain then acts as a template for the formation on to itself of a new companion chain, so that eventually we shall have two pairs of chains, where we only had one before. Moreover, the sequence of the pairs of bases will have been duplicated exactly.*

Figure 11 illustrates Watson and Crick's basic idea. To make it easier to follow, we show only a short section of double helix, in untwisted form. Notice that if you cover one of the two DNA strands of Figure 11a, you can still determine its linear sequence of nucleotides by referring to the uncovered strand and applying the base-pairing rules. The two strands are complementary; each stores the information necessary to reconstruct the other. When a cell copies a DNA molecule, each strand serves as a template for ordering nucleotides into a new, complementary strand. Nucleotides line up along the template strand according to the base-pairing rules and are linked to form the new strands. Where there was one double-stranded DNA molecule at the beginning of the process, there are soon two, each an exact replica of the "parental" molecule.

This model of DNA replication remained untested for several years following publication of the DNA structure. The requisite experiments were simple in concept but difficult to perform. Watson and Crick's model predicts that when a double helix replicates, each of the two daughter molecules will have one old strand, from the parental molecule, and one newly made strand. This **semiconservative model** can be distinguished from a conservative model of replication, in which the two parental strands somehow come back together after the process (that is, the parental molecule is conserved). In yet a third model, called the dispersive model, all four strands of DNA following replication have a mixture of old and new DNA. These three models are shown in **Figure 12**. Although mechanisms for conservative or dispersive DNA replication are not easy to come up with, these models remained possibilities until they could be ruled out. After two years of preliminary work at the California Institute of Technology in the late 1950s, Matthew Meselson and Franklin Stahl devised a clever experiment that distinguished between the three models, described in detail in **Figure 13**. Their results supported the semiconservative model of DNA replication, predicted by Watson and Crick, and their experiment is widely recognized among biologists as a classic example of elegant design.

*J. D. Watson and F. H. C. Crick, Genetical implications of the structure of deoxyribonucleic acid, *Nature* 171:964–967 (1953).

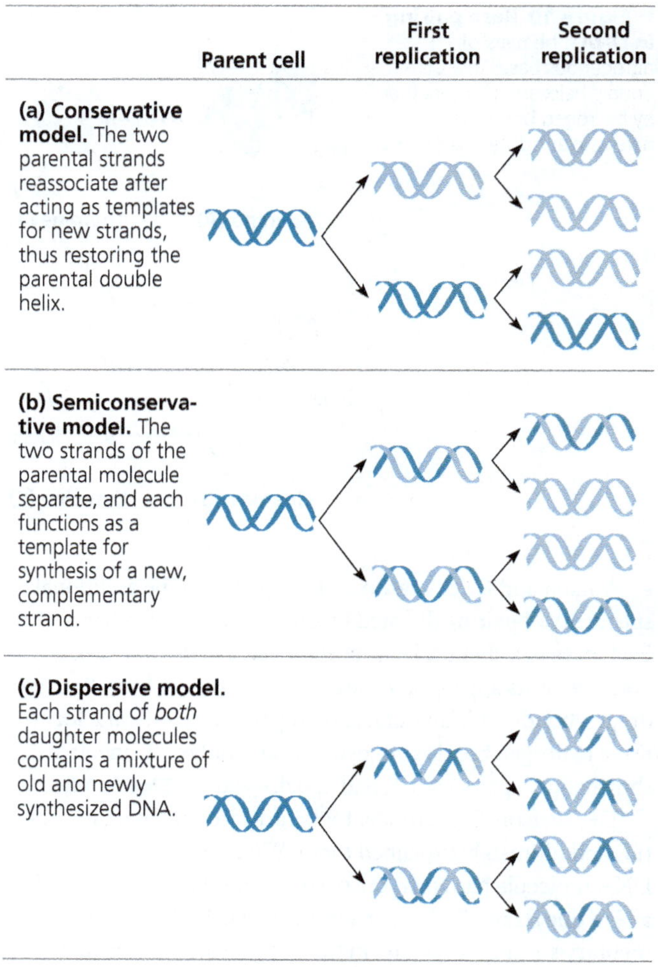

▲ **Figure 12 Three alternative models of DNA replication.** Each segment of double helix symbolizes the DNA within a cell. Beginning with a parent cell, we follow the DNA for two more generations of cells—two rounds of DNA replication. Parental DNA is dark blue; newly made DNA is light blue.

The basic principle of DNA replication is conceptually simple. However, the actual process involves some complicated biochemical gymnastics, as we will now see.

DNA Replication: *A Closer Look*

The bacterium *E. coli* has a single chromosome of about 4.6 million nucleotide pairs. In a favorable environment, an *E. coli* cell can copy all this DNA and divide to form two genetically identical daughter cells in less than an hour. Each of *your* cells has 46 DNA molecules in its nucleus, one long double-helical molecule per chromosome. In all, that represents about 6 billion nucleotide pairs, or over a thousand times more DNA than is found in most bacterial cells. If we were to print the one-letter symbols for these bases (A, G, C, and T) the size of the type you are now reading, the 6 billion nucleotide pairs of information in a diploid human cell would fill about 1,400 biology textbooks. Yet it takes one of your cells just a few hours to copy all of this DNA. This replication of an enormous amount

▼ **Figure 13 Inquiry**

Does DNA replication follow the conservative, semiconservative, or dispersive model?

Experiment Matthew Meselson and Franklin Stahl cultured *E. coli* for several generations in a medium containing nucleotide precursors labeled with a heavy isotope of nitrogen, ^{15}N. They then transferred the bacteria to a medium with only ^{14}N, a lighter isotope. They took one sample after the first DNA replication and another after the second replication. They extracted DNA from the bacteria in the samples and then centrifuged each DNA sample to separate DNA of different densities.

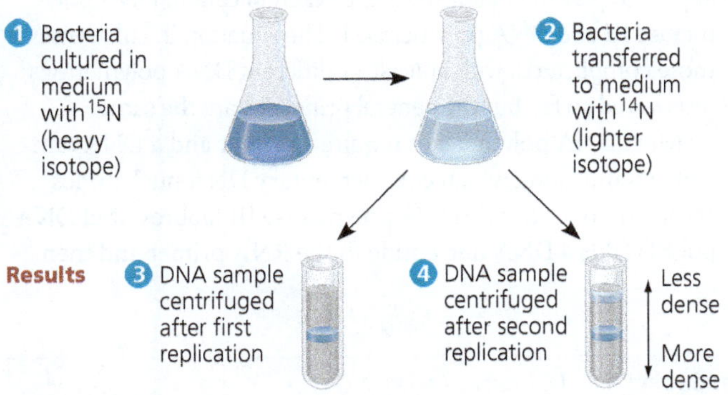

Conclusion Meselson and Stahl compared their results to those predicted by each of the three models in Figure 12, as shown below. The first replication in the ^{14}N medium produced a band of hybrid (^{15}N -^{14}N) DNA. This result eliminated the conservative model. The second replication produced both light and hybrid DNA, a result that refuted the dispersive model and supported the semiconservative model. They therefore concluded that DNA replication is semiconservative.

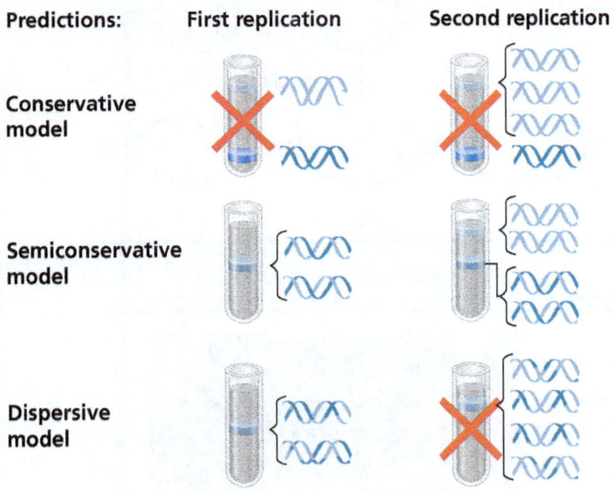

Data from M. Meselson and F. W. Stahl, The replication of DNA in *Escherichia coli, Proceedings of the National Academy of Sciences USA* 44:671–682 (1958).

Inquiry in Action Read and analyze the original paper in *Inquiry in Action: Interpreting Scientific Papers.*

(MB) A related Experimental Inquiry Tutorial can be assigned in MasteringBiology.

WHAT IF? If Meselson and Stahl had first grown the cells in ^{14}N-containing medium and then moved them into ^{15}N-containing medium before taking samples, what would have been the result?

of genetic information is achieved with very few errors—only about one per 10 billion nucleotides. The copying of DNA is remarkable in its speed and accuracy.

More than a dozen enzymes and other proteins participate in DNA replication. Much more is known about how this "replication machine" works in bacteria (such as *E. coli*) than in eukaryotes, and we will describe the basic steps of the process for *E. coli*, except where otherwise noted. What scientists have learned about eukaryotic DNA replication suggests, however, that most of the process is fundamentally similar for prokaryotes and eukaryotes.

Getting Started

The replication of chromosomal DNA begins at particular sites called **origins of replication**, short stretches of DNA that have a specific sequence of nucleotides. Proteins that initiate DNA replication recognize this sequence and attach to the DNA, separating the two strands and opening up a replication "bubble." At each end of a bubble is a **replication fork**, a Y-shaped region where the parental strands of DNA are being unwound. Several kinds of proteins participate in the unwinding (**Figure 14**). **Helicases** are enzymes that untwist the double helix at the replication forks, separating the two parental strands and making them available as template strands. After the parental strands separate, **single-strand binding proteins** bind to the unpaired DNA strands, keeping them from re-pairing. The untwisting of the double helix causes tighter twisting and strain ahead of the replication fork. **Topoisomerase** helps relieve this strain by breaking, swiveling, and rejoining DNA strands.

The *E. coli* chromosome, like many other bacterial chromosomes, is circular and has a single origin of replication, forming

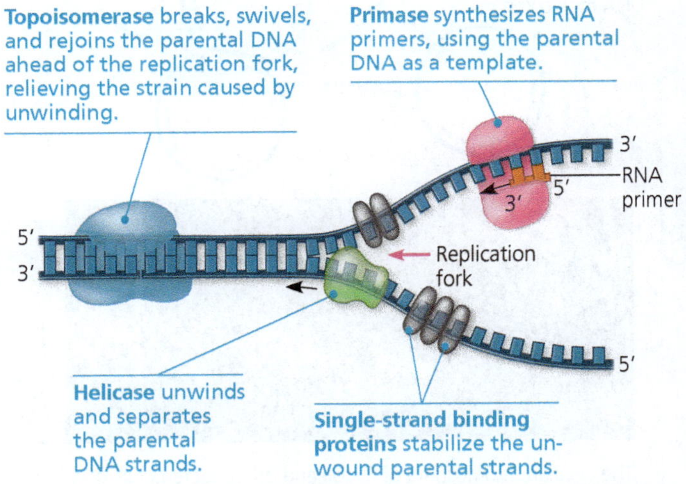

▲ **Figure 14 Some of the proteins involved in the initiation of DNA replication.** The same proteins function at both replication forks in a replication bubble. For simplicity, only the left-hand fork is shown, and the DNA bases are drawn much larger in relation to the proteins than they are in reality.

one replication bubble **(Figure 15a)**. Replication of DNA then proceeds in both directions until the entire molecule is copied. In contrast to a bacterial chromosome, a eukaryotic chromosome may have hundreds or even a few thousand replication origins. Multiple replication bubbles form and eventually fuse, thus speeding up the copying of the very long DNA molecules **(Figure 15b)**. As in bacteria, eukaryotic DNA replication proceeds in both directions from each origin.

Synthesizing a New DNA Strand

Within a bubble, the unwound sections of parental DNA strands are available to serve as templates for the synthesis of new complementary DNA strands. However, the enzymes that synthesize DNA cannot *initiate* the synthesis of a polynucleotide; they can only add nucleotides to the end of an already existing chain that is base-paired with the template strand. The initial nucleotide chain that is produced during DNA synthesis is actually a short stretch of RNA, not DNA. This RNA chain is called a **primer** and is synthesized by the enzyme **primase** (see Figure 14). Primase starts a complementary RNA chain with a single RNA nucleotide and adds RNA nucleotides one at a time, using the parental DNA strand as a template. The completed primer, generally 5–10 nucleotides long, is thus base-paired to the template strand. The new DNA strand will start from the 3′ end of the RNA primer.

Enzymes called **DNA polymerases** catalyze the synthesis of new DNA by adding nucleotides at the 3′ end of a preexisting chain. In *E. coli*, there are several DNA polymerases, but two appear to play the major roles in DNA replication: DNA polymerase III and DNA polymerase I. The situation in eukaryotes is more complicated, with at least 11 different DNA polymerases discovered so far, but the general principles are the same.

Most DNA polymerases require a primer and a DNA template strand along which complementary DNA nucleotides are lined up. In *E. coli*, DNA polymerase III (abbreviated DNA pol III) adds a DNA nucleotide to the RNA primer and then

▼ Figure 15 Origins of replication in *E. coli* and eukaryotes. The red arrows indicate the movement of the replication forks and thus the overall directions of DNA replication within each bubble.

(a) Origin of replication in an *E. coli* cell

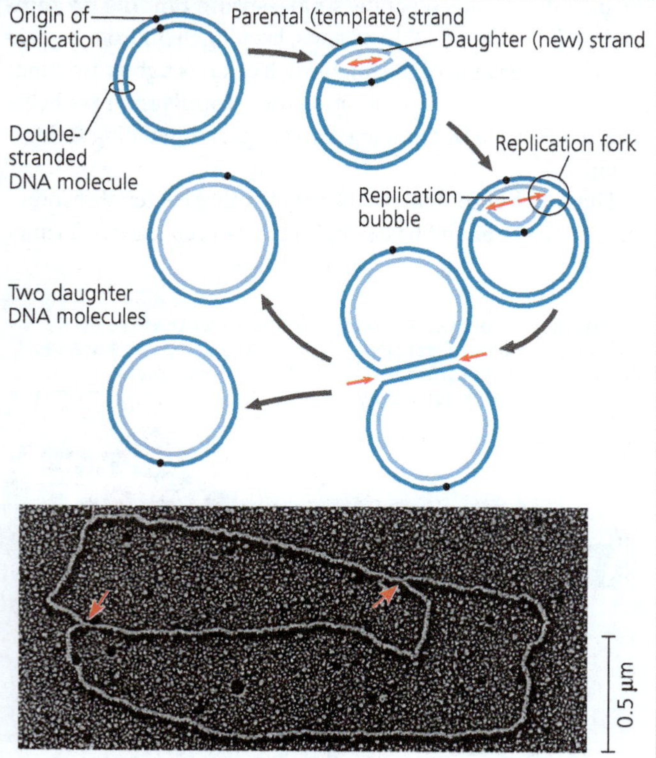

The circular chromosome of *E. coli* and other bacteria has only one origin of replication. The parental strands separate there, forming a replication bubble with two forks (red arrows). Replication proceeds in both directions until the forks meet on the other side, resulting in two daughter DNA molecules. The TEM shows a bacterial chromosome with a replication bubble.

(b) Origins of replication in a eukaryotic cell

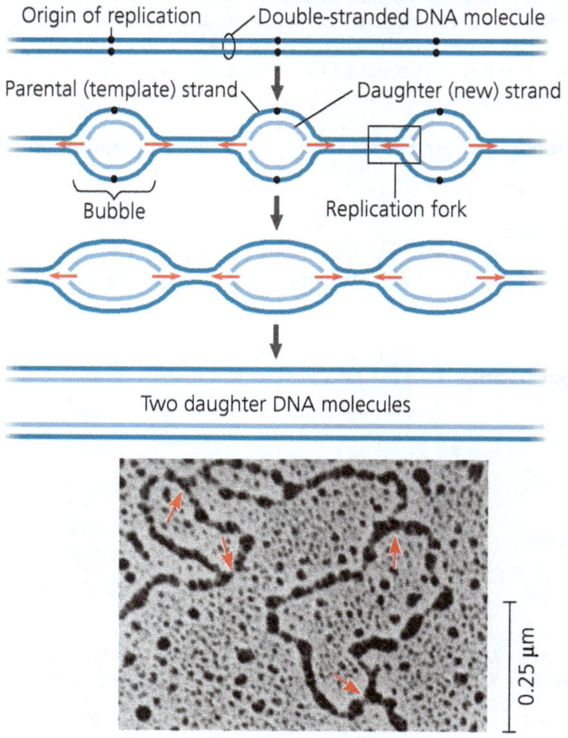

In a linear chromosome of a eukaryote, replication bubbles form at many sites along the giant DNA molecule. The bubbles expand as replication proceeds in both directions (red arrows). Eventually, the bubbles fuse and synthesis of the daughter strands is complete. The TEM shows three replication bubbles along the DNA of a cultured Chinese hamster cell.

DRAW IT *In the TEM above, add arrows for the third bubble.*

continues adding DNA nucleotides, complementary to the parental DNA template strand, to the growing end of the new DNA strand. The rate of elongation is about 500 nucleotides per second in bacteria and 50 per second in human cells.

Each nucleotide to be added to a growing DNA strand consists of a sugar attached to a base and to three phosphate groups. You have already encountered such a molecule—ATP (adenosine triphosphate). The only difference between the ATP of energy metabolism and dATP, the adenine nucleotide used to make DNA, is the sugar component, which is deoxyribose in the building block of DNA but ribose in ATP. Like ATP, the nucleotides used for DNA synthesis are chemically reactive, partly because their triphosphate tails have an unstable cluster of negative charge. As each monomer joins the growing end of a DNA strand in a dehydration reaction catalyzed by DNA polymerase, two phosphate groups are lost as a molecule of pyrophosphate (Ⓟ—$Ⓟ_i$). Subsequent hydrolysis of the pyrophosphate to two molecules of inorganic phosphate ($Ⓟ_i$) is a coupled exergonic reaction that helps drive the polymerization reaction **(Figure 16)**.

Antiparallel Elongation

As we have noted previously, the two ends of a DNA strand are different, giving each strand directionality, like a one-way street (see Figure 6). In addition, the two strands of DNA in a double helix are antiparallel, meaning that they are oriented in opposite directions to each other, like the two sides of a divided street (see Figure 16). Therefore, the two new strands formed during DNA replication must also be antiparallel to their template strands.

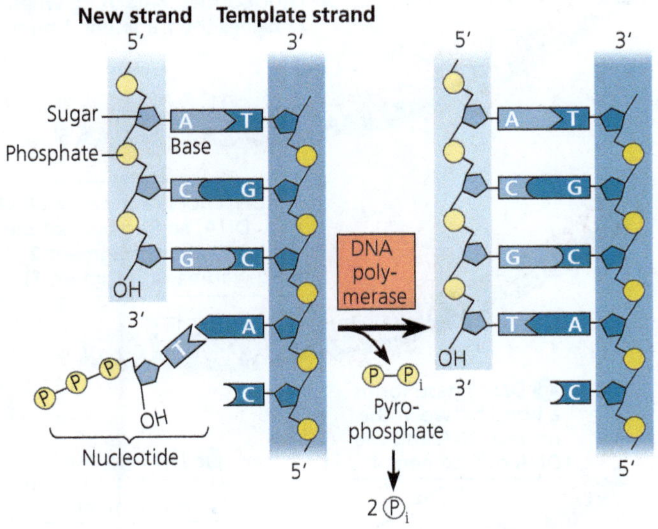

▲ **Figure 16 Addition of a nucleotide to a DNA strand.** DNA polymerase catalyzes the addition of a nucleotide to the 3′ end of a growing DNA strand, with the release of two phosphates.

? *Use this diagram to explain what we mean when we say that each DNA strand has directionality.*

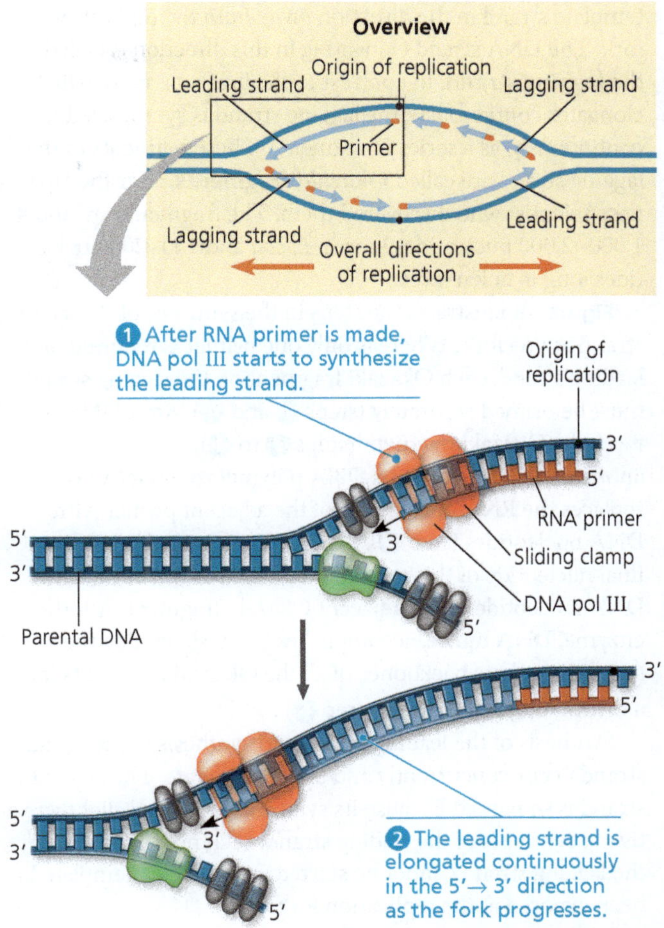

▲ **Figure 17 Synthesis of the leading strand during DNA replication.** This diagram focuses on the left replication fork shown in the overview box. DNA polymerase III (DNA pol III), shaped like a cupped hand, is shown closely associated with a protein called the "sliding clamp" that encircles the newly synthesized double helix like a doughnut. The sliding clamp moves DNA pol III along the DNA template strand.

How does the antiparallel arrangement of the double helix affect replication? Because of their structure, DNA polymerases can add nucleotides only to the free 3′ end of a primer or growing DNA strand, never to the 5′ end (see Figure 16). Thus, a new DNA strand can elongate only in the 5′ → 3′ direction. With this in mind, let's examine one of the two replication forks in a bubble **(Figure 17)**. Along one template strand, DNA polymerase III can synthesize a complementary strand continuously by elongating the new DNA in the mandatory 5′ → 3′ direction. DNA pol III remains in the replication fork on that template strand and continuously adds nucleotides to the new complementary strand as the fork progresses. The DNA strand made by this mechanism is called the **leading strand**. Only one primer is required for DNA pol III to synthesize the entire leading strand.

To elongate the other new strand of DNA in the mandatory 5′ → 3′ direction, DNA pol III must work along the other template strand in the direction *away from* the replication fork. The DNA strand elongating in this direction is called the **lagging strand**. In contrast to the leading strand, which elongates continuously, the lagging strand is synthesized discontinuously, as a series of segments. These segments of the lagging strand are called **Okazaki fragments**, after the Japanese scientist who discovered them. The fragments are about 1,000–2,000 nucleotides long in *E. coli* and 100–200 nucleotides long in eukaryotes.

Figure 18 illustrates the steps in the synthesis of the lagging strand at one fork. Whereas only one primer is required on the leading strand, each Okazaki fragment on the lagging strand must be primed separately (steps 1 and 4). After DNA pol III forms an Okazaki fragment (steps 2 to 4), another DNA polymerase, DNA polymerase I (DNA pol I), replaces the RNA nucleotides of the adjacent primer with DNA nucleotides (step 5). But DNA pol I cannot join the final nucleotide of this replacement DNA segment to the first DNA nucleotide of the adjacent Okazaki fragment. Another enzyme, **DNA ligase**, accomplishes this task, joining the sugar-phosphate backbones of all the Okazaki fragments into a continuous DNA strand (step 6).

Synthesis of the leading strand and synthesis of the lagging strand occur concurrently and at the same rate. The lagging strand is so named because its synthesis is delayed slightly relative to synthesis of the leading strand; each new fragment of the lagging strand cannot be started until enough template has been exposed at the replication fork.

Figure 19 summarizes DNA replication. Please study it carefully before proceeding.

The DNA Replication Complex

It is traditional—and convenient—to represent DNA polymerase molecules as locomotives moving along a DNA railroad track, but such a model is inaccurate in two important ways. First, the various proteins that participate in DNA replication actually form a single large complex, a "DNA replication machine." Many protein-protein interactions facilitate the efficiency of this complex. For example, by interacting with other proteins at the fork, primase apparently acts as a molecular brake, slowing progress of the replication fork and coordinating the placement of primers and the rates of replication on the leading and lagging strands. Second, the DNA replication complex may not move along the DNA; rather, the DNA may move through the complex during the replication process. In eukaryotic cells, multiple copies of the complex, perhaps grouped into "factories," may be anchored to the nuclear matrix, a framework of fibers extending through the interior of the nucleus. Experimental evidence supports a model in which two DNA polymerase molecules, one on each template strand, "reel in" the parental DNA and extrude newly made daughter DNA molecules. In this so-called trombone model, the lagging strand is also looped back through the complex **(Figure 20)**.

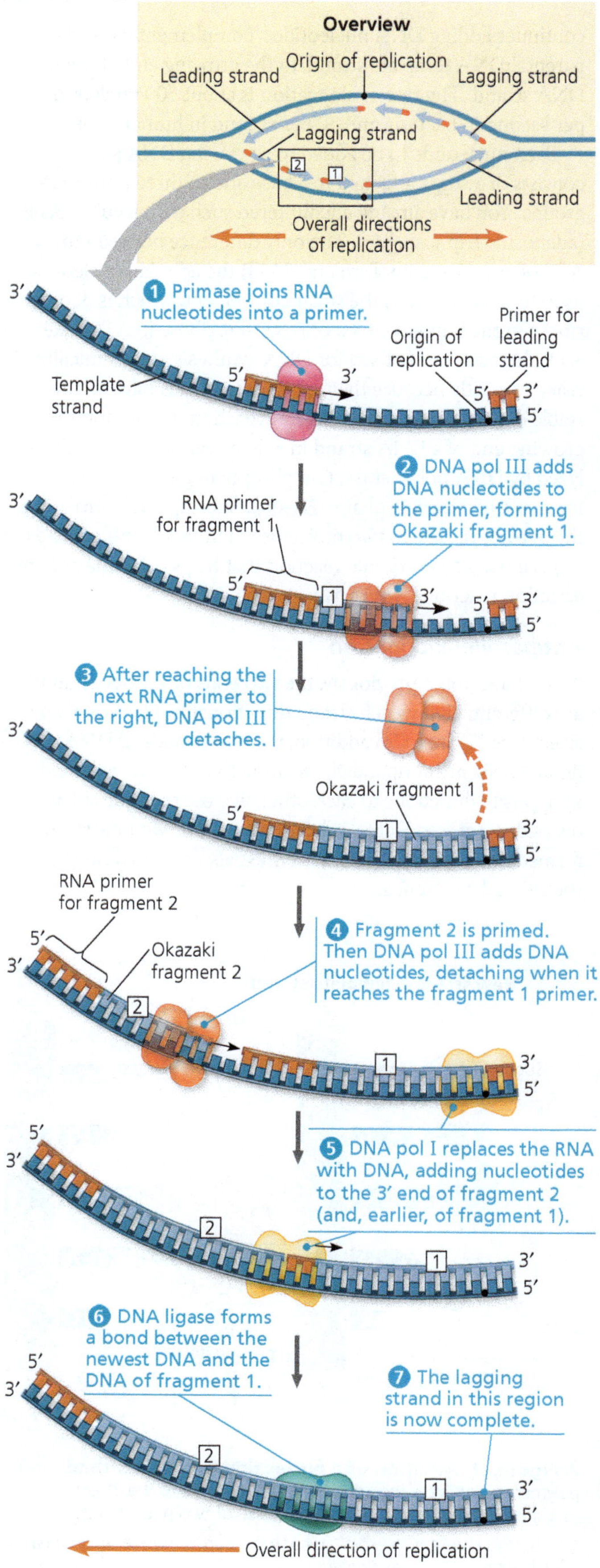

▲ **Figure 18 Synthesis of the lagging strand.**

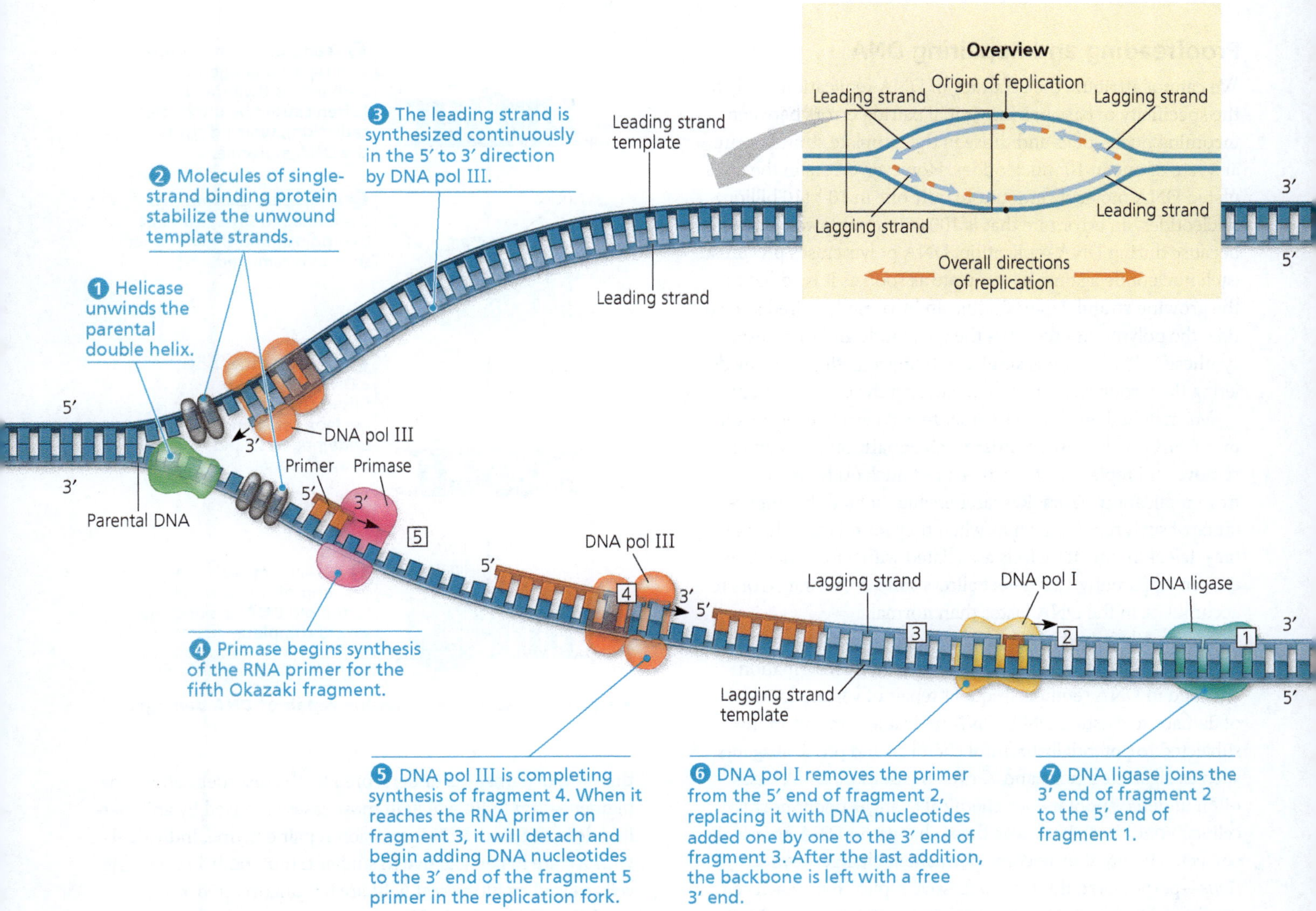

▲ **Figure 19 A summary of bacterial DNA replication.** The detailed diagram shows the left-hand replication fork of the replication bubble in the overview (upper right). Viewing each daughter strand in its entirety in the overview, you can see that half of it is made continuously as the leading strand, while the other half (on the other side of the origin) is synthesized in fragments as the lagging strand.

DRAW IT *Draw a diagram showing the right-hand fork of the bubble in Figure 19. Number the Okazaki fragments and label all 5′ and 3′ ends.*

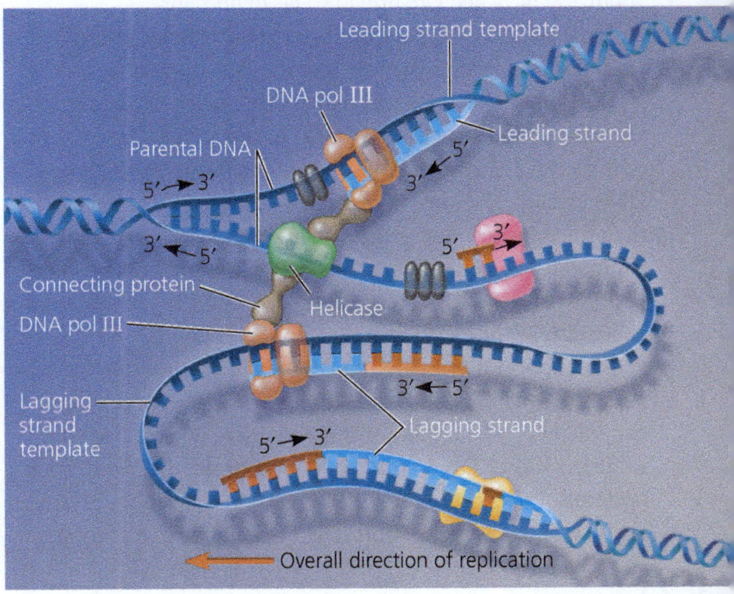

► **Figure 20 The "trombone" model of the DNA replication complex.** Two DNA polymerase III molecules work together in a complex, one on each template strand. The lagging strand template DNA loops through the complex, resembling the slide of a trombone.

DRAW IT *Draw a line tracing the lagging strand template along the entire stretch of DNA shown here.*

ANIMATION

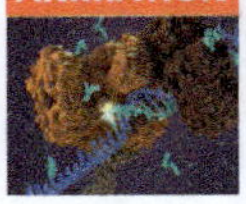

Visit the Study Area in **MasteringBiology** for the BioFlix® 3-D Animation on DNA Replication.

Proofreading and Repairing DNA

We cannot attribute the accuracy of DNA replication solely to the specificity of base pairing. Initial pairing errors between incoming nucleotides and those in the template strand occur at a rate of one in 10^5 nucleotides. However, errors in the completed DNA molecule amount to only one in 10^{10} (10 billion) nucleotides, an error rate that is 100,000 times lower. This is because during DNA replication, DNA polymerases proofread each nucleotide against its template as soon as it is added to the growing strand. Upon finding an incorrectly paired nucleotide, the polymerase removes the nucleotide and then resumes synthesis. (This action is similar to fixing a texting error by deleting the wrong letter and then entering the correct letter.)

Mismatched nucleotides sometimes do evade proofreading by a DNA polymerase. In **mismatch repair**, other enzymes remove and replace incorrectly paired nucleotides resulting from replication errors. Researchers highlighted the importance of such repair enzymes when they found that a hereditary defect in one of them is associated with a form of colon cancer. Apparently, this defect allows cancer-causing errors to accumulate in the DNA faster than normal.

Incorrectly paired or altered nucleotides can also arise after replication. In fact, maintenance of the genetic information encoded in DNA requires frequent repair of various kinds of damage to existing DNA. DNA molecules are constantly subjected to potentially harmful chemical and physical agents, such as cigarette smoke and X-rays. In addition, DNA bases often undergo spontaneous chemical changes under normal cellular conditions. However, these changes in DNA are usually corrected before they become permanent changes—*mutations*—perpetuated through successive replications. Each cell continuously monitors and repairs its genetic material. Because repair of damaged DNA is so important to the survival of an organism, it is no surprise that many different DNA repair enzymes have evolved. Almost 100 are known in *E. coli*, and about 130 have been identified so far in humans.

Most cellular systems for repairing incorrectly paired nucleotides, whether they are due to DNA damage or to replication errors, use a mechanism that takes advantage of the base-paired structure of DNA. In many cases, a segment of the strand containing the damage is cut out (excised) by a DNA-cutting enzyme—a **nuclease**—and the resulting gap is then filled in with nucleotides, using the undamaged strand as a template. The enzymes involved in filling the gap are a DNA polymerase and DNA ligase. One such DNA repair system is called **nucleotide excision repair** (Figure 21).

An important function of DNA repair enzymes in our skin cells is to repair genetic damage caused by the ultraviolet rays of sunlight. One example of this damage is when adjacent thymine bases on a DNA strand become covalently linked. Such *thymine dimers* cause the DNA to buckle (see Figure 21), interfering with DNA replication. The importance of repairing this kind of damage is underscored by the disorder xeroderma pigmentosum (XP), which in most cases is caused by an inherited defect in a nucleotide excision repair enzyme. Individuals with XP are hypersensitive to sunlight; mutations in their skin cells caused by ultraviolet light are left uncorrected, often resulting in skin cancer. The effects are extreme: Without sun protection, children who have XP can develop skin cancer by age 10.

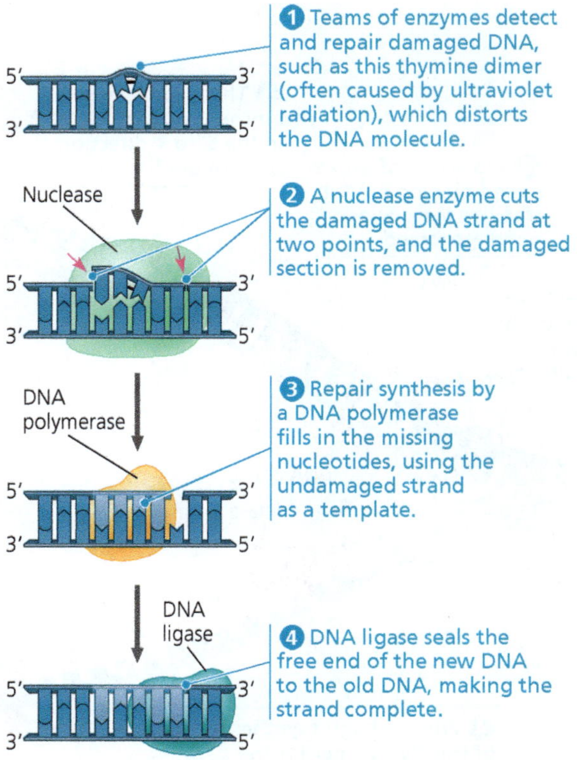

▲ **Figure 21 Nucleotide excision repair of DNA damage.**

Evolutionary Significance of Altered DNA Nucleotides

EVOLUTION Faithful replication of the genome and repair of DNA damage are important for the functioning of the organism and for passing on a complete, accurate genome to the next generation. The error rate after proofreading and repair is extremely low, but rare mistakes do slip through. Once a mismatched nucleotide pair is replicated, the sequence change is permanent in the daughter molecule that has the incorrect nucleotide as well as in any subsequent copies. As you know, a permanent change in the DNA sequence is called a mutation.

Mutations can change the phenotype of an organism. And if they occur in germ cells, which give rise to gametes, mutations can be passed on from generation to generation. The vast majority of such changes are harmful, but a very small percentage can be beneficial. In either case, mutations are the source of the variation on which natural selection operates during evolution and are ultimately responsible for the appearance of new species. The balance between complete fidelity of DNA replication and repair and a low mutation rate has, over long periods of time, allowed the evolution of the rich diversity of species we see on Earth today.

Replicating the Ends of DNA Molecules

For linear DNA, such as the DNA of eukaryotic chromosomes, the usual replication machinery cannot complete the 5′ ends of daughter DNA strands. (This is a consequence of the fact that a DNA polymerase can add nucleotides only to the 3′ ends of a pre-existing polynucleotide.) As a result, repeated rounds of replication produce shorter and shorter DNA molecules with uneven ends.

What protects the genes near the ends of eukaryotic chromosomes from being eroded away during successive replications? Eukaryotic chromosomal DNA molecules have special nucleotide sequences called **telomeres** at their ends **(Figure 22)**. Telomeres do not contain genes; instead, the DNA typically consists of multiple repetitions of one short nucleotide sequence. In each human telomere, for example, the sequence TTAGGG is repeated 100 to 1,000 times. Telomeric DNA acts as a buffer zone that protects the organism's genes.

Telomeres do not prevent the erosion of genes near ends of chromosomes; they merely postpone it. As you would expect, telomeres tend to be shorter in cultured cells that have divided many times and in dividing somatic cells of older individuals. Shortening of telomeres is proposed to play a role in the aging process of some tissues and even of the organism as a whole.

If the chromosomes of germ cells became shorter in every cell cycle, essential genes would eventually be missing from the gametes they produce. However, this does not occur: An enzyme called *telomerase* catalyzes the lengthening of telomeres in eukaryotic germ cells, thus restoring their original length and compensating for the shortening that occurs during DNA replication. Telomerase is not active in most human somatic cells, but shows inappropriate activity in some cancer cells that may remove limits to a cell's normal life span. Thus, telomerase is under study as a target for cancer therapies.

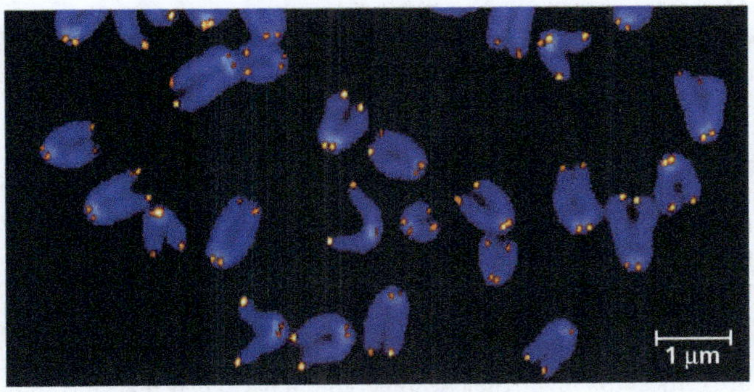

▲ **Figure 22 Telomeres.** Eukaryotes have repetitive, noncoding sequences called telomeres at the ends of their DNA. Telomeres are stained orange in these mouse chromosomes (LM).

CONCEPT CHECK 2

1. What role does base pairing play in the replication of DNA?
2. Make a table listing the functions of seven proteins involved in DNA replication in *E. coli*.
3. MAKE CONNECTIONS What is the relationship between DNA replication and the S phase of the cell cycle?

CONCEPT 3

A chromosome consists of a DNA molecule packed together with proteins

Now that you have learned about the structure and replication of DNA, let's take a step back and examine how DNA is packaged into chromosomes, the structures that carry genetic information. The main component of the genome in most bacteria is one double-stranded, circular DNA molecule that is associated with a small amount of protein. Although we refer to this structure as a bacterial chromosome, it is very different from a eukaryotic chromosome, which consists of one linear DNA molecule associated with a large amount of protein. In *E. coli*, the chromosomal DNA consists of about 4.6 million nucleotide pairs, representing about 4,400 genes. This is 100 times more DNA than is found in a typical virus, but only about one-thousandth as much DNA as in a human somatic cell. Still, that is a lot of DNA to be packaged in such a small container.

Stretched out, the DNA of an *E. coli* cell would measure about a millimeter in length, 500 times longer than the cell. Within a bacterium, however, certain proteins cause the chromosome to coil and "supercoil," densely packing it so that it fills only part of the cell. Unlike the nucleus of a eukaryotic cell, this dense region of DNA in a bacterium, called the **nucleoid**, is not surrounded by membrane.

Each eukaryotic chromosome contains a single linear DNA double helix that, in humans, averages about 1.5×10^8 nucleotide pairs. This is an enormous amount of DNA relative to a chromosome's condensed length. If completely stretched out, such a DNA molecule would be about 4 cm long, thousands of times the diameter of a cell nucleus—and that's not even considering the DNA of the other 45 human chromosomes!

In the cell, eukaryotic DNA is precisely combined with a large amount of protein. Together, this complex of DNA and protein, called **chromatin**, fits into the nucleus through an elaborate, multilevel system of packing.

Chromatin undergoes striking changes in its degree of packing during the course of the cell cycle. In interphase cells stained for light microscopy, the chromatin usually appears as a diffuse mass within the nucleus, suggesting that the chromatin is highly extended. As a cell prepares for mitosis, its chromatin coils and folds up (condenses), eventually forming a characteristic number of short, thick metaphase chromosomes that are distinguishable from each other with the light microscope. Our current view of the successive levels of DNA packing in a chromosome is outlined in **Figure 23**.

Though interphase chromatin is generally much less condensed than the chromatin of mitotic chromosomes, it shows several of the same levels of higher-order packing. Some of the chromatin comprising a chromosome seems to be present as a 10-nm fiber, but much is compacted into a 30-nm fiber, which in some regions is further folded into looped domains. Even during interphase, the centromeres of chromosomes, as well as other chromosomal regions in some cells, exist in a highly condensed state similar to that seen in a metaphase chromosome. This type of interphase chromatin, visible as irregular clumps with a light

▼ Figure 23 Exploring Chromatin Packing in a Eukaryotic Chromosome

This illustration, accompanied by transmission electron micrographs, depicts a current model for the progressive levels of DNA coiling and folding. The illustration zooms out from a single molecule of DNA to a metaphase chromosome, which is large enough to be seen with a light microscope.

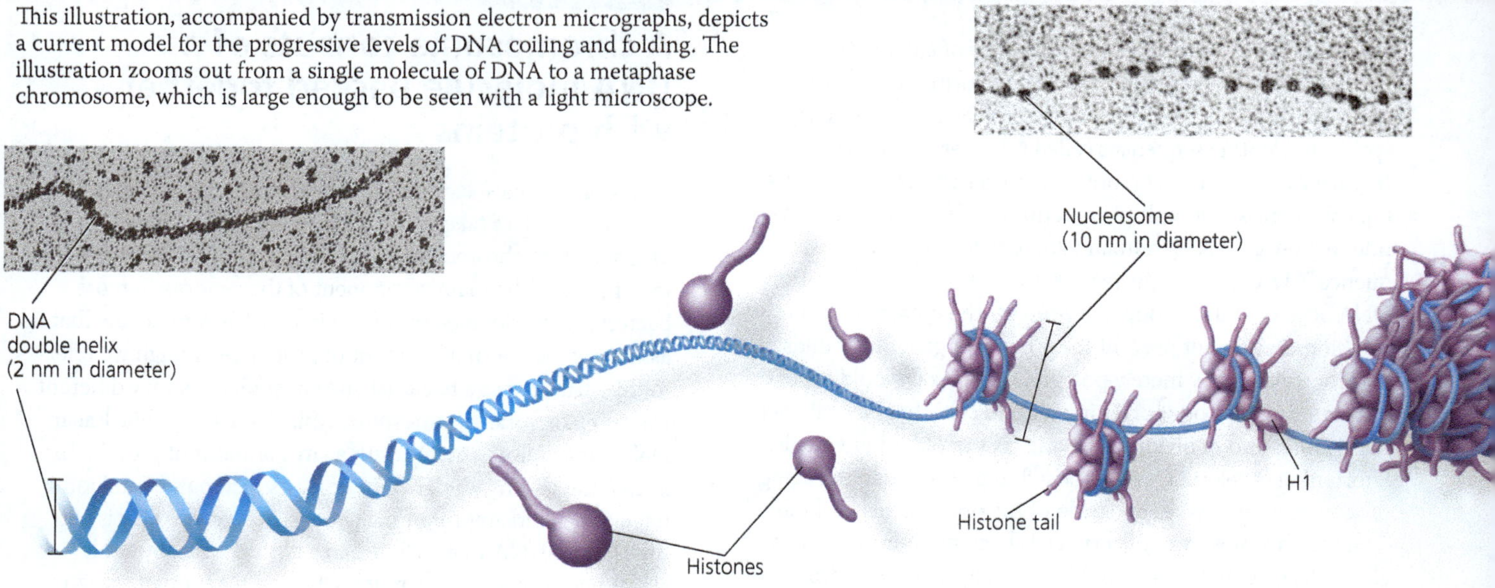

DNA, the double helix

Shown above is a ribbon model of DNA, with each ribbon representing one of the polynucleotide strands. Recall that the phosphate groups along the backbone contribute a negative charge along the outside of each strand. The TEM shows a molecule of naked (protein-free) DNA; the double helix alone is 2 nm across.

Histones

Proteins called **histones** are responsible for the first level of DNA packing in chromatin. Although each histone is small—containing only about 100 amino acids—the total mass of histone in chromatin roughly equals the mass of DNA. More than a fifth of a histone's amino acids are positively charged (lysine or arginine) and therefore bind tightly to the negatively charged DNA.

Four types of histones are most common in chromatin: H2A, H2B, H3, and H4. The histones are very similar among eukaryotes; for example, all but two of the amino acids in cow H4 are identical to those in pea H4. The apparent conservation of histone genes during evolution probably reflects the important role of histones in organizing DNA within cells.

These four types of histones are critical to the next level of DNA packing. (A fifth type of histone, called H1, is involved in a further stage of packing.)

Nucleosomes, or "beads on a string" (10-nm fiber)

In electron micrographs, unfolded chromatin is 10 nm in diameter (the *10-nm fiber*). Such chromatin resembles beads on a string (see the TEM). Each "bead" is a **nucleosome**, the basic unit of DNA packing; the "string" between beads is called *linker* DNA.

A nucleosome consists of DNA wound twice around a protein core of eight histones, two each of the main histone types (H2A, H2B, H3, and H4). The amino end (N-terminus) of each histone (the *histone tail*) extends outward from the nucleosome.

In the cell cycle, the histones leave the DNA only briefly during DNA replication. Generally, they do the same during transcription, another process that requires access to the DNA by the cell's molecular machinery. Nucleosomes, and in particular their histone tails, are involved in the regulation of gene expression.

microscope, is called **heterochromatin**, to distinguish it from the less compacted, more dispersed **euchromatin** ("true chromatin"). Because of its compaction, heterochromatic DNA is largely inaccessible to the machinery in the cell responsible for transcribing the genetic information coded in the DNA, a crucial early step in gene expression. In contrast, the looser packing of euchromatin makes its DNA accessible to this machinery, so the genes present in euchromatin can be transcribed.

The chromosome is a dynamic structure that is condensed, loosened, modified, and remodeled as necessary for various cell processes, including mitosis, meiosis, and gene activity. Certain chemical modifications of histones affect the state of chromatin condensation and also have multiple effects on gene activity.

CONCEPT CHECK 3

1. Describe the structure of a nucleosome, the basic unit of DNA packing in eukaryotic cells.
2. What two properties, one structural and one functional, distinguish heterochromatin from euchromatin?

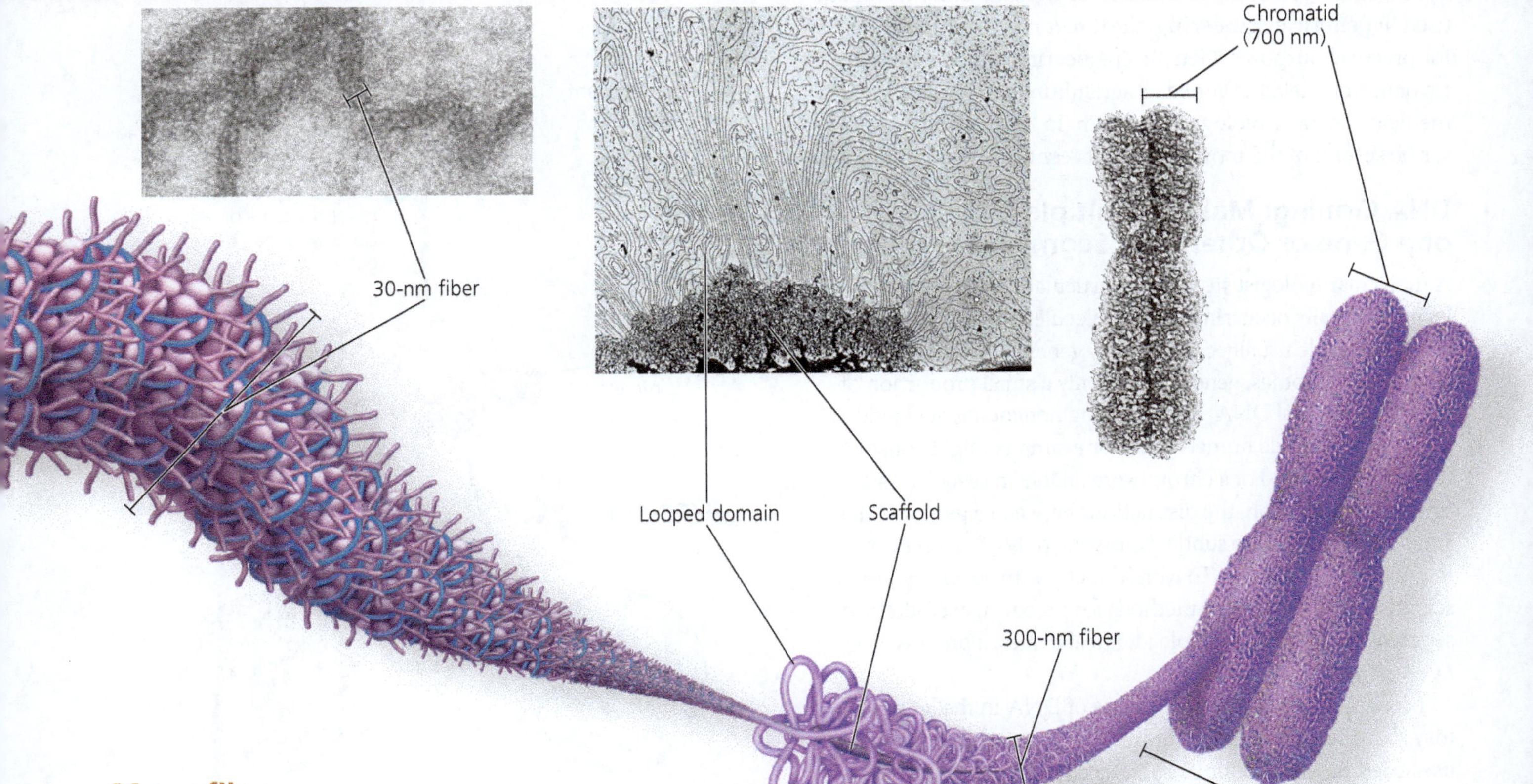

30-nm fiber

The next level of packing results from interactions between the histone tails of one nucleosome and the linker DNA and nucleosomes on either side. The fifth histone, H1, is involved at this level. These interactions cause the extended 10-nm fiber to coil or fold, forming a chromatin fiber roughly 30 nm in thickness, the *30-nm fiber*. Although the 30-nm fiber is quite prevalent in the interphase nucleus, the packing arrangement of nucleosomes in this form of chromatin is still a matter of some debate.

Looped domains (300-nm fiber)

The 30-nm fiber, in turn, forms loops called *looped* domains attached to a chromosome scaffold composed of proteins, thus making up a *300-nm fiber*. The scaffold is rich in one type of topoisomerase, and H1 molecules also appear to be present.

Metaphase chromosome

In a mitotic chromosome, the looped domains themselves coil and fold in a manner not yet fully understood, further compacting all the chromatin to produce the characteristic metaphase chromosome (also shown in the micrograph above). The width of one chromatid is 700 nm. Particular genes always end up located at the same places in metaphase chromosomes, indicating that the packing steps are highly specific and precise.

CONCEPT 4

Understanding DNA structure and replication makes genetic engineering possible

The discovery of the structure of DNA marked a milestone in biology and changed the course of biological research. Most notable was the realization that the two strands of a DNA molecule are complementary to each other. This fundamental structural property of DNA is the basis for **nucleic acid hybridization**, the base pairing of one strand of a nucleic acid to a complementary sequence on another strand. Nucleic acid hybridization forms the foundation of virtually every technique used in **genetic engineering**, the direct manipulation of genes for practical purposes. Genetic engineering has launched a revolution in fields as varied as agriculture, criminal law, and medical and basic biological research. In this section, we'll describe several of the most important techniques and their uses.

DNA Cloning: Making Multiple Copies of a Gene or Other DNA Segment

A molecular biologist studying a particular gene faces a challenge. Naturally occurring DNA molecules are very long, and a single molecule usually carries many genes. Moreover, in many eukaryotic genomes, genes occupy only a small proportion of the chromosomal DNA, the rest being noncoding nucleotide sequences. A single human gene, for example, might constitute only 1/100,000 of a chromosomal DNA molecule. As a further complication, the distinctions between a gene and the surrounding DNA are subtle, consisting only of differences in nucleotide sequence. To work directly with specific genes, scientists have developed methods for preparing well-defined segments of DNA in multiple identical copies, a process called *DNA cloning*.

Most methods for cloning pieces of DNA in the laboratory share certain general features. One common approach uses bacteria, most often *E. coli*. Recall from Figure 15 that the *E. coli* chromosome is a large circular molecule of DNA. In addition, *E. coli* and many other bacteria have **plasmids**, small circular DNA molecules that are replicated separately. A plasmid has only a small number of genes; these genes may be useful when the bacterium is in a particular environment but may not be required for survival or reproduction under most conditions.

To clone pieces of DNA using bacteria, researchers first obtain a plasmid (originally isolated from a bacterial cell and genetically engineered for efficient cloning) and insert DNA from another source ("foreign" DNA) into it **(Figure 24)**. The resulting plasmid is now a **recombinant DNA molecule**, a molecule containing DNA from two different sources, very often different species. The plasmid is then returned to a bacterial cell, producing a *recombinant bacterium*. This single cell

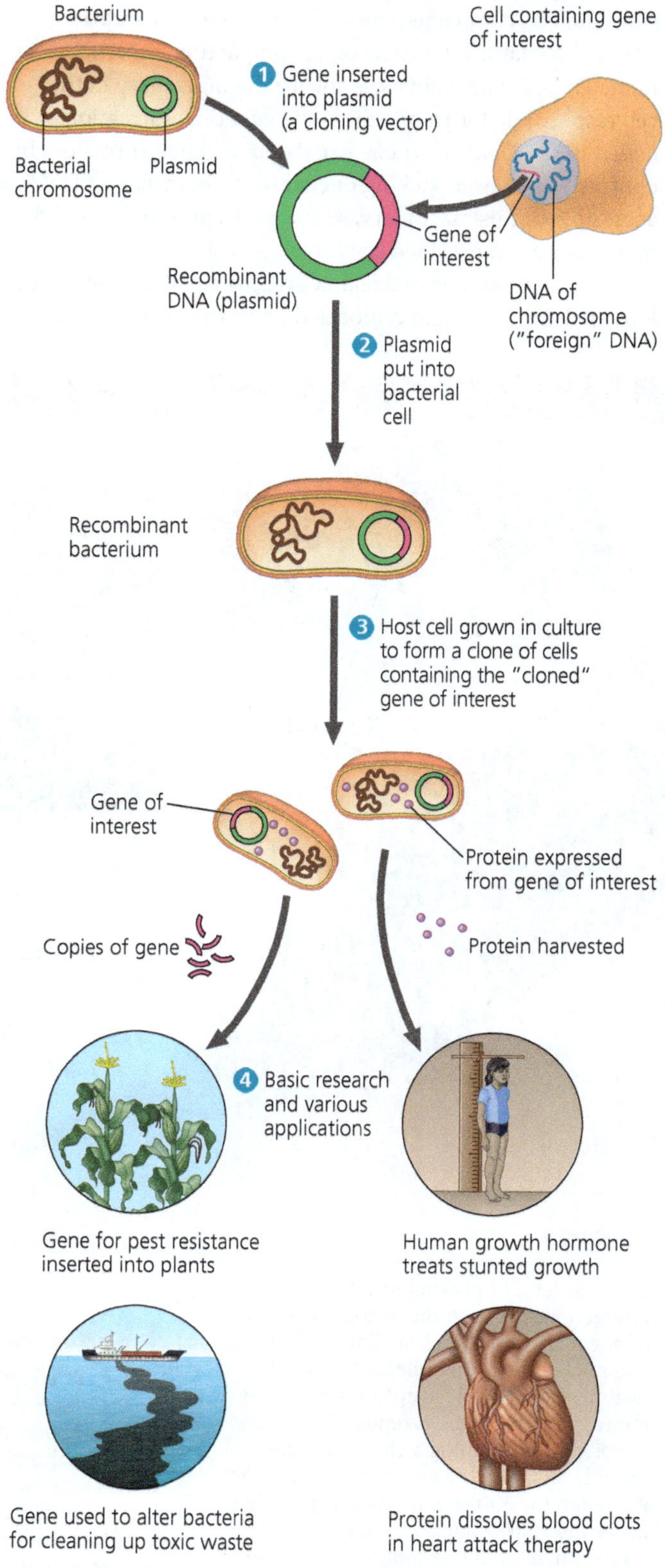

▲ **Figure 24 An overview of gene cloning and some uses of cloned genes.** In this simplified diagram of gene cloning, we start with a plasmid (originally isolated from a bacterial cell) and a gene of interest from another organism. Only one plasmid and one copy of the gene of interest are shown at the top of the figure, but the starting materials would include many of each.

reproduces through repeated cell divisions to form a clone of cells, a population of genetically identical cells. Because the dividing bacteria replicate the recombinant plasmid and pass it on to their descendants, the foreign DNA and any genes it carries are cloned at the same time. The production of multiple copies of a single gene is a type of DNA cloning called **gene cloning**. In our example in Figure 24, the plasmid acts as a **cloning vector**, a DNA molecule that can carry foreign DNA into a cell and be replicated there. The foreign DNA segment could be a gene from a eukaryotic cell; we will describe in more detail how it was obtained later in this section.

Gene cloning is useful for two basic purposes: to make many copies of, or *amplify*, a particular gene and to produce a protein product from it. Researchers can isolate copies of a cloned gene from bacteria for use in basic research or to endow another organism with a new metabolic trait, such as pest resistance. For example, a resistance gene present in one crop species might be cloned and transferred into plants of another species. (Such organisms are called Genetically Modified Organisms, or GMOs.) Alternatively, a protein with medical uses, such as human growth hormone, can be harvested in large quantities from cultures of bacteria carrying a cloned gene for the protein. Since a single gene is usually a very small part of the total DNA in a cell, the ability to amplify such rare DNA fragments is crucial for any application involving a single gene.

Using Restriction Enzymes to Make a Recombinant DNA Plasmid

Gene cloning and genetic engineering generally rely on the use of enzymes that cut DNA molecules at a limited number of specific locations. These enzymes, called restriction endonucleases, or **restriction enzymes**, were discovered in the late 1960s by biologists doing basic research on bacteria. Restriction enzymes protect the bacterial cell by cutting up foreign DNA from other organisms or phages.

Hundreds of different restriction enzymes have been identified and isolated. Each restriction enzyme is very specific, recognizing a particular short DNA sequence, or **restriction site**, and cutting both DNA strands at precise points within this restriction site. The DNA of a bacterial cell is protected from the cell's own restriction enzymes by the addition of methyl groups (—CH_3) to adenines or cytosines within the sequences recognized by the enzymes.

Figure 25 shows how restriction enzymes are used to clone a foreign DNA fragment into a bacterial plasmid. At the top is a plasmid (like the one in Figure 24) that has a single restriction site recognized by a particular restriction enzyme from *E. coli*. (Various such plasmids are available from commercial suppliers.) As shown in this example, most restriction sites are symmetrical. That is, the sequence of nucleotides is the same on both strands when read in the 5′ → 3′ direction. The most commonly used restriction enzymes recognize sequences containing four to eight nucleotide pairs. Because any sequence this short

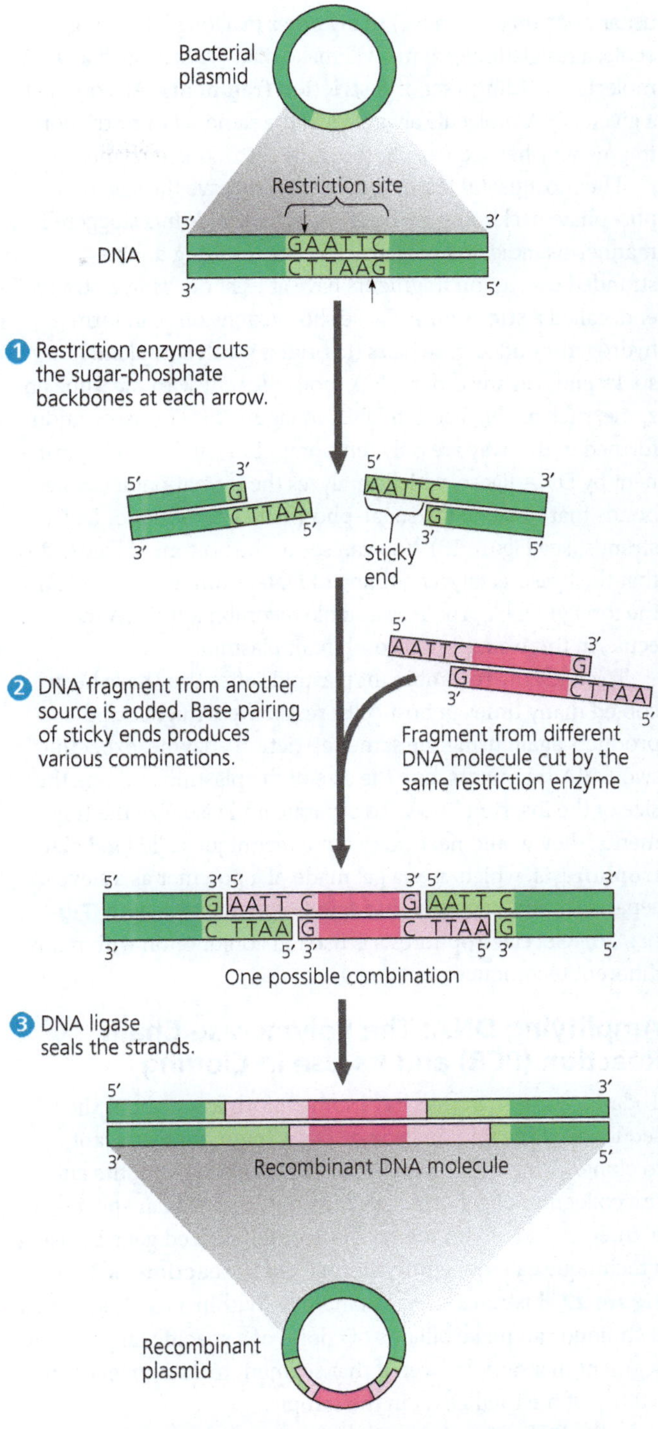

▲ **Figure 25 Using a restriction enzyme and DNA ligase to make a recombinant DNA plasmid.** The restriction enzyme in this example (called *Eco*RI) recognizes a single six-base-pair restriction site present in the plasmid. It makes staggered cuts in the sugar-phosphate backbones, producing fragments with "sticky ends." Foreign DNA fragments with complementary sticky ends can base-pair with the plasmid ends; the ligated product is a recombinant plasmid. (If the two plasmid sticky ends base-pair, the original non-recombinant plasmid would re-form.)

DRAW IT *The restriction enzyme Hind III recognizes the sequence 5′-AAGCTT-3′, cutting between the two A's. Draw the double-stranded sequence before and after the enzyme cuts.*

usually occurs (by chance) many times in a long DNA molecule, a restriction enzyme will make many cuts in such a DNA molecule, yielding a set of **restriction fragments**. All copies of a given DNA molecule always yield the same set of restriction fragments when exposed to the same restriction enzyme.

The most useful restriction enzymes cleave the sugar-phosphate backbones in the two DNA strands in a staggered manner, as indicated in Figure 25. The resulting double-stranded restriction fragments have at least one single-stranded end, called a **sticky end**. These short extensions can form hydrogen-bonded base pairs (hybridize) with complementary sticky ends on any other DNA molecules cut with the same enzyme, such as the inserted DNA in Figure 25. The associations formed in this way are only temporary but can be made permanent by DNA ligase, which catalyzes the formation of covalent bonds that close up the sugar-phosphate backbones of DNA strands (see Figure 18). You can see at the bottom of Figure 25 that the ligase-catalyzed joining of DNA from the plasmid and the foreign DNA produces a stable recombinant DNA molecule, in this example a recombinant plasmid.

To check the recombinant plasmids after they have been copied many times in host cells, researchers might cut the products again using the same restriction enzyme, expecting two DNA fragments, one the size of the plasmid and one the size of the inserted DNA. To separate and visualize the fragments, they would next carry out a technique called **gel electrophoresis**, which uses a gel made of a polymer as a sieve to separate a mixture of nucleic acid fragments by length **(Figure 26)**. Gel electrophoresis is used in conjunction with many different techniques in molecular biology.

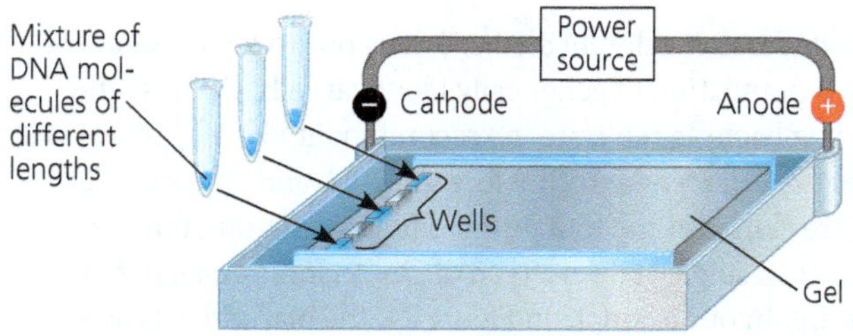

(a) Each sample, a mixture of different DNA molecules, is placed in a separate well near one end of a thin slab of agarose gel. The gel is set into a small plastic support and immersed in an aqueous, buffered solution in a tray with electrodes at each end. The current is then turned on, causing the negatively charged DNA molecules to move toward the positive electrode.

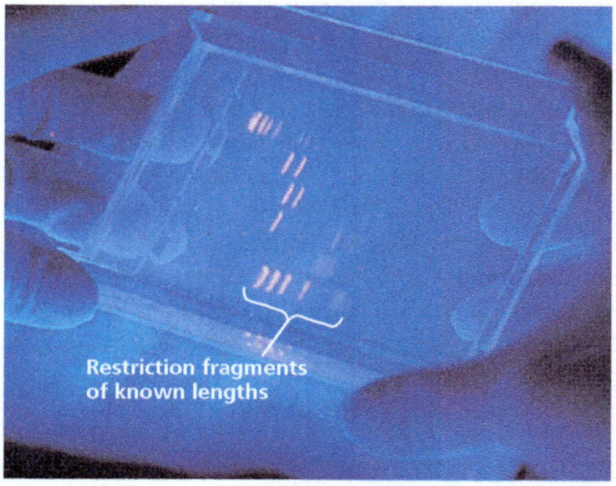

(b) Shorter molecules are slowed down less than longer ones, so they move faster through the gel. After the current is turned off, a DNA-binding dye is added that fluoresces pink in UV light. Each pink band corresponds to many thousands of DNA molecules of the same length. The horizontal ladder of bands at the bottom of the gel is a set of restriction fragments of known lengths for comparison with samples of unknown length.

▲ **Figure 26 Gel electrophoresis.** A gel made of a polymer acts as a molecular sieve to separate nucleic acids or proteins differing in size, electrical charge, or other physical properties as they move in an electric field. In the example shown here, DNA molecules are separated by length in a gel made of a polysaccharide called agarose.

Amplifying DNA: The Polymerase Chain Reaction (PCR) and Its Use in Cloning

Today, most researchers have some information about the sequence of the foreign gene or DNA fragment they want to clone. Using this information, they can start with the entire collection of genomic DNA from the particular species of interest and obtain many copies of the desired gene by using a technique called the **polymerase chain reaction**, or **PCR**. **Figure 27** illustrates the steps in PCR. Within a few hours, this technique can make billions of copies of a specific target DNA segment in a sample, even if that segment makes up less than 0.001% of the total DNA in the sample.

In the PCR procedure, a three-step cycle brings about a chain reaction that produces an exponentially growing population of identical DNA molecules. During each cycle, the reaction mixture is heated to denature (separate) the DNA strands and then cooled to allow annealing (hybridization) of short, single-stranded DNA primers complementary to sequences on opposite strands at each end of the target segment; finally, a DNA polymerase extends the primers in the $5' \rightarrow 3'$ direction. If a standard DNA polymerase were used, the protein would be denatured along with the DNA during the first heating step and would have to be replaced after each cycle.

The key to automating PCR was the discovery of an unusually heat-stable DNA polymerase called *Taq* polymerase, named after the bacterial species from which it was first isolated. This bacterial species, *Thermus aquaticus*, lives in hot springs, and the stability of its DNA polymerase at high temperatures is an evolutionary adaptation that enables the enzyme to function at temperatures up to 95°C. Today, researchers also use a DNA polymerase from the archaean species *Pyrococcus furiosus*. This enzyme, called *Pfu* polymerase, is more accurate, more stable, and costlier than *Taq* polymerase.

PCR is speedy and very specific. Only minuscule amounts of DNA need be present in the starting material, and this DNA can be partially degraded, as long as a few molecules contain the complete target segment. The key to this high specificity is the primers, the sequences of which are chosen so they

▼ Figure 27 Research Method

The Polymerase Chain Reaction (PCR)

Application With PCR, any specific segment—the so-called target sequence—in a DNA sample can be copied many times (amplified) within a test tube.

Technique PCR requires double-stranded DNA containing the target sequence, a heat-resistant DNA polymerase, all four nucleotides, and two 15- to 20-nucleotide DNA strands that serve as primers. One primer is complementary to one end of the target sequence on one strand; the second primer is complementary to the other end of the sequence on the other strand.

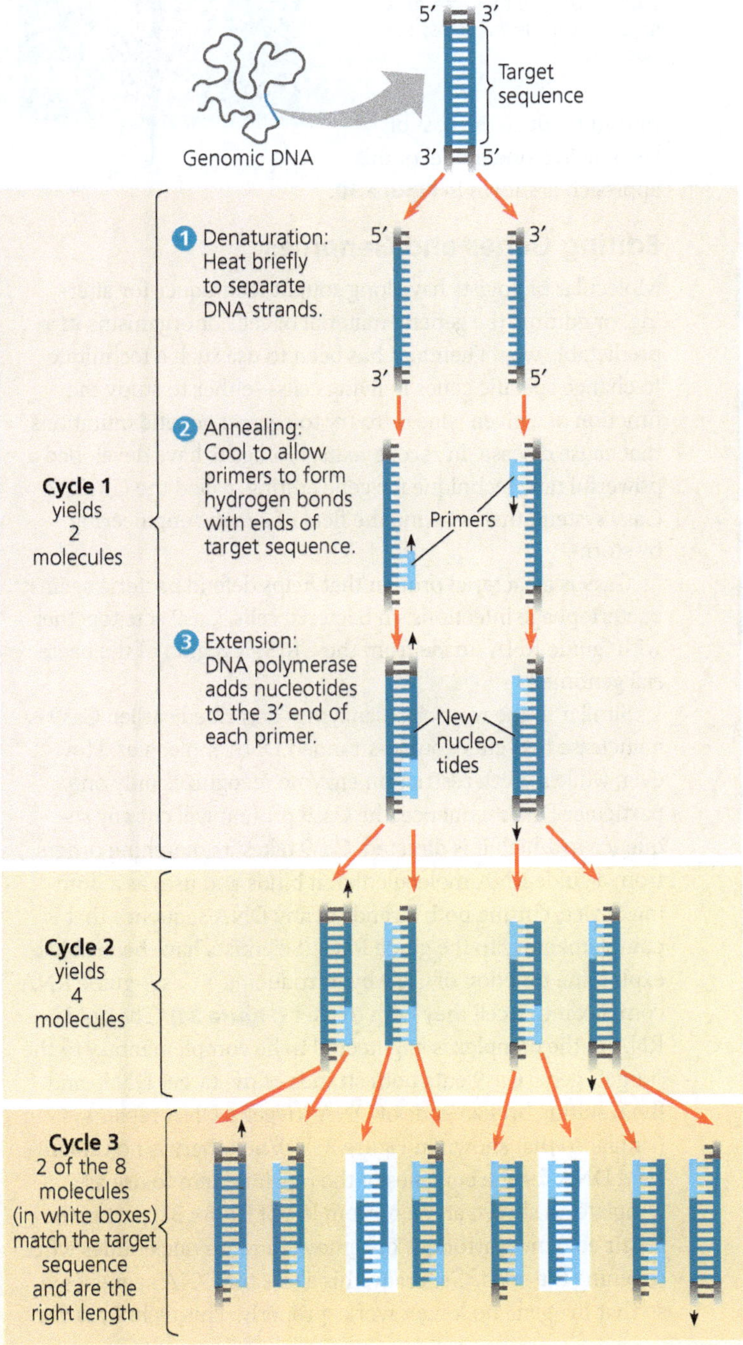

Results After three cycles, two molecules match the target sequence exactly. After 30 more cycles, over 1 billion (10^9) molecules match the target sequence.

hybridize *only* with complementary sequences at opposite ends of the target segment. (For high specificity, the primers must be at least 15 or so nucleotides long.) With each successive cycle, the number of target segment molecules of the correct length doubles, so the number of molecules equals $2n$, where n is the number of cycles. After 30 or so cycles, about a billion copies of the target sequence are present!

Despite its speed and specificity, PCR amplification alone cannot substitute for gene cloning in cells to make large amounts of a gene. This is because occasional errors during PCR replication limit the number of good copies and the length of DNA fragments that can be copied. Instead, PCR is used to provide the specific DNA fragment for cloning. PCR primers are synthesized to include a restriction site at each end of the DNA fragment that matches the site in the cloning vector, and the fragment and vector are cut and ligated together (**Figure 28**). The resulting plasmids are sequenced so that those with error-free inserts can be selected.

Devised in 1985, PCR has had a major impact on biological research and genetic engineering. PCR has been used to amplify DNA from a wide variety of sources: a 40,000-year-old frozen woolly mammoth; fingerprints or tiny amounts of blood, tissue, or semen found at crime scenes; single embryonic cells for rapid prenatal diagnosis of genetic disorders; and cells infected with viruses that are difficult to detect, such as HIV (in the latter case, viral genes are amplified).

DNA Sequencing

Once a gene is cloned, researchers can exploit the principle of complementary base pairing to determine the gene's complete nucleotide sequence, a process called **DNA sequencing**. In the last 15 years, "next-generation" sequencing techniques have been developed that are rapid and inexpensive. In machines that carry out next-generation

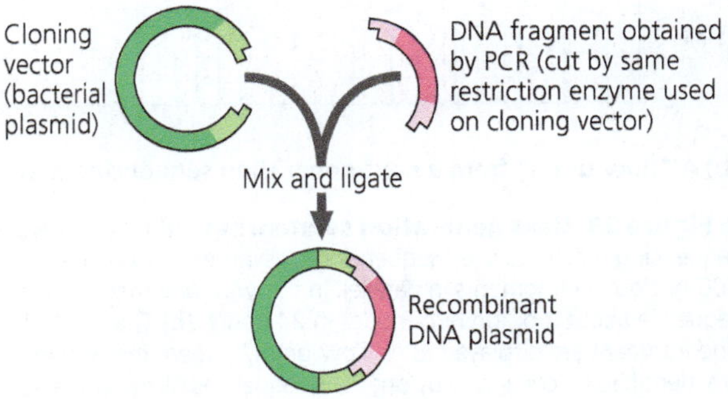

▲ Figure 28 Use of a restriction enzyme and PCR in gene cloning. In a closer look at the process shown at the top of Figure 24, PCR is used to produce the DNA fragment or gene of interest that will be ligated into a cloning vector, in this case a bacterial plasmid. Both the plasmid and the DNA fragments are cut with the same restriction enzyme, combined so the sticky ends can hybridize, and ligated together. The resulting plasmids will then be introduced into bacterial cells.

sequencing **(Figure 29a)**, a single template strand is immobilized, and DNA polymerase and other reagents are added that allow so-called *sequencing by synthesis* of the complementary strand, one nucleotide at a time. A specialized chemical technique enables electronic monitors to identify which of the four nucleotides is being added, allowing determination of the sequence **(Figure 29b)**.

Next-generation sequencing is rapidly being complemented or even replaced by "third-generation" sequencing techniques, with each new technique being faster and less expensive than the previous one. In some of these new methods, the DNA is neither cut into fragments nor amplified. Instead, a single, very long DNA molecule is sequenced on its own. Several groups have been working on moving a single strand of a DNA molecule through a very small pore (a *nanopore*) in a membrane. This is one of the many approaches to further increase the rate and reduce the cost of sequencing; one model of this approach is shown in **Figure 30**.

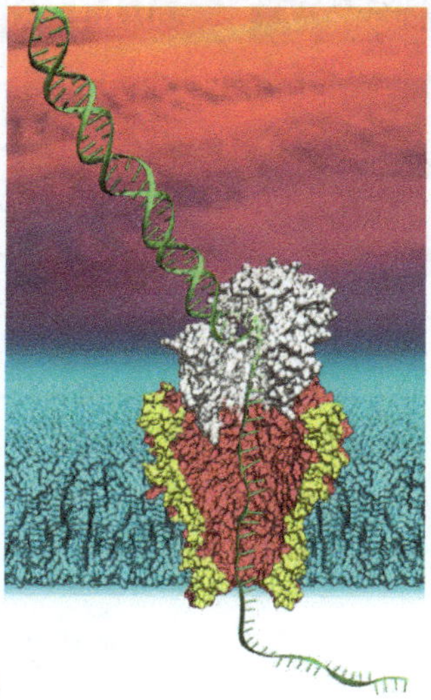

▶ **Figure 30 An example of a third-generation sequencing technique.** In this approach, a single strand of an uncut DNA molecule would be passed, nucleotide by nucleotide, through a nanopore in a membrane. Here, the pore is a protein channel embedded in a lipid membrane. The nucleotides are identified by slight differences in the amount of time they interrupt an electrical current across the opening.

(a) Next-generation sequencing machines

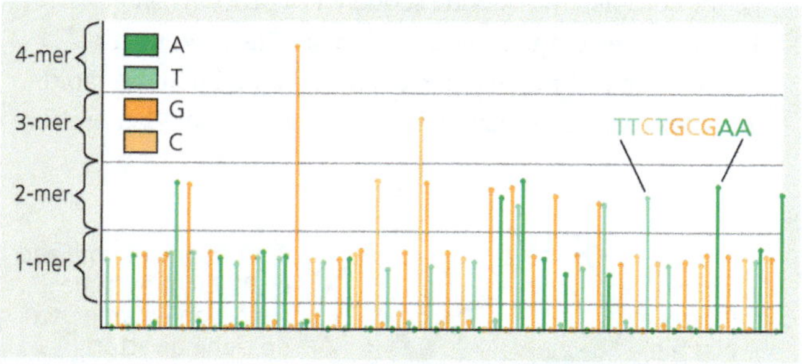

(b) A "flow-gram" from a next-generation sequencing machine

▲ **Figure 29 Next-generation sequencing. (a)** Next-generation sequencing machines use "sequencing by synthesis" to sequence many 300-nucleotide fragments in parallel. In this way, one machine can sequence about 2 billion nucleotides in 24 hours. **(b)** The results for one fragment are displayed as a "flow-gram," where the nucleotides are identified by color, one by one. The sequences of the entire set of fragments are analyzed using computer software, which "stitches" them together into a whole sequence—often, an entire genome.

INTERPRET THE DATA *If the template strand has two or more identical nucleotides in a row, the complementary nucleotides on the synthesized strand will be added, one after the other, in the same step. How are two or more of the same nucleotide in a row displayed in the flow-gram? (See example sequence on the right.) Write out the sequence of the first 25 nucleotides in the flow-gram, starting from the left as the 5′ end. (Ignore the very short lines.)*

Editing Genes and Genomes

Molecular biologists have long sought techniques for altering, or editing, the genetic material of cells or organisms in a predictable way. Their aim has been to use such a technique to change specific genes in living cells—either to study the function of a given gene or to try to correct genetic mutations that cause disease. In recent years, biologists have developed a powerful new technique for gene editing, called the CRISPR-Cas9 system, that is taking the field of genetic engineering by storm.

Cas9 is a bacterial protein that helps defend bacteria against bacteriophage infections. In bacterial cells, Cas9 acts together with "guide RNA" made from the CRISPR region of the bacterial genome.

Similar to the restriction enzymes described earlier, Cas9 is a nuclease that cuts double-stranded DNA molecules. However, while a given restriction enzyme recognizes only one particular DNA sequence, the Cas9 protein will cut any sequence to which it is directed. Cas9 takes its marching orders from a guide RNA molecule that it binds and uses as a homing device, cutting both strands of any DNA sequence that is complementary to the guide RNA. Scientists have been able to exploit the function of Cas9 by introducing a Cas9–guide RNA complex into a cell they wish to alter **(Figure 31)**. The guide RNA in the complex is engineered to be complementary to the "target" gene. Cas9 cuts both strands of the target DNA, and the resulting broken ends of DNA trigger a DNA repair system (similar to that shown in Figure 21). When there is no undamaged DNA for the enzymes of the repair system to use as a template, as shown at the bottom left of Figure 31, part a, the repair enzymes introduce or remove random nucleotides while rejoining the ends. Generally, this alters the DNA sequence so that the gene no longer works properly. This technique is a

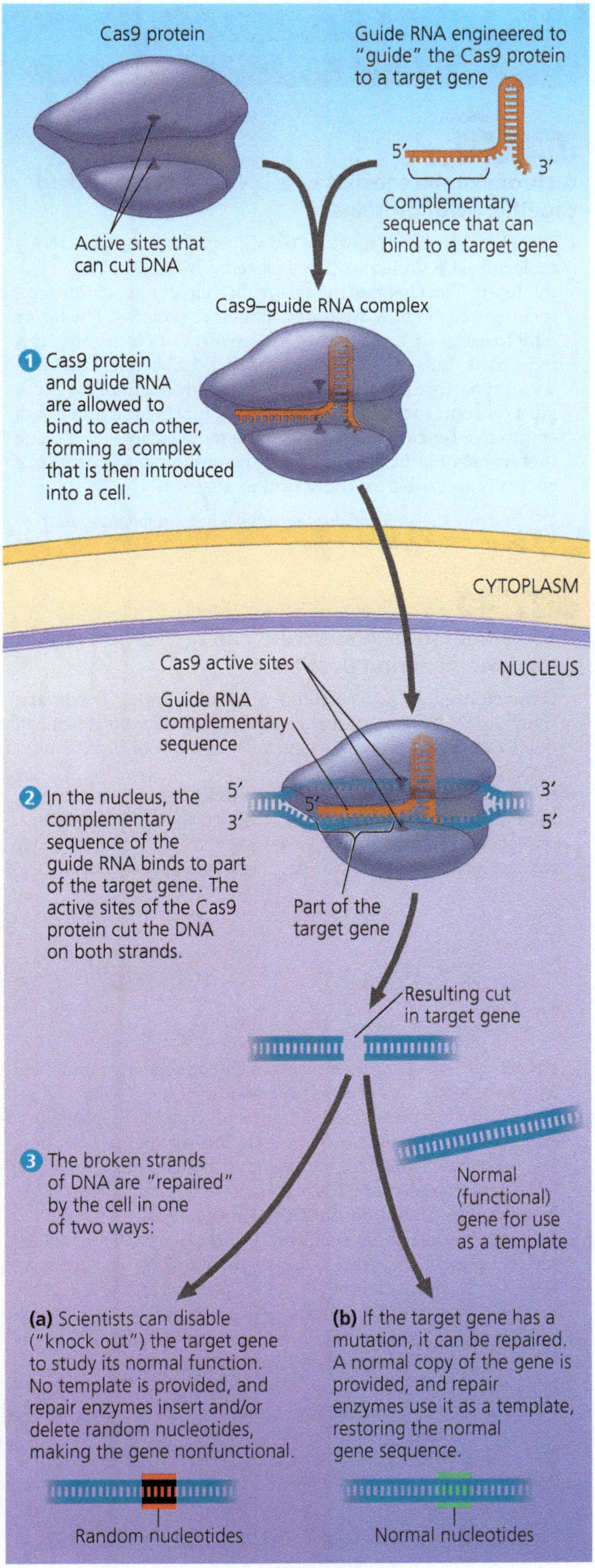

▲ **Figure 31 Gene editing using the CRISPR-Cas9 system.**

highly successful way for researchers to "knock out" (disable) a given gene to study what that gene does in an organism.

But what about using this system to help treat genetic diseases? Researchers have modified the technique so that the CRISPR-Cas9 system can be used to repair a gene that has a mutation. They introduce a segment from the normal (functional) gene along with the CRISPR-Cas9 system. After Cas9 cuts the target DNA, repair enzymes can use the normal DNA as a template to repair the target DNA at the break point. In this way, the CRISPR-Cas9 system edits the defective gene so that it is corrected (see the bottom right of Figure 31, part b).

In 2014, a group of researchers reported success in correcting a genetic defect in mice using CRISPR technology. The lab mice had been genetically engineered to have a mutation in a gene encoding a liver enzyme that metabolizes the amino acid tyrosine, mimicking a fatal genetic disorder in humans called tyrosinemia. A guide RNA molecule complementary to the mutated region of the gene was introduced into the mouse along with the Cas9 protein and a segment of DNA from the same region of the normal gene for use as a template. Subsequent analysis indicated that the faulty gene had been corrected in enough of the liver cells that the amount of functional enzyme made was sufficient to alleviate the disease symptoms. There are still many hurdles to overcome before this approach can be tried in humans, but the CRISPR technology is sparking widespread excitement among researchers and physicians alike.

In this section, you have learned how understanding the elegant structure of DNA has led to powerful techniques for genetic engineering. Earlier in the chapter, you also saw how DNA molecules are arranged in chromosomes and how DNA replication provides the copies of genes that parents pass to offspring. However, it is not enough that genes be copied and transmitted; the information they carry must be used by the cell. In other words, genes must also be "expressed."

CONCEPT CHECK 4

1. **MAKE CONNECTIONS** The restriction site for an enzyme called *Pvu*I is the following sequence:

   5′-C G A T C G-3′
   3′-G C T A G C-5′

   Staggered cuts are made between the T and C on each strand. What type of bonds are being cleaved?

2. **DRAW IT** One strand of a DNA molecule has the following sequence: 5′-CCTTGACGATCGTTACCG-3′. Draw the other strand. Will *Pvu*I (see question 1) cut this molecule? If so, draw the products.

3. Describe the role of complementary base pairing during cloning, PCR, DNA sequencing, and gene editing using the CRISPR-Cas9 system.

Chapter Review

SUMMARY OF KEY CONCEPTS

VOCAB SELF-QUIZ
goo.gl/gbai8v

CONCEPT 1

DNA is the genetic material

- Experiments with bacteria and **phages** provided the first strong evidence that the genetic material is DNA.
- Watson and Crick deduced that DNA is a **double helix** and built a structural model. Two **antiparallel** sugar-phosphate chains wind around the outside of the molecule; the nitrogenous bases project into the interior, where they hydrogen-bond in specific pairs: A with T, G with C.

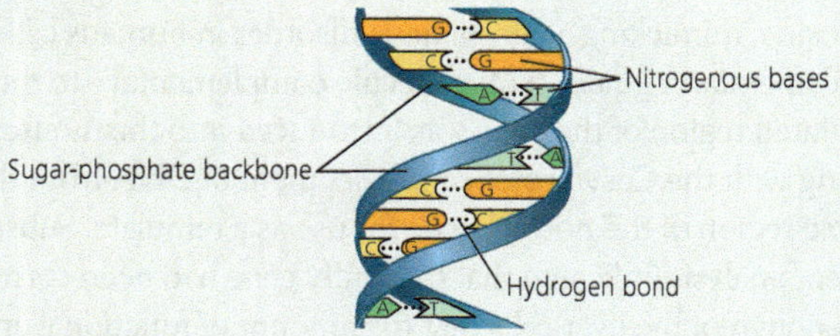

? *What does it mean when we say that the two DNA strands in the double helix are antiparallel? What would an end of the double helix look like if the strands were parallel?*

CONCEPT 2

Many proteins work together in DNA replication and repair

- The Meselson-Stahl experiment showed that **DNA replication** is **semiconservative**: The parental molecule unwinds, and each strand then serves as a template for the synthesis of a new strand according to base-pairing rules.
- DNA replication at one **replication fork** is summarized here:

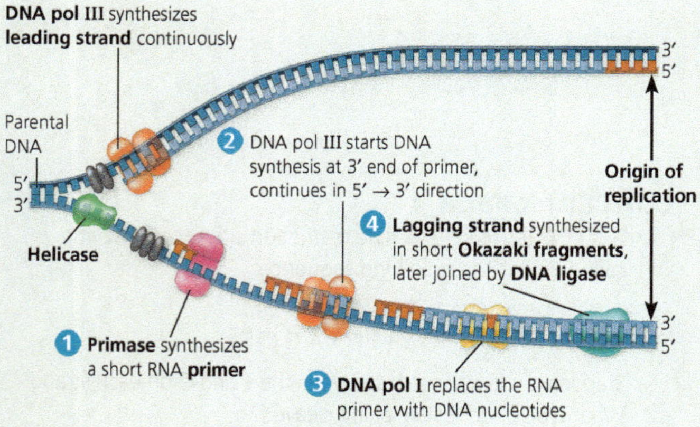

- **DNA polymerases** proofread new DNA, replacing incorrect nucleotides. In **mismatch repair**, enzymes correct errors that persist. **Nucleotide excision repair** is a general process by which **nucleases** cut out and replace damaged stretches of DNA.

? *Compare DNA replication on the leading and lagging strands, including both similarities and differences.*

CONCEPT 3

A chromosome consists of a DNA molecule packed together with proteins

- The chromosome of most bacterial species is a circular DNA molecule with some associated proteins, making up the **nucleoid**. The **chromatin** making up a eukaryotic chromosome is composed of DNA, **histones**, and other proteins. The histones bind to each other and to the DNA to form **nucleosomes**, the most basic units of DNA packing. Additional coiling and folding lead ultimately to the highly condensed chromatin of the metaphase chromosome. In interphase cells, most chromatin is less compacted (**euchromatin**), but some remains highly condensed (**heterochromatin**). Euchromatin, but not heterochromatin, is generally accessible for transcription of genes.

? *Describe the levels of chromatin packing you would expect to see in an interphase nucleus.*

CONCEPT 4

Understanding DNA structure and replication makes genetic engineering possible

- **Gene cloning** (or DNA cloning) produces multiple copies of a gene (or DNA segment) that can be used to manipulate and analyze DNA and to produce useful new products or organisms with beneficial traits.
- In **genetic engineering**, bacterial **restriction enzymes** are used to cut DNA molecules within short, specific nucleotide sequences (**restriction sites**), yielding a set of double-stranded **restriction fragments** with single-stranded **sticky ends**.

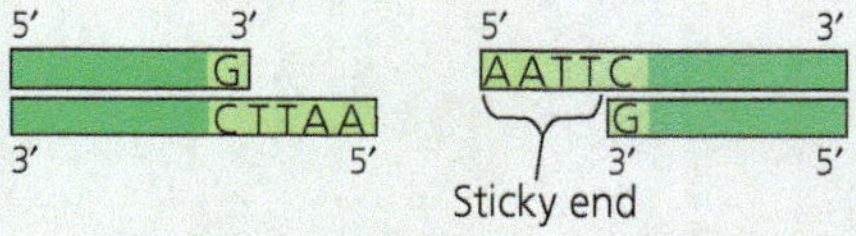

- DNA fragments of different lengths can be separated and their lengths assessed by **gel electrophoresis**.
- The sticky ends on restriction fragments from one DNA source—such as a bacterial **plasmid** or other **cloning vector**—can base-pair with complementary sticky ends on fragments from other DNA molecules; sealing the base-paired fragments with DNA ligase produces **recombinant DNA molecules**.
- The **polymerase chain reaction (PCR)** can amplify (produce many copies of) a specific target segment of DNA for use as a DNA fragment for cloning. PCR uses primers that bracket the desired segment and requires a heat-resistant DNA polymerase.
- The rapid development of fast, inexpensive techniques for **DNA sequencing** is based on *sequencing by synthesis*: DNA polymerase is used to replicate a stretch of DNA from a single-stranded template, and the order in which nucleotides are added reveals the sequence.
- The CRISPR-Cas9 system allows researchers to edit genes in a specific, desired way. This may ultimately be used for treatment of genetic diseases.

? *Describe how the process of gene cloning results in a cell clone containing a recombinant plasmid.*

TEST YOUR UNDERSTANDING

PRACTICE TEST
goo.gl/CRZjvS

Level 1: Knowledge/Comprehension

1. In his work with pneumonia-causing bacteria and mice, Griffith found that
   (A) the protein coat from pathogenic cells was able to transform nonpathogenic cells.
   (B) heat-killed pathogenic cells caused pneumonia.
   (C) some substance from pathogenic cells was transferred to nonpathogenic cells, making them pathogenic.
   (D) the polysaccharide coat of bacteria caused pneumonia.

2. What is the basis for the difference in how the leading and lagging strands of DNA molecules are synthesized?
   (A) DNA polymerase can join new nucleotides only to the 3′ end of a preexisting strand.
   (B) Helicases and single-strand binding proteins work at the 5′ end.
   (C) The origins of replication occur only at the 5′ end.
   (D) DNA ligase works only in the 3′ → 5′ direction.

3. In analyzing the number of different bases in a DNA sample, which result would be consistent with the base-pairing rules?
   (A) A = G
   (B) A + G = C + T
   (C) A + T = G + C
   (D) A = C

4. The elongation of the leading strand during DNA synthesis
   (A) progresses away from the replication fork.
   (B) occurs in the 3′ → 5′ direction.
   (C) produces Okazaki fragments.
   (D) depends on the action of DNA polymerase.

5. In a nucleosome, the DNA is wrapped around
   (A) polymerase molecules.
   (B) ribosomes.
   (C) histones.
   (D) a thymine dimer.

6. Which of the following sequences in double-stranded DNA is most likely to be recognized as a cutting site for a restriction enzyme?
   (A) AAGG / TTCC
   (B) GGCC / CCGG
   (C) ACCA / TGGT
   (D) AAAA / TTTT

Level 2: Application/Analysis

7. *E. coli* cells grown on ^{15}N medium are transferred to ^{14}N medium and allowed to grow for two more generations (two rounds of DNA replication). DNA extracted from these cells is centrifuged. What density distribution of DNA would you expect in this experiment?
   (A) one high-density and one low-density band
   (B) one intermediate-density band
   (C) one high-density and one intermediate-density band
   (D) one low-density and one intermediate-density band

8. A student isolates, purifies, and combines in a test tube a variety of molecules needed for DNA replication. After adding some DNA to the mixture, replication occurs, but each DNA molecule consists of a normal strand paired with numerous segments of DNA a few hundred nucleotides long. What has the student probably left out of the mixture?
   (A) DNA polymerase
   (B) DNA ligase
   (C) Okazaki fragments
   (D) primase

9. The spontaneous loss of amino groups from adenine in DNA results in hypoxanthine, an uncommon base, opposite thymine. What combination of proteins could repair such damage?
   (A) nuclease, DNA polymerase, DNA ligase
   (B) topoisomerase, primase, DNA polymerase
   (C) topoisomerase, helicase, single-strand binding protein
   (D) DNA ligase, replication fork proteins, adenylyl cyclase

10. **MAKE CONNECTIONS** Although the proteins that cause the *E. coli* chromosome to coil are not histones, identify a property you would expect them to share with histones, given their ability to bind to DNA.

Level 3: Synthesis/Evaluation

11. **SCIENTIFIC INQUIRY**

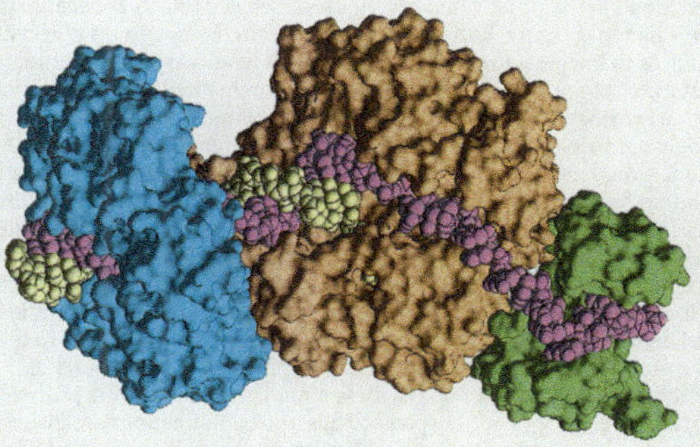

DRAW IT Model building can be an important part of the scientific process. The illustration shown here is a computer-generated model of a DNA replication complex. The parental and newly synthesized DNA strands are color-coded differently, as are each of the following three proteins: DNA pol III, the sliding clamp, and single-strand binding protein. (a) Using what you've learned in this chapter to clarify this model, label each DNA strand and each protein. (b) Draw an arrow to indicate the overall direction of DNA replication.

12. **FOCUS ON EVOLUTION**
Some bacteria may be able to respond to environmental stress by increasing the rate at which mutations occur during cell division. How might this be accomplished? Might there be an evolutionary advantage of this ability? Explain.

13. **FOCUS ON ORGANIZATION**
The continuity of life is based on heritable information in the form of DNA, and structure and function are correlated at all levels of biological organization. In a short essay (100–150 words), describe how the structure of DNA is correlated with its role as the molecular basis of inheritance.

14. **SYNTHESIZE YOUR KNOWLEDGE**

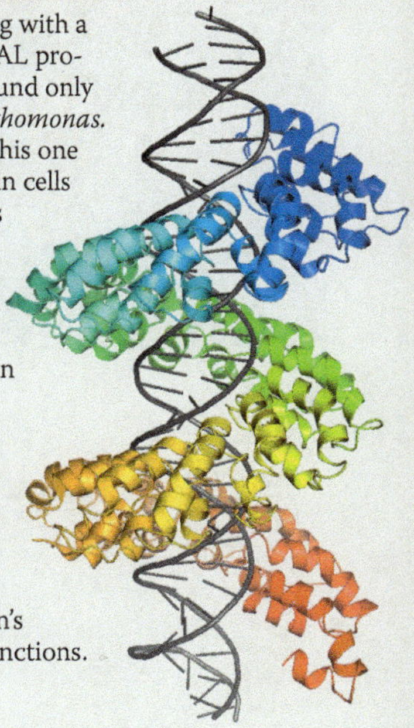

This image shows DNA interacting with a computer-generated model of a TAL protein, one of a family of proteins found only in a species of the bacterium *Xanthomonas*. The bacterium uses proteins like this one to find particular gene sequences in cells of the organisms it infects, such as tomatoes, rice, and citrus fruits. Researchers are excited about working with this family of proteins. Their goal is to generate modified versions that can home in on specific gene sequences. Such proteins could then be used in an approach called gene therapy to "fix" mutated genes in individuals with genetic diseases. Given what you know about DNA structure and considering the image shown here, discuss what the TAL protein's structure suggests about how it functions.

Answers

Figure Questions

Figure 3 The living S cells found in the blood sample were able to reproduce to yield more S cells, indicating that the S trait is a permanent, heritable change, rather than just a one-time use of the dead S cells' capsules. **Figure 5** The radioactivity would have been found in the pellet when proteins were labeled (batch 1) because proteins would have had to enter the bacterial cells to program them with genetic instructions. It's hard for us to imagine now, but the DNA might have played a structural role that allowed some of the proteins to be injected while it remained outside the bacterial cell (thus no radioactivity in the pellet in batch 2). **Figure 13** The tube from the first replication would look the same, with a middle band of hybrid ^{15}N-^{14}N DNA, but the second tube would not have the upper band of two light blue strands. Instead, it would have a bottom band of two dark blue strands, like the bottom band in the result predicted after one replication in the conservative model. **Figure 15** In the bubble at the top of the micrograph in (b), arrows should be drawn pointing left and right to indicate the two replication forks. **Figure 16** Looking at any of the DNA strands, we see that one end is called the 5′ end and the other the 3′ end. If we proceed from the 5′ end to the 3′ end on the left-most strand, for example, we list the components in this order: phosphate group → 5′ C of the sugar → 3′ C → phosphate → 5′ C → 3′ C. Going in the opposite direction on the same strand, the components proceed in the reverse order: 3′ C → 5′ C → phosphate. Thus, the two directions are distinguishable, which is what we mean when we say that the strands have directionality. (Review Figure 6 if necessary.)

Figure 19

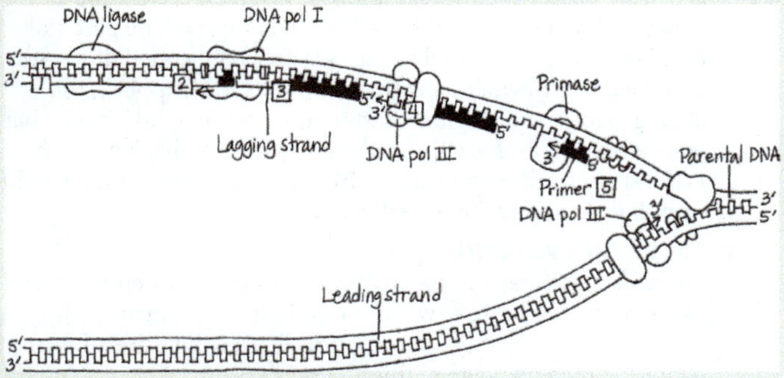

Figure 20

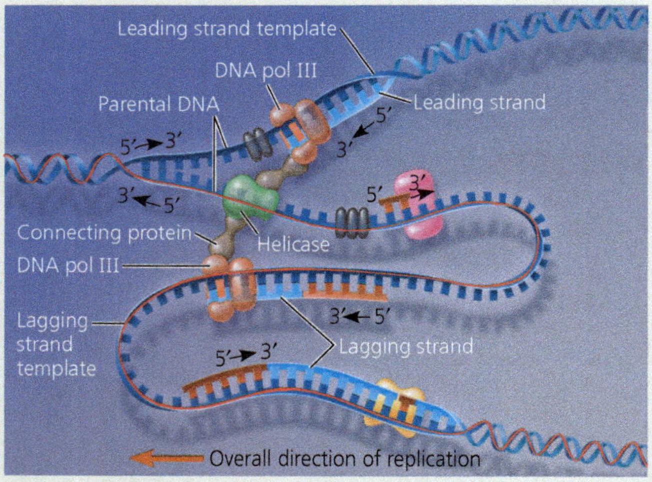

Figure 25

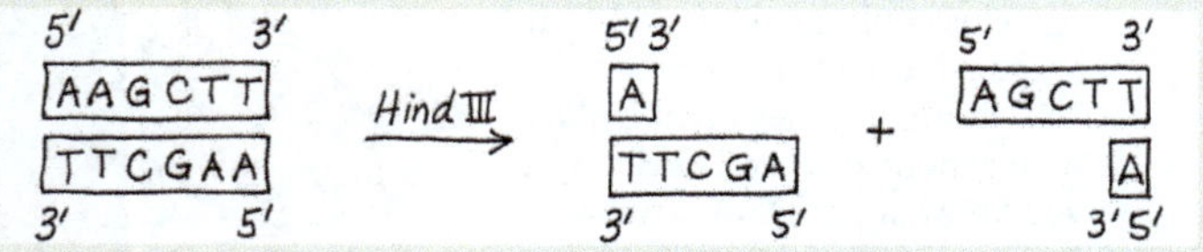

Concept Check 1

1. In order to tell which end is the 5′ end, you need to know which end has a phosphate group on the 5′ carbon (the 5′ end) or which end has an —OH group on the 3′ carbon (the 3′ end). **2.** Griffith expected that the mouse injected with the mixture of heat-killed S cells and living R cells would survive, since neither type of cell alone could kill the mouse.

Concept Check 2

1. Complementary base pairing ensures that the two daughter molecules are exact copies of the parental molecule. When the two strands of the parental molecule separate, each serves as a template on which nucleotides are arranged, by the base-pairing rules, into new complementary strands.

2.

Protein	Function
Helicase	Unwinds parental double helix at replication forks
Single-strand binding protein	Binds to and stabilizes single-stranded DNA until it can be used as a template
Topoisomerase	Relieves "overwinding" strain ahead of replication forks by breaking, swiveling, and rejoining DNA strands
Primase	Synthesizes an RNA primer at 5′ end of leading strand and at 5′ end of each Okazaki fragment of lagging strand
DNA pol III	Using parental DNA as a template, synthesizes new DNA strand by covalently adding nucleotides to 3′ end of a preexisting DNA strand or RNA primer
DNA pol I	Removes RNA nucleotides of previous fragment's primer from its 5′ end and replaces them with DNA nucleotides attached to 3′ end of next fragment
DNA ligase	Joins 3′ end of DNA that replaces primer to rest of leading strand and joins Okazaki fragments of lagging strand

3. In the cell cycle, DNA synthesis occurs during the S phase, between the G_1 and G_2 phases of interphase. DNA replication is therefore complete before the mitotic phase begins.

Concept Check 3

1. A nucleosome is made up of eight histone proteins, two each of four different types, around which DNA is wound. Linker DNA runs from one nucleosome to the next. **2.** Euchromatin is chromatin that becomes less compacted during interphase and is accessible to the cellular machinery responsible for gene activity. Heterochromatin, on the other hand, remains quite condensed during interphase and contains genes that are largely inaccessible to this machinery.

Concept Check 4

1. The covalent sugar-phosphate bonds of the DNA strands **2.** Yes, *PvuI* will cut the molecule (at the position indicated by the dashed red line).

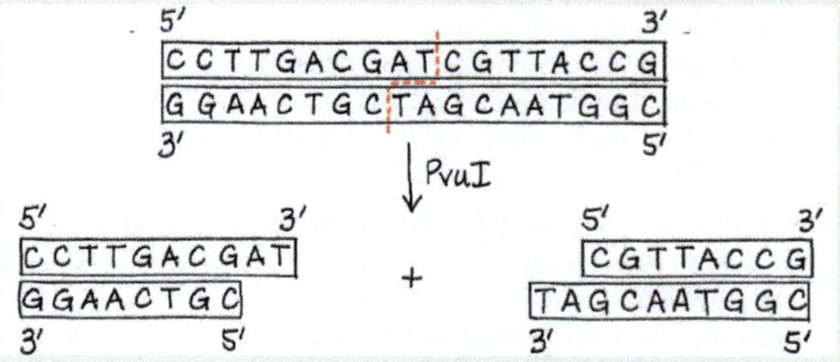

3. Cloning requires joining two pieces of DNA—a cloning vector, such as a bacterial plasmid, and a gene or DNA fragment from another source. Both pieces must be cut with the same restriction enzyme, creating sticky ends that will base-pair with complementary ends on other fragments. (The sugar-phosphate backbones will then be ligated together by ligase.) In PCR, the primers must base-pair with their target sequences in the DNA mixture, bracketing one specific region among many, and complementary base pairing is the basis for the building of the new strand during the extension step. In DNA sequencing, primers base-pair to the template, allowing DNA synthesis to start, and then nucleotides are added to the growing strand based on complementarity of base pairing. Using the CRISPR-Cas9 system to edit genes involves an RNA used as a guide that is complementary to the sequence to be edited. Complementary base pairing anchors the Cas9 protein to the target DNA sequence it will then cut.

Summary of Key Concepts Questions

1 Each strand in the double helix has polarity, the end with a phosphate group on the 5′ carbon of the sugar being called the 5′ end, and the end with an —OH group on the 3′ carbon of the sugar being called the 3′ end. The two strands run in opposite directions, one running 5′ → 3′ and the other running 3′ → 5′. Thus, each end of the molecule has both a 5′ and a 3′ end. This arrangement is called "antiparallel." If the strands were parallel, they would both run 5′ → 3′ in the same direction, so an end of the molecule would have either two 5′ ends or two 3′ ends. **2** On both the leading and lagging strands, DNA polymerase adds onto the 3′ end of an RNA primer synthesized by primase, synthesizing DNA in the 5′ → 3′ direction. Because the parental strands are antiparallel, however, only on the leading strand does synthesis proceed continuously into the replication fork. The lagging strand is synthesized bit by bit in the direction away from the fork as a series of shorter Okazaki fragments, which are later joined together by DNA ligase. Each fragment is initiated by synthesis of an RNA primer by primase as soon as a given stretch of single-stranded template strand is opened up. Although both strands are synthesized at the same rate, synthesis of the lagging strand is delayed because initiation of each fragment begins only when sufficient template strand is available. **3** Most of the chromatin in an interphase nucleus is fairly uncondensed. Much is present as the 30-nm fiber, with some in the form of the 10-nm fiber and some as looped domains of the 30-nm fiber. (These different levels of chromatin packing may reflect differences in gene expression occurring in these regions.) Also, a small percentage of the chromatin, such as that at the centromeres and telomeres, is highly condensed heterochromatin.

4 A plasmid vector and a source of foreign DNA to be cloned are both cut with the same restriction enzyme, generating restriction fragments with sticky ends. These fragments are mixed together, ligated, and reintroduced into bacterial cells, which can then make many copies of the foreign DNA or its product.

Test Your Understanding

1. C **2.** A **3.** B **4.** D **5.** C **6.** B **7.** D **8.** B **9.** A **10.** Like histones, the *E. coli* proteins would be expected to contain many basic (positively charged) amino acids, such as lysine and arginine, which can form weak bonds with the negatively charged phosphate groups on the sugar-phosphate backbone of the DNA molecule.

11.

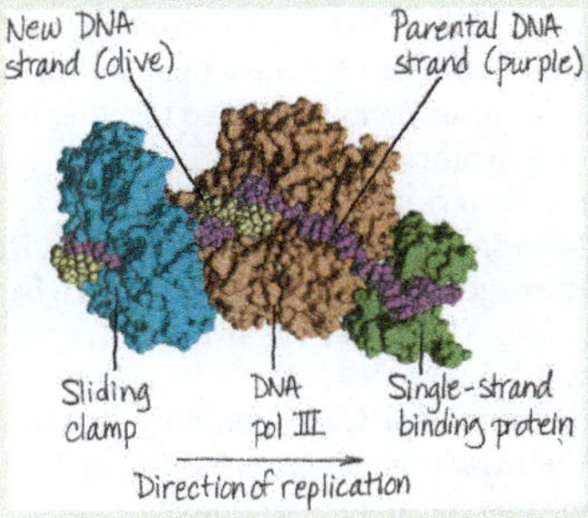

Credits

Credits are listed in the order of appearance.

Photo Credits

1 Andrey Prokhorov/Getty Images; **2** A. Barrington Brown/Science Source; **4** Oliver Meckes/Science Source; **13 SSE** Marevision/age fotostock; **7a** Library of Congress; **7b** From: *The Double Helix* by James D. Watson, Atheneum Press, N. Y., p. 215. © 1968. Courtesy CSHL Archive; **8c** Photodisc/Getty Images; **15a** Micrograph by Jerome Vinograd. From: *Molecular Biology of the Cell*. 4th edition. DNA Replication Mechanisms. Figure 5-6; **15b** From: Enrichment and visualization of small replication units from cultured mammalian cells. D. J. Burks et al. *J Cell Biol*. 1978 Jun; 77(3):762–73. Fig. 6A; **22** Peter Lansdorp; **23 left to right** Gopal Murti/Science Source; Victoria E. Foe; Barbara Hamkalo; U. Laemmli/Science Source; Biophoto/Science Source; **26b** Repligen Corporation; **29a** P. Morris/Garvan Institute of Medical Research; **30** Ian Derrington; **Chapter Review DNA model** Thomas A. Steitz, Yale University; **SYK** Barry Stoddard.

Illustration and Text Credits

Quote ("It has not . . .") J. D. Watson and F. H. C. Crick. A Structure for Deoxyribose Nucleic Acid. *Nature*, 2 April 1953, VOL 171:737 1953; **Quote** ("Now our model . . .") J. D. Watson and F. H. C. Crick, Genetical implications of the structure of deoxyribonucleic acid, *Nature* 171:964–967 (1953); **3** Data from F. Griffith, The significance of pneumococcal types, *Journal of Hygiene* 27:113–159 (1928); **5** Data from A. D. Hershey and M. Chase, Independent functions of viral protein and nucleic acid in growth of bacteriophage, *Journal of General Physiology* 36:39–56 (1952); **SSE.1** Data from several papers by Chargaff: for example, E. Chargaff et al., Composition of the desoxypentose nucleic acids of four genera of sea-urchin, *Journal of Biological Chemistry* 195:155–160 (1952); **13** Data from M. Meselson and F. W. Stahl, The replication of DNA in *Escherichia coli*, *Proceedings of the National Academy of Sciences USA* 44:671–682 (1958); **27** Adapted from *The World of the Cell*, 3rd Edition, by W.M. Becker, J.B. Reece, and M.F. Poenie, 1996. Copyright © 1996 by Pearson Education, Inc. Reprinted and electronically reproduced by permission of Pearson Education, Inc. Upper Saddle River, NJ.

Glossary

bacteriophage (bak-tēr′-ē-ō-fāj) A virus that infects bacteria; also called a phage.

cloning vector In genetic engineering, a DNA molecule that can carry foreign DNA into a host cell and replicate there. Cloning vectors include plasmids.

DNA ligase (lī′-gās) A linking enzyme essential for DNA replication; catalyzes the covalent bonding of the 3′ end of one DNA fragment (such as an Okazaki fragment) to the 5′ end of another DNA fragment (such as a growing DNA chain).

DNA polymerase (puh-lim′-er-ās) An enzyme that catalyzes the elongation of new DNA (for example, at a replication fork) by the addition of nucleotides to the 3′ end of an existing chain. There are several different DNA polymerases; DNA polymerase III and DNA polymerase I play major roles in DNA replication in *E. coli*.

DNA replication The process by which a DNA molecule is copied; also called DNA synthesis.

DNA sequencing Determining the order of nucleotide bases in a gene or DNA fragment.

euchromatin (yū-krō′-muh-tin) The less condensed form of eukaryotic chromatin that is available for transcription.

gel electrophoresis (ē-lek′-trō-fōr-ē′-sis) A technique for separating nucleic acids or proteins on the basis of their size and electrical charge, both of which affect their rate of movement through an electric field in a gel made of agarose or another polymer.

gene cloning The production of multiple copies of a gene.

genetic engineering The direct manipulation of genes for practical purposes.

helicase An enzyme that untwists the double helix of DNA at replication forks, separating the two strands and making them available as template strands.

heterochromatin (het′-er-ō-krō′-muh-tin) Eukaryotic chromatin that remains highly compacted during interphase and is generally not transcribed.

lagging strand A discontinuously synthesized DNA strand that elongates by means of Okazaki fragments, each synthesized in a 5′ → 3′ direction away from the replication fork.

leading strand The new complementary DNA strand synthesized continuously along the template strand toward the replication fork in the mandatory 5′ → 3′ direction.

mismatch repair The cellular process that uses specific enzymes to remove and replace incorrectly paired nucleotides.

nuclease An enzyme that cuts DNA or RNA, either removing one or a few bases or hydrolyzing the DNA or RNA completely into its component nucleotides.

nucleotide excision repair A repair system that removes and then correctly replaces a damaged segment of DNA using the undamaged strand as a guide.

Okazaki fragment (ō'-kah-zah'-kē) A short segment of DNA synthesized away from the replication fork on a template strand during DNA replication. Many such segments are joined together to make up the lagging strand of newly synthesized DNA.

PCR *See* polymerase chain reaction.

plasmid (plaz'-mid) A small, circular, double-stranded DNA molecule that carries accessory genes separate from those of a bacterial chromosome; in DNA cloning, can be used as a vector carrying up to about 10,000 base pairs (10 kb) of DNA.

polymerase chain reaction (PCR) (puh-lim'-uh-rās) A technique for amplifying DNA *in vitro* by incubating it with specific primers, a heat-resistant DNA polymerase, and nucleotides.

primase An enzyme that joins RNA nucleotides to make a primer during DNA replication, using the parental DNA strand as a template.

primer A short stretch of RNA with a free 3′ end, bound by complementary base pairing to the template strand and elongated with DNA nucleotides during DNA replication.

replication fork A Y-shaped region on a replicating DNA molecule where the parental strands are being unwound and new strands are being synthesized.

restriction enzyme An endonuclease (type of enzyme) that recognizes and cuts DNA molecules foreign to a bacterium (such as phage genomes). The enzyme cuts at specific nucleotide sequences (restriction sites).

restriction fragment A DNA segment that results from the cutting of DNA by a restriction enzyme.

restriction site A specific sequence on a DNA strand that is recognized and cut by a restriction enzyme.

semiconservative model Type of DNA replication in which the replicated double helix consists of one old strand, derived from the parental molecule, and one newly made strand.

single-strand binding protein A protein that binds to the unpaired DNA strands during DNA replication, stabilizing them and holding them apart while they serve as templates for the synthesis of complementary strands of DNA.

sticky end A single-stranded end of a double-stranded restriction fragment.

telomere (tel'-uh-mēr) The tandemly repetitive DNA at the end of a eukaryotic chromosome's DNA molecule. Telomeres protect the organism's genes from being eroded during successive rounds of replication.

topoisomerase A protein that breaks, swivels, and rejoins DNA strands. During DNA replication, topoisomerase helps to relieve strain in the double helix ahead of the replication fork.

Gene Expression: From Gene to Protein

From Chapter 14 of *Campbell Biology in Focus*, Second Edition. Lisa A. Urry, Michael L. Cain, Steven A. Wasserman, Peter V. Minorsky, and Jane B. Reece.

Gene Expression: From Gene to Protein

KEY CONCEPTS

1. Genes specify proteins via transcription and translation
2. Transcription is the DNA-directed synthesis of RNA: *a closer look*
3. Eukaryotic cells modify RNA after transcription
4. Translation is the RNA-directed synthesis of a polypeptide: *a closer look*
5. Mutations of one or a few nucleotides can affect protein structure and function

▲ **Figure 1** How does a single faulty gene result in the dramatic appearance of an albino donkey?

The Flow of Genetic Information

The island of Asinara lies off the coast of Sardinia, an Italian island. The name Asinara probably originated from the Latin word *sinuaria*, which means "sinus-shaped." A second meaning of Asinara is "donkey-inhabited," which is particularly appropriate because Asinara is home to a wild population of albino donkeys (**Figure 1**). The donkeys were brought to Asinara in the early 1800s and abandoned there in 1885 when the 500 residents were forced to leave the island so it could be used as a penal colony. What is responsible for the phenotype of the albino donkey, strikingly different from its pigmented relative?

Inherited traits are determined by genes, and the trait of albinism is caused by a recessive allele of a pigmentation gene. The information content of genes is in the form of specific sequences of nucleotides along strands of DNA, the genetic material. But how does this information determine an organism's traits? Put another way, what does a gene actually say? And how is its message translated by cells into a specific trait, such as brown hair, type A blood, or, in the case of an albino donkey, a total lack of pigment? The albino donkey has a faulty version of a key protein, an enzyme required for pigment synthesis, and this protein is faulty because the gene that codes for it contains incorrect information.

This example illustrates the main point of this chapter: The DNA inherited by an organism leads to specific traits by dictating the synthesis of proteins and of RNA molecules involved in protein synthesis. In other words, proteins are the link between genotype and phenotype. **Gene expression** is the process by which DNA directs the synthesis of proteins (or, in some cases, just RNAs). The expression of genes that code for proteins includes two stages: transcription and translation. This chapter describes the flow of information from gene to protein and explains how genetic mutations affect organisms through their proteins. Understanding the processes of gene expression, which are similar in all three domains of life, will allow us to revisit the concept of the gene in more detail at the end of the chapter.

CONCEPT 1

Genes specify proteins via transcription and translation

Before going into the details of how genes direct protein synthesis, let's step back and examine how the fundamental relationship between genes and proteins was discovered.

Evidence from the Study of Metabolic Defects

In 1902, British physician Archibald Garrod was the first to suggest that genes dictate phenotypes through enzymes, proteins that catalyze specific chemical reactions in the cell. He postulated that the symptoms of an inherited disease reflect an inability to make a particular enzyme. For example, people with a disease called alkaptonuria have black urine because it contains a chemical called alkapton. Garrod reasoned that these people cannot make an enzyme that breaks down alkapton, so the chemical is expelled in their urine.

Several decades later, research supported Garrod's hypothesis that a gene dictates the production of a specific enzyme. Biochemists learned that cells synthesize and degrade most organic molecules via metabolic pathways, in which each chemical reaction in a sequence is catalyzed by a specific enzyme. Such metabolic pathways lead, for instance, to the synthesis of the pigments that give the brown donkey in Figure 1 its fur color or fruit flies (*Drosophila*) their eye color. In the 1930s, the American geneticist George Beadle and his French colleague Boris Ephrussi speculated that in *Drosophila* each mutation affecting eye color blocks pigment synthesis at a specific step by preventing production of the enzyme that catalyzes that step. But neither the chemical reactions nor the enzymes that catalyze them were known.

Nutritional Mutants: Scientific Inquiry

A breakthrough in demonstrating the relationship between genes and enzymes came a few years later at Stanford University, where Beadle and Edward Tatum began experimenting with the bread mold *Neurospora crassa*. Unlike the diploid organisms studied by Mendel (peas) and T. H. Morgan (fruit flies), *Neurospora* is a haploid species. In order to observe a change in a mutant's phenotype, Beadle and Tatum needed to disable just one allele (rather than two) of a protein-coding gene required for a specific metabolic activity. By cleverly electing to work with *Neurospora*, they could cause a mutation in one allele in a cell and directly deduce the function of the wild-type gene.

Another advantage was that *Neurospora* has modest food requirements. For this species, the *minimal medium*, the culture containing minimal nutrients for growth of wild-type cells, is a simple solution of inorganic salts, glucose, and the vitamin biotin. On a gel-like substance called agar, saturated with minimal medium, single wild-type cells can synthesize all the nutrients they need for growth, dividing repeatedly and forming visible colonies of genetically identical cells.

Beadle and Tatum designed an experiment in which they generated different "nutritional mutants" of *Neurospora* cells, each of which was unable to synthesize a particular essential nutrient. Such cells could not grow on minimal medium but could grow on *complete medium,* which contains all nutrients needed for growth, including any that a mutant cell can't synthesize. For *Neurospora*, the complete medium consists of the minimal medium supplemented with all 20 amino acids and a few other nutrients. Beadle and Tatum hypothesized that in each nutritional mutant, the gene for the enzyme that synthesizes a particular nutrient had been disabled. **Figure 2** summarizes their experimental approach, in which they tested each type of mutant to determine which nutrient it was unable to synthesize.

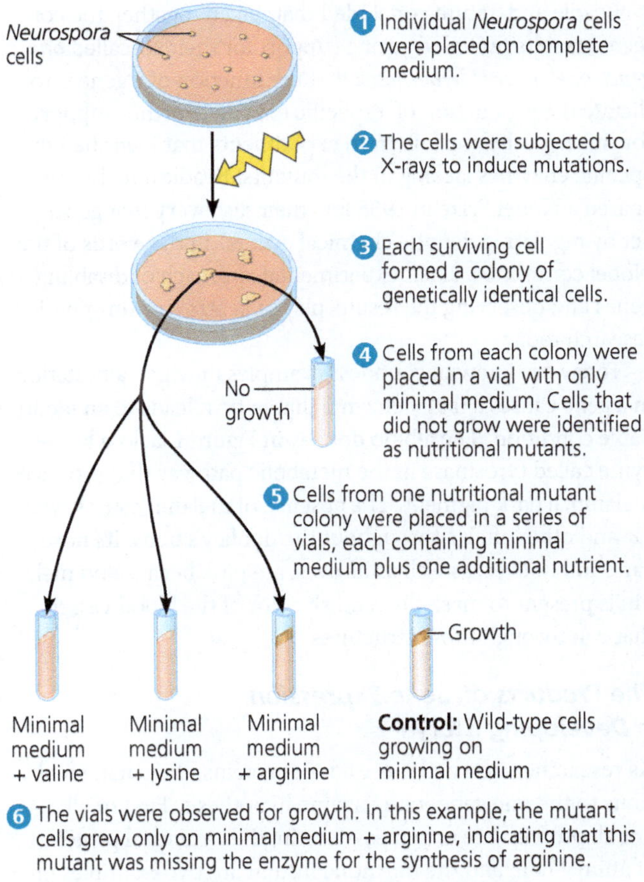

▲ **Figure 2 The experimental approach of Beadle and Tatum.** To obtain nutritional mutants, Beadle and Tatum exposed *Neurospora* cells to X-rays to induce mutations. They then screened mutants with new nutritional requirements, such as arginine, as shown here.

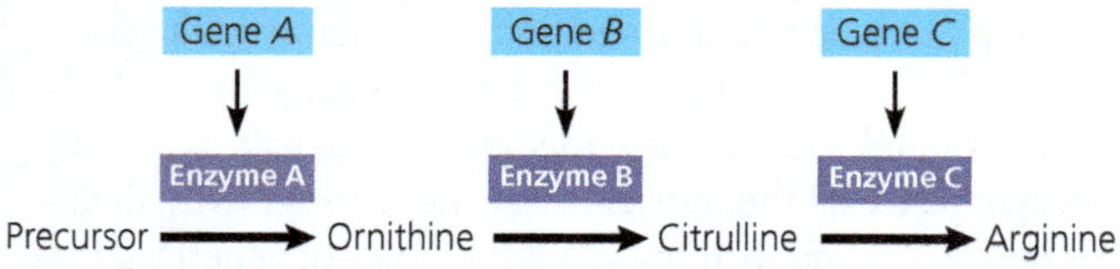

▲ **Figure 3 The one gene–one protein hypothesis.** Based on results from work in their lab on nutritional mutants, Beadle and Tatum proposed that the function of a specific gene is to dictate production of a specific enzyme that catalyzes a particular reaction. The model shown here for the arginine-synthesizing pathway illustrates their hypothesis.

Beadle and Tatum amassed a valuable collection of mutant strains of *Neurospora*, catalogued not only by their defects in a particular metabolic pathway but also at different steps in that pathway. For example, a series of experiments on mutants requiring the amino acid arginine revealed that they could be grouped into distinct classes, each corresponding to a particular blocked step in the biochemical pathway for arginine synthesis. Presumably, the enzyme that normally catalyzes that step was unable to do so as a result of a mutation in its encoding gene **(Figure 3)**.

Beadle and Tatum concluded that, taken together, the collected results provided strong support for their so-called *one gene–one enzyme hypothesis*: that the function of a gene is to dictate the production of a specific enzyme. Further support for this hypothesis came from experiments that identified the specific enzymes lacking in the mutants. Beadle and Tatum shared a Nobel Prize in 1958 for "their discovery that genes act by regulating definite chemical events," in the words of the Nobel committee. Their experimental approach of disabling genes and observing the results plays a central role in genetic research today.

Today, we know of countless examples in which a mutation in a gene causes a faulty enzyme that in turn leads to an identifiable condition. The albino donkey in Figure 1 lacks a key enzyme called tyrosinase in the metabolic pathway that produces melanin, a dark pigment. The absence of melanin causes white fur and other effects throughout the donkey's body. Its nose, ears, and hooves, as well as its eyes, are pink because no melanin is present to mask the reddish color of the blood vessels that run through those structures.

The Products of Gene Expression: A Developing Story

As researchers learned more about proteins, they made revisions to the one gene–one enzyme hypothesis. First of all, not all proteins are enzymes. Keratin, the structural protein of animal hair, and the hormone insulin are two examples of nonenzyme proteins. Because proteins that are not enzymes are nevertheless gene products, molecular biologists began to think in terms of one gene–one protein. However, many proteins are constructed from two or more different polypeptide chains, and each polypeptide is specified by its own gene. For example, hemoglobin, the oxygen-transporting protein of vertebrate red blood cells, contains two kinds of polypeptides, and thus two genes code for this protein. Beadle and Tatum's idea was therefore restated as the *one gene–one polypeptide hypothesis*. Even this description is not entirely accurate, though. First, in many cases, a eukaryotic gene can code for a set of closely related polypeptides via a process called alternative splicing, which you will learn about later in this chapter. Second, quite a few genes code for RNA molecules that have important functions in cells even though they are never translated into protein. For now, we will focus on genes that do code for polypeptides. (Note that it is common to refer to these gene products as proteins—a practice you'll encounter in this text—rather than more precisely as polypeptides.)

Basic Principles of Transcription and Translation

Genes provide the instructions for making specific proteins. But a gene does not build a protein directly. The bridge between DNA and protein synthesis is the nucleic acid RNA. RNA is chemically similar to DNA except that it contains ribose instead of deoxyribose as its sugar and has the nitrogenous base uracil rather than thymine. Thus, each nucleotide along a DNA strand has A, G, C, or T as its base, and each nucleotide along an RNA strand has A, G, C, or U as its base. An RNA molecule usually consists of a single strand.

It is customary to describe the flow of information from gene to protein in linguistic terms. Just as specific sequences of letters communicate information in a language like English, both nucleic acids and proteins are polymers with specific sequences of monomers that convey information. In DNA or RNA, the monomers are the four types of nucleotides, which differ in their nitrogenous bases. Genes are typically hundreds or thousands of nucleotides long, each gene having a specific sequence of nucleotides. Each polypeptide of a protein also has monomers arranged in a particular linear order (the protein's primary structure, but its monomers are amino acids. Thus, nucleic acids and proteins contain information written in two different chemical languages. Getting from DNA to protein requires two major stages: transcription and translation.

Transcription is the synthesis of RNA using information in the DNA. The two nucleic acids are written in different forms of the same language, and the information is simply transcribed, or "rewritten," from DNA to RNA. Just as a DNA strand provides a template for making a new complementary strand during DNA replication, it also can serve as a template for assembling a complementary sequence of RNA nucleotides. For a protein-coding gene, the resulting RNA molecule is a faithful transcript of the gene's protein-building instructions.

This type of RNA molecule is called **messenger RNA (mRNA)** because it carries a genetic message from the DNA to the protein-synthesizing machinery of the cell. (Transcription is the general term for the synthesis of *any* kind of RNA on a DNA template. Later, you'll learn about some other types of RNA produced by transcription.)

Translation is the synthesis of a polypeptide using the information in the mRNA. During this stage, there is a change in language: The cell must translate the nucleotide sequence of an mRNA molecule into the amino acid sequence of a polypeptide. The sites of translation are **ribosomes**, molecular complexes that facilitate the orderly linking of amino acids into polypeptide chains.

Transcription and translation occur in all organisms. Because most studies of transcription and translation have used bacteria and eukaryotic cells, they are our main focus in this chapter. While our understanding of transcription and translation in archaea lags behind, we do know that archaeal cells share some features of gene expression with bacteria and others with eukaryotes.

The basic mechanics of transcription and translation are similar for bacteria and eukaryotes, but there is an important difference in the flow of genetic information within the cells. Bacteria do not have nuclei. Therefore, nuclear membranes do not separate bacterial DNA and mRNA from ribosomes and the other protein-synthesizing equipment **(Figure 4a)**. This lack of compartmentalization allows translation of an mRNA to begin while its transcription is still in progress, as you'll see later. By contrast, eukaryotic cells have nuclei. The presence of a nuclear envelope separates transcription from translation in space and time **(Figure 4b)**. Transcription occurs in the nucleus, but the mRNA must be transported to the cytoplasm for translation. Before eukaryotic RNA transcripts from protein-coding genes can leave the nucleus, they are modified in various ways to produce the final, functional mRNA. The transcription of a protein-coding eukaryotic gene results in *pre-mRNA*, and further processing yields the finished mRNA. The initial RNA transcript from any gene, including those specifying RNA that is not translated into protein, is more generally called a **primary transcript**.

To summarize: Genes program protein synthesis via genetic messages in the form of messenger RNA. Put another way, cells are governed by a molecular chain of command with a directional flow of genetic information, shown here by arrows:

DNA → RNA → Protein

This concept was dubbed the *central dogma* by Francis Crick in 1956. How has the concept held up over time? In the 1970s, scientists were surprised to discover that some enzymes exist that use RNA molecules as templates for DNA synthesis. However, these exceptions do not invalidate the idea that, in general, genetic information flows from DNA to RNA to protein. Now let's discuss how the instructions for assembling amino acids into a specific order are encoded in nucleic acids.

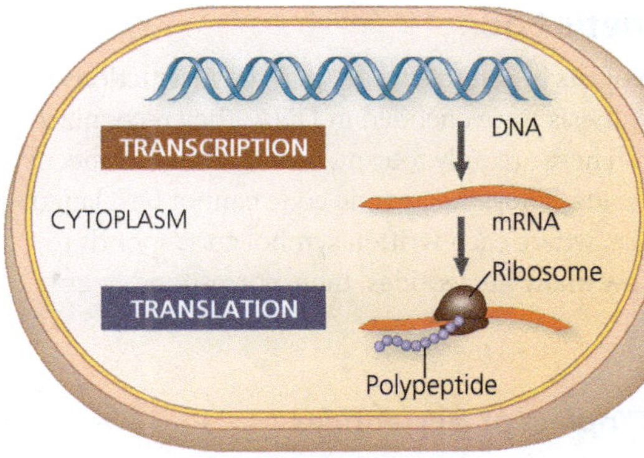

(a) Bacterial cell. In a bacterial cell, which lacks a nucleus, mRNA produced by transcription is immediately translated without additional processing.

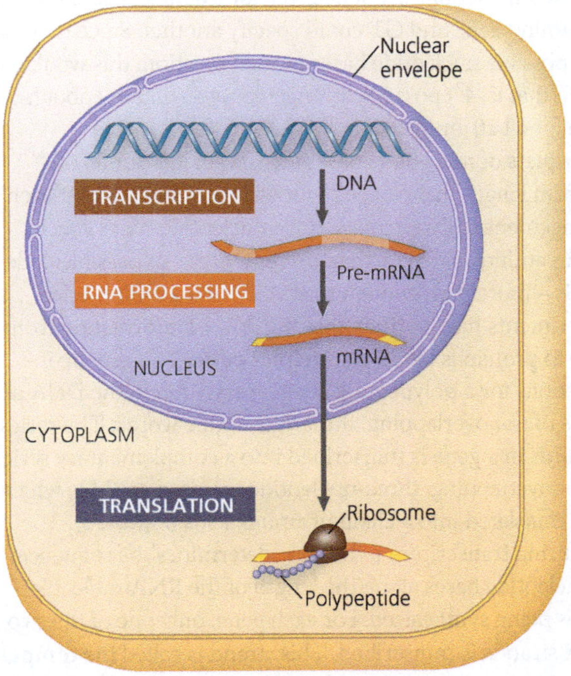

(b) Eukaryotic cell. The nucleus provides a separate compartment for transcription. The original RNA transcript, called pre-mRNA, is processed in various ways before leaving the nucleus as mRNA.

▲ **Figure 4 Overview: the roles of transcription and translation in the flow of genetic information.** In a cell, inherited information flows from DNA to RNA to protein. The two main stages of information flow are transcription and translation. A miniature version of part (a) or (b) accompanies several figures later in the chapter as an orientation diagram to help you see where a particular figure fits into the overall scheme of gene expression.

The Genetic Code

When biologists began to suspect that the instructions for protein synthesis were encoded in DNA, they recognized a problem: There are only four nucleotide bases to specify 20 amino acids. Thus, the genetic code cannot be a language like Chinese, where each written symbol corresponds to a word. How many nucleotides, then, correspond to an amino acid?

Codons: Triplets of Nucleotides

If each kind of nucleotide base were translated into an amino acid, only four amino acids could be specified, one per nucleotide base. Would a language of two-letter code words suffice? The two-nucleotide sequence AG, for example, could specify one amino acid, and GT could specify another. Since there are four possible nucleotide bases in each position, this would give us 16 (that is, 4^2) possible arrangements—still not enough to code for all 20 amino acids.

Triplets of nucleotide bases are the smallest units of uniform length that can code for all the amino acids. If each arrangement of three consecutive nucleotide bases specifies an amino acid, there can be 64 (that is, 4^3) possible code words—more than enough to specify all the amino acids. Experiments have verified that the flow of information from gene to protein is based on a **triplet code**: The genetic instructions for a polypeptide chain are written in the DNA as a series of nonoverlapping, three-nucleotide words. The series of words in a gene is transcribed into a complementary series of nonoverlapping, three-nucleotide words in mRNA, which is then translated into a chain of amino acids **(Figure 5)**.

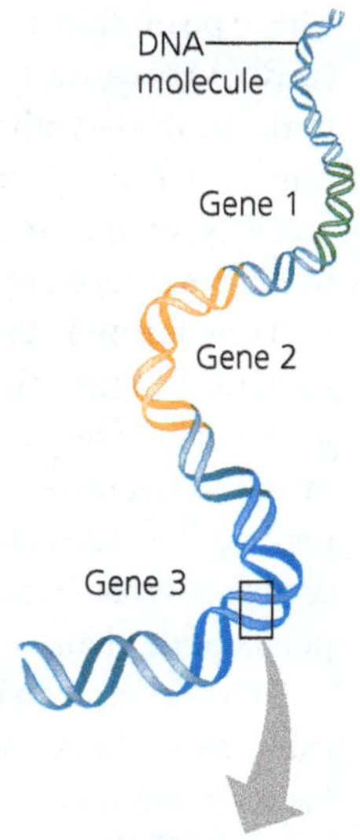

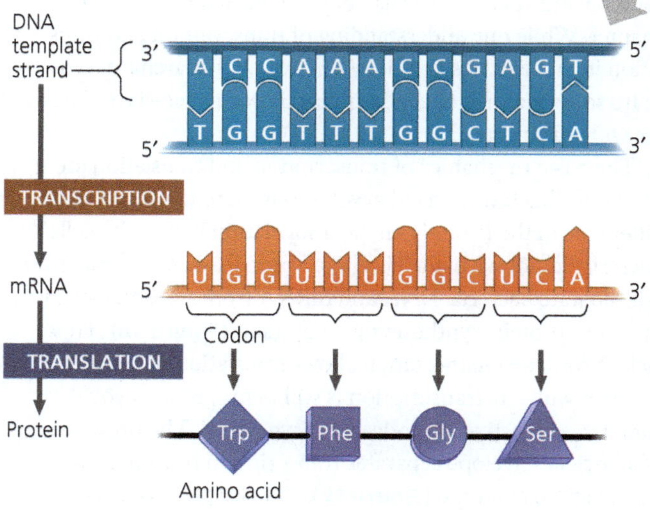

► **Figure 5 The triplet code.** For each gene, one DNA strand functions as a template for transcription of RNAs, such as mRNA. The base-pairing rules for DNA synthesis also guide transcription, except that uracil (U) takes the place of thymine (T) in RNA. During translation, the mRNA is read as a sequence of nucleotide triplets, called codons. Each codon specifies an amino acid to be added to the growing polypeptide chain. The mRNA is read in the 5′ → 3′ direction.

? *Compare the sequence of the mRNA to that of the nontemplate DNA strand, in both cases reading from 5′ → 3′.*

During transcription, the gene determines the sequence of nucleotide bases along the length of the RNA molecule that is being synthesized. For each gene, only one of the two DNA strands is transcribed. This strand is called the **template strand** because it provides the pattern, or template, for the sequence of nucleotides in an RNA transcript. For any given gene, the same strand is used as the template every time the gene is transcribed. However, on the same DNA molecule farther along the chromosome, the opposite strand may be the one that functions as the template for another gene.

An mRNA molecule is complementary rather than identical to its DNA template because RNA nucleotides are assembled on the template according to base-pairing rules (see Figure 5). The pairs are similar to those that form during DNA replication, except that U (the RNA substitute for T) pairs with A and the mRNA nucleotides contain ribose instead of deoxyribose. Like a new strand of DNA, the RNA molecule is synthesized in an antiparallel direction to the template strand of DNA. In the example in Figure 5, the nucleotide triplet ACC along the DNA template strand (written as 3′-ACC-5′) provides a template for 5′-UGG-3′ in the mRNA molecule. The mRNA nucleotide triplets are called **codons**, and they are customarily written in the 5′ → 3′ direction. In our example, UGG is the codon for the amino acid tryptophan (abbreviated Trp). The term *codon* is also used for the DNA nucleotide triplets along the *nontemplate* strand. These codons are complementary to the template strand and thus identical in sequence to the mRNA, except that they have T wherever there is a U in the mRNA. (For this reason, the nontemplate DNA strand is often called the *coding strand*.)

During translation, the sequence of codons along an mRNA molecule is decoded, or translated, into a sequence of amino acids making up a polypeptide chain. The codons are read by the translation machinery in the 5′ → 3′ direction along the mRNA. Each codon specifies which one of the 20 amino acids will be incorporated at the corresponding position along a polypeptide. Because codons are nucleotide triplets, the number of nucleotides making up a genetic message must be three times the number of amino acids in the protein product. For example, it takes 300 nucleotides along an mRNA strand to code for the amino acids in a polypeptide that is 100 amino acids long.

Cracking the Code

Molecular biologists cracked the genetic code of life in the early 1960s when a series of elegant experiments disclosed the amino acid translations of each of the RNA codons. The first codon was deciphered in 1961 by Marshall Nirenberg, of the National Institutes of Health, along with his colleagues. Nirenberg synthesized an artificial mRNA by linking together many identical RNA nucleotides containing uracil as their base. No matter where the genetic message started or stopped, it could contain only one codon (UUU) over and over. Nirenberg added this "poly-U" polynucleotide to a test-tube mixture that contained all 20 amino acids, ribosomes, and the other components required for protein synthesis. His artificial system translated the poly-U mRNA into a polypeptide containing many units of the amino acid phenylalanine (Phe), strung together as a polyphenylalanine chain. Thus, Nirenberg determined that the mRNA codon UUU specifies the amino acid phenylalanine. Soon, the amino acids specified by the codons AAA, GGG, and CCC were determined in the same way.

Although more elaborate techniques were required to decode mixed triplets such as AUA and CGA, all 64 codons were deciphered by the mid-1960s. As **Figure 6** shows, 61 of the 64 triplets code for amino acids. The three codons that do not designate amino acids are "stop" signals, or termination codons, marking the end of translation. Notice that the codon AUG has a dual function: It codes for the amino acid methionine (Met) and also functions as a "start" signal, or initiation codon. Genetic messages usually begin with the mRNA codon AUG, which signals the protein-synthesizing machinery to begin translating the mRNA at that location. (Because AUG also stands for methionine, polypeptide chains begin with methionine when they are synthesized. However, an enzyme may subsequently remove this starter amino acid from the chain.)

Notice in Figure 6 that there is redundancy in the genetic code, but no ambiguity. For example, although codons GAA and GAG both specify glutamic acid (redundancy), neither of them ever specifies any other amino acid (no ambiguity). The redundancy in the code is not altogether random. In many cases, codons that are synonyms for a particular amino acid differ only in the third nucleotide base of the triplet. We'll consider the significance of this redundancy later in the chapter.

Our ability to extract the intended message from a written language depends on reading the symbols in the correct groupings—that is, in the correct **reading frame**. Consider this statement: "The red dog ate the bug." Group the letters

First mRNA base (5′ end of codon)	Second mRNA base: U	Second mRNA base: C	Second mRNA base: A	Second mRNA base: G	Third mRNA base (3′ end of codon)
U	UUU Phe	UCU Ser	UAU Tyr	UGU Cys	U
U	UUC Phe	UCC Ser	UAC Tyr	UGC Cys	C
U	UUA Leu	UCA Ser	UAA Stop	UGA Stop	A
U	UUG Leu	UCG Ser	UAG Stop	UGG Trp	G
C	CUU Leu	CCU Pro	CAU His	CGU Arg	U
C	CUC Leu	CCC Pro	CAC His	CGC Arg	C
C	CUA Leu	CCA Pro	CAA Gln	CGA Arg	A
C	CUG Leu	CCG Pro	CAG Gln	CGG Arg	G
A	AUU Ile	ACU Thr	AAU Asn	AGU Ser	U
A	AUC Ile	ACC Thr	AAC Asn	AGC Ser	C
A	AUA Ile	ACA Thr	AAA Lys	AGA Arg	A
A	AUG Met or start	ACG Thr	AAG Lys	AGG Arg	G
G	GUU Val	GCU Ala	GAU Asp	GGU Gly	U
G	GUC Val	GCC Ala	GAC Asp	GGC Gly	C
G	GUA Val	GCA Ala	GAA Glu	GGA Gly	A
G	GUG Val	GCG Ala	GAG Glu	GGG Gly	G

▲ **Figure 6 The codon table for mRNA.** The three nucleotide bases of an mRNA codon are designated here as the first, second, and third bases, reading in the 5′ → 3′ direction along the mRNA. (Practice using this table by finding the codons in Figure 5.) The codon AUG not only stands for the amino acid methionine (Met) but also functions as a "start" signal for ribosomes to begin translating the mRNA at that point. Three of the 64 codons function as "stop" signals, marking where ribosomes end translation.

incorrectly by starting at the wrong point, and the result will probably be gibberish: for example, "her edd oga tet heb ug." The reading frame is also important in the molecular language of cells. The short stretch of polypeptide shown in Figure 5, for instance, will be made correctly only if the mRNA nucleotides are read from left to right (5′ → 3′) in the groups of three shown in the figure: UGG UUU GGC UCA. Although a genetic message is written with no spaces between the codons, the cell's protein-synthesizing machinery reads the message as a series of nonoverlapping three-letter words. The message is *not* read as a series of overlapping words—UGGUUU, and so on—which would convey a very different message.

Evolution of the Genetic Code

EVOLUTION The genetic code is nearly universal, shared by organisms from the simplest bacteria to the most complex plants and animals. The mRNA codon CCG, for instance, is translated as the amino acid proline in all organisms whose genetic code has been examined. In laboratory experiments, genes can be transcribed and translated after being transplanted from one species to another, sometimes with

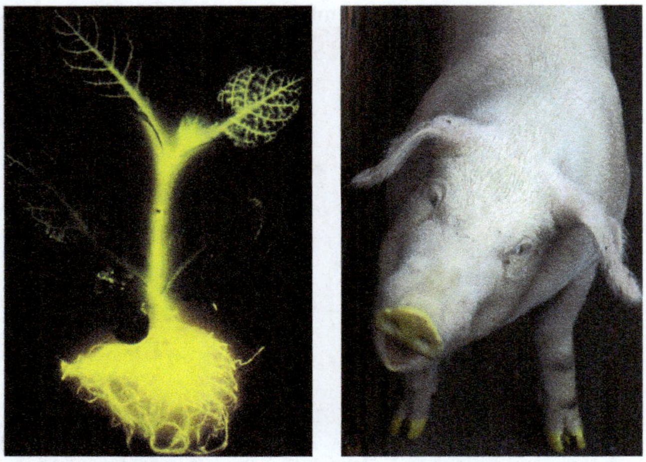

(a) **Tobacco plant expressing a firefly gene.** The yellow glow is produced by a chemical reaction catalyzed by the protein product of the firefly gene.

(b) **Pig expressing a jellyfish gene.** Researchers injected the gene for a fluorescent protein into fertilized pig eggs. One of the eggs developed into this fluorescent pig.

▲ **Figure 7 Expression of genes from different species.** Because diverse forms of life share a common genetic code, one species can be programmed to produce proteins characteristic of a second species by introducing DNA from the second species into the first.

quite striking results, as shown in **Figure 7**. Bacteria can be programmed by the insertion of human genes to synthesize certain human proteins for medical use, such as insulin. Such applications have produced many exciting developments in the area of genetic engineering.

Despite a small number of exceptions in which a few codons differ from the standard ones, the evolutionary significance of the code's *near* universality is clear. A language shared by all living things must have been operating very early in the history of life—early enough to be present in the common ancestor of all present-day organisms. A shared genetic vocabulary is a reminder of the kinship that bonds all life on Earth.

CONCEPT CHECK 1

1. MAKE CONNECTIONS In a research article about alkaptonuria published in 1902, Garrod suggested that humans inherit two "characters" (alleles) for a particular enzyme and that both parents must contribute a faulty version for the offspring to have the disorder. Today, would this disorder be called dominant or recessive?
2. What polypeptide product would you expect from a poly-G mRNA that is 30 nucleotides long?
3. DRAW IT The template strand of a gene contains the sequence 3′-TTCAGTCGT-5′. Suppose that the nontemplate sequence could be transcribed instead of the template sequence. Draw the nontemplate sequence in 3′ to 5′ order. Then draw the mRNA sequence and translate it using Figure 6. (Be sure to pay attention to the 5′ and 3′ ends, remembering that the mRNA is antiparallel to the DNA strand.) Predict how well the protein synthesized from the nontemplate strand would function, if at all.

CONCEPT 2

Transcription is the DNA-directed synthesis of RNA: *a closer look*

Now that we have considered the linguistic logic and evolutionary significance of the genetic code, we are ready to reexamine transcription, the first stage of gene expression, in more detail.

Molecular Components of Transcription

Messenger RNA, the carrier of information from DNA to the cell's protein-synthesizing machinery, is transcribed from the template strand of a gene. An enzyme called an **RNA polymerase** pries the two strands of DNA apart and joins together RNA nucleotides complementary to the DNA template strand, thus elongating the RNA polynucleotide **(Figure 8)**. Like the DNA polymerases that function in DNA replication, RNA polymerases can assemble a polynucleotide only in its $5' \rightarrow 3'$ direction. Unlike DNA polymerases, however, RNA polymerases are able to start a chain from scratch; they don't need a primer.

Specific sequences of nucleotides along the DNA mark where transcription of a gene begins and ends. The DNA sequence where RNA polymerase attaches and initiates transcription is known as the **promoter**; in bacteria, the sequence that signals the end of transcription is called the **terminator**. (The termination mechanism is different in eukaryotes; we'll describe it later.) Molecular biologists refer to the direction of transcription as "downstream" and the other direction as "upstream." These terms are also used to describe the positions of nucleotide sequences within the DNA or RNA. Thus, the promoter sequence in DNA is said to be upstream from the terminator. The stretch of DNA downstream from the promoter that is transcribed into an RNA molecule is called a **transcription unit**.

Bacteria have a single type of RNA polymerase that synthesizes not only mRNA but also other types of RNA that function in protein synthesis, such as ribosomal RNA. In contrast, eukaryotes have at least three types of RNA polymerase in their nuclei; the one used for pre-mRNA synthesis is called RNA polymerase II. In the discussion that follows, we start with the features of mRNA synthesis common to both bacteria and eukaryotes and then describe some key differences.

Synthesis of an RNA Transcript

The three stages of transcription, as shown in Figure 8 and described next, are initiation, elongation, and termination of the RNA chain. Study Figure 8 to familiarize yourself with the stages and the terms used to describe them.

RNA Polymerase Binding and Initiation of Transcription

The promoter of a gene includes within it the transcription **start point** (the nucleotide where RNA synthesis actually begins) and typically extends several dozen or more nucleotide pairs upstream from the start point. RNA polymerase binds in a precise location and orientation on the promoter, thereby determining where transcription starts and which of the two strands of the DNA helix is used as the template.

Certain sections of a promoter are especially important for binding RNA polymerase. In bacteria, part of the RNA polymerase itself specifically recognizes and binds to the promoter. In eukaryotes, a collection of proteins called **transcription factors** mediate the binding of RNA polymerase and the initiation of transcription. Only after transcription factors are attached to the promoter does RNA polymerase II bind to it. The whole complex of transcription factors and RNA polymerase II bound to the promoter is called a **transcription initiation complex**. **Figure 9** shows the role of transcription factors and

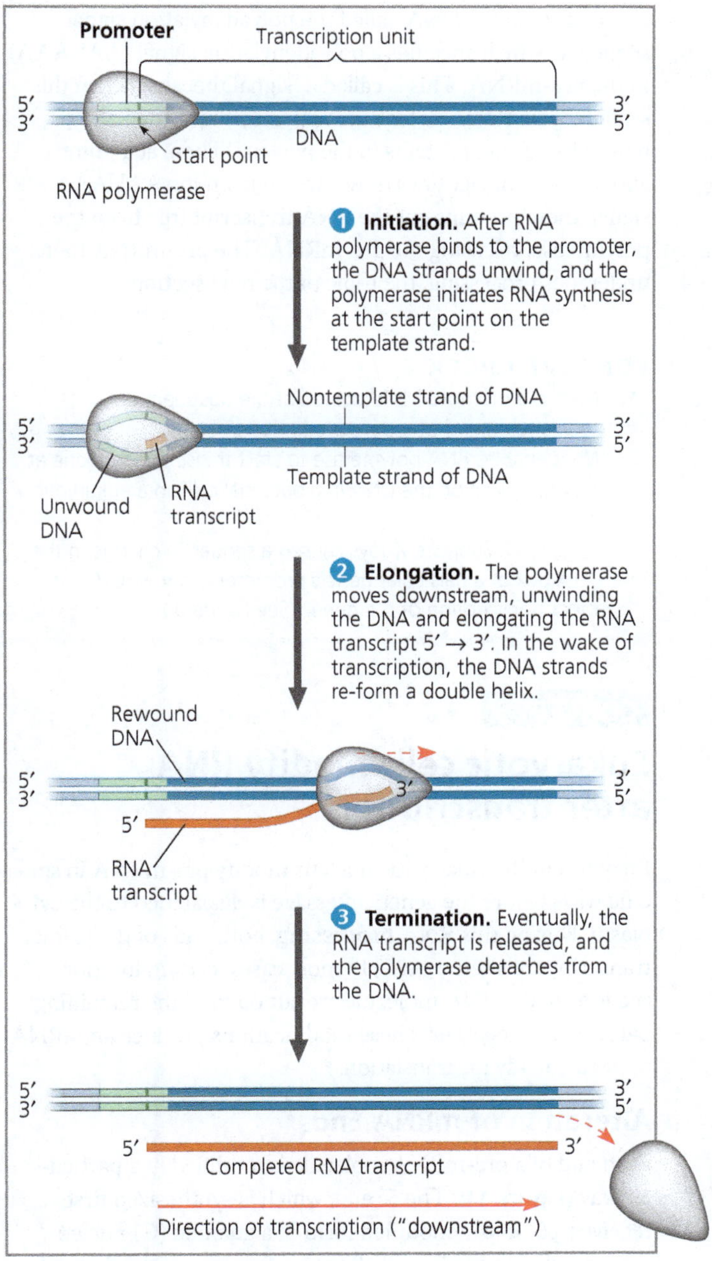

▲ **Figure 8 The stages of transcription: initiation, elongation, and termination.** This general depiction of transcription applies to both bacteria and eukaryotes, but the details of termination differ, as described in the text. Also, in a bacterium, the RNA transcript is immediately usable as mRNA; in a eukaryote, the RNA transcript must first undergo processing.

MAKE CONNECTIONS *Compare the use of a template strand during transcription and replication.*

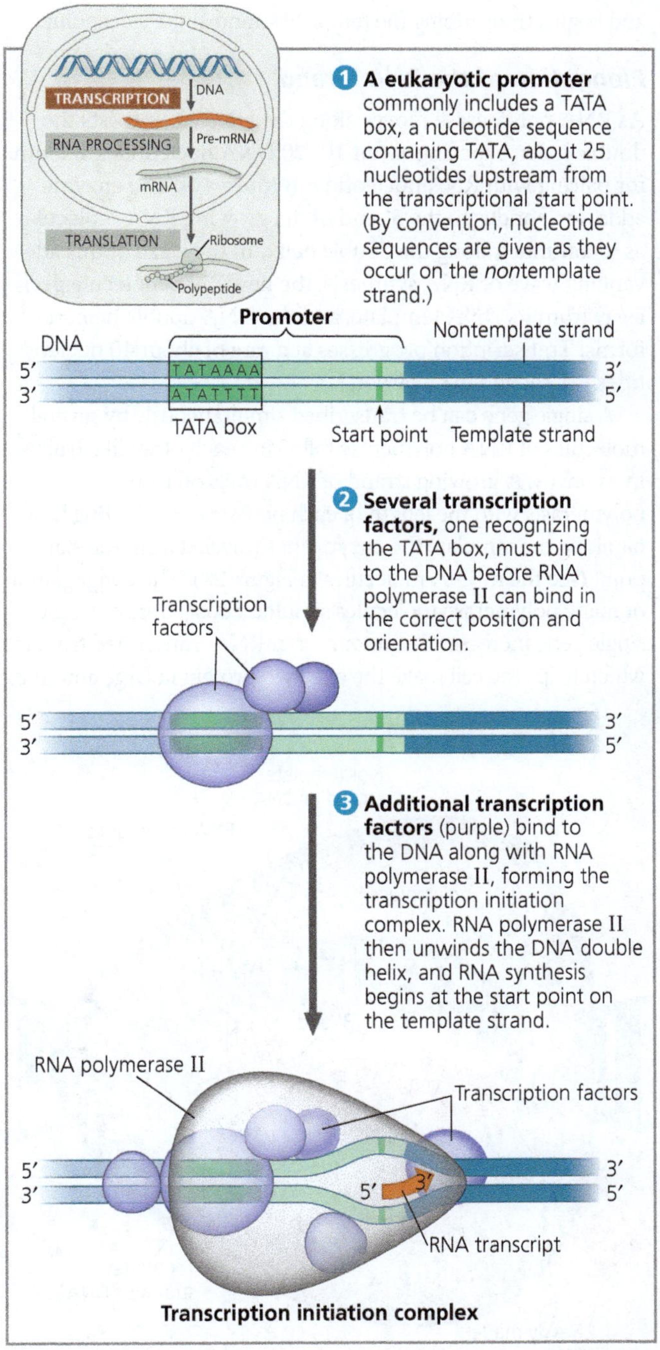

▲ **Figure 9 The initiation of transcription at a eukaryotic promoter.** In eukaryotic cells, proteins called transcription factors mediate the initiation of transcription by RNA polymerase II.

? *Explain how the interaction of RNA polymerase with the promoter would differ if the figure showed transcription initiation for bacteria.*

a crucial promoter DNA sequence called the **TATA box** in forming the initiation complex at a eukaryotic promoter.

The interaction between eukaryotic RNA polymerase II and transcription factors is an example of the importance of protein-protein interactions in controlling eukaryotic transcription. Once the appropriate transcription factors are firmly attached to the promoter DNA and the polymerase is bound in the correct orientation, the enzyme unwinds the two DNA strands and begins transcribing the template strand at the start point.

Elongation of the RNA Strand

As RNA polymerase moves along the DNA, it untwists the double helix, exposing about 10–20 DNA nucleotides at a time for pairing with RNA nucleotides **(Figure 10)**. The enzyme adds nucleotides to the 3′ end of the growing RNA molecule as it continues along the double helix. In the wake of this advancing wave of RNA synthesis, the new RNA molecule peels away from its DNA template, and the DNA double helix re-forms. Transcription progresses at a rate of about 40 nucleotides per second in eukaryotes.

A single gene can be transcribed simultaneously by several molecules of RNA polymerase following each other like trucks in a convoy. A growing strand of RNA trails off from each polymerase, with the length of each new strand reflecting how far along the template the enzyme has traveled from the start point (see the mRNA molecules in Figure 23). The congregation of many polymerase molecules simultaneously transcribing a single gene increases the amount of mRNA transcribed from it, which helps the cell make the encoded protein in large amounts.

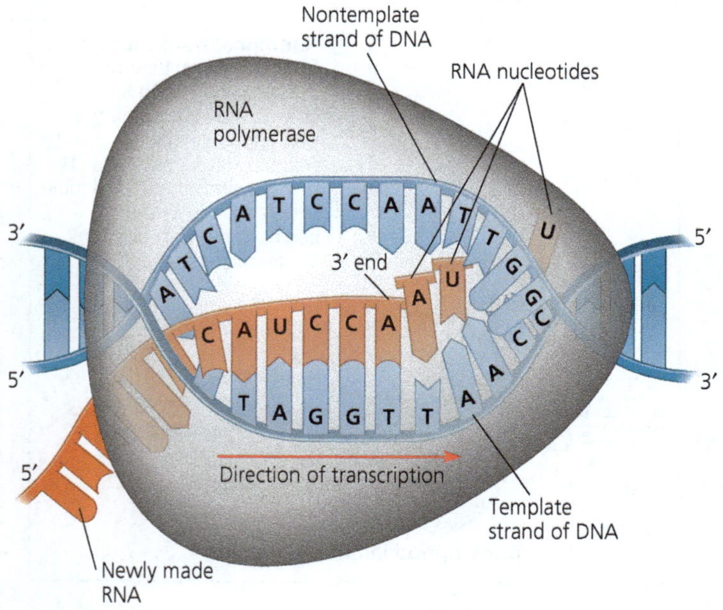

▲ **Figure 10 Transcription elongation.** RNA polymerase moves along the DNA template strand, joining complementary RNA nucleotides to the 3′ end of the growing RNA transcript. Behind the polymerase, the new RNA peels away from the template strand, which re-forms a double helix with the nontemplate strand.

Termination of Transcription

The mechanism of termination differs between bacteria and eukaryotes. In bacteria, transcription proceeds through a terminator sequence in the DNA. The transcribed terminator (an RNA sequence) functions as the termination signal, causing the polymerase to detach from the DNA and release the transcript, which requires no further modification before translation. In eukaryotes, RNA polymerase II transcribes a sequence on the DNA called the polyadenylation signal sequence, which specifies a polyadenylation signal (AAUAAA) in the pre-mRNA. This is called a "signal" because when this stretch of six RNA nucleotides appears, it is immediately bound by certain proteins in the nucleus. Then, at a point about 10–35 nucleotides downstream from the AAUAAA signal, these proteins cut the RNA transcript free from the polymerase, releasing the pre-mRNA. The pre-mRNA then undergoes processing, the topic of the next section.

CONCEPT CHECK 2

1. What is a promoter? Is it located at the upstream or downstream end of a transcription unit?
2. What enables RNA polymerase to start transcribing a gene at the right place on the DNA in a bacterial cell? In a eukaryotic cell?
3. **WHAT IF?** Suppose X-rays caused a sequence change in the TATA box of a particular gene's promoter. How would that affect transcription of the gene? (See Figure 9.)

CONCEPT 3

Eukaryotic cells modify RNA after transcription

Enzymes in the eukaryotic nucleus modify pre-mRNA in specific ways before the genetic message is dispatched to the cytoplasm. During this **RNA processing**, both ends of the primary transcript are altered. Also, in most cases, certain interior sections of the RNA molecule are cut out and the remaining parts spliced together. These modifications produce an mRNA molecule ready for translation.

Alteration of mRNA Ends

Each end of a pre-mRNA molecule is modified in a particular way **(Figure 11)**. The 5′ end, which is synthesized first, receives a **5′ cap**, a modified form of a guanine (G) nucleotide added onto the 5′ end after transcription of the first 20–40 nucleotides. The 3′ end of the pre-mRNA molecule is also modified before the mRNA exits the nucleus. Recall that the pre-mRNA is released soon after the polyadenylation signal, AAUAAA, is transcribed. At the 3′ end, an enzyme then adds 50–250 more adenine (A) nucleotides, forming a **poly-A tail**. The 5′ cap and poly-A tail share several important functions. First, they seem to facilitate the export of the mature

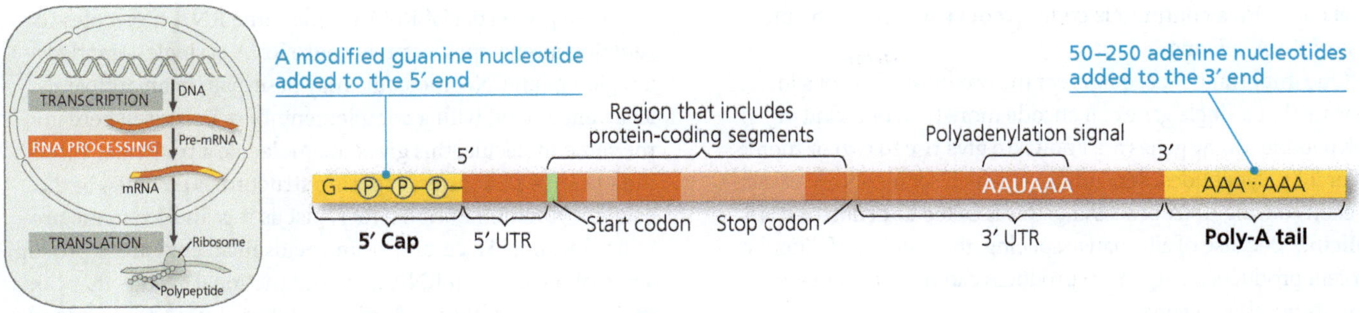

▲ **Figure 11 RNA processing: Addition of the 5′ cap and poly-A tail.** Enzymes modify the two ends of a eukaryotic pre-mRNA molecule. The modified ends may promote the export of mRNA from the nucleus, and they help protect the mRNA from degradation. When the mRNA reaches the cytoplasm, the modified ends, in conjunction with certain cytoplasmic proteins, facilitate ribosome attachment. The 5′ cap and poly-A tail are not translated into protein, nor are the regions called the 5′ untranslated region (5′ UTR) and 3′ untranslated region (3′ UTR). The pink segments will be described shortly.

mRNA from the nucleus. Second, they help protect the mRNA from degradation by hydrolytic enzymes. And third, they help ribosomes attach to the 5′ end of the mRNA once the mRNA reaches the cytoplasm. Figure 11 shows a diagram of a eukaryotic mRNA molecule with cap and tail. The figure also shows the untranslated regions (UTRs) at the 5′ and 3′ ends of the mRNA (referred to as the 5′ UTR and 3′ UTR). The UTRs are parts of the mRNA that will not be translated into protein, but they have other functions, such as ribosome binding.

Split Genes and RNA Splicing

A remarkable stage of RNA processing in the eukaryotic nucleus is the removal of large portions of the RNA molecule and reconnection of the remaining portions. This cut-and-paste job, called **RNA splicing**, is similar to editing a video. The average length of a transcription unit along a human DNA molecule is about 27,000 nucleotide pairs, so the primary RNA transcript is also that long. However, the average-sized protein of 400 amino acids requires only 1,200 nucleotides in RNA to code for it. (Remember, each amino acid is encoded by a *triplet* of nucleotides.) This is because most eukaryotic genes and their RNA transcripts have long noncoding stretches of nucleotides, regions that are not translated. Even more surprising is that most of these noncoding sequences are interspersed between coding segments of the gene and thus between coding segments of the pre-mRNA. In other words, the sequence of DNA nucleotides that codes for a eukaryotic polypeptide is usually not continuous; it is split into segments. The noncoding segments of nucleic acid that lie between coding regions are called *in*tervening sequences, or **introns**. The other regions are called **exons**, because they are eventually *ex*pressed, usually by being translated into amino acid sequences. (Exceptions include the UTRs of the exons at the ends of the RNA, which make up part of the mRNA but are not translated into protein. Because of these exceptions, you may prefer to think of exons as sequences of RNA that *exit* the nucleus.) The terms *intron* and *exon* are used for both RNA sequences and the DNA sequences that specify them.

In making a primary transcript from a gene, RNA polymerase II transcribes both introns and exons from the DNA, but the mRNA molecule that enters the cytoplasm is an abridged version **(Figure 12)**. The introns are cut out from the molecule and the exons joined together, forming an mRNA

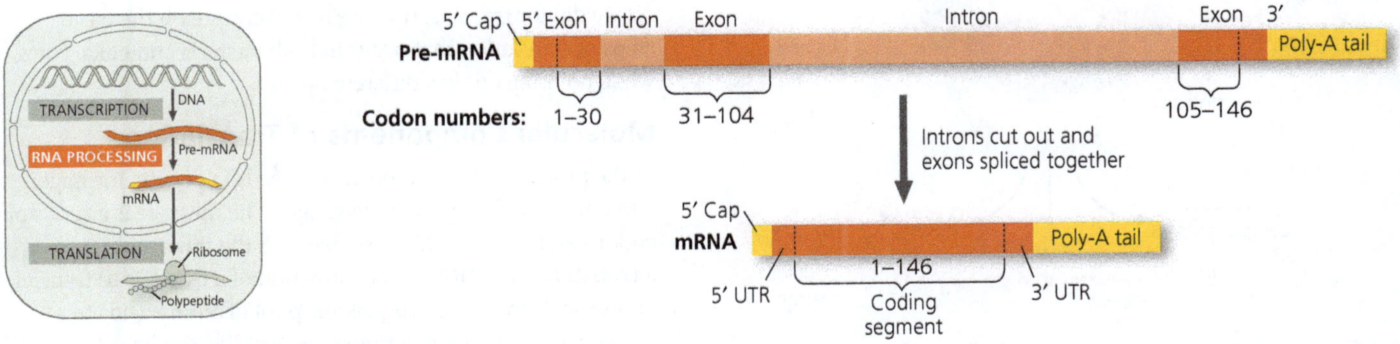

▲ **Figure 12 RNA processing: RNA splicing.** The RNA molecule shown here codes for β-globin, one of the polypeptides of hemoglobin. The numbers under the RNA refer to codons; β-globin is 146 amino acids long. The β-globin gene and its pre-mRNA transcript have three exons, corresponding to sequences that will leave the nucleus as mRNA. (The 5′ UTR and 3′ UTR are parts of exons because they are included in the mRNA; however, they do not code for protein.) During RNA processing, the introns are cut out and the exons spliced together. In many genes, the introns are much larger than the exons.

? *On the mRNA, indicate the sites of the start and stop codons.*

molecule with a continuous coding sequence. This is the process of RNA splicing.

One important consequence of the presence of introns in genes is that a single gene can encode more than one kind of polypeptide. Many genes are known to give rise to two or more different polypeptides, depending on which segments are treated as exons during RNA processing; this is called **alternative RNA splicing**. Because of alternative splicing, the number of different protein products an organism produces can be much greater than its number of genes.

How is pre-mRNA splicing carried out? The removal of introns is accomplished by a large complex made of proteins and small RNAs called a **spliceosome**. This complex binds to several short nucleotide sequences along the intron, including key sequences at each end **(Figure 13)**. The intron is then released (and rapidly degraded), and the spliceosome joins together the two exons that flanked the intron. It turns out that the small RNAs in the spliceosome not only participate in the assembly of the spliceosome and recognition of the splice site but also catalyze the splicing process.

Ribozymes

The idea of a catalytic role for the RNAs in the spliceosome arose from the discovery of **ribozymes**, RNA molecules that function as enzymes. In some organisms, RNA splicing can occur without proteins or even additional RNA molecules: The intron RNA functions as a ribozyme and catalyzes its own excision! For example, in the ciliate protist *Tetrahymena*, self-splicing occurs in the production of ribosomal RNA (rRNA), a component of the organism's ribosomes. The pre-rRNA actually removes its own introns. The discovery of ribozymes rendered obsolete the idea that all biological catalysts are proteins.

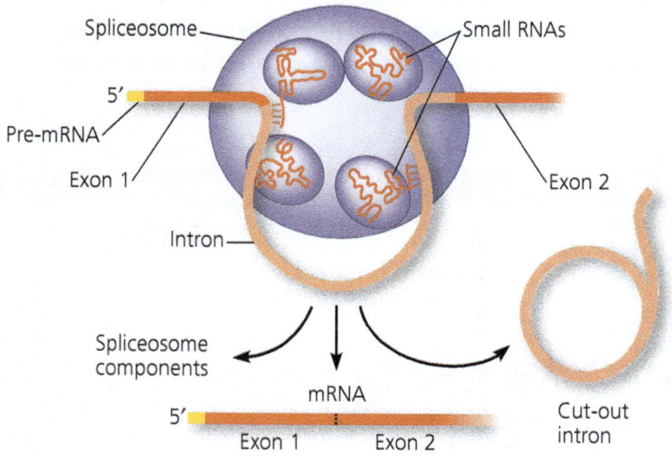

▲ **Figure 13 A spliceosome splicing a pre-mRNA.** The diagram shows a portion of a pre-mRNA transcript, with an intron (pink) flanked by two exons (red). Small RNAs within the spliceosome base-pair with nucleotides at specific sites along the intron. Next, small spliceosome RNAs also catalyze cutting of the pre-mRNA and the splicing together of the exons, releasing the intron for rapid degradation. (These RNAs are *ribozymes*, discussed in the next section.)

Three properties of RNA enable some RNA molecules to function as enzymes. First, because RNA is single-stranded, a region of an RNA molecule may base-pair, in an antiparallel arrangement, with a complementary region elsewhere in the same molecule; this gives the molecule a particular three-dimensional structure. A specific structure is essential to the catalytic function of ribozymes, just as it is for enzymatic proteins. Second, like certain amino acids in an enzymatic protein, some of the bases in RNA contain functional groups that can participate in catalysis. Third, the ability of RNA to hydrogen-bond with other nucleic acid molecules (either RNA or DNA) adds specificity to its catalytic activity. For example, complementary base pairing between the RNA of the spliceosome and the RNA of a primary RNA transcript precisely locates the region where the ribozyme catalyzes splicing. Later in this chapter, you'll see how these properties of RNA also allow it to perform important noncatalytic roles in the cell, such as recognition of the three-nucleotide codons on mRNA.

CONCEPT CHECK 3

1. Given that there are about 20,000 human genes, how can human cells make 75,000–100,000 different proteins?
2. How is RNA splicing similar to how you would watch a TV show recorded earlier using a DVR? In what ways is it different? What would introns correspond to in this analogy?
3. **WHAT IF?** What would be the effect of treating cells with an agent that removed the cap from mRNAs?

CONCEPT 4

Translation is the RNA-directed synthesis of a polypeptide: *a closer look*

We will now examine in greater detail how genetic information flows from mRNA to protein—the process of translation. As we did for transcription, we'll concentrate on the basic steps of translation that occur in both bacteria and eukaryotes, while pointing out key differences.

Molecular Components of Translation

In the process of translation, a cell "reads" a genetic message and builds a polypeptide accordingly. The message is a series of codons along an mRNA molecule, and the translator is called a **transfer RNA (tRNA)**. The function of a tRNA is to transfer amino acids from the cytoplasmic pool of amino acids to a growing polypeptide in a ribosome. A cell keeps its cytoplasm stocked with all 20 amino acids, either by synthesizing them from other compounds or by taking them up from the surrounding solution. The ribosome, a structure made of proteins and RNAs, adds each amino acid brought to it by a tRNA to the growing end of a polypeptide chain **(Figure 14)**.

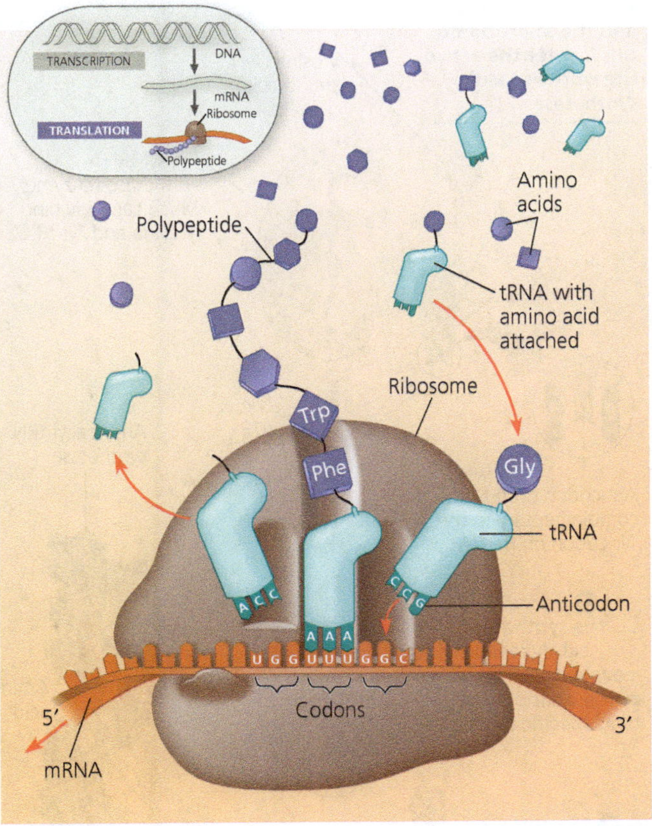

▲ **Figure 14 Translation: the basic concept.** As a molecule of mRNA is moved through a ribosome, codons are translated into amino acids, one by one. The translators, or interpreters, are tRNA molecules, each type with a specific nucleotide triplet called an anticodon at one end and a corresponding amino acid at the other end. A tRNA adds its amino acid cargo to a growing polypeptide chain after the anticodon hydrogen-bonds to the complementary codon on the mRNA. The figures that follow show some of the details of translation in a bacterial cell.

Visit the Study Area in **MasteringBiology** for the BioFlix® 3-D Animation on Protein Synthesis.

Translation is simple in principle but complex in its biochemistry and mechanics. In dissecting translation, we'll focus on the slightly less complicated version of the process that occurs in bacteria. We'll first look at the major players in this cellular process.

The Structure and Function of Transfer RNA

The key to translating a genetic message into a specific amino acid sequence is the fact that each tRNA enables translation of a given mRNA codon into a certain amino acid. This is possible because a tRNA bears a specific amino acid at one end of its three-dimensional structure, while at the other end is a nucleotide triplet that can base-pair with the complementary codon on mRNA.

A tRNA molecule consists of a single RNA strand that is only about 80 nucleotides long (compared to hundreds of

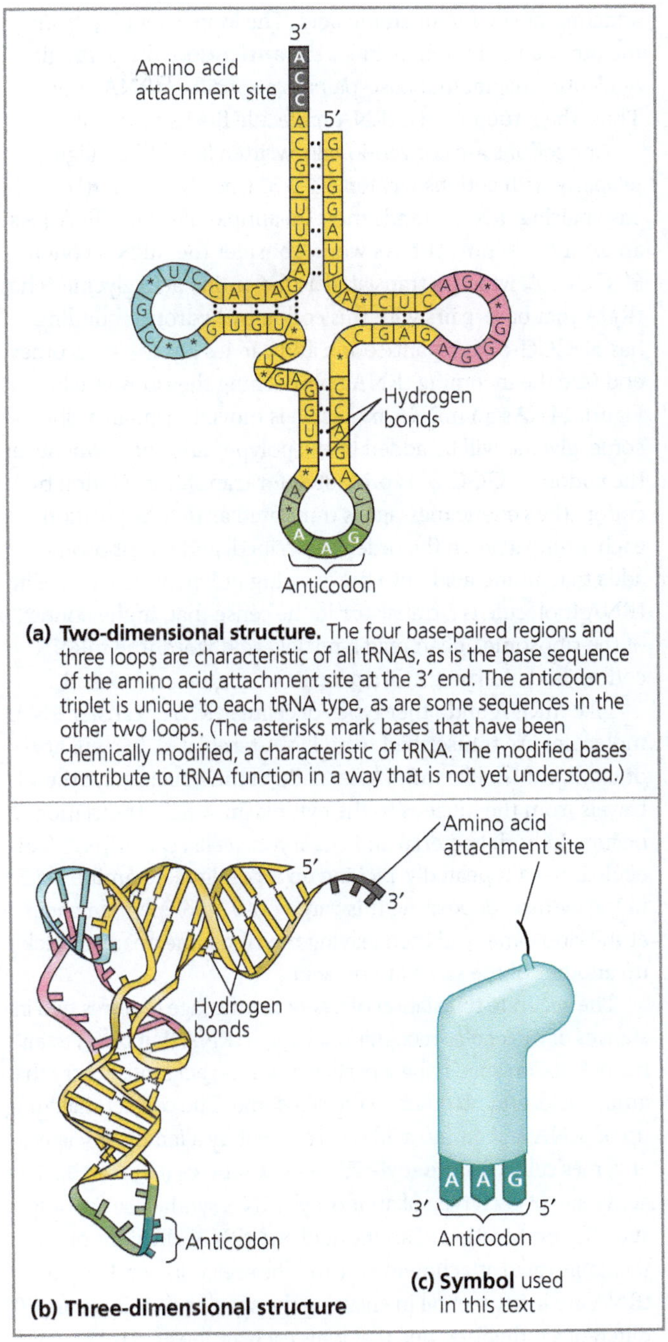

(a) Two-dimensional structure. The four base-paired regions and three loops are characteristic of all tRNAs, as is the base sequence of the amino acid attachment site at the 3′ end. The anticodon triplet is unique to each tRNA type, as are some sequences in the other two loops. (The asterisks mark bases that have been chemically modified, a characteristic of tRNA. The modified bases contribute to tRNA function in a way that is not yet understood.)

(b) Three-dimensional structure

(c) Symbol used in this text

▲ **Figure 15 The structure of transfer RNA (tRNA).**

nucleotides for most mRNA molecules). Because of the presence of complementary stretches of nucleotide bases that can hydrogen-bond to each other, this single strand can fold back on itself and form a molecule with a three-dimensional structure. Flattened into one plane to clarify this base pairing, a tRNA molecule looks like a cloverleaf **(Figure 15a)**. The tRNA actually twists and folds into a compact three-dimensional structure that is roughly L-shaped, with the 5′ and 3′ ends of the linear tRNA both located near one end of the structure **(Figure 15b)**. The protruding 3′ end acts as the attachment site for an amino acid. The loop extending from the other end of the L includes the **anticodon**, the particular nucleotide triplet that base-pairs to a specific mRNA codon. Thus, the structure of a tRNA molecule fits its function.

Anticodons are conventionally written $3' \rightarrow 5'$ to align properly with codons written $5' \rightarrow 3'$ (see Figure 14). (For base pairing, RNA strands must be antiparallel, like DNA.) As an example of how tRNAs work, consider the mRNA codon 5′-GGC-3′, which is translated as the amino acid glycine. The tRNA that base-pairs with this codon by hydrogen bonding has 3′-CCG-5′ as its anticodon and carries glycine at its other end (see the incoming tRNA approaching the ribosome in Figure 14). As an mRNA molecule is moved through a ribosome, glycine will be added to the polypeptide chain whenever the codon 5′-GGC-3′ is presented for translation. Codon by codon, the genetic message is translated as tRNAs position each amino acid, in the order prescribed, and the ribosome adds that amino acid onto the growing polypeptide chain. The tRNA molecule is a translator in the sense that, in the context of the ribosome, it can read a nucleic acid word (the mRNA codon) and interpret it as a protein word (the amino acid).

Like mRNA and other types of cellular RNA, transfer RNA molecules are transcribed from DNA templates. In a eukaryotic cell, tRNA, like mRNA, is made in the nucleus and then travels from the nucleus to the cytoplasm, where translation occurs. In both bacterial and eukaryotic cells, each tRNA molecule is used repeatedly, picking up its designated amino acid in the cytosol, depositing this cargo onto a polypeptide chain at the ribosome, and then leaving the ribosome, ready to pick up another of the same amino acid.

The accurate translation of a genetic message requires two instances of molecular recognition. First, a tRNA that binds to an mRNA codon specifying a particular amino acid must carry that amino acid, and no other, to the ribosome. The correct matching up of tRNA and amino acid is carried out by a family of related enzymes called **aminoacyl-tRNA synthetases** (**Figure 16**). The active site of each type of aminoacyl-tRNA synthetase fits only a specific combination of amino acid and tRNA. (Regions of both the amino acid attachment end and the anticodon end of the tRNA are instrumental in ensuring the specific fit.) There are 20 different synthetases, one that joins each amino acid to the right tRNA; each synthetase is able to bind to all the different tRNAs that code for its particular amino acid. The synthetase catalyzes the covalent attachment of the amino acid to its tRNA in a process driven by the hydrolysis of ATP. The resulting aminoacyl tRNA, also called a charged tRNA, is released from the enzyme and is then available to deliver its amino acid to a growing polypeptide chain on a ribosome.

The second instance of molecular recognition is the pairing of the tRNA anticodon with the appropriate mRNA codon. If one tRNA variety existed for each mRNA codon specifying an amino acid, there would be 61 tRNAs (see Figure 6). In fact, there are only about 45, signifying that some tRNAs must be able to bind to more than one codon. Such versatility is possible because the rules for base pairing between the third nucleotide base of a codon and the corresponding base of a tRNA anticodon are relaxed compared to those at other codon positions. For example, the nucleotide base U at the 5′ end of a tRNA anticodon can pair with A or G in the third position (at the 3′ end) of an mRNA codon. The flexible base pairing at this codon position is called **wobble**. Wobble explains why synonymous codons for an amino acid most often differ in their third nucleotide base but not in the other bases. One such case is that a tRNA with the anticodon 3′-UCU-5′ can base-pair with either the mRNA codon 5′-AGA-3′ or 5′-AGG-3′, both of which code for arginine (see Figure 6).

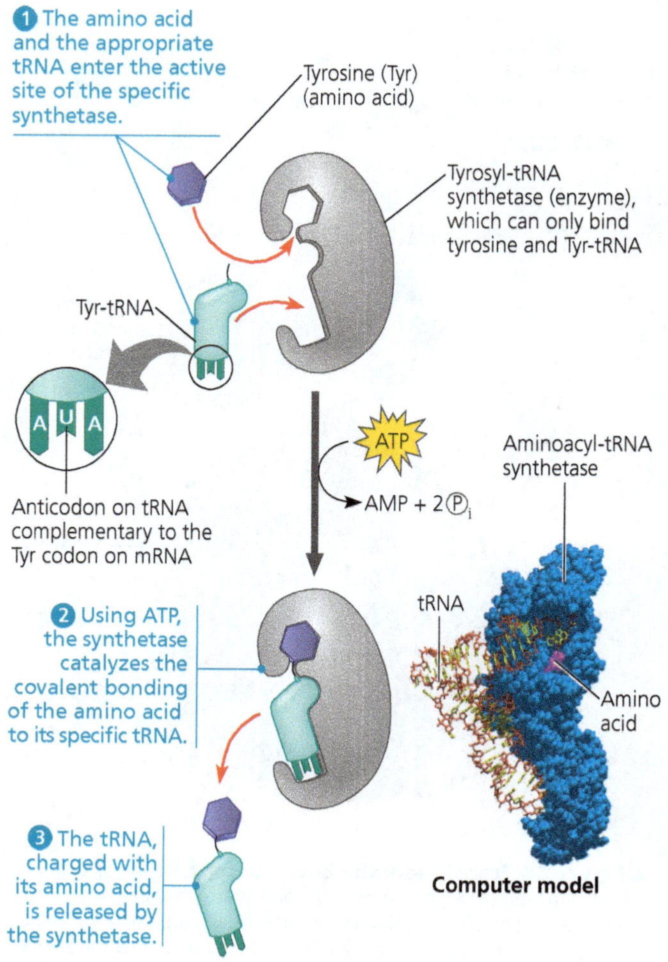

▲ **Figure 16 Aminoacyl-tRNA synthetases provide specificity in joining amino acids to their tRNAs.** Linkage of a tRNA to its amino acid is an endergonic process that occurs at the expense of ATP (which loses two phosphate groups, becoming AMP).

Ribosomes

Ribosomes facilitate the specific coupling of tRNA anticodons with mRNA codons during protein synthesis. A ribosome consists of a large subunit and a small subunit, each made up of proteins and one or more **ribosomal RNAs (rRNAs)** (**Figure 17**). In eukaryotes, the subunits are made in the nucleolus. Ribosomal RNA genes are transcribed, and the RNA is

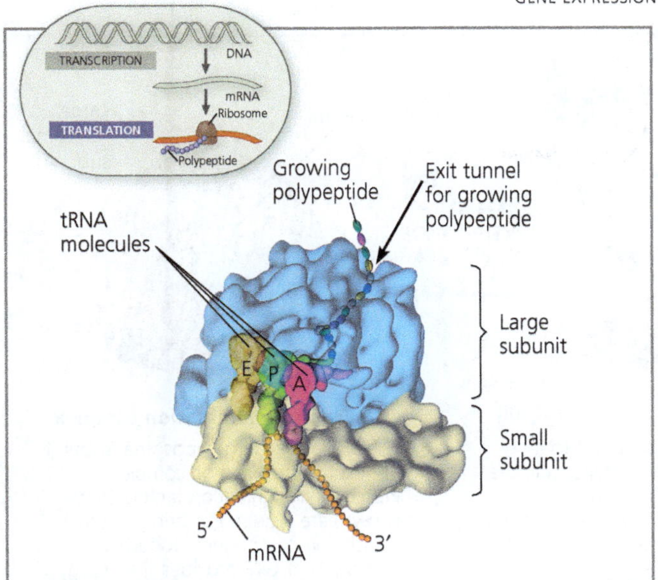

(a) Computer model of functioning ribosome. This is a model of a bacterial ribosome, showing its overall shape. The eukaryotic ribosome is roughly similar. A ribosomal subunit is a complex of ribosomal RNA molecules and proteins.

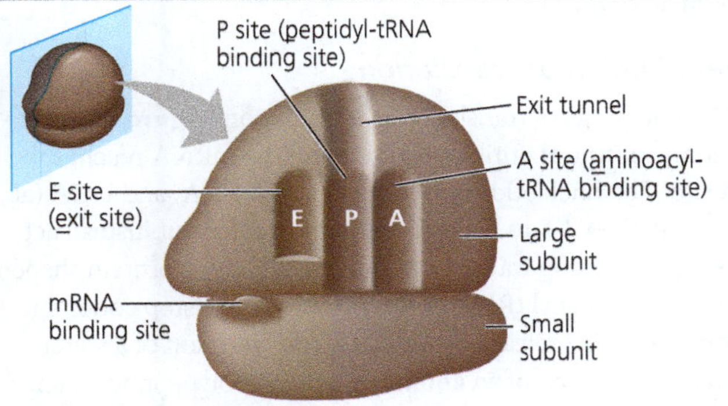

(b) Schematic model showing binding sites. A ribosome has an mRNA binding site and three tRNA binding sites, known as the A, P, and E sites. This schematic ribosome will appear in later diagrams.

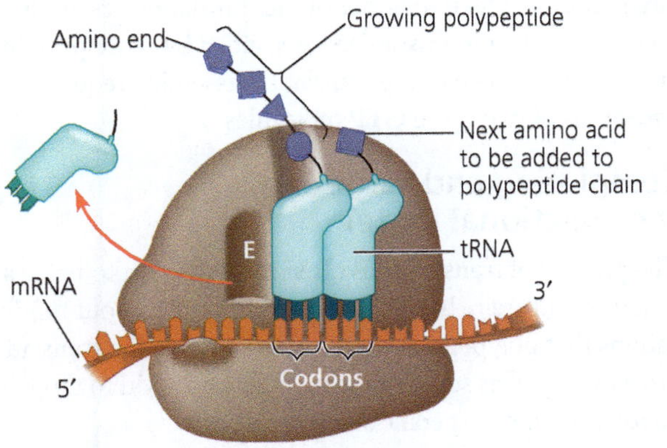

(c) Schematic model with mRNA and tRNA. A tRNA fits into a binding site when its anticodon base-pairs with an mRNA codon. The P site holds the tRNA attached to the growing polypeptide. The A site holds the tRNA carrying the next amino acid to be added to the polypeptide chain. Discharged tRNAs leave from the E site. The polypeptide grows at its carboxyl end.

▲ **Figure 17 The anatomy of a functioning ribosome.**

processed and assembled with proteins imported from the cytoplasm. Completed ribosomal subunits are then exported via nuclear pores to the cytoplasm. In both bacteria and eukaryotes, a large and a small subunit join to form a functional ribosome only when attached to an mRNA molecule. About one-third of the mass of a ribosome is made up of proteins; the rest consists of rRNAs, either three molecules (in bacteria) or four (in eukaryotes). Because most cells contain thousands of ribosomes, rRNA is the most abundant type of cellular RNA.

Although the ribosomes of bacteria and eukaryotes are very similar in structure and function, eukaryotic ribosomes are slightly larger, as well as differing somewhat from bacterial ribosomes in their molecular composition. The differences are medically significant. Certain antibiotic drugs can inactivate bacterial ribosomes without affecting eukaryotic ribosomes to make proteins. These drugs, including tetracycline and streptomycin, are used to combat bacterial infections.

The structure of a ribosome reflects its function of bringing mRNA together with tRNAs carrying amino acids. In addition to a binding site for mRNA, each ribosome has three binding sites for tRNA, as described in Figure 17. The **P site** (peptidyl-tRNA binding site) holds the tRNA carrying the growing polypeptide chain, while the **A site** (aminoacyl-tRNA binding site) holds the tRNA carrying the next amino acid to be added to the chain. Discharged tRNAs leave the ribosome from the **E site** (exit site). The ribosome holds the tRNA and mRNA in close proximity and positions the new amino acid so that it can be added to the carboxyl end of the growing polypeptide. It then catalyzes the formation of the peptide bond. As the polypeptide becomes longer, it passes through an *exit tunnel* in the ribosome's large subunit. When the polypeptide is complete, it is released through the exit tunnel.

The widely accepted model is that rRNA, not the proteins, is primarily responsible for both the structure and the function of the ribosome. The proteins, which are largely on the exterior, support the shape changes of the rRNA molecules as they carry out catalysis during translation. Ribosomal RNA is the main constituent of the A and P sites and of the interface between the two ribosomal subunits; it also acts as the catalyst of peptide bond formation. Thus, a ribosome could actually be considered one colossal ribozyme!

Building a Polypeptide

We can divide translation, the synthesis of a polypeptide chain, into three stages: initiation, elongation, and termination. All three stages require protein "factors" that aid in the translation process. For certain aspects of chain initiation and elongation, energy is also required. It is provided by the hydrolysis of guanosine triphosphate (GTP).

Ribosome Association and Initiation of Translation

The initiation stage of translation brings together mRNA, a tRNA bearing the first amino acid of the polypeptide, and the two subunits of a ribosome. First, a small ribosomal subunit binds to both mRNA and a specific initiator tRNA, which carries the amino acid methionine. In bacteria, the small subunit can bind these two in either order; it binds the mRNA at a specific RNA sequence, just upstream of the start codon, AUG. In eukaryotes, the small subunit, with the initiator tRNA already bound, binds to the 5′ cap of the mRNA and then moves, or *scans*, downstream along the mRNA until it reaches the AUG start codon, where the initiator tRNA hydrogen-bonds, as shown in step 1 of **Figure 18**. In either case, the start codon signals the start of translation; this is important because it establishes the codon reading frame for the mRNA. In the **Scientific Skills Exercise**, you can work with DNA sequences encoding the ribosomal binding sites on the mRNAs of a group of *Escherichia coli* (*E. coli*) genes.

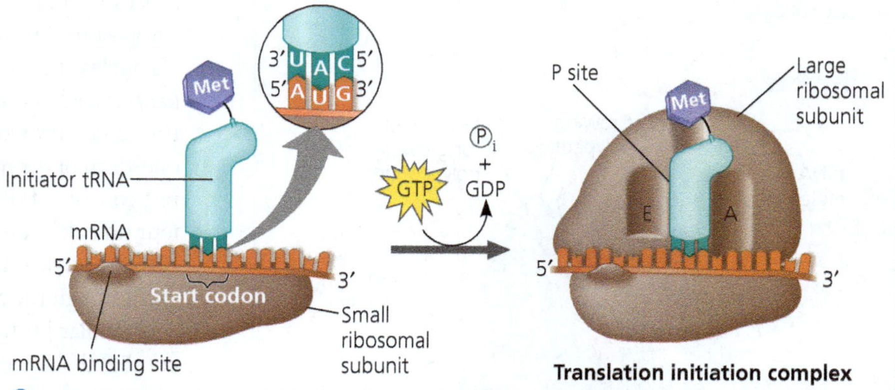

1 A small ribosomal subunit binds to a molecule of mRNA. In a bacterial cell, the mRNA binding site on this subunit recognizes a specific nucleotide sequence on the mRNA just upstream of the start codon. An initiator tRNA, with the anticodon UAC, base-pairs with the start codon, AUG. This tRNA carries the amino acid methionine (Met).

2 The arrival of a large ribosomal subunit completes the initiation complex. Proteins called initiation factors (not shown) are required to bring all the translation components together. Hydrolysis of GTP provides the energy for the assembly. The initiator tRNA is in the P site; the A site is available to the tRNA bearing the next amino acid.

▲ **Figure 18 The initiation of translation.**

The union of mRNA, initiator tRNA, and a small ribosomal subunit is followed by the attachment of a large ribosomal subunit, completing the *translation initiation complex*. Proteins called *initiation factors* are required to bring all these components together. The cell also expends energy obtained by hydrolysis of a GTP molecule to form the initiation complex. At the completion of the initiation process, the initiator tRNA sits in the P site of the ribosome, and the vacant A site is ready for the next aminoacyl tRNA. Note that a polypeptide is always synthesized in one direction, from the initial methionine at the amino end, also called the N-terminus, toward the final amino acid at the carboxyl end, also called the C-terminus.

Elongation of the Polypeptide Chain

In the elongation stage of translation, amino acids are added one by one to the previous amino acid at the C-terminus of the growing chain. Each addition involves the participation of several proteins called *elongation factors* and occurs in a three-step cycle described in **Figure 19**. Energy expenditure occurs in the first and third steps. Codon recognition requires hydrolysis of one molecule of GTP, which increases the accuracy and efficiency of this step. One more GTP is hydrolyzed to provide energy for the translocation step.

The mRNA is moved through the ribosome in one direction only, 5′ end first; this is equivalent to the ribosome moving 5′ → 3′ on the mRNA. The important point is that the ribosome and the mRNA move relative to each other, unidirectionally, codon by codon. The elongation cycle takes less than a tenth of a second in bacteria and is repeated as each amino acid is added to the chain until the polypeptide is completed. The empty tRNAs that are released from the E site return to the cytoplasm, where they will be reloaded with the proper amino acid (see Figure 16).

Termination of Translation

The final stage of translation is termination **(Figure 20)**. Elongation continues until a stop codon in the mRNA reaches the A site. The nucleotide base triplets UAG, UAA, and UGA (all written 5′ → 3′) do not code for amino acids but instead act as signals to stop translation. A *release factor*, a protein shaped like an aminoacyl tRNA, binds directly to the stop codon in the A site. The release factor causes the addition of a water molecule instead of an amino acid to the polypeptide chain. (Water molecules are abundant in the aqueous cellular environment.) This reaction breaks (hydrolyzes) the bond between the completed polypeptide and the tRNA in the P site, releasing the polypeptide through the exit tunnel of the ribosome's large subunit. The remainder of the translation assembly then comes apart in a multistep process, aided by other protein factors. Breakdown of the translation assembly requires the hydrolysis of two more GTP molecules.

Completing and Targeting the Functional Protein

The process of translation is often not sufficient to make a functional protein. In this section, you'll learn about modifications that polypeptide chains undergo after the translation process as well as some of the mechanisms used to target completed proteins to specific sites in the cell.

Protein Folding and Post-Translational Modifications

During its synthesis, a polypeptide chain begins to coil and fold spontaneously as a consequence of its amino acid sequence (primary structure), forming a protein with a specific shape: a three-dimensional molecule with secondary and tertiary

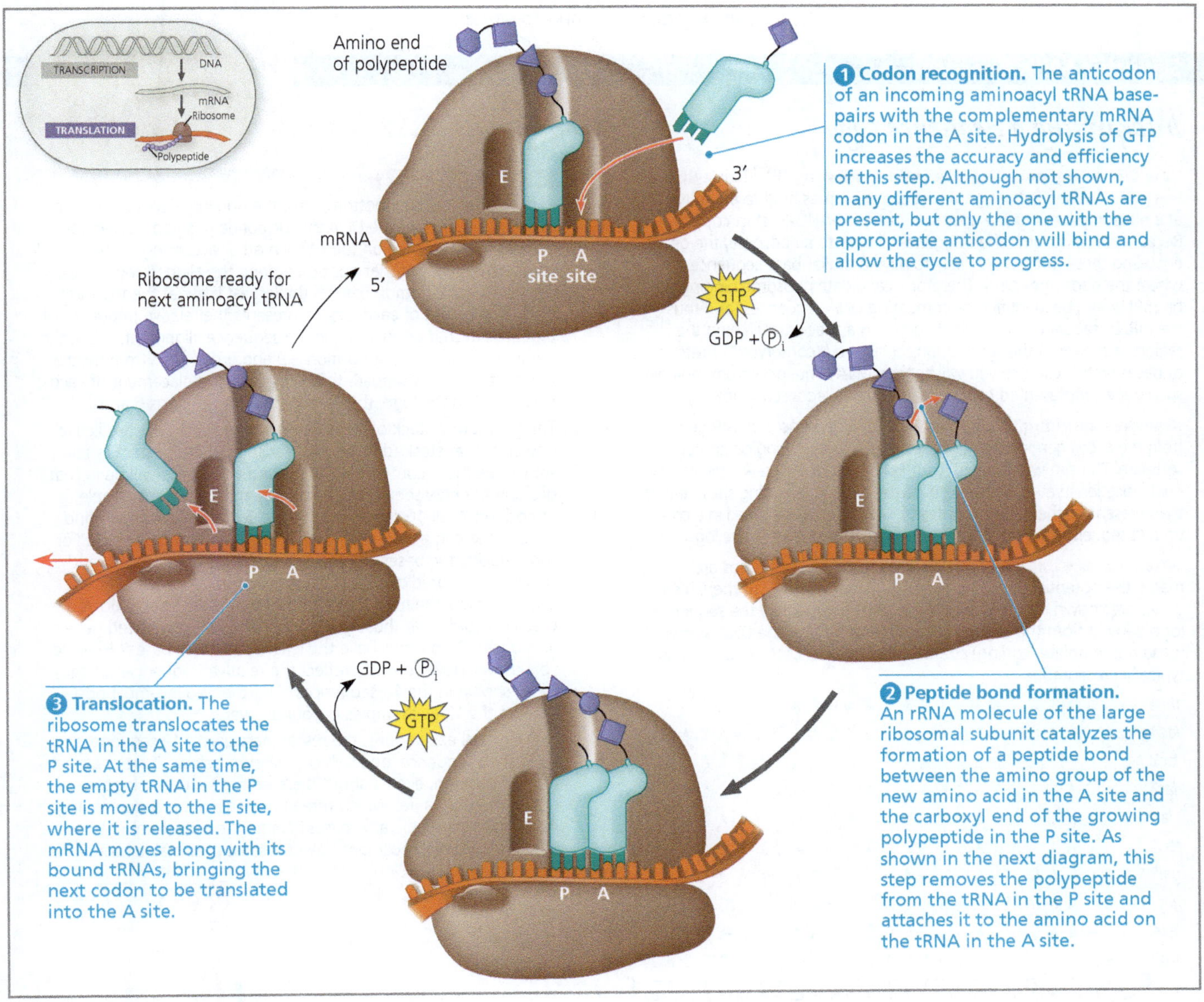

▲ **Figure 19 The elongation cycle of translation.** The hydrolysis of GTP plays an important role in the elongation process. Not shown are the proteins called elongation factors.

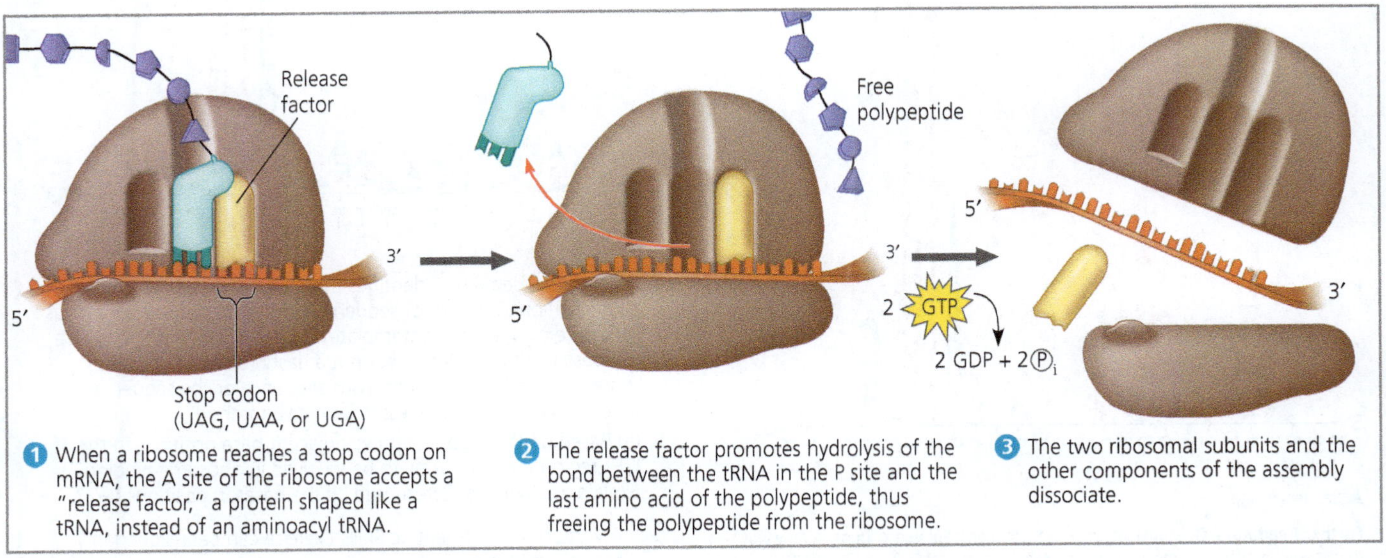

▲ **Figure 20 The termination of translation.** Like elongation, termination requires GTP hydrolysis as well as additional protein factors, which are not shown here.

Scientific Skills Exercise

Interpreting a Sequence Logo

How Can a Sequence Logo Be Used to Identify Ribosome Binding Sites? When initiating translation, ribosomes bind to an mRNA at a ribosome binding site upstream of the 5′-AUG-3′ start codon. Because mRNAs from different genes all bind to a ribosome, the genes encoding these mRNAs are likely to have a similar base sequence where the ribosomes bind. Therefore, candidate ribosome binding sites on mRNA can be identified by comparing DNA sequences (and thus the mRNA sequences) of multiple genes in a species, searching the region upstream of the start codon for shared ("conserved") stretches of bases. In this exercise you will analyze DNA sequences from multiple such genes, represented by a visual graphic called a sequence logo.

How the Experiment Was Done The DNA sequences of 149 genes from the *E. coli* genome were aligned and analyzed using computer software. The aim was to identify similar base sequences—at the appropriate location in each gene—as potential ribosome binding sites. Rather than presenting the data as a series of 149 sequences aligned in a column (a sequence alignment), the researchers used a sequence logo.

Data from the Experiment To show how sequence logos are made, the potential ribosome binding regions from 10 of the *E. coli* genes are shown in a sequence alignment, followed by the sequence logo derived from the aligned sequences. Note that the DNA shown is the nontemplate (coding) strand, which is how DNA sequences are typically presented.

thrA	G G T A A C G A G G T A A C A A C C A T G C G A G T G
lacA	C A T A A C G G A G T G A T C G C A T T G A A C A T G
lacY	C G C G T A A G G A A A T C C A T T A T G T A C T A T
lacZ	T T C A C A C A G G A A A C A G C T A T G A C C A T G
lacI	C A A T T C A G G G T G G T G A A T G T G A A A C C A
recA	G G C A T G A C A G G A G T A A A A A T G G C T A T C
galR	A C C C A C T A A G G T A T T T T C A T G G C G A C C
metJ	A A G A G G A T T A A G T A T C T C A T G G C T G A A
lexA	A T A C A C C C A G G G G G C G G A A T G A A A G C G
trpR	T A A C A A T G G C G A C A T A T T A T G G C C C A A
	5′ −18 −17 −16 −15 −14 −13 −12 −11 −10 −9 −8 −7 −6 −5 −4 −3 −2 −1 0 1 2 3 4 5 6 7 8 3′

▲ Sequence alignment

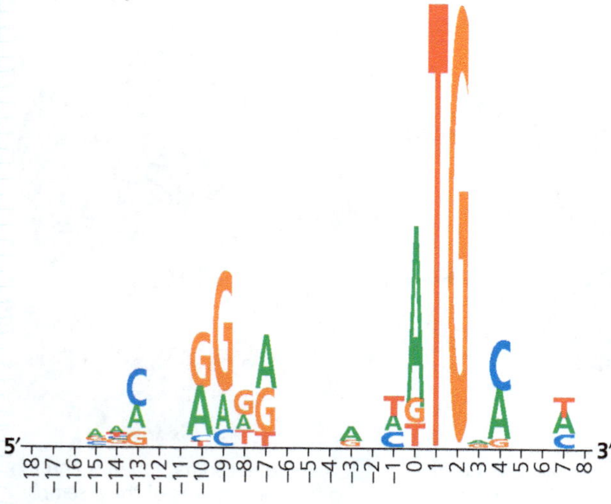

▲ Sequence logo

Further Reading T. D. Schneider and R. M. Stephens, Sequence logos: A new way to display consensus sequences, *Nucleic Acids Research* 18:6097–6100 (1990).

INTERPRET THE DATA

1. In the sequence logo (bottom, left), the horizontal axis shows the primary sequence of the DNA by nucleotide position. Letters for each base are stacked on top of each other according to their relative frequency at that position among the aligned sequences, with the most common base as the largest letter at the top of the stack. The height of each letter represents the relative frequency of that base *at that position*. (a) In the sequence alignment, count the number of each base at position −9 and order them from most to least frequent. Compare this to the size and placement of each base at −9 in the logo. (b) Do the same for positions 0 and 1.
2. The height of a stack of letters in a logo indicates the predictive power of that stack (determined statistically). If the stack is tall, we can be more confident in predicting what base will be in that position if a new sequence is added to the logo. For example, at position 2, all 10 sequences have a G; the probability of finding a G there in a new sequence is very high, as is the stack. For short stacks, the bases all have about the same frequency, and so it's hard to predict a base at those positions. (a) Which two positions have the most predictable bases? What bases do you predict would be at those positions in a newly sequenced gene? (b) Which 12 positions have the least predictable bases? How do you know? How does this reflect the relative frequencies of the bases shown in the 10 sequences? Use the two left-most positions of the 12 as examples in your answer.
3. In the actual experiment, the researchers used 149 sequences to build their sequence logo, which is shown below. There is a stack at each position, even if short, because the sequence logo includes more data. (a) Which three positions in this sequence logo have the most predictable bases? Name the most frequent base at each. (b) Which positions have the least predictable bases? How can you tell?

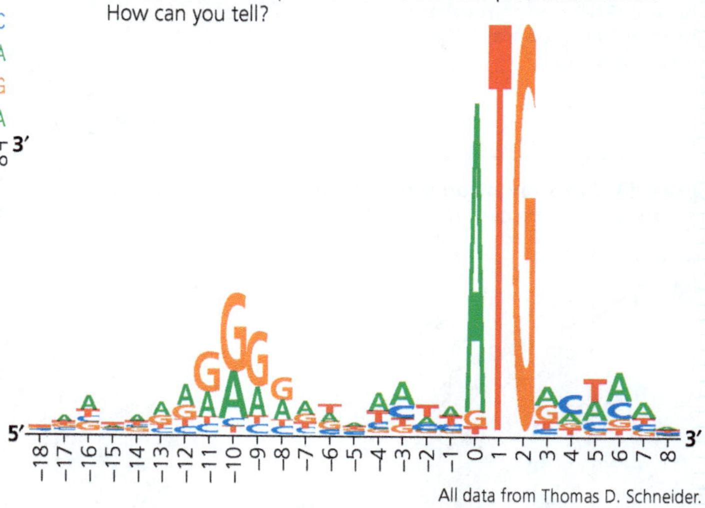

All data from Thomas D. Schneider.

4. A consensus sequence identifies the base occurring most often at each position in the set of sequences. (a) Write out the consensus sequence of this (the nontemplate) strand. In any position where the base can't be determined, put a dash. (b) Which provides more information—the consensus sequence or the sequence logo? What is lost in the less informative method?
5. (a) Based on the logo, what five adjacent base positions in the 5′ UTR region are most likely to be involved in ribosome binding? Explain. (b) What is represented by the bases in positions 0–2?

MB A version of this Scientific Skills Exercise can be assigned in MasteringBiology.

structure. Thus, a gene determines primary structure, and primary structure in turn determines shape.

Additional steps—*post-translational modifications*—may be required before the protein can begin doing its particular job in the cell. Certain amino acids may be chemically modified by the attachment of sugars, lipids, phosphate groups, or other additions. Enzymes may remove one or more amino acids from the leading (amino) end of the polypeptide chain. In some cases, a polypeptide chain may be enzymatically cleaved into two or more pieces. For example, the protein insulin is first synthesized as a single polypeptide chain but becomes active only after an enzyme cuts out a central part of the chain, leaving a protein made up of two shorter polypeptide chains connected by disulfide bridges. In other cases, two or more polypeptides that are synthesized separately may come together, becoming the subunits of a protein that has quaternary structure. A familiar example is hemoglobin.

Targeting Polypeptides to Specific Locations

In electron micrographs of eukaryotic cells active in protein synthesis, two populations of ribosomes are evident: free and bound. Free ribosomes are suspended in the cytosol and mostly synthesize proteins that stay in the cytosol and function there. In contrast, bound ribosomes are attached to the cytosolic side of the endoplasmic reticulum (ER) or to the nuclear envelope. Bound ribosomes make proteins of the endomembrane system (the nuclear envelope, ER, Golgi apparatus, lysosomes, vacuoles, and plasma membrane) as well as proteins secreted from the cell, such as insulin. It is important to note that the ribosomes themselves are identical and can alternate between being free ribosomes one time they are used and being bound ribosomes the next time.

What determines whether a ribosome is free in the cytosol or bound to rough ER? Polypeptide synthesis always begins in the cytosol as a free ribosome starts to translate an mRNA molecule. There the process continues to completion—*unless* the growing polypeptide itself cues the ribosome to attach to the ER. The polypeptides of proteins destined for the endomembrane system or for secretion are marked by a **signal peptide**, which targets the protein to the ER (**Figure 21**). The signal peptide, a sequence of about 20 amino acids at or near the leading end (N-terminus) of the polypeptide, is recognized as it emerges from the ribosome by a protein-RNA complex called a **signal-recognition particle (SRP)**. This particle functions as an escort that brings the ribosome to a receptor protein built into the ER membrane. The receptor is part of a multiprotein translocation complex. Polypeptide synthesis continues there, and the growing polypeptide snakes across the membrane into the ER lumen via a protein pore. The signal peptide is usually removed by an enzyme. The rest of the completed polypeptide, if it is to be secreted from the cell, is released into solution within the ER lumen (as in Figure 21). Alternatively, if the polypeptide is to be a membrane protein, it remains partially embedded in the ER membrane. In either

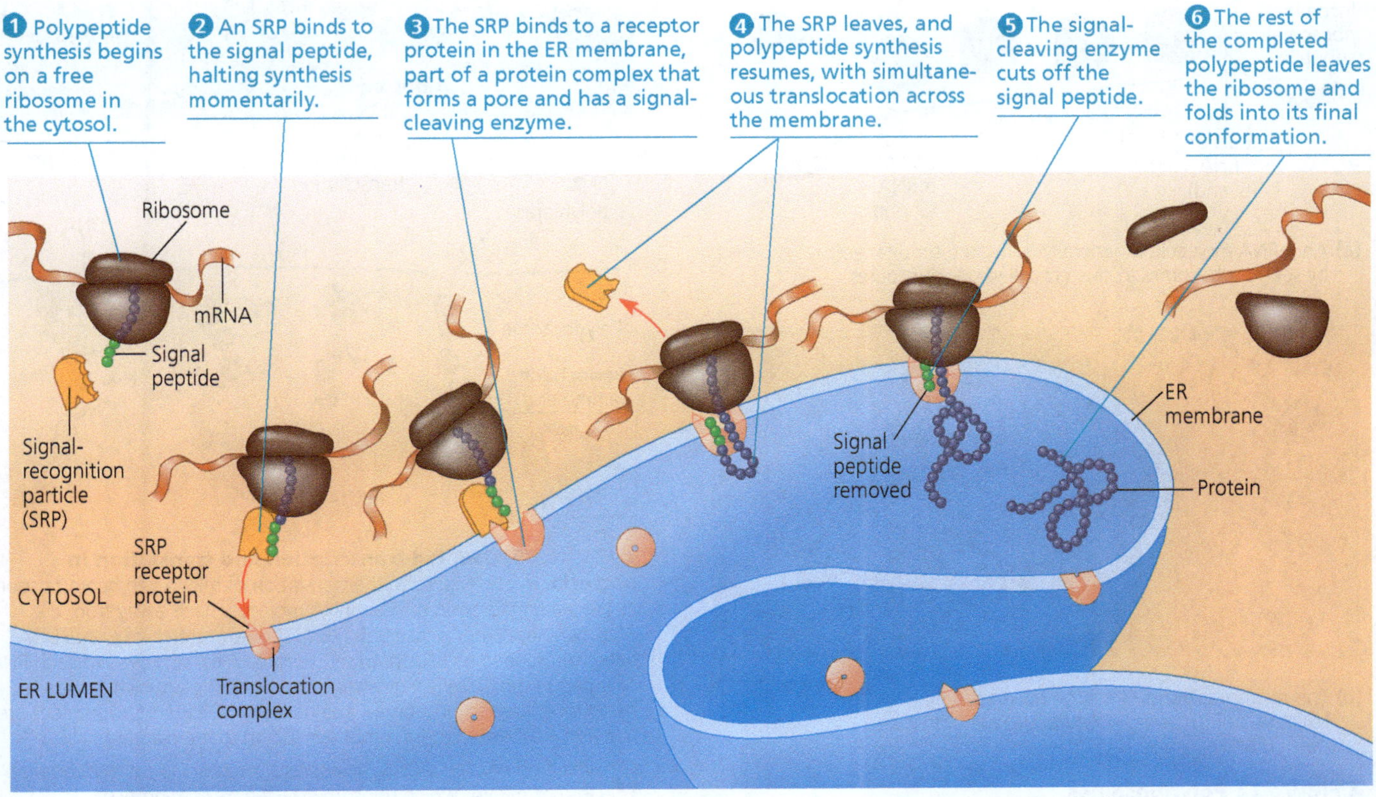

▲ **Figure 21 The signal mechanism for targeting proteins to the ER.**

case, it travels in a transport vesicle to the plasma membrane.

Other kinds of signal peptides are used to target polypeptides to mitochondria, chloroplasts, the interior of the nucleus, and other organelles that are not part of the endomembrane system. The critical difference in these cases is that translation is completed in the cytosol before the polypeptide is imported into the organelle. Translocation mechanisms also vary, but in all cases studied to date, the "postal zip codes" that address proteins for secretion or to cellular locations are signal peptides of some sort. Bacteria also employ signal peptides to target proteins to the plasma membrane for secretion.

Making Multiple Polypeptides in Bacteria and Eukaryotes

In previous sections, you have learned how a single polypeptide is synthesized using the information encoded in an mRNA molecule. When a polypeptide is required in a cell, though, the need is for many copies, not just one.

In both bacteria and eukaryotes, multiple ribosomes translate an mRNA at the same time **(Figure 22)**; that is, a single mRNA is used to make many copies of a polypeptide simultaneously. Once a ribosome is far enough past the start codon, a second ribosome can attach to the mRNA, eventually resulting in a number of ribosomes trailing along the mRNA. Such strings of ribosomes, called polyribosomes (or polysomes), can be seen with an electron microscope (see Figure 22); they can be either free or bound. They enable a cell to make many copies of a polypeptide very quickly.

Another way both bacteria and eukaryotes augment the number of copies of a polypeptide is by transcribing multiple mRNAs from the same gene, as we mentioned earlier. However, the coordination of the two processes—transcription and translation—differs in the two groups. The most important differences between bacteria and eukaryotes arise from the bacterial cell's lack of compartmental organization. Like a one-room workshop, a bacterial cell ensures a streamlined operation by coupling the two processes. In the absence of a nucleus, it can simultaneously transcribe and translate the same gene **(Figure 23)**, and the newly made protein can quickly diffuse to its site of function.

In contrast, the eukaryotic cell's nuclear envelope segregates transcription from translation and provides a compartment for extensive RNA processing. This processing stage includes additional steps whose regulation can help coordinate the eukaryotic cell's elaborate activities. **Figure 24** summarizes the path from gene to polypeptide in a eukaryotic cell.

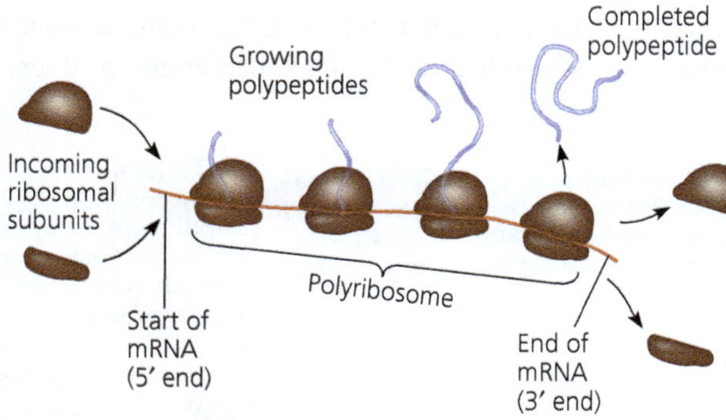

(a) An mRNA molecule is generally translated simultaneously by several ribosomes in clusters called polyribosomes.

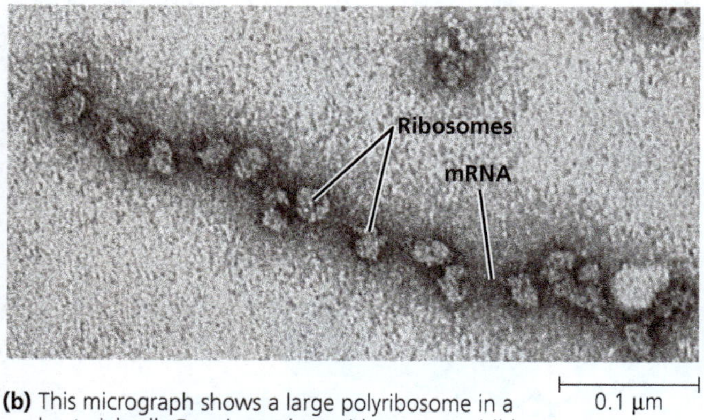

(b) This micrograph shows a large polyribosome in a bacterial cell. Growing polypeptides are not visible here (TEM).

▲ **Figure 22 Polyribosomes.**

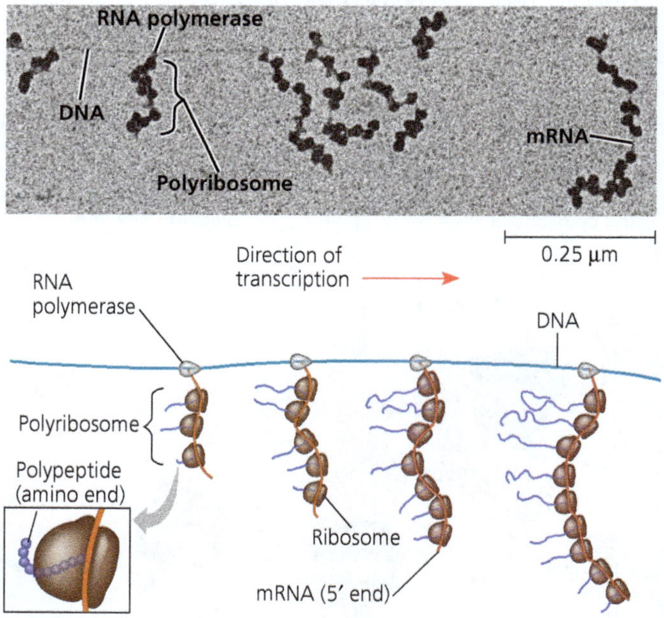

▲ **Figure 23 Coupled transcription and translation in bacteria.** In bacterial cells, the translation of mRNA can begin as soon as the leading (5′) end of the mRNA molecule peels away from the DNA template. The micrograph (TEM) shows a stretch of *E. coli* DNA being transcribed by RNA polymerase molecules. Attached to each RNA polymerase molecule is a growing strand of mRNA, which is already being translated by ribosomes. The newly synthesized polypeptides are not visible in the micrograph but are shown in the diagram.

? *Which one of the mRNA molecules started being transcribed first? On that mRNA, which ribosome started translating first?*

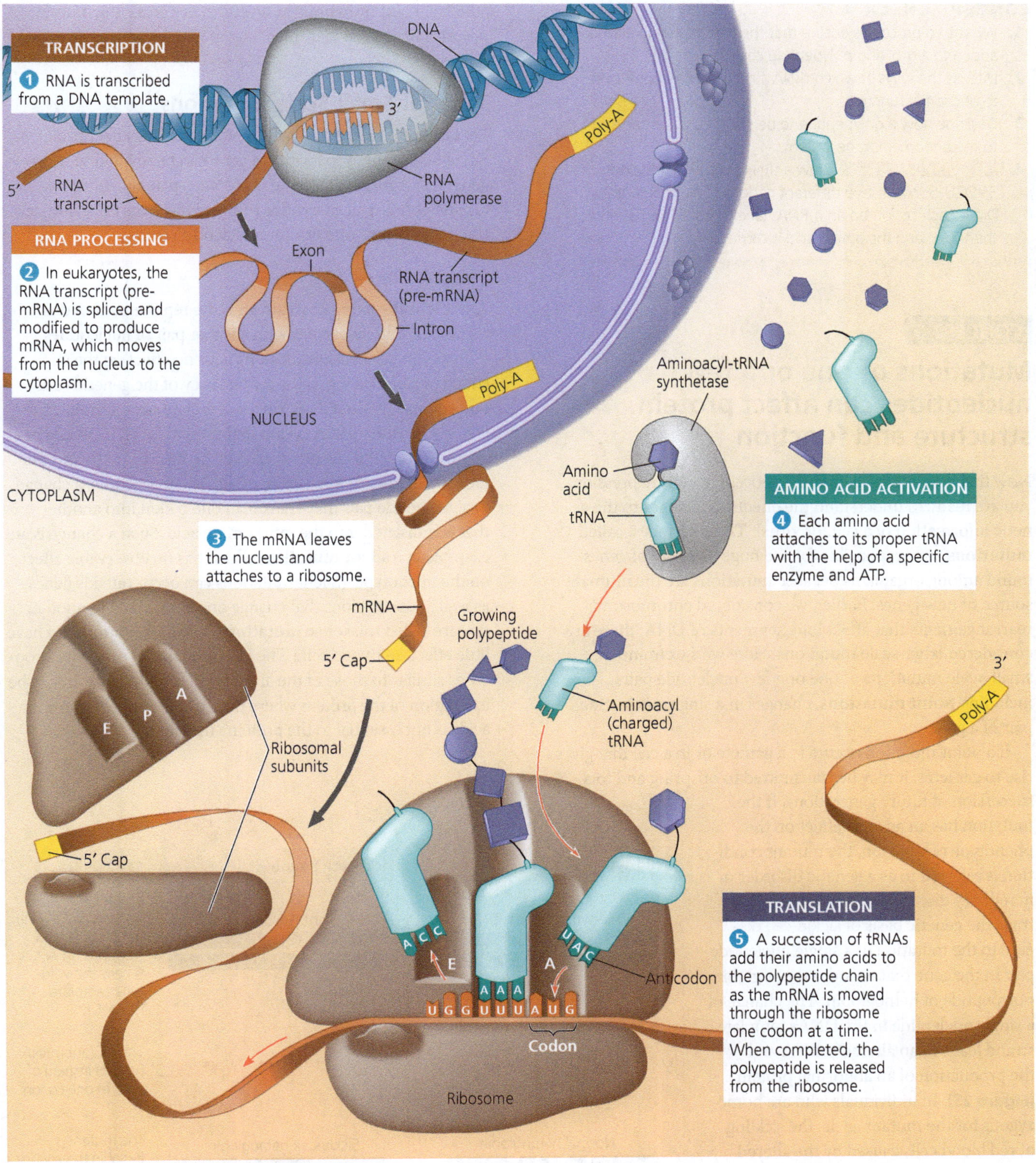

▲ **Figure 24 A summary of transcription and translation in a eukaryotic cell.** This diagram shows the path from one gene to one polypeptide. Keep in mind that each gene in the DNA can be transcribed repeatedly into many identical RNA molecules and that each mRNA can be translated repeatedly to yield many identical polypeptide molecules. (Also, remember that the final products of some genes are not polypeptides but RNA molecules that don't get translated, including tRNA and rRNA.) In general, the steps of transcription and translation are similar in bacterial, archaeal, and eukaryotic cells. The major difference is the occurrence of RNA processing in the eukaryotic nucleus. Other significant differences are found in the initiation stages of both transcription and translation and in the termination of transcription.

CONCEPT CHECK 4

1. What two processes ensure that the correct amino acid is added to a growing polypeptide chain?
2. Discuss the ways in which rRNA structure likely contributes to ribosomal function.
3. Describe how a polypeptide to be secreted is transported to the endomembrane system.
4. WHAT IF? DRAW IT Draw a tRNA with the anticodon 3′-CGU-5′. What two different codons could it bind to? Draw each codon on an mRNA, labeling all 5′ and 3′ ends, the tRNA, and the amino acid it carries.

CONCEPT 5

Mutations of one or a few nucleotides can affect protein structure and function

Now that you have explored the process of gene expression, you are ready to understand the effects of changes to the genetic information of a cell (or virus). These changes, called **mutations**, are responsible for the huge diversity of genes found among organisms because mutations are the ultimate source of new genes. Earlier, we considered chromosomal rearrangements that affect long segments of DNA; these are considered large-scale mutations. Here we'll examine small-scale mutations of one or a few nucleotide pairs, including **point mutations**, changes in a single nucleotide pair of a gene.

If a point mutation occurs in a gamete or in a cell that gives rise to gametes, it may be transmitted to offspring and to a succession of future generations. If the mutation has an adverse effect on the phenotype of a person, the mutant condition is referred to as a genetic disorder or hereditary disease. For example, we can trace the genetic basis of sickle-cell disease to the mutation of a single nucleotide pair in the gene that encodes the β-globin polypeptide of hemoglobin. The change of a single nucleotide in the DNA's template strand leads to an altered mRNA and the production of an abnormal protein (**Figure 25**). In individuals who are homozygous for the mutant allele, the sickling of red blood cells caused by the altered hemoglobin produces the multiple symptoms associated with sickle-cell disease. Another disorder caused by a point mutation is a heart condition called familial cardiomyopathy, which is responsible for some incidents of sudden death in young athletes. Point mutations in several genes encoding muscle proteins have been identified, any of which can lead to this disorder.

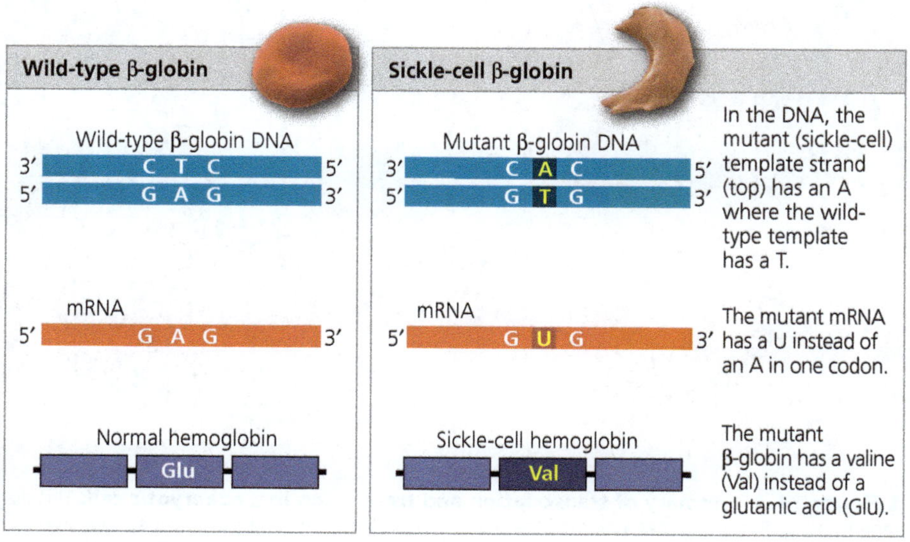

▲ **Figure 25 The molecular basis of sickle-cell disease: a point mutation.** The allele that causes sickle-cell disease differs from the wild-type (normal) allele by a single DNA nucleotide pair. The micrographs are SEMs of a normal red blood cell (on the left) and a sickled red blood cell (right) from individuals homozygous for wild-type and mutant alleles, respectively.

Types of Small-Scale Mutations

Let's now consider how small-scale mutations affect proteins. Small-scale mutations within a gene can be divided into two general categories: (1) single nucleotide-pair substitutions and (2) nucleotide-pair insertions or deletions. Insertions and deletions can involve one or more nucleotide pairs.

Substitutions

A **nucleotide-pair substitution** is the replacement of one nucleotide and its partner with another pair of nucleotides (**Figure 26a**). Some substitutions have no effect on the encoded protein, owing to the redundancy of the genetic code. For example, if 3′-CCG-5′ on the template strand mutated to 3′-CCA-5′, the mRNA codon that used to be GGC would become GGU, but a glycine would still be inserted at the proper location in the protein (see Figure 6). In other words, a change in a nucleotide pair may transform one codon into another that is translated into the same amino acid. Such a change is an example of a **silent mutation**, which has no observable effect on the phenotype. (Silent mutations can occur outside genes as well.) Substitutions that change one amino acid to another one are called **missense mutations**. Such a mutation may have little effect on the protein: The new amino acid may have properties similar to those of the amino acid it replaces, or it may be in a region of the protein where the exact sequence of amino acids is not essential to the protein's function.

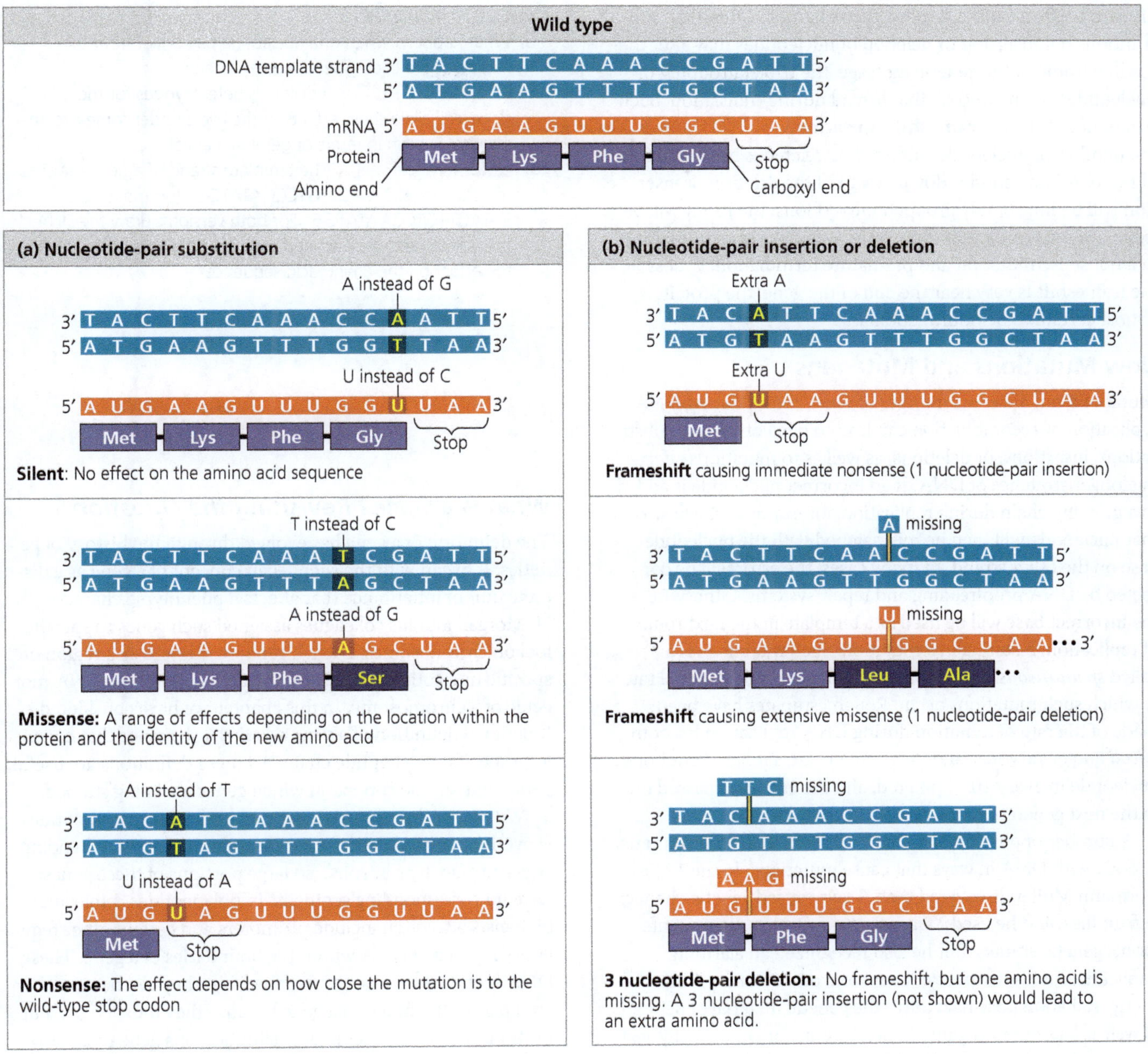

▲ **Figure 26** **Types of small-scale mutations that affect mRNA sequence.** All but one of the types shown here also affect the amino acid sequence of the encoded polypeptide.

However, the nucleotide-pair substitutions of greatest interest are those that cause a major change in a protein. The alteration of a single amino acid in a crucial area of a protein—such as in the part of the β-globin subunit of hemoglobin shown in Figure 25 or in the active site of an enzyme—can significantly alter protein activity. Occasionally, such a mutation leads to an improved protein or one with novel capabilities, but much more often such mutations are neutral or detrimental, leading to a useless or less active protein that impairs cellular function.

Substitution mutations are usually missense mutations; that is, the altered codon still codes for an amino acid and thus makes sense, although not necessarily the *right* sense. But a point mutation can also change a codon for an amino acid into a stop codon. This is called a **nonsense mutation**, and it causes translation to be terminated prematurely; the resulting polypeptide will be shorter than the polypeptide encoded by the normal gene. Nearly all nonsense mutations lead to nonfunctional proteins.

Insertions and Deletions

Insertions and **deletions** are additions or losses of nucleotide pairs in a gene **(Figure 26b)**. These mutations have a disastrous effect on the resulting protein more often than substitutions do. Insertion or deletion of nucleotides may alter the reading frame of the genetic message, the triplet grouping of nucleotides on the mRNA that is read during translation. Such a mutation, called a **frameshift mutation**, occurs whenever the number of nucleotides inserted or deleted is not a multiple of three. All nucleotides downstream of the deletion or insertion will be improperly grouped into codons; the result will be extensive missense mutations, usually ending sooner or later in a nonsense mutation and premature termination. Unless the frameshift is very near the end of the gene, the protein is almost certain to be nonfunctional.

New Mutations and Mutagens

Mutations can arise in a number of ways. Errors during DNA replication or recombination can lead to nucleotide-pair substitutions, insertions, or deletions, as well as to mutations affecting longer stretches of DNA. If an incorrect nucleotide is added to a growing chain during replication, for example, the base on that nucleotide will then be mismatched with the nucleotide base on the other strand. In many cases, the error will be corrected by DNA proofreading and repair systems. Otherwise, the incorrect base will be used as a template in the next round of replication, resulting in a mutation. Such mutations are called *spontaneous mutations*. It is difficult to calculate the rate at which such mutations occur. Rough estimates have been made of the rate of mutation during DNA replication for both *E. coli* and eukaryotes, and the numbers are similar: About one nucleotide in every 10^{10} is altered, and the change is passed on to the next generation of cells.

A number of physical and chemical agents, called **mutagens**, interact with DNA in ways that cause mutations. In the 1920s, Hermann Muller discovered that X-rays caused genetic changes in fruit flies, and he used X-rays to make *Drosophila* mutants for his genetic studies. But he also recognized an alarming implication of his discovery: X-rays and other forms of high-energy radiation pose hazards to the genetic material of people as well as laboratory organisms. Mutagenic radiation, a physical mutagen, includes ultraviolet (UV) light, which can cause disruptive thymine dimers in DNA.

Chemical mutagens fall into several categories. Nucleotide analogs are chemicals that are similar to normal DNA nucleotides but that pair incorrectly during DNA replication. Some other chemical mutagens interfere with correct DNA replication by inserting themselves into the DNA and distorting the double helix. Still other mutagens cause chemical changes in bases that change their pairing properties.

Researchers have developed a variety of methods to test the mutagenic activity of chemicals. A major application of these tests is the preliminary screening of chemicals to identify those that may cause cancer. This approach makes sense because most carcinogens (cancer-causing chemicals) are mutagenic, and conversely, most mutagens are carcinogenic.

CONCEPT CHECK 5

1. What happens when one nucleotide pair is lost from the middle of the coding sequence of a gene?
2. MAKE CONNECTIONS Individuals heterozygous for the sickle-cell allele show effects of the allele under some circumstances. Explain in terms of gene expression.
3. WHAT IF? DRAW IT The template strand of a gene includes this sequence: 3′-TACTTGTCCGATATC-5′. It is mutated to 3′-TACTTGTCCAATATC-5′. For both versions, draw the DNA, the mRNA, and the encoded amino acid sequence. What is the effect on the amino acid sequence?

What Is a Gene? *Revisiting the Question*

The definition of a gene has evolved through the history of genetics. It began with the Mendelian concept of a gene as a discrete unit of inheritance that affects a phenotypic character. T. H. Morgan and his colleagues assigned such genes to specific loci on chromosomes. Later, a gene was viewed as a region of specific nucleotide sequence along the length of the DNA molecule of a chromosome. In this chapter, we have considered a functional definition of a gene as a DNA sequence that codes for a specific polypeptide chain. All these definitions are useful, depending on the context in which genes are being studied.

We now realize that saying a gene codes for a polypeptide is overly simplistic. Most eukaryotic genes contain noncoding segments (such as introns), so large portions of these genes have no corresponding segments in polypeptides. Molecular biologists also often include promoters and certain other regulatory regions of DNA within the boundaries of a gene. These DNA sequences are not transcribed, but they can be considered part of the functional gene because they must be present for transcription to occur. Our definition of a gene must also be broad enough to include the DNA that is transcribed into rRNA, tRNA, and other RNAs that are not translated. These genes have no polypeptide products but play crucial roles in the cell. Thus, we arrive at the following definition: *A gene is a region of DNA that can be expressed to produce a final functional product that is either a polypeptide or an RNA molecule.*

When considering phenotypes, however, it is useful to focus on genes that code for polypeptides. In this chapter, you have learned how a typical gene is expressed—by transcription into RNA and then translation into a polypeptide that forms a protein of specific structure and function. Proteins, in turn, bring about an organism's observable phenotype.

A given type of cell expresses only a subset of its genes. This is an essential feature in multicellular organisms: Gene expression is precisely regulated.

Go to MasteringBiology® for Assignments, the eText, and the Study Area with Animations, Activities, Vocab Self-Quiz, and Practice Tests.

Chapter Review

SUMMARY OF KEY CONCEPTS

VOCAB SELF-QUIZ
goo.gl/gbai8v

CONCEPT 1

Genes specify proteins via transcription and translation

- Beadle and Tatum's studies of mutant strains of *Neurospora* led to the one gene–one polypeptide hypothesis. During **gene expression**, the information encoded in genes is used to make specific polypeptide chains (enzymes and other proteins) or RNA molecules.
- **Transcription** is the synthesis of RNA complementary to a **template strand** of DNA. **Translation** is the synthesis of a polypeptide whose amino acid sequence is specified by the nucleotide sequence in **mRNA**.
- Genetic information is encoded as a sequence of nonoverlapping nucleotide triplets, or **codons**. A codon in messenger RNA (mRNA) either is translated into an amino acid (61 of the 64 codons) or serves as a stop signal (3 codons). Codons must be read in the correct **reading frame**.

? *Describe the process of gene expression, by which a gene affects the phenotype of an organism.*

CONCEPT 2

Transcription is the DNA-directed synthesis of RNA: *a closer look*

- RNA synthesis is catalyzed by **RNA polymerase**, which links together RNA nucleotides complementary to a DNA template strand. This process follows the same base-pairing rules as DNA replication, except that in RNA, uracil substitutes for thymine.

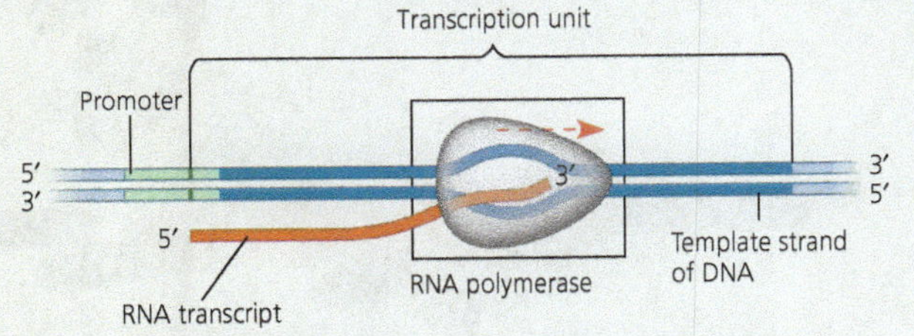

- The three stages of transcription are initiation, elongation, and termination. A **promoter**, often including a **TATA box** in eukaryotes, establishes where RNA synthesis is initiated. **Transcription factors** help eukaryotic RNA polymerase recognize promoter sequences, forming a **transcription initiation complex**. Termination differs in bacteria and eukaryotes.

? *What are the similarities and differences in the initiation of gene transcription in bacteria and eukaryotes?*

CONCEPT 3

Eukaryotic cells modify RNA after transcription

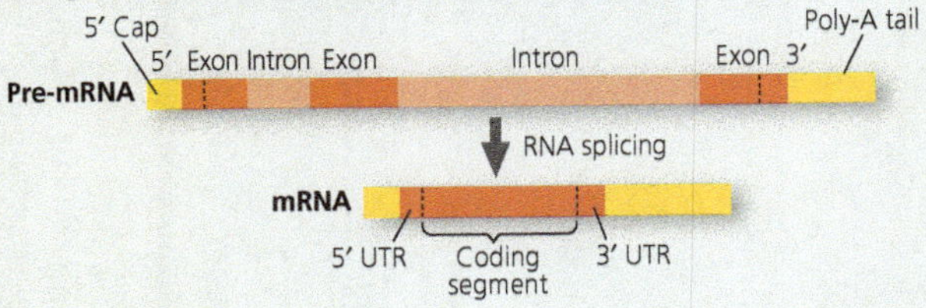

- Eukaryotic pre-mRNAs undergo **RNA processing**, which includes RNA splicing, the addition of a modified nucleotide **5′ cap** to the 5′ end, and the addition of a **poly-A tail** to the 3′ end. The processed mRNA includes an untranslated region (5′ UTR or 3′ UTR) at each end of the coding segment.
- Most eukaryotic genes are split into segments: They have **introns** interspersed among the **exons** (regions included in the mRNA). In **RNA splicing**, introns are removed and exons joined. RNA splicing is typically carried out by **spliceosomes**, but in some cases, RNA alone catalyzes its own splicing. The catalytic ability of some RNA molecules, called **ribozymes**, derives from the properties of RNA. The presence of introns allows for **alternative RNA splicing**.

? *What function do the 5′ cap and the poly-A tail serve on a eukaryotic mRNA?*

CONCEPT 4

Translation is the RNA-directed synthesis of a polypeptide: *a closer look*

- A cell translates an mRNA message into protein using **transfer RNAs (tRNAs)**. After being bound to a specific amino acid by an **aminoacyl-tRNA synthetase**, a tRNA lines up via its **anticodon** at the complementary codon on mRNA. A ribosome, made up of **ribosomal RNAs (rRNAs)** and proteins, facilitates this coupling with binding sites for mRNA and tRNA.
- Ribosomes coordinate the three stages of translation: initiation, elongation, and termination. The formation of peptide bonds between amino acids is catalyzed by ribosomal RNAs as tRNAs move through the **A** and **P sites** and exit through the **E site**.
- After translation, modifications to proteins can affect their shape. Free ribosomes in the cytosol initiate synthesis of all proteins, but proteins with a **signal peptide** are synthesized on the ER.
- A gene can be transcribed by multiple RNA polymerases simultaneously. Also, a single mRNA molecule can be translated simultaneously by a number of ribosomes, forming a polyribosome. In bacteria, these processes are coupled, but in eukaryotes they are separated in time and space by the nuclear membrane.

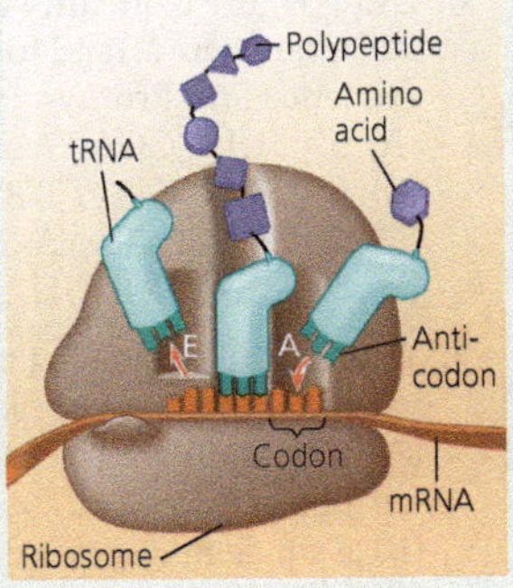

? *What function do tRNAs serve in the process of translation?*

CONCEPT 5

Mutations of one or a few nucleotides can affect protein structure and function

- Small-scale **mutations** include **point mutations**, changes in one DNA nucleotide pair, which may lead to production of nonfunctional proteins. **Nucleotide-pair substitutions** can cause **missense** or **nonsense mutations**. Nucleotide-pair **insertions** or **deletions** may produce **frameshift mutations**.
- Spontaneous mutations can occur during DNA replication, recombination, or repair. Chemical and physical **mutagens** cause DNA damage that can alter genes.

? *What will be the results of chemically modifying one nucleotide base of a gene? What role is played by DNA repair systems in the cell?*

TEST YOUR UNDERSTANDING

PRACTICE TEST

goo.gl/CRZjvS

Level 1: Knowledge/Comprehension

1. In eukaryotic cells, transcription cannot begin until
(A) the two DNA strands have completely separated and exposed the promoter.
(B) several transcription factors have bound to the promoter.
(C) the 5′ caps are removed from the mRNA.
(D) the DNA introns are removed from the template.

2. Which of the following is *not* true of a codon?
(A) It may code for the same amino acid as another codon.
(B) It never codes for more than one amino acid.
(C) It extends from one end of a tRNA molecule.
(D) It is the basic unit of the genetic code.

3. The anticodon of a particular tRNA molecule is
(A) complementary to the corresponding mRNA codon.
(B) complementary to the corresponding triplet in rRNA.
(C) the part of tRNA that bonds with a specific amino acid.
(D) catalytic, making the tRNA a ribozyme.

4. Which of the following is *not* true of RNA processing?
(A) Exons are cut out before mRNA leaves the nucleus.
(B) Nucleotides may be added at both ends of the RNA.
(C) Ribozymes may function in RNA splicing.
(D) RNA splicing can be catalyzed by spliceosomes.

5. Which component is *not* directly involved in translation?
(A) GTP (C) tRNA
(B) DNA (D) ribosomes

Level 2: Application/Analysis

6. Using Figure 6, identify a 5′ → 3′ sequence of nucleotides in the DNA template strand for an mRNA coding for the polypeptide sequence Phe-Pro-Lys.
(A) 5′-UUUCCCAAA-3′
(B) 5′-GAACCCCTT-3′
(C) 5′-CTTCGGGAA-3′
(D) 5′-AAACCCUUU-3′

7. Which of the following mutations would be *most* likely to have a harmful effect on an organism?
(A) a deletion of three nucleotides near the middle of a gene
(B) a single nucleotide deletion in the middle of an intron
(C) a single nucleotide deletion near the end of the coding sequence
(D) a single nucleotide insertion downstream of, and close to, the start of the coding sequence

8. Would the coupling of the processes shown in Figure 23 be found in a eukaryotic cell? Explain why or why not.

9. Fill in the following table:

Type of RNA	Functions
Messenger RNA (mRNA)	
Transfer RNA (tRNA)	
	Plays catalytic (ribozyme) roles and structural roles in ribosomes
Primary transcript	
Small RNAs in the spliceosome	

Level 3: Synthesis/Evaluation

10. **SCIENTIFIC INQUIRY**
Knowing that the genetic code is almost universal, a scientist uses molecular biological methods to insert the human β-globin gene (shown in Figure 12) into bacterial cells, hoping the cells will express it and synthesize functional β-globin protein. Instead, the protein produced is nonfunctional and is found to contain many fewer amino acids than does β-globin made by a eukaryotic cell. Explain why.

11. **FOCUS ON EVOLUTION**
Most amino acids are coded for by a set of similar codons (see Figure 6). What evolutionary explanation can you give for this pattern?

12. **FOCUS ON INFORMATION**
Evolution accounts for the unity and diversity of life, and the continuity of life is based on heritable information in the form of DNA. In a short essay (100–150 words), discuss how the fidelity with which DNA is inherited is related to the processes of evolution.

13. **SYNTHESIZE YOUR KNOWLEDGE**

Some mutations result in proteins that function well at one temperature but are nonfunctional at a different (usually higher) temperature. Siamese cats have such a "temperature-sensitive" mutation in a gene encoding an enzyme that makes dark pigment in the fur. The mutation results in the breed's distinctive point markings and lighter body color (see the photo). Using this information and what you learned in the chapter, explain the pattern of the cat's fur pigmentation.

Answers

Figure Questions

Figure 5 The mRNA sequence (5′-UGGUUUGGCUCA-3′) is the same as the nontemplate DNA strand sequence (5′-TGGTTTGGCTCA-3′), except there is a U in the mRNA wherever there is a T in the DNA. **Figure 8** The processes are similar in that polymerases form polynucleotides complementary to an antiparallel DNA template strand. In replication, however, both strands act as templates, whereas in transcription, only one DNA strand acts as a template. **Figure 9** The RNA polymerase would bind directly to the promoter, rather than being dependent on the previous binding of other factors.

Figure 12

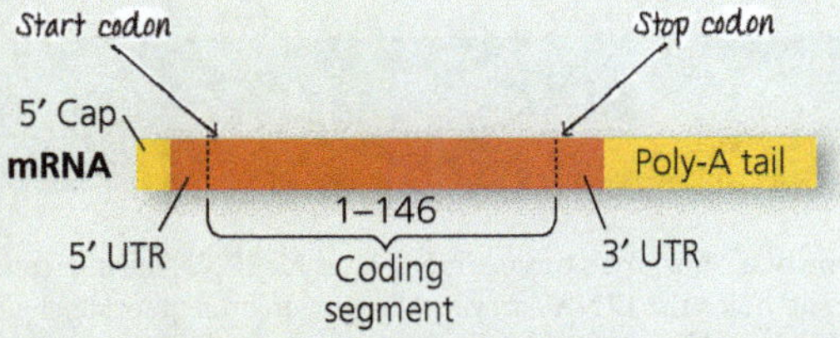

Figure 23 The mRNA farthest to the right (the longest one) started being transcribed first. The ribosome at the top, closest to the DNA, started translating first and thus has the longest polypeptide.

Concept Check 1

1. Recessive **2.** A polypeptide made up of 10 Gly (glycine) amino acids
3.

Nontemplate sequence in 3′ to 5′ order:	3′-ACGACTGAA-5′
mRNA sequence:	5′-UGCUGACUU-3′
Translated:	Cys-STOP

If the nontemplate sequence could have been used as a template for transcribing the mRNA, the protein translated from the mRNA would have a completely different amino acid sequence and would most likely be nonfunctional. (It would also be shorter because of the UGA stop signal shown in the mRNA sequence above—and possibly others earlier in the mRNA sequence.)

Concept Check 2

1. A promoter is the region of DNA to which RNA polymerase binds to begin transcription. It is at the upstream end of the gene (transcription unit). **2.** In a bacterial cell, part of the RNA polymerase recognizes the gene's promoter and binds to it. In a eukaryotic cell, transcription factors mediate the binding of RNA polymerase to the promoter. In both cases, sequences in the promoter bind precisely to the RNA polymerase, so the enzyme is in the right location and orientation. **3.** The transcription factor that recognizes the TATA sequence would be unable to bind, so RNA polymerase could not bind and transcription of that gene probably would not occur.

Concept Check 3

1. Due to alternative splicing of exons, each gene can result in multiple different mRNAs and can thus direct synthesis of multiple different proteins. **2.** In watching a show recorded with a DVR, you watch segments of the show itself (exons) and fast-forward through the commercials, which are thus like introns. However, unlike introns, commercials remain in the recording, while the introns are cut out of the RNA transcript during RNA processing. **3.** Once the mRNA has exited the nucleus, the cap prevents it from being degraded by hydrolytic enzymes and facilitates its attachment to ribosomes. If the cap were removed from all mRNAs, the cell would no longer be able to synthesize any proteins and would probably die.

Concept Check 4

1. First, each aminoacyl-tRNA synthetase specifically recognizes a single amino acid and attaches it only to an appropriate tRNA. Second, a tRNA charged with its specific amino acid binds only to an mRNA codon for that amino acid. **2.** The structure and function of the ribosome seem to depend more on the rRNAs than on the ribosomal proteins. Because it is single-stranded, an RNA molecule can hydrogen-bond with itself and with other RNA molecules. RNA molecules make up the interface between the two ribosomal subunits, so presumably RNA-RNA binding helps hold the ribosome together. The binding site for mRNA in the ribosome includes rRNA that can bind the mRNA. Also, complementary hydrogen bonding within an RNA molecule allows it to assume a particular three-dimensional shape and, along with the RNA's functional groups, presumably enables rRNA to catalyze peptide bond formation during translation. **3.** A signal peptide on the leading end of the polypeptide being synthesized is recognized by a signal-recognition particle that brings the ribosome to the ER membrane. There the ribosome attaches and continues to synthesize the polypeptide, depositing it in the ER lumen. **4.** Because of wobble, the tRNA could bind to either 5′-GCA-3′ or 5′-GCG-3′, both of which code for alanine (Ala). Alanine would be attached to the tRNA (see diagram, upper right).

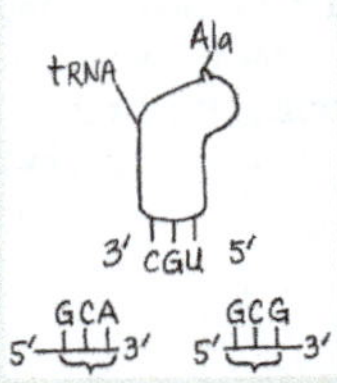

Concept Check 5

1. In the mRNA, the reading frame downstream from the deletion is shifted, leading to a long string of incorrect amino acids in the polypeptide, and in most cases, a stop codon will arise, leading to premature termination. The polypeptide will most likely be nonfunctional. **2.** Heterozygous individuals, said to have sickle-cell trait, have a copy each of the wild-type allele and the sickle-cell allele. Both alleles will be expressed, so these individuals will have both normal and sickle-cell hemoglobin molecules. Apparently, having a mix of the two forms of β-globin has no effect under most conditions, but during prolonged periods of low blood oxygen (such as at higher altitudes), these individuals can show some signs of sickle-cell disease.
3.

Normal DNA sequence (template strand is on top):	3′-TACTTGTCCGATATC-5′ 5′-ATGAACAGGCTATAG-3′
mRNA sequence:	5′-AUGAACAGGCUAUAG-3′
Amino acid sequence:	Met-Asn-Arg-Leu-STOP
Mutated DNA sequence (template strand is on top):	3′-TACTTGTCCAATATC-5′ 5′-ATGAACAGGTTATAG-3′
mRNA sequence:	5′-AUGAACAGGUUAUAG-3′
Amino acid sequence:	Met-Asn-Arg-Leu-STOP

No effect: The amino acid sequence is Met-Asn-Arg-Leu both before and after the mutation because the mRNA codons 5′-CUA-3′ and 5′-UUA-3′ both code for Leu. (The fifth codon is a stop codon.)

Summary of Key Concepts Questions

1 A gene contains genetic information in the form of a nucleotide sequence. The gene is first transcribed into an RNA molecule, and a messenger RNA molecule is ultimately translated into a polypeptide. The polypeptide makes up part or all of a protein, which performs a function in the cell and contributes to the phenotype of the organism. **2** Both bacterial and eukaryotic genes have promoters, regions where RNA polymerase ultimately binds and begins transcription. In bacteria, RNA polymerase binds directly to the promoter; in eukaryotes, transcription factors bind first to the promoter, and then RNA polymerase binds to the transcription factors and promoter together. **3** Both the 5′ cap and the 3′ poly-A tail help the mRNA exit from the nucleus and then, in the cytoplasm, help ensure mRNA stability and allow it to bind to ribosomes. **4** In the context of the ribosome, tRNAs function as translators between the nucleotide-based language of mRNA and the amino-acid-based language of polypeptides. A tRNA carries a specific amino acid, and the anticodon on the tRNA is complementary to the codon on the mRNA that codes for that amino acid. In the ribosome, the tRNA binds to the A site. Then the polypeptide being synthesized (currently on the tRNA in the P site) is joined to the new amino acid, which becomes the new (C-terminal) end of the polypeptide. Next, the tRNA in the A site moves to the P site. After the polypeptide is transferred to the new tRNA, thus adding the new amino acid, the now empty tRNA moves from the P site to the E site, where it exits the ribosome. **5** When a nucleotide base is altered chemically, its base-pairing characteristics may be changed. When that happens, an incorrect nucleotide is likely to be incorporated into the complementary strand during the next replication of the DNA, and successive rounds of replication will perpetuate the mutation. Once the gene is transcribed, the mutated codon may code for a different amino acid that inhibits or changes the function of a protein. If the chemical change in the base is detected and repaired by the DNA repair system before the next replication, no mutation will result.

Test Your Understanding

1. B **2.** C **3.** A **4.** A **5.** B **6.** C **7.** D **8.** No. Transcription and translation are separated in space and time in a eukaryotic cell, as a result of the eukaryotic cell's nuclear membrane.

9.

Type of RNA	Functions
Messenger RNA (mRNA)	Carries information specifying amino acid sequences of proteins from DNA to ribosomes
Transfer RNA (tRNA)	Serves as translator molecule in protein synthesis; translates mRNA codons into amino acids
Ribosomal RNA (rRNA)	Plays catalytic (ribozyme) roles and structural roles in ribosomes
Primary transcript	Is a precursor to mRNA, rRNA, or tRNA, before being processed; some intron RNA acts as a ribozyme, catalyzing its own splicing
Small RNAs in spliceosome	Play structural and catalytic roles in spliceosomes, the complexes of protein and RNA that splice pre-mRNA

Credits

Credits are listed in the order of appearance.

Photo Credits

4.1 Alessandro Bianchi/Reuters; **7a** Keith V. Wood, University of California, San Diego; **7b** Reuters; **22b** Barbara Hamkalo; **23** Reproduced with permission from O. L. Miller, Jr., B. A. Hamkalo, and C. A. Thomas, Jr., *Science* 169 (1970):392. Copyright © 1970 American Association for the Advancement of Science. Fig. 3; **25** Eye of Science/Science Source; **Chapter Review SYK** Vasiliy Koval/Shutterstock.

Illustration and Text Credits

12 Adapted from *The World of the Cell*, 3rd Edition, by W.M. Becker, J.B. Reece and M.F. Poenie, 1996. Copyright © 1996 by Pearson Education, Inc. Reprinted and electronically reproduced by permission of Pearson Education, Inc. Upper Saddle River, NJ; **17b-c** Adapted from *Biochemistry*, 2nd ed., by C.K. Matthews and K.E. van Holde, 1996. Copyright © 1996 by Pearson Education, Inc. Reprinted and electronically reproduced by permission of Pearson Education, Inc. Upper Saddle River, NJ; **SSE.1–3** Material provided courtesy of Dr. Thomas Schneider, National Cancer Institute, National Institutes of Health, 2012.

Glossary

5′ cap A modified form of guanine nucleotide added onto the end of a pre-mRNA molecule.

A site One of a ribosome's three binding sites for tRNA during translation. The A site holds the tRNA carrying the next amino acid to be added to the polypeptide chain. (A stands for aminoacyl tRNA.)

aminoacyl-tRNA synthetase An enzyme that joins each amino acid to the appropriate tRNA.

anticodon (an′-tī-kō′-don) A nucleotide triplet at one end of a tRNA molecule that base-pairs with a particular complementary codon on an mRNA molecule.

codon (kō′-don) A three-nucleotide sequence of DNA or mRNA that specifies a particular amino acid or termination signal; the basic unit of the genetic code.

E site One of a ribosome's three binding sites for tRNA during translation. The E site is the place where discharged tRNAs leave the ribosome. (E stands for exit.)

exon A sequence within a primary transcript that remains in the RNA after RNA processing; also refers to the region of DNA from which this sequence was transcribed.

frameshift mutation A mutation occurring when nucleotides are inserted in or deleted from a gene and the number inserted or deleted is not a multiple of three, resulting in the improper grouping of the subsequent nucleotides into codons.

gene expression The process by which information encoded in DNA directs the synthesis of proteins or, in some cases, RNAs that are not translated into proteins and instead function as RNAs.

insertion A mutation involving the addition of one or more nucleotide pairs to a gene.

intron (in′-tron) A noncoding, intervening sequence within a primary transcript that is removed from the transcript during RNA processing; also refers to the region of DNA from which this sequence was transcribed.

messenger RNA (mRNA) A type of RNA, synthesized using a DNA template, that attaches to ribosomes in the cytoplasm and specifies the primary structure of a protein. (In eukaryotes, the primary RNA transcript must undergo RNA processing to become mRNA.)

missense mutation A nucleotide-pair substitution that results in a codon that codes for a different amino acid.

mutagen (myū′-tuh-jen) A chemical or physical agent that interacts with DNA and can cause a mutation.

mutation (myū-tā′-shun) A change in the nucleotide sequence of an organism's DNA or in the DNA or RNA of a virus.

nonsense mutation A mutation that changes an amino acid codon to one of the three stop codons, resulting in a shorter and usually nonfunctional protein.

nucleotide-pair substitution A type of point mutation in which one nucleotide in a DNA strand and its partner in the complementary strand are replaced by another pair of nucleotides.

P site One of a ribosome's three binding sites for tRNA during translation. The P site holds the tRNA carrying the growing polypeptide chain. (P stands for peptidyl tRNA.)

point mutation A change in a single nucleotide pair of a gene.

poly-A tail A sequence of 50–250 adenine nucleotides added onto the 3′ end of a pre-mRNA molecule.

primary transcript An initial RNA transcript from any gene; also called pre-mRNA when transcribed from a protein-coding gene.

promoter A specific nucleotide sequence in the DNA of a gene that binds RNA polymerase, positioning it to start transcribing RNA at the appropriate place.

reading frame On an mRNA, the triplet grouping of ribonucleotides used by the translation machinery during polypeptide synthesis.

ribosomal RNA (rRNA) (rī′-buh-sō′-mul) RNA molecules that, together with proteins, make up ribosomes; the most abundant type of RNA.

ribozyme (rī′-buh-zīm) An RNA molecule that functions as an enzyme, such as an intron that catalyzes its own removal during RNA splicing.

RNA polymerase An enzyme that links ribonucleotides into a growing RNA chain during transcription, based on complementary binding to nucleotides on a DNA template strand.

RNA processing Modification of RNA primary transcripts, including splicing out of introns, joining together of exons, and alteration of the 5′ and 3′ ends.

RNA splicing After synthesis of a eukaryotic primary RNA transcript, the removal of portions of the transcript (introns) that will not be included in the mRNA and the joining together of the remaining portions (exons).

signal peptide A sequence of about 20 amino acids at or near the leading (amino) end of a polypeptide that targets it to the endoplasmic reticulum or other organelles in a eukaryotic cell.

signal-recognition particle (SRP) A protein-RNA complex that recognizes a signal peptide as it emerges from a ribosome and helps direct the ribosome to the endoplasmic reticulum (ER) by binding to a receptor protein on the ER.

silent mutation A nucleotide-pair substitution that has no observable effect on the phenotype; for example, within a gene, a mutation that results in a codon that codes for the same amino acid.

spliceosome (splī′-sō-sōm) A large complex made up of proteins and RNA molecules that splices RNA by interacting with the ends of an RNA intron, releasing the intron and joining the two adjacent exons.

start point In transcription, the nucleotide position on the promoter where RNA polymerase begins synthesis of RNA.

TATA box A DNA sequence in eukaryotic promoters crucial in forming the transcription initiation complex.

template strand The DNA strand that provides the pattern, or template, for ordering, by complementary base pairing, the sequence of nucleotides in an RNA transcript.

terminator In bacteria, a sequence of nucleotides in DNA that marks the end of a gene and signals RNA polymerase to release the newly made RNA molecule and detach from the DNA.

transcription The synthesis of RNA using a DNA template.

transcription factor A regulatory protein that binds to DNA and affects transcription of specific genes.

transcription initiation complex The completed assembly of transcription factors and RNA polymerase bound to a promoter.

transcription unit A region of DNA that is transcribed into an RNA molecule.

transfer RNA (tRNA) An RNA molecule that functions as a translator between nucleic acid and protein languages by picking up a specific amino acid and carrying it to the ribosome, where the tRNA recognizes the appropriate codon in the mRNA.

translation The synthesis of a polypeptide using the genetic information encoded in an mRNA molecule. There is a change of "language" from nucleotides to amino acids.

triplet code A genetic information system in which a series of three-nucleotide-long words specifies a sequence of amino acids for a polypeptide chain.

wobble Flexibility in the base-pairing rules in which the nucleotide at the 5′ end of a tRNA anticodon can form hydrogen bonds with more than one kind of base in the third position (3′ end) of a codon.

Regulation of Gene Expression

KEY CONCEPTS

1. Bacteria often respond to environmental change by regulating transcription
2. Eukaryotic gene expression is regulated at many stages
3. Noncoding RNAs play multiple roles in controlling gene expression
4. Researchers can monitor expression of specific genes

▲ **Figure 1** How can this fish's eyes see equally well in both air and water?

Beauty in the Eye of the Beholder

The fish shown in **Figure 1** is keeping an eye out for predators—or, more precisely, both halves of each eye! *Anableps anableps*, commonly known as "cuatro ojos" ("four eyes"), glides through freshwater lakes and ponds in Central and South America with the upper half of each eye protruding from the water. The eye's upper half is particularly well-suited for aerial vision and the lower half for aquatic vision. The molecular basis of this specialization has recently been revealed: The cells of the two parts of the eye express a slightly different set of genes involved in vision, even though these two groups of cells are quite similar and contain identical genomes.

A hallmark of prokaryotic and eukaryotic cells alike—from a bacterium to the cells of a fish—is their intricate and precise regulation of gene expression. Both prokaryotes and eukaryotes must alter their patterns of gene expression in response to changes in environmental conditions. Multicellular eukaryotes must also develop and maintain multiple cell types, each expressing a different subset of genes. This is a significant challenge in gene regulation.

In this chapter, we'll first explore how bacteria regulate expression of their genes in response to different environmental conditions. We'll then examine the general mechanisms by which eukaryotes regulate gene expression. Next, we'll look at the many roles played by RNA molecules in regulating gene expression in eukaryotes. Finally, we'll describe a few genetic engineering techniques that have been developed to investigate gene expression. Elucidating how gene expression is regulated in different cells is crucial to our understanding of living systems.

CONCEPT 1

Bacteria often respond to environmental change by regulating transcription

Bacterial cells that can conserve resources and energy have a selective advantage over cells that are unable to do so. Thus, natural selection has favored bacteria that express only the genes whose products are needed by the cell.

From Chapter 15 of *Campbell Biology in Focus*, Second Edition. Lisa A. Urry, Michael L. Cain, Steven A. Wasserman, Peter V. Minorsky, and Jane B. Reece.

Consider, for instance, an individual *Escherichia coli* (*E. coli*) cell living in the erratic environment of a human colon, dependent for its nutrients on the whimsical eating habits of its host. If the environment is lacking in the amino acid tryptophan, which the bacterium needs to survive, the cell responds by activating a metabolic pathway that makes tryptophan from another compound. If the human host later eats a tryptophan-rich meal, the bacterial cell stops producing tryptophan, thus avoiding wasting resources to produce a substance that is readily available from the surrounding solution.

A metabolic pathway can be controlled on two levels, as shown for the synthesis of tryptophan in **Figure 2**. First, cells can adjust the activity of enzymes already present. This is a fairly rapid response, which relies on the sensitivity of many enzymes to chemical cues that increase or decrease their catalytic activity. The activity of the first enzyme in the tryptophan synthesis pathway is inhibited by tryptophan, the pathway's end product **(Figure 2a)**. Thus, if tryptophan accumulates in a cell, it shuts down the synthesis of more tryptophan by inhibiting enzyme activity. Such *feedback inhibition*, typical of anabolic (biosynthetic) pathways, allows a cell to adapt to short-term fluctuations in the supply of a substance it needs.

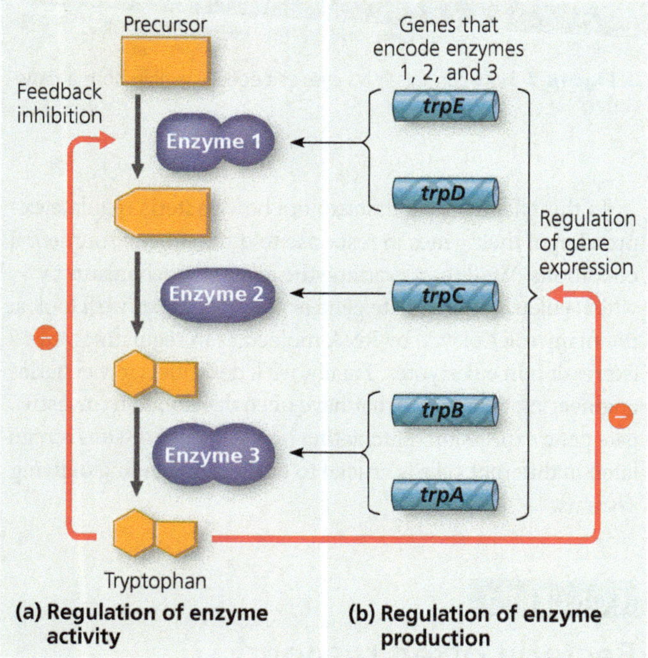

▲ **Figure 2 Regulation of a metabolic pathway.** In the pathway for tryptophan synthesis, an abundance of tryptophan can both **(a)** inhibit the activity of the first enzyme in the pathway (feedback inhibition), a rapid response, and **(b)** repress expression of the genes encoding all subunits of the enzymes in the pathway, a longer-term response. Genes *trpE* and *trpD* encode the two subunits of enzyme 1, and genes *trpB* and *trpA* encode the two subunits of enzyme 3. (The genes were named before the order in which they functioned in the pathway was determined.) The ⊖ symbol stands for inhibition.

Second, cells can adjust the production of certain enzymes; that is, they can regulate the expression of the genes encoding the enzymes. If, in our example, the environment provides all the tryptophan the cell needs, the cell stops making the enzymes that catalyze the synthesis of tryptophan **(Figure 2b)**. In this case, the control of enzyme production occurs at the level of transcription, the synthesis of messenger RNA from the genes that code for these enzymes.

Regulation of the tryptophan synthesis pathway is just one example of how bacteria tune their metabolism to changing environments. Many genes of the bacterial genome are switched on or off by changes in the metabolic status of the cell. One basic mechanism for this control of gene expression in bacteria, described as the *operon model*, was discovered in 1961 by François Jacob and Jacques Monod at the Pasteur Institute in Paris. Let's see what an operon is and how it works, using the control of tryptophan synthesis as our first example.

Operons: The Basic Concept

E. coli synthesizes the amino acid tryptophan from a precursor molecule in the three-step pathway shown in Figure 2. Each reaction in the pathway is catalyzed by a specific enzyme, and the five genes that code for the subunits of these enzymes are clustered together on the bacterial chromosome. A single promoter serves all five genes, which together constitute a transcription unit. (Recall that a promoter is a site where RNA polymerase can bind to DNA and begin transcription.) Thus, transcription gives rise to one long mRNA molecule that codes for the five polypeptides making up the enzymes in the tryptophan pathway **(Figure 3a)**. The cell can translate this one mRNA into five separate polypeptides because the mRNA is punctuated with start and stop codons that signal where the coding sequence for each polypeptide begins and ends.

A key advantage of grouping genes of related function into one transcription unit is that a single "on-off switch" can control the whole cluster of functionally related genes; in other words, these genes are *coordinately controlled*. When an *E. coli* cell must make tryptophan for itself because its surrounding environment lacks this amino acid, all the enzymes for the metabolic pathway are synthesized at one time. The "on-off switch" is a segment of DNA called an **operator**. Both its location and name suit its function: Positioned within the promoter or, in some cases, between the promoter and the enzyme-coding genes, the operator controls the access of RNA polymerase to the genes. All together, the operator, the promoter, and the genes they control—the entire stretch of DNA required for enzyme production for the tryptophan pathway—constitute an **operon**. The *trp* operon (*trp* for tryptophan) is one of many operons in the *E. coli* genome (see Figure 3a).

If the operator is the operon's switch for controlling transcription, how does this switch work? By itself, the *trp* operon is turned on; that is, RNA polymerase can bind to the

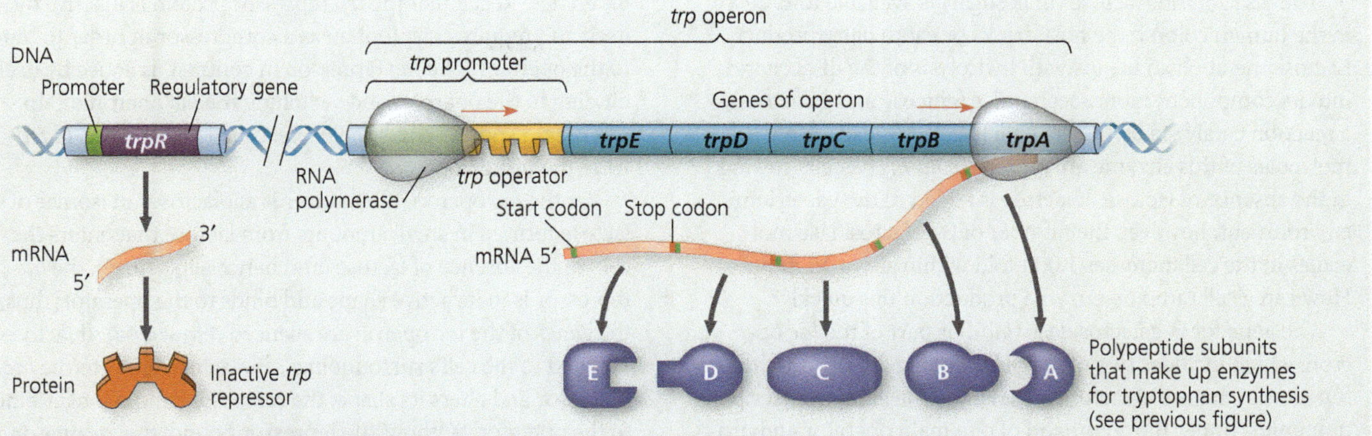

(a) Tryptophan absent, repressor inactive, operon on. RNA polymerase attaches to the DNA at the operon's promoter and transcribes the operon's genes. Enzymes for tryptophan synthesis are made.

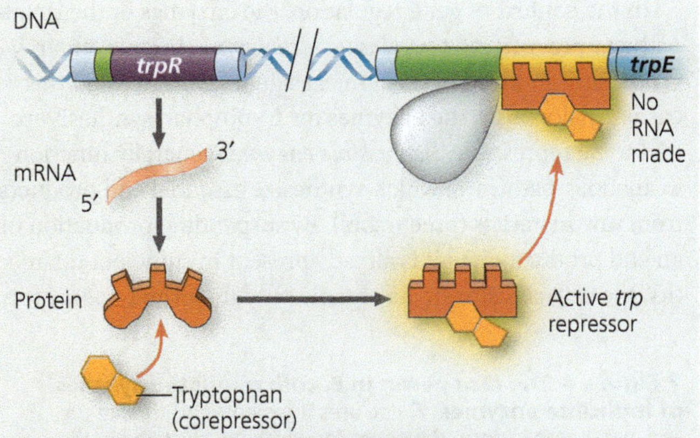

(b) Tryptophan present, repressor active, operon off. As tryptophan accumulates, it inhibits its own production by activating the repressor protein, which binds to the operator, blocking transcription. Enzymes for tryptophan synthesis are not made.

▲ **Figure 3 The *trp* operon in *E. coli*: regulated synthesis of repressible enzymes.** Tryptophan is an amino acid, produced by an anabolic pathway catalyzed by three enzymes (see Figure 2). **(a)** The five genes encoding the polypeptide subunits of the enzymes in this pathway are grouped, along with a promoter, into the *trp* operon. The *trp* operator (the repressor binding site) is located within the *trp* promoter (the RNA polymerase binding site). **(b)** Accumulation of tryptophan, the end product of the pathway, represses transcription of the *trp* operon, thus blocking synthesis of all the enzymes in the pathway and shutting down tryptophan production.

? *Describe what happens to the* trp *operon as the cell uses up its store of tryptophan.*

promoter and transcribe the genes of the operon. The operon can be switched off by a protein called the *trp* repressor. A **repressor** binds to the operator and blocks the attachment of RNA polymerase to the promoter, preventing transcription of the genes. A repressor protein is specific for the operator of a particular operon. For example, the *trp* repressor, which switches off the *trp* operon by binding to the *trp* operator, has no effect on other operons in the *E. coli* genome.

The *trp* repressor is the protein product of a **regulatory gene** called *trpR*, which is located some distance from the *trp* operon and has its own promoter. Regulatory genes are expressed continuously, although at a low rate, and a few *trp* repressor molecules are always present in *E. coli* cells. Why, then, is the *trp* operon not switched off permanently? First, the binding of repressors to operators is reversible. An operator alternates between two states: one with the repressor bound and one without. The relative duration of the repressor-bound state is higher when more active repressor molecules are present. Second, the *trp* repressor, like most regulatory proteins, is an allosteric protein, with two alternative shapes: active and inactive. The *trp* repressor is synthesized in the inactive form, which has little affinity for the *trp* operator. Only when tryptophan binds to the *trp* repressor at an allosteric site does the repressor protein change to the active form that can attach to the operator, turning the operon off (**Figure 3b**).

Tryptophan functions in this system as a **corepressor**, a small molecule that cooperates with a repressor protein to switch an operon off. As tryptophan accumulates, more tryptophan molecules associate with *trp* repressor molecules, which can then bind to the *trp* operator and shut down production of the tryptophan pathway enzymes. If the cell's tryptophan level drops, transcription of the operon's genes resumes. The *trp* operon is one example of how gene expression can respond to changes in the cell's internal and external environment.

Repressible and Inducible Operons: Two Types of Negative Gene Regulation

The *trp* operon is said to be a *repressible operon* because its transcription is usually on but can be inhibited (repressed) when a specific small molecule (in this case, tryptophan) binds allosterically to a regulatory protein. In contrast, an *inducible operon* is usually off but can be stimulated (induced) when a specific small molecule interacts with a regulatory protein. The classic example of an inducible operon is the *lac* operon (*lac* for lactose).

The disaccharide lactose (milk sugar) is available to *E. coli* in the human colon if the host drinks or eats a dairy product. Lactose metabolism begins with hydrolysis of the disaccharide into its component monosaccharides (glucose and galactose), a reaction catalyzed by the enzyme β-galactosidase. Only a few molecules of this enzyme are present in an *E. coli* cell growing in the absence of lactose. If lactose is added to the bacterium's environment, however, the number of β-galactosidase molecules in the cell increases 1,000-fold within about 15 minutes. How can a cell ramp up enzyme production this quickly?

The gene for β-galactosidase (*lacZ*) is part of the *lac* operon, which includes two other genes coding for enzymes that function in the use of lactose **(Figure 4)**. The entire transcription unit is under the command of one main operator and promoter. The regulatory gene, *lacI*, located outside the operon, codes for an allosteric repressor protein that can switch off the *lac* operon by binding to the operator. So far, this sounds just like regulation of the *trp* operon, but there is one important difference. Recall that the *trp* repressor protein is inactive by itself and requires tryptophan as a corepressor in order to bind to the operator. The *lac* repressor, in contrast, is active by itself, binding to the operator and switching the *lac* operon off. In this case, a specific small molecule, called an **inducer**, *inactivates* the repressor.

For the *lac* operon, the inducer is allolactose, an isomer of lactose formed in small amounts from lactose that enters the cell. In the absence of lactose (and hence allolactose), the *lac* repressor is in its active shape and binds to the operator; thus, the genes of the *lac* operon are silenced **(Figure 4a)**. If lactose is added to the cell's surroundings, allolactose binds to the *lac* repressor and alters its shape; the repressor can no longer bind to the operator. Without the repressor bound, the *lac* operon is transcribed into mRNA, and the enzymes for using lactose are made **(Figure 4b)**.

In the context of gene regulation, the enzymes of the lactose pathway are referred to as *inducible enzymes* because their synthesis is induced by a chemical signal (allolactose, in this case). Analogously, the enzymes for tryptophan synthesis are said to be repressible. *Repressible enzymes* generally function in anabolic pathways, which synthesize essential end products from raw materials (precursors). By suspending production of an end product when it is already present in sufficient quantity, the cell can allocate its organic precursors and energy for

▼ Figure 4 The *lac* operon in *E. coli*: regulated synthesis of inducible enzymes. *E. coli* uses three enzymes to take up and metabolize lactose, the genes for which are clustered in the *lac* operon. The first gene, *lacZ*, codes for β-galactosidase, which hydrolyzes lactose to glucose and galactose. The second gene, *lacY*, codes for a permease, a membrane protein that transports lactose into the cell. The third gene, *lacA*, codes for an enzyme that detoxifies other molecules entering the cell via the permease. Unusually, the gene for the *lac* repressor, *lacI*, is adjacent to the *lac* operon. The function of the teal region within the promoter will be revealed in Figure 5.

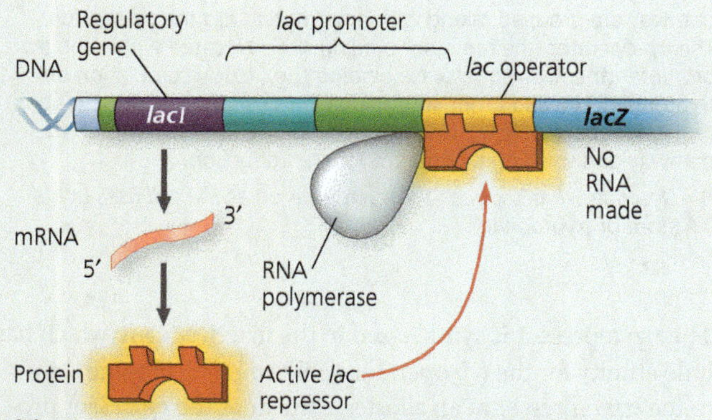

(a) Lactose absent, repressor active, operon off. The *lac* repressor is innately active, and in the absence of lactose it switches off the operon by binding to the operator. The enzymes for using lactose are not made.

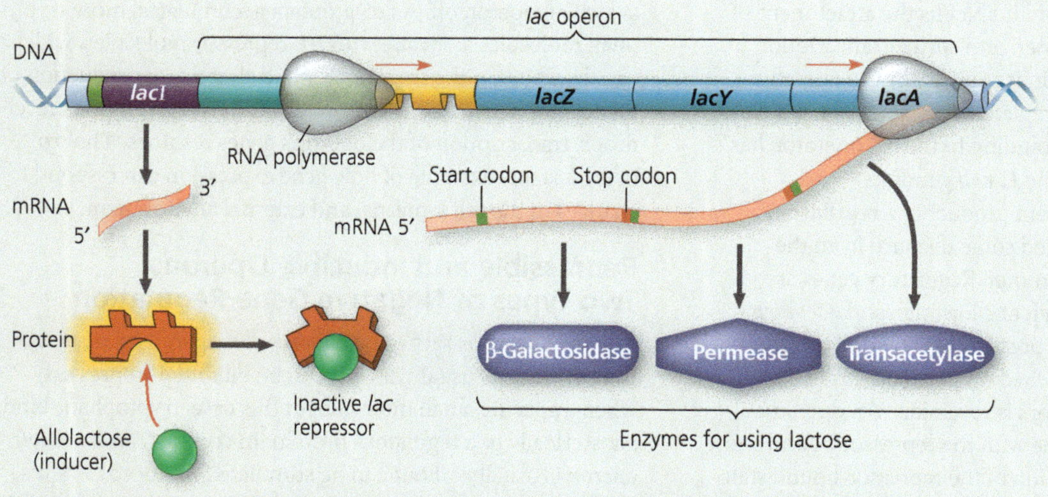

(b) Lactose present, repressor inactive, operon on. Allolactose, an isomer of lactose, binds to the repressor, inactivating it and "derepressing" the operon. The inactive repressor cannot bind to the operator, and so the genes of the *lac* operon are transcribed, and the enzymes for using lactose are made.

other uses. In contrast, inducible enzymes usually function in catabolic pathways, which break down a nutrient to simpler molecules. By producing the appropriate enzymes only when the nutrient is available, the cell avoids wasting energy and precursors making proteins that are not needed.

Regulation of both the *trp* and *lac* operons involves the *negative* control of genes, because the operons are switched off by the active form of the repressor protein (see Figures 3b and 4a). It may be easier to see this for the *trp* operon, but it is also true for the *lac* operon. In the case of the *lac* operon, allolactose induces enzyme synthesis not by directly activating the *lac* operon, but by freeing it from the negative effect of the repressor (see Figure 4b). Gene regulation is said to be *positive* only when a regulatory protein interacts directly with the genome to switch transcription on. Let's look at an example of the positive control of genes, again involving the *lac* operon.

Positive Gene Regulation

When glucose and lactose are both present in its environment, *E. coli* preferentially uses glucose. The enzymes for glucose breakdown in glycolysis are continually present. Only when lactose is present *and* glucose is in short supply does *E. coli* use lactose as an energy source, and only then does it synthesize appreciable quantities of the enzymes for lactose breakdown.

How does the *E. coli* cell sense the glucose concentration and relay this information to the genome? Again, the mechanism depends on the interaction of an allosteric regulatory protein with a small organic molecule, in this case **cyclic AMP (cAMP)**, which accumulates when glucose is scarce. The regulatory protein, called *cAMP receptor protein (CRP)*, is an **activator**, a protein that binds to DNA and stimulates transcription of a gene. When cAMP binds to this regulatory protein, CRP assumes its active shape and can attach to a specific site at the upstream end of the *lac* promoter **(Figure 5a)**. This attachment increases the affinity of RNA polymerase for the *lac* promoter, which is actually rather low even when no *lac* repressor is bound to the operator. By facilitating the binding of RNA polymerase to the promoter and thereby increasing the rate of transcription of the *lac* operon, the attachment of CRP to the promoter directly stimulates gene expression. Therefore, this mechanism qualifies as positive regulation.

If the amount of glucose in the cell increases, the cAMP concentration falls, and without cAMP, CRP detaches from the operon. Because CRP is inactive, RNA polymerase binds less efficiently to the promoter, and transcription of the *lac* operon proceeds at only a low level, even in the presence of lactose **(Figure 5b)**. Thus, the *lac* operon is under dual control: negative control by the *lac* repressor and positive control by CRP. The state of the *lac* repressor (with or without bound allolactose) determines whether or not transcription of the *lac* operon's genes occurs at all; the state of CRP (with or without bound cAMP) controls the *rate* of transcription if the operon is repressor-free. It is as though the operon has both an on-off switch and a volume control.

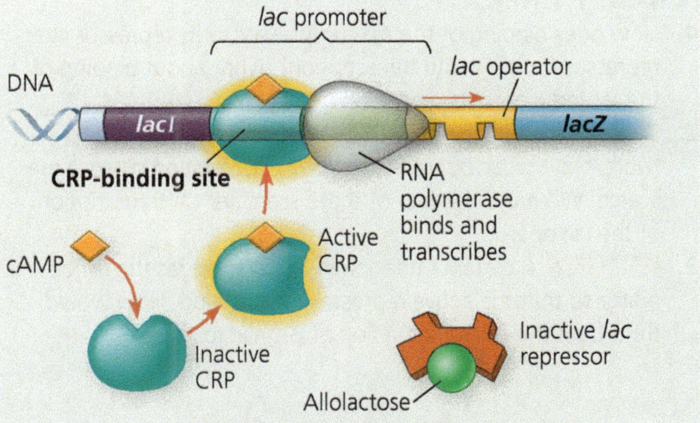

(a) Lactose present, glucose scarce (cAMP level high): abundant *lac* mRNA synthesized. If glucose is scarce, the high level of cAMP activates CRP, which binds to the promoter and increases RNA polymerase binding there. The *lac* operon produces large amounts of mRNA coding for the enzymes that the cell needs for use of lactose.

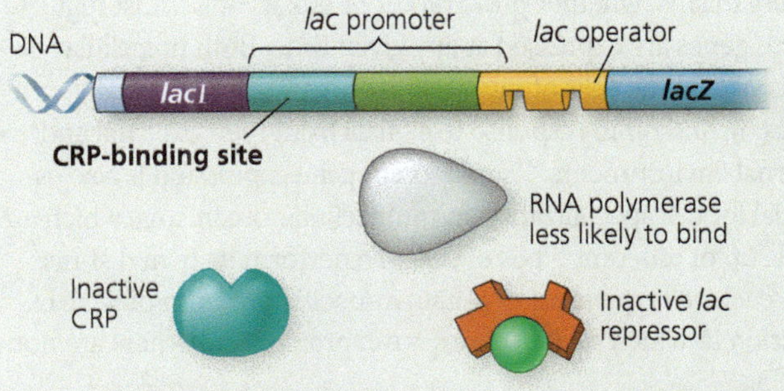

(b) Lactose present, glucose present (cAMP level low): little *lac* mRNA synthesized. When glucose is present, cAMP is scarce, and CRP is unable to stimulate transcription at a significant rate, even though no repressor is bound.

▲ **Figure 5 Positive control of the *lac* operon by cAMP receptor protein (CRP).** RNA polymerase has high affinity for the *lac* promoter only when CRP is bound to a DNA site at the upstream end of the promoter. CRP, in turn, attaches to its DNA site only when associated with cyclic AMP (cAMP), whose concentration in the cell rises when the glucose concentration falls. Thus, when glucose is present, even if lactose also is available, the cell preferentially catabolizes glucose and makes very little of the enzymes for using lactose.

In addition to regulating the *lac* operon, CRP helps regulate other operons that encode enzymes used in catabolic pathways. All told, it may affect the expression of more than 100 genes in *E. coli*. When glucose is plentiful and CRP is inactive, the synthesis of enzymes that catabolize compounds other than glucose generally slows down. The ability to catabolize other compounds, such as lactose, enables a cell deprived of glucose to survive. The compounds present in the cell at any given moment determine which operons are switched on—the result of simple interactions of activator and repressor proteins with the promoters of the genes in question.

CONCEPT CHECK 1

1. How does binding of the *trp* corepressor to its repressor alter repressor function and transcription? What about binding of the *lac* inducer to its repressor?
2. Describe the binding of RNA polymerase, repressors, and activators to the *lac* operon when both lactose and glucose are scarce. What is the effect of these scarcities on transcription of the *lac* operon?
3. **WHAT IF?** A certain mutation in *E. coli* changes the *lac* operator so that the active repressor cannot bind. How would this affect the cell's production of β-galactosidase?

CONCEPT 2

Eukaryotic gene expression is regulated at many stages

All organisms, whether prokaryotes or eukaryotes, must regulate which genes are expressed at any given time. Both unicellular organisms and the cells of multicellular organisms continually turn genes on and off in response to signals from their external and internal environments. Regulation of gene expression is also essential for cell specialization in multicellular organisms, which are made up of different types of cells. To perform its own distinct role, each cell type must maintain a specific program of gene expression in which certain genes are expressed and others are not.

Differential Gene Expression

A typical human cell might express about 20% of its protein-coding genes at any given time. Highly differentiated cells, such as muscle or nerve cells, express an even smaller fraction of their genes. Almost all the cells in a multicellular organism contain an identical genome. (Cells of the immune system are one exception.) A subset of genes is expressed in each cell type; some of those are "housekeeping" genes, expressed by many cell types, while others are unique to that cell type. The uniquely expressed genes allow these cells to carry out their specific function. The differences between cell types, therefore, are due not to different genes being present, but to **differential gene expression**, the expression of different genes by cells with the same genome.

The function of any cell, whether a single-celled eukaryote or a particular cell type in a multicellular organism, depends on the appropriate set of genes being expressed. The transcription factors of a cell must locate the right genes at the right time, a task on a par with finding a needle in a haystack. When gene expression proceeds abnormally, serious imbalances and diseases, including cancer, can arise.

Figure 6 summarizes the process of gene expression in a eukaryotic cell, highlighting key stages in the expression of a protein-coding gene. Each stage depicted in Figure 6 is a potential control point at which gene expression can be turned on or off, accelerated, or slowed down.

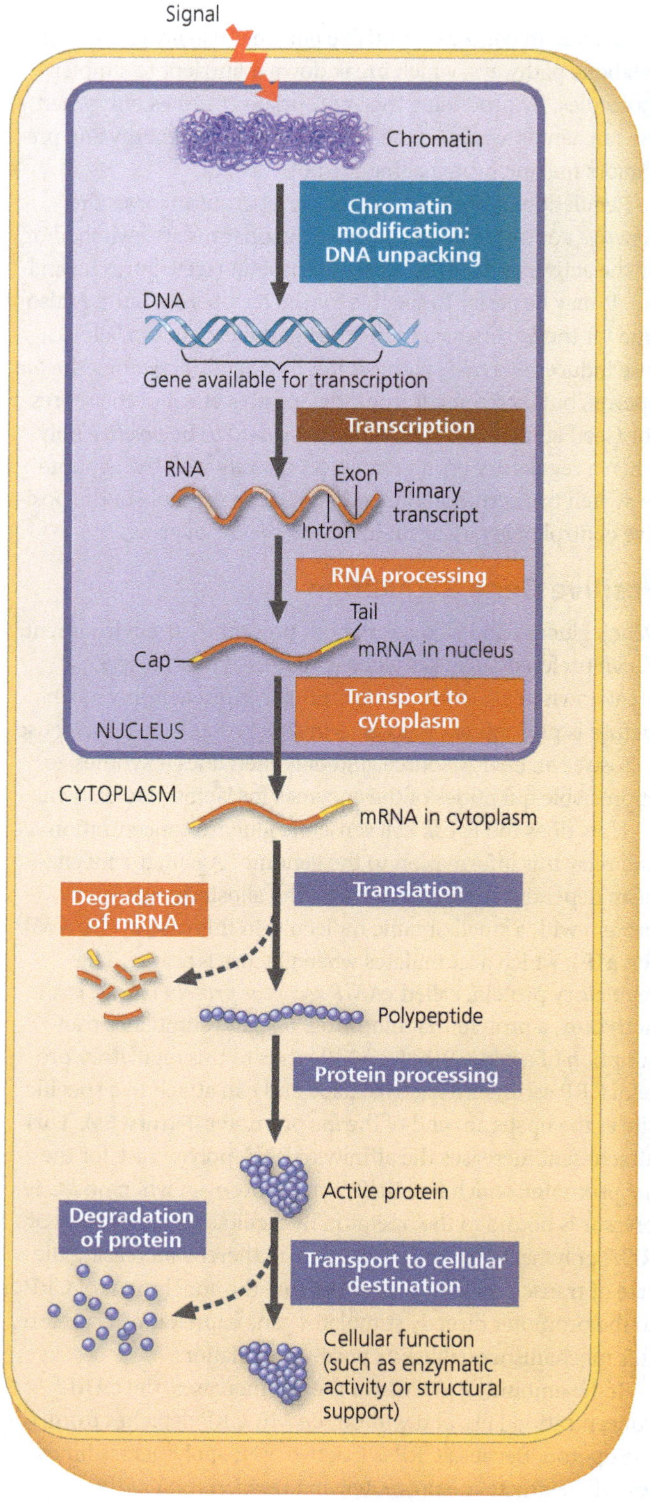

▲ **Figure 6 Stages in gene expression that can be regulated in eukaryotic cells.** In this diagram, the colored boxes indicate the processes most often regulated; each color indicates the type of molecule that is affected (blue = DNA, red/orange = RNA, purple = protein). The nuclear envelope separating transcription from translation in eukaryotic cells offers an opportunity for post-transcriptional control in the form of RNA processing that is absent in prokaryotes. In addition, eukaryotes have a greater variety of control mechanisms operating before transcription and after translation. A miniature version of this figure accompanies several figures later in the chapter as an orientation diagram.

Fifty years ago, an understanding of the mechanisms that control gene expression in eukaryotes seemed almost hopelessly out of reach. Since then, new research methods, notably advances in DNA technology that we will discuss in Concept 4, have enabled molecular biologists to uncover many details of eukaryotic gene regulation. In all organisms, gene expression is commonly controlled at transcription; regulation at this stage often occurs in response to signals coming from outside the cell, such as hormones or other signaling molecules. For this reason, the term *gene expression* is often equated with transcription, for both bacteria and eukaryotes. While this may often be the case for bacteria, the greater complexity of eukaryotic cell structure and function provides opportunities for regulating gene expression at many additional stages (see Figure 6). In the remainder of this section, we'll examine some of the important control points of eukaryotic gene expression more closely.

Regulation of Chromatin Structure

Recall that the DNA of eukaryotic cells is packaged with proteins in an elaborate complex known as chromatin, the basic unit of which is the nucleosome. A nucleosome consists of a cluster of eight histone proteins around which the DNA double helix is wrapped. The structural organization of chromatin not only packs a cell's DNA into a compact form that fits inside the nucleus, but also helps regulate gene expression in several ways. Whether or not a gene is transcribed is affected by the location of nucleosomes along a gene's promoter and also the sites where the promoter DNA attaches to the protein scaffolding of the chromosome. In addition, genes within highly condensed chromatin (heterochromatin) are usually not expressed. Lastly, certain chemical modifications to the histone proteins and to the DNA of chromatin can influence both chromatin structure and gene expression. Here we examine the effects of these modifications, which are catalyzed by specific enzymes.

Histone Modifications and DNA Methylation

Chemical modifications to histones play a direct role in the regulation of gene transcription. The N-terminus of each histone protein in a nucleosome protrudes outward from the nucleosome. These so-called *histone tails* are accessible to various modifying enzymes that catalyze the addition or removal of specific chemical groups, such as acetyl (— $COCH_3$), methyl, and phosphate groups. Generally, **histone acetylation** appears to promote transcription by opening up the chromatin structure **(Figure 7)**, while addition of methyl groups can lead to the condensation of chromatin and reduced transcription.

Rather than modifying proteins associated with DNA, a different set of enzymes can methylate the DNA itself on certain bases, usually cytosine. Such **DNA methylation** occurs in most plants, animals, and fungi. Long stretches of inactive DNA, such as that of inactivated mammalian X chromosomes, are generally more methylated than regions of actively transcribed DNA (although there are exceptions). On a smaller

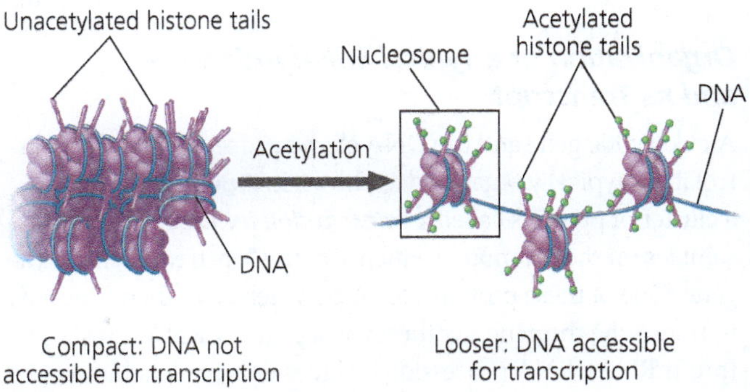

▲ **Figure 7 A simple model of the effect of histone acetylation.** Amino acids in histone tails may be chemically modified. For example, acetyl groups (green balls) may be added, causing the chromatin to be less compact and making the DNA accessible for transcription.

scale, the DNA of individual genes is usually more heavily methylated in cells in which those genes are not expressed. Once methylated, genes usually stay that way through successive cell divisions in a given individual. At DNA sites where one strand is already methylated, enzymes methylate the correct daughter strand after each round of DNA replication. In this way, methylation patterns can be inherited.

Epigenetic Inheritance

The chromatin modifications we just discussed do not change the DNA sequence, yet they still may be passed along to future generations of cells. Inheritance of traits transmitted by mechanisms not involving the nucleotide sequence itself is called **epigenetic inheritance**. Whereas mutations in DNA are permanent, modifications to the chromatin can be reversed. For example, DNA methylation patterns are largely erased and reestablished during gamete formation.

Researchers are amassing more and more evidence for the importance of epigenetic information in the regulation of gene expression. Epigenetic variations might help explain why one identical twin acquires a genetically based disease, such as schizophrenia, but the other does not, despite their identical genomes. Alterations in normal patterns of DNA methylation are seen in some cancers, where they are associated with inappropriate gene expression. Evidently, enzymes that modify chromatin structure are integral parts of the eukaryotic cell's machinery for regulating transcription.

Regulation of Transcription Initiation

Chromatin-modifying enzymes provide initial control of gene expression by making a region of DNA either more or less able to bind the transcription machinery. Once the chromatin of a gene is optimally modified for expression, the initiation of transcription is the next major step at which gene expression is regulated. As in bacteria, the regulation of transcription initiation in eukaryotes involves proteins that bind to DNA and either facilitate or inhibit binding of RNA polymerase. The process is more complicated in eukaryotes, however. Before looking at how eukaryotic cells control their transcription, let's review the structure of a typical eukaryotic gene and its transcript.

Organization of a Typical Eukaryotic Gene and Its Transcript

A eukaryotic gene and the DNA elements (segments) that control it are typically organized as shown in **Figure 8**. Recall that a cluster of proteins called a *transcription initiation complex* assembles on the promoter sequence at the "upstream" end of the gene. One of these proteins, RNA polymerase II, then proceeds to transcribe the gene, synthesizing a primary RNA transcript (pre-mRNA). RNA processing includes enzymatic addition of a 5′ cap and a poly-A tail, as well as splicing out of introns, to yield a mature mRNA. Associated with most eukaryotic genes are multiple **control elements**, segments of noncoding DNA having particular nucleotide sequences that serve as binding sites for the proteins called transcription factors. Control elements on the DNA and the transcription factors that bind to them are critical to the precise regulation of gene expression seen in different cell types.

The Roles of General and Specific Transcription Factors

Transcription factors are of two types: General transcription factors act at the promoter of all genes, while some genes require specific transcription factors that bind to control elements close to or further away from the promoter.

General Transcription Factors at the Promoter To initiate transcription, eukaryotic RNA polymerase requires the assistance of transcription factors. Some transcription factors are essential for the transcription of *all* protein-coding genes; therefore, they are often called *general transcription factors.* A few general transcription factors bind to a DNA sequence such as the TATA box within the promoter, but many bind to proteins, including other transcription factors and RNA polymerase II. Protein-protein interactions are crucial to the initiation of eukaryotic transcription. Only when the complete initiation complex has assembled can the polymerase begin to move along the DNA template strand, producing a complementary strand of RNA.

The interaction of general transcription factors and RNA polymerase II with a promoter usually leads to only a low rate of initiation and production of few RNA transcripts from genes that are not expressed all the time, but, instead, are regulated. In eukaryotes, high levels of transcription of these particular genes at the appropriate time and place depend on the interaction of control elements with another set of proteins, which can be thought of as *specific transcription factors.*

Enhancers and Specific Transcription Factors As you can see in Figure 8, some control elements, named *proximal control elements*, are located close to the promoter. (Although some biologists consider proximal control elements part of the promoter, in this text we do not.) The more distant *distal control elements*, groupings of which are called **enhancers**, may be thousands of nucleotides upstream or downstream of a gene or even within an intron. A given gene may have multiple enhancers, each active at a different time or in a different cell

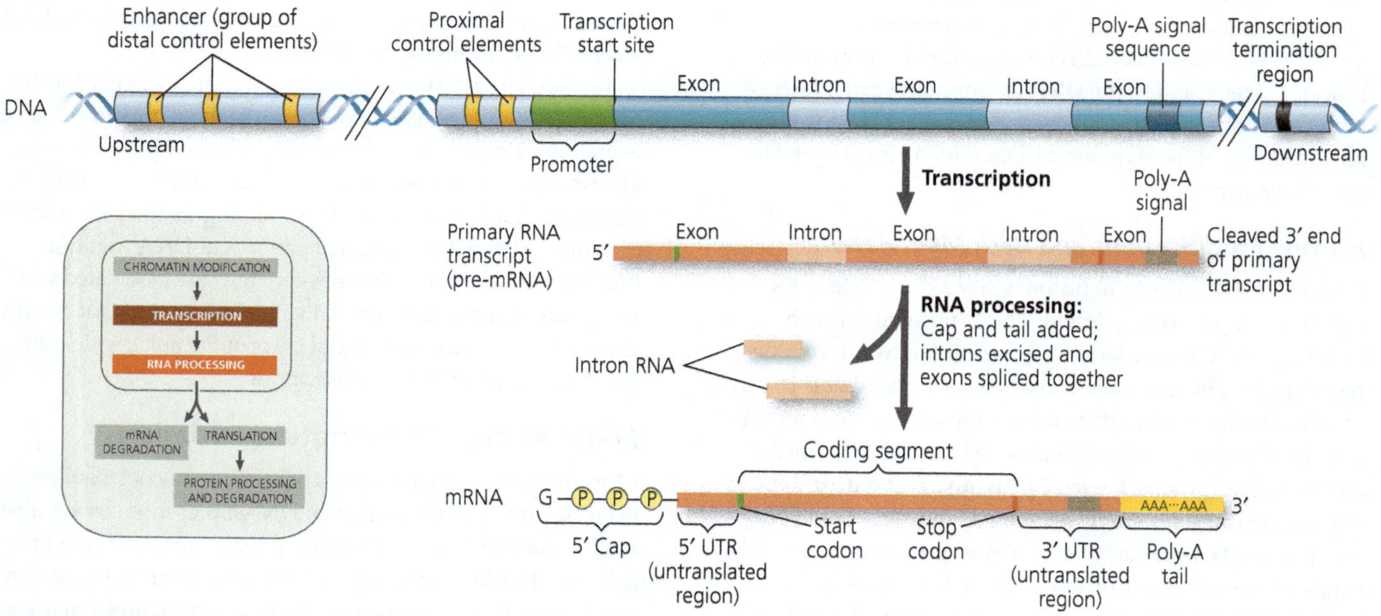

▲ **Figure 8 A eukaryotic gene and its transcript.** Each eukaryotic gene has a promoter, a DNA sequence where RNA polymerase binds and starts transcription, proceeding "downstream." A number of control elements (gold) are involved in regulating the initiation of transcription; these are DNA sequences located near (proximal to) or far from (distal to) the promoter. Distal control elements can be grouped together as enhancers, one of which is shown for this gene. At the other end of the gene, a polyadenylation (poly-A) signal sequence in the last exon of the gene is transcribed into an RNA sequence that signals where the transcript is cleaved and the poly-A tail added. Transcription may continue for hundreds of nucleotides beyond the poly-A signal before terminating. RNA processing of the primary transcript into a functional mRNA involves three steps: addition of the 5′ cap, addition of the poly-A tail, and splicing. In the cell, the 5′ cap is added soon after transcription is initiated, and splicing occurs while transcription is still under way.

type or location in the organism. Each enhancer, however, is generally associated with only that gene and no other.

In eukaryotes, the rate of gene expression can be strongly increased or decreased by the binding of specific transcription factors, either activators or repressors, to the control elements of enhancers. Hundreds of transcription activators have been discovered in eukaryotes; the structure of one example is shown in **Figure 9**. Researchers have identified two types of structural domains that are commonly found in a large number of activator proteins: a *DNA-binding domain*—a part of the protein's three-dimensional structure that binds to DNA—and one or more *activation domains*. Activation domains bind other regulatory proteins or components of the transcription machinery, facilitating a series of protein-protein interactions that result in enhanced transcription of a given gene.

Figure 10 shows the currently accepted model for how binding of activators to an enhancer located far from the promoter can influence transcription. Protein-mediated bending of the DNA brings the bound activators into contact with a

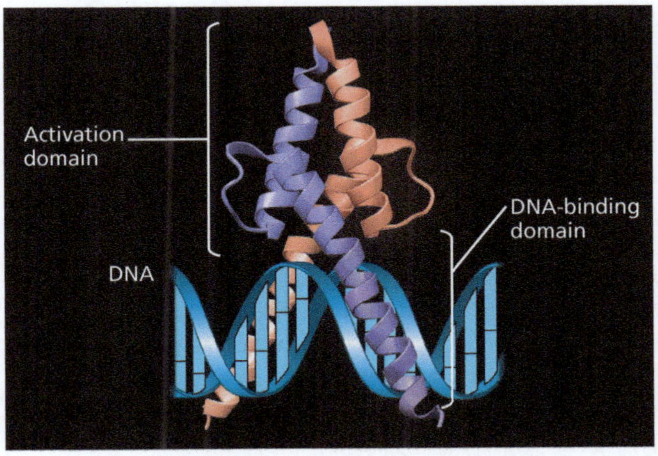

▲ **Figure 9 MyoD, a transcriptional activator.** The MyoD protein is made up of two subunits (purple and salmon) with extensive regions of α helix. Each subunit has one DNA-binding domain and one activation domain. The latter includes binding sites for the other subunit and for other proteins. MyoD is involved in muscle development in vertebrate embryos.

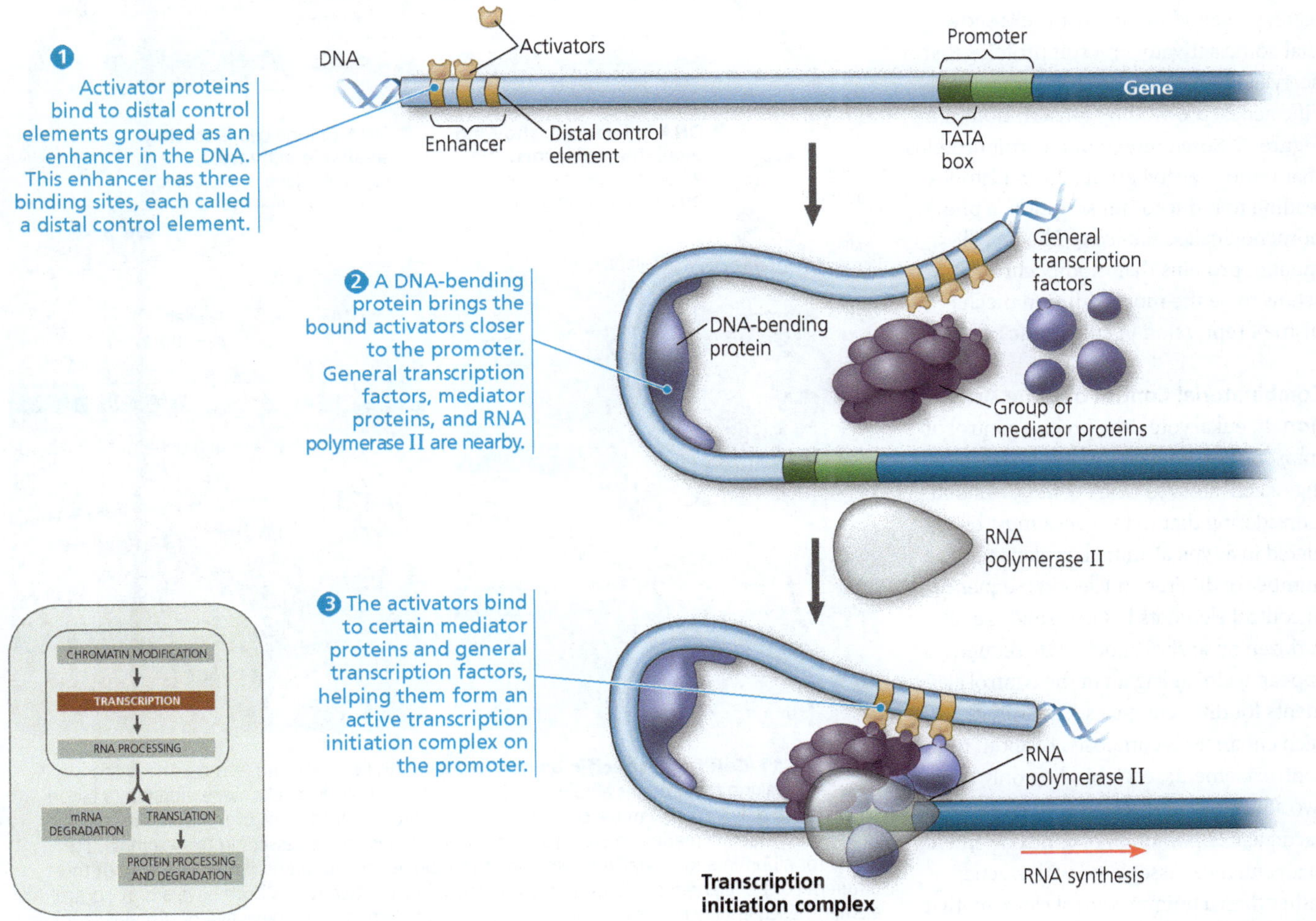

▲ **Figure 10 A model for the action of enhancers and transcription activators.** Bending of the DNA by a protein enables enhancers to influence a promoter hundreds or even thousands of nucleotides away. Specific transcription factors called activators bind to the enhancer DNA sequences and then to a group of mediator proteins, which in turn bind to general transcription factors and then RNA polymerase II, assembling the transcription initiation complex. These protein-protein interactions lead to correct positioning of the complex on the promoter and the initiation of RNA synthesis. Only one enhancer (with three gold control elements) is shown here, but a gene may have several enhancers that act at different times or in different cell types.

group of *mediator proteins*, which interact with proteins at the promoter. These interactions help assemble and position the initiation complex on the promoter. One of the studies supporting this model shows that proteins regulating a mouse globin gene contact both the gene's promoter and an enhancer about 50,000 nucleotides upstream. Protein interactions allow these two DNA regions to come together in a very specific fashion, in spite of the many nucleotide pairs between them. In the **Scientific Skills Exercise**, you can work with data from an experiment that identified the control elements in the enhancer of a particular human gene.

Specific transcription factors that function as repressors can inhibit gene expression in several different ways. Some repressors bind directly to control element DNA, blocking activator binding. Other repressors interfere with the activator itself so it can't bind the DNA.

In addition to influencing transcription directly, some activators and repressors indirectly affect chromatin structure. Studies using yeast and mammalian cells show that some activators recruit proteins that acetylate histones near promoters of specific genes, promoting transcription (see Figure 7). Some repressors recruit proteins that remove acetyl groups from histones, leading to reduced transcription, a phenomenon called *silencing*. The recruitment of proteins that modify chromatin seems to be the most common mechanism of repression in eukaryotic cells.

Combinatorial Control of Gene Activation In eukaryotes, the precise control of transcription depends largely on the binding of activators to DNA control elements. Considering that many genes must be regulated in a typical animal or plant cell, the number of different nucleotide sequences in control elements is surprisingly small. A dozen or so short nucleotide sequences appear again and again in the control elements for different genes. On average, each enhancer is composed of about ten control elements, each binding only one or two specific transcription factors. It is the particular *combination* of control elements in an enhancer associated with a gene, rather than a unique control element, that is important in regulating transcription of the gene.

Even with only a dozen control element sequences available, many combinations are possible. Each combination can activate transcription only when the appropriate activator proteins are present, which may occur at a precise time during development or in a particular cell type. **Figure 11** shows how different combinations of a few control elements can allow differential regulation of transcription in two cell types—liver and lens cells. This can occur because each cell type contains a different group of activator proteins.

Coordinately Controlled Genes in Eukaryotes

How does the eukaryotic cell deal with a group of genes of related function that need to be turned on or off at the same time? You have learned that in bacteria such *coordinately*

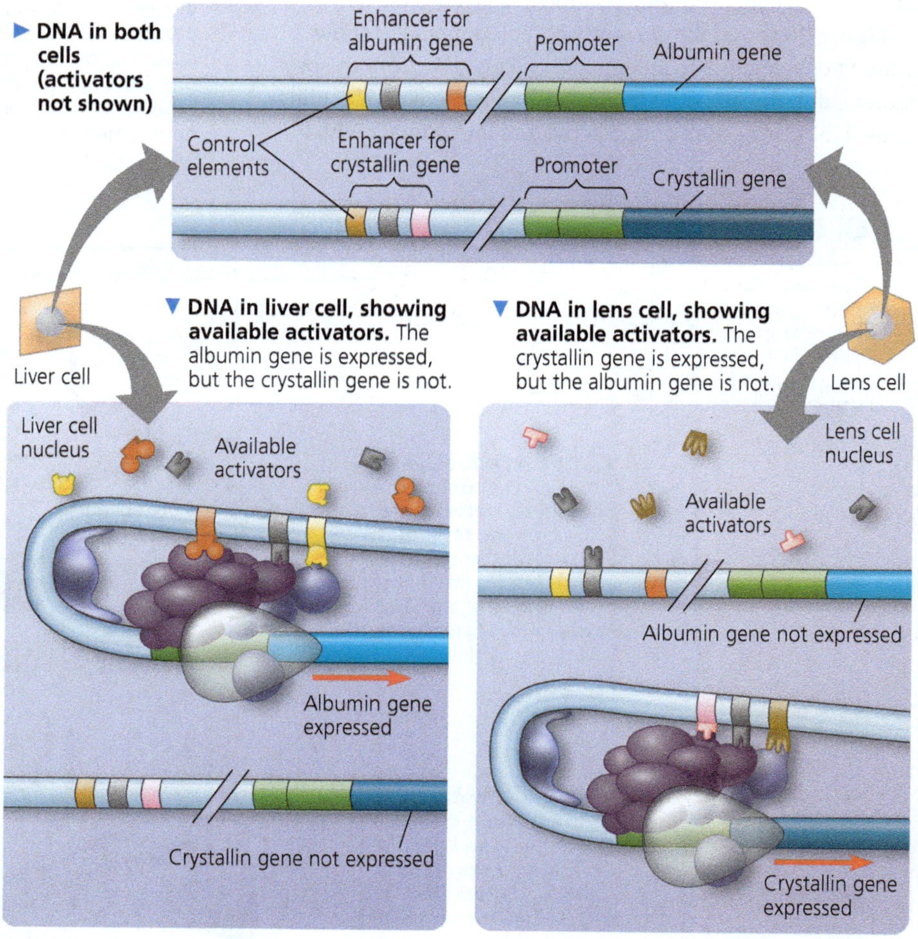

▲ **Figure 11 Cell type–specific transcription.** Both liver cells and lens cells have the genes for making the proteins albumin and crystallin, but only liver cells make albumin (a blood protein) and only lens cells make crystallin (the main protein of the lens of the eye). The specific transcription factors made in a cell determine which genes are expressed. In this example, the genes for albumin and crystallin are shown at the top, each with an enhancer made up of three different control elements. Although the enhancers for the two genes both have a gray control element, each enhancer has a unique *combination* of elements. All the activator proteins required for high-level expression of the albumin gene are present in liver cells only **(left)**, whereas the activators needed for expression of the crystallin gene are present in lens cells only **(right)**. For simplicity, we consider only the role of activators here, although the presence or absence of repressors may also influence transcription in certain cell types.

? *Describe the enhancer for the albumin gene in each type of cell. How would the nucleotide sequence of this enhancer in the liver cell compare with that in the lens cell?*

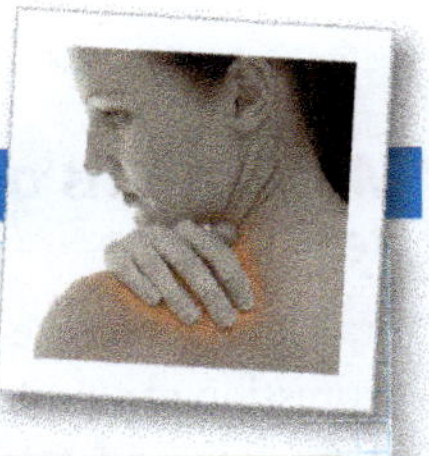

Scientific Skills Exercise

Analyzing DNA Deletion Experiments

What Control Elements Regulate Expression of the *mPGES-1* Gene? The promoter of a gene includes the DNA immediately upstream of the transcription start site, but the control elements regulating the level of transcription of the gene (grouped in an enhancer) may be thousands of base pairs upstream of the promoter. Because the distance and spacing of control elements make them difficult to identify, scientists begin by deleting possible control elements and measuring the effect on gene expression. In this exercise, you will analyze data obtained from DNA deletion experiments that tested possible control elements for the human gene *mPGES-1*. This gene codes for an enzyme that synthesizes a type of prostaglandin, a chemical made during tissue inflammation.

How the Experiment Was Done The researchers hypothesized that there were three possible control elements in an enhancer region located 8–9 kilobases upstream of the *mPGES-1* gene. Control elements regulate whatever gene is in the appropriate downstream location. Thus, to test the activity of the possible elements, researchers first synthesized molecules of DNA ("constructs") that had the intact enhancer region upstream of a "reporter gene," a gene whose mRNA product could be easily measured experimentally. Next, they made three more DNA constructs with one of the three proposed control elements deleted in each (see left side of figure). The researchers then introduced each DNA construct into a separate human cell culture, where the cells took up the DNA constructs. After 48 hours, the amount of reporter gene mRNA made by the cells was measured. Comparing these amounts allowed researchers to determine if any of the deletions had an effect on expression of the reporter gene, mimicking the effect of the deletions on *mPGES-1* gene expression. (The *mPGES-1* gene itself couldn't be used to measure expression levels because the cells express their own *mPGES-1* gene. The mRNA from that gene would confuse the results.)

Data from the Experiment The diagrams on the left side of the figure show the intact DNA sequence (top) and the three experimental DNA constructs. A red X is located on the possible control element (1, 2, or 3) that was deleted in each experimental DNA construct. The area between the slashes represents the approximately 8 kilobases of DNA located between the promoter and the enhancer region. The horizontal bar graph on the right shows the amount of reporter gene mRNA that was present in each cell culture after 48 hours relative to the amount that was in the culture containing the intact enhancer region (top bar = 100%).

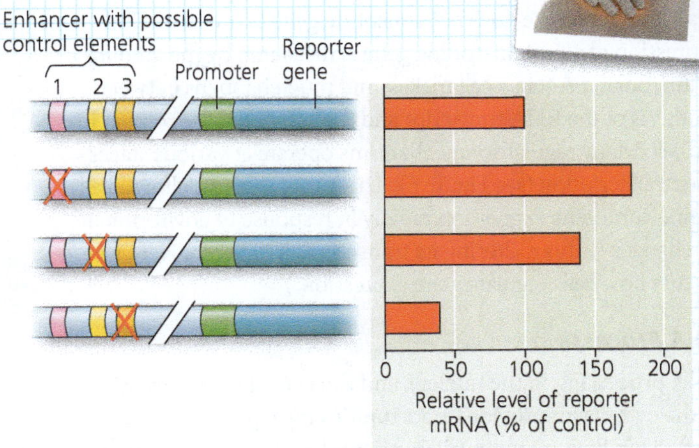

Data from J. N. Walters et al., Regulation of human microsomal prostaglandin E synthase-1 by IL-1β requires a distal enhancer element with a unique role for C/EBPβ, *Biochemical Journal* 443:561–571 (2012).

INTERPRET THE DATA

1. (a) What is the independent variable in the graph? (b) What is the dependent variable? (c) What was the control treatment in this experiment? Label it on the diagram.
2. Do the data suggest that any of these possible control elements are actual control elements? Explain.
3. (a) Did deletion of any of the possible control elements cause a *reduction* in reporter gene expression? If so, which one(s), and how can you tell? (b) If loss of a control element causes a reduction in gene expression, what must be the normal role of that control element? Provide a biological explanation for how the loss of such a control element could lead to a reduction in gene expression.
4. (a) Did deletion of any of the possible control elements cause an *increase* in reporter gene expression relative to the control? If so, which one(s), and how can you tell? (b) If loss of a control element causes an increase in gene expression, what must be the normal role of that control element? Propose a biological explanation for how the loss of such a control element could lead to an increase in gene expression.

MB A version of this Scientific Skills Exercise can be assigned in MasteringBiology.

controlled genes are often clustered into an operon regulated by a promoter and transcribed into an mRNA molecule. Thus, the genes are expressed together, and the encoded proteins are produced concurrently. With few exceptions, operons that work in this way have *not* been found in eukaryotic cells.

Eukaryotic genes that are co-expressed are typically scattered over different chromosomes. Thus, coordinate gene expression depends on every gene of a dispersed group having a specific combination of control elements. Activator proteins bind to the control elements, promoting simultaneous transcription of the genes, no matter where they are in the genome.

Coordinate control of dispersed genes in a eukaryotic cell often occurs in response to chemical signals from outside the cell. A steroid hormone, for example, enters a cell and binds to a receptor protein, forming a hormone-receptor complex that acts as a transcription activator. Every gene whose transcription is stimulated by a given steroid hormone, on any chromosome, has a control element recognized by that hormone-receptor complex. This is how estrogen activates a group of genes that stimulate cell division in uterine cells, preparing the uterus for pregnancy.

Many signaling molecules, such as nonsteroid hormones and growth factors, bind to receptors on a cell's surface and never enter the cell. Such molecules can control gene expression indirectly by triggering signal transduction pathways that activate particular transcription factors. Coordinate regulation in such pathways is the same as for steroid hormones: Genes with the same sets of control elements are activated by the same chemical signals. Because this system for coordinating gene regulation is so widespread, biologists think that it probably arose early in evolutionary history.

Mechanisms of Post-transcriptional Regulation

Transcription alone does not constitute gene expression. The expression of a protein-coding gene is measured by the amount of functional protein a cell makes, and much happens between synthesis of the RNA transcript and the activity of the protein in the cell. Many regulatory mechanisms operate at various stages after transcription (see Figure 6). These mechanisms allow a cell to fine-tune gene expression rapidly in response to environmental changes without altering its transcription patterns. Here we discuss how cells regulate gene expression.

RNA Processing

RNA processing in the nucleus and the export of mature RNA to the cytoplasm provide opportunities for regulating gene expression that are not available in prokaryotes. One example of regulation at the RNA-processing level is **alternative RNA splicing**, in which different mRNA molecules are produced from the same primary transcript, depending on which RNA segments are treated as exons and which as introns. Regulatory proteins specific to a cell type control intron-exon choices by binding to RNA sequences within the primary transcript.

A simple example of alternative RNA splicing is shown in **Figure 12** for the troponin T gene, which encodes two different (though related) proteins. Other genes code for many more possible products. For instance, researchers have found a gene in *Drosophila* with enough alternatively spliced exons to generate about 19,000 membrane proteins with different extracellular domains. At least 17,500 (94%) of the alternative mRNAs are actually synthesized. Each developing nerve cell in the fly appears to synthesize a different form of the protein, which acts as a unique identifier on the cell surface.

It is clear that alternative RNA splicing can significantly expand the repertoire of a eukaryotic genome. In fact, alternative splicing was proposed as one explanation for the surprisingly low number of human genes counted when the human genome was sequenced. The number of human genes was found to be similar to that of a soil worm (nematode), mustard plant, or sea anemone. Scientists wondered what, if not the total number of genes, accounts for the more complex morphology (external form) of humans. It turns out that more than 90% of human protein-coding genes probably undergo alternative splicing. Thus, the extent of alternative splicing greatly multiplies the number of possible human proteins, which may be better correlated with complexity of form than the number of genes.

▼ Figure 12 Alternative RNA splicing of the troponin T gene. The primary transcript of this gene can be spliced in more than one way, generating different mRNA molecules. Notice that one mRNA molecule has ended up with exon 3 (green) and the other with exon 4 (purple). These two mRNAs are translated into different but related muscle proteins.

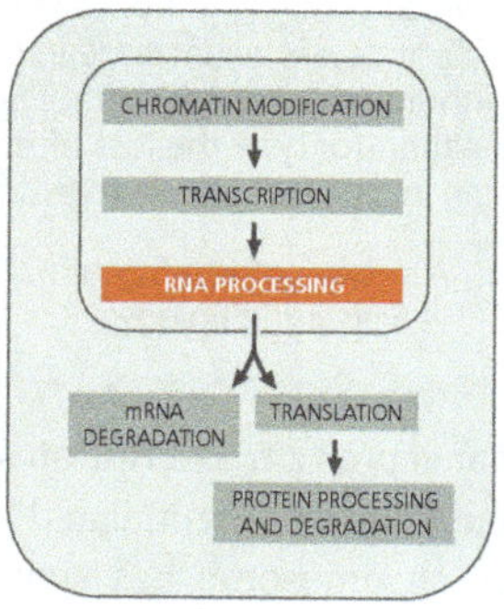

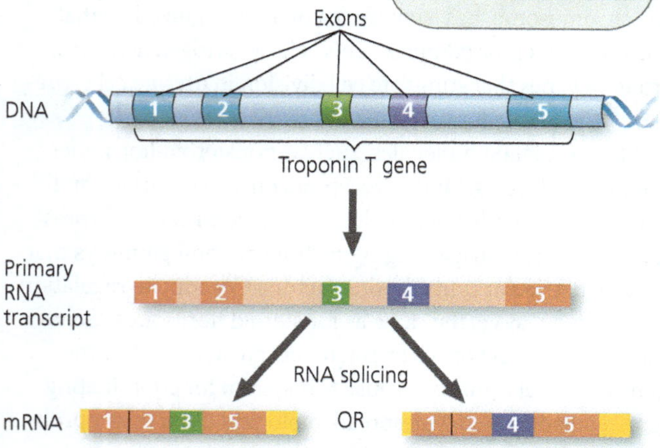

Initiation of Translation and mRNA Degradation

Translation is another occasion when gene expression is regulated, most commonly at the initiation stage. For some mRNAs, the initiation of translation can be blocked by regulatory proteins that bind to specific sequences or structures within the untranslated region (UTR) at the 5′ or 3′ end, preventing the attachment of ribosomes. (Recall that both the 5′ cap and the poly-A tail of an mRNA molecule are important for ribosome binding.)

Alternatively, translation of *all* the mRNAs in a cell may be regulated simultaneously. In a eukaryotic cell, such "global" control usually involves activation or inactivation of one or more protein factors required to initiate translation. This mechanism plays a role in starting translation of mRNAs stored in eggs. Just after fertilization, translation is triggered by sudden activation of translation initiation factors. The response is a burst of synthesis of proteins encoded by the stored mRNAs. Some plants and algae store mRNAs during periods of darkness; light triggers reactivation of the translational apparatus.

The life span of mRNA molecules in the cytoplasm is important in determining the pattern of protein synthesis in a cell. Bacterial mRNA molecules typically are degraded by enzymes within a few minutes. This short life span of mRNAs is one reason bacteria can change their patterns of protein synthesis so quickly in response to environmental changes. In contrast, mRNAs in multicellular eukaryotes typically survive for hours, days, or even weeks. For instance, the mRNAs for the hemoglobin polypeptides (α-globin and β-globin) in developing red blood cells are unusually stable, and these mRNAs are translated repeatedly in these cells. Nucleotide sequences that affect how long an mRNA remains intact are often found in the untranslated region (UTR) at the 3′ end of the molecule (see Figure 8).

Other mechanisms that degrade or block expression of mRNA molecules have come to light. They involve a group of newly discovered RNA molecules that regulate gene expression at several levels, as we'll discuss shortly.

Protein Processing and Degradation

The final opportunities for controlling gene expression occur after translation. Often, eukaryotic polypeptides must be processed to yield functional protein molecules. For instance, cleavage of the initial insulin polypeptide (pro-insulin) forms the active hormone. In addition, many proteins undergo chemical modifications that make them functional. Regulatory proteins are commonly activated or inactivated by the reversible addition of phosphate groups, and proteins destined for the surface of animal cells acquire sugars. Cell-surface proteins and many others must also be transported to target destinations in the cell in order to function. Regulation might occur at any of the steps involved in modifying or transporting a protein.

Finally, the length of time each protein functions in the cell is strictly regulated by selective degradation. Many proteins, such as the cyclins involved in regulating the cell cycle, must be short-lived if the cell is to function appropriately. To mark a protein for destruction, the cell commonly attaches molecules of a small protein called ubiquitin to the protein, which triggers its destruction by protein complexes.

CONCEPT CHECK 2

1. In general, what are the effects of histone acetylation and DNA methylation on gene expression?
2. Compare the roles of general and specific transcription factors in regulating gene expression.
3. **WHAT IF?** Suppose you compared the nucleotide sequences of the distal control elements in the enhancers of three genes that are expressed only in muscle cells. What would you expect to find? Why?

CONCEPT 3

Noncoding RNAs play multiple roles in controlling gene expression

Genome sequencing has revealed that protein-coding DNA accounts for only 1.5% of the human genome and a similarly small percentage of the genomes of many other multicellular eukaryotes. A very small fraction of the non-protein-coding DNA consists of genes for RNAs such as ribosomal RNA and transfer RNA. Scientists assumed until recently that most of the remaining DNA was not transcribed, thinking that since it didn't specify proteins or the few known types of RNA, such DNA didn't contain meaningful genetic information—in fact, it was called "junk DNA." However, a flood of recent data has contradicted this idea. For example, a massive study of the entire human genome showed that roughly 75% of the genome is transcribed at some point in any given cell. Introns account for only a fraction of this transcribed RNA, most of which is untranslated. These and other results suggest that a significant amount of the genome may be transcribed into non-protein-coding RNAs—also called *noncoding RNAs*, including a variety of small RNAs. Researchers are uncovering more evidence of the biological roles of these RNAs every day.

These discoveries have revealed a large, diverse population of RNA molecules in the cell that play crucial roles in regulating gene expression—and that have gone largely unnoticed until recently. Our long-standing view, that because mRNAs code for proteins they are the most important RNAs functioning in the cell, demands revision. This represents a major shift in the thinking of biologists, one that you are witnessing as students entering this field of study.

Effects on mRNAs by MicroRNAs and Small Interfering RNAs

Regulation by both small and large noncoding RNAs occurs at several points in the pathway of gene expression, including mRNA translation and chromatin modification. We'll examine two types of small noncoding RNAs, the importance of which was acknowledged when their discovery was the focus of the 2006 Nobel Prize in Physiology or Medicine.

Since 1993, a number of research studies have uncovered small, single-stranded RNA molecules, called **microRNAs (miRNAs)**, that are capable of binding to complementary sequences in mRNA molecules. A longer RNA precursor is processed by cellular enzymes into an miRNA, a single-stranded RNA of about 22 nucleotides that forms a complex with one or more proteins **(Figure 13)**. The miRNA allows the complex

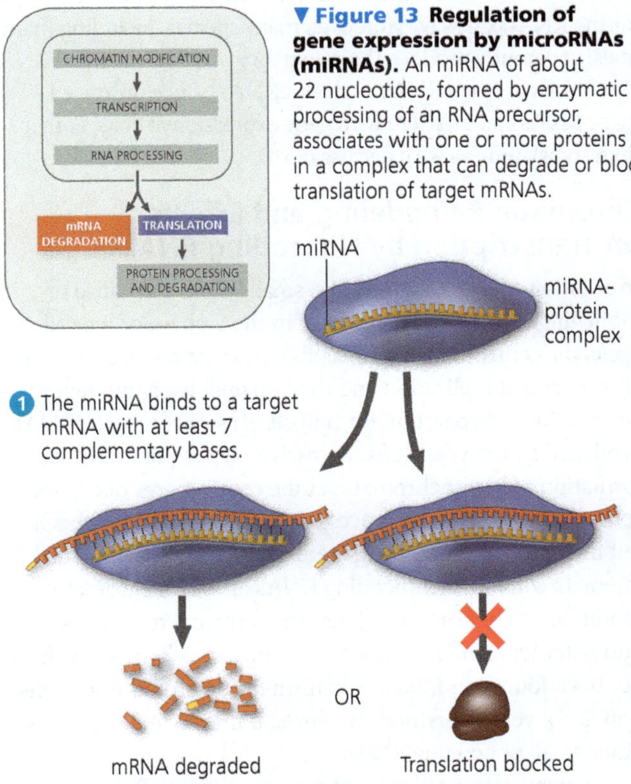

▼ Figure 13 Regulation of gene expression by microRNAs (miRNAs). An miRNA of about 22 nucleotides, formed by enzymatic processing of an RNA precursor, associates with one or more proteins in a complex that can degrade or block translation of target mRNAs.

2 If miRNA and mRNA bases are complementary all along their length, the mRNA is degraded (left); if the match is less complete, translation is blocked (right).

to bind to any mRNA molecule with at least 7 or 8 nucleotides of complementary sequence. The miRNA-protein complex then either degrades the target mRNA or blocks its translation. Biologists estimate that expression of at least one-half of all human genes may be regulated by miRNAs, a remarkable figure given that the existence of miRNAs was unknown 25 years ago.

Another class of small RNAs are called **small interfering RNAs (siRNAs)**. These are similar in size and function to miRNAs—both can associate with the same proteins, producing similar results. In fact, if siRNA precursor RNA molecules are injected into a cell, the cell's machinery can process them into siRNAs that turn off expression of genes with related sequences, similarly to how miRNAs function. The distinction between miRNAs and siRNAs is based on subtle differences in the structure of their precursors, which in both cases are RNA molecules that are mostly double-stranded. The blocking of gene expression by siRNAs is called **RNA interference (RNAi)**, and it is used in the laboratory as a means of disabling specific genes to investigate their function.

EVOLUTION How did the RNAi pathway evolve? Some viruses have double-stranded RNA genomes. Given that the cellular RNAi pathway can process double-stranded RNAs into homing devices that lead to the destruction of related RNAs, some researchers hypothesize that this pathway may have evolved as a natural defense against infection by such viruses. However, the fact that RNAi can also affect the expression of nonviral cellular genes may reflect a different evolutionary origin for the RNAi pathway. Moreover, many species, including mammals, apparently produce their own long, double-stranded RNA precursors to small RNAs such as siRNAs. Once produced, these RNAs can interfere with gene expression at stages other than translation, as we'll discuss next.

Chromatin Remodeling and Effects on Transcription by Noncoding RNAs

In addition to affecting mRNAs, small RNAs can cause remodeling of chromatin structure. In the S phase of the cell cycle, the centromeric regions of DNA must be loosened for chromosomal replication and then recondensed into heterochromatin in preparation for mitosis. In some yeasts, siRNAs produced by the yeast cells themselves are required for this formation of heterochromatin at the centromeres of chromosomes. Exactly how the process occurs is still under debate, but biologists agree on the general idea: The siRNA system interacts with other noncoding RNAs and with chromatin-modifying enzymes to condense the centromere chromatin into heterochromatin. In most mammalian cells, siRNAs have not been found, and the mechanism for centromere condensation is not yet understood. However, it may turn out to involve other small noncoding RNAs.

A newly discovered class of small noncoding RNAs called *piwi-associated RNAs* (*piRNAs*) also induces formation of heterochromatin, blocking expression of some parasitic DNA elements in the genome known as transposons. Usually 24–31 nucleotides in length, piRNAs are probably processed from a longer, single-stranded RNA precursor. They play an indispensable role in the germ cells of many animal species, where they appear to help reestablish appropriate methylation patterns in the genome during gamete formation.

The role of noncoding RNAs in regulation of gene expression adds yet another layer to the complex and intricate process described in the previous section. As more is learned about the multiple interacting ways a cell can fine-tune expression of its genes, the goal is to understand how a specific set of genes is expressed in a particular cell. In the next section, we'll describe a few methods that researchers use to monitor expression of specific genes, such as in different cell types.

CONCEPT CHECK 3

1. **WHAT IF?** Suppose the mRNA being degraded in Figure 13 coded for a protein that promotes cell division in a multicellular organism. What would happen if a mutation disabled the gene for the miRNA that triggers this degradation?
2. **MAKE CONNECTIONS** Inactivation of one of the X chromosomes in female mammals results in a Barr body. Suggest a model for how noncoding RNA functions to cause Barr body formation.

CONCEPT 4

Researchers can monitor expression of specific genes

The diverse mechanisms of regulating gene expression discussed in this chapter underlie one basic generality: Cells of a given multicellular organism differ from each other because they express different genes from an identical genome. Biologists driven to understand the assorted cell types of a multicellular organism, cancer cells, or the developing tissues of an embryo first try to discover which genes are expressed by the cells of interest. The most straightforward way to do this is usually to identify the mRNAs being made. Techniques related to those developed for genetic engineering are widely used to track expression of mRNAs. In this section we'll first examine techniques that look for patterns of expression of specific individual genes. Next, we'll explore techniques that characterize groups of genes being expressed by cells or tissues of interest. As you will see, all of these techniques depend in some way on base pairing between complementary nucleotide sequences.

Studying the Expression of Single Genes

Suppose we have cloned a gene that may play an important role in the embryonic development of *Drosophila* (the fruit fly). The first thing we might want to know is which embryonic cells express the gene—in other words, where in the embryo is the corresponding mRNA found? We can detect the mRNA using the technique of **nucleic acid hybridization**, the base pairing of one strand of a nucleic acid to the complementary sequence on another strand. The complementary molecule, a short, single-stranded nucleic acid that can be either RNA or DNA, is called a **nucleic acid probe**. Using our cloned gene as a template, we can synthesize a probe complementary to the mRNA. For example, if part of the sequence on the mRNA were

5′ ···CUCAUCACCGGC··· 3′

then we would synthesize this single-stranded DNA probe:

3′ GAGTAGTGGCCG 5′

Each probe molecule is labeled with a fluorescent tag so we can follow it. A solution with the probe is applied to *Drosophila* embryos, allowing probe molecules to hybridize to any complementary sequences on the many mRNAs in embryonic cells in which the gene is being transcribed **(Figure 14)**.

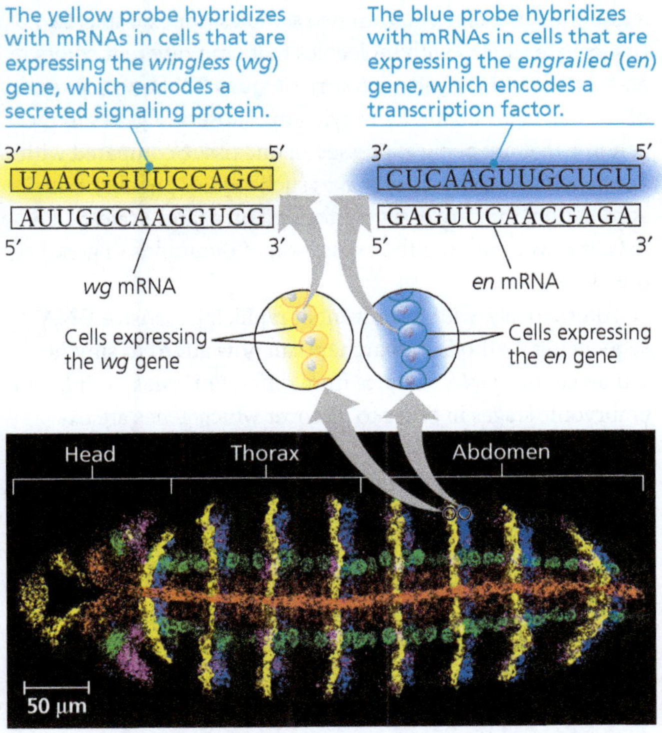

▲ **Figure 14 Determining where genes are expressed by *in situ* hybridization analysis.** This *Drosophila* embryo was incubated in a solution containing probes for five different mRNAs, each probe labeled with a different fluorescently colored tag. The embryo was then viewed using fluorescence microscopy. Each color marks cells in which a specific gene is expressed as mRNA. The arrows from the groups of yellow and blue cells above the micrograph point to a magnified view of nucleic acid hybridization of the appropriately colored probe to the mRNA.

Because this technique allows us to see the mRNA in place (or *in situ*, in Latin) in the intact organism, it is called ***in situ* hybridization**. Different probes can be labeled with different fluorescent dyes, sometimes with strikingly beautiful results. (see Figure 14).

Other mRNA detection techniques may be preferable for comparing the amounts of a specific mRNA in several samples at the same time—for example, in different cell types or in embryos of different stages. One method that is widely used is called the **reverse transcriptase–polymerase chain reaction (RT-PCR)**. RT-PCR begins by turning sample sets of mRNAs into double-stranded DNAs with the corresponding sequences. This feat is accomplished by an enzyme called *reverse transcriptase*, isolated in the late 1980s from a type of virus called a retrovirus. Reverse transcriptase is able to synthesize a complementary DNA copy of an mRNA, thus making a *reverse transcript* **(Figure 15)**. Following enzymatic degradation of the mRNA, a second DNA strand, complementary to the first, is synthesized by DNA polymerase. The resulting double-stranded DNA is called **complementary DNA (cDNA)**, and the reverse transcription step accounts for the "RT" in the name RT-PCR. To analyze the timing of expression of the *Drosophila* gene of interest, for example, we would first isolate all

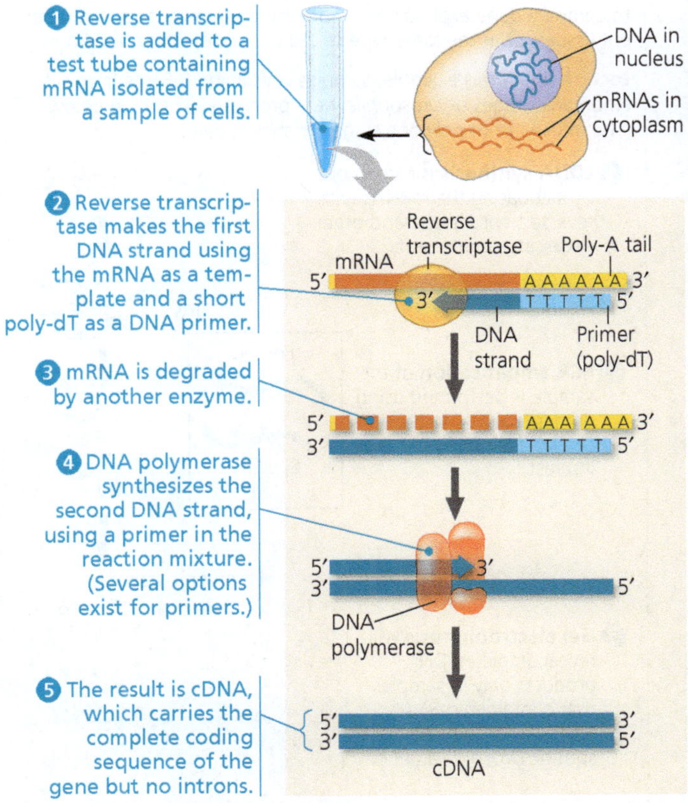

▲ **Figure 15 Making complementary DNA (cDNA) from eukaryotic genes.** Complementary DNA is DNA made in a test tube using mRNA as a template for the first strand. Only one mRNA is shown here, but the final collection of cDNAs would reflect all the mRNAs that were present in the cell.

the mRNAs from different stages of *Drosophila* embryos and then make cDNA from each stage **(Figure 16)**.

Next in RT-PCR is the PCR step. As you may recall, PCR is a way of rapidly making many copies of one specific stretch of double-stranded DNA, using primers that hybridize to the opposite ends of the region of interest. In our case, we would add primers corresponding to a region of our *Drosophila* gene, using the cDNA from each sample as a template for PCR amplification. When the products are run on a gel, copies of the amplified region will be observed as bands only in samples that originally contained mRNA from the gene we're focusing on. A recent enhancement involves using a fluorescent dye that fluoresces only when bound to a double-stranded PCR product. The newer PCR machines can detect the light and measure the PCR product, thus avoiding the need for electrophoresis while also providing quantitative data, a distinct advantage. RT-PCR can also be carried out with mRNAs collected from different tissues at one time to discover which tissue is producing a specific mRNA.

▼ Figure 16 Research Method

RT-PCR Analysis of the Expression of Single Genes

Application RT-PCR uses the enzyme reverse transcriptase (RT) in combination with PCR and gel electrophoresis. RT-PCR can be used to compare gene expression in different embryonic stages, in different tissues, or in the same type of cell under different conditions.

Technique In this example, samples containing mRNAs from six embryonic stages of *Drosophila* were processed as shown below. (The mRNAs from only one stage are shown here.)

1. **cDNA synthesis** is carried out by incubating the mRNAs with reverse transcriptase and other necessary components.
2. **PCR amplification** of the sample is performed using primers specific to the *Drosophila* gene of interest.
3. **Gel electrophoresis** will reveal amplified DNA products only in samples that contained mRNA transcribed from the specific *Drosophila* gene.

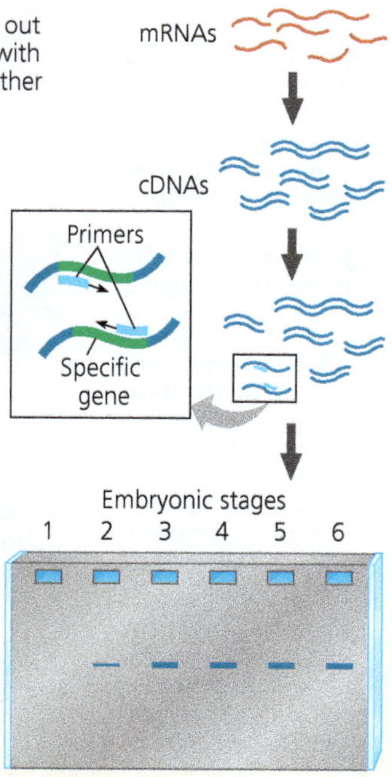

Results The mRNA for this gene is expressed from stage 2 through stage 6. The size of the amplified fragment (shown by its position on the gel) depends on the distance between the primers that were used.

Studying the Expression of Groups of Genes

A major goal of biologists is to learn how genes act together to produce and maintain a functioning organism. Now that the genomes of a number of species have been sequenced, it is possible to study the expression of large groups of genes—the so-called *systems approach*. Researchers use what is known about the whole genome to investigate which groups of genes are transcribed in different tissues or at different stages of development. One aim is to identify networks of gene expression across an entire genome.

Genome-wide expression studies can be carried out using **DNA microarray assays**. A DNA microarray consists of tiny amounts of a large number of single-stranded DNA fragments representing different genes fixed to a glass slide in a tightly spaced array, or grid. (The microarray is also called a *DNA chip* by analogy to a computer chip.) Ideally, these fragments represent all the genes in the genome of an organism.

The basic strategy in such studies is to isolate the mRNAs made in a cell of interest and use these mRNAs as templates for making the corresponding cDNAs by reverse transcription. In microarray assays, these cDNAs are labeled with fluorescent molecules and then allowed to hybridize to a DNA microarray. Most often, the cDNAs from two samples (for example, two tissues) are labeled with molecules that emit different colors and tested on the same microarray. **Figure 17** shows the result of such an experiment, identifying the subsets of genes in the genome that are being expressed in one tissue compared with another. DNA technology makes such studies possible; with automation, they are easily performed on a large scale. Scientists can now measure the expression of thousands of genes at one time.

Alternatively, with the advent of rapid, inexpensive DNA-sequencing methods, researchers can now afford to simply sequence the cDNA samples from different tissues or different embryonic stages in order to discover which genes are expressed. This straightforward method is called *RNA sequencing* or *RNA-seq*, even though it is the cDNA that is actually sequenced. As the price of sequencing plummets, this method is growing more widespread.

By characterizing sets of genes that are expressed together in some tissues but not others, genome-wide gene expression studies may contribute to a better understanding of diseases and suggest new diagnostic techniques or therapies. For instance, comparing patterns of gene expression in breast cancer tumors and noncancerous breast tissue has already resulted in more informed and effective treatment protocols. Ultimately, information from these methods should provide more of a big-picture view of how ensembles of genes interact to form an organism.

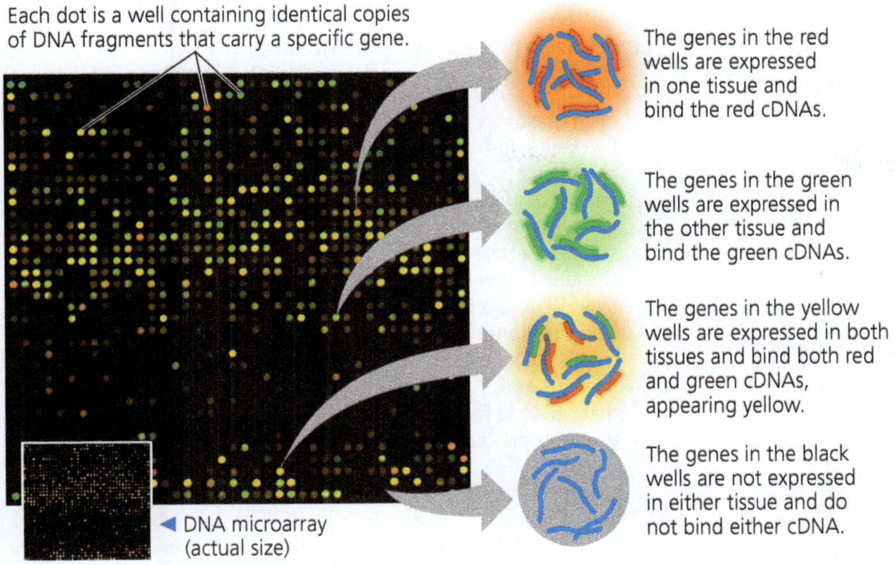

▲ **Figure 17 DNA microarray assay of gene expression levels.** Researchers synthesized two sets of cDNAs, fluorescently labeled red or green, from mRNAs from two different human tissues. These cDNAs were hybridized with a microarray containing 5,760 human genes (about 25% of human genes), resulting in the pattern shown here. The intensity of fluorescence at each spot measures the relative expression in the two samples of the gene represented by that spot: Red indicates expression in one sample, green in the other, yellow in both, and black in neither.

CONCEPT CHECK 4

1. Describe the role of complementary base pairing during RT-PCR and DNA microarray analysis.
2. **WHAT IF?** Study the microarray in Figure 17. If a sample from normal tissue is labeled with a green fluorescent dye, and a sample from cancerous tissue is labeled red, what color spots would represent genes that you would be interested in if you were studying cancer? Explain.

Go to **MasteringBiology**® for Assignments, the eText, and the Study Area with Animations, Activities, Vocab Self-Quiz, and Practice Tests.

Chapter Review

SUMMARY OF KEY CONCEPTS

VOCAB SELF-QUIZ
goo.gl/gbai8v

CONCEPT 1

Bacteria often respond to environmental change by regulating transcription

- In bacteria, genes are often clustered into an operon with a single promoter. An **operator** site on the DNA switches the **operon** on or off, resulting in coordinate regulation of the genes.
- Both repressible and inducible operons are examples of negative gene regulation: Binding of a specific **repressor** protein to the operator shuts off transcription. (The repressor is encoded by a separate **regulatory gene**.) In a repressible operon (usually encoding anabolic enzymes), the repressor is active when bound to a **corepressor**:

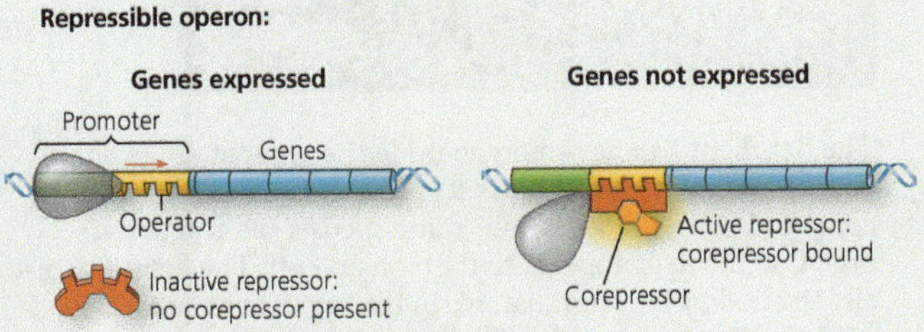

In an inducible operon (usually encoding catabolic enzymes), binding of an **inducer** to an innately active repressor inactivates the repressor and turns on transcription.
- Some operons have positive gene regulation. A stimulatory **activator** protein (such as CRP, when activated by **cyclic AMP**), binds to a site within the promoter and stimulates transcription.

? *Compare and contrast the roles of a corepressor and an inducer in negative regulation of an operon.*

CONCEPT 2

Eukaryotic gene expression is regulated at many stages

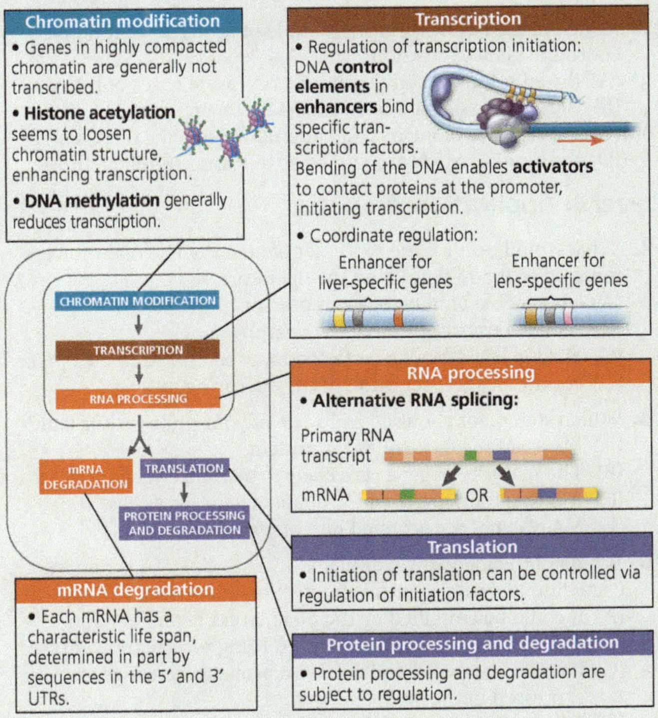

? *Describe what must happen in a cell before a gene specific to that type of cell can be transcribed.*

CONCEPT 3

Noncoding RNAs play multiple roles in controlling gene expression

- Noncoding RNAs (for example, **miRNAs** and **siRNAs**) can block translation or cause degradation of mRNAs.

? *Why are miRNAs called noncoding RNAs? Explain how they participate in gene regulation.*

CONCEPT 4

Researchers can monitor expression of specific genes

- In **nucleic acid hybridization**, a **nucleic acid probe** is used to detect the presence of a specific mRNA.
- ***In situ* hybridization** and **RT-PCR** can detect the presence of a given mRNA in a tissue or an RNA sample, respectively.
- **DNA microarrays** are used to identify sets of genes co-expressed by a group of cells. Their **cDNAs** can also be sequenced (RNA-seq).

? *What does detecting expression of specific genes tell a researcher?*

TEST YOUR UNDERSTANDING

PRACTICE TEST

goo.gl/CRZjvS

Level 1: Knowledge/Comprehension

1. If a particular operon encodes enzymes for making an essential amino acid and is regulated like the *trp* operon, then
(A) the amino acid inactivates the repressor.
(B) the repressor is active in the absence of the amino acid.
(C) the amino acid acts as a corepressor.
(D) the amino acid turns on transcription of the operon.

2. The functioning of enhancers is an example of
(A) a eukaryotic equivalent of prokaryotic promoter function.
(B) transcriptional control of gene expression.
(C) the stimulation of translation by initiation factors.
(D) post-translational control that activates certain proteins.

3. Which of the following is an example of post-transcriptional control of gene expression?
(A) the addition of methyl groups to cytosine bases of DNA
(B) the binding of transcription factors to a promoter
(C) the removal of introns and alternative splicing of exons
(D) the binding of RNA polymerase to transcription factors

Level 2: Application/Analysis

4. What would occur if the repressor of an inducible operon were mutated so it could not bind the operator?
(A) irreversible binding of the repressor to the promoter
(B) reduced transcription of the operon's genes
(C) buildup of a substrate for the pathway controlled by the operon
(D) continuous transcription of the operon's genes

5. Which statement about DNA in one of your brain cells is true?
(A) Most of the DNA codes for protein.
(B) The majority of genes are likely to be transcribed.
(C) It is the same as the DNA in one of your liver cells.
(D) Many genes are grouped into operon-like clusters.

6. Which of the following would *not* be true of cDNA produced using human brain tissue as the starting material?
(A) It could be amplified by the polymerase chain reaction.
(B) It was produced from pre-mRNA using reverse transcriptase.
(C) It could be labeled and used as a probe to detect genes expressed in the brain.
(D) It lacks the introns of the pre-mRNA.

Level 3: Synthesis/Evaluation

7. DRAW IT The diagram shows five genes, including their enhancers, from the genome of a certain species. Imagine that orange, blue, green, black, red, and purple activator proteins exist that can bind to the appropriately color-coded control elements in the enhancers of these genes. (a) Draw an X above enhancer elements (of all the genes) that would have activators bound in a cell in which only gene 5 is transcribed. Identify the colored activators that would have to be present. (b) Draw a dot above all enhancer elements that would have activators bound in a cell in which the green, blue, and orange activators are present. Identify the gene(s) that would be transcribed. (c) Imagine that genes 1, 2, and 4 code for nerve-specific proteins and genes 3 and 5 are skin-specific. Identify the activators that would ensure transcription of the appropriate genes.

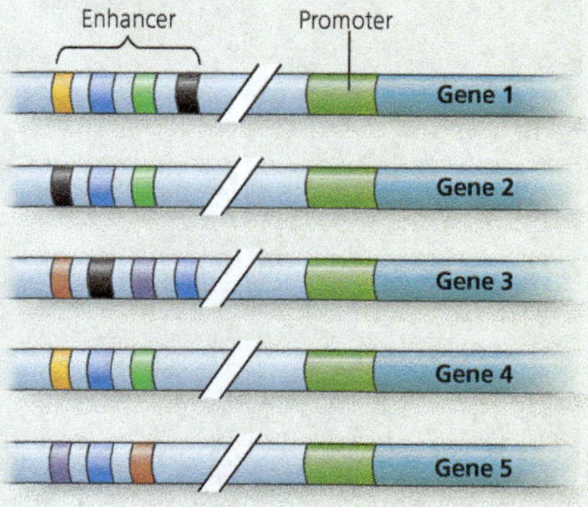

8. SCIENTIFIC INQUIRY
Imagine you want to study one of the mouse crystallins, proteins present in the lens of the eye. Assuming that the gene has been cloned, describe two ways to study its expression in the embryo.

9. FOCUS ON EVOLUTION
DNA sequences can act as "tape measures of evolution." Some highly conserved regions of the human genome (similar to comparable regions in other species) don't code for proteins. Propose a possible explanation.

10. FOCUS ON INTERACTIONS
In a short essay (100–150 words), discuss how the processes shown in Figure 2 are examples of feedback mechanisms regulating biological systems in bacterial cells.

11. SYNTHESIZE YOUR KNOWLEDGE

The flashlight fish has an organ under its eye that emits light, which startles predators, attracts prey, and allows the fish to communicate with other fish. Some species can rotate the organ, so the light appears to flash on and off. The light is actually emitted by bacteria that live in the organ in a mutualistic relationship with the fish. (The bacteria receive nutrients from the fish.) The bacteria must multiply until they reach a certain density in the organ, at which point they all begin emitting light at the same time. There is a group of six or so genes, called *lux* genes, whose gene products are necessary for light formation. Given that these bacterial genes are regulated together, propose a hypothesis for how the genes are organized and regulated.

Answers

Figure Questions

Figure 3 As the concentration of tryptophan in the cell falls, eventually there will be none bound to *trp* repressor molecules. These will then change into their inactive shapes and dissociate from the operator, allowing transcription of the operon to resume. The enzymes for tryptophan synthesis will be made, and they will again synthesize tryptophan in the cell. **Figure 11** In both types of cell, the albumin gene enhancer has the three control elements colored yellow, gray, and red. The sequences in the liver and lens cells would be identical, since the cells are in the same organism.

Concept Check 1

1. Binding by the *trp* corepressor (tryptophan) activates the *trp* repressor, allowing it to bind to the *trp* operator, shutting off transcription of the *trp* operon. Binding by the *lac* inducer (allolactose) inactivates the *lac* repressor so it can no longer bind to the *lac* operator, leading to transcription of the *lac* operon. **2.** When glucose is scarce, cAMP is bound to CRP and CRP is bound to the *lac* promoter, favoring the binding of RNA polymerase. However, in the absence of lactose, the *lac* repressor is bound to the operator, blocking RNA polymerase binding to the promoter. Therefore, the *lac* operon genes are not transcribed. **3.** The cell would continuously produce β-galactosidase and the two other enzymes for using lactose, even in the absence of lactose, thus wasting cell resources.

Concept Check 2

1. Histone acetylation is generally associated with gene expression, while DNA methylation is generally associated with lack of expression. **2.** General transcription factors function in assembling the transcription initiation complex at the promoters for all genes. Specific transcription factors bind to control elements associated with a particular gene and, once bound, either increase (activators) or decrease (repressors) transcription of that gene. **3.** The three genes should have some similar or identical sequences in the control elements of their enhancers. Because of this similarity, the same specific transcription factors in muscle cells could bind to the enhancers of all three genes and stimulate their expression coordinately.

Concept Check 3

1. The mRNA would persist and be translated into the cell division–promoting protein, and the cell would probably divide. If the intact miRNA is necessary for inhibition of cell division, then division of this cell might be inappropriate. Uncontrolled cell division could lead to formation of a mass of cells (tumor) that prevents proper functioning of the organism and could contribute to the development of cancer. **2.** The *XIST* RNA is transcribed from the *XIST* gene on the X chromosome that will be inactivated. It then binds to that chromosome and induces heterochromatin formation. A likely model is that the *XIST* RNA somehow recruits chromatin modification enzymes that lead to formation of heterochromatin.

Concept Check 4

1. In RT-PCR, the primers must base-pair with their target sequences in the DNA mixture, locating one specific region among many. Also, the DNA polymerase (e.g., Taq polymerase) used in PCR relies on complementary base pairing to the template strand to add new nucleotides during synthesis of the fragments. In DNA microarray analysis, the labeled probe binds only to the specific target sequence due to complementary nucleic acid hybridization (DNA-DNA hybridization). **2.** As a researcher interested in cancer, you would want to study genes represented by spots that are green or red because these are genes for which the expression level differs between the two types of tissues. Some of these genes may be expressed differently as a result of cancer, while others might play a role in causing cancer, so both would be of interest.

Summary of Key Concepts Questions

1 A corepressor and an inducer are both small molecules that bind to the repressor protein in an operon, causing the repressor to change shape. In the case of a corepressor (like tryptophan), this shape change allows the repressor to bind to the operator, blocking transcription. In contrast, an inducer causes the repressor to dissociate from the operator, allowing transcription to begin. **2** In that specific type of cell, the chromatin must not be tightly condensed because it must be accessible to transcription factors. The appropriate specific transcription factors (activators), which are made in that type of cell, must bind to the control elements in the enhancer of the gene, while repressors must not be bound. The DNA must be bent by a bending protein so the activators can contact the mediator proteins and form a complex with general transcription factors at the promoter. Then RNA polymerase must bind and begin transcription. **3** miRNAs do not "code" for the amino acids of a protein—they are never translated. Each miRNA associates with a group of proteins to form a complex. Binding of the complex to an mRNA with a complementary sequence causes that mRNA to be degraded or blocks its translation. This is considered gene regulation because it controls the amount of a particular mRNA that can be translated into a functional protein. **4** The genes that are expressed in a given tissue or cell type determine the proteins (and noncoding RNAs) that are the basis of the structure and functions of that tissue or cell type. Understanding which groups of interacting genes establish particular structures and carry out certain functions will help us learn how the parts of an organism form and are maintained. We will also be better able to treat diseases that occur when faulty gene expression leads to malfunctioning tissues.

Test Your Understanding

1. C **2.** B **3.** C **4.** D **5.** C **6.** B

7.

(a)

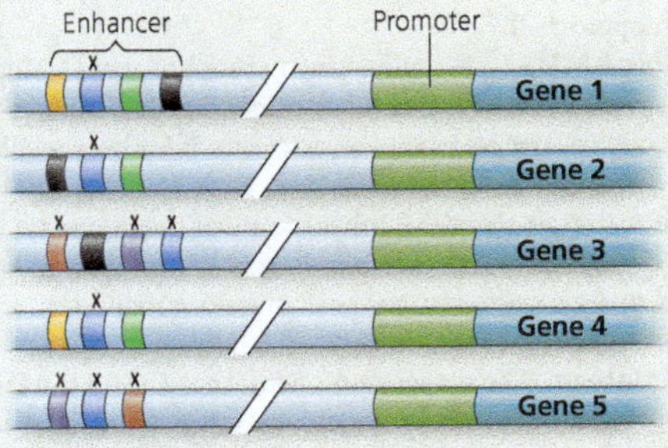

The purple, blue, and red activator proteins would be present.

(b)

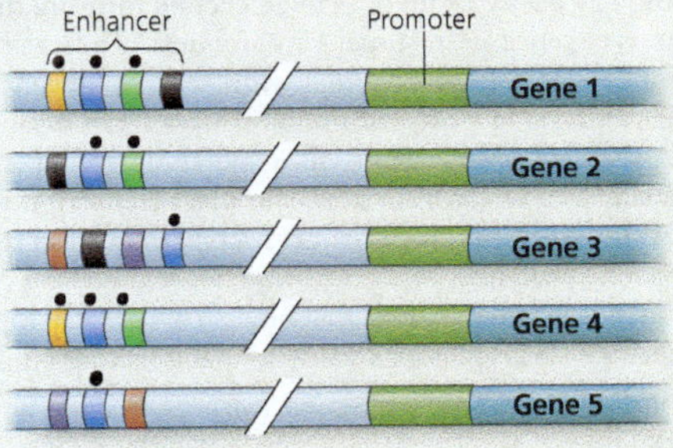

Only gene 4 would be transcribed. (c) In nerve cells, the orange, blue, green, and black activators would have to be present, thus activating transcription of genes 1, 2, and 4. In skin cells, the red, black, purple, and blue activators would have to be present, thus activating genes 3 and 5.

Credits

Credits are listed in the order of appearance.

Photo Credits

1 Andreas Werth; **15 SSE** Amanda Rohde/Getty Images; **14** Ethan Bier; **17** George S. Watts and Bernard W. Futscher, University of Arizona Cancer Center; **Chapter Review SYK** Peter Herring/Image Quest Marine.

Illustration and Text Credits

9 PDB ID 1MDY: P. C. Ma et al. Crystal structure of MyoD bHLH domain-DNA complex: perspectives on DNA recognition and implications for transcriptional activation, *Cell* 77: 451–459 (1994); **SSE.2** Adapted from "Regulation of human microsomal prostaglandin E synthase-1 by IL-1β requires a distal enhancer element with a unique role for C/EBPβ" by J.N. Walters et al., *Biochemical Journal*, 443(2): 561–571 (4/15/12). Copyright © 2012 by The Biochemical Society. Reprinted with permission of Portland Press Ltd.

Glossary

activator A protein that binds to DNA and stimulates gene transcription. In prokaryotes, activators bind in or near the promoter; in eukaryotes, activators generally bind to control elements in enhancers.

alternative RNA splicing A type of eukaryotic gene regulation at the RNA-processing level in which different mRNA molecules are produced from the same primary transcript, depending on which RNA segments are treated as exons and which as introns.

complementary DNA (cDNA) A double-stranded DNA molecule made *in vitro* using mRNA as a template and the enzymes reverse transcriptase and DNA polymerase. A cDNA molecule corresponds to the exons of a gene.

control element A segment of noncoding DNA that helps regulate transcription of a gene by serving as a binding site for a transcription factor. Multiple control elements are present in a eukaryotic gene's enhancer.

corepressor A small molecule that binds to a bacterial repressor protein and changes the protein's shape, allowing it to bind to the operator and switch an operon off.

cyclic AMP (cAMP) Cyclic adenosine monophosphate, a ring-shaped molecule made from ATP that is a common intracellular signaling molecule (second messenger) in eukaryotic cells. It is also a regulator of some bacterial operons.

cyclostome (sī′-cluh-stōm) Member of one of the two main clades of vertebrates; cyclostomes lack jaws and include lampreys and hagfishes.

differential gene expression The expression of different sets of genes by cells with the same genome.

DNA methylation The presence of methyl groups on the DNA bases (usually cytosine) of plants, animals, and fungi. (The term also refers to the process of adding methyl groups to DNA bases.)

DNA microarray assay A method to detect and measure the expression of thousands of genes at one time. Tiny amounts of a large number of single-stranded DNA fragments representing different genes are fixed to a glass slide and tested for hybridization with samples of labeled cDNA.

enhancer A segment of eukaryotic DNA containing multiple control elements, usually located far from the gene whose transcription it regulates.

histone acetylation The attachment of acetyl groups to certain amino acids of histone proteins.

inducer A specific small molecule that binds to a bacterial repressor protein and changes the repressor's shape so that it cannot bind to an operator, thus switching an operon on.

microRNA (miRNA) A small, single-stranded RNA molecule, generated from a hairpin structure on a precursor RNA transcribed from a particular gene. The miRNA associates with one or more proteins in a complex that can degrade or prevent translation of an mRNA with a complementary sequence.

nucleic acid hybridization The process of base pairing between a gene and a complementary sequence on another nucleic acid molecule.

nucleic acid probe In DNA technology, a labeled single-stranded nucleic acid molecule used to locate a specific nucleotide sequence in a nucleic acid sample. Molecules of the probe hydrogen-bond to the complementary sequence wherever it occurs; radioactive, fluorescent, or other labeling of the probe allows its location to be detected.

operator In bacterial and phage DNA, a sequence of nucleotides near the start of an operon to which an active repressor can attach. The binding of the repressor prevents RNA polymerase from attaching to the promoter and transcribing the genes of the operon.

operon (op′-er-on) A unit of genetic function found in bacteria and phages, consisting of a promoter, an operator, and a coordinately regulated cluster of genes whose products function in a common pathway.

regulatory gene A gene that codes for a protein, such as a repressor, that controls the transcription of another gene or group of genes.

repressor A protein that inhibits gene transcription. In prokaryotes, repressors bind to the DNA in or near the promoter. In eukaryotes, repressors may bind to control elements within enhancers, to activators, or to other proteins in a way that blocks activators from binding to DNA.

reverse transcriptase–polymerase chain reaction (RT-PCR) A technique for determining expression of a particular gene. It uses reverse transcriptase and DNA polymerase to synthesize cDNA from all the mRNA in a sample and then subjects the cDNA to PCR amplification using primers specific for the gene of interest.

RNA interference (RNAi) A technique used to silence the expression of selected genes. RNAi uses synthetic double-stranded RNA molecules that match the sequence of a particular gene to trigger the breakdown of the gene's messenger RNA.

small interfering RNA (siRNA) One of multiple small, single-stranded RNA molecules generated by cellular machinery from a long, linear, double-stranded RNA molecule. The siRNA associates with one or more proteins in a complex that can degrade or prevent translation of an mRNA with a complementary sequence. In some cases, siRNA can also block transcription by promoting chromatin modification.

Development, Stem Cells, and Cancer

KEY CONCEPTS

1 A program of differential gene expression leads to the different cell types in a multicellular organism
2 Cloning of organisms showed that differentiated cells could be "reprogrammed" and ultimately led to the production of stem cells
3 Abnormal regulation of genes that affect the cell cycle can lead to cancer

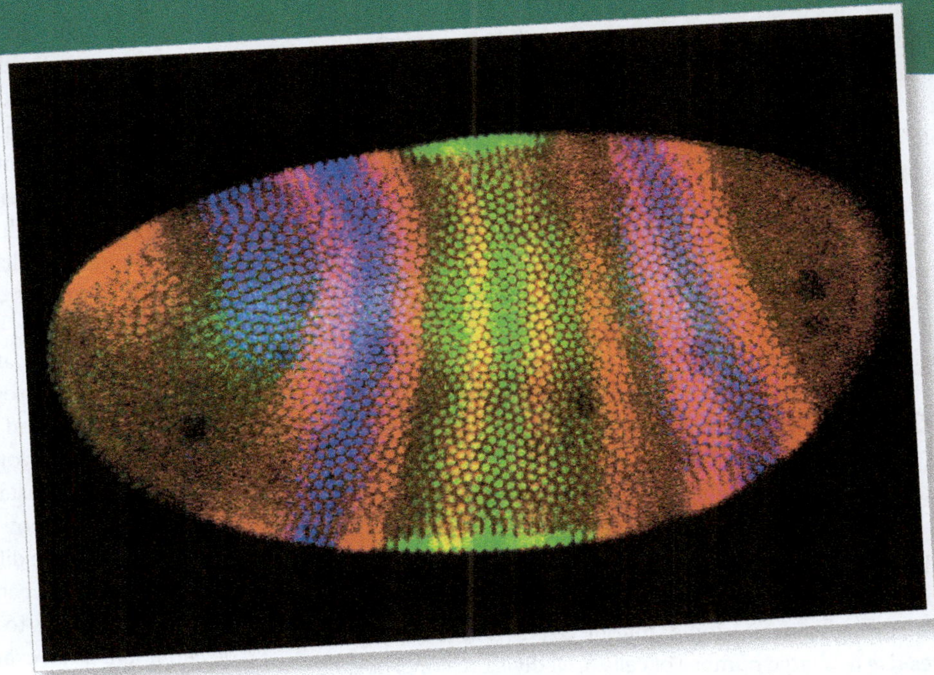

▲ **Figure 1** What regulates the precise pattern of gene expression in the developing fly embryo?

Orchestrating Life's Processes

The development of the fertilized egg, a single cell, into an embryo and later an adult is an astounding transformation that requires a precisely regulated program of gene expression. All of the levels of eukaryotic gene regulation come into play during embryonic development. The elaborate sequence of genes being turned on and off in different cells is the ultimate example of regulation of gene expression.

Understanding the genetic underpinnings of development has progressed mainly by studying the process in **model organisms**, species that are easy to raise in the lab and use in experiments. A prime example is the fruit fly *Drosophila melanogaster*. An adult fruit fly develops from a fertilized egg into an embryo then a wormlike stage called a larva that metamorphoses into the adult fly. At every stage, gene expression is carefully regulated, ensuring that the right genes are expressed only at the correct time and place. The embryo shown in **Figure 1** has been analyzed by *in situ* hybridization to reveal the mRNA for three genes—labeled red, blue, and green. (Red and green together appear yellow; red and blue together appear purple.) The intricate pattern of expression for each gene is the same from embryo to embryo at this stage, and it provides a graphic display of the precision of gene regulation. But what is the molecular basis for this pattern? Why is one particular gene expressed only in the few hundred cells that appear blue or purple in this image and not in the other cells?

Part of the answer involves transcription factors and other regulatory molecules. But how do they come to be different in distinct cell types? In this chapter, we'll first explain the mechanisms that send cells down diverging genetic pathways to adopt different fates. Then we'll take a closer look at *Drosophila* development. Next, we'll describe the discovery of stem cells, a powerful cell type that is key to the developmental process. Understanding these cells offers hope for medical treatments as well. Finally, after having explored embryonic development and stem cells, we will underscore the crucial role played by regulation of gene expression by investigating how cancer can result when this regulation goes awry. Orchestrating proper gene expression by all cells is crucial to the functions of life.

From Chapter 16 of *Campbell Biology in Focus*, Second Edition. Lisa A. Urry, Michael L. Cain, Steven A. Wasserman, Peter V. Minorsky, and Jane B. Reece.

CONCEPT 1

A program of differential gene expression leads to the different cell types in a multicellular organism

In the embryonic development of multicellular organisms, a fertilized egg (a zygote) gives rise to cells of many different types, each with a different structure and corresponding function. Typically, cells are organized into tissues, tissues into organs, organs into organ systems, and organ systems into the whole organism. Thus, any developmental program must produce cells of different types that form higher-level structures arranged in a particular way in three dimensions. Here, we'll focus on the program of regulation of gene expression that orchestrates development using a few animal species as examples.

A Genetic Program for Embryonic Development

The photos in **Figure 2** illustrate the dramatic difference between a frog zygote (fertilized egg) and the tadpole it becomes. This remarkable transformation results from three interrelated processes: cell division, cell differentiation, and morphogenesis. Through a succession of mitotic cell divisions, the zygote gives rise to a large number of cells. Cell division alone, however, would merely produce a great ball of identical cells, nothing like a tadpole. During embryonic development, cells not only increase in number but also undergo cell **differentiation**, the process by which cells become specialized in structure and function. Moreover, the different kinds of cells are not randomly distributed but are organized into tissues and organs in a particular three-dimensional arrangement. The physical processes that give an organism its shape constitute **morphogenesis**, the development of the form of an organism and its structures.

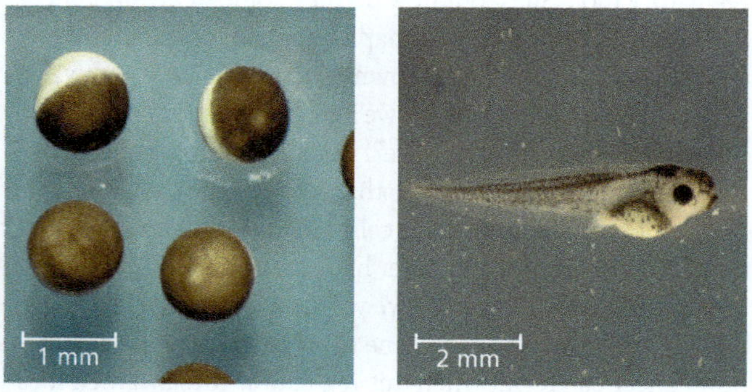

(a) Fertilized eggs of a frog **(b) Newly hatched tadpole**

▲ **Figure 2 From fertilized egg to animal: What a difference four days makes.** It takes just four days for cell division, differentiation, and morphogenesis to transform each of the fertilized frog eggs shown in **(a)** into a tadpole like the one in **(b)**.

All three processes are rooted in cellular behavior. Even morphogenesis, the shaping of the organism, can be traced to changes in the motility, shape, and other characteristics of the cells that make up the regions of the embryo. As you know, the activities of a cell depend on the genes it expresses and the proteins it produces. Almost all cells in an organism have the same genome; therefore, differential gene expression results from the genes being regulated differently in each cell type.

For example, differential gene expression occurs in two cell types, a liver cell and a lens cell. Each of these fully differentiated cells has a particular mix of specific activators that turn on the collection of genes whose products are required in the cell. The fact that both cells arose through a series of mitoses from a common fertilized egg inevitably leads to a question: How do different sets of activators come to be present in the two cells?

It turns out that materials placed into the egg by the mother set up a sequential program of gene regulation that is carried out as cells divide, and this program coordinates cell differentiation during embryonic development. To understand how this works, we'll consider two basic developmental processes: First, we'll explore how cells that arise from early embryonic mitoses develop the differences that start each cell along its own differentiation pathway. Second, we'll see how cellular differentiation leads to one particular cell type, using muscle development as an example.

Cytoplasmic Determinants and Inductive Signals

What generates the first differences among cells in an early embryo? And what controls the differentiation of all the various cell types as development proceeds? You can probably deduce the answer: The specific genes expressed in any particular cell of a developing organism determine its path. Two sources of information, used to varying extents in different species, "tell" a cell which genes to express at any given time during embryonic development.

One important source of information early in development is the egg's cytoplasm, which contains both RNA and proteins encoded by the mother's DNA. The cytoplasm of an unfertilized egg is not homogeneous. Messenger RNA, proteins, other substances, and organelles are distributed unevenly in the unfertilized egg, and this unevenness has a profound impact on the development of the future embryo in many species. Maternal substances in the egg that influence the course of early development are called **cytoplasmic determinants** (**Figure 3a**). After fertilization, early mitotic divisions distribute the zygote's cytoplasm into separate cells. The nuclei of these cells may thus be exposed to different cytoplasmic determinants, depending on which portions of the zygotic cytoplasm a cell received. The combination of cytoplasmic determinants in a cell helps determine its developmental fate by regulating gene expression during cell differentiation.

▼ **Figure 3 Sources of developmental information for the early embryo.**

(a) Cytoplasmic determinants in the egg

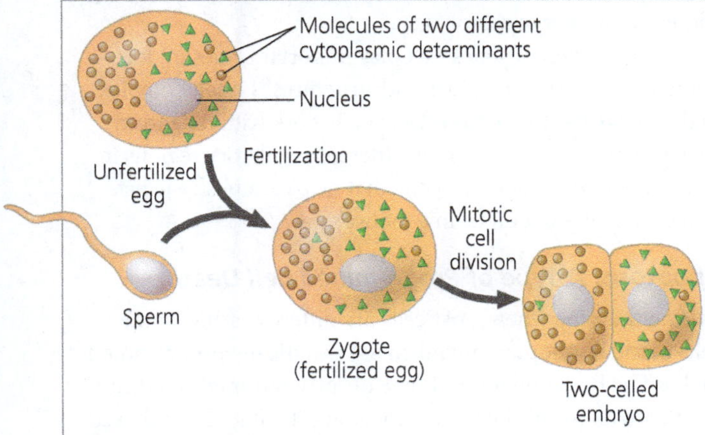

The unfertilized egg has molecules in its cytoplasm, encoded by the mother's genes, that influence development. Many of these cytoplasmic determinants, like the two shown here, are unevenly distributed in the egg. After fertilization and mitotic division, the cell nuclei of the embryo are exposed to different sets of cytoplasmic determinants and, as a result, express different genes.

(b) Induction by nearby cells

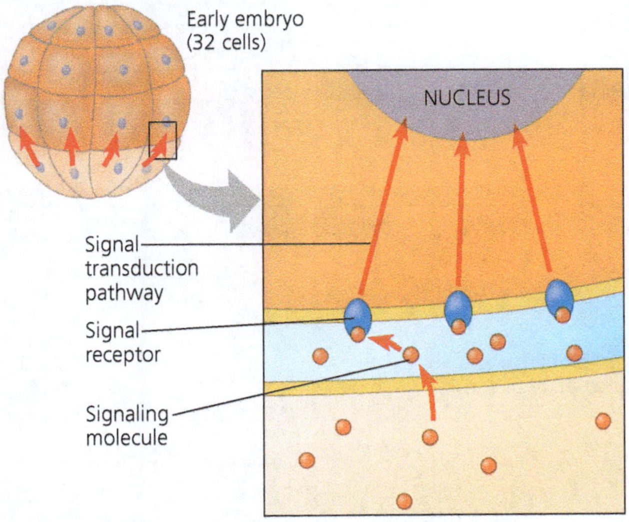

Cells at the bottom of the early embryo are releasing molecules that signal (induce) nearby cells to change their gene expression.

The other major source of developmental information, which becomes increasingly important as the number of embryonic cells increases, is the environment around a particular cell. Most influential are the signals conveyed to an embryonic cell from other embryonic cells in the vicinity, including contact with cell-surface molecules on neighboring cells and the binding of growth factors secreted by neighboring cells. Such signals cause changes in the target cells, a process called **induction** (Figure 3b). The molecules passing along these signals within the target cell are cell-surface receptors and other signaling pathway proteins. In general, the signaling molecules send a cell down a specific developmental path by causing changes in its gene expression that eventually result in observable cellular changes. Thus, interactions between embryonic cells help induce differentiation into the many specialized cell types making up a new organism.

Sequential Regulation of Gene Expression during Cellular Differentiation

The earliest changes that set a cell on a path to specialization are subtle ones, showing up only at the molecular level. Before biologists knew much about the molecular changes occurring in embryos, they coined the term **determination** to refer to the point at which an embryonic cell is irreversibly committed to becoming a particular cell type. Once an embryonic cell has undergone determination, it can be experimentally placed at another site in the embryo and will still differentiate into the cell type that is its normal fate. Differentiation, then, is the process by which a cell attains its determined fate. As the tissues and organs of an embryo develop, their cells differentiate, becoming more noticeably different in structure and function.

Differentiation of Cell Types

Today we understand determination in terms of molecular changes. The outcome of determination, observable cell differentiation, is marked by the expression of genes for *tissue-specific proteins*. These are found only in a specific cell type and give the cell its characteristic structure and function. The first sign of differentiation is the appearance of mRNAs for these proteins. Eventually, differentiation is observable with a microscope as changes in cellular structure. On the molecular level, different sets of genes are sequentially expressed in a regulated manner as new cells arise from division of their precursors. Some steps in gene expression may be regulated during differentiation, transcription being the most common. In the fully differentiated cell, transcription is the principal regulatory point for maintaining appropriate gene expression.

Differentiated cells are specialists at making tissue-specific proteins. For example, as a result of transcriptional regulation, liver cells specialize in making albumin, and lens cells specialize in making crystallin. Skeletal muscle cells in vertebrates are another instructive example. Each of these cells is a long fiber containing many nuclei within a single plasma membrane. Skeletal muscle cells have high concentrations of muscle-specific versions of the contractile proteins myosin and actin, as well as membrane receptor proteins that detect signals from nerve cells.

Muscle cells develop from embryonic precursor cells that have the potential to become a number of cell types, including cartilage cells and fat cells, but particular conditions commit them to becoming muscle cells. Although the committed cells appear unchanged under the microscope, determination has occurred, and they are now *myoblasts*. Eventually, myoblasts start to churn out large amounts of muscle-specific proteins and fuse to form mature, elongated, multinucleate skeletal muscle cells.

Researchers have worked out what happens at the molecular level during muscle cell determination. To do so, they grew embryonic precursor cells in culture and analyzed them using molecular techniques. In a series of experiments, they isolated different genes, caused each to be expressed in a separate precursor cell, and then looked for differentiation into myoblasts and muscle cells. In this way, they identified several so-called "master regulatory genes" whose protein products commit the cells to becoming skeletal muscle. Thus, in the case of muscle cells, the molecular basis of determination is the expression of one or more of these master regulatory genes.

To understand more about how determination occurs in muscle cell differentiation, let's focus on the master regulatory gene called *myoD* **(Figure 4)**. This gene encodes MyoD protein, a transcription factor that binds to specific control elements in the enhancers of various target genes and stimulates their expression. Some target genes for MyoD encode still other muscle-specific transcription factors. MyoD also stimulates expression of the *myoD* gene itself, an example of positive feedback that perpetuates MyoD's effect in maintaining the cell's differentiated state. Since all the genes activated by MyoD have enhancer control elements recognized by MyoD, they are coordinately controlled. Finally, the secondary transcription factors activate the genes for proteins such as myosin and actin that confer the unique properties of skeletal muscle cells. The MyoD protein deserves its designation as a master regulatory gene.

The regulation of genes that play important roles in development of embryonic tissues and structures is often complex. In the **Scientific Skills Exercise**, you'll work with data from an experiment that tested how different regulatory regions in DNA affect expression of a gene that helps establish the pattern of the different digits in a mouse's paw.

Apoptosis: A Type of Programmed Cell Death

During the time when most cells are differentiating, some cells in the developing organism are genetically programmed to die. The best-understood type of "programmed cell death" is **apoptosis** (from the Greek, meaning "falling off," and used in a classic Greek poem to refer to leaves falling from a tree). Apoptosis also occurs in cells of the mature organism that are infected, damaged, or have reached the end of their functional life span. During this process, cellular agents chop up the DNA and fragment the organelles and other cytoplasmic

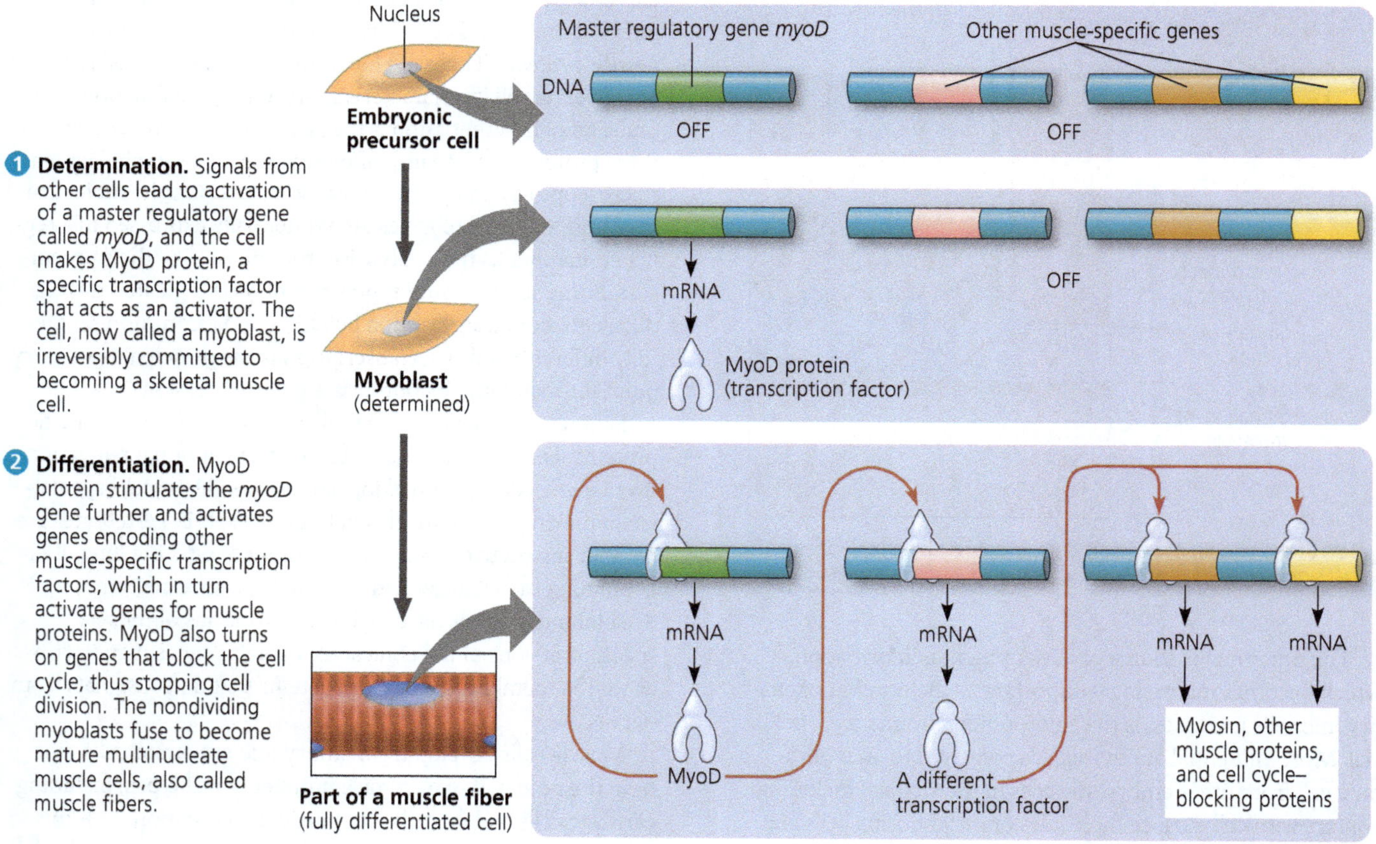

▲ **Figure 4 Determination and differentiation of muscle cells.** Skeletal muscle cells arise from embryonic cells as a result of changes in gene expression. (In this depiction, the process of gene activation is greatly simplified.)

WHAT IF? *What would happen if a mutation in the* myoD *gene resulted in a MyoD protein that could not activate the* myoD *gene?*

Scientific Skills Exercise

Analyzing Quantitative and Spatial Gene Expression Data

How Is a Particular *Hox* Gene Regulated During Paw Development? *Hox* genes code for transcription factor proteins, which in turn control sets of genes important for animal development. One group of *Hox* genes, the *Hoxd* genes, plays a role in establishing the pattern of the different digits (fingers and toes) at the end of a limb. *Hox* genes have very large, complicated regulatory regions, including control elements that may be hundreds of kilobases (kb; thousands of nucleotides) away from the gene.

In cases like this, how do biologists locate the DNA segments that contain important elements? They begin by removing (deleting) large segments of DNA and studying the effect on gene expression. In this exercise, you'll compare data from two different but complementary approaches that look at the expression of a specific *Hoxd* gene (*Hoxd13*). One approach quantifies overall expression; the other approach is less quantitative but gives important spatial localization information.

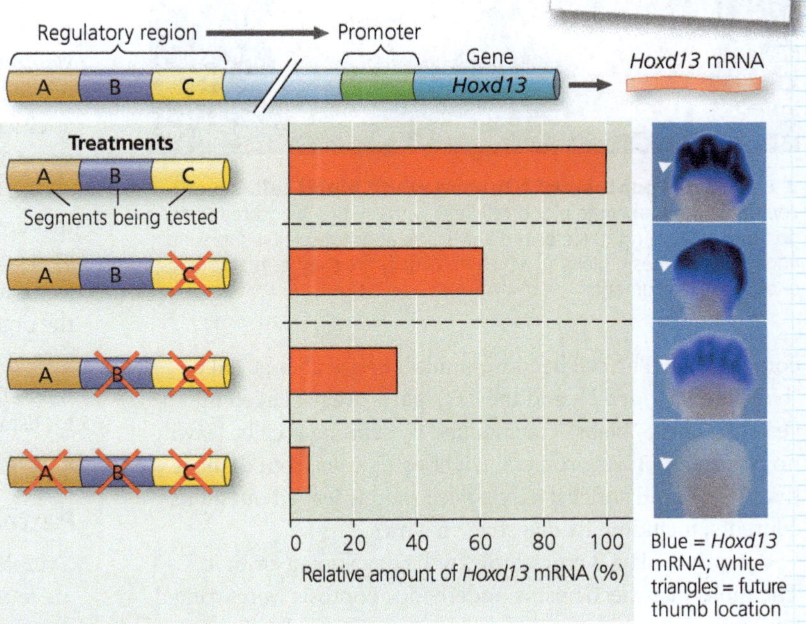

Data from T. Montavon et al., A regulatory archipelago controls *Hox* genes transcription in digits, *Cell* 147:1132–1145 (2011). doi:10.1016/j.cell.2011.10.023

How the Experiment Was Done Researchers interested in the regulation of *Hoxd13* gene expression genetically engineered a set of mice that had different segments of DNA deleted upstream of the gene. They then compared levels and patterns of *Hoxd13* gene expression in developing paws of 12.5-day-old mouse embryos that had the DNA deletions with gene expression seen in wild-type mouse embryos of the same age.

They used two different approaches: In some mice, they extracted the mRNA from the embryonic paws and quantified the overall level of *Hoxd13* mRNA in the whole paw. In another set of the same mice, they used *in situ* hybridization to pinpoint exactly where in the paws the *Hoxd13* gene was expressed as mRNA. The particular technique that was used causes the *Hoxd13* mRNA to appear blue or black, which indicates the highest mRNA levels.

Data from the Experiment The topmost diagram depicts the very large regulatory region upstream of the *Hoxd13* gene. The area between the slashes represents the long stretch of DNA located between the promoter and the regulatory region.

The diagrams to the left of the bar graph show, first, the intact DNA (830 kb) and, next, the three altered DNA sequences. (Each is called a "deletion" because a particular section of DNA has been deleted from it.) A red X indicates the segment (A, B, and/or C) that was deleted in each experimental treatment.

The horizontal bar graph shows the amount of *Hoxd13* mRNA that was present in the digit formation zone of each 12.5-day-old embryo with a deletion, relative to the amount that was in the digit formation zone of the mouse with the intact regulatory region (top bar = 100%).

The images to the right of the graph are fluorescent micrographs of the embryo paws showing the location of the *Hoxd13* mRNA (the stain appears blue or black). The white triangles show the location where the thumb will form.

INTERPRET THE DATA

1. The researchers hypothesized that all three regulatory segments (A, B, and C) were required for full expression of the *Hoxd13* gene. By measuring the amount of *Hoxd13* mRNA in the embryo paw zones where digits develop, they could measure the effect of the regulatory segments individually and in combination. Refer to the graph to answer these questions, noting that the segments being tested are shown on the vertical axis and the relative amount of *Hoxd13* mRNA is shown on the horizontal axis. (a) Which of the four treatments was used as a control for the experiment? (b) The hypothesis is that all three segments together are required for highest expression of the *Hoxd13* gene. Is this supported by the results? Explain your answer.
2. (a) What is the effect on the amount of *Hoxd13* mRNA when segments B and C are both deleted, compared with the control? (b) Is this effect visible in the blue-stained regions of the *in situ* hybridizations? How would you describe the spatial pattern of gene expression in the embryo paws that lack segments B and C? (You'll need to look carefully at different regions of each paw and how they differ.)
3. (a) What is the effect on the amount of *Hoxd13* mRNA when just segment C is deleted, compared with the control? (b) Is this effect visible in the *in situ* hybridizations? How would you describe the spatial pattern of gene expression in embryo paws that lack just segment C, compared with the control and with the paws that lack segments B and C?
4. If the researchers had only measured the amount of *Hoxd13* mRNA and not done the *in situ* hybridizations, what important information about the role of the regulatory segments in *Hoxd13* gene expression during paw development would have been missed? Conversely, if the researchers had only done the *in situ* hybridizations, what information would have been inaccessible?

MB A version of this Scientific Skills Exercise can be assigned in MasteringBiology.

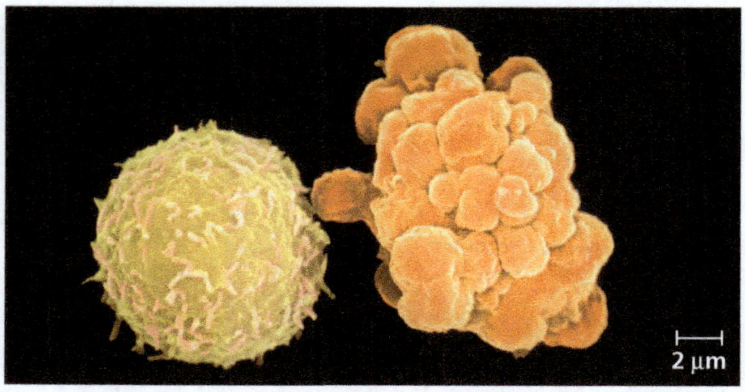

▲ **Figure 5 Apoptosis of a human white blood cell.** We can compare a normal white blood cell (left) with a white blood cell undergoing apoptosis (right) (colorized SEM). The apoptotic cell is shrinking and forming lobes ("blebs"), which eventually are shed as membrane-enclosed cell fragments.

components. The cell becomes multilobed, a change called "blebbing" **(Figure 5)**, and the cell's parts are packaged in vesicles. These "blebs" are engulfed by scavenger cells, leaving no trace. Apoptosis protects neighboring cells from damage that they would suffer if a dying cell merely leaked out all its contents, including its digestive enzymes.

Apoptosis plays a crucial role in the developing embryo. The molecular mechanisms underlying apoptosis were worked out by studying embryonic development of a small soil worm, a nematode called *Caenorhabditis elegans* that has now become a popular model organism for genetic studies. Because the adult worm has only about a thousand cells, researchers worked out the complete ancestry of each cell. The timely suicide of cells occurs exactly 131 times during normal development of *C. elegans*, at precisely the same points in the cell lineage of each worm. In worms and other species, apoptosis is triggered by signal transduction pathways. These activate a cascade of apoptotic "suicide" proteins in the cells destined to die, including the enzymes that break down and package cellular molecules in the "blebs."

Apoptosis is essential to development and maintenance in all animals. There are similarities in genes encoding apoptotic proteins in nematodes and mammals, and apoptosis is known to occur as well in multicellular fungi and single-celled yeasts, evidence that the basic mechanism evolved early among eukaryotes. In vertebrates, apoptosis is essential for normal development of the nervous system and for normal morphogenesis of hands and feet in humans and paws in other mammals **(Figure 6)**. The level of apoptosis between the developing digits is lower in the webbed feet of ducks and other water birds than in the nonwebbed feet of land birds, such as chickens. In the case of humans, the failure of appropriate apoptosis can result in webbed fingers and toes.

We have seen how different programs of gene expression that are activated in the fertilized egg can result in differentiated cells and tissues as well as the death of some cells. But for tissues to function effectively in the organism as a whole, the organism's *body plan*—its overall three-dimensional arrangement—must be established and superimposed on the differentiation process. Next we'll look at the molecular basis for establishing the body plan, using the well-studied fruit fly *Drosophila* as an example.

Pattern Formation: Setting Up the Body Plan

Cytoplasmic determinants and inductive signals both contribute to the development of a spatial organization in which the tissues and organs of an organism are all in their characteristic places. This process is called **pattern formation**.

Just as the locations of the front, back, and sides of a new building are determined before construction begins, pattern formation in animals begins in the early embryo, when the major axes of an animal are established. In a bilaterally symmetrical animal, the relative positions of head and tail, right and left sides, and back and front—the three major body axes—are set up before the organs appear. The molecular cues that control pattern formation, collectively called **positional information**, are provided by cytoplasmic determinants and inductive signals (see Figure 3). These cues tell a cell its location relative to the body axes and to neighboring cells and determine how the cell and its descendants will respond to future molecular signals.

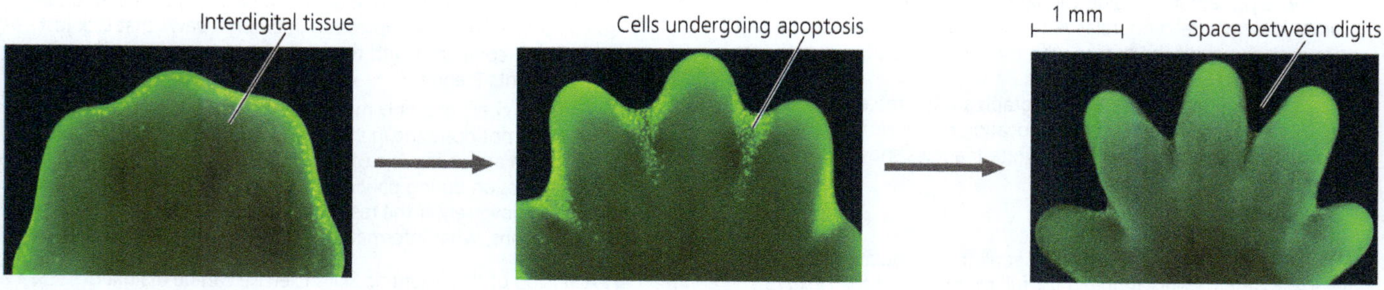

▲ **Figure 6 Effect of apoptosis during paw development in the mouse.** In mice, humans, other mammals, and land birds, the embryonic region that develops into feet or hands initially has a solid, platelike structure. Apoptosis eliminates the cells in the interdigital regions, thus forming the digits. The embryonic mouse paws shown in these fluorescence light micrographs are stained so that cells undergoing apoptosis appear bright yellow. Apoptosis of cells begins at the margin of each interdigital region (left), peaks as the tissue in these regions is reduced (middle), and is no longer visible when the interdigital tissue has been eliminated (right). (Note that the Scientific Skills Exercise shows a different genetic process involved in mouse paw development.)

During the first half of the 20th century, classical embryologists made detailed anatomical observations of embryonic development in a number of species and performed experiments in which they manipulated embryonic tissues. This research laid the groundwork for understanding the mechanisms of development, but it did not reveal the specific molecules that guide development or determine how patterns are established.

In the 1940s, scientists began using the genetic approach—the study of mutants—to investigate *Drosophila* development. That approach has had spectacular success. Genetic studies have established that genes control development and have led to an understanding of the key roles that specific molecules play in defining position and directing differentiation. By combining anatomical, genetic, and biochemical approaches to the study of *Drosophila* development, researchers have discovered developmental principles common to many other species, including humans.

The Life Cycle of Drosophila

Fruit flies and other arthropods have a modular construction, an ordered series of segments. These segments make up the body's three major parts: the head, the thorax (the mid-body, from which the wings and legs extend), and the abdomen **(Figure 7a)**. Like other bilaterally symmetrical animals, *Drosophila* has an anterior-posterior (head-to-tail) axis, a dorsal-ventral (back-to-belly) axis, and a right-left axis. In *Drosophila*, cytoplasmic determinants that are localized in the unfertilized egg provide positional information for the placement of anterior-posterior and dorsal-ventral axes even before fertilization. We'll focus here on the molecules involved in establishing the anterior-posterior axis as a case in point.

The *Drosophila* egg develops in one of the female's ovaries, surrounded by ovarian cells called nurse cells and follicle cells (**Figure 7b**, top). These support cells supply the egg with nutrients, mRNAs, and other substances needed for development, and make the egg shell. After fertilization and laying of the egg, embryonic development results in the formation of a segmented larva, which goes through three larval stages. Then, in a process much like that by which a caterpillar becomes a butterfly, the larva forms a pupa in which it metamorphoses into the adult fly pictured in Figure 7a.

Genetic Analysis of Early Development: *Scientific Inquiry*

Edward B. Lewis was a visionary American biologist who, in the 1940s, first showed the value of the genetic approach to studying embryonic development in *Drosophila*. Lewis studied bizarre mutant flies with developmental defects that led to extra wings or legs in the wrong place **(Figure 8)**. He located the mutations on the fly's genetic map, connecting the abnormalities to specific genes. This research supplied the first evidence that genes direct the developmental processes studied by embryologists. The genes Lewis discovered, called

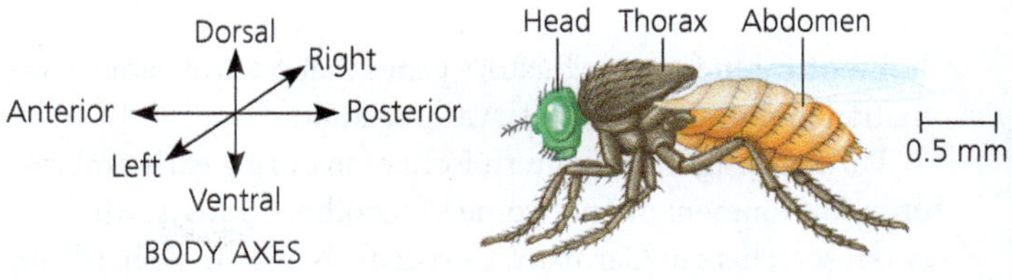

(a) Adult. The adult fly is segmented, and multiple segments make up each of the three main body parts—head, thorax, and abdomen. The body axes are shown by arrows.

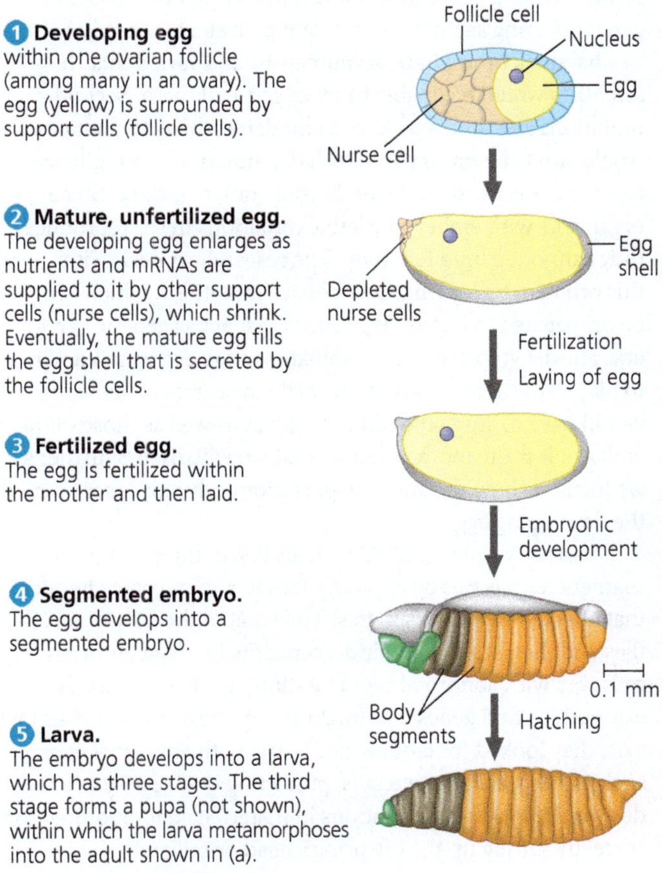

(b) Development from egg to larva.

▲ **Figure 7 Key events in *Drosophila* development.**

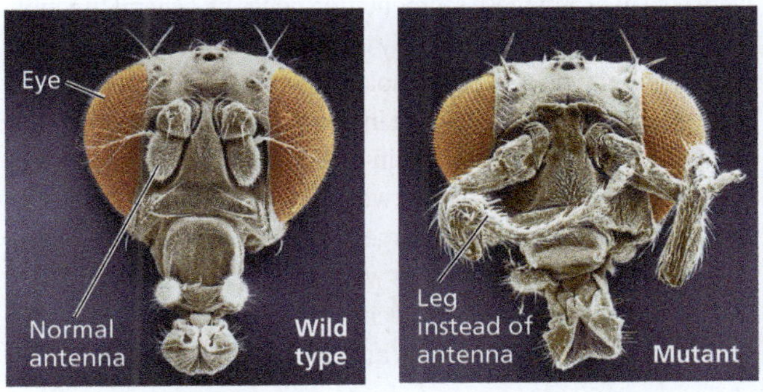

▲ **Figure 8 Abnormal pattern formation in *Drosophila*.** Mutations in homeotic genes cause misplacement of structures. These scanning electron micrographs contrast the head of a wild-type fly with that of a homeotic mutant (a fly with a mutation in a homeotic gene) bearing a pair of legs in place of antennae.

homeotic genes, are regulatory genes that control pattern formation in the late embryo, larva, and adult.

Further insight into pattern formation during early embryonic development did not come for another 30 years, when two researchers in Germany, Christiane Nüsslein-Volhard and Eric Wieschaus, set out to identify *all* the genes that affect segment formation in *Drosophila*. The project was daunting for three reasons. The first was the sheer number of *Drosophila* protein-coding genes, now known to total about 14,000. The genes affecting segmentation might be just a few needles in a haystack or might be so numerous and varied that the scientists would be unable to make sense of them. Second, mutations affecting a process as fundamental as segmentation would surely be **embryonic lethals**, mutations with phenotypes causing death at the embryonic or larval stage. Since organisms with embryonic lethal mutations never reproduce, they cannot be bred for study. The researchers dealt with this problem by looking for recessive mutations, which can be propagated in heterozygous flies that act as genetic carriers. Third, cytoplasmic determinants in the egg were known to play a role in axis formation, so the researchers knew they would have to study the mother's genes as well as those of the embryo. It is the mother's genes that we will discuss further as we focus on how the anterior-posterior body axis is set up in the developing egg.

Nüsslein-Volhard and Wieschaus began their search for segmentation genes by exposing flies to a mutagenic chemical that affected the flies' gametes. They mated the mutagenized flies and then scanned their descendants for dead embryos or larvae with abnormal segmentation or other defects. For example, to find genes that might set up the anterior-posterior axis, they looked for embryos or larvae with abnormal ends, such as two heads or two tails, predicting that such abnormalities would arise from mutations in maternal genes required for correctly setting up the offspring's head or tail end.

Using this approach, Nüsslein-Volhard and Wieschaus eventually identified about 1,200 genes essential for pattern formation during embryonic development. Of these, about 120 were essential for normal segmentation. Over several years, the researchers were able to group these segmentation genes by general function, to map them, and to clone many of them for further study in the lab. The result was a detailed molecular understanding of the early steps in pattern formation in *Drosophila*.

When the results of Nüsslein-Volhard and Wieschaus were combined with Lewis's earlier work, a coherent picture of *Drosophila* development emerged. In recognition of their discoveries, the three researchers were awarded a Nobel Prize in 1995. Next, let's consider a specific example of the genes that Nüsslein-Volhard, Wieschaus, and co-workers found.

Axis Establishment

As we mentioned earlier, cytoplasmic determinants in the egg are the substances that initially establish the axes of the *Drosophila* body. These substances are encoded by genes of the mother, fittingly called maternal effect genes. A **maternal effect gene** is a gene that, when mutant in the mother, results in a mutant phenotype in the offspring, regardless of the offspring's own genotype. In fruit fly development, the mRNA or protein products of maternal effect genes are placed in the egg while it is still in the mother's ovary. When the mother has a mutation in such a gene, she makes a defective gene product (or none at all), and her eggs are defective; when these eggs are fertilized, they fail to develop properly.

Because maternal effect genes control the orientation (polarity) of the egg and consequently of the fly, they are also called *egg-polarity genes*. One group of these genes sets up the anterior-posterior axis of the embryo, while a second group establishes the dorsal-ventral axis. Like mutations in segmentation genes, mutations in maternal effect genes are generally embryonic lethals.

Bicoid: A Morphogen That Determines Head Structures

To see how maternal effect genes determine the body axes of the offspring, we'll focus on one such gene, called ***bicoid***, a term meaning "two-tailed." An embryo whose mother has two mutant *bicoid* alleles lacks the front half of its body and has posterior structures at both ends **(Figure 9)**. This phenotype suggested to Nüsslein-Volhard and her colleagues that the product of the mother's *bicoid* gene is essential for setting up the anterior end of the fly and might be concentrated at the future anterior end of the embryo. This hypothesis is an example of the *morphogen gradient hypothesis* first proposed by embryologists a century ago, in which gradients of substances called **morphogens** establish an embryo's axes and other features of its form.

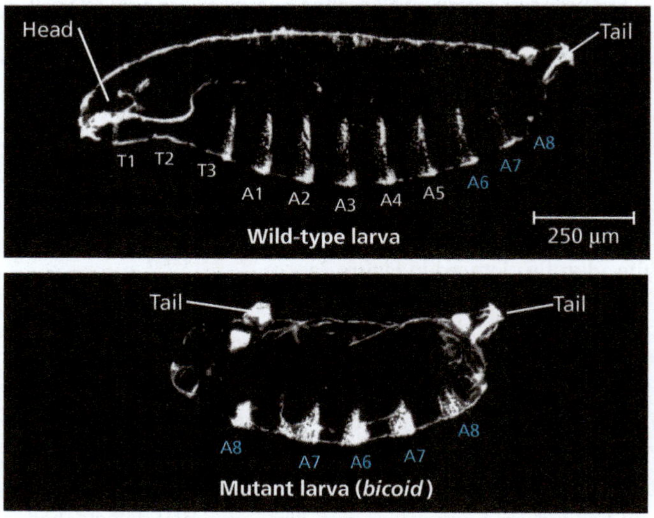

▲ **Figure 9 Effect of the *bicoid* gene on *Drosophila* development.** A wild-type fruit fly larva has a head, three thoracic (T) segments, eight abdominal (A) segments, and a tail. A larva whose mother has two mutant alleles of the *bicoid* gene has two tails and lacks all anterior structures (LMs).

DNA technology and other modern biochemical methods enabled the researchers to test whether the *bicoid* product, a protein called Bicoid, is in fact a morphogen that determines the anterior end of the fly. The first question they asked was whether the mRNA and protein products of this gene are located in the egg in a position consistent with the hypothesis. They found that *bicoid* mRNA is highly concentrated at the extreme anterior end of the mature egg, as predicted by the hypothesis **(Figure 10)**. After the egg is fertilized, the mRNA is translated into protein. The Bicoid protein then diffuses from the anterior end toward the posterior, resulting in a gradient of protein within the early embryo, with the highest concentration at the anterior end. These results are consistent with the hypothesis that Bicoid protein specifies the fly's anterior end. To test the hypothesis more specifically, scientists injected pure *bicoid* mRNA into various regions of early embryos. The protein that resulted from its translation caused anterior structures to form at the injection sites.

▼ Figure 10 Inquiry

Could Bicoid be a morphogen that determines the anterior end of a fruit fly?

Experiment Using a genetic approach to study *Drosophila* development, Christiane Nüsslein-Volhard and colleagues at two research institutions in Germany analyzed expression of the *bicoid* gene. The researchers hypothesized that *bicoid* normally codes for a morphogen that specifies the head (anterior) end of the embryo. To begin to test this hypothesis, they used molecular techniques to determine whether the mRNA and protein encoded by this gene were found in the anterior end of the fertilized egg and early embryo of wild-type flies.

Results *Bicoid* mRNA (dark blue in the light micrographs and drawings) was confined to the anterior end of the unfertilized egg. Later in development, Bicoid protein (dark orange) was seen to be concentrated in cells at the anterior end of the embryo.

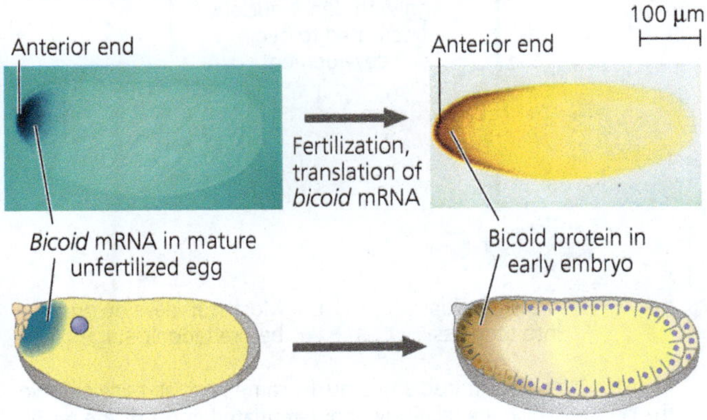

Conclusion The location of *bicoid* mRNA and the diffuse gradient of Bicoid protein seen later are consistent with the hypothesis that Bicoid protein is a morphogen specifying formation of head-specific structures.

Further Reading C. Nüsslein-Volhard et al., Determination of anteroposterior polarity in *Drosophila*, *Science* 238:1675–1681 (1987); W. Driever and C. Nüsslein-Volhard, A gradient of Bicoid protein in *Drosophila* embryos, *Cell* 54:83–93 (1988); T. Berleth et al., The role of localization of *bicoid* RNA in organizing the anterior pattern of the *Drosophila* embryo, *EMBO Journal* 7:1749–1756 (1988).

WHAT IF? The researchers needed further evidence, so they injected *bicoid* mRNA into the anterior end of an egg from a female with a mutation disabling the *bicoid* gene. Given that the hypothesis was supported, what must their results have been?

The *bicoid* research was groundbreaking for several reasons. First, it led to the identification of a specific protein required for some of the earliest steps in pattern formation. It thus helped us understand how different regions of the egg can give rise to cells that go down different developmental pathways. Second, it increased our understanding of the mother's critical role in the initial phases of embryonic development. Finally, the principle that a gradient of morphogens can determine polarity and position has proved to be a key developmental concept for a number of species, just as early embryologists had hypothesized.

Maternal mRNAs are crucial during development of many species. In *Drosophila*, gradients of specific proteins encoded by maternal mRNAs not only determine the posterior and anterior ends, but also establish the dorsal-ventral axis. As the fly embryo grows, it reaches a point when the embryonic program of gene expression takes over, and the maternal mRNAs must be destroyed. (This process involves miRNAs in *Drosophila* and other species.) Later, positional information encoded by the embryo's genes, operating on an ever finer scale, establishes a specific number of correctly oriented segments and triggers the formation of each segment's characteristic structures. When the genes operating in this final step are abnormal, the pattern of the adult is abnormal, as you saw in Figure 8.

EVOLUTION The fly with legs emerging from its head in Figure 8 is the result of a single mutation in one gene. The gene does not encode any antenna protein, however. Instead, it encodes a transcription factor that regulates other genes, and its malfunction leads to misplaced structures like legs instead of antennae. The observation that a change in gene regulation during development could lead to such a fantastic change in body form prompted some scientists to consider whether these types of mutations could contribute to evolution by generating novel body shapes. Ultimately this line of inquiry gave rise to the field of evolutionary developmental biology, so-called "evo-devo."

CONCEPT CHECK 1

1. **MAKE CONNECTIONS** Mitosis gives rise to two daughter cells that are genetically identical to the parent cell. Yet you, the product of many mitotic divisions, are not composed of identical cells. Why?
2. **MAKE CONNECTIONS** Explain how the signaling molecules released by an embryonic cell can induce changes in a neighboring cell without entering the cell.
3. How do fruit fly maternal effect genes determine the polarity of the egg and the embryo?

CONCEPT 2

Cloning of organisms showed that differentiated cells could be "reprogrammed" and ultimately led to the production of stem cells

When the field of developmental biology (then called embryology) was first taking shape in the early 1900s, a major question was whether all the cells of an organism have the same genes or whether cells lose genes during the process of differentiation. Today, we know that genes are not lost—but the question is whether each cell can express all of its genes.

One way to answer this question is to see whether a differentiated cell has the potential to generate a whole organism. Because an organism generated in this way develops from a single cell without either meiosis or fertilization, this is known as "cloning." In this context, cloning produces one or more organisms genetically identical to the "parent" that donated the single cell. This is often called *organismal cloning* to differentiate it from gene cloning and, more significantly, from cell cloning—the division of an asexually reproducing cell such as a bacterium into a group of genetically identical cells. (The common theme for all types of cloning is that the product is genetically identical to the parent. In fact, the word *clone* comes from the Greek *klon*, meaning "twig.")

The current interest in organismal cloning arises primarily from its potential to generate **stem cells**, relatively unspecialized cells that can both reproduce themselves and, under the right conditions, differentiate into many different tissues. We'll return to stem cells later. For now, we'll discuss a series of experiments that provides a conceptual framework for thinking about what determines the potential of a cell—in other words, what genes that cell can express at any given time.

Cloning Plants: Single-Cell Cultures

The successful cloning of whole plants from single differentiated cells was accomplished during the 1950s by F. C. Steward and his students at Cornell University, who worked with carrot plants. They found that single differentiated cells taken from the root (the carrot) and incubated in culture medium could grow into normal adult plants, each genetically identical to the parent plant. These results showed that differentiation does not necessarily involve irreversible changes in the DNA. In plants, at least, mature cells can "dedifferentiate" and then give rise to all the specialized cell types of the organism. Any cell with this potential is said to be **totipotent**.

Cloning Animals: Nuclear Transplantation

Differentiated cells from animals generally do not divide in culture, much less develop into the multiple cell types of a new organism. Therefore, early researchers had to use a different approach to the question of whether differentiated animal cells are totipotent. Their approach was to remove the nucleus of an egg (creating an *enucleated* egg) and replace it with the nucleus of a differentiated cell, a procedure called *nuclear transplantation*. If the nucleus from the differentiated donor cell retains its full genetic capability, then it should be able to direct development of the recipient cell into all the tissues and organs of an organism.

Such experiments were conducted on one species of frog (*Rana pipiens*) by Robert Briggs and Thomas King in the 1950s and on another frog species (*Xenopus laevis*) by John Gurdon in the 1970s (**Figure 11**). These researchers transplanted a

▼ Figure 11 Inquiry

Can the nucleus from a differentiated animal cell direct development of an organism?

Experiment John Gurdon and colleagues at Oxford University, in England, destroyed the nuclei of frog (*Xenopus laevis*) eggs by exposing the eggs to ultraviolet light. They then transplanted nuclei from cells of frog embryos and tadpoles into the enucleated eggs.

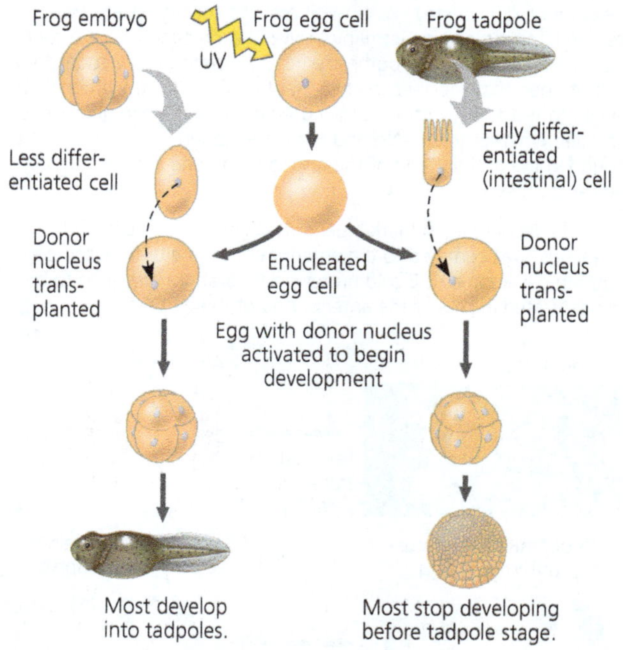

Results When the transplanted nuclei came from an early embryo, the cells of which are relatively undifferentiated, most of the recipient eggs developed into tadpoles. But when the nuclei came from the fully differentiated intestinal cells of a tadpole, fewer than 2% of the eggs developed into normal tadpoles, and most of the embryos stopped developing at a much earlier stage.

Conclusion The nucleus from a differentiated frog cell can direct development of a tadpole. However, its ability to do so decreases as the donor cell becomes more differentiated, presumably because of changes in the nucleus.

Data from J. B. Gurdon et al., The developmental capacity of nuclei transplanted from keratinized cells of adult frogs, *Journal of Embryology and Experimental Morphology* 34:93–112 (1975).

WHAT IF? If each cell in a four-cell embryo were already so specialized that it was not totipotent, what results would you predict for the experiment on the left side of the figure?

nucleus from an embryonic or tadpole cell into an enucleated egg of the same species. In Gurdon's experiments, the transplanted nucleus was often able to support normal development of the egg into a tadpole. However, he found that the potential of a transplanted nucleus to direct normal development was inversely related to the age of the donor: The older the donor nucleus, the lower the percentage of normal tadpoles.

From these results, Gurdon concluded that something in the nucleus *does* change as animal cells differentiate. In frogs and most other animals, nuclear potential tends to be restricted more and more as embryonic development and cell differentiation progress. These were foundational experiments that ultimately led to stem cell technology, and Gurdon received the 2012 Nobel Prize in Medicine for this work.

Reproductive Cloning of Mammals

In addition to cloning frogs, researchers had long been able to clone mammals using early embryonic cells as a source of donor nuclei. Until about 20 years ago, though, it was not known whether a nucleus from a fully differentiated cell could be reprogrammed successfully to act as a donor nucleus. In 1997, researchers in Scotland announced the birth of Dolly, a lamb cloned from an adult sheep by nuclear transplantation from a differentiated mammary gland cell **(Figure 12)**. Using a technique related to that shown in Figure 11, the researchers implanted early embryos into surrogate mothers. Out of several hundred embryos, one successfully completed normal development, and Dolly was born, a genetic clone of the nucleus donor. At the age of 6, Dolly suffered complications from a lung disease usually seen only in much older sheep and was euthanized. Dolly's premature death, as well as an arthritic condition, led to speculation that her cells were in some way not quite as healthy as those of a normal sheep, possibly reflecting incomplete reprogramming of the original transplanted nucleus.

Since that time, researchers have cloned numerous other mammals, including mice, cats, cows, horses, pigs, dogs, and monkeys. In most cases, their goal has been the production of new individuals; this is known as *reproductive cloning*. We have already learned a lot from such experiments. For example, cloned animals of the same species do *not* always look or behave identically. In a herd of cows cloned from the same line

▶ Figure 12 Dolly, the first mammal cloned by nuclear transplantation. Dolly, shown here as a lamb, has a very different appearance from her surrogate mother, standing behind her.

▶ Figure 13 CC ("Carbon Copy"), the first cloned cat (right), and her single parent. Rainbow (left) donated the nucleus in a cloning procedure that resulted in CC. However, the two cats are not identical: Rainbow has orange patches on her fur, but CC does not.

of cultured cells, certain cows are dominant in behavior and others are more submissive. Another example of nonidentity in clones is the first cloned cat, named CC for Carbon Copy **(Figure 13)**. She has a calico coat, like her single female parent, but the color and pattern are different because of random X chromosome inactivation, which is a normal occurrence during embryonic development. And identical human twins, which are naturally occurring "clones," are always slightly different. Evidently, environmental influences and random phenomena can play a significant role during development.

Faulty Gene Regulation in Cloned Animals

In most nuclear transplantation studies thus far, only a small percentage of cloned embryos develop normally to birth. And like Dolly, many cloned animals exhibit defects. Cloned mice, for instance, are prone to obesity, pneumonia, liver failure, and premature death. Scientists assert that even cloned animals that appear normal are likely to have subtle defects.

Researchers have uncovered some reasons for the low efficiency of cloning and the high incidence of abnormalities. In the nuclei of fully differentiated cells, a small subset of genes is turned on and expression of the rest of the genes is repressed. This regulation often is the result of epigenetic changes in chromatin, such as acetylation of histones or methylation of DNA. During the nuclear transfer procedure, many of these changes must be reversed in the later-stage nucleus from a donor animal for genes to be expressed or repressed appropriately in early stages of development. Researchers have found that the DNA in cells from cloned embryos, like that of differentiated cells, often has more methyl groups than does the DNA in equivalent cells from normal embryos of the same species. This finding suggests that the reprogramming of donor nuclei requires more accurate and complete chromatin restructuring than occurs during cloning procedures. Because DNA methylation helps regulate gene expression, misplaced or extra methyl groups in the DNA of donor nuclei may interfere with the pattern of gene expression necessary for normal embryonic development. In fact, the success of a cloning attempt may depend in large part on whether or not the chromatin in the donor nucleus can be artificially "rejuvenated" to resemble that of a newly fertilized egg.

Stem Cells of Animals

Progress in cloning mammalian embryos, including primates, has heightened speculation about the cloning of humans, which has not yet been achieved. The main reason researchers are trying to clone human embryos is not for reproduction, but for the production of stem cells to treat human diseases. Recall that a stem cell is a relatively unspecialized cell that can both reproduce itself indefinitely and, under appropriate conditions, differentiate into specialized cells of one or more types **(Figure 14)**. Thus, stem cells are able to both replenish their own population and generate cells that travel down specific differentiation pathways.

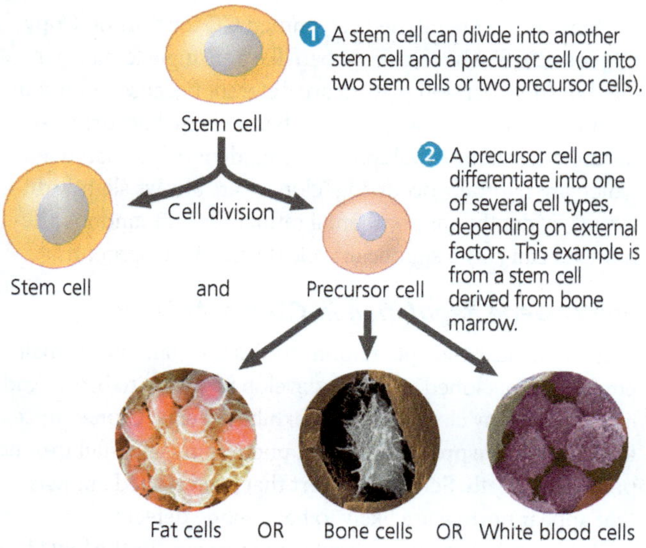

▲ **Figure 14 How stem cells maintain their own population and generate differentiated cells.**

Embryonic and Adult Stem Cells

Many early animal embryos contain stem cells capable of giving rise to differentiated embryonic cells of any type. Stem cells can be isolated from early embryos at a stage called the blastula stage or its human equivalent, the blastocyst stage. In culture, these *embryonic stem (ES) cells* reproduce indefinitely, and depending on culture conditions, they can be made to differentiate into a wide variety of specialized cells **(Figure 15)**, including even eggs and sperm.

The adult body also has stem cells, which serve to replace nonreproducing specialized cells as needed. In contrast to ES cells, *adult stem cells* are not able to give rise to all cell types in the organism, though they can generate multiple types. For example, one of the several types of stem cells in bone marrow can generate all the different kinds of blood cells (see Figure 15), and another type of bone marrow stem cell can differentiate into bone, cartilage, fat, muscle, and the linings of blood vessels. To the surprise of many, the adult brain has been found to contain stem cells that continue to produce certain kinds of nerve cells there. Researchers have also reported

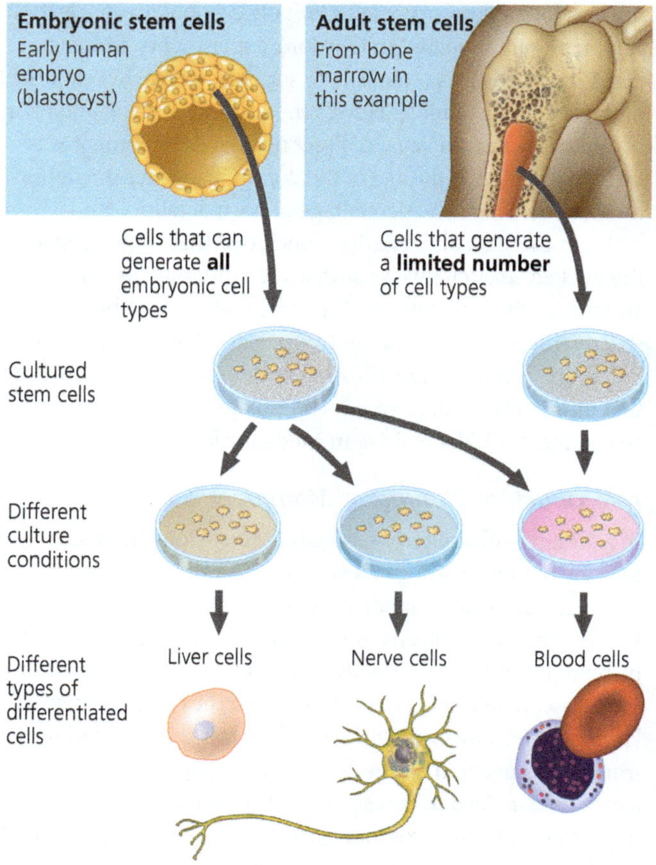

▲ **Figure 15 Working with stem cells.** Animal stem cells, which can be isolated from early embryos or adult tissues and grown in culture, are self-perpetuating, relatively undifferentiated cells. Embryonic stem cells are easier to grow than adult stem cells and can theoretically give rise to *all* types of cells in an organism. The range of cell types that can arise from adult stem cells is not yet fully understood.

finding stem cells in skin, hair, eyes, and dental pulp. Although adult animals have only tiny numbers of stem cells, scientists are learning to identify and isolate these cells from various tissues and, in some cases, to grow them in culture. With the right culture conditions (for instance, the addition of specific growth factors), cultured stem cells from adult animals have been made to differentiate into multiple types of specialized cells, although none are as versatile as ES cells.

Research with embryonic or adult stem cells is a source of valuable data about differentiation and has enormous potential for medical applications. The ultimate aim is to supply cells for the repair of damaged or diseased organs: for example, insulin-producing pancreatic cells for people with type 1 diabetes or certain kinds of brain cells for people with Parkinson's disease or Huntington's disease. Adult stem cells from bone marrow have long been used as a source of immune system cells in patients whose own immune systems are nonfunctional because of genetic disorders or radiation treatments for cancer.

The developmental potential of adult stem cells is limited to certain tissues. ES cells hold more promise than adult stem cells for most medical applications because ES cells are

pluripotent, capable of differentiating into many different cell types. Until recently, ES cells were obtained only from embryos donated (with informed consent) by patients undergoing infertility treatment or from long-term cell cultures originally established with cells isolated from donated embryos. However, in 2013, researchers in Oregon were able to obtain human ES cells from cloned human blastocysts and to establish several ES cell lines from them. When the main aim of cloning is to produce ES cells to treat disease, the process is called *therapeutic cloning*. Although most people believe that reproductive cloning of humans is unethical, opinions vary about the morality of therapeutic cloning.

Induced Pluripotent Stem (iPS) Cells

Resolving the debate about therapeutic cloning now seems less urgent because researchers have learned to turn back the clock in fully differentiated cells, reprogramming them to act like ES cells. This major feat was announced in 2007, first by labs using mouse skin cells and then by additional groups using cells from human skin and other organs or tissues. In all these cases, researchers transformed the differentiated cells into a type of ES cell by using viruses called retroviruses to introduce extra cloned copies of four "stem cell" master regulatory genes. The "deprogrammed" cells are known as *induced pluripotent stem (iPS) cells* because, in using this fairly simple laboratory technique to return them to their undifferentiated state, pluripotency has been restored. The experiments that first transformed human differentiated cells into iPS cells are described in **Figure 16**. Shinya Yamanaka received the 2012 Nobel Prize in Medicine for this work, shared with John Gurdon (see Figure 11).

By many criteria, iPS cells can perform most of the functions of ES cells, but there are some differences in gene expression and other cellular functions, such as cell division. At least until these differences are fully understood, the study of ES cells will continue to make important contributions to the development of stem cell therapies. In the meantime, work is proceeding using iPS cells that have been experimentally produced.

There are two major potential uses for human iPS cells. First, cells from patients suffering from diseases can be reprogrammed to become iPS cells, which can act as model cells for studying the disease and potential treatments. Human iPS cell lines have been developed from people with type 1 diabetes, Parkinson's disease, and at least a dozen other diseases. Second, in the field of regenerative medicine, a patient's cells could be reprogrammed into iPS cells and then used to replace nonfunctional tissues, such as insulin-producing cells of the pancreas or cells of the retina in the eye. In fact, Dr. Masayo Takahashi, of the RIKEN Center for Developmental Biology in Japan, has been leading a study on patients with age-related macular degeneration, a disease of the retina. She obtains skin cells from each patient and turns them into iPS cells, and then into retinal cells, ultimately transplanting them into the patient's eye. While this technique is currently too expensive to be used as a standard treatment for the disorder, it holds promise for the future.

▼ Figure 16 Inquiry

Can a fully differentiated human cell be "deprogrammed" to become a stem cell?

Experiment Shinya Yamanaka and colleagues at Kyoto University, in Japan, used a virus called a retrovirus as a vector to introduce four regulatory genes into fully differentiated human skin fibroblast cells. The cells were then cultured in a medium that would support growth of stem cells.

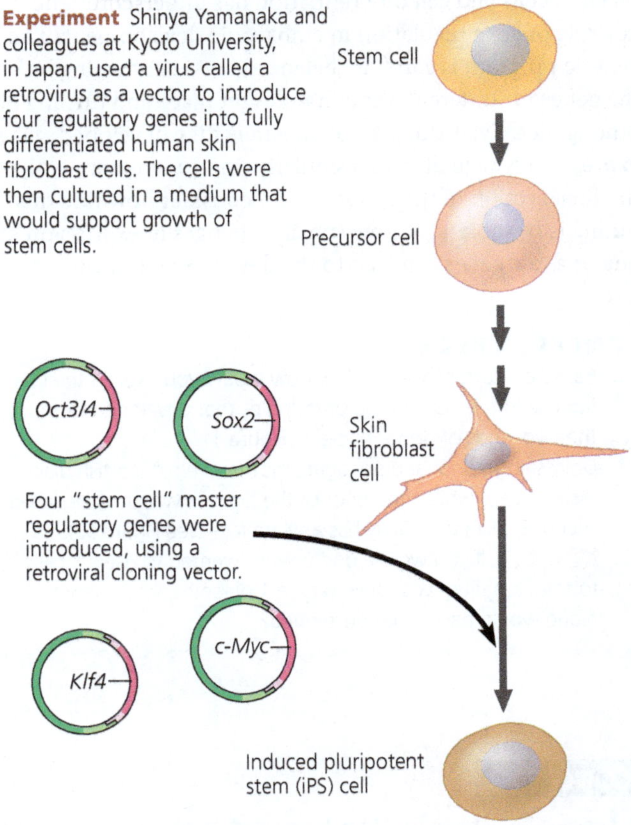

Results Two weeks later, the cells resembled embryonic stem cells in appearance and were actively dividing. Their gene expression patterns, gene methylation patterns, and other characteristics were also consistent with those of embryonic stem cells. The iPS cells were able to differentiate into heart muscle cells, as well as other cell types.

Conclusion The four regulatory genes induced differentiated skin cells to become pluripotent stem cells, with characteristics of embryonic skin cells.

Data from K. Takahashi et al., Induction of pluripotent stem cells from adult human fibroblasts by defined factors, *Cell* 131:861–872 (2007).

WHAT IF? When organs are transplanted from a donor to a diseased recipient, the recipient's immune system may reject the transplant, a condition with serious and often fatal consequences. Would the same risk be faced if the patients' own cells were used in an iPS-based procedure? Explain.

In another surprising recent development, researchers have identified genes that can reprogram a differentiated cell into another type of differentiated cell without passing through a pluripotent state. In the first reported example, one type of cell in the pancreas was transformed into another type. However, the two types of cells do not need to be very closely related: Another research group has been able to directly reprogram a skin fibroblast into a nerve cell. Developing techniques that direct iPS cells or even differentiated cells to become cell types for regenerative medicine is an area of intense research that has seen some success. The iPS cells created in this way could eventually provide tailor-made "replacement" cells for patients without using human eggs or embryos, thus avoiding most ethical objections.

The research described in this and the preceding section on stem cells and cell differentiation has underscored the key role of gene regulation in embryonic development. The genetic program is carefully balanced between turning on the genes for differentiation in the right place and turning off other genes. Even when an organism is fully developed, gene expression is regulated in a similarly fine-tuned manner. In the final section of the chapter, we'll consider how fine this tuning is by looking at how specific changes in expression of one or a few genes can lead to the development of cancer.

CONCEPT CHECK 2

1. Based on current knowledge, how would you explain the difference in the percentage of tadpoles that developed from the two kinds of donor nuclei in Figure 11?
2. Dolly's egg donor and surrogate mother were "Scottish blackface" sheep, while the donor of the nucleus was a white-faced sheep. Explain why Dolly has a white face (see Figure 12).
3. **WHAT IF?** If you were a doctor who wanted to use iPS cells to treat a patient with severe type 1 diabetes, what new technique would have to be developed?

CONCEPT 3

Abnormal regulation of genes that affect the cell cycle can lead to cancer

Cancer is a type of disease in which cells escape from the control mechanisms that normally limit their growth. Now that we have discussed the molecular basis of gene expression and its regulation, we are ready to look at cancer more closely. The gene regulation systems that go wrong during cancer turn out to be the very same systems that play important roles in embryonic development, the maintenance of stem cell populations, and many other biological processes. Thus, research into the molecular basis of cancer has both benefited from and informed many other fields of biology.

Types of Genes Associated with Cancer

The genes that normally regulate cell growth and division during the cell cycle include genes for growth factors, their receptors, and the intracellular molecules of signaling pathways. Mutations that alter any of these genes in somatic cells can lead to cancer. The agent of such change can be random spontaneous mutation. However, it is also likely that many cancer-causing mutations result from environmental influences, such as chemical carcinogens, X-rays and other high-energy radiation, and some viruses.

Cancer research led to the discovery of cancer-causing genes called **oncogenes** (from the Greek *onco*, tumor) in certain types of viruses. Subsequently, close counterparts of viral oncogenes were found in the genomes of humans and other animals. The normal versions of the cellular genes, called **proto-oncogenes**, code for proteins that stimulate normal cell growth and division.

How might a proto-oncogene—a gene that has an essential function in normal cells—become an oncogene, a cancer-causing gene? In general, an oncogene arises from a genetic change that leads to an increase either in the amount of the proto-oncogene's protein product or in the intrinsic activity of each protein molecule. The genetic changes that convert proto-oncogenes to oncogenes fall into three main categories: movement of DNA within the genome, amplification of a proto-oncogene, and point mutations in a control element or in the proto-oncogene itself (Figure 17).

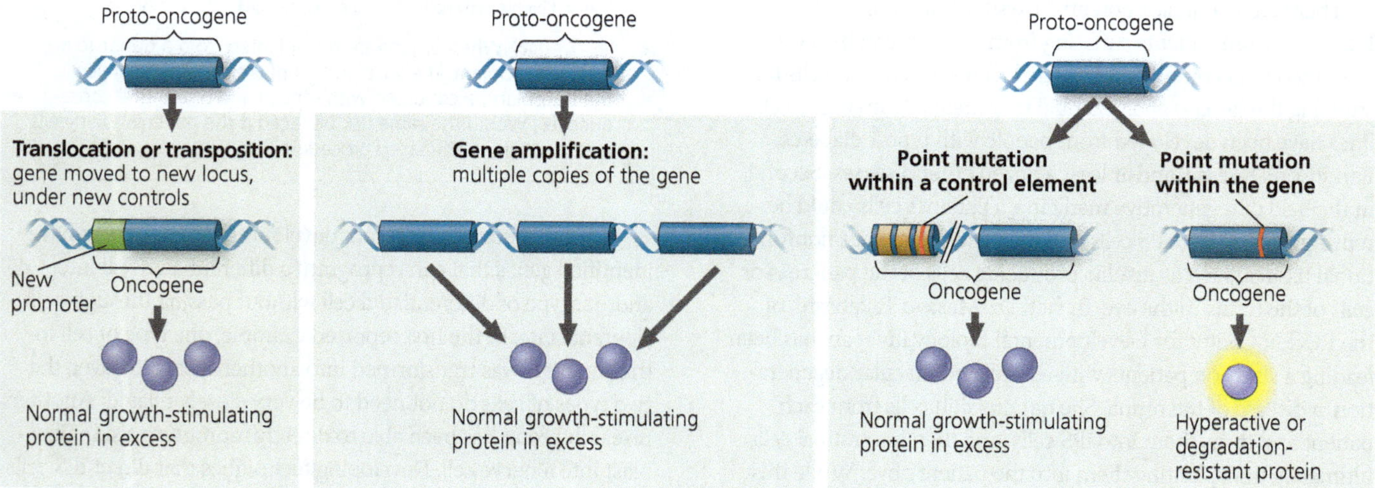

▲ **Figure 17** **Genetic changes that can turn proto-oncogenes into oncogenes.**

Cancer cells are frequently found to contain chromosomes that have broken and rejoined incorrectly, translocating fragments from one chromosome to another. Having learned how gene expression is regulated, you can now see the possible consequences of such translocations. If a translocated proto-oncogene ends up near an especially active promoter (or other control element), its transcription may increase, making it an oncogene. The second main type of genetic change, amplification, increases the number of copies of the proto-oncogene in the cell through repeated gene duplication. The third possibility is a point mutation either in the promoter or an enhancer that controls a proto-oncogene, causing an increase in its expression, or in the coding sequence of the proto-oncogene, changing the gene's product to a protein that is more active or more resistant to degradation than the normal protein. These mechanisms can lead to abnormal stimulation of the cell cycle and put the cell on the path to becoming a cancer cell.

In addition to genes whose products normally promote cell division, cells contain genes whose normal products *inhibit* cell division. Such genes are called **tumor-suppressor genes**, since the proteins they encode help prevent uncontrolled cell growth. Any mutation that decreases the normal activity of a tumor-suppressor protein may contribute to the onset of cancer, in effect stimulating growth through the absence of suppression.

Proteins produced from tumor-suppressor genes have various functions. Some repair damaged DNA, a function that prevents the cell from accumulating cancer-causing mutations. Other tumor-suppressor proteins control the adhesion of cells to each other or to the extracellular matrix; proper cell anchorage is crucial in normal tissues—and is often absent in cancers. Still other tumor-suppressor proteins are components of cell-signaling pathways that inhibit the cell cycle.

Interference with Cell-Signaling Pathways

The proteins encoded by many proto-oncogenes and tumor-suppressor genes are components of cell-signaling pathways. Let's take a closer look at how such proteins function in normal cells and what goes wrong with their function in cancer cells. We'll focus on the products of two key genes, the *ras* proto-oncogene and the *p53* tumor-suppressor gene. Mutations in *ras* occur in about 30% of human cancers, and mutations in *p53* in more than 50%.

The Ras protein, encoded by the ***ras* gene** (named for rat sarcoma, a connective tissue cancer), is a G protein that relays a signal from a growth factor receptor on the plasma membrane to a cascade of protein kinases. The cellular response at the end of the pathway is the synthesis of a protein that stimulates the cell cycle (**Figure 18a**). Normally, such a pathway will not operate unless triggered by the appropriate growth factor. But certain mutations in the *ras* gene can lead to production of a hyperactive Ras protein that triggers the kinase cascade even in the absence of growth factor, resulting in increased cell division (**Figure 18b**). In fact, hyperactive versions or excess amounts of any of the pathway's components can have the same outcome: excessive cell division.

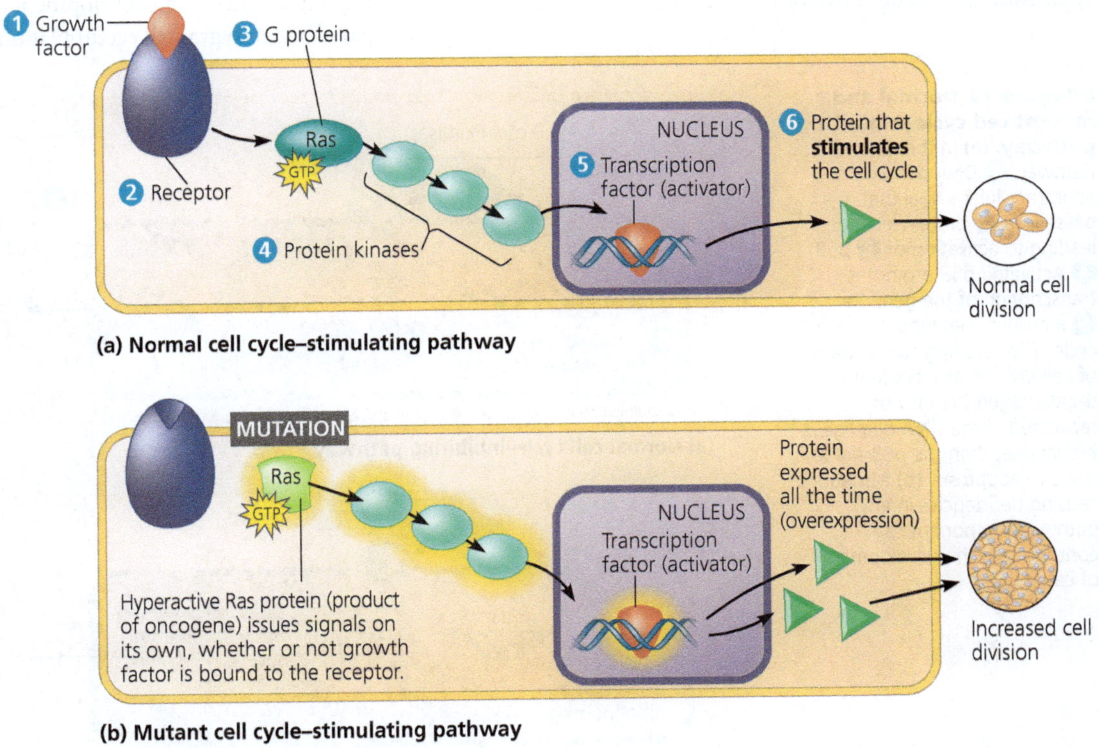

▶ **Figure 18 Normal and mutant cell cycle–stimulating pathway.** **(a)** The normal pathway is triggered by 1 a growth factor that binds to 2 its receptor in the plasma membrane. The signal is relayed to 3 a G protein called Ras. Like all G proteins, Ras is active when GTP is bound to it. Ras passes the signal to 4 a series of protein kinases. The last kinase activates 5 a transcription factor (activator) that turns on one or more genes for 6 a protein that stimulates the cell cycle. **(b)** If a mutation makes Ras or any other pathway component abnormally active, excessive cell division and cancer may result.

Figure 19a shows a pathway in which an intracellular signal leads to the synthesis of a protein that suppresses the cell cycle. In this case, the signal is damage to the cell's DNA, perhaps as the result of exposure to ultraviolet light. Operation of this signaling pathway blocks the cell cycle until the damage has been repaired. Otherwise, the damage might contribute to tumor formation by causing mutations or chromosomal abnormalities. Thus, the genes for the components of the pathway act as tumor-suppressor genes. The ***p53* gene**, named for the 53,000-dalton molecular weight of its protein product, is a tumor-suppressor gene. The protein it encodes is a specific transcription factor that promotes the synthesis of cell cycle–inhibiting proteins. That is why a mutation that knocks out the *p53* gene, like a mutation that leads to a hyperactive Ras protein, can lead to excessive cell growth and cancer **(Figure 19b)**.

The *p53* gene has been called the "guardian angel of the genome." Once the gene is activated—for example, by DNA damage—the p53 protein functions as an activator for several other genes. Often it activates a gene called *p21*, whose product halts the cell cycle by binding to cyclin-dependent kinases, allowing time for the cell to repair the DNA. Researchers recently showed that p53 also activates expression of a group of miRNAs, which in turn inhibit the cell cycle. In addition, the p53 protein can turn on genes directly involved in DNA repair. Finally, when DNA damage is irreparable, p53 activates "suicide" genes, whose protein products bring about apoptosis, as described in the first section of this chapter. Thus, p53 acts in several ways to prevent a cell from passing on mutations due to DNA damage. If mutations do accumulate and the cell survives through many divisions—as is more likely if the *p53* tumor-suppressor gene is defective or missing—cancer may ensue.

The many functions of p53 suggest a complex picture of regulation in normal cells, one that we do not yet fully understand.

For the present, the diagrams in Figures 18 and 19 are an accurate view of how mutations contribute to cancer, but we don't know exactly how a cell becomes a cancer cell. As we discover previously unknown aspects of gene regulation, it is informative to study their role in the onset of cancer. Such studies have shown, for instance, that DNA methylation and histone modification patterns differ in normal and cancer cells and that miRNAs probably participate in cancer development. While we've learned a lot about cancer by studying cell-signaling pathways, there is still a lot to learn.

The Multistep Model of Cancer Development

More than one somatic mutation is generally needed to produce all the changes characteristic of a full-fledged cancer cell. This may help explain why the incidence of cancer increases greatly with age. If cancer results from an accumulation of mutations and if mutations occur throughout life, then the longer we live, the more likely we are to develop cancer.

The model of a multistep path to cancer is well supported by studies of one of the best-understood types of human cancer, colorectal cancer, which affects the colon and/or rectum. About 140,000 new cases of colorectal cancer are diagnosed each year in the United States, and the disease causes 50,000 deaths each year. Like most cancers, colorectal cancer develops gradually **(Figure 20)**. The first sign is often a polyp, a small, benign growth in the colon lining. The cells of the polyp look normal, although they divide unusually frequently. The tumor grows and may eventually become malignant, invading other tissues. The development of a malignant tumor is paralleled by a gradual accumulation of mutations that convert

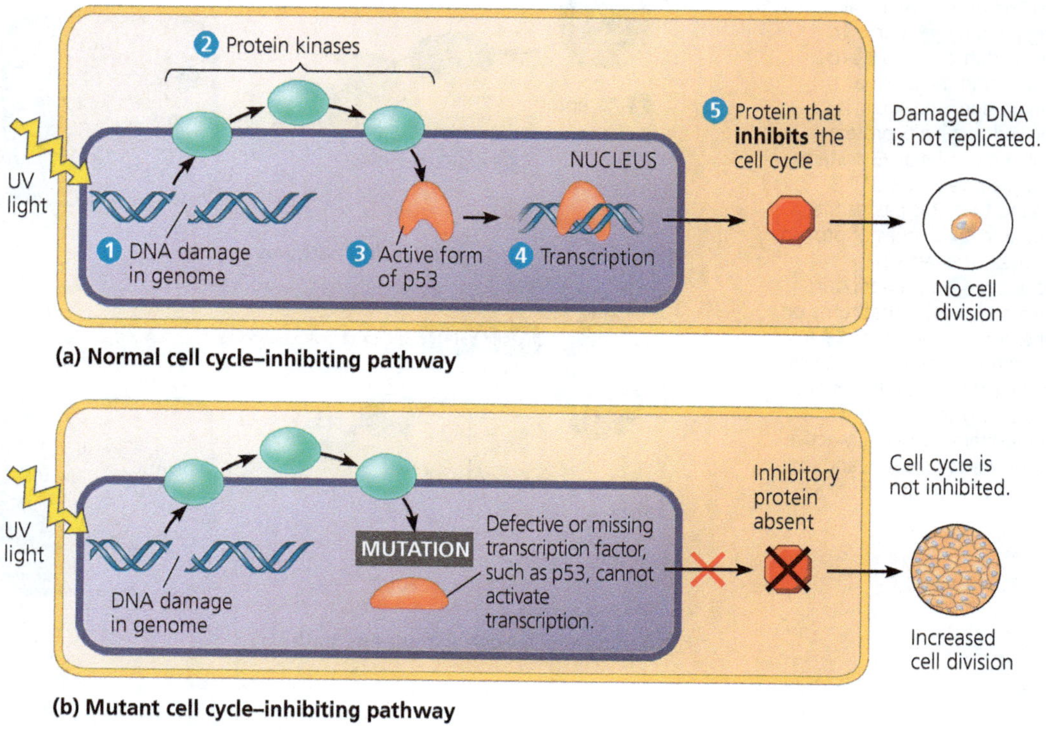

▶ **Figure 19 Normal and mutant cell cycle–inhibiting pathway. (a)** In the normal pathway, (1) DNA damage is an intracellular signal that is passed via (2) protein kinases, leading to activation of (3) p53. (4) Activated p53 promotes transcription of the gene for (5) a protein that inhibits the cell cycle. The resulting suppression of cell division ensures that the damaged DNA is not replicated. If the DNA damage is irreparable, then the p53 signal leads to apoptosis. **(b)** Mutations causing deficiencies in any pathway component can contribute to the development of cancer.

proto-oncogenes to oncogenes and knock out tumor-suppressor genes. A *ras* oncogene and a mutated *p53* tumor-suppressor gene are often involved.

About half a dozen changes must occur at the DNA level for a cell to become fully cancerous. These changes usually include the appearance of at least one active oncogene and the mutation or loss of several tumor-suppressor genes. Furthermore, since mutant tumor-suppressor alleles are usually recessive, in most cases mutations must knock out *both* alleles in a cell's genome to block tumor suppression. (Most oncogenes, on the other hand, behave as dominant alleles.)

Since we understand the progression of this type of cancer, routine screenings (colonoscopies, for example) are recommended to identify and remove any suspicious polyps. The colorectal cancer rate has been declining for the past 20 years, due to increased screening and improved treatments. Treatments for other cancers have improved as well. Advances in the sequencing of DNA and mRNA allow medical researchers to compare the genes expressed by different types of tumors and by the same type in different people. These comparisons have led to personalized treatments based on the molecular characteristics of a person's tumor.

Breast cancer is the second most common form of cancer in the United States, and the most common form of cancer among women. Each year, this cancer strikes over 230,000 women (and some men) in the United States and kills 40,000 (450,000 worldwide). A major problem with understanding breast cancer is its heterogeneity: Tumors differ in significant ways. Identifying differences between types of breast cancer is expected to improve treatment and decrease the mortality rate. In 2012, The Cancer Genome Atlas Network, sponsored by the National Institutes of Health, published the results of a genomics approach to profile subtypes of breast cancer based on their molecular signatures. Four major types of breast cancer were identified **(Figure 21)**.

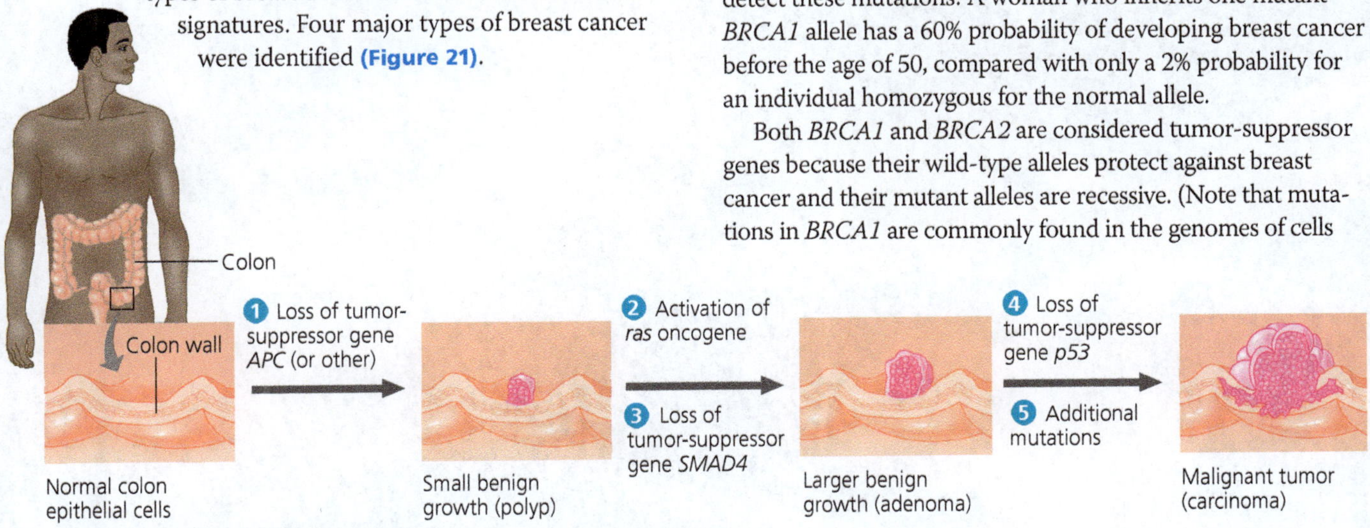

▲ **Figure 20 A multistep model for the development of colorectal cancer.** This type of cancer is one of the best understood. Changes in a tumor parallel a series of genetic changes, including mutations affecting several tumor-suppressor genes (such as *p53*) and the *ras* proto-oncogene. Mutations of tumor-suppressor genes often entail loss (deletion) of the gene. *APC* stands for "adenomatous polyposis coli," and *SMAD4* is a gene involved in signaling that results in apoptosis.

Inherited Predisposition and Other Factors Contributing to Cancer

The fact that multiple genetic changes are required to produce a cancer cell helps explain the observation that cancers can run in families. An individual inheriting an oncogene or a mutant allele of a tumor-suppressor gene is one step closer to accumulating the necessary mutations for cancer to develop than is an individual without any such mutations.

Geneticists are identifying inherited cancer alleles so that predisposition to certain cancers can be detected early in life. About 15% of colorectal cancers, for example, involve inherited mutations. Many affect the tumor-suppressor gene called *adenomatous polyposis coli*, or *APC* (see Figure 20). This gene has multiple functions in the cell, including regulation of cell migration and adhesion. Even in patients with no family history of the disease, the *APC* gene is mutated in 60% of colorectal cancers. In these people, new mutations must occur in both *APC* alleles before the gene's function is lost. Since only 15% of colorectal cancers are associated with known inherited mutations, researchers are trying to identify "markers" that could predict the risk of developing this type of cancer.

Given the prevalence and significance of breast cancer, it is not surprising that it was one of the first cancers for which the role of inheritance was investigated. It turns out that for 5–10% of patients with breast cancer, there is evidence of a strong inherited predisposition. Geneticist Mary-Claire King began working on this problem in the mid-1970s. After 16 years of research, she demonstrated that mutations in one gene—*BRCA1*—were associated with increased susceptibility to breast cancer, a finding that flew in the face of medical opinion at the time. (*BRCA* stands for <u>br</u>east <u>ca</u>ncer.) Mutations in that gene or a gene called *BRCA2* are found in at least half of inherited breast cancers, and tests using DNA sequencing can detect these mutations. A woman who inherits one mutant *BRCA1* allele has a 60% probability of developing breast cancer before the age of 50, compared with only a 2% probability for an individual homozygous for the normal allele.

Both *BRCA1* and *BRCA2* are considered tumor-suppressor genes because their wild-type alleles protect against breast cancer and their mutant alleles are recessive. (Note that mutations in *BRCA1* are commonly found in the genomes of cells

▼ **Figure 21**

MAKE CONNECTIONS

Genomics, Cell Signaling, and Cancer

Modern medicine that melds genome-wide molecular studies with cell-signaling research is transforming the treatment of many diseases, such as breast cancer. Using microarray analysis and other techniques, researchers measured the relative levels of mRNA transcripts for every gene in hundreds of breast cancer tumor samples. They identified four major subtypes of breast cancer that differ in their expression of three signal receptors involved in regulating cell growth and division:

- Estrogen receptor alpha (ERα)
- Progesterone receptor (PR)
- HER2, a type of receptor called a receptor tyrosine kinase

(ERα and PR are steroid receptor.) The absence or excess expression of these receptors can cause aberrant cell signaling, leading in some cases to inappropriate cell division, which may contribute to cancer (see Figure 18).

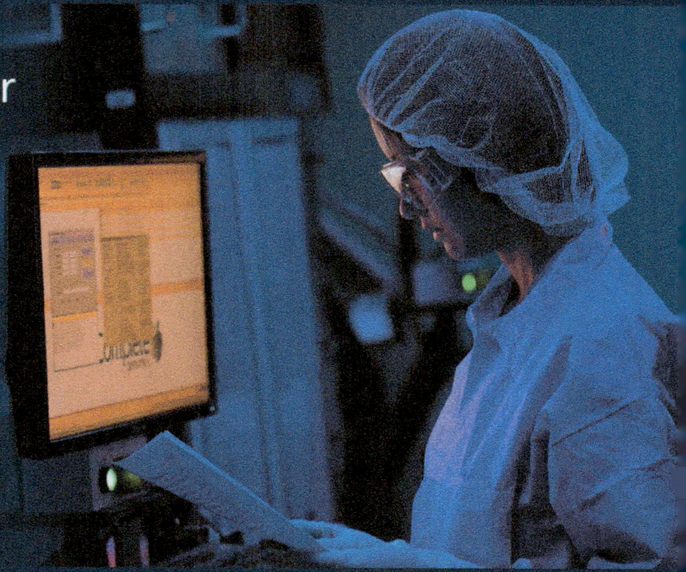

▲ **A research scientist examines DNA sequencing data from breast cancer samples.**

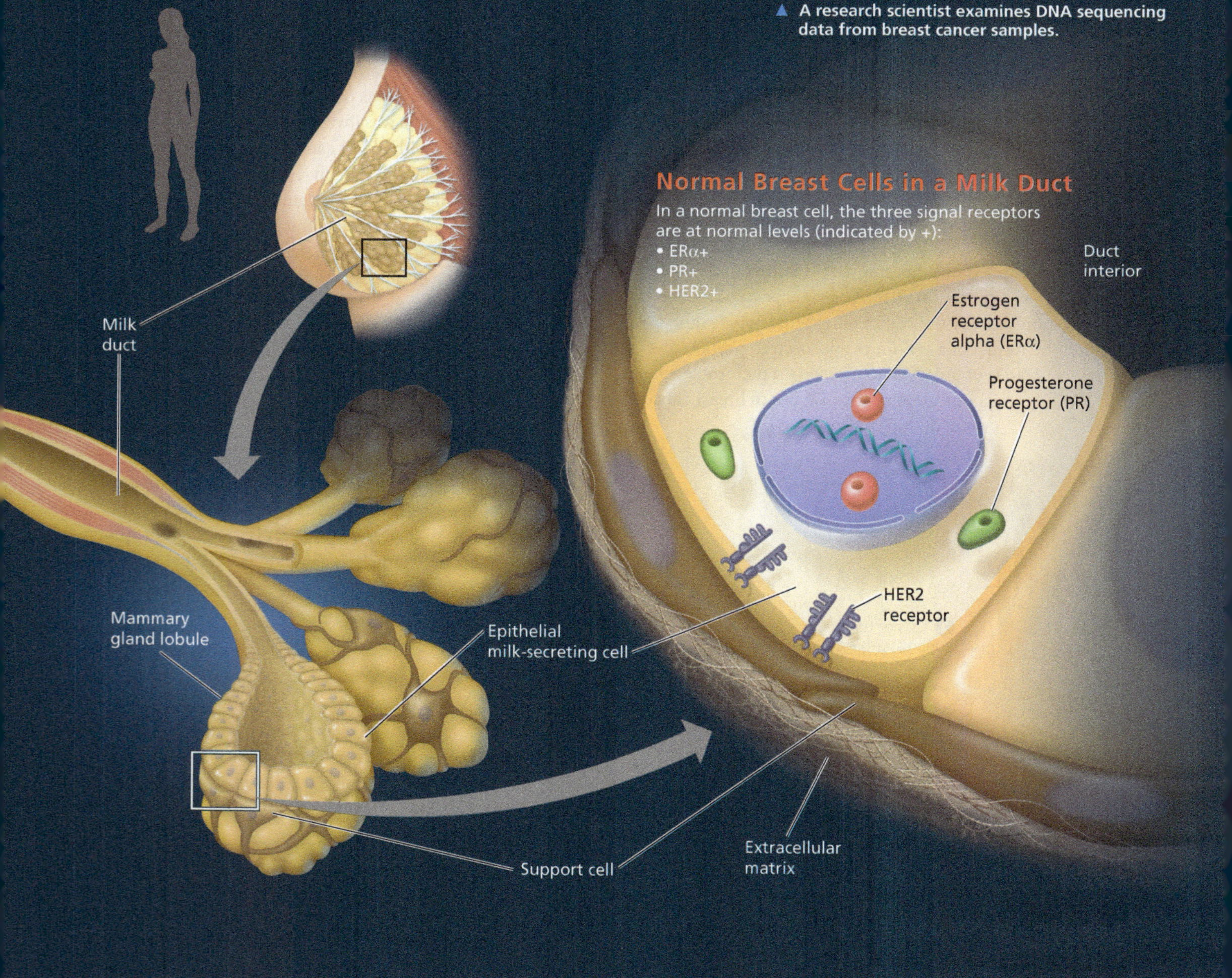

Breast Cancer Subtypes

Each breast cancer subtype is characterized by the overexpression (indicated by ++ or +++) or absence (–) of three signal receptors: ERα, PR, and HER2. Breast cancer treatments are becoming more effective because they can be tailored to the specific cancer subtype.

Luminal A

Luminal B

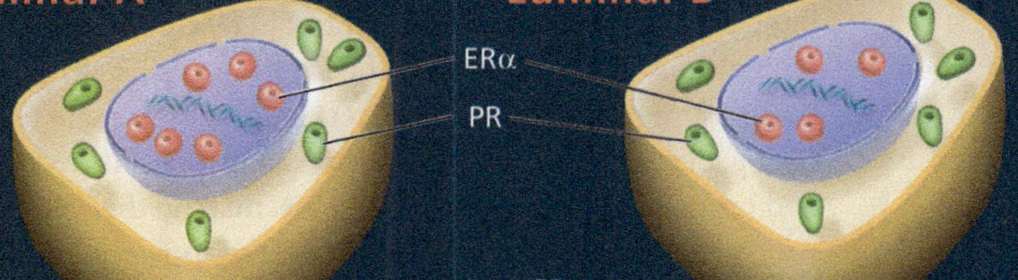

Luminal A
- ERα+++
- PR++
- HER2–
- 40% of breast cancers
- Best prognosis

Luminal B
- ERα++
- PR++
- HER2– (shown here); some HER2++
- 15–20% of breast cancers
- Poorer prognosis than luminal A subtype

Both luminal subtypes overexpress ERα (luminal A more than luminal B) and PR, and usually lack expression of HER2. Both can be treated with drugs that target ERα and inactivate it, the most well-known drug being tamoxifen. These subtypes can also be treated with drugs that inhibit estrogen synthesis.

Basal-like

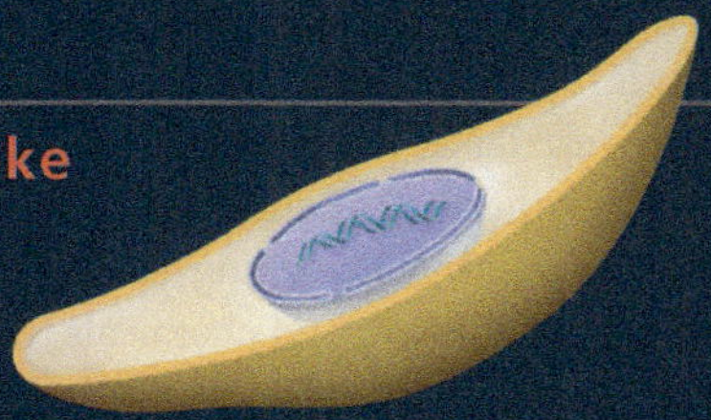

- ERα–
- PR–
- HER2–
- 15–20% of breast cancers
- More aggressive; poorer prognosis than other subtypes

The basal-like subtype is "triple negative"—it does not express ERα, PR, or HER2. It often has a mutation in the tumor-suppressor gene *BRCA1* (see Concept 3). Treatments that target ER, PR, or HER2 are not effective, but new treatments are being developed. Currently, patients are treated with cytotoxic chemotherapy, which selectively kills fast-growing cells.

HER2

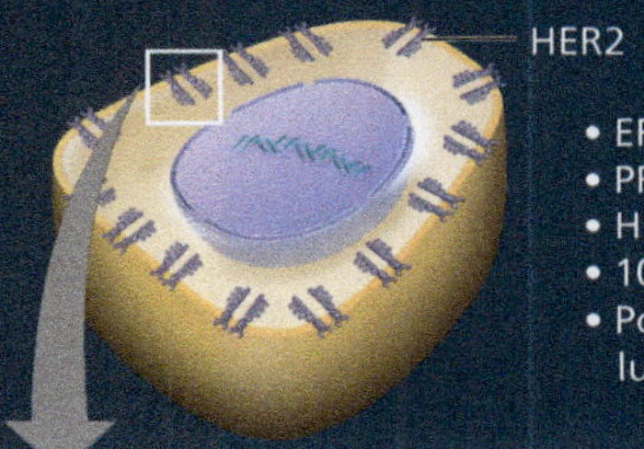

- ERα–
- PR–
- HER2++
- 10–15% of breast cancers
- Poorer prognosis than luminal A subtype

The HER2 subtype overexpresses HER2. Because it does not express either ERα or PR at normal levels, the cells are unresponsive to therapies that target those two receptors. However, patients with the HER2 subtype can be treated with Herceptin, an antibody protein that inactivates HER2.

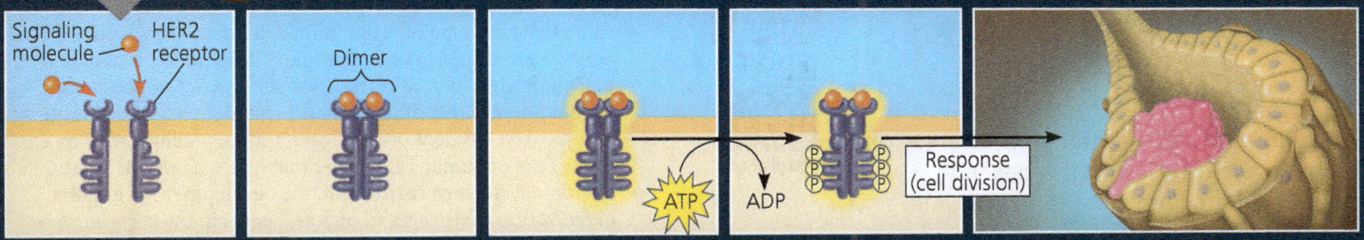

1 Signaling molecules (such as a growth factor) bind to HER2 receptor monomers (single receptor proteins).

2 Binding of signaling molecules causes two receptor monomers to associate closely with each other, forming a dimer.

3 Formation of a dimer activates each monomer.

4 Each monomer adds phosphate from ATP to the other monomer, triggering a signal transduction pathway.

5 The signal is transduced through the cell, which leads to a cellular response—in this case, turning on genes that trigger cell division. HER2 cells have up to 100 times as many HER2 receptors as normal cells, so they undergo uncontrolled cell division.

Treatment with Herceptin for the HER2 subtype

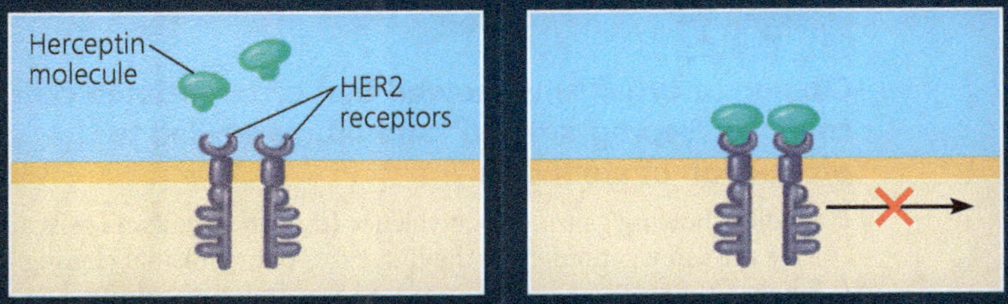

1 The drug Herceptin binds to HER2 receptors in place of the usual signaling molecules, preventing signaling.

2 In certain patients with the HER2 subtype, signaling is blocked and excessive cell division does not occur.

MAKE CONNECTIONS *When researchers compared gene expression in normal breast cells and cells from breast cancers, they found that the genes showing the most significant differences in expression encoded signal receptors, as shown here. Given what you learned in this chapter, explain why this result is not surprising.*

from basal-like breast cancers; see Figure 21.) The BRCA1 and BRCA2 proteins appear to function in the cell's DNA damage repair pathway. More is known about BRCA2. Along with another protein, BRCA2 helps repair breaks that occur in both strands of DNA, crucial for maintaining undamaged DNA.

Because DNA breakage can contribute to cancer, it makes sense that the risk of cancer can be lowered by minimizing exposure to DNA-damaging agents, such as the ultraviolet radiation in sunlight and chemicals found in cigarette smoke. Novel genomics-based analyses of specific cancers, such as the approach described in Figure 21, are contributing to both early diagnosis and development of treatments that interfere with expression of key genes in tumors. Ultimately, such approaches are expected to lower the death rate from cancer.

The study of genes associated with cancer, inherited or not, increases our basic understanding of how disruption of normal gene regulation can result in this disease. In addition to the mutations and other genetic alterations described in this section, a number of *tumor viruses* can cause cancer in various animals, including humans. In fact, one of the earliest breakthroughs in understanding cancer came in 1911, when Peyton Rous, an American pathologist, discovered a virus that causes cancer in chickens. The Epstein-Barr virus, which causes infectious mononucleosis, has been linked to several types of cancer in humans, notably Burkitt's lymphoma. Papillomaviruses are associated with cancer of the cervix, and a virus called HTLV-1 causes a type of adult leukemia. Viruses play a role in about 15% of the cases of human cancer.

Viruses may at first seem very different from mutations as a cause of cancer. However, we now know that viruses can interfere with gene regulation in several ways if they integrate their genetic material into the DNA of a cell. Viral integration may donate an oncogene to the cell, disrupt a tumor-suppressor gene, or convert a proto-oncogene to an oncogene. Some viruses produce proteins that inactivate p53 and other tumor-suppressor proteins, making the cell more prone to becoming cancerous. Viruses are powerful biological agents. You'll.

CONCEPT CHECK 3

1. The p53 protein can activate genes involved in apoptosis. Review Concept 1 and discuss how mutations in genes coding for proteins that function in apoptosis could contribute to cancer.
2. Under what circumstances is cancer considered to have a hereditary component?
3. **WHAT IF?** Cancer-promoting mutations are likely to have different effects on the activities of proteins encoded by proto-oncogenes compared with proteins encoded by tumor-suppressor genes. Explain.

Chapter Review

Go to **MasteringBiology**® for Assignments, the eText, and the Study Area with Animations, Activities, Vocab Self-Quiz, and Practice Tests.

SUMMARY OF KEY CONCEPTS

VOCAB SELF-QUIZ
goo.gl/gbai8v

CONCEPT 1
A program of differential gene expression leads to the different cell types in a multicellular organism

- Embryonic cells undergo **differentiation**, becoming specialized in structure and function. **Morphogenesis** encompasses the processes that give shape to the organism and its various structures. Cells differ in structure and function not because they contain different genomes but because they express different genes.
- Localized **cytoplasmic determinants** in the unfertilized egg are distributed differentially to daughter cells, where they regulate the expression of genes that affect those cells' developmental fates. In the process called **induction**, signaling molecules from embryonic cells cause transcriptional changes in nearby target cells.

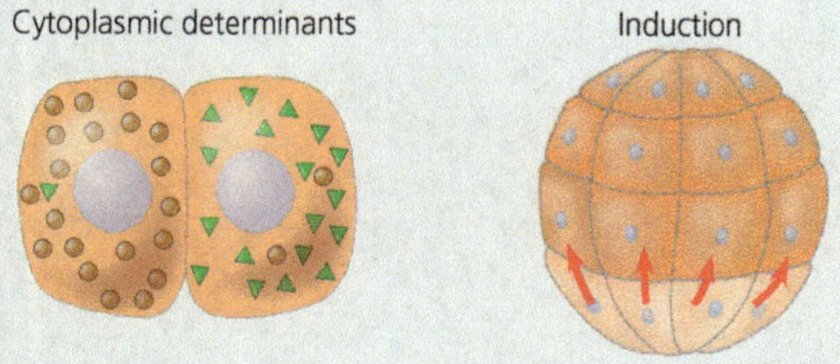

- Differentiation is heralded by the appearance of tissue-specific proteins, which enable differentiated cells to carry out their roles.
- **Apoptosis** is a type of programmed cell death in which cell components are disposed of in an orderly fashion, without damage to neighboring cells. Studies of the soil worm *Caenorhabditis elegans* showed that apoptosis occurs at defined times during embryonic development. Related apoptotic signaling pathways exist in the cells of humans and other mammals, as well as yeasts.
- In animals, **pattern formation**, the development of a spatial organization of tissues and organs, begins in the early embryo. **Positional information**, the molecular cues that control pattern formation, tells a cell its location relative to the body's axes and to other cells. In *Drosophila melanogaster*, gradients of **morphogens** encoded by **maternal effect genes** determine the body axes. For example, the gradient of **Bicoid** protein determines the anterior-posterior axis.

? *Describe the two main processes that cause embryonic cells to head down different pathways to their final fates.*

CONCEPT 2
Cloning of organisms showed that differentiated cells could be "reprogrammed" and ultimately led to the production of stem cells

- Studies showing genomic equivalence (that an organism's cells all have the same genome) are the first cases of organismal cloning.
- Single differentiated cells from plants are often **totipotent**: capable of generating all the tissues of a complete new plant.
- Transplantation of the nucleus from a differentiated animal cell into an enucleated egg can sometimes give rise to a new animal.

- Certain embryonic **stem cells** (ES cells) from animal embryos or adult stem cells from adult tissues can reproduce and differentiate in a test tube as well as in the organism, offering the potential for medical use. ES cells are **pluripotent** but difficult to acquire. Induced pluripotent stem (iPS) cells resemble ES cells in their capacity to differentiate; they can be generated by reprogramming differentiated cells. iPS cells hold promise for medical research.

? *Describe how a researcher could carry out organismal cloning, production of ES cells, and generation of iPS cells, focusing on how the cells are reprogrammed and using mice as an example. (The procedures are basically the same in humans and mice.)*

CONCEPT 3

Abnormal regulation of genes that affect the cell cycle can lead to cancer

- The products of **proto-oncogenes** and **tumor-suppressor genes** control cell division. A DNA change that makes a proto-oncogene excessively active converts it to an **oncogene**, which then may promote excessive cell division and cancer. A tumor-suppressor gene encodes a protein that inhibits abnormal cell division. A mutation that reduces activity of its protein product may lead to excessive cell division and cancer.
- Many proto-oncogenes and tumor-suppressor genes encode components of growth-stimulating and growth-inhibiting signaling pathways, respectively, and mutations in these genes can interfere with normal cell-signaling pathways. A hyperactive version of a protein in a stimulatory pathway, such as **Ras** (a G protein), functions as an oncogene protein. A defective version of a protein in an inhibitory pathway, such as **p53** (a transcription activator), fails to function as a tumor suppressor.

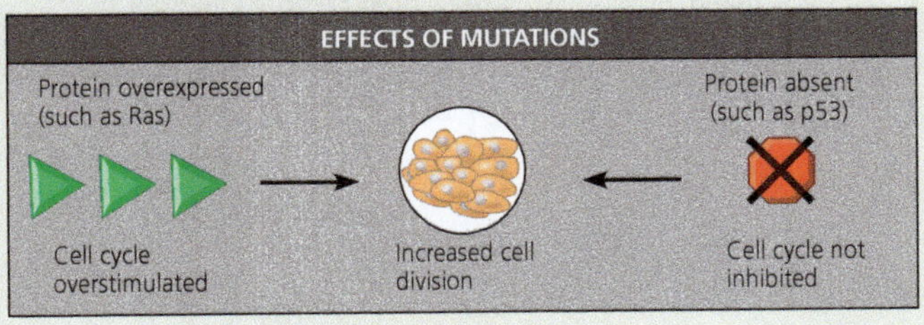

- In the multistep model of cancer development, normal cells are converted to cancer cells by the accumulation of mutations affecting proto-oncogenes and tumor-suppressor genes. Technical advances in DNA and mRNA sequencing are enabling cancer treatments that are more individually based
- Genomics studies have resulted in four proposed subtypes of breast cancer, based on expression of genes by tumor cells.
- Individuals who inherit a mutant allele of a proto-oncogene or tumor-suppressor gene have a predisposition to develop a particular cancer. Certain viruses promote cancer by integration of viral DNA into a cell's genome.

? *Compare the usual functions of proteins encoded by proto-oncogenes with the functions of proteins encoded by tumor-suppressor genes.*

TEST YOUR UNDERSTANDING

PRACTICE TEST

goo.gl/CRZjvS

Level 1: Knowledge/Comprehension

1. Muscle cells differ from nerve cells mainly because they
   (A) express different genes.
   (B) contain different genes.
   (C) use different genetic codes.
   (D) have unique ribosomes.

2. Cell differentiation always involves
   (A) the transcription of the *myoD* gene.
   (B) the movement of cells.
   (C) the production of tissue-specific proteins.
   (D) the selective loss of certain genes from the genome.

Level 2: Application/Analysis

3. Apoptosis involves all but which of the following?
   (A) fragmentation of the DNA
   (B) cell-signaling pathways
   (C) lysis of the cell
   (D) digestion of cellular contents by scavenger cells

4. Absence of *bicoid* mRNA from a *Drosophila* egg leads to the absence of anterior larval body parts and mirror-image duplication of posterior parts. This is evidence that the product of the *bicoid* gene
   (A) normally leads to formation of head structures.
   (B) normally leads to formation of tail structures.
   (C) is transcribed in the early embryo.
   (D) is a protein present in all head structures.

5. Proto-oncogenes can change into oncogenes that cause cancer. Which of the following best explains the presence of these potential time bombs in eukaryotic cells?
   (A) Proto-oncogenes first arose from viral infections.
   (B) Proto-oncogenes are mutant versions of normal genes.
   (C) Proto-oncogenes are genetic "junk."
   (D) Proto-oncogenes normally help regulate cell division.

Level 3: Synthesis/Evaluation

6. **SCIENTIFIC INQUIRY**
   Prostate cells usually require testosterone and other androgens to survive. But some prostate cancer cells thrive despite treatments that eliminate androgens. One hypothesis is that estrogen, often considered a female hormone, may be activating genes normally controlled by an androgen in these cancer cells. Describe one or more experiments to test this hypothesis.

7. **FOCUS ON EVOLUTION**
   Cancer cells can be considered a population that undergoes evolutionary processes such as random mutation and natural selection. Apply what you learned about cancer in this chapter to discuss this concept.

8. **FOCUS ON ORGANIZATION**
   The property of life emerges at the biological level of the cell. The highly regulated process of apoptosis is not simply the destruction of a cell; it is also an emergent property. In a short essay (about 100–150 words), briefly explain the role of apoptosis in the development and proper functioning of an animal and describe how this form of programmed cell death is a process that emerges from the orderly integration of signaling pathways.

9. **SYNTHESIZE YOUR KNOWLEDGE**

Recently, new anti-aging skin creams have been developed that claim to harness the power of stem cells. One such skin cream contains stem cells from red grape plants, and the manufacturer claims that it will "restore our skin's stem cells." Do you think this cream will fulfill the manufacturer's promise? Explain.

Answers

Figure Questions

Figure 4 Even if the mutant MyoD protein couldn't activate the *myoD* gene, it could still turn on genes for the other proteins in the pathway (other transcription factors, which would turn on the genes for muscle-specific proteins, for example). Therefore, some differentiation would occur. But unless there were other activators that could compensate for the loss of the MyoD protein's activation of the *myoD* gene, the cell would not be able to maintain its differentiated state. **Figure 10** Normal Bicoid protein would be made in the anterior end and compensate for the presence of mutant *bicoid* mRNA put into the egg by the mother. Development should be normal, with a head present. (This is what was observed.) **Figure 11** None of the eggs with the transplanted nuclei from the four-cell embryo at the upper left would have developed into a tadpole. Also, the resulting samples might include only some of the tissues of a tadpole. The tissues that develop might differ from treatment to treatment, depending on which of the four nuclei was transplanted. (This assumes that there was some way to tell the four cells apart, as one can in some frog species.) **Figure 16** Using converted iPS cells would not carry the same risk, which is its major advantage. Because the donor cells would come from the patient, they would be perfectly matched. The patient's immune system would recognize them as "self" cells and would not mount an attack (which is what leads to rejection). **Figure 21** Cancer is a disease in which cell division occurs without its usual regulation. Cell division can be stimulated by growth factors, which bind to cell-surface receptors. Cancer cells evade these normal controls and can often divide in the absence of growth factors. This suggests that the receptor proteins or some other components in a signaling pathway are abnormal in some way (see, for example, the mutant Ras protein in Figure 18) or are expressed at abnormal levels, as seen for the receptors in this figure. Under some circumstances in the mammalian body, steroid hormones such as estrogen and progesterone can also promote cell division. These molecules also use cell-signaling pathways. Because signaling receptors are involved in triggering cells to undergo cell division, it is not surprising that altered genes encoding these proteins might play a significant role in the development of cancer. Genes might be altered either through a mutation that changes the function of the protein product or through a mutation that causes the gene to be expressed at abnormal levels that disrupt the overall regulation of the signaling pathway.

Concept Check 1

1. Cells undergo differentiation during embryonic development, becoming different from each other. Therefore, the adult organism is made up of many highly specialized cell types. **2.** By binding to a receptor on the receiving cell's surface and triggering a signal transduction pathway involving intracellular molecules such as second messengers and transcription factors that affect gene expression **3.** The products of maternal effect genes, made and deposited into the egg by the mother, determine the head and tail ends, as well as the back and belly, of the egg and embryo (and eventually the adult fly).

Concept Check 2

1. The state of chromatin modification in the nucleus from the intestinal cell was undoubtedly less similar to that of a nucleus from a fertilized egg, explaining why many fewer of these nuclei were able to be reprogrammed. In contrast, the chromatin in a nucleus from a cell at the four-cell stage would have been much more like that of a nucleus in a fertilized egg and therefore much more easily programmed to direct development. **2.** Dolly's face matched that of the donor of the nucleus, because the nucleus contained the genetic information that determined Dolly's phenotype. No genetic information was contributed by either the egg donor (the egg was enucleated) or the surrogate mother. **3.** A technique would have to be worked out for turning a human iPS cell into a pancreatic cell (probably by inducing expression of pancreas-specific regulatory genes in the cell).

Concept Check 3

1. Apoptosis is signaled by p53 protein when a cell has extensive DNA damage, so apoptosis plays a protective role in eliminating a cell that might contribute to cancer. If mutations in the genes in the apoptotic pathway blocked apoptosis, a cell with such damage could continue to divide and might lead to tumor formation. **2.** When an individual has inherited an oncogene or a mutant allele of a tumor-suppressor gene **3.** A cancer-causing mutation in a proto-oncogene usually makes the gene product overactive, whereas a cancer-causing mutation in a tumor-suppressor gene usually makes the gene product nonfunctional.

Summary of Key Concepts Questions

1 The first process involves cytoplasmic determinants, including mRNAs and proteins, placed into specific locations in the egg by the mother. The cells that are formed from different regions in the egg during early cell divisions will have different proteins in them, which will direct different programs of gene expression. The second process involves the cell in question responding to signaling molecules secreted by neighboring cells. The signaling pathway in the responding cells also leads to a different pattern of gene expression. The coordination of these two processes results in each cell following a unique pathway in the developing embryo. **2** Cloning a mouse involves transplanting a nucleus from a differentiated mouse cell into a mouse egg cell that has had its own nucleus removed. Activating the egg cell and promoting its development into an embryo in a surrogate mother results in a mouse that is genetically identical to the mouse that donated the nucleus. In this case, the differentiated nucleus has been reprogrammed by factors in the egg cytoplasm. Mouse ES cells are generated from inner cells in mouse blastocysts, so in this case the cells are "naturally" reprogrammed by the process of reproduction and development. (Cloned mouse embryos can also be used as a source of ES cells.) iPS cells can be generated without the use of embryos from a differentiated adult mouse cell by adding certain transcription factors into the cell. In this case, the transcription factors are reprogramming the cells to become pluripotent. **3** The protein product of a proto-oncogene is usually involved in a pathway that stimulates cell division. The protein product of a tumor-suppressor gene is usually involved in a pathway that inhibits cell division.

Test Your Understanding

1. A **2.** C **3.** C **4.** A **5.** D

Credits

Credits are listed in the order of appearance.

Photo Credits

1 Stephen W. Paddock; **2** Mike Wu; **16 SSE mouse** Joseph Zellner/E+/Getty Images; **paws** Thomas Montavon; **5** Gopal Murti/Science Source; **6** From: Mesenchymal cells engulf and clear apoptotic footplate cells in macrophageless PU. 1 null mouse embryos. William Wood, et al. *Development* 127, 5245–5252 (2000). Fig. 2abc; **8** F. Rudolf Turner, Indiana University; **9** Wolfgang Driever, University of Freiburg, Freiburg, Germany; **10** Ruth Lehmann, The Whitehead Institution; **12** The Roslin Institute, The University of Edinburgh, Roslin, Scotland, UK; **13** Pat Sullivan/AP Images; **14a** Steve Gschmeissner/Science Source; **14b** SPL/Science Source; **14c** Steve Gschmeissner/Science Source; **21** David Paul Morris/Bloomberg via Getty Images; **Chapter Review SYK** i love images/Cultura/Getty Images.

Illustration and Text Credits

SSE.1 Data from T. Montavon et al., A regulatory archipelago controls *Hox* genes transcription in digits, *Cell* 147:1132–1145 (2011). doi:10.1016/j.cell.2011.10.023;
10 C. Nüsslein-Volhard et al., Determination of anteroposterior polarity in *Drosophila*, *Science* 238:1675–1681 (1987); W. Driever and C. Nüsslein-Volhard, A gradient of Bicoid protein in *Drosophila* embryos, *Cell* 54:83–93 (1988); T. Berleth et al., The role of localization of bicoid RNA in organizing the anterior pattern of the *Drosophila* embryo, *EMBO Journal* 7:1749–1756 (1988); **11** Data from J. B. Gurdon et al., The developmental capacity of nuclei transplanted from keratinized cells of adult frogs, *Journal of Embryology and Experimental Morphology* 34:93–112 (1975); **16** Data from K. Takahashi et al., Induction of pluripotent stem cells from adult human fibroblasts by defined factors, *Cell* 131:861–872 (2007); **20** Adapted from *The World of the Cell*, 3rd Edition, by W.M. Becker, J.B. Reece, and M.F. Poenie, 1996. Copyright © 1996 by Pearson Education, Inc. Reprinted and electronically reproduced by permission of Pearson Education, Inc. Upper Saddle River, NJ.

Glossary

apoptosis (ā-puh-tō′-sus) A type of programmed cell death that is brought about by activation of enzymes that break down many chemical components in the cell.

bicoid A maternal effect gene that codes for a protein responsible for specifying the anterior end in *Drosophila melanogaster*.

cytoplasmic determinant A maternal substance, such as a protein or RNA, that when placed into an egg influences the course of early development by regulating the expression of genes that affect the developmental fate of cells.

determination The progressive restriction of developmental potential whereby the possible fate of each cell becomes more limited as an embryo develops. At the end of determination, a cell is committed to its fate.

differentiation The process by which a cell or group of cells becomes specialized in structure and function.

embryonic lethal A mutation with a phenotype leading to death of an embryo or larva.

homeotic gene (hō-mē-o′-tik) Any of the master regulatory genes that control placement and spatial organization of body parts in animals, plants, and fungi by controlling the developmental fate of groups of cells.

induction The process in which one group of embryonic cells influences the development of another, usually by causing changes in gene expression.

maternal effect gene A gene that, when mutant in the mother, results in a mutant phenotype in the offspring, regardless of the offspring's genotype. Maternal effect genes, also called egg-polarity genes, were first identified in *Drosophila melanogaster*.

model organism A particular species chosen for research into broad biological principles because it is representative of a larger group and usually easy to grow in a lab.

morphogen A substance, such as Bicoid protein in *Drosophila*, that provides positional information in the form of a concentration gradient along an embryonic axis.

oncogene (on′-kō-jēn) A gene found in viral or cellular genomes that is involved in triggering molecular events that can lead to cancer.

pattern formation The development of a multicellular organism's spatial organization, the arrangement of organs and tissues in their characteristic places in three-dimensional space.

pluripotent Describing a cell that can give rise to many, but not all, parts of an organism.

positional information Molecular cues that control pattern formation in an animal or plant embryonic structure by indicating a cell's location relative to the organism's body axes. These cues elicit a response by genes that regulate development.

proto-oncogene (prō′-tō-on′-kō-jēn) A normal cellular gene that has the potential to become an oncogene.

stem cell Any relatively unspecialized cell that can produce, during a single division, one identical daughter cell and one more specialized daughter cell that can undergo further differentiation.

tumor-suppressor gene A gene whose protein product inhibits cell division, thereby preventing the uncontrolled cell growth that contributes to cancer.

Viruses

From Chapter 17 of *Campbell Biology in Focus*, Second Edition. Lisa A. Urry, Michael L. Cain, Steven A. Wasserman, Peter V. Minorsky, and Jane B. Reece.

Viruses

KEY CONCEPTS

1. A virus consists of a nucleic acid surrounded by a protein coat
2. Viruses replicate only in host cells
3. Viruses and prions are formidable pathogens in animals and plants

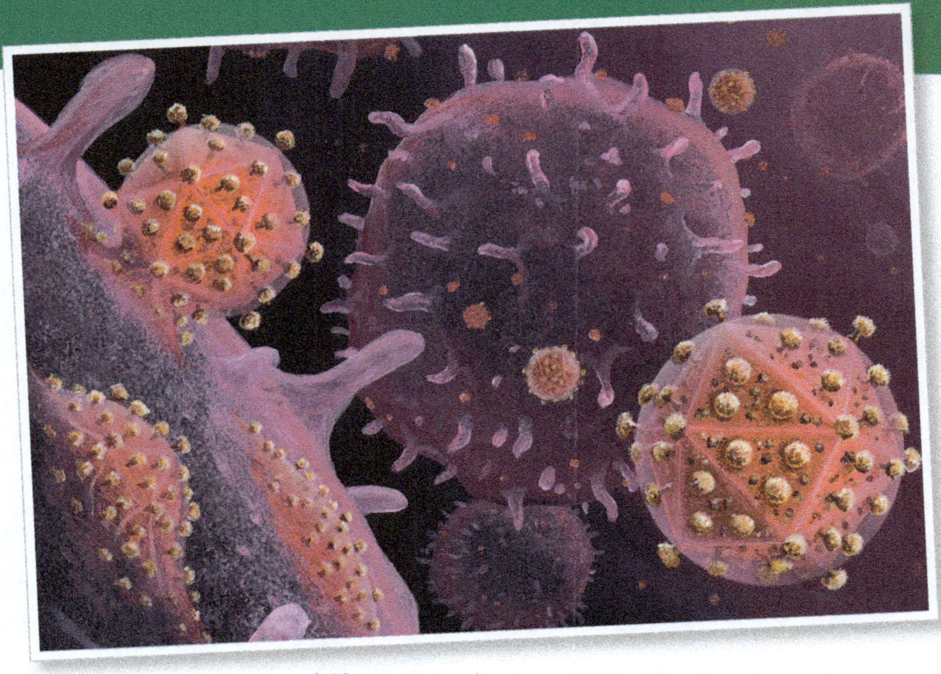

▲ **Figure 1** Are the viruses (red) budding from these cells alive?

A Borrowed Life

The illustration in **Figure 1** shows a remarkable event: Human immune cells (purple) infected by human immunodeficiency viruses (HIV) are releasing more HIV viruses. These viruses (red, surrounded by purple membrane from the immune cell) will infect other cells. By injecting its genetic information into a cell, a virus hijacks the cell, recruiting cellular machinery to manufacture many new viruses and promote further infection. Left untreated, HIV causes acquired immunodeficiency syndrome (AIDS) by destroying vital immune system cells.

Compared with eukaryotic and even prokaryotic cells, viruses are much smaller and simpler in structure. Lacking the structures and metabolic machinery found in a cell, a **virus** is an infectious particle consisting of little more than genes packaged in a protein coat.

Are viruses living or nonliving? Because viruses are capable of causing a wide variety of diseases, researchers in the late 1800s saw a parallel with bacteria and proposed that viruses were the simplest of living forms. However, viruses cannot reproduce or carry out metabolic activities outside of a host cell. Most biologists studying viruses today would probably agree that they are not alive but exist in a shady area between life-forms and chemicals. The simple phrase used recently by two researchers describes them aptly enough: Viruses lead "a kind of borrowed life."

In this chapter, we'll explore the biology of viruses, beginning with their structure and then describing how they replicate. We'll next look at the role of viruses as disease-causing agents, or pathogens, of plants and animals. Finally, we'll consider some even simpler infectious agents called prions.

CONCEPT 1

A virus consists of a nucleic acid surrounded by a protein coat

The tiniest viruses are only 20 nm in diameter—smaller than a ribosome. Millions could easily fit on a pinhead. Even the largest known virus, which has a diameter of several hundred nanometers, is barely visible under the light microscope. An early discovery that some viruses could be crystallized was exciting and puzzling news. Not even the simplest of cells can aggregate into regular crystals. But if viruses are not cells, then what are they? Examining the structure of a virus more closely reveals that it is an infectious particle consisting of nucleic acid enclosed in a protein coat and, for some viruses, surrounded by a membranous envelope.

Viral Genomes

We usually think of genes as being made of double-stranded DNA, but many viruses defy this convention. Their genomes may consist of double-stranded DNA, single-stranded DNA, double-stranded RNA, or single-stranded RNA, depending on the type of virus. A virus is called a DNA virus or an RNA virus, based on the kind of nucleic acid that makes up its genome. In either case, the genome is usually organized as a single linear or circular molecule of nucleic acid, although the genomes of some viruses consist of multiple molecules of nucleic acid. The smallest viruses known have only three genes in their genome, while the largest have several hundred to a thousand. For comparison, bacterial genomes contain about 200 to a few thousand genes.

Capsids and Envelopes

The protein shell enclosing the viral genome is called a **capsid**. Depending on the type of virus, the capsid may be rod-shaped, polyhedral, or more complex in shape. Capsids are built from a large number of protein subunits called *capsomeres*, but the number of different *kinds* of proteins in a capsid is usually small. Tobacco mosaic virus (TMV), for example, has a rigid, rod-shaped capsid made from over a thousand molecules of a single type of protein arranged in a helix; rod-shaped viruses are commonly called *helical viruses* for this reason **(Figure 2a)**. Adenoviruses, which infect the respiratory tracts of animals, have 252 identical protein molecules arranged in a polyhedral capsid with 20 triangular facets—an icosahedron; thus, these and other similarly shaped viruses are referred to as *icosahedral viruses* **(Figure 2b)**.

Some viruses have accessory structures that help them infect their hosts. For instance, a membranous envelope surrounds the capsids of influenza viruses and many other viruses found in animals **(Figure 2c)**. These **viral envelopes**, which are derived from the membranes of the host cell, contain host cell phospholipids and membrane proteins. They also contain proteins and glycoproteins of

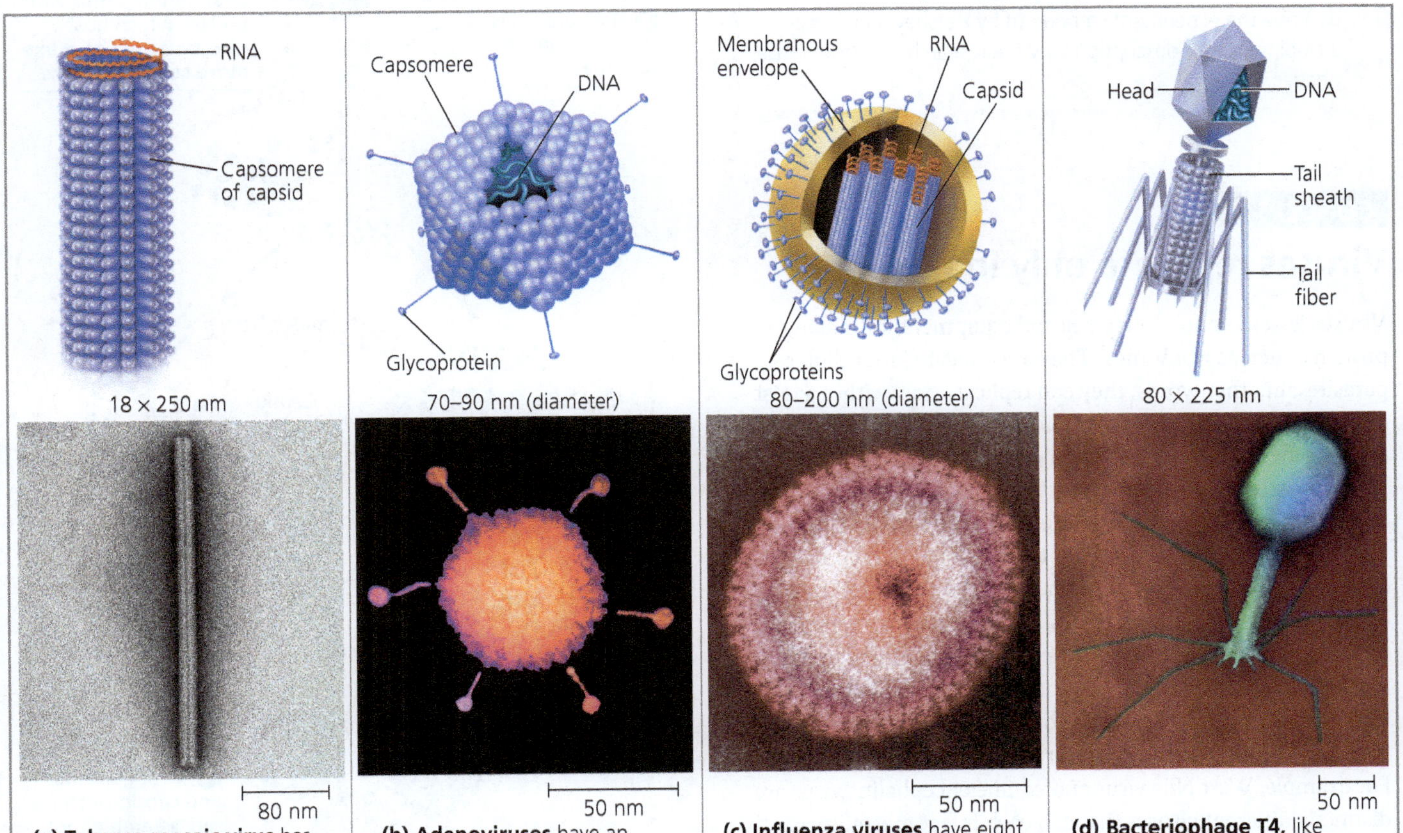

(a) Tobacco mosaic virus has a helical capsid with the overall shape of a rigid rod.

(b) Adenoviruses have an icosahedral capsid with a glycoprotein spike at each vertex (corner).

(c) Influenza viruses have eight different RNA molecules, each wrapped in a helical capsid, and an outer envelope studded with glycoprotein spikes.

(d) Bacteriophage T4, like other "T-even" phages, has a complex capsid consisting of an icosahedral head and a tail apparatus.

▲ **Figure 2 Viral structure.** Viruses are made up of nucleic acid (DNA or RNA) enclosed in a protein coat (the capsid) and sometimes further wrapped in a membranous envelope. The individual protein subunits making up the capsid are called capsomeres. Although diverse in size and shape, viruses have many common structural features. (All micrographs are colorized TEMs.)

viral origin. (Glycoproteins are proteins with carbohydrates covalently attached.) Some viruses carry a few viral enzyme molecules within their capsids.

Many of the most complex capsids are found among the viruses that infect bacteria, called **bacteriophages**, or simply **phages**. The first phages studied included seven that infect *Escherichia coli*. These seven phages were named type 1 (T1), type 2 (T2), and so forth, in the order of their discovery. The three "T-even" phages (T2, T4, and T6) turned out to be very similar in structure. Their capsids have elongated icosahedral heads enclosing their DNA. Attached to the head is a protein tail piece with fibers by which the phages attach to a bacterium **(Figure 2d)**. In the next section, we'll examine how these few viral parts function together with cellular components to produce large numbers of viral progeny.

CONCEPT CHECK 1

1. Compare the structures of tobacco mosaic virus and influenza virus (see Figure 2).
2. **MAKE CONNECTIONS** Bacteriophages were used to provide evidence that DNA carries genetic information. Briefly describe the experiment carried out by Hershey and Chase, including in your description why the researchers chose to use phages.

CONCEPT 2

Viruses replicate only in host cells

Viruses lack metabolic enzymes and equipment for making proteins, such as ribosomes. They are obligate intracellular parasites; in other words, they can replicate only within a host cell. It is fair to say that viruses in isolation are merely packaged sets of genes in transit from one host cell to another.

Each particular virus can infect cells of only a limited number of host species, called the **host range** of the virus. This host specificity results from the evolution of recognition systems by the virus. Viruses usually identify host cells by a "lock-and-key" fit between viral surface proteins and specific receptor molecules on the outside of cells. According to one model, such receptor molecules originally carried out functions that benefited the host cell but were co-opted later by viruses as portals of entry. Some viruses have broad host ranges. For example, West Nile virus and equine encephalitis virus are distinctly different viruses that can each infect mosquitoes, birds, horses, and humans. Other viruses have host ranges so narrow that they infect only a single species. Measles virus, for instance, can infect only humans. Furthermore, viral infection of multicellular eukaryotes is usually limited to particular tissues. Human cold viruses infect only the cells lining the upper respiratory tract, and the HIV seen in Figure 1 binds to receptors present only on certain types of immune cells.

General Features of Viral Replicative Cycles

A viral infection begins when a virus binds to a host cell and the viral genome makes its way inside **(Figure 3)**. The mechanism of genome entry depends on the type of virus and the type of host cell. For example, T-even phages use their elaborate tail apparatus to inject DNA into a bacterium (see Figure 2d). Other viruses are taken up by endocytosis or, in the case of enveloped viruses, by fusion of the viral envelope with the host's plasma membrane. Once the viral genome is inside, the proteins it encodes can commandeer the host, reprogramming the cell to copy the viral genome and manufacture viral proteins. The host provides the nucleotides for making viral nucleic acids, as well as enzymes, ribosomes, tRNAs,

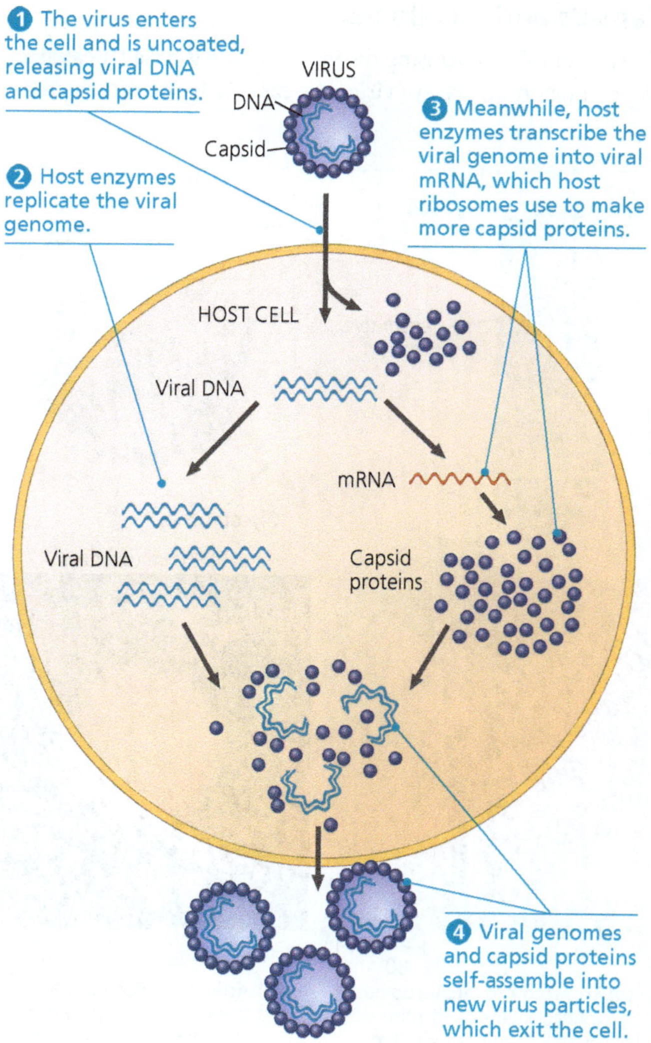

▲ **Figure 3 A simplified viral replicative cycle.** A virus is an intracellular parasite that uses the equipment and small molecules of its host cell to replicate. In this simplest of viral cycles, the parasite is a DNA virus with a capsid consisting of a single type of protein.

MAKE CONNECTIONS *Label each of the straight gray arrows with one word representing the name of the process that is occurring.*

amino acids, ATP, and other components needed for making the viral proteins. Many DNA viruses use the DNA polymerases of the host cell to synthesize new genomes along the templates provided by the viral DNA. In contrast, to replicate their genomes, RNA viruses use virally encoded RNA polymerases that can use RNA as a template. (Uninfected cells generally make no enzymes for carrying out this process.)

After the viral nucleic acid molecules and capsomeres are produced, they spontaneously self-assemble into new viruses. In fact, researchers can separate the RNA and capsomeres of TMV and then reassemble complete viruses simply by mixing the components together under the right conditions. The simplest type of viral replicative cycle ends with the exit of hundreds or thousands of viruses from the infected host cell, a process that often damages or destroys the cell. Such cellular damage and death, as well as the body's responses to this destruction, cause many of the symptoms associated with viral infections. The viral progeny that exit a cell have the potential to infect additional cells, spreading the viral infection.

There are many variations on the simplified viral replicative cycle we have just described. We'll now take a look at some of these variations in bacterial viruses (phages) and animal viruses; later in the chapter, we'll consider plant viruses.

Replicative Cycles of Phages

Phages are the best understood of all viruses, although some of them are also among the most complex. Research on phages led to the discovery that some double-stranded DNA viruses can replicate by two alternative mechanisms: the lytic cycle and the lysogenic cycle.

The Lytic Cycle

A phage replicative cycle that culminates in death of the host cell is known as a **lytic cycle**. The term refers to the last stage of infection, during which the bacterium lyses (breaks open) and releases the phages that were produced within the cell. Each of these phages can then infect a healthy cell, and a few successive lytic cycles can destroy an entire bacterial population in just a few hours. A phage that replicates only by a lytic cycle is called a **virulent phage**. **Figure 4** illustrates the major steps in the lytic cycle of T4, a typical virulent phage. Study this figure before proceeding.

The Lysogenic Cycle

Instead of lysing their host cells, many phages coexist with them in a state called lysogeny. In contrast to the lytic cycle,

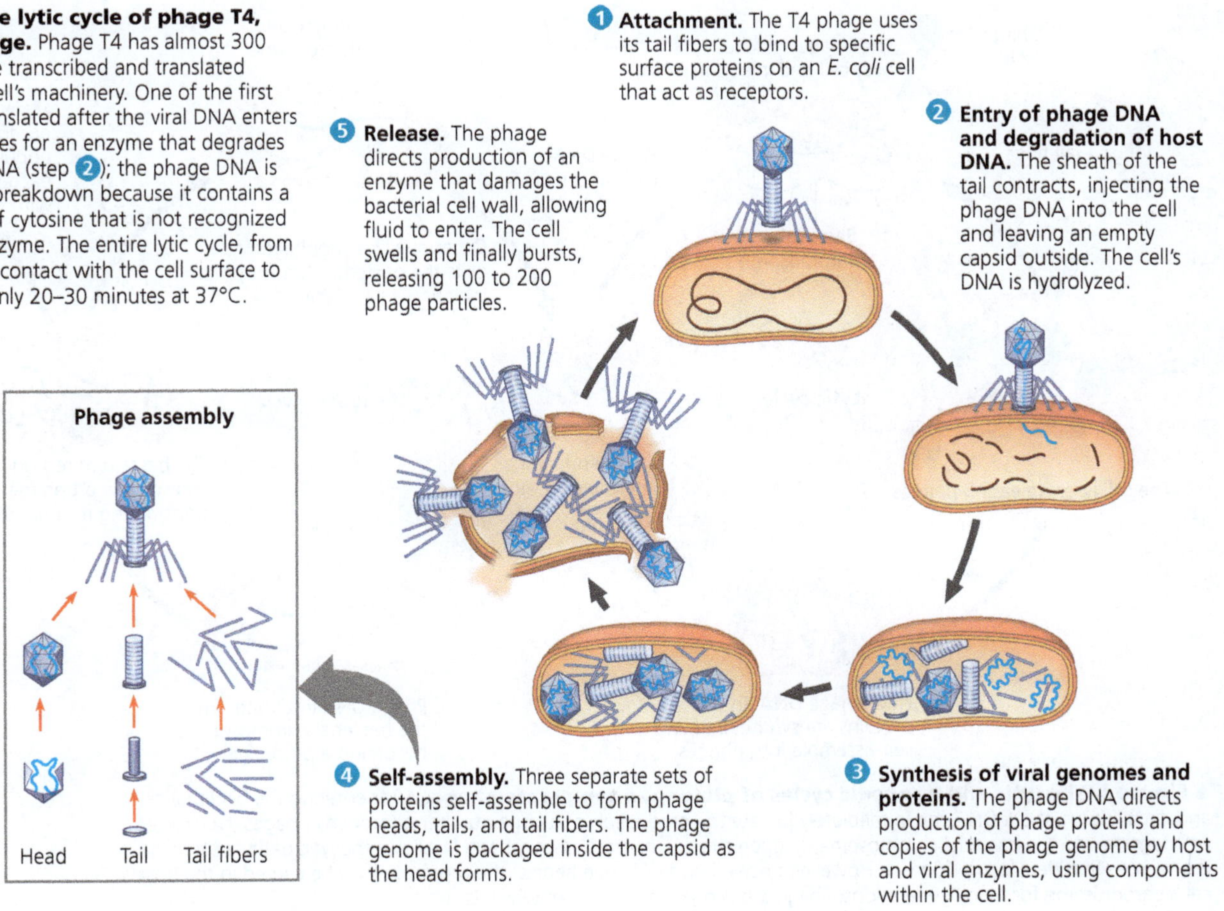

▶ **Figure 4 The lytic cycle of phage T4, a virulent phage.** Phage T4 has almost 300 genes, which are transcribed and translated using the host cell's machinery. One of the first phage genes translated after the viral DNA enters the host cell codes for an enzyme that degrades the host cell's DNA (step 2); the phage DNA is protected from breakdown because it contains a modified form of cytosine that is not recognized by the phage enzyme. The entire lytic cycle, from the phage's first contact with the cell surface to cell lysis, takes only 20–30 minutes at 37°C.

which kills the host cell, the **lysogenic cycle** allows replication of the phage genome without destroying the host. Phages capable of using both modes of replicating within a bacterium are called **temperate phages**. A temperate phage called lambda, written with the Greek letter λ, has been widely used in biological research. Phage λ resembles T4, but its tail has only one short tail fiber.

Infection of an *E. coli* cell by phage λ begins when the phage binds to the surface of the cell and injects its linear DNA genome **(Figure 5)**. Within the host, the λ DNA molecule forms a circle. What happens next depends on the replicative mode: lytic cycle or lysogenic cycle. During a lytic cycle, the viral genes immediately turn the host cell into a λ-producing factory, and the cell soon lyses and releases its virus progeny. During a lysogenic cycle, however, the λ DNA molecule is incorporated into a specific site on the *E. coli* chromosome by viral proteins that break both circular DNA molecules and join them to each other. When integrated into the bacterial chromosome in this way, the viral DNA is known as a **prophage**. One prophage gene codes for a protein that prevents transcription of most of the other prophage genes. Thus, the phage genome is mostly silent within the bacterium. Every time the *E. coli* cell prepares to divide, it replicates the phage DNA along with its own chromosome such that each daughter cell inherits a prophage. A single infected cell can quickly give rise to a large population of bacteria carrying the virus in prophage form. This mechanism enables viruses to propagate without killing the host cells on which they depend.

The term *lysogenic* signifies that prophages are capable of generating active phages that lyse their host cells. This occurs when the λ genome (or that of another temperate phage) is induced to exit the bacterial chromosome and initiate a lytic cycle. An environmental signal, such as a certain chemical or high-energy radiation, usually triggers the switchover from the lysogenic to the lytic mode.

In addition to the gene for the viral protein that prevents transcription, a few other prophage genes may be expressed during lysogeny. Expression of these genes may alter the host's phenotype, a phenomenon that can have important medical significance. For example, the three species of bacteria that cause the human diseases diphtheria, botulism, and scarlet fever would not be so harmful to humans without certain prophage genes that cause the host bacteria to make toxins. And the difference between the *E. coli* strain in our intestines and the O157:H7 strain that has caused several deaths by food poisoning appears to be the presence of toxin genes of prophages in the O157:H7 strain.

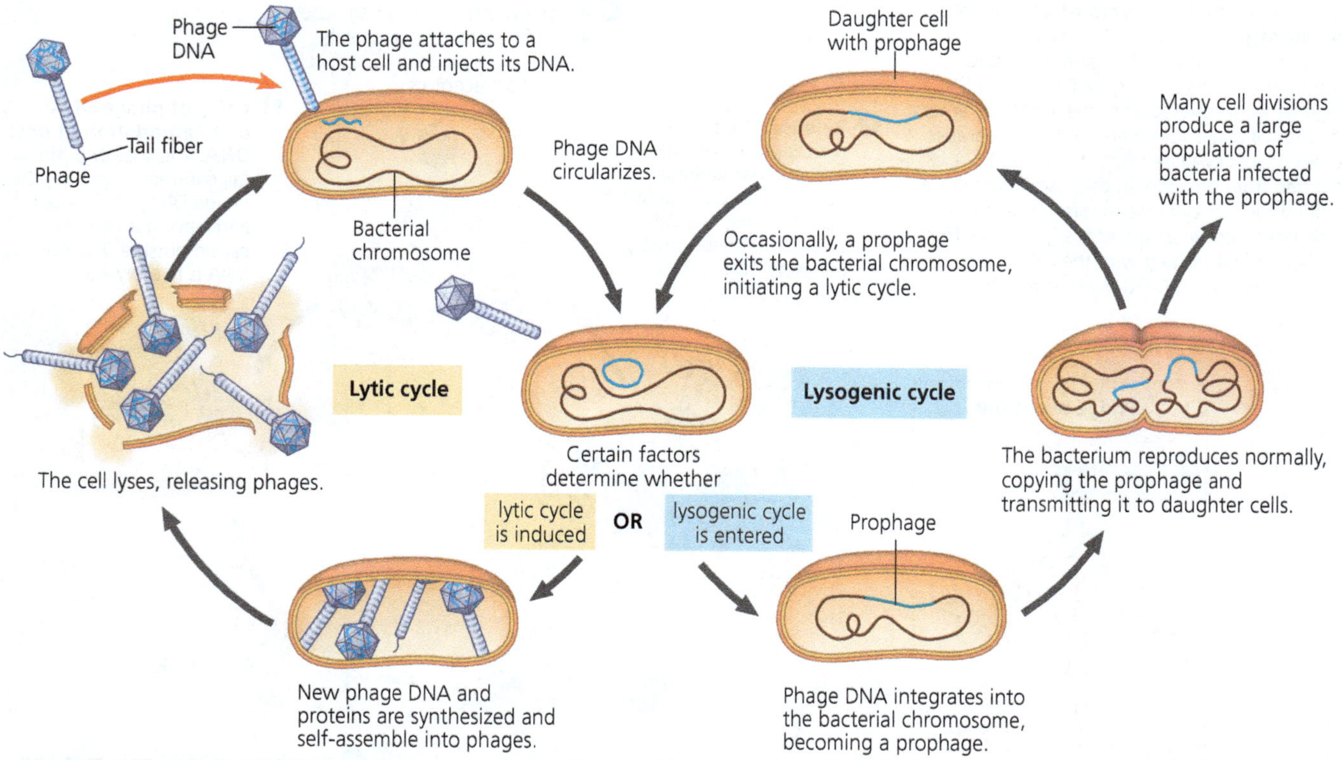

▲ **Figure 5 The lytic and lysogenic cycles of phage λ, a temperate phage.** After entering the bacterial cell and circularizing, the λ DNA can immediately initiate the production of a large number of progeny phages (lytic cycle) or integrate into the bacterial chromosome (lysogenic cycle). In most cases, phage λ follows the lytic pathway, which is similar to that detailed in Figure 4. However, once a lysogenic cycle begins, the prophage may be carried in the host cell's chromosome for many generations. Phage λ has one main tail fiber, which is short.

Bacterial Defenses Against Phages

After reading about the lytic cycle, you may have wondered why phages haven't exterminated all bacteria. Lysogeny is one major reason why bacteria have been spared from extinction caused by phages. Bacteria also have their own defenses against phages. First, natural selection favors bacterial mutants with surface proteins that are no longer recognized as receptors by a particular type of phage. Second, when phage DNA does enter a bacterium, the DNA often is identified as foreign and cut up by **restriction enzymes**, so named because they *restrict* a phage's ability to replicate within the bacterium. (Restriction enzymes are used in molecular biology and DNA cloning techniques.) The bacterium's own DNA is methylated in a way that prevents attack by its own restriction enzymes. A third defense is the CRISPR-Cas system, which exists in both bacteria and archaea. Here we'll focus on how it works within bacterial cells.

The CRISPR-Cas system was discovered during a study of repetitive DNA sequences present in the genomes of many prokaryotes. These sequences, which puzzled scientists, were named clustered regularly interspaced short palindromic repeats (CRISPRs) because each sequence read the same forward and backward (a palindrome), with different stretches of "spacer DNA" in between the repeats. At first, scientists assumed the spacer DNA sequences were random and meaningless, but analysis by several research groups showed that each spacer sequence corresponded to DNA from a particular phage that had infected the cell. Further studies revealed that particular nuclease proteins interact with the CRISPR region. These nucleases, called Cas (CRISPR-associated) proteins, can identify and cut phage DNA, thereby defending the bacterium against phage infection.

When a phage infects a bacterial cell that has the CRISPR-Cas system, the DNA of the invading phage is integrated into the genome between two repeat sequences. If the cell survives the infection, any further attempt by the same type of phage to infect this cell (or its offspring) triggers transcription of the CRISPR region into RNA molecules (**Figure 6**). These RNAs are cut into pieces and then bound by Cas proteins, such as the Cas9 protein. A Cas protein uses a portion of the phage-related RNA as a homing device to identify the invading phage DNA and cut it, leading to its destruction.

Just as natural selection favors bacteria that have receptors altered by mutation or that have enzymes that cut phage DNA, it also favors phage mutants that can bind to altered receptors or that are resistant to enzymes. Thus, the bacterium-phage relationship is in constant evolutionary flux.

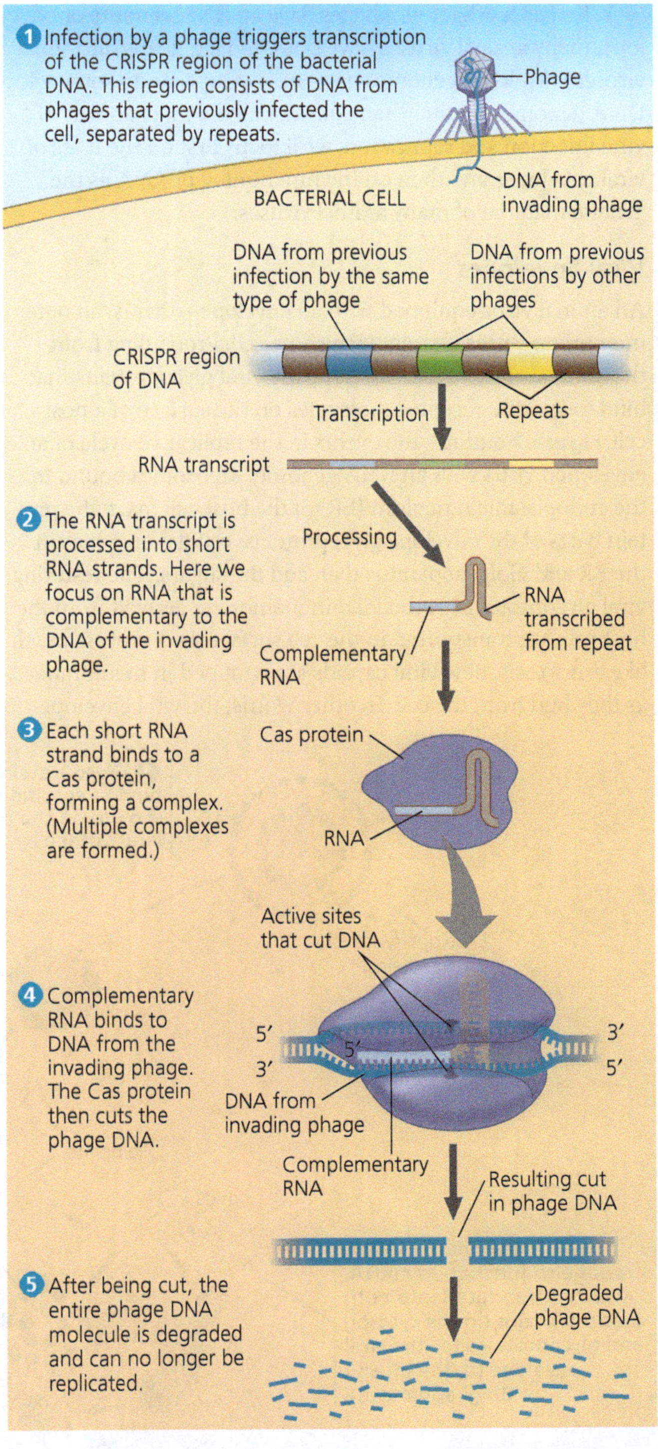

▲ **Figure 6 The CRISPR system: a type of bacterial immune system.**

Replicative Cycles of Animal Viruses

Everyone has suffered from viral infections, whether cold sores, influenza, or the common cold. Like all viruses, those that cause illness in humans and other animals can replicate only inside host cells. Many variations on the basic scheme of viral infection and replication are represented among the animal viruses. Key variables are the nature of the viral genome (double- or single-stranded DNA or RNA) and the presence or absence of an envelope.

Whereas few bacteriophages have an RNA genome or envelope, many animal viruses have both. In fact, nearly all animal viruses with RNA genomes have an envelope, as do some with DNA genomes. Rather than consider all the mechanisms of viral infection and replication, we'll focus first on the roles of viral envelopes and then on the functioning of RNA as the genetic material of many animal viruses.

Viral Envelopes

An animal virus equipped with an envelope—that is, an outer membrane—uses it to enter the host cell. Protruding from the outer surface of this envelope are viral glycoproteins that bind to specific receptor molecules on the surface of a host cell. **Figure 7** outlines the events in the replicative cycle of an enveloped virus with an RNA genome. Ribosomes bound to the endoplasmic reticulum (ER) of the host cell make the protein parts of the envelope glycoproteins; cellular enzymes in the ER and Golgi apparatus then add the sugars. The resulting viral glycoproteins, embedded in membrane derived from the host cell, are transported to the cell surface. In a process much like exocytosis, new viral capsids are wrapped in membrane as they bud from the cell. In other words, the viral envelope is usually derived from the host cell's plasma membrane, although all or most of the molecules of this membrane are specified by viral genes. The enveloped viruses are now free to infect other cells. This replicative cycle does not necessarily kill the host cell, in contrast to the lytic cycles of phages.

Some viruses have envelopes that are not derived from plasma membrane. Herpesviruses, for example, are temporarily cloaked in membrane derived from the nuclear envelope of the host; they then shed this membrane in the cytoplasm and acquire a new envelope made from membrane of the Golgi apparatus. These viruses have a double-stranded DNA genome and replicate within the host cell nucleus, using a combination of viral and cellular enzymes to replicate and transcribe their DNA. In the case of herpesviruses, copies of the viral DNA can remain behind as mini-chromosomes in the nuclei of certain nerve cells. There they remain latent until some sort of physical or emotional stress triggers a new round of active virus production. The infection of other cells by these new viruses causes the blisters characteristic of herpes, such as cold sores or genital sores. Once someone acquires a herpesvirus infection, flare-ups may recur throughout the person's life.

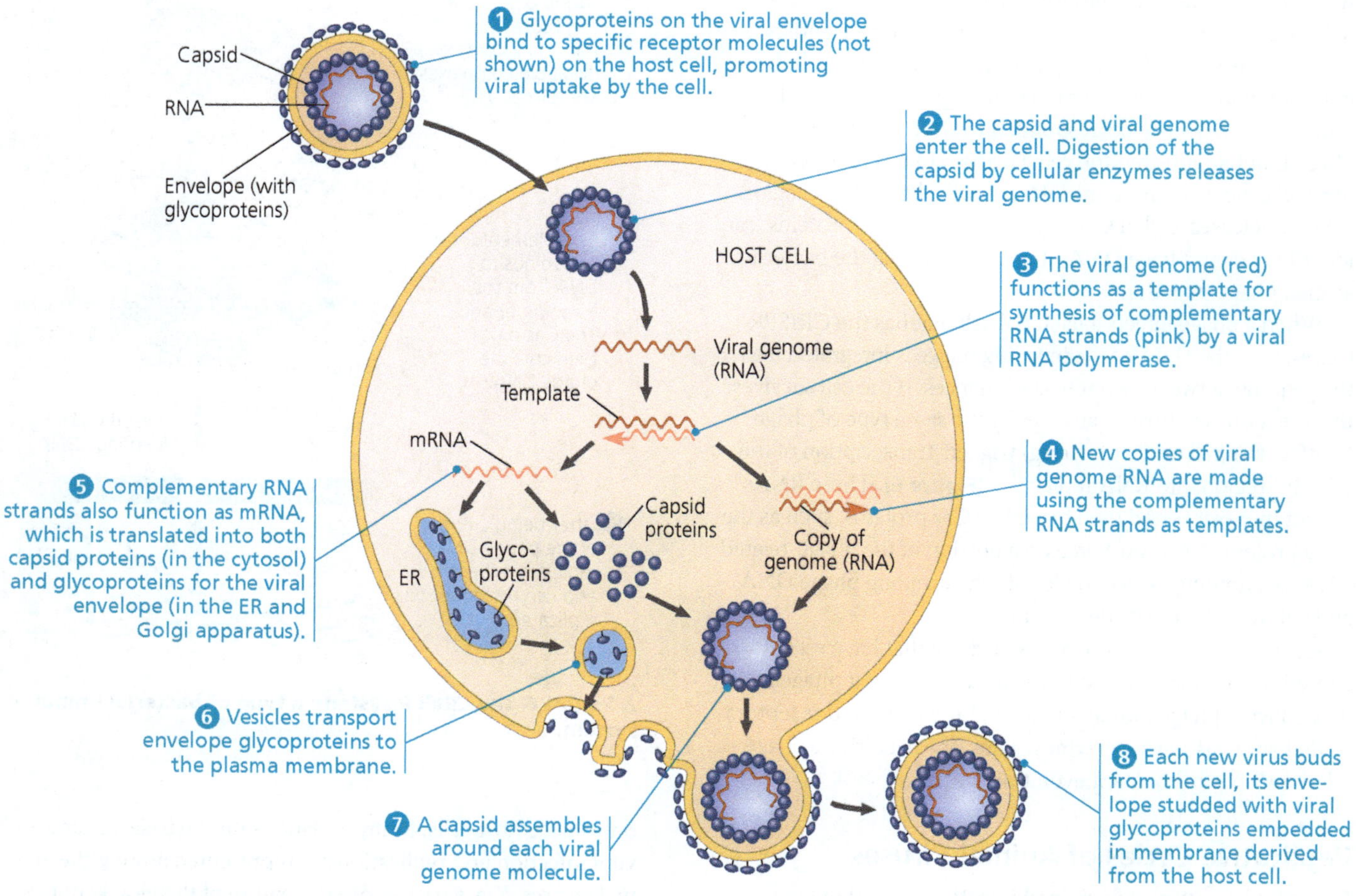

▲ **Figure 7 The replicative cycle of an enveloped RNA virus.** Shown here is a virus with a single-stranded RNA genome that functions as a template for synthesis of mRNA. Some enveloped viruses enter the host cell by fusion of the envelope with the cell's plasma membrane; others enter by endocytosis. For all enveloped RNA viruses, formation of new envelopes for progeny viruses occurs by the mechanism depicted in this figure.

RNA as Viral Genetic Material

Although some phages and most plant viruses are RNA viruses, the broadest variety of RNA genomes is found among the viruses that infect animals. There are three types of single-stranded RNA genomes found in animal viruses. In the first type, the viral genome can directly serve as mRNA and thus can be translated into viral protein immediately after infection. In a second type, the RNA genome serves instead as a *template* for mRNA synthesis. The RNA genome is transcribed into complementary RNA strands, which function both as mRNA and as templates for the synthesis of additional copies of genomic RNA. All viruses that use an RNA genome as a template for mRNA transcription require RNA → RNA synthesis. These viruses use a viral enzyme capable of carrying out this process; there are no such enzymes in most cells. The viral enzyme used in this process is packaged during viral self-assembly with the genome inside the viral capsid.

The RNA animal viruses with the most complicated replicative cycles are the third type, the **retroviruses**. These viruses are equipped with an enzyme called **reverse transcriptase**, which transcribes an RNA template into DNA, providing an RNA → DNA information flow, the opposite of the usual direction. (Reverse transcriptase is the enzyme used in the technique called RT-PCR.) This unusual phenomenon is the source of the name retroviruses (*retro* means "backward"). Of particular medical importance is **HIV (human immunodeficiency virus)**, the retrovirus that causes **AIDS (acquired immunodeficiency syndrome)**. HIV and other retroviruses are enveloped viruses that contain two identical molecules of single-stranded RNA and two molecules of reverse transcriptase.

The HIV replicative cycle (traced in **Figure 8**) is typical of a retrovirus. After HIV enters a host cell, its reverse transcriptase molecules are released into the cytoplasm, where they catalyze synthesis of viral DNA. The newly made viral DNA then enters the cell's nucleus and integrates into the DNA of a chromosome. The integrated viral DNA, called a **provirus**, never leaves the host's genome, remaining a permanent resident of the cell. (Recall that a prophage, in contrast, leaves the host's genome at the start of a lytic cycle.) The host's RNA polymerase transcribes the proviral DNA into RNA molecules, which can function both as mRNA for the synthesis of viral proteins and as genomes for the new viruses that will be assembled and released from the cell.

Evolution of Viruses

EVOLUTION We began this chapter by asking whether or not viruses are alive. Viruses do not really fit our definition of living organisms. An isolated virus is biologically inert, unable to replicate its genes or regenerate its own ATP. Yet it has a genetic program written in the universal language of life. Do we think of viruses as nature's most complex associations of molecules or as the simplest forms of life? Either way, we must bend our usual definitions. Although viruses cannot replicate or carry out metabolic activities independently, their use of the genetic code makes it hard to deny their evolutionary connection to the living world.

How did viruses originate? Viruses have been found that infect every form of life—not just bacteria, animals, and plants, but also archaea, fungi, and algae and other protists. Because they depend on cells for their own propagation, it seems likely that viruses are not the descendants of precellular forms of life, but evolved—possibly multiple times—*after* the first cells appeared. Most molecular biologists favor the hypothesis that viruses originated from naked bits of cellular nucleic acids that moved from one cell to another, perhaps via injured cell surfaces. The evolution of genes coding for capsid proteins may have allowed viruses to bind cell membranes, thus facilitating the infection of uninjured cells.

Candidates for the original sources of viral genomes include plasmids and transposons. *Plasmids* are small, circular DNA molecules found in bacteria and in the unicellular eukaryotes called yeasts. Plasmids exist apart from and can replicate independently of the genome and are occasionally transferred between cells. *Transposons* are DNA segments that can move from one location to another within a cell's genome. Thus, plasmids, transposons, and viruses all share an important feature: They are *mobile genetic elements.* Consistent with this notion of pieces of DNA shuttling from cell to cell is the observation that a viral genome can have more in common with the genome of its host than with the genomes of viruses that infect other hosts.

Debate about the origin of viruses was revived in 2003 when an extremely large virus was discovered: Mimivirus is a double-stranded DNA (dsDNA) virus with an icosahedral capsid that is 400 nm in diameter, the size of a small bacterium. Its genome has 1.2 million bases (Mb)—approximately 100 times as many as influenza virus—and an estimated 1,000 genes. Surprisingly, its genome included genes previously found only in cellular genomes. In 2013 an even larger virus was discovered, unrelated to any known virus. This virus is 1 μm (1,000 nm) in diameter, with a dsDNA genome of around 2–2.5 Mb, larger than that of some small eukaryotes. What's more, over 90% of its 2,000 or so genes are not related to cellular genes, so it was fittingly named pandoravirus. How these and all other viruses fit in the tree of life is an intriguing question.

The ongoing evolutionary relationship between viruses and the genomes of their host cells is an association that makes viruses very useful experimental systems in molecular biology. Knowledge about viruses also allows many practical applications, since viruses have a tremendous impact on all organisms through their ability to cause disease.

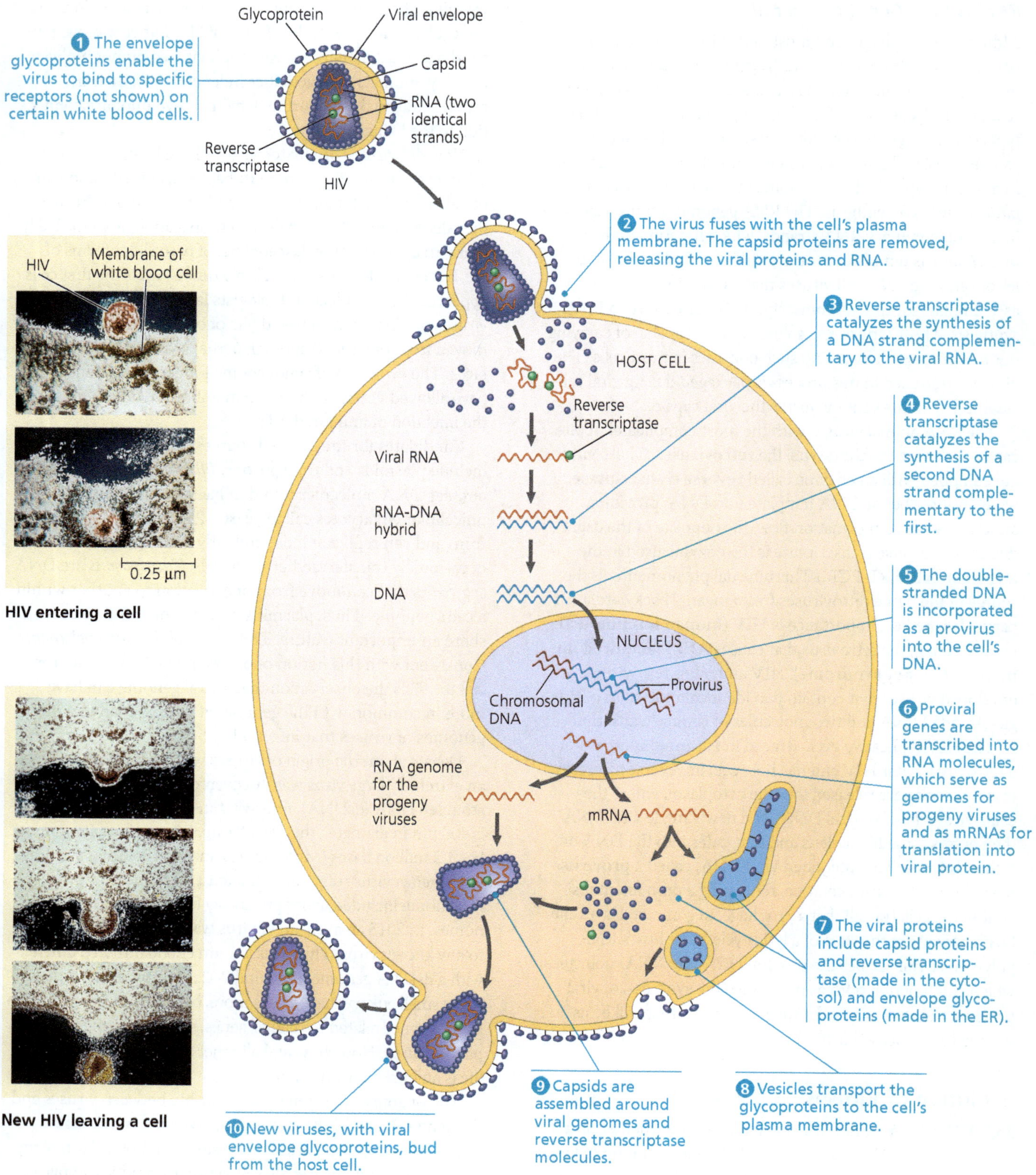

▲ **Figure 8 The replicative cycle of HIV, the retrovirus that causes AIDS.** Note in step 5 that DNA synthesized from the viral RNA genome is integrated as a provirus into the host cell chromosomal DNA, a characteristic unique to retroviruses. For simplicity, the cell-surface proteins that act as receptors for HIV are not shown (see Figure 3). The photos on the left (artificially colored TEMs) show HIV entering and leaving a human white blood cell.

WHAT IF? *If you were a researcher trying to combat HIV infection, what molecular processes could you attempt to block?*

CONCEPT CHECK 2

1. Compare the effect on the host cell of a lytic (virulent) phage and a lysogenic (temperate) phage.
2. MAKE CONNECTIONS The RNA virus in Figure 7 has a viral RNA polymerase that functions in step 3 of the virus's replicative cycle. Compare this with a cellular RNA polymerase in terms of template and overall function.
3. Why is HIV called a retrovirus?
4. MAKE CONNECTIONS Compare the CRISPR system to miRNAs , including their mechanisms and their functions.

CONCEPT 3

Viruses and prions are formidable pathogens in animals and plants

Diseases caused by viral infections afflict humans, crops, and livestock worldwide. Other smaller, less complex entities known as prions also cause disease in animals. We'll first discuss animal viruses.

Viral Diseases in Animals

A viral infection can produce symptoms by a number of different routes. Viruses may damage or kill cells by causing the release of hydrolytic enzymes from lysosomes. Some viruses cause infected cells to produce toxins that lead to disease symptoms, and some have molecular components that are toxic, such as envelope proteins. How much damage a virus causes depends partly on the ability of the infected tissue to regenerate by cell division. People usually recover completely from colds because the epithelium of the respiratory tract, which the viruses infect, can efficiently repair itself. In contrast, damage inflicted by poliovirus to mature nerve cells is permanent because these cells do not divide and usually cannot be replaced. Many of the temporary symptoms associated with viral infections, such as fever and body aches, actually result from the body's own efforts to defend itself against infection rather than from cell death caused by the virus.

The immune system is a complex and critical part of the body's natural defenses. It is also the basis for the major medical tool for preventing viral infections—vaccines. A **vaccine** is a harmless variant or derivative of a pathogen that stimulates the immune system to mount defenses against the harmful pathogen. Smallpox, a viral disease that was at one time a devastating scourge in many parts of the world, was eradicated by a vaccination program carried out by the World Health Organization (WHO). The very narrow host range of the smallpox virus—it infects only humans—was a critical factor in the success of this program. Similar worldwide vaccination campaigns are currently under way to eradicate polio and measles. Effective vaccines are also available to protect against rubella, mumps, hepatitis A and B, and a number of other viral diseases.

Although vaccines can prevent certain viral illnesses, medical technology can do little, at present, to cure most viral infections once they occur. The antibiotics that help us recover from bacterial infections are powerless against viruses. Antibiotics kill bacteria by inhibiting enzymes specific to bacteria but have no effect on eukaryotic or virally encoded enzymes. However, the few enzymes that are encoded only by viruses have provided targets for other drugs. Most antiviral drugs resemble nucleosides and as a result interfere with viral nucleic acid synthesis. One such drug is acyclovir, which impedes herpesvirus replication by inhibiting the viral polymerase that synthesizes viral DNA but not the eukaryotic one. Similarly, azidothymidine (AZT) curbs HIV replication by interfering with the synthesis of DNA by reverse transcriptase. In the past two decades, much effort has gone into developing drugs to treat HIV. Currently, multidrug treatments, sometimes called "cocktails," have been found to be most effective. Such treatments commonly include a combination of two nucleoside mimics and a protease inhibitor, which interferes with an enzyme required for assembly of the viruses.

Receptor proteins on the surface of cells are important factors to consider in the development of ways to prevent or treat viral infection. For example, medical researchers knew that a protein called CD4 helped HIV infect immune cells. They also noticed, however, that despite multiple exposures to HIV, a small number of people failed to develop AIDS and showed no evidence of harboring HIV-infected cells. Comparing their genes with the genes of infected individuals, the researchers learned that resistant people have an unusual form of a gene that codes for an immune cell-surface protein called CCR5. Further work showed that although CD4 is the main HIV receptor, HIV must also bind to CCR5 as a "co-receptor" to infect most cells **(Figure 9a)**. An absence of CCR5 on the cells of resistant individuals, due to the gene alteration, prevents the virus from entering the cells **(Figure 9b)**.

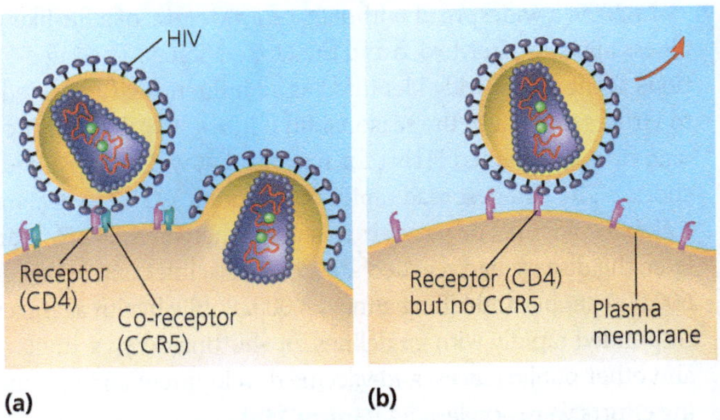

▲ **Figure 9 HIV resistance due to differences in cell-surface proteins. (a)** HIV can infect a cell with CCR5 on its surface, as in most people. **(b)** HIV cannot infect a cell lacking CCR5 on its surface, as in resistant individuals.

MAKE CONNECTIONS *What would you predict about CCR5 that would allow HIV to bind to it? How could a drug molecule interfere with this binding?*

This information has been key to developing a treatment for HIV infection. Interfering with CD4 could cause dangerous side effects because CD4 performs many vital functions in the cell (and was probably co-opted later for use as an HIV entry gate). Discovery of the CCR5 co-receptor provided a safer target for development of drugs that mask this protein and block HIV entry. One such drug, maraviroc (brand name Selzentry), was approved for treatment of HIV in 2007 and is still being used today. A clinical trial began in 2012 to test whether this drug might also work to prevent HIV infection in uninfected, at-risk patients.

Emerging Viruses

Viruses that suddenly become apparent are often referred to as *emerging viruses*. HIV, the AIDS virus, is a classic example: This virus appeared in San Francisco in the early 1980s, seemingly out of nowhere, although later studies uncovered a case in the Belgian Congo in 1959. The deadly Ebola virus **(Figure 10)**, recognized initially in 1976 in central Africa, is one of several emerging viruses that cause *hemorrhagic fever*, an often fatal syndrome (set of symptoms) characterized by fever, vomiting, massive bleeding, and circulatory system collapse. In 2014, a widespread outbreak of Ebola virus caused the World Health Organization to declare an international health emergency. A number of other dangerous emerging viruses cause encephalitis, inflammation of the brain. One example is the West Nile virus, which appeared in North America for the first time in 1999 and has now spread to all 48 contiguous states in the United States. By 2013, there had been about 40,000 cases total, with over 1,600 deaths.

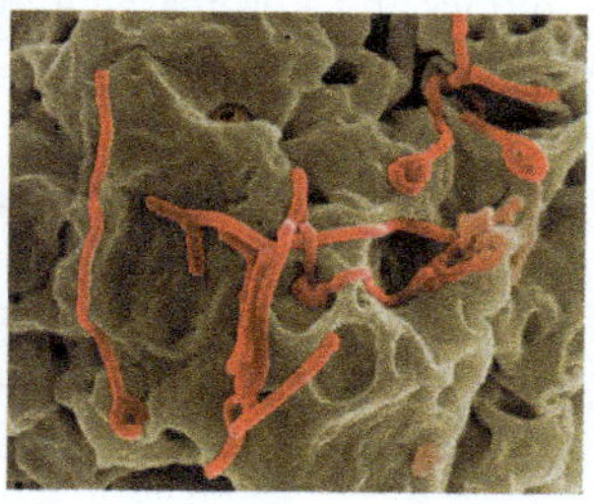

▲ **Figure 10** **Ebola viruses infecting a monkey cell.**

In 2009, a widespread outbreak, or **epidemic**, of a flu-like illness appeared in Mexico and the United States. The infectious agent was quickly identified as an influenza virus related to viruses that cause the seasonal flu **(Figure 11a)**. This particular virus was named H1N1 for reasons that will be explained shortly. The illness spread rapidly, prompting WHO to declare a global epidemic, or **pandemic**, shortly thereafter. Half a year later, the disease had reached 207 countries, infecting over 600,000 people and killing almost 8,000. Public health agencies responded rapidly with guidelines for shutting down schools and other public places, and vaccine development and screening efforts were accelerated **(Figure 11b)**.

How do such viruses burst on the human scene, giving rise to harmful diseases that were previously rare or even unknown? Three processes contribute to the emergence of viral diseases. The first, and perhaps most important, is the mutation of existing viruses. RNA viruses tend to have an unusually high rate of mutation because viral RNA polymerases do not proofread and correct errors in replicating their RNA genomes. Some mutations change existing viruses into new genetic varieties (strains) that can cause disease, even in individuals who are immune to the ancestral virus. For instance, seasonal flu epidemics are caused by new strains of influenza virus genetically different enough from earlier strains that people have little immunity to them. You'll see an example of this process in the **Scientific Skills Exercise**, where you'll analyze genetic changes in variants of the H1N1 flu virus and correlate them with spread of the disease.

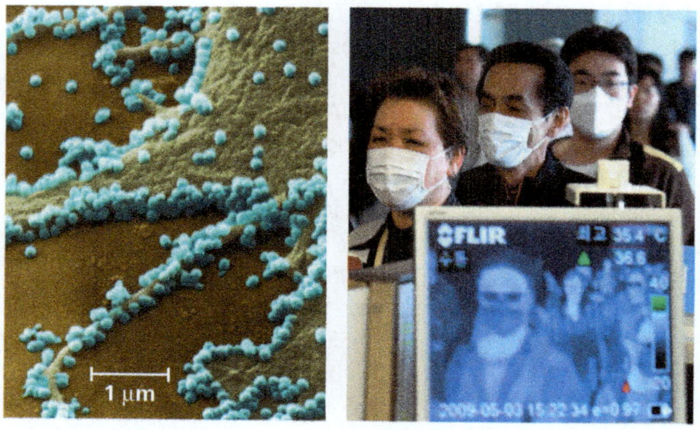

(a) 2009 pandemic H1N1 influenza A virus. Viruses (blue) are seen on an infected cell (green) in this colorized SEM.

(b) 2009 pandemic screening. At a South Korean airport, thermal scans were used to detect passengers with a fever who might have the H1N1 flu.

▲ **Figure 11** **Influenza in humans.**

A second process that can lead to the emergence of viral diseases is the dissemination of a viral disease from a small, isolated human population. For instance, AIDS went unnamed and virtually unnoticed for decades before it began to spread around the world. In this case, technological and social factors, including affordable international travel, blood transfusions, sexual promiscuity, and the abuse of intravenous drugs, allowed a previously rare human disease to become a global scourge.

A third source of new viral diseases in humans is the spread of existing viruses from other animals. Scientists estimate that about three-quarters of new human diseases originate in this way. Animals that harbor and can transmit a particular virus but are generally unaffected by it are said to act as a natural reservoir for that virus. For example, the H1N1 virus that caused the 2009 flu pandemic mentioned earlier was likely passed to humans from pigs; for this reason, the disease it caused was originally called "swine flu."

In general, flu epidemics provide an instructive example of the effects of viruses moving between species. There are three types of influenza virus: types B and C, which infect only humans and have never caused an epidemic, and type A, which infects a wide range of animals, including birds, pigs, horses, and humans. Influenza A strains have caused four major flu epidemics among humans in the last 100 years. The worst was the first one, the "Spanish flu" pandemic of 1918–1919, which killed 40–50 million people, including many World War I soldiers.

Scientific Skills Exercise

Analyzing a Sequence-Based Phylogenetic Tree to Understand Viral Evolution

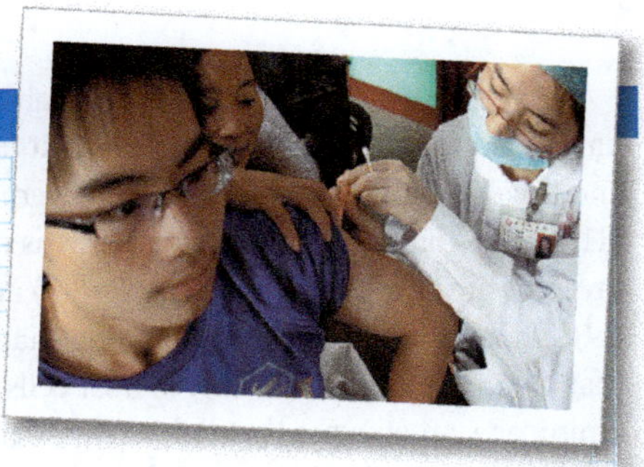

▶ H1N1 flu vaccination

How Can Sequence Data Be Used to Track Flu Virus Evolution? In 2009, an influenza A H1N1 virus caused a pandemic, and the virus has continued to resurface in outbreaks across the world. Researchers in Taiwan were curious about why the virus kept appearing despite widespread flu vaccine initiatives. They hypothesized that newly evolved variant strains of the H1N1 virus were able to evade human immune system defenses. To test this hypothesis, they needed to determine if each wave of flu infection was caused by a different H1N1 variant strain.

How the Experiment Was Done Scientists obtained the genome sequences for 4,703 virus isolates collected from patients with H1N1 flu in Taiwan. They compared the sequences in different strains for the viral hemagglutinin (HA) gene, and based on mutations that had occurred, arranged the isolates into a phylogenetic tree.

Data from the Experiment In the phylogenetic tree, each branch tip is one variant strain of the H1N1 virus with a unique HA gene sequence. The tree is a way to visualize a working hypothesis about the evolutionary relationships between H1N1 variants.

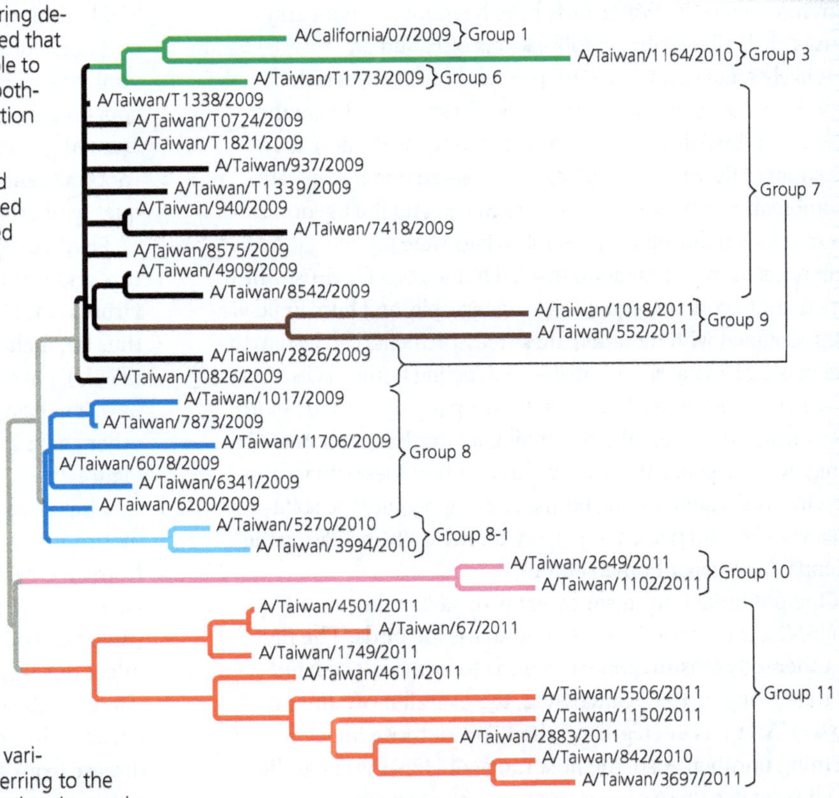

INTERPRET THE DATA

1. The phylogenetic tree shows the hypothesized evolutionary relationship between the variant strains of H1N1 virus. The more closely connected two variants are, the more alike they are in terms of HA gene sequence. Each fork in a branch, called a node, shows where two lineages separate due to different accumulated mutations. The length of the branches is a measure of how many sequence differences there are between the variants, indicating how distantly related they are. Referring to the phylogenetic tree, which variants are more closely related to each other: A/Taiwan/1018/2011 and A/Taiwan/552/2011 or A/Taiwan/1018/2011 and A/Taiwan/8542/2009? Explain your answer.
2. The scientists arranged the branches into groups made up of one ancestral variant and all of its descendant, mutated variants. They are color-coded in the figure. Using Group 11 as an example, trace the lineage of its variants. (a) Do all of the nodes have the same number of branches or branch tips? (b) Are all of the branches in the group the same length? (c) What do these results indicate?
3. The graph shows the number of isolates collected (each from an ill patient) on the *y*-axis and the month and year that the isolates were collected on the *x*-axis. Each group of variants is plotted separately with a line color that matches the tree diagram. (a) Which group of variants was the earliest to cause the first wave of H1N1 flu in over 100 patients in Taiwan? (b) After a group of variants had a peak number of infections, did members of that same group cause another (later) wave of infection? (c) One variant in Group 1 (green, uppermost branch) was used to make a vaccine that was distributed very early in the pandemic. Based on the graphed data, does it look like the vaccine was effective?
4. Groups 9, 10, and 11 all had H1N1 variants that caused a large number of infections at the same time in Taiwan. Does this mean that the scientists' hypothesis, that new variants cause new waves of infection, was incorrect? Explain your answer.

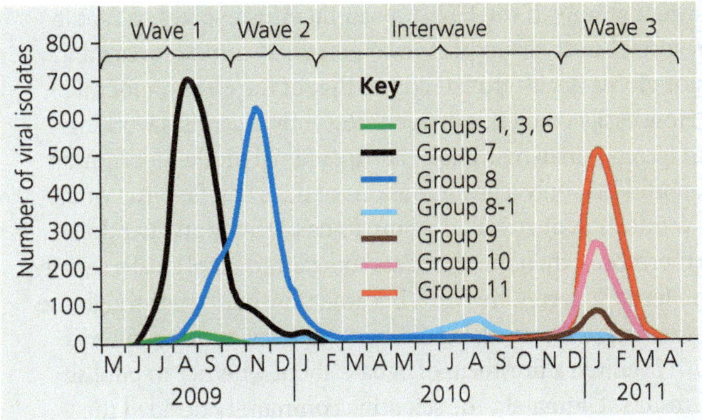

▲ Scientists graphed the number of isolates by the month and year of isolate collection to show the period in which each viral variant was actively causing illness in people.

Data from J.-R. Yang et al., New variants and age shift to high fatality groups contribute to severe successive waves in the 2009 influenza pandemic in Taiwan, *PLoS ONE* 6(11):e28288 (2011).

MB A version of this Scientific Skills Exercise can be assigned in MasteringBiology.

Different strains of influenza A are given standardized names; for example, both the strain that caused the 1918 flu and the one that caused the 2009 pandemic flu are called H1N1. The name identifies which forms of two viral surface proteins are present: hemagglutinin (H) and neuraminidase (N). There are 16 different types of hemagglutinin, a protein that helps the flu virus attach to host cells, and 9 types of neuraminidase, an enzyme that helps release new virus particles from infected cells. Waterbirds have been found that carry viruses with all possible combinations of H and N.

A likely scenario for the 1918 pandemic and others is that the virus mutated as it passed from one host species to another. When an animal like a pig or a bird is infected with more than one strain of flu virus, the different strains can undergo genetic recombination if the RNA molecules making up their genomes mix and match during viral assembly. Pigs were probably the main hosts for recombination that led to the 2009 flu virus, which turned out to contain sequences from bird, pig, and human flu viruses. Coupled with mutation, these reassortments can lead to the emergence of a viral strain capable of infecting human cells. People who have never been exposed to that particular strain before will lack immunity, and the recombinant virus has the potential to be highly pathogenic. If such a flu virus recombines with viruses that circulate widely among humans, it may acquire the ability to spread easily from person to person, dramatically increasing the potential for a major human outbreak.

One potential long-term threat is the avian flu caused by an H5N1 virus carried by wild and domestic birds. The first documented transmission from birds to humans occurred in Hong Kong in 1997. Since then, the overall mortality rate due to H5N1 has been greater than 50% of those infected—an alarming number. Also, the host range of H5N1 is expanding, which provides increasing chances for reassortment between different strains. If the H5N1 avian flu virus evolves so that it can spread easily from person to person, it could represent a major global health threat akin to that of the 1918 pandemic.

How easily could this happen? Recently, scientists working with ferrets, small mammals that are animal models for human flu, found that only a few mutations of the avian flu virus would allow infection of cells in the human nasal cavity and windpipe. Furthermore, when the scientists transferred nasal swabs serially from ferret to ferret, the virus became transmissible through the air. Reports of this startling discovery at a scientific conference in 2011 ignited a firestorm of debate about whether to publish the results. Ultimately, the scientific community decided the benefits of potentially understanding how to prevent pandemics would outweigh the risks of the information being used for harmful purposes, and the work was published in 2012.

As we have seen, emerging viruses are generally not new; rather, they are existing viruses that mutate, disseminate more widely in the current host species, or spread to new host species. Changes in host behavior or environmental changes can increase the viral traffic responsible for emerging diseases. For instance, new roads built through remote areas can allow viruses to spread between previously isolated human populations. Also, the destruction of forests to expand cropland can bring humans into contact with other animals that may host viruses capable of infecting humans.

Viral Diseases in Plants

More than 2,000 types of viral diseases of plants are known, and together they account for an estimated annual loss of $15 billion worldwide due to their destruction of agricultural and horticultural crops. Common signs of viral infection include bleached or brown spots on leaves and fruits **(Figure 12)**, stunted growth, and damaged flowers or roots, all tending to diminish the yield and quality of crops.

▲ Figure 12 Virus-infected tomato.

Plant viruses have the same basic structure and mode of replication as animal viruses. Most plant viruses discovered thus far, including tobacco mosaic virus (TMV), have an RNA genome. Many have a helical capsid, like TMV, while others have an icosahedral capsid (see Figure 2b).

Viral diseases of plants spread by two major routes. In the first route, called *horizontal transmission*, a plant is infected from an external source of the virus. Because the invading virus must get past the plant's outer protective layer of cells (the epidermis), a plant becomes more susceptible to viral infections if it has been damaged by wind, injury, or herbivores. Herbivores, especially insects, pose a double threat because they can also act as carriers of viruses, transmitting disease from plant to plant. Moreover, farmers and gardeners may transmit plant viruses inadvertently on pruning shears and other tools. The other route of viral infection is *vertical transmission*, in which a plant inherits a viral infection from a parent. Vertical transmission can occur in asexual propagation (for example, through cuttings) or in sexual reproduction via infected seeds.

Once a virus enters a plant cell and begins replicating, viral genomes and associated proteins can spread throughout the plant by means of plasmodesmata, the cytoplasmic connections that penetrate the walls between adjacent plant cells. The passage of viral macromolecules from cell to cell is facilitated by virally encoded proteins that cause enlargement of plasmodesmata. Scientists have not yet devised cures for most viral plant diseases. Consequently, research efforts are focused largely on reducing the transmission of such diseases and on breeding resistant varieties of crop plants.

Earlier in this chapter, we mentioned the ongoing evolutionary relationship between viruses and the genomes of their host cells. In fact, the original source of viral genetic material may have been transposons, mobile genetic elements that are present in multiple copies in many genomes.

Prions: Proteins as Infectious Agents

The viruses we have discussed in this chapter are infectious agents that spread diseases, and their genetic material is composed of nucleic acids, whose ability to be replicated is well known. Surprisingly, there are also *proteins* that are known to be infectious. Proteins called **prions** appear to cause a number of degenerative brain diseases in various animal species. These diseases include scrapie in sheep; mad cow disease, which has plagued the European beef industry in recent years; and Creutzfeldt-Jakob disease in humans, which has caused the death of some 150 people in Great Britain. Prions can be transmitted in food, as may occur when people eat prion-laden beef from cattle with mad cow disease. Kuru, another human disease caused by prions, was identified in the early 1900s among the South Fore natives of New Guinea. A kuru epidemic peaked there in the 1960s, puzzling scientists, who at first thought the disease had a genetic basis. Eventually, however, anthropological investigations ferreted out how the disease was spread: ritual cannibalism, a widespread practice among South Fore natives at that time.

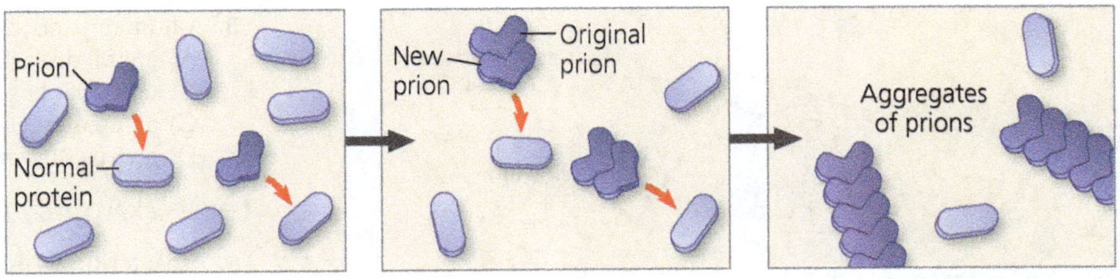

▲ **Figure 13 Model for how prions propagate.** Prions are misfolded versions of normal brain proteins. When a prion contacts a normally folded version of the same protein, it may induce the normal protein to assume the abnormal shape. The resulting chain reaction may continue until high levels of prion aggregation cause cellular malfunction and eventual degeneration of the brain.

Two characteristics of prions are especially alarming. First, prions act very slowly, with an incubation period of at least ten years before symptoms develop. The lengthy incubation period prevents sources of infection from being identified until long after the first cases appear, allowing many more infections to occur. Second, prions are virtually indestructible; they are not destroyed or deactivated by heating to normal cooking temperatures. To date, there is no known cure for prion diseases, and the only hope for developing effective treatments lies in understanding the process of infection.

How can a protein, which cannot replicate itself, be a transmissible pathogen? According to the leading model, a prion is a misfolded form of a protein normally present in brain cells. When the prion gets into a cell containing the normal form of the protein, the prion somehow converts normal protein molecules to the misfolded prion versions. Several prions then aggregate into a complex that can convert other normal proteins to prions, which join the chain **(Figure 13)**. Prion aggregation interferes with normal cellular functions and causes disease symptoms. This model was greeted with much skepticism when it was first proposed by Stanley Prusiner in the early 1980s, but it is now widely accepted. Prusiner was awarded the Nobel Prize in 1997 for his work on prions. He has recently proposed that prions are also involved in neurodegenerative disorders such as Alzheimer's and Parkinson's disease. There are many outstanding questions about these small infectious agents.

CONCEPT CHECK 3

1. Describe two ways in which a preexisting virus can become an emerging virus.
2. Contrast horizontal and vertical transmission of viruses in plants.
3. **WHAT IF?** TMV has been isolated from virtually all commercial tobacco products. Why, then, is TMV infection not an additional hazard for smokers?

Go to **MasteringBiology**® for Assignments, the eText, and the Study Area with Animations, Activities, Vocab Self-Quiz, and Practice Tests.

Chapter Review

SUMMARY OF KEY CONCEPTS

VOCAB SELF-QUIZ
goo.gl/gbai8v

CONCEPT 1

A virus consists of a nucleic acid surrounded by a protein coat

- A **virus** is a small nucleic acid genome enclosed in a protein **capsid** and sometimes a membranous **viral envelope** containing viral proteins that help the virus enter a cell. The genome may be single- or double-stranded DNA or RNA.

? *Are viruses generally considered living or nonliving? Explain.*

CONCEPT 2

Viruses replicate only in host cells

- Viruses use enzymes, ribosomes, and small molecules of host cells to synthesize progeny viruses during replication. Each type of virus has a characteristic **host range**, affected by whether cell-surface proteins that viral surface proteins can bind to are present.
- Phages (viruses that infect bacteria) can replicate by two alternative mechanisms: the **lytic cycle** and the **lysogenic cycle**.

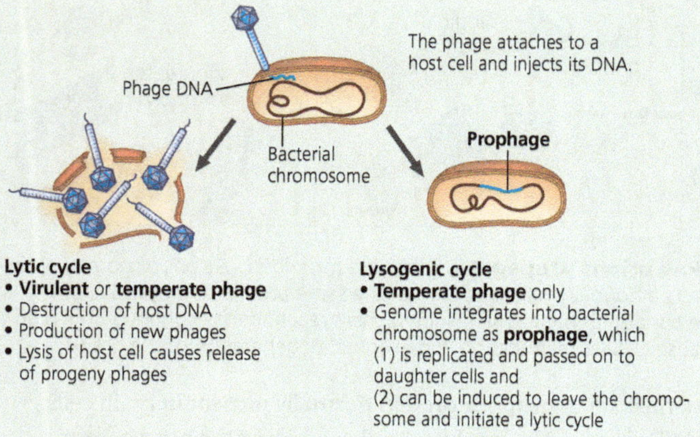

- Bacteria have various ways of defending themselves against phage infections, including the CRISPR system.
- Many animal viruses have an envelope. **Retroviruses** (such as **HIV**) use the enzyme **reverse transcriptase** to copy their RNA genome into DNA, which can be integrated into the host genome as a **provirus**.
- Since viruses can replicate only within cells, they probably evolved after the first cells appeared, perhaps as packaged fragments of cellular nucleic acid.

? *Describe enzymes that are not found in most cells but are necessary for the replication of certain types of viruses.*

CONCEPT 3

Viruses and prions are formidable pathogens in animals and plants

- Symptoms of viral diseases may be caused by direct viral harm to cells or by the body's immune response. **Vaccines** stimulate the immune system to defend the host against specific viruses.
- Outbreaks of emerging viral diseases in humans are usually caused by existing viruses that expand their host territory. The H1N1 2009 flu virus was a new combination of pig, human, and avian viral genes that caused a **pandemic**.
- Viruses enter plant cells through damaged cell walls (horizontal transmission) or are inherited from a parent (vertical transmission).
- **Prions** are slow-acting, virtually indestructible infectious proteins that cause brain diseases in mammals.

? *What aspect of an RNA virus makes it more likely than a DNA virus to become an emerging virus?*

TEST YOUR UNDERSTANDING

PRACTICE TEST

goo.gl/CRZjvS

Level 1: Knowledge/Comprehension

1. Which of the following characteristics, structures, or processes is common to both bacteria and viruses?
   (A) metabolism
   (B) ribosomes
   (C) genetic material composed of nucleic acid
   (D) cell division
2. Emerging viruses arise by
   (A) mutation of existing viruses.
   (B) the spread of existing viruses to new host species.
   (C) the spread of existing viruses more widely within their host species.
   (D) all of the above.
3. A human pandemic is
   (A) a viral disease that infects all humans.
   (B) a flu that kills more than 1 million people.
   (C) an epidemic that extends around the world.
   (D) a virus that increases in mortality rate as it spreads.

Level 2: Application/Analysis

4. A bacterium is infected with an experimentally constructed bacteriophage composed of the T2 phage protein coat and T4 phage DNA. The new phages produced would have
   (A) T2 protein and T4 DNA.
   (B) T4 protein and T2 DNA.
   (C) T2 protein and T2 DNA.
   (D) T4 protein and T4 DNA.
5. RNA viruses require their own supply of certain enzymes because
   (A) host cells rapidly destroy the viruses.
   (B) host cells lack enzymes that can replicate the viral genome.
   (C) these enzymes translate viral mRNA into proteins.
   (D) these enzymes penetrate host cell membranes.
6. **DRAW IT** Redraw Figure 8 to show the replicative cycle of a virus with a single-stranded genome that can function as mRNA.

Level 3: Synthesis/Evaluation

7. **SCIENTIFIC INQUIRY**
   When bacteria infect an animal, the number of bacteria in the body increases in an exponential fashion (graph A). After infection by a virulent animal virus with a lytic replicative cycle, there is no evidence of infection for a while. Then the number of viruses rises suddenly and subsequently increases in a series of steps (graph B). Explain the difference in the curves.

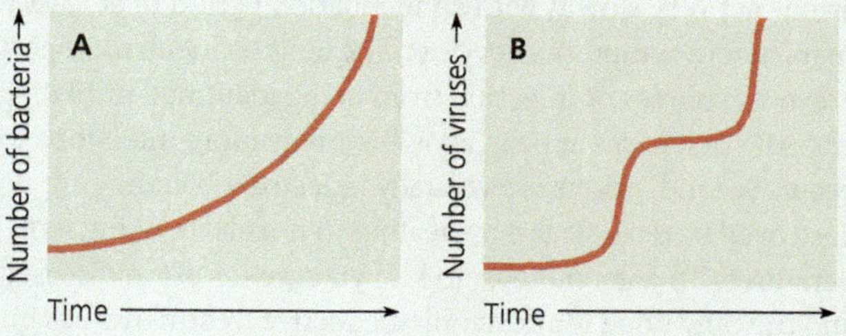

8. **FOCUS ON EVOLUTION**
   The success of some viruses lies in their ability to evolve rapidly within the host. Such viruses evade the host's defenses by mutating and producing many altered progeny viruses before the body can mount an attack. Thus, the viruses present late in infection differ from those that initially infected the body. Discuss this as an example of evolution in microcosm. Which viral lineages tend to predominate?
9. **FOCUS ON ORGANIZATION**
   While viruses are considered by most scientists to be nonliving, they do show some characteristics of life, including the correlation of structure and function. In a short essay (100–150 words), discuss how the structure of a virus correlates with its function.
10. **SYNTHESIZE YOUR KNOWLEDGE**

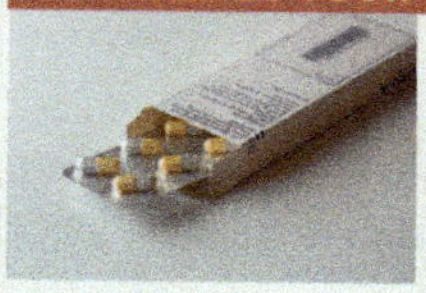

Oseltamivir (Tamiflu), an antiviral drug prescribed for the flu, inhibits the enzyme neuraminidase. Explain how this drug could prevent infection in someone who is exposed to the flu or could shorten the course of flu in an infected patient (the reasons for which it is prescribed).

Answers

Figure Questions

Figure 3 Top vertical arrow: Infection. Left upper arrow: Replication. Right upper arrow: Transcription. Right middle arrow: Translation. Lower left and right arrows: Self-assembly. Bottom middle arrow: Exit. **Figure 8** There are many steps that could be interfered with: binding of the virus to the cell, reverse transcriptase function, integration into the host cell chromosome, genome synthesis (in this case, transcription of RNA from the integrated provirus), assembly of the virus inside the cell, and budding of the virus. (Many of these, if not all, are targets of actual medical strategies to block progress of the infection in HIV-infected people.) **Figure 9** The shape of a protein on the HIV surface is likely to be complementary to the shape of the receptor (CD4) and also to that of the co-receptor (CCR5). A molecule with a shape similar to the HIV surface protein could bind CCR5, blocking HIV binding. (Another answer would be a molecule that bound to CCR5 and changed the shape of CCR5 so it could no longer bind to HIV.)

Concept Check 1

1. TMV consists of one molecule of RNA surrounded by a helical array of proteins. The influenza virus has eight molecules of RNA, each surrounded by a helical array of proteins, similar to the arrangement of the single RNA molecule in TMV. Another difference between the viruses is that the influenza virus has an outer envelope and TMV does not. **2.** The T2 phages were an excellent choice for use in the Hershey-Chase experiment because they consist of only DNA surrounded by a protein coat, and DNA and protein were the two candidates for macromolecules that carried genetic information. Hershey and Chase were able to radioactively label each type of molecule alone and follow it during separate infections of *E. coli* cells with T2. Only the DNA entered the bacterial cell during infection, and only labeled DNA showed up in some of the progeny phage. Hershey and Chase concluded that the DNA must carry the genetic information necessary for the phage to reprogram the cell and produce progeny phages.

Concept Check 2

1. Lytic phages can only carry out lysis of the host cell, whereas lysogenic phages may either lyse the host cell or integrate into the host chromosome. In the latter case, the viral DNA (prophage) is simply replicated along with the host chromosome. Under certain conditions, a prophage may exit the host chromosome and initiate a lytic cycle.
2. Both the viral RNA polymerase and the cellular RNA polymerase in Figure 10 synthesize an RNA molecule complementary to a template strand. However, the cellular RNA polymerase in Figure 10 uses one of the strands of the DNA double helix as a template, whereas the viral RNA polymerase uses the RNA of the viral genome as a template.
3. HIV is called a retrovirus because it synthesizes DNA using its RNA genome as a template. This is the reverse ("retro") of the usual DNA → RNA information flow.
4. Both the CRISPR system and miRNAs involve RNA molecules bound in a protein complex and acting as "homing devices" that enable the complex to bind a complementary sequence, but miRNAs are involved in regulating gene expression (by affecting mRNAs) and the CRISPR system protects bacterial cells from foreign invaders (infecting phages). Thus the CRISPR system is more like an immune system than are miRNAs.

Concept Check 3

1. Mutations can lead to a new strain of a virus that can no longer be effectively fought by the immune system, even if an animal had been exposed to the original strain; a virus can jump from one species to a new host; and a rare virus can spread if a host population becomes less isolated. **2.** In horizontal transmission, a plant is infected from an external source of the virus, which enters through a break in the plant's epidermis due to damage by herbivores or other agents. In vertical transmission, a plant inherits viruses from its parent either via infected seeds (sexual reproduction) or via an infected cutting (asexual reproduction). **3.** Humans are not within the host range of TMV, so they can't be infected by the virus. (TMV can't bind to receptors on human cells and infect them.)

Summary of Key Concepts Questions

1 Viruses are generally considered nonliving because they are not capable of replicating outside of a host cell and are unable to carry out the energy-transforming reactions of metabolism. To replicate and carry out metabolism, they depend completely on host enzymes and resources. **2** Single-stranded RNA viruses require an RNA polymerase that can make RNA using an RNA template. (Cellular RNA polymerases make RNA using a DNA template.) Retroviruses require reverse transcriptases to make DNA using an RNA template. (Once the first DNA strand has been made, the same enzyme can promote synthesis of the second DNA strand.) **3** The mutation rate of RNA viruses is higher than that of DNA viruses because RNA polymerase has no proofreading function, so errors in replication are not corrected. Their higher mutation rate means that RNA viruses change faster than DNA viruses, leading to their being able to have an altered host range and to evade immune defenses in possible hosts.

Test Your Understanding

1. C **2.** D **3.** C **4.** D **5.** B
6. As shown in the sketch, the viral genome would be translated into capsid proteins and envelope glycoproteins directly, rather than after a complementary RNA copy was made. A complementary RNA strand would still be made, however, that could be used as a template for many new copies of the viral genome.

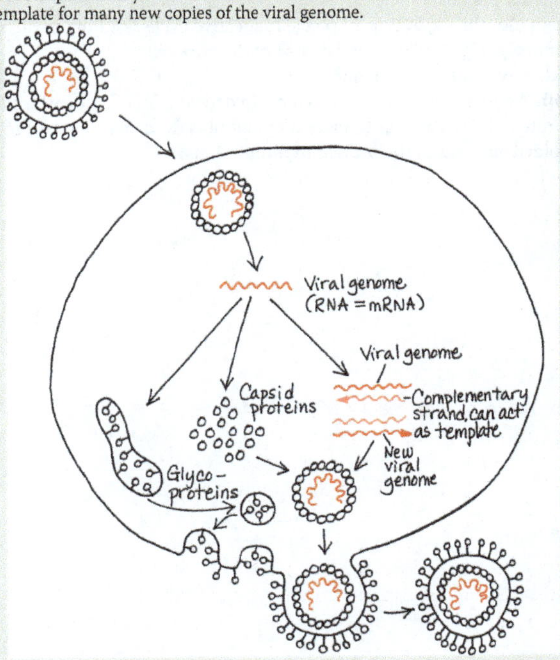

Credits

Credits are listed in the order of appearance.

Photo Credits

1 Richard Bizley/Science Source; **2a** Omikron/Science Source; **2b** Linda M. Stannard, University of Cape Town/Science Source; **2c** Hazel Appleton/Science Source; **2d** Ami Images/Science Source; **8** C. Dauguet/Institute Pasteur/Science Source;
10 National Institute of Allergy and Infectious Diseases (NIAID) and CDC; **11a** NIBSC/Science Source; **11b** Seo Myung-gon/AP Images; **17 SSE** AP Images; **12** Nigel Cattlin/Alamy; **Chapter Review SYK** Nelson Hale/Shutterstock.

Illustration and Text Credits

SSE.1 Data from J-R. Yang et al., New variants and age shift to high fatality groups contribute to severe successive waves in the 2009 influenza pandemic in Taiwan, *PLoS ONE* 6(11): e28288 (2011).

Glossary

AIDS (acquired immunodeficiency syndrome) The symptoms and signs present during the late stages of HIV infection, defined by a specified reduction in the number of T cells and the appearance of characteristic secondary infections.

capsid The protein shell that encloses a viral genome. It may be rod-shaped, polyhedral, or more complex in shape.

epidemic A general outbreak of a disease.

HIV (human immunodeficiency virus) The infectious agent that causes AIDS. HIV is a retrovirus.

host range The limited number of species whose cells can be infected by a particular virus.

lysogenic cycle (lī′-sō-jen′-ik) A type of phage replicative cycle in which the viral genome becomes incorporated into the bacterial host chromosome as a prophage, is replicated along with the chromosome, and does not kill the host.

lytic cycle (lit′-ik) A type of phage replicative cycle resulting in the release of new phages by lysis (and death) of the host cell.

pandemic A global epidemic.

prion An infectious agent that is a misfolded version of a normal cellular protein. Prions appear to increase in number by converting correctly folded versions of the protein to more prions.

prophage (prō′-fāj) A phage genome that has been inserted into a specific site on a bacterial chromosome.

provirus A viral genome that is permanently inserted into a host genome.

retrovirus (re′-trō-vī′-rus) An RNA virus that replicates by transcribing its RNA into DNA and then inserting the DNA into a cellular chromosome; an important class of cancer-causing viruses.

reverse transcriptase (tran-skrip′-tās) An enzyme encoded by certain viruses (retroviruses) that uses RNA as a template for DNA synthesis.

vaccine A harmless variant or derivative of a pathogen that stimulates a host's immune system to mount defenses against the pathogen.

viral envelope A membrane, derived from membranes of the host cell, that cloaks the capsid, which in turn encloses a viral genome.

virulent phage A phage that replicates only by a lytic cycle.

virus An infectious particle incapable of replicating outside of a cell, consisting of an RNA or DNA genome surrounded by a protein coat (capsid) and, for some viruses, a membranous envelope.

Index

D

J

K

L

M

N

O

P

Q

R

S

T

U

V

W

X

Y

Z